Classical Electrodynamics

by **JOHN DAVID JACKSON**

Professor of Physics, University of Illinois

CLASSICAL
ELECTRODYNAMICS

John Wiley & Sons, Inc., New York · London · Sydney

SIXTH PRINTING, MAY, 1967

Printed in the United States of America
Library of Congress Catalog Card Number: 62-8774

To the memory of my father,
Walter David Jackson

Preface

Classical electromagnetic theory, together with classical and quantum mechanics, forms the core of present-day theoretical training for undergraduate and graduate physicists. A thorough grounding in these subjects is a requirement for more advanced or specialized training.

Typically the undergraduate program in electricity and magnetism involves two or perhaps three semesters beyond elementary physics, with the emphasis on the fundamental laws, laboratory verification and elaboration of their consequences, circuit analysis, simple wave phenomena, and radiation. The mathematical tools utilized include vector calculus, ordinary differential equations with constant coefficients, Fourier series, and perhaps Fourier or Laplace transforms, partial differential equations, Legendre polynomials, and Bessel functions.

As a general rule a two-semester course in electromagnetic theory is given to beginning graduate students. It is for such a course that my book is designed. My aim in teaching a graduate course in electromagnetism is at least threefold. The first aim is to present the basic subject matter as a coherent whole, with emphasis on the unity of electric and magnetic phenomena, both in their physical basis and in the mode of mathematical description. The second, concurrent aim is to develop and utilize a number of topics in mathematical physics which are useful in both electromagnetic theory and wave mechanics. These include Green's theorems and Green's functions, orthonormal expansions, spherical harmonics, cylindrical and spherical Bessel functions. A third and perhaps most important purpose is the presentation of new material, especially on the interaction of relativistic charged particles with electromagnetic fields. In this last area personal preferences and prejudices enter strongly. My choice of topics is governed by what I feel is important and useful for students interested in theoretical physics, experimental nuclear and high-energy physics, and that as yet ill-defined field of plasma physics.

The book begins in the traditional manner with electrostatics. The first six chapters are devoted to the development of Maxwell's theory of electromagnetism. Much of the necessary mathematical apparatus is constructed along the way, especially in Chapters 2 and 3, where boundary-value problems are discussed thoroughly. The treatment is initially in terms of the electric field E and the magnetic induction B, with the derived macroscopic quantities, D and H, introduced by suitable averaging over ensembles of atoms or molecules. In the discussion of dielectrics, simple classical models for atomic polarizability are described, but for magnetic materials no such attempt is made. Partly this omission was a question of space, but truly classical models of magnetic susceptibility are not possible. Furthermore, elucidation of the interesting phenomenon of ferromagnetism needs almost a book in itself.

The next three chapters (7–9) illustrate various electromagnetic phenomena, mostly of a macroscopic sort. Plane waves in different media, including plasmas, as well as dispersion and the propagation of pulses, are treated in Chapter 7. The discussion of wave guides and cavities in Chapter 8 is developed for systems of arbitrary cross section, and the problems of attenuation in guides and the Q of a cavity are handled in a very general way which emphasizes the physical processes involved. The elementary theory of multipole radiation from a localized source and diffraction occupy Chapter 9. Since the simple scalar theory of diffraction is covered in many optics textbooks, as well as undergraduate books on electricity and magnetism, I have presented an improved, although still approximate, theory of diffraction based on vector rather than scalar Green's theorems.

The subject of magnetohydrodynamics and plasmas receives increasingly more attention from physicists and astrophysicists. Chapter 10 represents a survey of this complex field with an introduction to the main physical ideas involved.

The first nine or ten chapters constitute the basic material of classical electricity and magnetism. A graduate student in physics may be expected to have been exposed to much of this material, perhaps at a somewhat lower level, as an undergraduate. But he obtains a more mature view of it, understands it more deeply, and gains a considerable technical ability in analytic methods of solution when he studies the subject at the level of this book. He is then prepared to go on to more advanced topics. The advanced topics presented here are predominantly those involving the interaction of charged particles with each other and with electromagnetic fields, especially when moving relativistically.

The special theory of relativity had its origins in classical electrodynamics. And even after almost 60 years, classical electrodynamics still impresses and delights as a beautiful example of the covariance of physical laws under Lorentz transformations. The special theory of relativity is discussed in Chapter 11, where all the necessary formal apparatus is developed, various kinematic consequences are explored, and the covariance of electrodynamics is established. The next chapter is devoted to relativistic particle kinematics and dynamics. Although the dynamics of charged particles in electromagnetic fields can properly be considered electrodynamics, the reader may wonder whether such things as kinematic transformations of collision problems can. My reply is that these examples occur naturally once one has established the four-vector character of a particle's momentum and energy, that they serve as useful practice in manipulating Lorentz transformations, and that the end results are valuable and often hard to find elsewhere.

Chapter 13 on collisions between charged particles emphasizes energy loss and scattering and develops concepts of use in later chapters. Here for the first time in the book I use semiclassical arguments based on the uncertainty principle to obtain approximate quantum-mechanical expressions for energy loss, etc., from the classical results. This approach, so fruitful in the hands of Niels Bohr and E. J. Williams, allows one to see clearly how and when quantum-mechanical effects enter to modify classical considerations.

The important subject of emission of radiation by accelerated point charges is discussed in detail in Chapters 14 and 15. Relativistic effects are stressed, and expressions for the frequency and angular dependence of the emitted radiation are developed in sufficient generality for all applications. The examples treated range from synchrotron radiation to bremsstrahlung and radiative beta processes. Cherenkov radiation and the Weizsäcker-Williams method of virtual quanta are also discussed. In the atomic and nuclear collision processes semiclassical arguments are again employed to obtain approximate quantum-mechanical results. I lay considerable stress on this point because I feel that it is important for the student to see that radiative effects such as bremsstrahlung are almost entirely classical in nature, even though involving small-scale collisions. A student who meets bremsstrahlung for the first time as an example of a calculation in quantum field theory will not understand its physical basis.

Multipole fields form the subject matter of Chapter 16. The expansion of scalar and vector fields in spherical waves is developed from first principles with no restrictions as to the relative dimensions of source and wavelength. Then the properties of electric and magnetic multipole radiation fields are considered. Once the connection to the multipole moments of the source has been made, examples of atomic and nuclear multipole radiation are discussed, as well as a macroscopic source whose dimensions are comparable to a wavelength. The scattering of a plane electromagnetic wave by a spherical object is treated in some detail in order to illustrate a boundary-value problem with vector spherical waves.

In the last chapter the difficult problem of radiative reaction is discussed. The treatment is physical, rather than mathematical, with the emphasis on delimiting the areas where approximate radiative corrections are adequate and on finding where and why existing theories fail. The original Abraham-Lorentz theory of the self-force is presented, as well as more recent classical considerations.

The book ends with an appendix on units and dimensions and a bibliography. In the appendix I have attempted to show the logical steps involved in setting up a system of units, without haranguing the reader as to the obvious virtues of *my* choice of units. I have provided two tables which I hope will be useful, one for converting equations and symbols and the other for converting a given quantity of something from so many Gaussian units to so many mks units, and vice versa. The bibliography lists books which I think the reader may find pertinent and useful for reference or additional study. These books are referred to by author's name in the reading lists at the end of each chapter.

This book is the outgrowth of a graduate course in classical electrodynamics which I have taught off and on over the past eleven years, at both the University of Illinois and McGill University. I wish to thank my colleagues and students at both institutions for countless helpful remarks and discussions. Special mention must be made of Professor P. R. Wallace of McGill, who gave me the opportunity and encouragement to teach what was then a rather unorthodox course in electromagnetism, and Professors H. W. Wyld and G. Ascoli of Illinois, who have been particularly free with many helpful suggestions on the treatment of various topics. My thanks are also extended to Dr. A. N. Kaufman for reading and commenting on a preliminary version of the manuscript, and to Mr. G. L. Kane for his zealous help in preparing the index.

J. D. JACKSON

Urbana, Illinois
January, 1962

Contents

1

Introduction to
Electrostatics

Although amber and lodestone were known by the ancient Greeks, electrodynamics developed as a quantitative subject in about 80 years. Coulomb's observations on the forces between charged bodies were made around 1785. About 50 years later, Faraday was studying the effects of currents and magnetic fields. By 1864, Maxwell had published his famous paper on a dynamical theory of the electromagnetic field.

We will begin our discussion with the subject of electrostatics—problems involving time-independent electric fields. Much of the material will be covered rather rapidly because it is in the nature of a review. We will use electrostatics as a testing ground to develop and use mathematical techniques of general applicability.

1.1   Coulomb's Law

All of electrostatics stems from the quantitative statement of Coulomb's law concerning the force acting between charged bodies at rest with respect to each other. Coulomb (and, even earlier, Cavendish) showed experimentally that the force between two small charged bodies separated a distance large compared to their dimensions

(1) varied directly as the magnitude of each charge,

(2) varied inversely as the square of the distance between them,

(3) was directed along the line joining the charges,

(4) was attractive if the bodies were oppositely charged and repulsive if the bodies had the same type of charge.

Furthermore it was shown experimentally that the total force produced

1

on one small charged body by a number of the other small charged bodies placed around it was the *vector* sum of the individual two-body forces of Coulomb.

1.2 Electric Field

Although the thing that eventually gets measured is a force, it is useful to introduce a concept one step removed from the forces, the concept of an electric field due to some array of charged bodies. At the moment, the electric field can be defined as the force per unit charge acting at a given point. It is a vector function of position, denoted by **E**. One must be careful in its definition, however. It is not necessarily the force that one would observe by placing one unit of charge on a pith ball and placing it in position. The reason is that one unit of charge (e.g., 100 strokes of cat's fur on an amber rod) may be so large that its presence alters appreciably the field configuration of the array. Consequently one must use a limiting process whereby the ratio of the force on the small test body to the charge on it is measured for smaller and smaller amounts of charge. Experimentally, this ratio and the direction of the force will become constant as the amount of test charge is made smaller and smaller. These limiting values of magnitude and direction define the magnitude and direction of the electric field **E** at the point in question. In symbols we may write

$$\mathbf{F} = q\mathbf{E} \qquad (1.1)$$

where **F** is the force, **E** the electric field, and q the charge. In this equation it is assumed that the charge q is located at a point, and the force and the electric field are evaluated at that point.

Coulomb's law can be written down similarly. If **F** is the force on a point charge q_1, located at $\mathbf{x}_1$, due to another point charge q_2, located at $\mathbf{x}_2$, then Coulomb's law is

$$\mathbf{F} = kq_1q_2\frac{(\mathbf{x}_1 - \mathbf{x}_2)}{|\mathbf{x}_1 - \mathbf{x}_2|^3} \qquad (1.2)$$

Note that q_1 and q_2 are algebraic quantities which can be positive or negative. The constant of proportionality k depends on the system of units used.

The electric field at the point **x** due to a point charge q_1 at the point $\mathbf{x}_1$ can be obtained directly:

$$\mathbf{E}(\mathbf{x}) = kq_1\frac{(\mathbf{x} - \mathbf{x}_1)}{|\mathbf{x} - \mathbf{x}_1|^3} \qquad (1.3)$$

as indicated in Fig. 1.1. The constant k is determined by the unit of charge

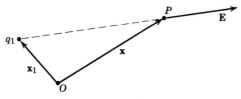

Fig. 1.1

chosen. In electrostatic units (esu), unit charge is chosen as that charge which exerts a force of one dyne on an equal charge located one centimeter away. Thus, with cgs units, $k = 1$ and the unit of charge is called the "stat-coulomb." In the mks system, $k = (4\pi\epsilon_0)^{-1}$, where ϵ_0 (= 8.854 × 10^{-12} farad/meter) is the permittivity of free space. We will use esu.*

The experimentally observed linear superposition of forces due to many charges means that we may write the electric field at $\mathbf{x}$ due to a system of point charges q_i, located at $\mathbf{x}_i$, $i = 1, 2, \ldots, n$, as the vector sum:

$$\mathbf{E}(\mathbf{x}) = \sum_{i=1}^{n} q_i \frac{(\mathbf{x} - \mathbf{x}_i)}{|\mathbf{x} - \mathbf{x}_i|^3} \tag{1.4}$$

If the charges are so small and so numerous that they can be described by a charge density $\rho(\mathbf{x}')$ [if Δq is the charge in a small volume $\Delta x\,\Delta y\,\Delta z$ at the point $\mathbf{x}'$, then $\Delta q = \rho(\mathbf{x}')\,\Delta x\,\Delta y\,\Delta z$], the sum is replaced by an integral:

$$\mathbf{E}(\mathbf{x}) = \int \rho(\mathbf{x}') \frac{(\mathbf{x} - \mathbf{x}')}{|\mathbf{x} - \mathbf{x}'|^3}\, d^3x' \tag{1.5}$$

where $d^3x' = dx'\,dy'\,dz'$ is a three-dimensional volume element at $\mathbf{x}'$.

At this point it is worth while to introduce the *Dirac delta function*. In one dimension, the delta function, written $\delta(x - a)$, is a mathematically improper function having the properties:

(1) $\delta(x - a) = 0$ for $x \neq a$, and

(2) $\int \delta(x - a)\, dx = 1$ if the region of integration includes $x = a$, and is zero otherwise.

The delta function can be given rigorous meaning as the limit of a peaked curve such as a Gaussian which becomes narrower and narrower, but higher and higher, in such a way that the area under the curve is always constant. L. Schwartz's theory of distributions is a comprehensive rigorous mathematical approach to delta functions and their manipulations.†

---

* The question of units is discussed in detail in the Appendix.

† A useful, rigorous account of the Dirac delta function is given by Lighthill. (Full references for items cited in the text or footnotes by author only will be found in the Bibliography.)

From the definitions above it is evident that, for an arbitrary function $f(x)$,

(3) $\int f(x)\, \delta(x - a)\, dx = f(a)$, and

(4) $\int f(x)\, \delta'(x - a)\, dx = -f'(a)$,

where a prime denotes differentiation with respect to the argument.

If the delta function has as argument a function $f(x)$ of the independent variable x, it can be transformed according to the rule,

(5) $\delta(f(x)) = \dfrac{1}{\left|\dfrac{df}{dx}\right|}\, \delta(x - x_0)$,

where $f(x_0) = 0$. This can be proved by noting that $\delta(f)\, df = \delta(x)\, dx$.

In more than one dimension, we merely take products of delta functions in each dimension. In three dimensions, for example,

(6) $\delta(\mathbf{x} - \mathbf{X}) = \delta(x_1 - X_1)\, \delta(x_2 - X_2)\, \delta(x_3 - X_3)$

is a function which vanishes everywhere except at $\mathbf{x} = \mathbf{X}$, and is such that

(7) $\int_{\Delta V} \delta(\mathbf{x} - \mathbf{X})\, d^3x = \begin{cases} 1 & \text{if } \Delta V \text{ contains } \mathbf{x} = \mathbf{X}, \\ 0 & \text{if } \Delta V \text{ does not contain } \mathbf{x} = \mathbf{X}. \end{cases}$

Note that a delta function has the dimensions of an inverse volume in whatever number of dimensions the space has.

A discrete set of point charges can be described with a charge density by means of delta functions. For example,

$$\rho(\mathbf{x}) = \sum_{i=1}^{n} q_i\, \delta(\mathbf{x} - \mathbf{x}_i) \tag{1.6}$$

represents a distribution of n point charges q_i, located at the points $\mathbf{x}_i$. Substitution of this charge density (1.6) into (1.5) and integration, using the properties of the delta function, yields the discrete sum (1.4).

1.3 Gauss's Law

The integral (1.5) is not the most suitable form for the evaluation of electric fields. There is another integral result, called *Gauss's law*, which is often more useful and which furthermore leads to a differential equation for $\mathbf{E}(\mathbf{x})$. To obtain Gauss's law we first consider a point charge q and a *closed* surface S, as shown in Fig. 1.2. Let r be the distance from the charge to a point on the surface, $\mathbf{n}$ be the outwardly directed unit normal to the surface at that point, da be an element of surface area. If the electric field $\mathbf{E}$ at the point on the surface due to the charge q makes an angle θ with the unit normal, then the normal component of $\mathbf{E}$ times the area element is:

$$\mathbf{E} \cdot \mathbf{n}\, da = q\, \frac{\cos\theta}{r^2}\, da \tag{1.7}$$

Since $\mathbf{E}$ is directed along the line from the surface element to the charge q,

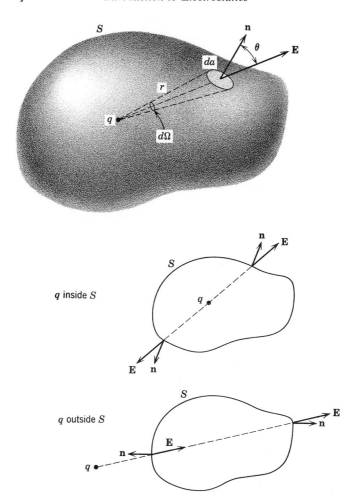

Fig. 1.2 Gauss's law. The normal component of electric field is integrated over the closed surface S. If the charge is inside (outside) S, the total solid angle subtended at the charge by the inner side of the surface is 4π (zero).

$\cos\theta\,da = r^2\,d\Omega$, where $d\Omega$ is the element of solid angle subtended by da at the position of the charge. Therefore

$$\mathbf{E}\cdot\mathbf{n}\,da = q\,d\Omega \tag{1.8}$$

If we now integrate the normal component of $\mathbf{E}$ over the whole surface, it is easy to see that

$$\oint_S \mathbf{E}\cdot\mathbf{n}\,da = \begin{cases} 4\pi q & \text{if } q \text{ lies } \textit{inside } S \\ 0 & \text{if } q \text{ lies } \textit{outside } S \end{cases} \tag{1.9}$$

This result is Gauss's law for a single point charge. For a discrete set of charges, it is immediately apparent that

$$\oint_S \mathbf{E} \cdot \mathbf{n} \, da = 4\pi \sum_i q_i \qquad (1.10)$$

where the sum is over only those charges *inside* the surface *S*. For a continuous charge density $\rho(\mathbf{x})$, Gauss's law becomes:

$$\oint_S \mathbf{E} \cdot \mathbf{n} \, da = 4\pi \int_V \rho(\mathbf{x}) \, d^3x \qquad (1.11)$$

where *V* is the volume enclosed by *S*.

Equation (1.11) is one of the basic equations of electrostatics. Note that it depends upon

(1) the inverse square law for the force between charges,
(2) the central nature of the force,
(3) the linear superposition of the effects of different charges.

Clearly, then, Gauss's law holds for Newtonian gravitational force fields, with matter density replacing charge density.

It is interesting to observe that before Coulomb's observations Cavendish, by what amounted to a direct application of Gauss's law, did an experiment with two concentric conducting spheres and deduced that the power law of the force was inverse *n*th power, where $n = 2.00 \pm 0.02$. By a refinement of the technique, Maxwell showed that $n = 2.0 \pm 0.00005$. (See Jeans, p. 37, or Maxwell, Vol. 1, p. 80.)

1.4 Differential Form of Gauss's Law

Gauss's law can be thought of as being an integral formulation of the law of electrostatics. We can obtain a differential form (i.e., a differential equation) by using the divergence theorem. The divergence theorem states that for any vector field $\mathbf{A}(\mathbf{x})$ defined within a volume *V* surrounded by the closed surface *S* the relation

$$\oint_S \mathbf{A} \cdot \mathbf{n} \, da = \int_V \mathbf{\nabla} \cdot \mathbf{A} \, d^3x$$

holds between the volume integral of the divergence of **A** and the surface integral of the outwardly directed normal component of **A**. The equation in fact can be used as the definition of the divergence (see Stratton, p. 4).

To apply the divergence theorem we consider the integral relation expressed in Gauss's theorem:

$$\oint_S \mathbf{E} \cdot \mathbf{n} \, da = 4\pi \int_V \rho(\mathbf{x}) \, d^3x$$

Now the divergence theorem allows us to write this as:

$$\int_V (\nabla \cdot \mathbf{E} - 4\pi\rho)\, d^3x = 0 \qquad (1.12)$$

for an arbitrary volume V. We can, in the usual way, put the integrand equal to zero to obtain

$$\nabla \cdot \mathbf{E} = 4\pi\rho \qquad (1.13)$$

which is the differential form of Gauss's law of electrostatics. This equation can itself be used to solve problems in electrostatics. However, it is often simpler to deal with scalar rather than vector functions of position, and then to derive the vector quantities at the end if necessary (see below).

1.5   Another Equation of Electrostatics and the Scalar Potential

The single equation (1.13) is not enough to specify completely the three components of the electric field $\mathbf{E}(\mathbf{x})$. Perhaps some readers know that a vector field can be specified completely if its divergence and curl are given everywhere in space. Thus we look for an equation specifying curl $\mathbf{E}$ as a function of position. Such an equation, namely,

$$\nabla \times \mathbf{E} = 0 \qquad (1.14)$$

follows directly from our generalized Coulomb's law (1.5):

$$\mathbf{E}(\mathbf{x}) = \int \rho(\mathbf{x}') \frac{(\mathbf{x} - \mathbf{x}')}{|\mathbf{x} - \mathbf{x}'|^3}\, d^3x'$$

The vector factor in the integrand, viewed as a function of $\mathbf{x}$, is the negative gradient of the scalar $1/|\mathbf{x} - \mathbf{x}'|$:

$$\frac{(\mathbf{x} - \mathbf{x}')}{|\mathbf{x} - \mathbf{x}'|^3} = -\nabla\left(\frac{1}{|\mathbf{x} - \mathbf{x}'|}\right)$$

Since the gradient operation involves $\mathbf{x}$, but not the integration variable $\mathbf{x}'$, it can be taken outside the integral sign. Then the field can be written

$$\mathbf{E}(\mathbf{x}) = -\nabla \int \frac{\rho(\mathbf{x}')}{|\mathbf{x} - \mathbf{x}'|}\, d^3x' \qquad (1.15)$$

Since the curl of the gradient of *any* scalar function of position vanishes ($\nabla \times \nabla\psi = 0$, for all ψ), (1.14) follows immediately from (1.15).

Note that $\nabla \times \mathbf{E} = 0$ depends on the central nature of the force between charges, and on the fact that the force is a function of relative distances only, but does not depend on the inverse square nature.

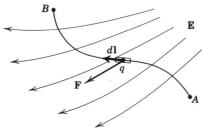

Fig. 1.3

In (1.15) the electric field (a vector) is derived from a scalar by the gradient operation. Since one function of position is easier to deal with than three, it is worth while concentrating on the scalar function and giving it a name. Consequently we define the "scalar potential" $\Phi(\mathbf{x})$ by the equation:

$$\mathbf{E} = -\nabla\Phi \qquad (1.16)$$

Then (1.15) shows that the scalar potential is given in terms of the charge density by

$$\Phi(\mathbf{x}) = \int \frac{\rho(\mathbf{x}')}{|\mathbf{x} - \mathbf{x}'|}\, d^3x' \qquad (1.17)$$

where the integration is over all charges in the universe, and Φ is arbitrary to the extent that a constant can be added to the right side of (1.17).

The scalar potential has a physical interpretation when we consider the work done on a test charge q in transporting it from one point (A) to another point (B) in the presence of an electric field $\mathbf{E}(\mathbf{x})$, as shown in Fig. 1.3. The force acting on the charge at any point is

$$\mathbf{F} = q\mathbf{E}$$

so that the work done in moving the charge from A to B is

$$W = -\int_A^B \mathbf{F} \cdot d\mathbf{l} = -q\int_A^B \mathbf{E} \cdot d\mathbf{l} \qquad (1.18)$$

The minus sign appears because we are calculating the work done *on* the charge against the action of the field. With definition (1.16) the work can be written

$$W = q\int_A^B \nabla\Phi \cdot d\mathbf{l} = q\int_A^B d\Phi = q[\Phi_B - \Phi_A] \qquad (1.19)$$

which shows that $q\Phi$ can be interpreted as the potential energy of the test charge in the electrostatic field.

From (1.18) and (1.19) it can be seen that the line integral of the electric field between two points is independent of the path and is the negative of the potential difference between the points:

$$\int_A^B \mathbf{E} \cdot d\mathbf{l} = -(\Phi_B - \Phi_A) \qquad (1.20)$$

This follows directly, of course, from definition (1.16). If the path is closed, the line integral is zero,

$$\oint \mathbf{E} \cdot d\mathbf{l} = 0 \qquad (1.21)$$

a result that can also be obtained directly from Coulomb's law. Then application of Stokes's theorem [if $\mathbf{A}(\mathbf{x})$ is a vector field, S is an open surface, and C is the closed curve bounding S,

$$\oint_C \mathbf{A} \cdot d\mathbf{l} = \int_S (\nabla \times \mathbf{A}) \cdot \mathbf{n} \, da$$

where $d\mathbf{l}$ is a line element of C, $\mathbf{n}$ is the normal to S, and the path C is traversed in a right-hand screw sense relative to $\mathbf{n}$] leads immediately back to $\nabla \times \mathbf{E} = 0$.

1.6 Surface Distributions of Charges and Dipoles and Discontinuities in the Electric Field and Potential

One of the common problems in electrostatics is the determination of electric field or potential due to a given surface distribution of charges. Gauss's law (1.11) allows us to write down a partial result directly. If a surface S, with a unit normal $\mathbf{n}$, has a surface-charge density of $\sigma(\mathbf{x})$ (measured in statcoulombs per square centimeter) and electric fields $\mathbf{E}_1$ and $\mathbf{E}_2$ on either side of the surface, as shown in Fig. 1.4, then Gauss's law tells us immediately that

$$(\mathbf{E}_2 - \mathbf{E}_1) \cdot \mathbf{n} = 4\pi\sigma \qquad (1.22)$$

This does not determine $\mathbf{E}_1$ and $\mathbf{E}_2$ unless there are no other sources of field and the geometry and form σ are especially simple. All that (1.22) says is that there is a discontinuity of $4\pi\sigma$ in the normal component of electric field in crossing a surface with a surface-charge density σ, the crossing being made from the "inner" to the "outer" side of the surface.

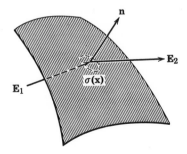

Fig. 1.4 Discontinuity in the normal component of electric field across a surface layer of charge.

The tangential component of electric field can be shown to be continuous across a boundary surface by using (1.21) for the line integral of **E** around a closed path. It is only necessary to take a rectangular path with negligible ends and one side on either side of the boundary.

A general result for the potential (and hence the field, by differentiation) at any point in space (not just at the surface) can be obtained from (1.17) by replacing $\rho \, d^3x$ by $\sigma \, da$:

$$\Phi(\mathbf{x}) = \int_S \frac{\sigma(\mathbf{x}')}{|\mathbf{x} - \mathbf{x}'|} \, da' \qquad (1.23)$$

Another problem of interest is the potential due to a dipole-layer distribution on a surface S. A dipole layer can be imagined as being formed by letting the surface S have a surface-charge density $\sigma(\mathbf{x})$ on it, and another surface S', lying close to S, have an equal and opposite surface-charge density on it at neighboring points, as shown in Fig. 1.5. The dipole-layer distribution of strength $D(\mathbf{x})$ is formed by letting S' approach infinitesimally close to S while the surface-charge density $\sigma(\mathbf{x})$ becomes infinite in such a manner that the product of $\sigma(\mathbf{x})$ and the local separation $d(\mathbf{x})$ of S and S' approaches the limit $D(\mathbf{x})$:

$$\lim_{d(\mathbf{x}) \to 0} \sigma(\mathbf{x}) \, d(\mathbf{x}) = D(\mathbf{x}) \qquad (1.24)$$

The direction of the dipole moment of the layer is normal to the surface S and in the direction going from negative to positive charge.

To find the potential due to a dipole layer we can consider a single dipole and then superpose a surface density of them, or we can obtain the same result by performing mathematically the limiting process described in words above on the surface-density expression (1.23). The first way is perhaps simpler, but the second gives useful practice in vector calculus. Consequently we proceed with the limiting process. With **n**, the unit normal to

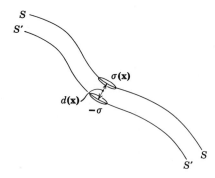

Fig. 1.5 Limiting process involved in creating a dipole layer.

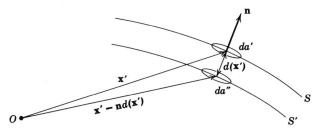

Fig. 1.6   Dipole-layer geometry.

the surface S, directed away from S', as shown in Fig. 1.6, the potential due to the two close surfaces is

$$\Phi(\mathbf{x}) = \int_S \frac{\sigma(\mathbf{x}')}{|\mathbf{x} - \mathbf{x}'|}\, da' - \int_{S'} \frac{\sigma(\mathbf{x}')}{|\mathbf{x} - \mathbf{x}' + \mathbf{n}d|}\, da''$$

For small d we can expand $|\mathbf{x} - \mathbf{x}' + \mathbf{n}d|^{-1}$. Consider the general expression $|\mathbf{x} + \mathbf{a}|^{-1}$, where $|\mathbf{a}| \ll |\mathbf{x}|$. Then we write

$$\frac{1}{|\mathbf{x} + \mathbf{a}|} \equiv \frac{1}{\sqrt{x^2 + a^2 + 2\mathbf{a}\cdot\mathbf{x}}}$$

$$= \frac{1}{x}\left(1 - \frac{\mathbf{a}\cdot\mathbf{x}}{x^2} + \cdots\right)$$

$$= \frac{1}{x} + \mathbf{a}\cdot\nabla\left(\frac{1}{x}\right) + \cdots$$

This is, of course, just a Taylor's series expansion in three dimensions. In this way we find that the potential becomes [upon taking the limit (1.24)]:

$$\Phi(\mathbf{x}) = \int_S D(\mathbf{x}')\mathbf{n}\cdot\nabla'\left(\frac{1}{|\mathbf{x} - \mathbf{x}'|}\right) da' \qquad (1.25)$$

Equation (1.25) has a simple geometrical interpretation. We note that

$$\mathbf{n}\cdot\nabla'\left(\frac{1}{|\mathbf{x} - \mathbf{x}'|}\right) da' = -\frac{\cos\theta\, da'}{|\mathbf{x} - \mathbf{x}'|^2} = -d\Omega$$

where $d\Omega$ is the element of solid angle subtended at the observation point by the area element da', as indicated in Fig. 1.7. Note that $d\Omega$ has a positive sign if θ is an acute angle, i.e., when the observation point views the "inner" side of the dipole layer. The potential can be written:

$$\Phi(\mathbf{x}) = -\int_S D(\mathbf{x}')\, d\Omega \qquad (1.26)$$

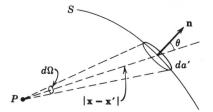

Fig. 1.7 The potential at P due to the dipole layer D on the area element da' is just the negative product of D and the solid angle element $d\Omega$ subtended by da' at P.

For a constant surface-dipole-moment density D, the potential is just the product of the moment and the solid angle subtended at the observation point by the surface, regardless of its shape.

There is a discontinuity in potential in crossing a double layer. This can be seen by letting the observation point come infinitesimally close to the double layer. The double layer is now imagined to consist of two parts, one being a small disc directly under the observation point. The disc is sufficiently small that it is sensibly flat and has constant surface-dipole-moment density D. Evidently the total potential can be obtained by linear superposition of the potential of the disc and that of the remainder. From (1.26) it is clear that the potential of the disc alone has a discontinuity of $4\pi D$ in crossing from the inner to the outer side, being $-2\pi D$ on the inner side and $+2\pi D$ on the outer. The potential of the remainder alone, with its hole where the disc fits in, is continuous across the plane of the hole. Consequently the total potential jump in crossing the surface is:

$$\Phi_2 - \Phi_1 = 4\pi D \qquad (1.27)$$

This result is analogous to (1.22) for the discontinuity of electric field in crossing a surface-charge density. Equation (1.27) can be interpreted "physically" as a potential drop occurring "inside" the dipole layer, and can be calculated as the product of the field between the two layers of surface charge times the separation before the limit is taken.

1.7 Poisson's and Laplace's Equations

In Sections 1.4 and 1.5 it was shown that the behavior of an electrostatic field can be described by the two differential equations:

$$\nabla \cdot \mathbf{E} = 4\pi\rho \qquad (1.13)$$

and

$$\nabla \times \mathbf{E} = 0 \qquad (1.14)$$

the latter equation being equivalent to the statement that $\mathbf{E}$ is the gradient of a scalar function, the scalar potential Φ:

$$\mathbf{E} = -\nabla\Phi \qquad (1.16)$$

Equations (1.13) and (1.16) can be combined into one partial differential equation for the single function $\Phi(\mathbf{x})$:

$$\nabla^2\Phi = -4\pi\rho \qquad (1.28)$$

This equation is called *Poisson's equation*. In regions of space where there is no charge density, the scalar potential satisfies Laplace's equation:

$$\nabla^2\Phi = 0 \qquad (1.29)$$

We already have a solution for the scalar potential in expression (1.17):

$$\Phi(\mathbf{x}) = \int \frac{\rho(\mathbf{x}')}{|\mathbf{x} - \mathbf{x}'|}\, d^3x' \qquad (1.17)$$

To verify that this does indeed satisfy Poisson's equation (1.28) we operate with the Laplacian on both sides:

$$\nabla^2\Phi = \nabla^2 \int \frac{\rho(\mathbf{x}')}{|\mathbf{x} - \mathbf{x}'|}\, d^3x' = \int \rho(\mathbf{x}')\nabla^2\left(\frac{1}{|\mathbf{x} - \mathbf{x}'|}\right) d^3x' \qquad (1.30)$$

We must now calculate the value of $\nabla^2(1/|\mathbf{x} - \mathbf{x}'|)$. It is convenient (and allowable) to translate the origin to $\mathbf{x}'$ and so consider $\nabla^2(1/r)$, where r is the magnitude of $\mathbf{x}$. By direct calculation we find that $\nabla^2(1/r) = 0$ for $r \neq 0$:

$$\nabla^2\left(\frac{1}{r}\right) = \frac{1}{r}\frac{d^2}{dr^2}\left(r \cdot \frac{1}{r}\right) = \frac{1}{r}\frac{d^2}{dr^2}(1) = 0$$

At $r = 0$, however, the expression is undefined. Hence we must use a limiting process. Since we anticipate something like a Dirac delta function, we integrate $\nabla^2(1/r)$ over a small volume V containing the origin. Then we use the divergence theorem to obtain a surface integral:

$$\int_V \nabla^2\left(\frac{1}{r}\right) d^3x \equiv \int_V \nabla \cdot \nabla\left(\frac{1}{r}\right) d^3x = \int_S \mathbf{n} \cdot \nabla\left(\frac{1}{r}\right) da$$

$$= \int_S \frac{\partial}{\partial r}\left(\frac{1}{r}\right) r^2\, d\Omega = -4\pi$$

It has now been established that $\nabla^2(1/r) = 0$ for $r \neq 0$, and that its volume integral is -4π. Consequently we can write the improper (but mathematically justifiable) equation, $\nabla^2(1/r) = -4\pi\delta(\mathbf{x})$, or, more generally,

$$\nabla^2\left(\frac{1}{|\mathbf{x} - \mathbf{x}'|}\right) = -4\pi\delta(\mathbf{x} - \mathbf{x}') \qquad (1.31)$$

Having established the singular nature of the Laplacian of $1/r$, we can now complete our check on (1.17) as a solution of Poisson's equation.

Equation (1.30) becomes

$$\nabla^2\Phi = \int \rho(\mathbf{x}')[-4\pi\delta(\mathbf{x}-\mathbf{x}')]\,d^3x' = -4\pi\rho(\mathbf{x})$$

verifying the correctness of our solution (1.17).

1.8   Green's Theorem

If electrostatic problems always involved localized discrete or continuous distributions of charge with no boundary surfaces, the general solution (1.17) would be the most convenient and straightforward solution to any problem. There would be no need of Poisson's or Laplace's equation. In actual fact, of course, many, if not most, of the problems of electrostatics involve finite regions of space, with or without charge inside, and with prescribed boundary conditions on the bounding surfaces. These boundary conditions may be simulated by an appropriate distribution of charges outside the region of interest (perhaps at infinity), but (1.17) becomes inconvenient as a means of calculating the potential, except in simple cases (e.g., method of images).

To handle the boundary conditions it is necessary to develop some new mathematical tools, namely, the identities or theorems due to George Green (1824). These follow as simple applications of the divergence theorem. The divergence theorem:

$$\int_V \nabla \cdot \mathbf{A}\,d^3x = \oint_S \mathbf{A} \cdot \mathbf{n}\,da$$

applies to any vector field $\mathbf{A}$ defined in the volume V bounded by the closed surface S. Let $\mathbf{A} = \phi\nabla\psi$, where ϕ and ψ are arbitrary scalar fields. Now

$$\nabla \cdot (\phi\nabla\psi) = \phi\nabla^2\psi + \nabla\phi \cdot \nabla\psi \tag{1.32}$$

and

$$\phi\nabla\psi \cdot \mathbf{n} = \phi\frac{\partial\psi}{\partial n} \tag{1.33}$$

where $\partial/\partial n$ is the normal derivative at the surface S (directed outwards from inside the volume V). When (1.32) and (1.33) are substituted into the divergence theorem, there results *Green's first identity*:

$$\int_V (\phi\nabla^2\psi + \nabla\phi \cdot \nabla\psi)\,d^3x = \oint_S \phi\frac{\partial\psi}{\partial n}\,da \tag{1.34}$$

If we write down (1.34) again with ϕ and ψ interchanged, and then subtract it from (1.34), the $\nabla\phi \cdot \nabla\psi$ terms cancel, and we obtain Green's second identity or *Green's theorem*:

$$\int_V (\phi \nabla^2 \psi - \psi \nabla^2 \phi) \, d^3x = \oint_S \left[\phi \frac{\partial \psi}{\partial n} - \psi \frac{\partial \phi}{\partial n} \right] da \qquad (1.35)$$

Poisson's differential equation for the potential can be converted into an integral equation if we choose a particular ψ, namely $1/R \equiv 1/|\mathbf{x} - \mathbf{x}'|$, where $\mathbf{x}$ is the observation point and $\mathbf{x}'$ is the integration variable. Further, we put $\phi = \Phi$, the scalar potential, and make use of $\nabla^2 \Phi = -4\pi \rho$. From (1.31) we know that $\nabla^2(1/R) = -4\pi \delta(\mathbf{x} - \mathbf{x}')$, so that (1.35) becomes

$$\int_V \left[-4\pi \Phi(\mathbf{x}') \, \delta(\mathbf{x} - \mathbf{x}') + \frac{4\pi}{R} \rho(\mathbf{x}') \right] d^3x' = \oint_S \left[\Phi \frac{\partial}{\partial n'} \left(\frac{1}{R} \right) - \frac{1}{R} \frac{\partial \Phi}{\partial n'} \right] da'$$

If the point $\mathbf{x}$ lies within the volume V, we obtain:

$$\Phi(\mathbf{x}) = \int_V \frac{\rho(\mathbf{x}')}{R} \, d^3x' + \frac{1}{4\pi} \oint_S \left[\frac{1}{R} \frac{\partial \Phi}{\partial n'} - \Phi \frac{\partial}{\partial n'} \left(\frac{1}{R} \right) \right] da' \qquad (1.36)$$

If $\mathbf{x}$ lies outside the surface S, the left-hand side of (1.36) is zero. [Note that this is consistent with the interpretation of the surface integral as being the potential due to a surface-charge density $\sigma = (1/4\pi)(\partial \Phi / \partial n')$ and a dipole layer $D = -(1/4\pi)\Phi$. The discontinuities in electric field and potential (1.22) and (1.27) across the surface then lead to zero field and zero potential outside the volume V.]

Two remarks are in order about result (1.36). First, if the surface S goes to infinity and the electric field on S falls off faster than R^{-1}, then the surface integral vanishes and (1.36) reduces to the familiar result (1.17). Second, for a charge-free volume the potential anywhere inside the volume (a solution of Laplace's equation) is expressed in (1.36) in terms of the potential and its normal derivative only on the surface of the volume. This rather surprising result is not a solution to a boundary-value problem, but only an integral equation, since the specification of both Φ and $\partial \Phi / \partial n$ (*Cauchy boundary conditions*) is an overspecification of the problem. This will be discussed in detail in the next sections, where techniques yielding solutions for appropriate boundary conditions will be developed using Green's theorem (1.35).

1.9 Uniqueness of the Solution with Dirichlet or Neumann Boundary Conditions

The question arises as to what are the boundary conditions appropriate for Poisson's (or Laplace's) equation in order that a unique and well-behaved (i.e., physically reasonable) solution exist inside the bounded region. Physical experience leads us to believe that specification of the potential on a closed surface (e.g., a system of conductors held at different potentials) defines a unique potential problem. This is called a *Dirichlet problem,* or *Dirichlet boundary conditions.* Similarly it is plausible that specification of the electric field (normal derivative of the potential) everywhere on the surface (corresponding to a given surface-charge density) also defines a unique problem. Specification of the normal derivative is known as the *Neumann boundary condition.* We now proceed to prove these expectations by means of Green's first identity (1.34).

We want to show the uniqueness of the solution of Poisson's equation, $\nabla^2 \Phi = -4\pi\rho$, inside a volume V subject to either Dirichlet or Neumann boundary conditions on the closed bounding surface S. We suppose, to the contrary, that there exist two solutions Φ_1 and Φ_2 satisfying the same boundary conditions. Let

$$U = \Phi_2 - \Phi_1 \tag{1.37}$$

Then $\nabla^2 U = 0$ inside V, and $U = 0$ or $\partial U/\partial n = 0$ on S for Dirichlet and Neumann boundary conditions, respectively. From Green's first identity (1.34), with $\phi = \psi = U$, we find

$$\int_V (U\nabla^2 U + \nabla U \cdot \nabla U)\, d^3x = \oint_S U \frac{\partial U}{\partial n}\, da \tag{1.38}$$

With the specified properties of U, this reduces (for both types of boundary conditions) to:

$$\int_V |\nabla U|^2\, d^3x = 0$$

which implies $\nabla U = 0$. Consequently, inside V, U is constant. For Dirichlet boundary conditions, $U = 0$ on S so that, inside V, $\Phi_1 = \Phi_2$ and the solution is unique. Similarly, for Neumann boundary conditions, the solution is unique, apart from an unimportant arbitrary additive constant.

From the right-hand side of (1.38) it is clear that there is also a unique solution to a problem with mixed boundary conditions (i.e., Dirichlet over part of the surface S, and Neumann over the remaining part).

It should be clear that a solution to Poisson's equation with both Φ and $\partial\Phi/\partial n$ specified on a closed boundary (Cauchy boundary conditions) does not exist, since there are unique solutions for Dirichlet and Neumann conditions separately and these will in general not be consistent. The question of whether Cauchy boundary conditions on an *open* surface define a unique electrostatic problem requires more discussion than is warranted here. The reader may refer to Morse and Feshbach, Section 6.2, pp. 692–706, or to Sommerfeld, *Partial Differential Equations in Physics,* Chapter

II, for a detailed discussion of these questions. Morse and Feshbach base their treatment on the replacement of the partial differential equation by appropriate difference equations which they then solve by an iterative procedure. On the other hand, Sommerfeld bases his discussion on the method of characteristics where possible. The result of these investigations on which boundary conditions are appropriate is summarized in the table below (based on one given in Morse and Feshbach), where different types

Type of Equation

Type of Boundary Condition	Elliptic (Poisson's eq.)	Hyperbolic (wave eq.)	Parabolic (heat-conduction eq.)
Dirichlet Open surface	Not enough	Not enough	Unique, stable solution in one direction
Closed surface	Unique, stable solution	Too much	Too much
Neumann Open surface	Not enough	Not enough	Unique, stable solution in one direction
Closed surface	Unique, stable solution in general	Too much	Too much
Cauchy Open surface	Unphysical results	Unique, stable solution	Too much
Closed surface	Too much	Too much	Too much

A stable solution is one for which small changes in the boundary conditions cause appreciable changes in the solution only in the neighborhood of the boundary.

of partial differential equations and different kinds of boundary conditions are listed.

Study of the table shows that electrostatic problems are specified only by Dirichlet *or* Neumann boundary conditions on a closed surface (part or all of which may be at infinity, of course).

1.10   Formal Solution of Electrostatic Boundary-Value Problem with Green's Function

The solution of Poisson's or Laplace's equation in a finite volume V with either Dirichlet or Neumann boundary conditions on the bounding surface S can be obtained by means of Green's theorem (1.35) and so-called "Green's functions."

In obtaining result (1.36)—not a solution—we chose the function ψ to be $1/|\mathbf{x} - \mathbf{x}'|$, it being the potential of a unit point charge, satisfying the equation:

$$\nabla'^2 \left(\frac{1}{|\mathbf{x} - \mathbf{x}'|} \right) = -4\pi\delta(\mathbf{x} - \mathbf{x}') \tag{1.31}$$

The function $1/|\mathbf{x} - \mathbf{x}'|$ is only one of a class of functions depending on the variables $\mathbf{x}$ and $\mathbf{x}'$, and called *Green's functions*, which satisfy (1.31). In general,

$$\nabla'^2 G(\mathbf{x}, \mathbf{x}') = -4\pi\delta(\mathbf{x} - \mathbf{x}') \tag{1.39}$$

where

$$G(\mathbf{x}, \mathbf{x}') = \frac{1}{|\mathbf{x} - \mathbf{x}'|} + F(\mathbf{x}, \mathbf{x}') \tag{1.40}$$

with the function F satisfying Laplace's equation inside the volume V:

$$\nabla'^2 F(\mathbf{x}, \mathbf{x}') = 0 \tag{1.41}$$

In facing the problem of satisfying the prescribed boundary conditions on Φ *or* $\partial\Phi/\partial n$, we can find the key by considering result (1.36). As has been pointed out already, this is not a solution satisfying the correct type of boundary conditions because both Φ *and* $\partial\Phi/\partial n$ appear in the surface integral. It is at best an integral equation for Φ. With the generalized concept of a Green's function and its additional freedom [via the function $F(\mathbf{x}, \mathbf{x}')$], there arises the possibility that we can use Green's theorem with $\psi = G(\mathbf{x}, \mathbf{x}')$ and choose $F(\mathbf{x}, \mathbf{x}')$ to eliminate one or the other of the two surface integrals, obtaining a result which involves only Dirichlet or Neumann boundary conditions. Of course, if the necessary $G(\mathbf{x}, \mathbf{x}')$ depended in detail on the exact form of the boundary conditions, the method would have little generality. As will be seen immediately, this is not required, and $G(\mathbf{x}, \mathbf{x}')$ satisfies rather simple boundary conditions on S.

With Green's theorem (1.35), $\phi = \Phi$, $\psi = G(\mathbf{x}, \mathbf{x}')$, and the specified properties of G (1.39), it is simple to obtain the generalization of (1.36):

$$\Phi(\mathbf{x}) = \int_V \rho(\mathbf{x}') G(\mathbf{x}, \mathbf{x}') \, d^3x' + \frac{1}{4\pi} \oint_S \left[G(\mathbf{x}, \mathbf{x}') \frac{\partial\Phi}{\partial n'} - \Phi(\mathbf{x}') \frac{\partial G(\mathbf{x}, \mathbf{x}')}{\partial n'} \right] da'$$

$$\tag{1.42}$$

The freedom available in the definition of G (1.40) means that we can make the surface integral depend only on the chosen type of boundary conditions. Thus, for *Dirichlet boundary conditions* we demand:

$$G_D(\mathbf{x}, \mathbf{x}') = 0 \quad \text{for } \mathbf{x}' \text{ on } S \tag{1.43}$$

Then the first term in the surface integral in (1.42) vanishes and the solution is

$$\Phi(\mathbf{x}) = \int_V \rho(\mathbf{x}')G_D(\mathbf{x}, \mathbf{x}') \, d^3x' - \frac{1}{4\pi} \oint_S \Phi(\mathbf{x}')\frac{\partial G_D}{\partial n'} \, da' \tag{1.44}$$

For *Neumann boundary conditions* we must be more careful. The obvious choice of boundary condition on $G(\mathbf{x}, \mathbf{x}')$ seems to be

$$\frac{\partial G_N}{\partial n'}(\mathbf{x}, \mathbf{x}') = 0 \quad \text{for } \mathbf{x}' \text{ on } S$$

since that makes the second term in the surface integral in (1.42) vanish, as desired. But an application of Gauss's theorem to (1.39) shows that

$$\oint_S \frac{\partial G}{\partial n'} \, da' = -4\pi$$

Consequently the simplest allowable boundary condition on G_N is

$$\frac{\partial G_N}{\partial n'}(\mathbf{x}, \mathbf{x}') = -\frac{4\pi}{S} \quad \text{for } \mathbf{x}' \text{ on } S \tag{1.45}$$

where S is the total area of the boundary surface. Then the solution is

$$\Phi(\mathbf{x}) = \langle\Phi\rangle_S + \int_V \rho(\mathbf{x}')G_N(\mathbf{x}, \mathbf{x}') \, d^3x' + \frac{1}{4\pi} \oint_S \frac{\partial \Phi}{\partial n'} G_N \, da' \tag{1.46}$$

where $\langle\Phi\rangle_S$ is the average value of the potential over the whole surface. The customary Neumann problem is the so-called "exterior problem" in which the volume V is bounded by two surfaces, one closed and finite, the other at infinity. Then the surface area S is infinite; the boundary condition (1.45) becomes homogeneous; the average value $\langle\Phi\rangle_S$ vanishes.

We note that the Green's functions satisfy simple boundary conditions (1.43) or (1.45) which do not depend on the detailed form of the Dirichlet (or Neumann) boundary values. Even so, it is often rather involved (if not impossible) to determine $G(\mathbf{x}, \mathbf{x}')$ because of its dependence on the shape of the surface S. We will encounter such problems in Chapter 2 and 3.

The mathematical symmetry property $G(\mathbf{x}, \mathbf{x}') = G(\mathbf{x}', \mathbf{x})$ can be proved for the Green's functions satisfying the Dirichlet boundary condition (1.43) by means of Green's theorem with $\phi = G(\mathbf{x}, \mathbf{y})$ and $\psi = G(\mathbf{x}', \mathbf{y})$, where $\mathbf{y}$ is the integration variable. Since the Green's function, as a function of one of its variables, is a potential due to a unit point charge, this symmetry merely represents the physical interchangeability of the source and the observation points. For Neumann boundary conditions the symmetry is not automatic, but can be imposed as a separate requirement.

As a final, important remark we note the physical meaning of $F(\mathbf{x}, \mathbf{x}')$. It is a solution of Laplace's equation inside V and so represents the potential of a system of charges external to the volume V. It can be thought of as the potential due to an external distribution of charges so chosen as to satisfy the homogeneous boundary conditions of zero potential (or zero normal derivative) on the surface S when combined with the potential of a point charge at the source point $\mathbf{x}'$. Since the potential at a point $\mathbf{x}$ on the surface due to the point charge depends on the position of the source point, the external distribution of charge $F(\mathbf{x}, \mathbf{x}')$ must also depend on the "parameter" $\mathbf{x}'$. From this point of view, we see that the method of images (to be discussed in Chapter 2) is a physical equivalent of the determination of the appropriate $F(\mathbf{x}, \mathbf{x}')$ to satisfy the boundary conditions (1.43) or (1.45). For the Dirichlet problem with conductors, $F(\mathbf{x}, \mathbf{x}')$ can also be interpreted as the potential due to the surface-charge distribution induced on the conductors by the presence of a point charge at the source point $\mathbf{x}'$.

1.11 Electrostatic Potential Energy and Energy Density

In Section 1.5 it was shown that the product of the scalar potential and the charge of a point object could be interpreted as potential energy. More precisely, if a point charge q_i is brought from infinity to a point $\mathbf{x}_i$ in a region of localized electric fields described by the scalar potential Φ (which vanishes at infinity), the work done on the charge (and hence its potential energy) is given by

$$W_i = q_i \Phi(\mathbf{x}_i) \tag{1.47}$$

The potential Φ can be viewed as produced by an array of $(n-1)$ charges $q_j (j = 1, 2, \ldots, n-1)$ at positions $\mathbf{x}_j$. Then

$$\Phi(\mathbf{x}_i) = \sum_{j=1}^{n-1} \frac{q_j}{|\mathbf{x}_i - \mathbf{x}_j|} \tag{1.48}$$

so that the potential energy of the charge q_i is

$$W_i = q_i \sum_{j=1}^{n-1} \frac{q_j}{|\mathbf{x}_i - \mathbf{x}_j|} \tag{1.49}$$

It is clear that the *total* potential energy of all the charges due to all the forces acting between them is:

$$W = \sum_{i=1}^{n} \sum_{j<i} \frac{q_i q_j}{|\mathbf{x}_i - \mathbf{x}_j|} \qquad (1.50)$$

as can be seen most easily by adding each charge in succession. A more symmetric form can be written by summing over i and j unrestricted, and then dividing by 2:

$$W = \frac{1}{2} \sum_{i} \sum_{j} \frac{q_i q_j}{|\mathbf{x}_i - \mathbf{x}_j|} \qquad (1.51)$$

It is understood that $i = j$ terms (infinite "self-energy" terms) are omitted in the double sum.

For a continuous charge distribution [or, in general, using the Dirac delta functions (1.6)] the potential energy takes the form:

$$W = \frac{1}{2} \int \int \frac{\rho(\mathbf{x})\rho(\mathbf{x}')}{|\mathbf{x} - \mathbf{x}'|} \, d^3x \, d^3x' \qquad (1.52)$$

Another expression, equivalent to (1.52), can be obtained by noting that one of the integrals in (1.52) is just the scalar potential (1.17). Therefore

$$W = \frac{1}{2} \int \rho(\mathbf{x}) \Phi(\mathbf{x}) \, d^3x \qquad (1.53)$$

Equations (1.51), (1.52), and (1.53) express the electrostatic potential energy in terms of the positions of the charges and so emphasize the interactions between charges via Coulomb forces. An alternative, and very fruitful, approach is to emphasize the electric field and to interpret the energy as being stored in the electric field surrounding the charges. To obtain this latter form, we make use of Poisson's equation to eliminate the charge density from (1.53):

$$W = \frac{-1}{8\pi} \int \Phi \nabla^2 \Phi \, d^3x$$

Integration by parts leads to the result:

$$W = \frac{1}{8\pi} \int |\nabla \Phi|^2 \, d^3x = \frac{1}{8\pi} \int |\mathbf{E}|^2 \, d^3x \qquad (1.54)$$

where the integration is over all space. In (1.54) all explicit reference to charges has gone, and the energy is expressed as an integral of the square of the electric field over all space. This leads naturally to the identification of the integrand as an energy density w:

$$w = \frac{1}{8\pi} |\mathbf{E}|^2 \qquad (1.55)$$

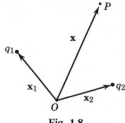

Fig. 1.8

This expression for energy density is intuitively reasonable, since regions of high fields "must" contain considerable energy.

There is perhaps one puzzling thing about (1.55). The energy density is positive definite. Consequently its volume integral is necessarily non-negative. This seems to contradict our impression from (1.51) that the potential energy of two charges of opposite sign is negative. The reason for this apparent contradiction is that (1.54) and (1.55) contain "self-energy" contributions to the energy density, whereas the double sum in (1.51) does not. To illustrate this, consider two point charges q_1 and q_2 located at $\mathbf{x}_1$ and $\mathbf{x}_2$, as in Fig. 1.8. The electric field at the point P with coordinate $\mathbf{x}$ is

$$\mathbf{E} = \frac{q_1(\mathbf{x} - \mathbf{x}_1)}{|\mathbf{x} - \mathbf{x}_1|^3} + \frac{q_2(\mathbf{x} - \mathbf{x}_2)}{|\mathbf{x} - \mathbf{x}_2|^3}$$

so that the energy density (1.55) is

$$w = \frac{q_1^{\,2}}{8\pi|\mathbf{x} - \mathbf{x}_1|^4} + \frac{q_2^{\,2}}{8\pi|\mathbf{x} - \mathbf{x}_2|^4} + \frac{q_1 q_2(\mathbf{x} - \mathbf{x}_1) \cdot (\mathbf{x} - \mathbf{x}_2)}{4\pi|\mathbf{x} - \mathbf{x}_1|^3 \, |\mathbf{x} - \mathbf{x}_2|^3} \quad (1.56)$$

Clearly the first two terms are self-energy contributions. To show that the third term gives the proper result for the interaction potential energy we integrate over all space:

$$W_{\text{int}} = \frac{q_1 q_2}{4\pi} \int \frac{(\mathbf{x} - \mathbf{x}_1) \cdot (\mathbf{x} - \mathbf{x}_2)}{|\mathbf{x} - \mathbf{x}_1|^3 \, |\mathbf{x} - \mathbf{x}_2|^3} \, d^3x \quad (1.57)$$

A change of integration variable to $\boldsymbol{\rho} = (\mathbf{x} - \mathbf{x}_1)/|\mathbf{x}_1 - \mathbf{x}_2|$ yields

$$W_{\text{int}} = \frac{q_1 q_2}{|\mathbf{x}_1 - \mathbf{x}_2|} \times \frac{1}{4\pi} \int \frac{\boldsymbol{\rho} \cdot (\boldsymbol{\rho} + \mathbf{n})}{\rho^3 |\boldsymbol{\rho} + \mathbf{n}|^3} \, d^3\rho \quad (1.58)$$

where $\mathbf{n}$ is a unit vector in the direction $(\mathbf{x}_1 - \mathbf{x}_2)$. By straightforward integration the dimensionless volume integral can be shown to have the value 4π, so that the interaction energy reduces to the expected value.

Forces acting between charged bodies can be obtained by calculating the change in the total electrostatic energy of the system under small virtual displacements. Examples of this are discussed in the problems. Care must be taken to exhibit the energy in a form showing clearly those factors which vary with a change in configuration and those which are kept constant.

As a simple illustration we calculate the force per unit area on the surface of a conductor with a surface-charge density $\sigma(\mathbf{x})$. In the immediate neighborhood of the surface the energy density is

$$w = \frac{1}{8\pi} |\mathbf{E}|^2 = 2\pi\sigma^2 \qquad (1.59)$$

If we now imagine a small outward displacement Δx of an elemental area Δa of the conducting surface, the electrostatic energy decreases by an amount which is the product of energy density w and the excluded volume $\Delta x\, \Delta a$:

$$\Delta W = -2\pi\sigma^2 \Delta a\, \Delta x \qquad (1.60)$$

This means that there is an outward force per unit area equal to $2\pi\sigma^2 = w$ at the surface of the conductor. This result is normally derived by taking the product of the surface-charge density and the electric field, with care taken to eliminate the electric field due to the element of surface-charge density itself.

REFERENCES AND SUGGESTED READING

On the mathematical side, the subject of delta functions is treated simply but rigorously by
   Lighthill.
For a discussion of different types of partial differential equations and the appropriate boundary conditions for each type, see
   Morse and Feshbach, Chapter 6,
   Sommerfeld, *Partial Differential Equations in Physics*, Chapter II,
   Courant and Hilbert, Vol. II, Chapters III-VI.
The general theory of Green's functions is treated in detail by
   Friedman, Chapter 3,
   Morse and Feshbach, Chapter 7.
The general theory of electrostatics is discussed extensively in many of the older books. Notable, in spite of some old-fashioned notation, are
   Maxwell, Vol. 1, Chapters II and IV,
   Jeans, Chapters II, VI, VII.
Of more recent books, mention may be made of the treatment of the general theory by
   Stratton, Chapter III, and parts of Chapter II.

PROBLEMS

1.1   Use Gauss's theorem to prove the following statements:
   (*a*) Any excess charge placed on a conductor must lie entirely on its surface. (A conductor by definition contains charges capable of moving freely under the action of applied electric fields.)

(*b*) A closed, hollow conductor shields its interior from fields due to charges outside, but does not shield its exterior from the fields due to charges placed inside it.

(*c*) The electric field at the surface of a conductor is normal to the surface and has a magnitude $4\pi\sigma$, where σ is the charge density per unit area on the surface.

1.2 Two infinite, conducting, plane sheets of uniform thicknesses t_1 and t_2, respectively, are placed parallel to one another with their adjacent faces separated by a distance L. The first sheet has a total charge per unit area (sum of the surface-charge densities on either side) equal to q_1, while the second has q_2. Use symmetry arguments and Gauss's law to prove that

(*a*) the surface-charge densities on the adjacent faces are equal and opposite;

(*b*) the surface-charge densities on the outer faces of the two sheets are the same;

(*c*) the magnitudes of the charge densities and the fields produced are independent of the thicknesses t_1 and t_2 and the separation L.

Find the surface-charge densities and fields explicitly in terms of q_1 and q_2, and apply your results to the special case $q_1 = -q_2 = Q$.

1.3 Each of three charged spheres of radius a, one conducting, one having a uniform charge density within its volume, and one having a spherically symmetric charge density which varies radially as r^n ($n > -3$), has a total charge Q. Use Gauss's theorem to obtain the electric fields both inside and outside each sphere. Sketch the behavior of the fields as a function of radius for the first two spheres, and for the third with $n = -2, +2$.

1.4 The time-average potential of a neutral hydrogen atom is given by

$$\Phi = q\,\frac{e^{-\alpha r}}{r}\left(1 + \frac{\alpha r}{2}\right)$$

where q is the magnitude of the electronic charge, and $\alpha^{-1} = a_0/2$. Find the distribution of charge (both continuous and discrete) which will give this potential and interpret your result physically.

1.5 A simple capacitor is a device formed by two insulated conductors adjacent to each other. If equal and opposite charges are placed on the conductors, there will be a certain difference of potential between them. The ratio of the magnitude of the charge on one conductor to the magnitude of the potential difference is called the capacitance (in electrostatic units it is measured in centimeters). Using Gauss's law, calculate the capacitance of

(*a*) two large, flat, conducting sheets of area A, separated by a small distance d;

(*b*) two concentric conducting spheres with radii a, b ($b > a$);

(*c*) two concentric conducting cylinders of length L, large compared to their radii a, b ($b > a$).

(*d*) What is the inner diameter of the outer conductor in an air-filled coaxial cable whose center conductor is B&S #20 gauge wire and whose capacitance is 0.5 micromicrofarad/cm? 0.05 micromicrofarad/cm?

1.6 Two long, cylindrical conductors of radii a_1 and a_2 are parallel and separated by a distance d which is large compared with either radius.

Show that the capacitance per unit length is given approximately by

$$C \simeq \left(4 \ln \frac{d}{a}\right)^{-1}$$

where a is the geometrical mean of the two radii.

Approximately what B&S gauge wire (state diameter in millimeters as well as gauge) would be necessary to make a two-wire transmission line with a capacitance of 0.1 $\mu\mu f/cm$ if the separation of the wires was 0.5 cm? 1.5 cm? 5.0 cm?

1.7 (a) For the three capacitor geometries in Problem 1.5 calculate the total electrostatic energy and express it alternatively in terms of the equal and opposite charges Q and $-Q$ placed on the conductors *and* the potential difference between them.

(b) Sketch the energy density of the electrostatic field in each case as a function of the appropriate linear coordinate.

1.8 Calculate the attractive force between conductors in the parallel plate capacitor (Problem 1.5a) and the parallel cylinder capacitor (Problem 1.6) for

(a) fixed charges on each conductor;

(b) fixed potential difference between conductors.

1.9 Prove the *mean value theorem*: For charge-free space the value of the electrostatic potential at any point is equal to the average of the potential over the surface of *any* sphere centered on that point.

1.10 Use Gauss's theorem to prove that at the surface of a curved charged conductor the normal derivative of the electric field is given by

$$\frac{1}{E}\frac{\partial E}{\partial n} = -\left(\frac{1}{R_1} + \frac{1}{R_2}\right)$$

where R_1 and R_2 are the principal radii of curvature of the surface.

1.11 Prove *Green's reciprocation theorem*: If Φ is the potential due to à volume-charge density ρ within a volume V and a surface-charge density σ on the surface S bounding the volume V, while Φ' is the potential due to another charge distribution ρ' and σ', then

$$\int_V \rho\Phi' \, d^3x + \int_S \sigma\Phi' \, da = \int_V \rho'\Phi \, d^3x + \int_S \sigma'\Phi \, da$$

1.12 Prove *Thomson's theorem*: If a number of conducting surfaces are fixed in position and a given total charge is placed on each surface, then the electrostatic energy in the region bounded by the surfaces is a minimum when the charges are placed so that every surface is an equipotential.

1.13 Prove the following theorem: If a number of conducting surfaces are fixed in position with a given total charge on each, the introduction of an uncharged, insulated conductor into the region bounded by the surfaces lowers the electrostatic energy.

2

Boundary-Value Problems
in Electrostatics: I

Many problems in electrostatics involve boundary surfaces on which either the potential or the surface-charge density is specified. The formal solution of such problems was presented in Section 1.10, using the method of Green's functions. In practical situations (or even rather idealized approximations to practical situations) the discovery of the correct Green's function is sometimes easy and sometimes not. Consequently a number of approaches to electrostatic boundary-value problems have been developed, some of which are only remotely connected to the Green's function method. In this chapter we will examine two of these special techniques: (1) the method of images, which is closely related to the use of Green's functions; (2) expansion in orthogonal functions, an approach directly through the differential equation and rather remote from the direct construction of a Green's function. Other methods of attack, such as the use of conformal mapping in two-dimensional problems, will be omitted. For a discussion of conformal mapping the interested reader may refer to the references cited at the end of the chapter.

2.1 Method of Images

The method of images concerns itself with the problem of one or more point charges in the presence of boundary surfaces, e.g., conductors either grounded or held at fixed potentials. Under favorable conditions it is possible to infer from the geometry of the situation that a small number of suitably placed charges of appropriate magnitudes, external to the region of interest, can simulate the required boundary conditions. These charges

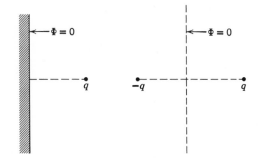

Fig. 2.1 Solution by method of images. The original potential problem is on the left, the equivalent-image problem on the right.

are called *image charges*, and the replacement of the actual problem with boundaries by an enlarged region with image charges but no boundaries is called the *method of images*. The image charges must be external to the volume of interest, since their potentials must be solutions of Laplace's equation inside the volume; the "particular integral" (i.e., solution of Poisson's equation) is provided by the sum of the potentials of the charges inside the volume.

A simple example is a point charge located in front of an infinite plane conductor at zero potential, as shown in Fig. 2.1. It is clear that this is equivalent to the problem of the original charge and an equal and opposite charge located at the mirror-image point behind the plane defined by the position of the conductor.

2.2 Point Charge in the Presence of a Grounded Conducting Sphere

As an illustration of the method of images we consider the problem illustrated in Fig. 2.2 of a point charge q located at $\mathbf{y}$ relative to the origin around which is centered a grounded conducting sphere of radius a.* We seek the potential $\Phi(\mathbf{x})$ such that $\Phi(|\mathbf{x}| = a) = 0$. By symmetry it is evident that the image charge q' (assuming that only one image is needed) will lie on the ray from the origin to the charge q. If we consider the charge q *outside* the sphere, the image position $\mathbf{y}'$ will lie inside the sphere. The

---

* The term *grounded* is used to imply that the surface or object is held at the same potential as the point at infinity by means of some fine conducting connector. The connection is assumed not to disturb the potential distribution. But arbitrary amounts of charge of either sign can flow onto the object from infinity in order to maintain its potential at "ground" (usually taken to be zero potential). A conductor held at a fixed potential is essentially the same situation, except that a voltage source is interposed between the object and "ground."

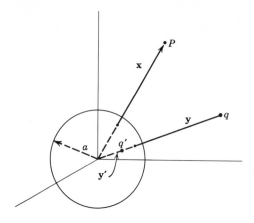

Fig. 2.2 Conducting sphere of radius *a*, with charge *q* and image charge *q'*.

potential due to the charges q and q' is:

$$\Phi(\mathbf{x}) = \frac{q}{|\mathbf{x} - \mathbf{y}|} + \frac{q'}{|\mathbf{x} - \mathbf{y}'|} \tag{2.1}$$

We now must try to choose q' and $|\mathbf{y}'|$ such that this potential vanishes at $|\mathbf{x}| = a$. If $\mathbf{n}$ is a unit vector in the direction $\mathbf{x}$, and $\mathbf{n}'$ a unit vector in the direction $\mathbf{y}$, then

$$\Phi(\mathbf{x}) = \frac{q}{|x\mathbf{n} - y\mathbf{n}'|} + \frac{q'}{|x\mathbf{n} - y'\mathbf{n}'|} \tag{2.2}$$

If x is factored out of the first term and y' out of the second, the potential at $x = a$ becomes:

$$\Phi(x = a) = \frac{q}{a\left|\mathbf{n} - \dfrac{y}{a}\mathbf{n}'\right|} + \frac{q'}{y'\left|\mathbf{n}' - \dfrac{a}{y'}\mathbf{n}\right|} \tag{2.3}$$

From the form of (2.3) it will be seen that the choices:

$$\frac{q}{a} = -\frac{q'}{y'}, \qquad \frac{y}{a} = \frac{a}{y'}$$

make $\Phi(x = a) = 0$, for all possible values of $\mathbf{n} \cdot \mathbf{n}'$. Hence the magnitude and position of the image charge are

$$q' = -\frac{a}{y}q, \qquad y' = \frac{a^2}{y} \tag{2.4}$$

We note that, as the charge q is brought closer to the sphere, the image charge grows in magnitude and moves out from the center of the sphere. When q is just outside the surface of the sphere, the image charge is equal and opposite in magnitude and lies just beneath the surface.

Now that the image charge has been found, we can return to the original problem of a charge q outside a grounded conducting sphere and consider various effects. The actual charge density induced on the surface of the sphere can be calculated from the normal component of Φ at the surface:

$$\sigma = -\frac{1}{4\pi}\frac{\partial\Phi}{\partial x}\bigg|_{x=a} = -\frac{q}{4\pi a^2}\left(\frac{a}{y}\right)\frac{\left(1 - \dfrac{a^2}{y^2}\right)}{\left(1 + \dfrac{a^2}{y^2} - 2\dfrac{a}{y}\cos\gamma\right)^{3/2}} \qquad (2.5)$$

where γ is the angle between **x** and **y**. This charge density in units of $-q/4\pi a^2$ is shown plotted in Fig. 2.3 as a function of γ for two values of y/a. The concentration of charge in the direction of the point charge q is evident, especially for $y/a = 2$. It is easy to show by direct integration that the total induced charge on the sphere is equal to the magnitude of the image charge, as it must according to Gauss's law.

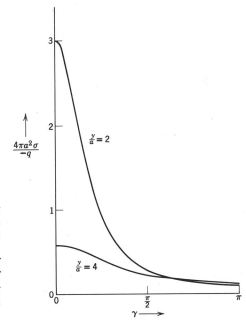

Fig. 2.3 Surface-charge density σ induced on the grounded sphere of radius a due to the presence of a point charge q located a distance y away from the center of the sphere. σ is plotted in units of $-q/4\pi a^2$ as function of the angular position γ away from the radius to the charge for $y = 2a, 4a$.

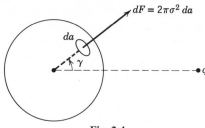

Fig. 2.4

The force acting on the charge q can be calculated in different ways. One (the easiest) way is to write down immediately the force between the charge q and the image charge q'. The distance between them is $y - y' = y(1 - a^2/y^2)$. Hence the attractive force, according to Coulomb's law, is:

$$|\mathbf{F}| = \frac{q^2}{a^2}\left(\frac{a}{y}\right)^3\left(1 - \frac{a^2}{y^2}\right)^{-2} \tag{2.6}$$

For large separations the force is an inverse cube law, but close to the sphere it is proportional to the inverse square of the distance away from the surface of the sphere.

The alternative method for obtaining the force is to calculate the total force acting on the surface of the sphere. The force on each element of area da is $2\pi\sigma^2\,da$, where σ is given by (2.5), as indicated in Fig. 2.4. But from symmetry it is clear that only the component parallel to the radius vector from the center of the sphere to q contributes to the total force. Hence the total force acting on the sphere (equal and opposite to the force acting on q) is given by the integral:

$$|\mathbf{F}| = \frac{q^2}{8\pi a^2}\left(\frac{a}{y}\right)^2\left(1 - \frac{a^2}{y^2}\right)^2 \int \frac{\cos\gamma}{\left(1 + \dfrac{a^2}{y^2} - \dfrac{2a}{y}\cos\gamma\right)^3}\,d\Omega \tag{2.7}$$

Integration immediately yields (2.6).

The whole discussion has been based on the understanding that the point charge q is *outside* the sphere. Actually, the results apply equally for the charge q *inside* the sphere. The only change necessary is in the surface-charge density (2.5), where the normal derivative out of the conductor is now radially inwards, implying a change in sign. The reader may transcribe all the formulas, remembering that now $y \le a$. The angular distributions of surface charge are similar to those of Fig. 2.3, but the total induced surface charge is evidently equal to $-q$, independent of y.

2.3   Point Charge in the Presence of a Charged, Insulated, Conducting Sphere

In the previous section we considered the problem of a point charge q near a grounded sphere and saw that a surface-charge density was induced on the sphere. This charge was of total amount $q' = -aq/y$, and was distributed over the surface in such a way as to be in equilibrium under all forces acting.

If we wish to consider the problem of an insulated conducting sphere with total charge Q in the presence of a point charge q, we can build up the solution for the potential by linear superposition. In an operational sense, we can imagine that we start with the grounded conducting sphere (with its charge q' distributed over its surface). We then disconnect the ground wire and add to the sphere an amount of charge $(Q - q')$. This brings the total charge on the sphere up to Q. To find the potential we merely note that the added charge $(Q - q')$ will distribute itself *uniformly* over the surface, since the electrostatic forces due to the point charge q are already balanced by the charge q'. Hence the potential due to the added charge $(Q - q')$ will be the same as if a point charge of that magnitude were at the origin, at least for points outside the sphere.

The potential is the superposition of (2.1) and the potential of a point charge $(Q - q')$ at the origin:

$$\Phi(\mathbf{x}) = \frac{q}{|\mathbf{x} - \mathbf{y}|} - \frac{aq}{y\left|\mathbf{x} - \frac{a^2}{y^2}\mathbf{y}\right|} + \frac{Q + \frac{a}{y}q}{|\mathbf{x}|} \qquad (2.8)$$

The force acting on the charge q can be written down directly from Coulomb's law. It is directed along the radius vector to q and has the magnitude:

$$\mathbf{F} = \frac{q}{y^2}\left[Q - \frac{qa^3(2y^2 - a^2)}{y(y^2 - a^2)^2}\right]\frac{\mathbf{y}}{y} \qquad (2.9)$$

In the limit of $y \gg a$, the force reduces to the usual Coulomb's law for two small charged bodies. But close to the sphere the force is modified because of the induced charge distribution on the surface of the sphere. Figure 2.5 shows the force as a function of distance for various ratios of Q/q. The force is expressed in units of q^2/y^2; positive (negative) values correspond to a repulsion (attraction). If the sphere is charged oppositely to q, or is

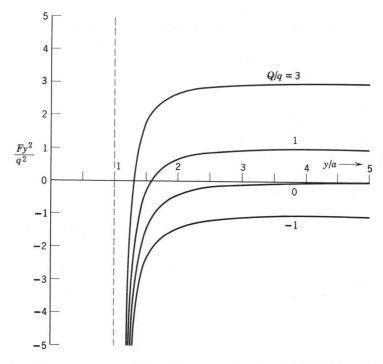

Fig. 2.5 The force on a point charge q due to an insulated, conducting sphere of radius a carrying a total charge Q. Positive values mean a repulsion, negative an attraction. The asymptotic dependence of the force has been divided out. Fy^2/q^2 is plotted versus y/a for $Q/q = -1, 0, 1, 3$. Regardless of the value of Q, the force is always attractive at close distances because of the induced surface charge.

uncharged, the force is attractive at all distances. Even if the charge Q is the same sign as q, however, the force becomes attractive at very close distances. In the limit of $Q \gg q$, the point of zero force (unstable equilibrium point) is very close to the sphere, namely, at $y \simeq a(1 + \frac{1}{2}\sqrt{q/Q})$. Note that the asymptotic value of the force is attained as soon as the charge q is more than a few radii away from the sphere.

This example exhibits a general property which explains why an excess of charge on the surface does not immediately leave the surface because of mutual repulsion of the individual charges. As soon as an element of charge is removed from the surface, the image force tends to attract it back. If sufficient work is done, of course, charge can be removed from the surface to infinity. The work function of a metal is in large part just the work done against the attractive image force in order to remove an electron from the surface.

2.4 Point Charge near a Conducting Sphere at Fixed Potential

Another problem which can be discussed easily is that of a point charge near a conducting sphere held at a fixed potential V. The potential is the same as for the charged sphere, except that the charge $(Q - q')$ at the center is replaced by a charge (Va). This can be seen from (2.8), since at $|\mathbf{x}| = a$ the first two terms cancel and the last term will be equal to V as required. Thus the potential is

$$\Phi(\mathbf{x}) = \frac{q}{|\mathbf{x} - \mathbf{y}|} - \frac{aq}{y\left|\mathbf{x} - \dfrac{a^2}{y^2}\mathbf{y}\right|} + \frac{Va}{|\mathbf{x}|} \tag{2.10}$$

The force on the charge q due to the sphere at fixed potential is

$$\mathbf{F} = \frac{q}{y^2}\left[Va - \frac{qay^3}{(y^2 - a^2)^2}\right]\frac{\mathbf{y}}{y} \tag{2.11}$$

For corresponding values of Va/q and Q/q this force is very similar to that of the charged sphere, shown in Fig. 2.5, although the approach to the asymptotic value (Vaq/y^2) is more gradual. For $Va \gg q$, the unstable equilibrium point has the equivalent location $y \simeq a(1 + \frac{1}{2}\sqrt{q/Va})$.

2.5 Conducting Sphere in a Uniform Electric Field by Method of Images

As a final example of the method of images we consider a conducting sphere of radius a in a uniform electric field E_0. A uniform field can be thought of as being produced by appropriate positive and negative charges at infinity. For example, if there are two charges $\pm Q$, located at positions $z = \mp R$, as shown in Fig. 2.6a, then in a region near the origin whose dimensions are very small compared to R there is an approximately constant electric field $E_0 \simeq 2Q/R^2$ parallel to the z axis. In the limit as $R, Q \to \infty$, with Q/R^2 constant, this approximation becomes exact.

If now a conducting sphere of radius a is placed at the origin, the potential will be that due to the charges $\pm Q$ at $\mp R$ and their images $\mp Qa/R$ at $z = \mp a^2/R$:

$$\Phi = \frac{Q}{(r^2 + R^2 + 2rR\cos\theta)^{\frac{1}{2}}} - \frac{Q}{(r^2 + R^2 - 2rR\cos\theta)^{\frac{1}{2}}}$$
$$- \frac{aQ}{R\left(r^2 + \dfrac{a^4}{R^2} + \dfrac{2a^2 r}{R}\cos\theta\right)^{\frac{1}{2}}} + \frac{aQ}{R\left(r^2 + \dfrac{a^4}{R^2} - \dfrac{2a^2 r}{R}\cos\theta\right)^{\frac{1}{2}}} \tag{2.12}$$

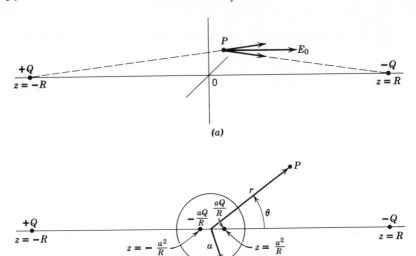

Fig. 2.6 Conducting sphere in a uniform electric field by the method of images.

where Φ has been expressed in terms of the spherical coordinates of the observation point. In the first two terms R is much larger than r by assumption. Hence we can expand the radicals after factoring out R^2. Similarly, in the third and fourth terms, we can factor out r^2 and then expand. The result is:

$$\Phi = \left[-\frac{2Q}{R^2} r \cos\theta + \frac{2Q}{R^2} \frac{a^3}{r^2} \cos\theta \right] + \cdots \qquad (2.13)$$

where the omitted terms vanish in the limit $R \to \infty$. In that limit $2Q/R^2$ becomes the applied uniform field, so that the potential is

$$\Phi = -E_0 \left(r - \frac{a^3}{r^2} \right) \cos\theta \qquad (2.14)$$

The first term $(-E_0 z)$ is, of course, just the potential of a uniform field E_0 which could have been written down directly instead of the first two terms in (2.12). The second is the potential due to the induced surface charge density or, equivalently, the image charges. Note that the image charges form a dipole of strength $D = Qa/R \times 2a^2/R = E_0 a^3$. The induced surface-charge density is

$$\sigma = -\frac{1}{4\pi} \frac{\partial\Phi}{\partial r}\bigg|_{r=a} = \frac{3}{4\pi} E_0 \cos\theta \qquad (2.15)$$

We note that the surface integral of this charge density vanishes, so that there is no difference between a grounded and an insulated sphere.

2.6 Method of Inversion

The method of images for a sphere and related topics discussed in the previous sections suggest that there is some sort of equivalence of solutions of potential problems under the reciprocal radius transformation,

$$r \rightarrow r' = \frac{a^2}{r} \qquad (2.16)$$

This equivalence forms the basis of the method of inversion, and transformation (2.16) is called *inversion in a sphere*. The radius of the sphere is called the *radius of inversion*, and the center of the sphere, the *center of inversion*. The mathematical equivalence is contained in the following theorem:

Let $\Phi(r, \theta, \phi)$ be the potential due to a set of point charges q_i at the points (r_i, θ_i, ϕ_i). Then the potential

$$\Phi'(r, \theta, \phi) = \frac{a}{r} \, \Phi\left(\frac{a^2}{r}, \theta, \phi\right) \qquad (2.17)$$

is the potential due to charges,

$$q_i' = \frac{a}{r_i} q_i \qquad (2.18)$$

located at the points $(a^2/r_i, \theta_i, \phi_i)$.

The proof of the theorem is as follows. The potential $\Phi(r, \theta, \phi)$ can be written as

$$\Phi = \sum_i \frac{q_i}{\sqrt{r^2 + r_i^2 - 2rr_i \cos \gamma_i}}$$

where γ_i is the angle between the radius vectors $\mathbf{x}$ and $\mathbf{x}_i$. Under transformation (2.16) the angles remain unchanged. Consequently the new potential Φ' is

$$\Phi'(r, \theta, \phi) = \frac{a}{r} \sum_i \frac{q_i}{\sqrt{\dfrac{a^4}{r^2} + r_i^2 - \dfrac{2a^2}{r} r_i \cos \gamma_i}}$$

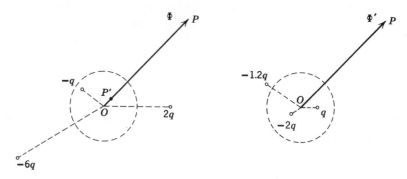

Fig. 2.7

By factoring $(r_i{}^2/r^2)$ out of the square root, this can be written

$$\Phi'(r, \theta, \phi) = \sum_i \frac{\left(\dfrac{aq_i}{r_i}\right)}{\sqrt{r^2 + \dfrac{a^4}{r_i{}^2} - 2r\,\dfrac{a^2}{r_i}\cos\gamma_i}}$$

This proves the theorem.

Figure 2.7 shows a simple configuration of charges before and after inversion. The potential Φ' at the point P due to the inverted distribution of charge is related by (2.17) to the original potential Φ at the point P' in the figure.

The inversion theorem has been stated and proved with discrete charges. It is left as an exercise for the reader to show that, if the potential Φ satisfies Poisson's equation,

$$\nabla^2\Phi = -4\pi\rho$$

the new potential Φ' (2.17) also satisfies Poisson's equation,

$$\nabla^2\Phi'(r, \theta, \phi) = -4\pi\rho'(r, \theta, \phi) \tag{2.19}$$

where the new charge density is given by

$$\rho'(r, \theta, \phi) = \left(\frac{a}{r}\right)^5 \rho\left(\frac{a^2}{r}, \theta, \phi\right) \tag{2.20}$$

The connection between this transformation law for charge densities and the law (2.18) for point charges can be established by considering the charge density as a sum of delta functions:

$$\rho(\mathbf{x}) = \sum_i q_i \delta(\mathbf{x} - \mathbf{x}_i)$$

In terms of spherical coordinates centered at the center of inversion the charge density can be written

$$\rho(r, \theta, \phi) = \sum_i q_i \delta(\Omega - \Omega_i) \frac{1}{r_i^2} \delta(r - r_i)$$

where $\delta(\Omega - \Omega_i)$ is the angular delta function whose integral over solid angle gives unity, and $\delta(r - r_i)$ is the radial delta function.* Under inversion the angular factor is unchanged. Consequently we have

$$\rho\left(\frac{a^2}{r}, \theta, \phi\right) = \sum_i q_i \delta(\Omega - \Omega_i) \frac{1}{r_i^2} \delta\left(\frac{a^2}{r} - r_i\right)$$

The radial delta function can be transformed according to rule 5 at the end of Section 1.2 as

$$\delta\left(\frac{a^2}{r} - r_i\right) = \frac{r^2}{a^2} \delta\left(r - \frac{a^2}{r_i}\right) = \frac{a^2}{r_i^2} \delta\left(r - \frac{a^2}{r_i}\right)$$

Then

$$\rho\left(\frac{a^2}{r}, \theta, \phi\right) = \sum_i q_i \delta(\Omega - \Omega_i) \frac{a^6}{r_i^6} \frac{\delta\left(r - \frac{a^2}{r_i}\right)}{\left(\frac{a^2}{r_i}\right)^2}$$

and the inverted charge density (2.20) becomes

$$\rho'(r, \theta, \phi) = \frac{a^5}{r^5} \sum_i q_i \left(\frac{a}{r_i}\right)^6 \delta(\mathbf{x} - \mathbf{x}_i') = \sum_i q_i' \delta(\mathbf{x} - \mathbf{x}_i')$$

where $\mathbf{x}_i' = (a^2/r_i, \theta, \phi)$ and $q_i' = (a/r_i)q_i$, as required by (2.18).

With the transformation laws for charges and volume-charge densities given by (2.18) and (2.20), it will not come as a great surprise that the transformation of *surface*-charge densities is according to

$$\sigma'(r, \theta, \phi) = \left(\frac{a}{r}\right)^3 \sigma\left(\frac{a^2}{r}, \theta, \phi\right) \tag{2.21}$$

Before treating any examples of inversion there are one or two physical and geometrical points which need discussion. First, in regard to the physical points, if the original potential problem is one where there are conducting surfaces at fixed potentials, the inverted problem will not in general involve the inversions of those surfaces held at fixed potentials. This is evident from (2.17), where the factor a/r shows that even if Φ is constant on the original surface the potential Φ' on the inverted surface is

* The factor r_i^{-2} multiplying the radial delta function is present to cancel out the r^2 which appears in the volume element $d^3x = r^2\, dr\, d\Omega$.

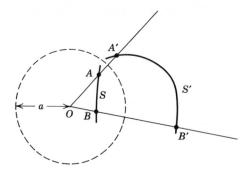

Fig. 2.8 Geometry of inversion. Center of inversion is at 0. Radius of inversion is a. The inversion of the surface S is the surface S', and vice versa.

not. The only exception occurs when Φ vanishes on some surface. Then Φ' also vanishes on the inverted surface.

One might think that, since Φ is arbitrary to the extent of an additive constant, we could make any surface in the original problem have zero potential and so also be at zero potential in the inverted problem. This brings us to the second physical point. The inverted potentials corresponding to two potential problems differing only by an added constant potential Φ_0 represent physically different charge configurations, namely, charge distributions which differ by a point charge $a\Phi_0$ located at the center of inversion. This can be seen from (2.17), where a constant term Φ_0 in Φ is transformed into $(a\Phi_0/r)$. Consequently care must be taken in applying the method of inversion to remember that the mapping of the point at infinity into the origin may introduce point charges there. If these are not wanted, they must be separately removed by suitable linear superposition.

The geometrical considerations involve only some elementary points which can be proved very simply. The notation is shown in Fig. 2.8. Let O be the center of inversion, and a the radius of inversion. The intersection of the sphere of inversion and the plane of the paper is shown as the dotted circle. A surface S intersects the page with the curve AB. The inverted surface S', obtained by transformation (2.16), intersects the page in the curve $A'B'$. The following facts are stated without proof:

(a) Angles of intersection are not altered by inversion.

(b) An element of area da on the surface S is related to an element of area da' on the inverted surface S' by $da/da' = r^2/r'^2$.

(c) The inverse of a sphere is always another sphere [perhaps of infinite radius; see (d)].

(d) The inverse of any plane is a sphere which passes through the center of inversion, and conversely.

Figure 2.9 illustrates the possibilities involved in (c) and (d) when the center of inversion lies outside, on the surface of, or inside the sphere.

As a very simple example of the solution of a potential problem by inversion we consider an isolated conducting sphere of radius R with a total charge Q on it. The potential has the constant value Q/R inside the sphere and falls off inversely with distance away from the center for points outside the sphere. By a suitable choice of center of inversion and associated parameters we can obtain the potential due to a point charge q a distance d away from an infinite, grounded, conducting plane. Evidently, if the center of inversion O is chosen to lie on the surface of the sphere of radius R, the sphere will invert into a plane. This geometric situation is shown in Fig. 2.10. Furthermore, if we choose the arbitrary additive constant potential Φ_0 to have the value $-Q/R$, the sphere and its inversion, the plane, will be at zero potential, while a point charge $-aQ/R$ will appear at the center of inversion. In order that we end up with a point charge q a distance d away from the plane it is necessary to choose the radius of inversion to be $a = (2Rd)^{1/2}$ and the initial charge, $Q = -(R/2d)^{1/2}q$. The surface-charge density induced on the plane can be found easily from (2.21). Since the charge density on the sphere is uniform over its surface, the induced charge density on the plane varies inversely as the cube of the distance away from the origin (as can be verified from the image solution; see Problem 2.1).

If the center of inversion is chosen to lie outside the isolated uniformly charged sphere, it is clear from Fig. 2.9 that the inverted problem can be

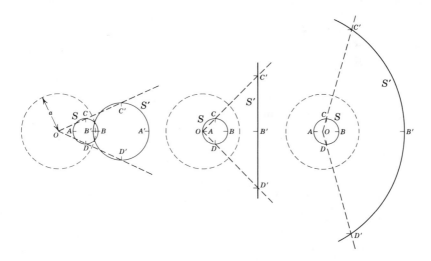

Fig. 2.9   Various possibilities for the inversion of a sphere. If the center of inversion O lies on the surface S of the sphere, the inverted surface S' is a plane; otherwise it is another sphere. The sphere of inversion is shown dotted.

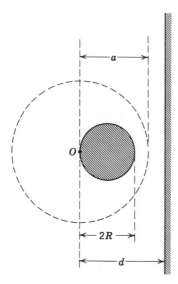

Fig. 2.10 Potential due to isolated, charged, conducting sphere of radius R is inverted to give the potential of a point charge a distance d away from an infinite, flat, conducting surface.

made that of a point charge near a grounded conducting sphere, handled by images in Section 2.2. The explicit verification of this is left to Problem 2.9.

A very interesting use of inversion was made by Lord Kelvin in 1847. He calculated the charge densities on the inner and outer surfaces of a thin, charged, conducting bowl made from a sphere with a cap cut out of it. The potential distribution which he inverted was that of a thin, flat, charged, circular disc (the charged disc is discussed in Section 3.12). As the shape of the bowl is varied from a shallow watch glass-like shape to an almost closed sphere, the charge densities go from those of the disc to those of a closed sphere, in the one limit being almost the same inside and out, but concentrated at the edges of the bowl, and in the other limit being almost zero on the inner surface and uniform over the outer surface. Numerical values are given in Kelvin's collected papers, p. 186, and in Jeans, pp. 250–251.

2.7   Green's Function for the Sphere;   General Solution for the Potential

In preceding sections the problem of a conducting sphere in the presence of a point charge has been discussed by the method of images. As was mentioned in Section 1.10, the potential due to a unit charge and its image (or images), chosen to satisfy homogeneous boundary conditions, is just the Green's function (1.43 or 1.45) appropriate for Dirichlet or Neumann boundary conditions. In $G(\mathbf{x}, \mathbf{x}')$ the variable $\mathbf{x}'$ refers to the location P' of the unit charge, while the variable $\mathbf{x}$ is the point P at which the potential is being evaluated. These coordinates and the sphere are shown in Fig. 2.11. For Dirichlet boundary conditions on the sphere of radius a the potential due to a unit charge and its image is given by (2.1) with $q = 1$ and relations (2.4). Transforming variables appropriately, we obtain the Green's function:

$$G(\mathbf{x}, \mathbf{x}') = \frac{1}{|\mathbf{x} - \mathbf{x}'|} - \frac{a}{x' \left| \mathbf{x} - \dfrac{a^2}{x'^2} \mathbf{x}' \right|} \qquad (2.22)$$

In terms of spherical coordinates this can be written:

$$G(\mathbf{x}, \mathbf{x}') = \frac{1}{(x^2 + x'^2 - 2xx' \cos \gamma)^{\frac{1}{2}}} - \frac{1}{\left(\dfrac{x^2 x'^2}{a^2} + a^2 - 2xx' \cos \gamma \right)^{\frac{1}{2}}}$$

$$(2.23)$$

where γ is the angle between $\mathbf{x}$ and $\mathbf{x}'$. The symmetry in the variables $\mathbf{x}$ and $\mathbf{x}'$ is obvious in the form (2.23), as is the condition that $G = 0$ if either $\mathbf{x}$ or $\mathbf{x}'$ is on the surface of the sphere.

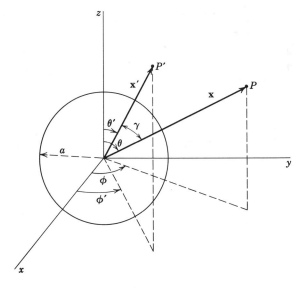

Fig. 2.11

For solution (1.44) of Poisson's equation we need not only G, but also $\partial G/\partial n'$. Remembering that $\mathbf{n}'$ is the unit normal outwards from the volume of interest, i.e., inwards along $\mathbf{x}'$ toward the origin, we have

$$\left.\frac{\partial G}{\partial n'}\right|_{x'=a} = -\frac{(x^2 - a^2)}{a(x^2 + a^2 - 2ax\cos\gamma)^{\frac{3}{2}}} \qquad (2.24)$$

[Note that this is essentially the induced surface-charge density (2.5).] Hence the solution of Laplace's equation *outside* a sphere with the potential specified on its surface is, according to (1.44),

$$\Phi(\mathbf{x}) = \frac{1}{4\pi}\int \Phi(a, \theta', \phi')\frac{a(x^2 - a^2)}{(x^2 + a^2 - 2ax\cos\gamma)^{\frac{3}{2}}}\, d\Omega' \qquad (2.25)$$

where $d\Omega'$ is the element of solid angle at the point (a, θ', ϕ') and $\cos\gamma = \cos\theta\cos\theta' + \sin\theta\sin\theta'\cos(\phi - \phi')$. For the *interior* problem, the normal derivative is radially outwards, so that the sign of $\partial G/\partial n'$ is opposite to (2.24). This is equivalent to replacing the factor $(x^2 - a^2)$ by $(a^2 - x^2)$ in (2.25). For a problem with a charge distribution, we must add to (2.25) the appropriate integral in (1.44), with the Green's function (2.23).

2.8 Conducting Sphere with Hemispheres at Different Potentials

As an example of general solution for the potential outside a sphere with prescribed values of potential on its surface, we consider the conducting sphere of radius a made up of two hemispheres separated by a small insulating ring. The hemispheres are kept at different potentials. It will suffice to consider the potentials as $\pm V$, since arbitrary potentials can be handled by superposition of the solution for a sphere at fixed potential over its whole surface. The insulating ring lies in the $z = 0$ plane, as shown in Fig. 2.12, with the upper (lower) hemisphere at potential $+V$ $(-V)$.

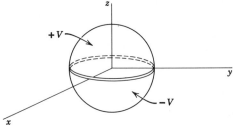

Fig. 2.12

From (2.25) the solution for $\Phi(x, \theta, \phi)$ is given by the integral:

$$\Phi(x, \theta, \phi) = \frac{V}{4\pi} \int_0^{2\pi} d\phi' \left\{ \int_0^1 d(\cos \theta') - \int_{-1}^0 d(\cos \theta') \right\} \frac{a(x^2 - a^2)}{(a^2 + x^2 - 2ax \cos \gamma)^{3/2}}$$

$$(2.26)$$

By a suitable change of variables in the second integral ($\theta' \to \pi - \theta'$, $\phi' \to \phi' + \pi$), this can be cast in the form:

$$\Phi(x, \theta, \phi) = \frac{Va(x^2 - a^2)}{4\pi} \int_0^{2\pi} d\phi' \int_0^1 d(\cos \theta')[(a^2 + x^2 - 2ax \cos \gamma)^{-3/2}$$

$$- (a^2 + x^2 + 2ax \cos \gamma)^{-3/2}] \quad (2.27)$$

Because of the complicated dependence of $\cos \gamma$ on the angles (θ', ϕ') and (θ, ϕ), equation (2.27) cannot in general be integrated in closed form.

As a special case we consider the potential on the positive z axis. Then $\cos \gamma = \cos \theta'$ since $\theta = 0$. The integration is elementary, and the potential can be shown to be

$$\Phi(z) = V \left[1 - \frac{(z^2 - a^2)}{z\sqrt{z^2 + a^2}} \right] \quad (2.28)$$

At $z = a$, this reduces to $\Phi = V$ as required, while at large distances it goes asymptotically as $\Phi \simeq 3Va^2/2z^2$.

In the absence of a closed expression for the integrals in (2.27), we can expand the denominator in power series and integrate term by term. Factoring out $(a^2 + x^2)$ from each denominator, we obtain

$$\Phi(x, \theta, \phi) = \frac{Va(x^2 - a^2)}{4\pi(x^2 + a^2)^{3/2}} \int_0^{2\pi} d\phi' \int_0^1 d(\cos \theta')[(1 - 2\alpha \cos \gamma)^{-3/2}$$

$$- (1 + 2\alpha \cos \gamma)^{-3/2}] \quad (2.29)$$

where $\alpha = ax/(a^2 + x^2)$. We observe that in the expansion of the radicals only odd powers of $\alpha \cos \gamma$ will appear:

$$[(1 - 2\alpha \cos \gamma)^{-3/2} - (1 + 2\alpha \cos \gamma)^{-3/2}] = 6\alpha \cos \gamma + 35\alpha^3 \cos^3 \gamma + \cdots$$

$$(2.30)$$

It is now necessary to integrate odd powers of $\cos \gamma$ over $d\phi' \, d(\cos \theta')$:

$$\left. \begin{array}{l} \int_0^{2\pi} d\phi' \int_0^1 d(\cos \theta') \cos \gamma = \pi \cos \theta \\[2mm] \int_0^{2\pi} d\phi' \int_0^1 d(\cos \theta') \cos^3 \gamma = \frac{\pi}{4} \cos \theta (3 - \cos^2 \theta) \end{array} \right\} \quad (2.31)$$

If (2.30) and (2.31) are inserted into (2.29), the potential becomes

$$\Phi(x, \theta, \phi) = \frac{3Va^2}{2x^2}\left(\frac{x^3(x^2 - a^2)}{(x^2 + a^2)^{5/2}}\right) \cos \theta$$

$$\times \left[1 + \frac{35}{24}\frac{a^2x^2}{(a^2 + x^2)^2}(3 - \cos^2 \theta) + \cdots\right] \quad (2.32)$$

We note that only odd powers of $\cos \theta$ appear, as required by the symmetry of the problem. If the expansion parameter is (a^2/x^2), rather than α^2, the series takes on the form:

$$\Phi(x, \theta, \phi) = \frac{3Va^2}{2x^2}\left[\cos \theta - \frac{7a^2}{12x^2}\left(\frac{5}{2}\cos^3 \theta - \frac{3}{2}\cos \theta\right) + \cdots\right] \quad (2.33)$$

For large values of x/a this expansion converges rapidly and so is a useful representation for the potential. Even for $x/a = 5$, the second term in the series is only of the order of 2 per cent. It is easily verified that, for $\cos \theta = 1$, expression (2.33) agrees with the expansion of (2.28) for the potential on the axis. [The particular choice of angular factors in (2.33) is dictated by the definitions of the Legendre polynomials. The two factors are, in fact, $P_1(\cos \theta)$ and $P_3(\cos \theta)$, and the expansion of the potential is one in Legendre polynomials of odd order. We shall establish this in a systematic fashion in Section 3.3.]

2.9 Orthogonal Functions and Expansions

The representation of solutions of potential problems (or any mathematical physics problem) by expansions in orthogonal functions forms a powerful technique that can be used in a large class of problems. The particular orthogonal set chosen depends on the symmetries or near symmetries involved. To recall the general properties of orthogonal functions and expansions in terms of them, we consider an interval (a, b) in a variable ξ with a set of real or complex functions $U_n(\xi)$, $n = 1, 2, \ldots$, orthogonal on the interval (a, b). The orthogonality condition on the functions $U_n(\xi)$ is expressed by

$$\int_a^b U_n^*(\xi)U_m(\xi)\,d\xi = 0, \quad m \neq n \quad (2.34)$$

If $n = m$, the integral is finite. We assume that the functions are normalized so that the integral is unity. Then the functions are said to be *orthonormal*, and they satisfy

$$\int_a^b U_n^*(\xi)U_m(\xi)\,d\xi = \delta_{nm} \quad (2.35)$$

An arbitrary function $f(\xi)$, square integrable on the interval (a, b), can be expanded in a series of the orthonormal functions $U_n(\xi)$. If the number of terms in the series is finite (say N),

$$f(\xi) \leftrightarrow \sum_{n=1}^{N} a_n U_n(\xi) \tag{2.36}$$

then we can ask for the "best" choice of coefficients a_n so that we get the "best" representation of the function $f(\xi)$. If "best" is defined as minimizing the mean square error M_N:

$$M_N = \int_a^b \left| f(\xi) - \sum_{n=1}^{N} a_n U_n(\xi) \right|^2 d\xi \tag{2.37}$$

it is easy to show that the coefficients are given by

$$a_n = \int_a^b U_n^*(\xi) f(\xi) \, d\xi \tag{2.38}$$

where the orthonormality condition (2.35) has been used. This is the standard result for the coefficients in an orthonormal function expansion.

If the number of terms N in series (2.36) is taken larger and larger, we intuitively expect that our series representation of $f(\xi)$ is "better" and "better." Our intuition will be correct provided the set of orthonormal functions is *complete*, completeness being defined by the requirement that there exist a finite number N_0 such that for $N > N_0$ the mean square error M_N can be made smaller than any arbitrarily small positive quantity. Then the series representation

$$\sum_{n=1}^{\infty} a_n U_n(\xi) = f(\xi) \tag{2.39}$$

with a_n given by (2.38) is said to *converge in the mean* to $f(\xi)$. Physicists generally leave the difficult job of proving completeness of a given set of functions to the mathematicians. All orthonormal sets of functions normally occurring in mathematical physics have been proved to be complete.

Series (2.39) can be rewritten with the explicit form (2.38) for the coefficients a_n:

$$f(\xi) = \int_a^b \left\{ \sum_{n=1}^{\infty} U_n^*(\xi') U_n(\xi) \right\} f(\xi') \, d\xi' \tag{2.40}$$

Since this represents any function $f(\xi)$ on the interval (a, b), it is clear that the sum of bilinear terms $U_n^*(\xi') U_n(\xi)$ must exist only in the neighborhood of $\xi' = \xi$. In fact, it must be true that

$$\sum_{n=1}^{\infty} U_n^*(\xi') U_n(\xi) = \delta(\xi' - \xi) \tag{2.41}$$

This is the so-called *completeness* or *closure relation*. It is analogous to the orthonormality condition (2.35), except that the roles of the continuous variable ξ and the discrete index n have been interchanged.

The most famous orthogonal functions are the sines and cosines, an expansion in terms of them being a Fourier series. If the interval in x is $(-a/2, a/2)$, the orthonormal functions are

$$\sqrt{\frac{2}{a}} \sin\left(\frac{2\pi m x}{a}\right), \quad \sqrt{\frac{2}{a}} \cos\left(\frac{2\pi m x}{a}\right)$$

where m is an integer. The series equivalent to (2.39) is customarily written in the form:

$$f(x) = \tfrac{1}{2}A_0 + \sum_{m=1}^{\infty} \left[A_m \cos\left(\frac{2\pi m x}{a}\right) + B_m \sin\left(\frac{2\pi m x}{a}\right) \right] \quad (2.42)$$

where

$$A_m = \frac{2}{a} \int_{-a/2}^{a/2} f(x) \cos\left(\frac{2\pi m x}{a}\right) dx$$

$$B_m = \frac{2}{a} \int_{-a/2}^{a/2} f(x) \sin\left(\frac{2\pi m x}{a}\right) dx$$

$$(2.43)$$

If the interval spanned by the orthonormal set has more than one dimension, formulas (2.34)–(2.39) have obvious generalizations. Suppose that the space is two dimensional, and that the variable ξ ranges over the interval (a, b) while the variable η has the interval (c, d). The orthonormal functions in each dimension are $U_n(\xi)$ and $V_m(\eta)$. Then the expansion of an arbitrary function $f(\xi, \eta)$ is

$$f(\xi, \eta) = \sum_n \sum_m a_{nm} U_n(\xi) V_m(\eta) \quad (2.44)$$

where

$$a_{nm} = \int_a^b d\xi \int_c^d d\eta\, U_n{}^*(\xi) V_m{}^*(\eta) f(\xi, \eta) \quad (2.45)$$

If the interval (a, b) becomes infinite, the set of orthogonal functions $U_n(\xi)$ may become a continuum of functions, rather than a denumerable set. Then the Kronecker delta symbol in (2.35) becomes a Dirac delta function. An important example is the Fourier integral. Start with the orthonormal set of complex exponentials,

$$U_m(x) = \frac{1}{\sqrt{a}} e^{i(2\pi m x/a)} \quad (2.46)$$

$m = 0, \pm 1, \pm 2, \ldots$, on the interval $(-a/2, a/2)$, with the expansion:

$$f(x) = \frac{1}{\sqrt{a}} \sum_{m=-\infty}^{\infty} A_m e^{i(2\pi m x/a)} \quad (2.47)$$

where

$$A_m = \frac{1}{\sqrt{a}} \int_{-a/2}^{a/2} e^{-i(2\pi m x'/a)} f(x')\, dx' \qquad (2.48)$$

Then let the interval become infinite $(a \to \infty)$, at the same time transforming

$$\left.\begin{array}{c} \dfrac{2\pi m}{a} \to k \\[2mm] \displaystyle\sum_m \to \int_{-\infty}^{\infty} dm = \dfrac{a}{2\pi} \int_{-\infty}^{\infty} dk \\[4mm] A_m \to \sqrt{\dfrac{2\pi}{a}}\, A(k) \end{array}\right\} \qquad (2.49)$$

The resulting expansion, equivalent to (2.47), is

$$f(x) = \frac{1}{\sqrt{2\pi}} \int_{-\infty}^{\infty} A(k) e^{ikx}\, dk \qquad (2.50)$$

where

$$A(k) = \frac{1}{\sqrt{2\pi}} \int_{-\infty}^{\infty} e^{-ikx} f(x)\, dx \qquad (2.51)$$

The orthogonality condition is

$$\frac{1}{2\pi} \int_{-\infty}^{\infty} e^{i(k-k')x}\, dx = \delta(k - k') \qquad (2.52)$$

while the completeness relation is

$$\frac{1}{2\pi} \int_{-\infty}^{\infty} e^{ik(x-x')}\, dk = \delta(x - x') \qquad (2.53)$$

These last integrals serve as convenient representations of a delta function. We note in (2.50)–(2.53) the complete equivalence of the two continuous variables x and k.

2.10 Separation of Variables; Laplace's Equation in Rectangular Coordinates

The partial differential equations of mathematical physics are often solved conveniently by a method called *separation of variables*. In the process, one often generates orthogonal sets of functions which are useful in their own right. Equations involving the three-dimensional Laplacian operator are known to be separable in eleven different coordinate systems (see Morse and Feshbach, pp. 509, 655). We will discuss only three of these in any detail—rectangular, spherical, and cylindrical—and will begin with the simplest, rectangular coordinates.

Laplace's equation in rectangular coordinates is

$$\frac{\partial^2 \Phi}{\partial x^2} + \frac{\partial^2 \Phi}{\partial y^2} + \frac{\partial^2 \Phi}{\partial z^2} = 0 \tag{2.54}$$

A solution of this *partial* differential equation can be found in terms of three *ordinary* differential equations, all of the same form, by the assumption that the potential can be represented by a product of three functions, one for each coordinate:

$$\Phi(x, y, z) = X(x)\,Y(y)Z(z) \tag{2.55}$$

Substitution into (2.54) and division of the result by (2.55) yields

$$\frac{1}{X(x)} \frac{d^2 X}{dx^2} + \frac{1}{Y(y)} \frac{d^2 Y}{dy^2} + \frac{1}{Z(z)} \frac{d^2 Z}{dz^2} = 0 \tag{2.56}$$

where total derivatives have replaced partial derivatives, since each term involves a function of one variable only. If (2.56) is to hold for arbitrary values of the independent coordinates, each of the three terms must be separately constant:

$$\left. \begin{aligned} \frac{1}{X} \frac{d^2 X}{dx^2} &= -\alpha^2 \\[2mm] \frac{1}{Y} \frac{d^2 Y}{dy^2} &= -\beta^2 \\[2mm] \frac{1}{Z} \frac{d^2 Z}{dz^2} &= \gamma^2 \end{aligned} \right\} \tag{2.57}$$

where
$$\alpha^2 + \beta^2 = \gamma^2$$

If we arbitrarily choose α^2 and β^2 to be positive, then the solutions of the three ordinary differential equations (2.57) are $\exp(\pm i\alpha x)$, $\exp(\pm i\beta y)$, $\exp(\pm \sqrt{\alpha^2 + \beta^2}\,z)$. The potential (2.55) can thus be built up from the product solutions:

$$\Phi = e^{\pm i\alpha x} e^{\pm i\beta y} e^{\pm \sqrt{\alpha^2 + \beta^2}\,z} \tag{2.58}$$

At this stage α and β are completely arbitrary. Consequently (2.58), by linear superposition, represents a very large class of solutions to Laplace's equation.

To determine α and β it is necessary to impose specific boundary conditions on the potential. As an example, consider a rectangular box, located as shown in Fig. 2.13, with dimensions (a, b, c) in the (x, y, z)

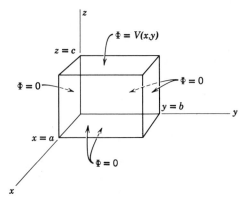

Fig. 2.13 Hollow, rectangular box with five sides at zero potential, while the sixth ($z = c$) has the specified potential $\Phi = V(x, y)$.

directions. All surfaces of the box are kept at zero potential, except the surface $z = c$, which is at a potential $V(x, y)$. It is required to find the potential everywhere inside the box. Starting with the requirement that $\Phi = 0$ for $x = 0$, $y = 0$, $z = 0$, it is easy to see that the required forms of X, Y, Z are

$$\left.\begin{aligned}
X &= \sin \alpha x \\
Y &= \sin \beta y \\
Z &= \sinh (\sqrt{\alpha^2 + \beta^2}z)
\end{aligned}\right\} \tag{2.59}$$

In order that $\Phi = 0$ at $x = a$ and $y = b$, it is necessary that $\alpha a = n\pi$ and $\beta b = m\pi$. With the definitions,

$$\left.\begin{aligned}
\alpha_n &= \frac{n\pi}{a} \\
\beta_m &= \frac{m\pi}{b} \\
\gamma_{nm} &= \pi\sqrt{\frac{n^2}{a^2} + \frac{m^2}{b^2}}
\end{aligned}\right\} \tag{2.60}$$

we can write the partial potential Φ_{nm}; satisfying all the boundary conditions except one,

$$\Phi_{nm} = \sin (\alpha_n x) \sin (\beta_m y) \sinh (\gamma_{nm}z) \tag{2.61}$$

The potential can be expanded in terms of these Φ_{nm} with initially arbitrary coefficients (to be chosen to satisfy the final boundary condition):

$$\Phi(x, y, z) = \sum_{n,m=1}^{\infty} A_{nm} \sin (\alpha_n x) \sin (\beta_m y) \sinh (\gamma_{nm}z) \tag{2.62}$$

There remains only the boundary condition $\Phi = V(x, y)$ at $z = c$:

$$V(x, y) = \sum_{n,m=1}^{\infty} A_{nm} \sin (\alpha_n x) \sin (\beta_m y) \sinh (\gamma_{nm} c) \qquad (2.63)$$

This is just a double Fourier series for the function $V(x, y)$. Consequently the coefficients A_{nm} are given by:

$$A_{nm} = \frac{4}{ab \sinh (\gamma_{nm} c)} \int_0^a dx \int_0^b dy V(x, y) \sin (\alpha_n x) \sin (\beta_m y) \qquad (2.64)$$

If the rectangular box has potentials different from zero on all six sides, the required solution for the potential inside the box can be obtained by a linear superposition of six solutions, one for each side, equivalent to (2.62) and (2.64). The problem of the solution of Poisson's equation, i.e., the potential inside the box with a charge distribution inside, as well as prescribed boundary conditions on the surface, requires the construction of the appropriate Green's function, according to (1.43) and (1.44). Discussion of this topic will be deferred until we have treated Laplace's equation in spherical and cylindrical coordinates. For the moment, we merely note that solution (2.62) and (2.64) is equivalent to the surface integral in the Green's function solution (1.44).

REFERENCES AND SUGGESTED READING

Images and inversion are treated in many books; among the better or more extensive discussions are those by
>   Jeans, Chapter VIII,
>   Maxwell, Vol. 1, Chapter XI,
>   Smythe, Chapters IV and V.

A truly encyclopedic source of examples with numerous diagrams is the book by
>   Durand, especially Chapters III and IV.

Durand discusses inversion on pp. 107–114.

Conformal mapping techniques for the solution of two-dimensional potential problems are discussed by
>   Durand, Chapter X,
>   Jeans, Chapter VIII, Sections 306–337,
>   Maxwell, Vol. 1, Chapter XII,
>   Smythe, Chapter IV, Sections 4.09–4.29.

There are, in addition, many engineering books devoted to the subject, e.g.,
>   Rothe, Ollendorff, and Polhausen.

Elementary, but clear, discussions of the mathematical theory of Fourier series and integrals, and orthogonal expansions, can be found in
>   Churchill,
>   Hildebrand, Chapter 5.

A somewhat old-fashioned treatment of Fourier series and integrals, but with many examples and problems, is given by
>   Byerly.

PROBLEMS

2.1 A point charge q is brought to a position a distance d away from an infinite plane conductor held at zero potential. Using the method of images, find:

(a) the surface-charge density induced on the plane, and plot it;

(b) the force between the plane and the charge by using Coulomb's law for the force between the charge and its image;

(c) the total force acting on the plane by integrating $2\pi\sigma^2$ over the whole plane;

(d) the work necessary to remove the charge q from its position to infinity;

(e) the potential energy between the charge q and its image [compare the answer to (d) and discuss].

(f) Find answer (d) in electron volts for an electron originally one angstrom from the surface.

2.2 Using the method of images, discuss the problem of a point charge q *inside* a hollow, grounded, conducting sphere of inner radius a. Find

(a) the potential inside the sphere;

(b) the induced surface-charge density;

(c) the magnitude and direction of the force acting on q.

Is there any change in the solution if the sphere is kept at a fixed potential V? If the sphere has a total charge Q on it?

2.3 Two infinite, grounded, conducting planes are located at $x = a/2$ and $x = -a/2$. A point charge q is placed between the planes at the point (x', y', z'), where $-(a/2) < x' < (a/2)$.

(a) Find the location and magnitude of all the image charges needed to satisfy the boundary conditions on the potential, and write down the Green's function $G(\mathbf{x}, \mathbf{x}')$.

(b) If the charge q is at $(x', 0, 0)$, find the surface-charge densities induced on each conducting plane and show that the sum of induced charge on the two planes is $-q$.

2.4 Consider a potential problem in the half-space defined by $z \geq 0$, with Dirichlet boundary conditions on the plane $z = 0$ (and at infinity).

(a) Write down the appropriate Green's function $G(\mathbf{x}, \mathbf{x}')$.

(b) If the potential on the plane $z = 0$ is specified to be $\Phi = V$ inside a circle of radius a centered at the origin, and $\Phi = 0$ outside that circle, find an integral expression for the potential at the point P specified in terms of cylindrical coordinates (ρ, ϕ, z).

(c) Show that, along the axis of the circle $(\rho = 0)$, the potential is given by

$$\Phi = V\left(1 - \frac{z}{\sqrt{a^2 + z^2}}\right)$$

(d) Show that at large distances $(\rho^2 + z^2 \gg a^2)$ the potential can be expanded in a power series in $(\rho^2 + z^2)^{-1}$, and that the leading terms are

$$\Phi = \frac{Va^2}{2}\frac{z}{(\rho^2 + z^2)^{3/2}}\left[1 - \frac{3a^2}{4(\rho^2 + z^2)} + \frac{5(3\rho^2a^2 + a^4)}{8(\rho^2 + z^2)^2} + \cdots\right]$$

Verify that the results of (*c*) and (*d*) are consistent with each other in their common range of validity.

2.5 An insulated, spherical, conducting shell of radius *a* is in a uniform electric field E_0. If the sphere is cut into two hemispheres by a plane perpendicular to the field, find the force required to prevent the hemispheres from separating

    (*a*) if the shell is uncharged;

    (*b*) if the total charge on the shell is *Q*.

2.6 A large parallel plate capacitor is made up of two plane conducting sheets, one of which has a small hemispherical boss of radius *a* on its inner surface. The conductor with the boss is kept at zero potential, and the other conductor is at a potential such that far from the boss the electric field between the plates is E_0.

    (*a*) Calculate the surface-charge densities at an arbitrary point on the plane and on the boss, and sketch their behavior as a function of distance (or angle).

    (*b*) Show that the total charge on the boss has the magnitude $3E_0a^2/4$.

    (*c*) If, instead of the other conducting sheet at a different potential, a point charge *q* is placed directly above the hemispherical boss at a distance *d* from its center, show that the charge induced on the boss is

$$q' = -q\left[1 - \frac{d^2 - a^2}{d\sqrt{d^2 + a^2}}\right]$$

2.7 A line charge with linear charge density τ is placed parallel to, and a distance *R* away from, the axis of a conducting cylinder of radius *b* held at fixed voltage such that the potential vanishes at infinity. Find

    (*a*) the magnitude and position of the image charge(s);

    (*b*) the potential at any point (expressed in polar coordinates with the line from the cylinder axis to the line charge as the *x* axis), including the asymptotic form far from the cylinder;

    (*c*) the induced surface-charge density, and plot it as a function of angle for $R/b = 2, 4$ in units of $\tau/2\pi b$;

    (*d*) the force on the charge.

2.8   (*a*) Find the Green's function for the two-dimensional potential problem with the potential specified on the surface of a cylinder of radius *b*, and show that the solution inside the cylinder is given by Poisson's integral:

$$\Phi(r, \theta) = \frac{1}{2\pi} \int_0^{2\pi} \Phi(b, \theta') \frac{b^2 - r^2}{b^2 + r^2 - 2br \cos(\theta' - \theta)} d\theta'$$

    (*b*) Two halves of a long conducting cylinder of radius *b* are separated by a small gap, and are kept at different potentials V_1 and V_2. Show that the potential inside is given by

$$\Phi(r, \theta) = \frac{V_1 + V_2}{2} + \frac{V_1 - V_2}{\pi} \tan^{-1}\left(\frac{2br}{b^2 - r^2} \cos \theta\right)$$

where θ is measured from a plane perpendicular to the plane through the gap.

    (*c*) Calculate the surface-charge density on each half of the cylinder.

    (*d*) What modification is necessary in (*a*) if the potential is desired in the region of space bounded by the cylinder and infinity?

2.9   (*a*) An isolated conducting sphere is raised to a potential V. Write down the (trivial) solution for the electrostatic potential everywhere in space.

(*b*) Apply the inversion theorem, choosing the center of inversion *outside* the conducting sphere. Show *explicitly* that the solution obtained for the potential is that of a grounded sphere in the presence of a point charge of magnitude $-VR$, where R is the inversion radius.

(*c*) What is the physical situation described by the inverted solution if the center of inversion lies inside the conducting sphere?

2.10 Knowing that the capacitance of a thin, flat, circular, conducting disc of radius a is $(2/\pi)a$ and that the surface-charge density on an isolated disc raised to a given potential is proportional to $(a^2 - r^2)^{-1/2}$, where r is the distance from the center of the disc,

(*a*) show that by inversion the potential can be found for the problem of an infinite, grounded, conducting plane with a circular hole in it and a point charge lying anywhere in the opening;

(*b*) show that, for a unit point charge at the center of the opening, the induced charge density on the plane is

$$\sigma(r, \theta, \phi) = -\frac{a}{\pi^2 r^2 \sqrt{r^2 - a^2}}$$

(*c*) show that (*a*) and (*b*) are a special case of the general problem, obtained by inversion of the disc, of a grounded, conducting, spherical bowl under the influence of a point charge located on the cap which is the complement of the bowl.

2.11 A hollow cube has conducting walls defined by six planes $x = y = z = 0$, and $x = y = z = a$. The walls $z = 0$ and $z = a$ are held at a constant potential V. The other four sides are at zero potential.

(*a*) Find the potential $\Phi(x, y, z)$ at any point inside the cube.

(*b*) Evaluate the potential at the center of the cube numerically, accurate to three significant figures. How many terms in the series is it necessary to keep in order to attain this accuracy? Compare your numerical result with the average value of the potential on the walls.

(*c*) Find the surface-charge density on the surface $z = a$.

3

Boundary-Value Problems in Electrostatics: II

In this chapter the discussion of boundary-value problems is continued. Spherical and cylindrical geometries are first considered, and solutions of Laplace's equation are represented by expansions in series of the appropriate orthonormal functions. Only an outline is given of the solution of the various ordinary differential equations obtained from Laplace's equation by separation of variables, but an adequate summary of the properties of the different functions is presented.

The problem of construction of Green's functions in terms of orthonormal functions arises naturally in the attempt to solve Poisson's equation in the various geometries. Explicit examples of Green's functions are obtained and applied to specific problems, and the equivalence of the various approaches to potential problems is discussed.

3.1 Laplace's Equation in Spherical Coordinates

In spherical coordinates (r, θ, ϕ), shown in Fig. 3.1, Laplace's equation can be written in the form:

$$\frac{1}{r} \frac{\partial^2}{\partial r^2} (r\Phi) + \frac{1}{r^2 \sin \theta} \frac{\partial}{\partial \theta} \left(\sin \theta \frac{\partial \Phi}{\partial \theta} \right) + \frac{1}{r^2 \sin^2 \theta} \frac{\partial^2 \Phi}{\partial \phi^2} = 0 \qquad (3.1)$$

If a product form for the potential is assumed, then it can be written:

$$\Phi = \frac{U(r)}{r} P(\theta)Q(\phi) \qquad (3.2)$$

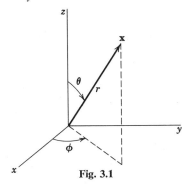

Fig. 3.1

When this is substituted into (3.1), there results the equation:

$$PQ \frac{d^2U}{dr^2} + \frac{UQ}{r^2 \sin \theta} \frac{d}{d\theta}\left(\sin \theta \frac{dP}{d\theta}\right) + \frac{UP}{r^2 \sin^2 \theta} \frac{d^2Q}{d\phi^2} = 0$$

If we multiply by $r^2 \sin^2 \theta / UPQ$, we obtain:

$$r^2 \sin^2 \theta \left[\frac{1}{U}\frac{d^2U}{dr^2} + \frac{1}{r^2 \sin \theta P}\frac{d}{d\theta}\left(\sin \theta \frac{dP}{d\theta}\right)\right] + \frac{1}{Q}\frac{d^2Q}{d\phi^2} = 0 \quad (3.3)$$

The ϕ dependence of the equation has now been isolated in the last term. Consequently that term must be a constant which we call $(-m^2)$:

$$\frac{1}{Q}\frac{d^2Q}{d\phi^2} = -m^2 \quad (3.4)$$

This has solutions

$$Q = e^{\pm im\phi} \quad (3.5)$$

In order that Q be single valued, m must be an integer. By similar considerations we find separate equations for $P(\theta)$ and $U(r)$:

$$\frac{1}{\sin \theta}\frac{d}{d\theta}\left(\sin \theta \frac{dP}{d\theta}\right) + \left[l(l+1) - \frac{m^2}{\sin^2 \theta}\right]P = 0 \quad (3.6)$$

$$\frac{d^2U}{dr^2} - \frac{l(l+1)}{r^2}U = 0 \quad (3.7)$$

where $l(l+1)$ is another real constant.

From the form of the radial equation it is apparent that a single power of r (rather than a power series) will satisfy it. The solution is found to be:

$$U = Ar^{l+1} + Br^{-l} \quad (3.8)$$

but l is as yet undetermined.

3.2   Legendre Equation and Legendre Polynomials

The θ equation for $P(\theta)$ is customarily expressed in terms of $x = \cos \theta$, instead of θ itself. Then it takes the form:

$$\frac{d}{dx}\left((1 - x^2)\frac{dP}{dx}\right) + \left(l(l + 1) - \frac{m^2}{1 - x^2}\right)P = 0 \qquad (3.9)$$

This equation is called the generalized Legendre equation, and its solutions are the associated Legendre functions. Before considering (3.9) we will outline the solution by power series of the ordinary Legendre differential equation with $m^2 = 0$:

$$\frac{d}{dx}\left((1 - x^2)\frac{dP}{dx}\right) + l(l + 1)P = 0 \qquad (3.10)$$

The desired solution should be single valued, finite, and continuous on the interval $-1 \le x \le 1$ in order that it represents a physical potential. The solution will be assumed to be represented by a power series of the form:

$$P(x) = x^\alpha \sum_{j=0}^{\infty} a_j x^j \qquad (3.11)$$

where α is a parameter to be determined. When this is substituted into (3.10), there results the series:

$$\sum_{j=0}^{\infty} \{(\alpha + j)(\alpha + j - 1)a_j x^{\alpha + j - 2}$$
$$- [(\alpha + j)(\alpha + j + 1) - l(l + 1)]a_j x^{\alpha + j}\} = 0 \qquad (3.12)$$

In this expansion the coefficient of each power of x must vanish separately. For $j = 0, 1$ we find that

$$\left.\begin{array}{ll} \text{if } a_0 \ne 0, & \text{then } \alpha(\alpha - 1) = 0 \\[2mm] \text{if } a_1 \ne 0, & \text{then } \alpha(\alpha + 1) = 0 \end{array}\right\} \qquad (3.13)$$

while for a general j value

$$a_{j+2} = \left[\frac{(\alpha + j)(\alpha + j + 1) - l(l + 1)}{(\alpha + j + 1)(\alpha + j + 2)}\right]a_j \qquad (3.14)$$

A moment's thought shows that the two relations (3.13) are equivalent and that it is sufficient to choose *either* a_0 *or* a_1 different from zero, but not both. Making the former choice, we have $\alpha = 0$ or $\alpha = 1$. From (3.14) we see that the power series has only even powers of $x(\alpha = 0)$ or only odd powers of $x(\alpha = 1)$.

For either of the series $\alpha = 0$ or $\alpha = 1$ it is possible to prove the following properties:

(*a*) the series converges for $x^2 < 1$, regardless of the value of *l*;

(*b*) the series diverges at $x = \pm 1$, unless it terminates.

Since we want a solution that is finite at $x = \pm 1$, as well as for $x^2 < 1$, we demand that the series terminate. Since α and j are positive integers or zero, the recurrence relation (3.14) will terminate only if *l is zero or a positive integer*. Even then only one of the two series converges at $x = \pm 1$. If *l* is even (odd), then only the $\alpha = 0$ ($\alpha = 1$) series terminates.* The polynomials in each case have x^l as their highest power of x, the next highest being x^{l-2}, and so on, down to x^0 (x) for *l* even (odd). By convention these polynomials are normalized to have the value unity at $x = +1$ and are called the *Legendre polynomials* of order *l*, $P_l(x)$. The first few Legendre polynomials are:

$$\begin{aligned}
P_0(x) &= 1 \\
P_1(x) &= x \\
P_2(x) &= \tfrac{1}{2}(3x^2 - 1) \\
P_3(x) &= \tfrac{1}{2}(5x^3 - 3x) \\
P_4(x) &= \tfrac{1}{8}(35x^4 - 30x^2 + 3)
\end{aligned} \qquad (3.15)$$

By manipulation of the power series solutions (3.11) and (3.14) it is possible to obtain a compact representation of the Legendre polynomials, known as *Rodrigues' formula*:

$$P_l(x) = \frac{1}{2^l l!} \frac{d^l}{dx^l} (x^2 - 1)^l \qquad (3.16)$$

[This can be obtained by other, more elegant means, or by direct *l*-fold integration of the differential equation (3.10).]

The Legendre polynomials form a complete orthogonal set of functions on the interval $-1 \le x \le 1$. To prove the orthogonality we can appeal directly to the differential equation (3.10). We write down the differential equation for $P_l(x)$, multiply by $P_{l'}(x)$, and then integrate over the interval:

$$\int_{-1}^{1} P_{l'}(x) \left[\frac{d}{dx} \left((1 - x^2) \frac{dP_l}{dx} \right) + l(l + 1)P_l(x) \right] dx = 0 \qquad (3.17)$$

---

* For example, if $l = 0$ the $\alpha = 1$ series has a general coefficient $a_j = a_0/j + 1$ for $j = 0, 2, 4, \ldots$. Thus the series is $a_0(x + \tfrac{1}{3}x^3 + \tfrac{1}{5}x^5 + \cdots.)$ This is just the power series expansion of a function $Q_0(x) = \tfrac{1}{2} \ln \left(\dfrac{1 + x}{1 - x} \right)$, which clearly diverges at $x = \pm 1$. For each *l* value there is a similar function $Q_l(x)$ with logarithms in it as the partner to the well-behaved polynomial solution. See Magnus and Oberhettinger, p. 59.

Integrating the first term by parts, we obtain

$$\int_{-1}^{1}\left[(x^2-1)\frac{dP_l}{dx}\frac{dP_{l'}}{dx}+l(l+1)P_{l'}(x)P_l(x)\right]dx=0 \qquad (3.18)$$

If we now write down (3.18) with l and l' interchanged and subtract it from (3.18), the result is the orthogonality condition:

$$[l(l+1)-l'(l'+1)]\int_{-1}^{1}P_{l'}(x)P_l(x)\,dx=0 \qquad (3.19)$$

For $l \neq l'$, the integral must vanish. For $l = l'$, the integral is finite. To determine its value it is necessary to use an explicit representation of the Legendre polynomials, e.g., Rodrigues' formula. Then the integral is explicitly:

$$\int_{-1}^{1}[P_l(x)]^2\,dx=\frac{1}{2^{2l}(l!)^2}\int_{-1}^{1}\frac{d^l}{dx^l}(x^2-1)^l\frac{d^l}{dx^l}(x^2-1)^l\,dx$$

Integration by parts l times yields the result:

$$\int_{-1}^{1}[P_l(x)]^2\,dx=\frac{(-1)^l}{2^{2l}(l!)^2}\int_{-1}^{1}(x^2-1)^l\frac{d^{2l}}{dx^{2l}}(x^2-1)^l\,dx$$

The differentiation of $(x^2-1)^l$ $2l$ times yields the constant $(2l)!$, so that

$$\int_{-1}^{1}[P_l(x)]^2\,dx=\frac{(2l)!}{2^{2l}(l!)^2}\int_{-1}^{1}(1-x^2)^l\,dx \qquad (3.20)$$

The remaining integral is easily shown to be $2^{2l+1}(l!)^2/(2l+1)!$ Consequently the orthogonality condition can be written:

$$\int_{-1}^{1}P_{l'}(x)P_l(x)\,dx=\frac{2}{2l+1}\delta_{l'l} \qquad (3.21)$$

and the orthonormal functions in the sense of Section 2.9 are

$$U_l(x)=\sqrt{\frac{2l+1}{2}}P_l(x) \qquad (3.22)$$

Since the Legendre polynomials form a complete set of orthogonal functions, any function $f(x)$ on the interval $-1 \leq x \leq 1$ can be expanded in

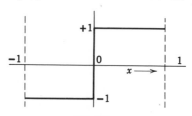

Fig. 3.2

terms of them. The Legendre series representation is:

$$f(x) = \sum_{l=0}^{\infty} A_l P_l(x) \tag{3.23}$$

where

$$A_l = \frac{2l+1}{2} \int_{-1}^{1} f(x) P_l(x)\, dx \tag{3.24}$$

As an example, consider the function shown in Fig. 3.2:

$$f(x) = +1 \text{ for } x > 0$$
$$= -1 \text{ for } x < 0$$

Then

$$A_l = \frac{2l+1}{2}\left[\int_{0}^{1} P_l(x)\, dx - \int_{-1}^{0} P_l(x)\, dx\right]$$

Since $P_l(x)$ is odd (even) about $x = 0$ if l is odd (even), only the odd l coefficients are different from zero. Thus, for l odd,

$$A_l = (2l+1)\int_{0}^{1} P_l(x)\, dx \tag{3.25}$$

By means of Rodrigues' formula the integral can be evaluated, yielding

$$A_l = \left(-\frac{1}{2}\right)^{(l-1)/2} \frac{(2l+1)(l-2)!!}{2\left(\frac{l+1}{2}\right)!} \tag{3.26}$$

where $(2n+1)!! \equiv (2n+1)(2n-1)(2n-3)\cdots \times 5 \times 3 \times 1$. Thus the series for $f(x)$ is:

$$f(x) = \tfrac{3}{2}P_1(x) - \tfrac{7}{8}P_3(x) + \tfrac{11}{16}P_5(x) - \cdots \tag{3.27}$$

Certain recurrence relations among Legendre polynomials of different order are useful in evaluating integrals, generating higher-order polynomials from lower-order ones, etc. From Rodrigues' formula it is a straightforward matter to show that

$$\frac{dP_{l+1}}{dx} - \frac{dP_{l-1}}{dx} - (2l+1)P_l = 0 \tag{3.28}$$

This result, combined with differential equation (3.10), can be made to yield various recurrence formulas, some of which are:

$$\left.\begin{array}{l}
(l+1)P_{l+1} - (2l+1)xP_l + lP_{l-1} = 0 \\[2mm]
\dfrac{dP_{l+1}}{dx} - x\dfrac{dP_l}{dx} - (l+1)P_l = 0 \\[2mm]
(x^2-1)\dfrac{dP_l}{dx} - lxP_l + lP_{l-1} = 0
\end{array}\right\} \tag{3.29}$$

As an illustration of the use of these recurrence formulas consider the evaluation of the integral:

$$I_1 = \int_{-1}^{1} x P_l(x) P_{l'}(x)\, dx \tag{3.30}$$

From the first of the recurrence formulas (3.29) we obtain an expression for $xP(x)$. Therefore (3.30) becomes

$$I_1 = \frac{1}{2l + 1} \int_{-1}^{1} P_{l'}(x)[(l + 1)P_{l+1}(x) + lP_{l-1}(x)]\, dx$$

The orthogonality integral (3.21) can now be employed to show that the integral vanishes unless $l' = l \pm 1$, and that, for those values,

$$\int_{-1}^{1} x P_l(x) P_{l'}(x)\, dx = \begin{cases} \dfrac{2(l + 1)}{(2l + 1)(2l + 3)}, & l' = l + 1 \\[3ex] \dfrac{2l}{(2l - 1)(2l + 1)}, & l' = l - 1 \end{cases} \tag{3.31}$$

These are really the same result with the roles of l and l' interchanged. In a similar manner it is easy to show that

$$\int_{-1}^{1} x^2 P_l(x) P_{l'}(x)\, dx = \begin{cases} \dfrac{2(l + 1)(l + 2)}{(2l + 1)(2l + 3)(2l + 5)}, & l' = l + 2 \\[3ex] \dfrac{2(2l^2 + 2l - 1)}{(2l - 1)(2l + 1)(2l + 3)}, & l' = l \end{cases} \tag{3.32}$$

where it is assumed that $l' \geq l$.

3.3 Boundary-Value Problems with Azimuthal Symmetry

From the form of the solution of Laplace's equation in spherical coordinates (3.2) it will be seen that, for a problem possessing azimuthal symmetry, $m = 0$ in (3.5). This means that the general solution for such a problem is:

$$\Phi(r, \theta) = \sum_{l=0}^{\infty} [A_l r^l + B_l r^{-(l+1)}] P_l(\cos \theta) \tag{3.33}$$

The coefficients A_l and B_l can be determined from the boundary conditions. Suppose that the potential is specified to be $V(\theta)$ on the surface of a sphere of radius a, and it is required to find the potential inside the sphere. If there are no charges at the origin, the potential must be finite there. Consequently $B_l = 0$ for all l. The coefficients A_l are found by evaluating

(3.33) on the surface of the sphere:

$$V(\theta) = \sum_{l=0}^{\infty} A_l a^l P_l(\cos \theta) \qquad (3.34)$$

This is just a Legendre series of the form (3.23), so that the coefficients A_l are:

$$A_l = \frac{2l + 1}{2a^l} \int_0^{\pi} V(\theta) P_l(\cos \theta) \sin \theta \, d\theta \qquad (3.35)$$

If, for example, $V(\theta)$ is that of Section 2.8, with two hemispheres at equal and opposite potentials,

$$V(\theta) = \begin{cases} +V, & 0 \le \theta < \dfrac{\pi}{2} \\[2mm] -V, & \dfrac{\pi}{2} < \theta \le \pi \end{cases} \qquad (3.36)$$

then the coefficients are proportional to those in (3.27). Thus the potential inside the sphere is:

$$\Phi(r, \theta) = V\left[\frac{3}{2}\frac{r}{a} P_1(\cos \theta) - \frac{7}{8}\left(\frac{r}{a}\right)^3 P_3(\cos \theta) + \frac{11}{16}\left(\frac{r}{a}\right)^5 P_5(\cos \theta) - \cdots\right] \qquad (3.37)$$

To find the potential outside the sphere we merely replace $(r/a)^l$ by $(a/r)^{l+1}$. The resulting potential can be seen to be the same as (2.33), obtained by another means.

Series (3.33), with its coefficients determined by the boundary conditions, is a unique expansion of the potential. This uniqueness provides a means of obtaining the solution of potential problems from a knowledge of the potential in a limited domain, namely on the symmetry axis. On the symmetry axis (3.33) becomes (with $z = r$):

$$\Phi(z = r) = \sum_{l=0}^{\infty} \left[A_l r^l + B_l r^{-(l+1)}\right] \qquad (3.38)$$

valid for positive z. For negative z each term must be multiplied by $(-1)^l$. Suppose that, by some means, we can evaluate the potential $\Phi(z)$ at an arbitrary point z on the symmetry axis. If this potential function can be expanded in a power series in $z = r$ of the form (3.38), with known coefficients, then the solution for the potential at any point in space is obtained by multiplying each power of r^l and $r^{-(l+1)}$ by $P_l(\cos \theta)$.

Classical Electrodynamics

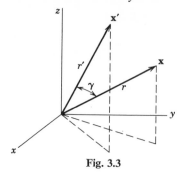

Fig. 3.3

At the risk of boring the reader we return to the problem of the hemi-spheres at equal and opposite potentials. We have already obtained the series solution in two different ways, (2.33) and (3.37). The method just stated gives a third way. For a point on the axis we have found the closed form (2.28):

$$\Phi(z = r) = V\left[1 - \frac{r^2 - a^2}{r\sqrt{r^2 + a^2}}\right]$$

This can be expanded in powers of a^2/r^2:

$$\Phi(z = r) = \frac{V}{\sqrt{\pi}} \sum_{j=1}^{\infty} (-1)^{j-1} \frac{(2j - \frac{1}{2})\Gamma(j - \frac{1}{2})}{j!} \left(\frac{a}{r}\right)^{2j} \qquad (3.39)$$

Comparison with expansion (3.38) shows that only odd l values ($l = 2j - 1$) enter. The solution, valid for all points outside the sphere, is consequently:

$$\Phi(r, \theta) = \frac{V}{\sqrt{\pi}} \sum_{j=1}^{\infty} (-1)^{j-1} \frac{(2j - \frac{1}{2})\Gamma(j - \frac{1}{2})}{j!} \left(\frac{a}{r}\right)^{2j} P_{2j-1}(\cos \theta) \qquad (3.40)$$

This is the same solution as already obtained, (2.33) and (3.37).

An important expansion is that of the potential at $\mathbf{x}$ due to a unit point charge at $\mathbf{x}'$:

$$\frac{1}{|\mathbf{x} - \mathbf{x}'|} = \sum_{l=0}^{\infty} \frac{r_<^l}{r_>^{l+1}} P_l(\cos \gamma) \qquad (3.41)$$

where $r_<$ ($r_>$) is the smaller (larger) of $|\mathbf{x}|$ and $|\mathbf{x}'|$, and γ is the angle between $\mathbf{x}$ and $\mathbf{x}'$, as shown in Fig. 3.3. This can be proved by rotating axes so that $\mathbf{x}'$ lies along the z axis. Then the potential satisfies Laplace's equation, possesses azimuthal symmetry, and can be expanded according to (3.33), except at the point $\mathbf{x} = \mathbf{x}'$:

$$\frac{1}{|\mathbf{x} - \mathbf{x}'|} = \sum_{l=0}^{\infty} (A_l r^l + B_l r^{-(l+1)}) P_l(\cos \gamma) \qquad (3.42)$$

If the point **x** is on the z axis, the right-hand side reduces to (3.38), while the left-hand side becomes:

$$\frac{1}{|\mathbf{x} - \mathbf{x'}|} \equiv \frac{1}{(r^2 + r'^2 - 2rr'\cos\gamma)^{\frac{1}{2}}} \rightarrow \frac{1}{|r - r'|} \tag{3.43}$$

Expanding (3.43), we find

$$\frac{1}{|\mathbf{x} - \mathbf{x'}|} = \frac{1}{r_>} \sum_{l=0}^{\infty} \left(\frac{r_<}{r_>}\right)^l \tag{3.44}$$

For points off the axis it is only necessary, according to (3.33) and (3.38), to multiply each term in (3.44) by $P_l(\cos\gamma)$. This proves the general result (3.41).

Another example is the potential due to a total charge q uniformly distributed around a circular ring of radius a, located as shown in Fig. 3.4, with its axis the z axis and its center at $z = b$. The potential at a point P on the axis of symmetry with $z = r$ is just q divided by the distance AP:

$$\Phi(z = r) = \frac{q}{(r^2 + c^2 - 2cr\cos\alpha)^{\frac{1}{2}}} \tag{3.45}$$

where $c^2 = a^2 + b^2$ and $\alpha = \tan^{-1}(a/b)$. The inverse distance AP can be expanded using (3.41). Thus, for $r > c$,

$$\Phi(z = r) = q\sum_{l=0}^{\infty} \frac{c^l}{r^{l+1}} P_l(\cos\alpha) \tag{3.46}$$

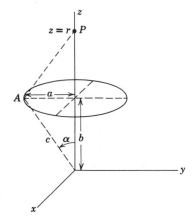

Fig. 3.4 Ring of charge of radius a and total charge q located on the z axis with center at $z = b$.

For $r < c$, the corresponding form is:

$$\Phi(z = r) = q \sum_{l=0}^{\infty} \frac{r^l}{c^{l+1}} P_l(\cos \alpha) \qquad (3.47)$$

The potential at *any point* in space is now obtained by multiplying each member of these series by $P_l(\cos \theta)$:

$$\Phi(r, \theta) = q \sum_{l=0}^{\infty} \frac{r^l_<}{r^{l+1}_>} P_l(\cos \alpha) P_l(\cos \theta) \qquad (3.48)$$

where $r_<$ $(r_>)$ is the smaller (larger) of r and c.

3.4 Associated Legendre Polynomials and the Spherical Harmonics $Y_{lm}(\theta,\phi)$

So far we have dealt with potential problems possessing azimuthal symmetry with solutions of the form (3.33). These involve only ordinary Legendre polynomials. The general potential problem can, however, have azimuthal variations so that $m \neq 0$ in (3.5) and (3.9). Then we need the generalization of $P_l(\cos \theta)$, namely, the solution of (3.9) with l and m both arbitrary. In essentially the same manner as for the ordinary Legendre functions it can be shown that in order to have finite solutions on the interval $-1 \leq x \leq 1$ the parameter l *must be zero or a positive integer* and that the *integer m* can take on only the values $-l$, $-(l - 1), \ldots, 0, \ldots,$ $(l - 1)$, l. The solution having these properties is called an associated Legendre function $P_l{}^m(x)$. For positive m it is defined by the formula*:

$$P_l{}^m(x) = (-1)^m (1 - x^2)^{m/2} \frac{d^m}{dx^m} P_l(x) \qquad (3.49)$$

If Rodrigues' formula is used to represent $P_l(x)$, a definition valid for both positive and negative m is obtained:

$$P_l{}^m(x) = \frac{(-1)^m}{2^l l!} (1 - x^2)^{m/2} \frac{d^{l+m}}{dx^{l+m}} (x^2 - 1)^l \qquad (3.50)$$

---

* The choice of phase for $P_l{}^m(x)$ is that of Magnus and Oberhettinger, and of E. U. Condon and G. H. Shortley in *Theory of Atomic Spectra*, Cambridge University Press (1953). For explicit expressions and recursion formulas, see Magnus and Oberhettinger, p. 54.

$P_l^{-m}(x)$ and $P_l^m(x)$ are proportional, since differential equation (3.9) depends only on m^2 and m is an integer. It can be shown that

$$P_l^{-m}(x) = (-1)^m \frac{(l-m)!}{(l+m)!} P_l^m(x) \qquad (3.51)$$

For fixed m the functions $P_l^m(x)$ form an orthogonal set in the index l on the interval $-1 \leq x \leq 1$. By the same means as for the Legendre functions the orthogonality relation can be obtained:

$$\int_{-1}^{1} P_{l'}^m(x) P_l^m(x)\, dx = \frac{2}{2l+1} \frac{(l+m)!}{(l-m)!} \delta_{l'l} \qquad (3.52)$$

The solution of Laplace's equation was decomposed into a product of factors for the three variables r, θ, and ϕ. It is convenient to combine the angular factors and construct orthonormal functions over the unit sphere. We will call these functions *spherical harmonics*, although this terminology is often reserved for solutions of the generalized Legendre equation (3.9). Our spherical harmonics are sometimes called "tesseral harmonics" in older books. The functions $Q_m(\phi) = e^{im\phi}$ form a complete set of ortho-gonal functions in the index m on the interval $0 \leq \phi \leq 2\pi$. The functions $P_l^m(\cos\theta)$ form a similar set in the index l for each m value on the interval $-1 \leq \cos\theta \leq 1$. Therefore their product $P_l^m Q_m$ will form a complete orthogonal set on the surface of the unit sphere in the two indices l, m. From the normalization condition (3.52) it is clear that the suitably normalized functions, denoted by $Y_{lm}(\theta, \phi)$, are:

$$Y_{lm}(\theta, \phi) = \sqrt{\frac{2l+1}{4\pi} \frac{(l-m)!}{(l+m)!}} P_l^m(\cos\theta) e^{im\phi} \qquad (3.53)$$

From (3.51) it can be seen that

$$Y_{l,-m}(\theta, \phi) = (-1)^m Y_{lm}^*(\theta, \phi) \qquad (3.54)$$

The normalization and orthogonality conditions are

$$\int_0^{2\pi} d\phi \int_0^{\pi} \sin\theta\, d\theta\, Y_{l'm'}^*(\theta, \phi) Y_{lm}(\theta, \phi) = \delta_{l'l}\delta_{m'm} \qquad (3.55)$$

The completeness relation, equivalent to (2.41), is

$$\sum_{l=0}^{\infty} \sum_{m=-l}^{l} Y_{lm}^*(\theta', \phi') Y_{lm}(\theta, \phi) = \delta(\phi - \phi')\delta(\cos\theta - \cos\theta') \qquad (3.56)$$

For a few small l values and $m \geq 0$ the table shows the explicit form of the $Y_{lm}(\theta, \phi)$. For negative m values (3.54) can be used.

Spherical harmonics $Y_{lm}(\theta, \phi)$

$l = 0$ $Y_{00} = \dfrac{1}{\sqrt{4\pi}}$

$l = 1$

$$Y_{11} = -\sqrt{\dfrac{3}{8\pi}}\, \sin\theta e^{i\phi}$$

$$Y_{10} = \sqrt{\dfrac{3}{4\pi}}\, \cos\theta$$

$l = 2$

$$Y_{22} = \dfrac{1}{4}\sqrt{\dfrac{15}{2\pi}}\, \sin^2\theta e^{2i\phi}$$

$$Y_{21} = -\sqrt{\dfrac{15}{8\pi}}\, \sin\theta \cos\theta e^{i\phi}$$

$$Y_{20} = \sqrt{\dfrac{5}{4\pi}}\left(\dfrac{3}{2}\cos^2\theta - \dfrac{1}{2}\right)$$

$l = 3$

$$Y_{33} = -\dfrac{1}{4}\sqrt{\dfrac{35}{4\pi}}\, \sin^3\theta e^{3i\phi}$$

$$Y_{32} = \dfrac{1}{4}\sqrt{\dfrac{105}{2\pi}}\, \sin^2\theta \cos\theta e^{2i\phi}$$

$$Y_{31} = -\dfrac{1}{4}\sqrt{\dfrac{21}{4\pi}}\, \sin\theta(5\cos^2\theta - 1)e^{i\phi}$$

$$Y_{30} = \sqrt{\dfrac{7}{4\pi}}\left(\dfrac{5}{2}\cos^3\theta - \dfrac{3}{2}\cos\theta\right)$$

Note that, for $m = 0$,

$$Y_{l0}(\theta, \phi) = \sqrt{\dfrac{2l + 1}{4\pi}}\, P_l(\cos\theta) \tag{3.57}$$

An arbitrary function $g(\theta, \phi)$ can be expanded in spherical harmonics:

$$g(\theta, \phi) = \sum_{l=0}^{\infty} \sum_{m=-l}^{l} A_{lm} Y_{lm}(\theta, \phi) \tag{3.58}$$

where the coefficients are

$$A_{lm} = \int d\Omega\, Y_{lm}^*(\theta, \phi) g(\theta, \phi)$$

A point of interest to us in the next section is the form of the expansion for $\theta = 0$. With definition (3.57), we find:

$$[g(\theta, \phi)]_{\theta=0} = \sum_{l=0}^{\infty} \sqrt{\frac{2l+1}{4\pi}} A_{l0} \tag{3.59}$$

where

$$A_{l0} = \sqrt{\frac{2l+1}{4\pi}} \int d\Omega \; P_l(\cos \theta) g(\theta, \phi) \tag{3.60}$$

All terms in the series with $m \neq 0$ vanish at $\theta = 0$.

The general solution for a boundary-value problem in spherical coordinates can be written in terms of spherical harmonics and powers of r in a generalization of (3.33):

$$\Phi(r, \theta, \phi) = \sum_{l=0}^{\infty} \sum_{m=-l}^{l} [A_{lm}r^l + B_{lm}r^{-(l+1)}]Y_{lm}(\theta, \phi) \tag{3.61}$$

If the potential is specified on a spherical surface, the coefficients can be determined by evaluating (3.61) on the surface and using (3.58).

3.5   Addition Theorem for Spherical Harmonics

A mathematical result of considerable interest and use is called the *addition theorem* for spherical harmonics. Two coordinate vectors **x** and **x′**, with spherical coordinates (r, θ, ϕ) and (r', θ', ϕ'), respectively, have an angle γ between them, as shown in Fig. 3.5. The addition theorem expresses a Legendre polynomial of order l in the angle γ in terms of

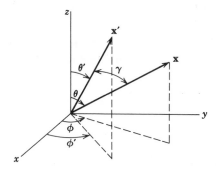

Fig. 3.5

products of the spherical harmonics of the angles θ, ϕ and θ', ϕ':

$$P_l(\cos \gamma) = \frac{4\pi}{2l + 1} \sum_{m=-l}^{l} Y_{lm}^*(\theta', \phi') Y_{lm}(\theta, \phi) \tag{3.62}$$

where $\cos \gamma = \cos \theta \cos \theta' + \sin \theta \sin \theta' \cos(\phi - \phi')$. To prove this theorem we consider the vector $\mathbf{x}'$ as fixed in space. Then $P_l(\cos \gamma)$ is a function of the angles θ, ϕ, with the angles θ', ϕ' as parameters. It may be expanded in a series (3.58):

$$P_l(\cos \gamma) = \sum_{l'=0}^{\infty} \sum_{m=-l'}^{l'} A_{l'm}(\theta', \phi') Y_{l'm}(\theta, \phi) \tag{3.63}$$

Comparison with (3.62) shows that only terms with $l' = l$ appear. To see why this is so, note that, if coordinate axes are chosen so that $\mathbf{x}'$ is on the z axis, then γ becomes the usual polar angle and $P_l(\cos \gamma)$ satisfies the equation:

$$\nabla'^2 P_l(\cos \gamma) + \frac{l(l + 1)}{r^2} P_l(\cos \gamma) = 0 \tag{3.64}$$

where ∇'^2 is the Laplacian referred to these new axes. If the axes are now rotated to the position shown in Fig. 3.5, $\nabla'^2 = \nabla^2$ and r is unchanged.* Consequently $P_l(\cos \gamma)$ still satisfies an equation of the form (3.64); i.e., it is a spherical harmonic of order l. This means that it is a linear combination of Y_{lm}'s of that order only:

$$P_l(\cos \gamma) = \sum_{m=-l}^{l} A_m(\theta', \phi') Y_{lm}(\theta, \phi) \tag{3.65}$$

The coefficients $A_m(\theta', \phi')$ are given by:

$$A_m(\theta', \phi') = \int Y_{lm}^*(\theta, \phi) P_l(\cos \gamma) \, d\Omega \tag{3.66}$$

To evaluate this coefficient we note that it may be viewed, according to (3.60), as the $m' = 0$ coefficient in an expansion of the function $\sqrt{4\pi/(2l + 1)} \; Y_{lm}^*(\theta, \phi)$ in a series of $Y_{lm'}(\gamma, \beta)$ referred to the primed axis of (3.64). From (3.59) it is then found that, since only one l value is present, coefficient (3.66) is

$$A_m(\theta', \phi') = \frac{4\pi}{2l + 1} [Y_{lm}^*(\theta(\gamma, \beta), \phi(\gamma, \beta))]_{\gamma=0} \tag{3.67}$$

In the limit $\gamma \to 0$, the angles (θ, ϕ), as functions of (γ, β), go over into

---

* The proof that $\nabla'^2 = \nabla^2$ under rotations follows most easily from noting that $\nabla^2 \psi = \nabla \cdot \nabla \psi$ is an operator scalar product, and that all scalar products are invariant under rotations.

(θ', ϕ'). Thus addition theorem (3.62) is proved. Sometimes the theorem is written in terms of $P_l^m(\cos \theta)$ rather than Y_{lm}. Then it has the form:

$$P_l(\cos \gamma) = P_l(\cos \theta)P_l(\cos \theta')$$

$$+2 \sum_{m=1}^{l} \frac{(l-m)!}{(l+m)!} P_l^m(\cos \theta)P_l^m(\cos \theta') \cos [m(\phi - \phi')] \quad (3.68)$$

It the angle γ goes to zero, there results a "sum rule" for the squares of Y_{lm}'s:

$$\sum_{m=-l}^{l} |Y_{lm}(\theta, \phi)|^2 = \frac{2l+1}{4\pi} \quad (3.69)$$

The addition theorem can be used to put expansion (3.41) of the potential at x due to a unit charge at x' into its most general form. Substituting (3.62) for $P_l(\cos \gamma)$ into (3.41) we obtain

$$\frac{1}{|\mathbf{x} - \mathbf{x}'|} = 4\pi \sum_{l=0}^{\infty} \sum_{m=-l}^{l} \frac{1}{2l+1} \frac{r_<^l}{r_>^{l+1}} Y_{lm}^*(\theta', \phi')Y_{lm}(\theta, \phi) \quad (3.70)$$

Equation (3.70) gives the potential in a completely factorized form in the coordinates x and x'. This is useful in any integrations over charge densities, etc., where one variable is the variable of integration and the other is the coordinate of the observation point. The price paid is that there is a double sum involved, rather than a single term.

3.6 Laplace's Equation in Cylindrical Coordinates; Bessel Functions

In cylindrical coordinates (ρ, ϕ, z), as shown in Fig. 3.6, Laplace's equation takes the form:

$$\frac{\partial^2 \Phi}{\partial \rho^2} + \frac{1}{\rho} \frac{\partial \Phi}{\partial \rho} + \frac{1}{\rho^2} \frac{\partial^2 \Phi}{\partial \phi^2} + \frac{\partial^2 \Phi}{\partial z^2} = 0 \quad (3.71)$$

The separation of variables is accomplished by the substitution:

$$\Phi(\rho, \phi, z) = R(\rho)Q(\phi)Z(z) \quad (3.72)$$

In the usual way this leads to the three ordinary differential equations:

$$\frac{d^2 Z}{dz^2} - k^2 Z = 0 \quad (3.73)$$

$$\frac{d^2 Q}{d\phi^2} + v^2 Q = 0 \quad (3.74)$$

$$\frac{d^2 R}{d\rho^2} + \frac{1}{\rho} \frac{dR}{d\rho} + \left(k^2 - \frac{v^2}{\rho^2}\right)R = 0 \quad (3.75)$$

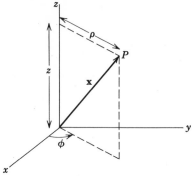

Fig. 3.6

The solutions of the first two equations are elementary:

$$Z(z) = e^{\pm kz}$$
$$Q(\phi) = e^{\pm i\nu\phi} \tag{3.76}$$

In order that the potential be single valued, ν must be an integer. But barring some boundary-condition requirement in the z direction, the parameter k is arbitrary. For the present we will assume that k is real.

The radial equation can be put in a standard form by the change of variable $x = k\rho$. Then it becomes

$$\frac{d^2R}{dx^2} + \frac{1}{x}\frac{dR}{dx} + \left(1 - \frac{\nu^2}{x^2}\right)R = 0 \tag{3.77}$$

This is Bessel's equation, and the solutions are called *Bessel functions* of order ν. If a power series solution of the form:

$$R(x) = x^\alpha \sum_{j=0}^{\infty} a_j x^j \tag{3.78}$$

is assumed, then it is found that

$$\alpha = \pm\nu \tag{3.79}$$

and

$$a_{2j} = -\frac{1}{4j(j+\alpha)}a_{2j-2} \tag{3.80}$$

for $j = 0, 1, 2, 3, \ldots$. All odd powers of x^j have vanishing coefficients. The recursion formula can be iterated to obtain

$$a_{2j} = \frac{(-1)^j\Gamma(\alpha+1)}{2^{2j}j!\,\Gamma(j+\alpha+1)}a_0 \tag{3.81}$$

It is conventional to choose the constant $a_0 = [2^\alpha \Gamma(\alpha + 1)]^{-1}$. Then the two solutions are

$$J_\nu(x) = \left(\frac{x}{2}\right)^\nu \sum_{j=0}^{\infty} \frac{(-1)^j}{j!\,\Gamma(j + \nu + 1)} \left(\frac{x}{2}\right)^{2j} \qquad (3.82)$$

$$J_{-\nu}(x) = \left(\frac{x}{2}\right)^{-\nu} \sum_{j=0}^{\infty} \frac{(-1)^j}{j!\,\Gamma(j - \nu + 1)} \left(\frac{x}{2}\right)^{2j} \qquad (3.83)$$

These solutions are called Bessel functions of the first kind of order $\pm\nu$. The series converge for all finite values of x. If ν *is not an integer*, these two solutions $J_{\pm\nu}(x)$ form a pair of linearly independent solutions to the second-order Bessel's equation. However, if ν is an integer, it is well known that the solutions are linearly dependent. In fact, for $\nu = m$, an integer, it can be seen from the series representation that

$$J_{-m}(x) = (-1)^m J_m(x) \qquad (3.84)$$

Consequently it is necessary to find another linearly independent solution when m is an integer. It is customary, even if ν is not an integer, to replace the pair $J_{\pm\nu}(x)$ by $J_\nu(x)$ and $N_\nu(x)$, the Neumann function (or Bessel's function of the second kind):

$$N_\nu(x) = \frac{J_\nu(x) \cos \nu\pi - J_{-\nu}(x)}{\sin \nu\pi} \qquad (3.85)$$

For ν not an integer, $N_\nu(x)$ is clearly linearly independent of $J_\nu(x)$. In the limit $\nu \to$ integer, it can be shown that $N_\nu(x)$ is still linearly independent of $J_\nu(x)$. As expected, it involves $\log x$. Its series representation is given in the reference books.

The Bessel functions of the third kind, called *Hankel functions*, are defined as linear combinations of $J_\nu(x)$ and $N_\nu(x)$:

$$\left. \begin{aligned} H_\nu^{(1)}(x) &= J_\nu(x) + iN_\nu(x) \\ H_\nu^{(2)}(x) &= J_\nu(x) - iN_\nu(x) \end{aligned} \right\} \qquad (3.86)$$

The Hankel functions form a fundamental set of solutions to Bessel's equation, just as do $J_\nu(x)$ and $N_\nu(x)$.

The functions J_ν, N_ν, $H_\nu^{(1)}$, $H_\nu^{(2)}$ all satisfy the recursion formulas:

$$\Omega_{\nu-1}(x) + \Omega_{\nu+1}(x) = \frac{2\nu}{x} \Omega_\nu(x) \qquad (3.87)$$

$$\Omega_{\nu-1}(x) - \Omega_{\nu+1}(x) = 2\frac{d\Omega_\nu(x)}{dx} \qquad (3.88)$$

where $\Omega_\nu(x)$ is any one of the cylinder functions of order ν. These may be verified directly from the series representation (3.82).

For reference purposes, the limiting forms of the various kinds of Bessel functions will be given for small and large values of their argument. Only the leading terms will be given for simplicity:

$$x \ll 1 \qquad J_\nu(x) \to \frac{1}{\Gamma(\nu+1)}\left(\frac{x}{2}\right)^\nu \tag{3.89}$$

$$N_\nu(x) \to \begin{cases} \dfrac{2}{\pi}\left(\ln\left(\dfrac{x}{2}\right) + 0.5772\cdots\right), & \nu = 0 \\[2ex] -\dfrac{\Gamma(\nu)}{\pi}\left(\dfrac{2}{x}\right)^\nu, & \nu \neq 0 \end{cases} \tag{3.90}$$

In these formulas ν is assumed to be real and nonnegative.

$$x \gg 1, \nu \qquad J_\nu(x) \to \sqrt{\frac{2}{\pi x}}\cos\left(x - \frac{\nu\pi}{2} - \frac{\pi}{4}\right)$$

$$N_\nu(x) \to \sqrt{\frac{2}{\pi x}}\sin\left(x - \frac{\nu\pi}{2} - \frac{\pi}{4}\right) \tag{3.91}$$

The transition from the small x behavior to the large x asymptotic form occurs in the region of $x \sim \nu$.

From the asymptotic forms (3.91) it is clear that each Bessel function has an infinite number of roots. We will be chiefly concerned with the roots of $J_\nu(x)$:

$$J_\nu(x_{\nu n}) = 0, \qquad n = 1, 2, 3, \ldots \tag{3.92}$$

$x_{\nu n}$ is the nth root of $J_\nu(x)$. For the first few integer values of ν, the first three roots are:

$$\nu = 0, \quad x_{0n} = 2.405,\ 5.520,\ 8.654,\ldots$$
$$\nu = 1, \quad x_{1n} = 3.832,\ 7.016,\ 10.173,\ldots$$
$$\nu = 2, \quad x_{2n} = 5.136,\ 8.417,\ 11.620,\ldots$$

For higher roots, the asymptotic formula

$$x_{\nu n} \simeq n\pi + (\nu - \tfrac{1}{2})\frac{\pi}{2}$$

gives adequate accuracy (to at least three figures). Tables of roots are given in Jahnke and Emde, pp. 166–168.

Having found the solution of the radial part of Laplace's equation in terms of Bessel functions, we can now ask in what sense the Bessel functions form an orthogonal, complete set of functions. We will consider only Bessel functions of the first kind, and will show that $\sqrt{\rho}\, J_\nu(x_{\nu n}\rho/a)$, for fixed $\nu \geq 0$, $n = 1, 2, \ldots$, form an orthogonal set on the interval $0 \leq \rho \leq a$. The demonstration starts with the differential equation satisfied by $J_\nu(x_{\nu n}\rho/a)$:

$$\frac{1}{\rho}\frac{d}{d\rho}\left(\rho\frac{dJ_\nu\left(x_{\nu n}\dfrac{\rho}{a}\right)}{d\rho}\right) + \left(\frac{x_{\nu n}^{\;2}}{a^2} - \frac{\nu^2}{\rho^2}\right)J_\nu\left(x_{\nu n}\frac{\rho}{a}\right) = 0 \qquad (3.93)$$

If we multiply the equation by $\rho J_\nu(x_{\nu n'}\rho/a)$ and integrate from 0 to a, we obtain

$$\int_0^a J_\nu\left(x_{\nu n'}\frac{\rho}{a}\right)\frac{d}{d\rho}\left(\rho\frac{dJ_\nu\left(x_{\nu n}\dfrac{\rho}{a}\right)}{d\rho}\right)d\rho$$

$$+\int_0^a\left(\frac{x_{\nu n}^{\;2}}{a^2} - \frac{\nu^2}{\rho^2}\right)\rho J_\nu\left(x_{\nu n'}\frac{\rho}{a}\right)J_\nu\left(x_{\nu n}\frac{\rho}{a}\right)d\rho = 0$$

Integration by parts, combined with the vanishing of $(\rho J_\nu J_\nu')$ at $\rho = 0$ (for $\nu \geq 0$) and $\rho = a$, leads to the result:

$$-\int_0^a \rho\,\frac{dJ_\nu\left(x_{\nu n'}\dfrac{\rho}{a}\right)}{d\rho}\frac{dJ_\nu\left(x_{\nu n}\dfrac{\rho}{a}\right)}{d\rho}\,d\rho$$

$$+\int_0^a\left(\frac{x_{\nu n}^2}{a^2} - \frac{\nu^2}{\rho^2}\right)\rho J_\nu\left(x_{\nu n'}\frac{\rho}{a}\right)J_\nu\left(x_{\nu n}\frac{\rho}{a}\right)d\rho = 0$$

If we now write down the same expression, with n and n' interchanged, and subtract, we obtain the orthogonality condition:

$$(x_{\nu n}^2 - x_{\nu n'}^2)\int_0^a \rho J_\nu\left(x_{\nu n'}\frac{\rho}{a}\right)J_\nu\left(x_{\nu n}\frac{\rho}{a}\right)d\rho = 0 \qquad (3.94)$$

By means of the recursion formulas (3.87) and (3.88) and the differential equation, the normalization integral can be found to be:

$$\int_0^a \rho J_\nu\left(x_{\nu n'}\frac{\rho}{a}\right)J_\nu\left(x_{\nu n}\frac{\rho}{a}\right)d\rho = \frac{a^2}{2}\left[J_{\nu+1}(x_{\nu n})\right]^2\delta_{n'n} \qquad (3.95)$$

Assuming that the set of Bessel functions is complete, we can expand an arbitrary function of ρ on the interval $0 \leq \rho \leq a$ in a Bessel-Fourier series:

$$f(\rho) = \sum_{n=1}^{\infty} A_{\nu n}J_\nu\left(x_{\nu n}\frac{\rho}{a}\right) \qquad (3.96)$$

where

$$A_{vn} = \frac{2}{a^2 J_{v+1}^2(x_{vn})} \int_0^a \rho f(\rho) J_v\left(\frac{x_{vn}\rho}{a}\right) d\rho \qquad (3.97)$$

Our derivation of (3.96) involved the restriction $v \geq 0$. Actually it can be proved to hold for all $v \geq -1$.

Expansion (3.96) and (3.97) is the conventional Fourier-Bessel series and is particularly appropriate to functions which vanish at $\rho = a$ (e.g., homogeneous Dirichlet boundary conditions on a cylinder; see the following section). But it will be noted that an alternative expansion is possible in a series of functions $\sqrt{\rho} J_v(y_{vn}\rho/a)$ where y_{vn} is the nth root of the equation $[dJ_v(x)]/dx = 0$. The reason is that, in proving the orthogonality of the functions, all that is demanded is that the quantity $[\rho J_v(\lambda\rho)(d/d\rho) J_v(\lambda'\rho)]$ vanish at the end points $\rho = 0$ and $\rho = a$. The requirement is met by *either* $\lambda = x_{vn}/a$ *or* $\lambda = y_{vn}/a$, where $J_v(x_{vn}) = 0$ and $J_v'(y_{vn}) = 0$. The expansion in terms of the set $\sqrt{\rho} J_v(y_{vn}\rho/a)$ is especially useful for functions with vanishing slope at $\rho = a$. (See Problem 3.8.)

A Fourier-Bessel series is only one type of expansion involving Bessel functions. Neumann series $\left[\sum_{n=0}^{\infty} a_n J_{v+n}(z)\right]$, Kapteyn series $\left[\sum_{n=0}^{\infty} a_n \times J_{v+n}((v + n)z)\right]$, and Schlömilch series $\left[\sum_{n=1}^{\infty} a_n J_v(nx)\right]$ are some of the other possibilities. The reader may refer to Watson, Chapters XVI–XIX, for a detailed discussion of the properties of these series. Kapteyn series occur in the discussion of the Kepler motion of planets and of radiation by rapidly moving charges (see Problems 14.7 and 14.8).

Before leaving the properties of Bessel functions it should be noted that if, in the separation of Laplace's equation, the separation constant k^2 in (3.73) had been taken as $-k^2$, then $Z(z)$ would have been sin kz or cos kz and the equation for $R(\rho)$ would have been:

$$\frac{d^2R}{d\rho^2} + \frac{1}{\rho}\frac{dR}{d\rho} - \left(k^2 + \frac{v^2}{\rho^2}\right)R = 0 \qquad (3.98)$$

With $k\rho = x$, this becomes

$$\frac{d^2R}{dx^2} + \frac{1}{x}\frac{dR}{dx} - \left(1 + \frac{v^2}{x^2}\right)R = 0 \qquad (3.99)$$

The solutions of this equation are called *modified Bessel functions*. It is evident that they are just Bessel functions of pure imaginary argument.

The usual choices of linearly independent solutions are denoted by $I_\nu(x)$ and $K_\nu(x)$. They are defined by

$$I_\nu(x) = i^{-\nu} J_\nu(ix) \tag{3.100}$$

$$K_\nu(x) = \frac{\pi}{2} i^{\nu+1} H_\nu^{(1)}(ix) \tag{3.101}$$

and are real functions for real x. Their limiting forms for small and large x are, assuming real $\nu \geq 0$:

$$x \ll 1 \quad I_\nu(x) \to \frac{1}{\Gamma(\nu + 1)} \left(\frac{x}{2}\right)^\nu \tag{3.102}$$

$$K_\nu(x) \to \begin{cases} -\left(\ln\left(\frac{x}{2}\right) + 0.5772 \cdots\right), & \nu = 0 \\ \frac{\Gamma(\nu)}{2} \left(\frac{2}{x}\right)^\nu, & \nu \neq 0 \end{cases} \tag{3.103}$$

$$x \gg 1, \nu \quad I_\nu(x) \to \frac{1}{\sqrt{2\pi x}} e^x \left[1 + 0\left(\frac{1}{x}\right)\right] \\ K_\nu(x) \to \sqrt{\frac{\pi}{2x}} e^{-x} \left[1 + 0\left(\frac{1}{x}\right)\right] \tag{3.104}$$

3.7 Boundary-Value Problems in Cylindrical Coordinates

The solution of Laplace's equation in cylindrical coordinates is $\Phi = R(\rho)Q(\phi)Z(z)$, where the separate factors are given in the previous section. Consider now the specific boundary-value problem shown in Fig. 3.7. The cylinder has a radius a and a height L, the top and bottom surfaces being at $z = L$ and $z = 0$. The potential on the side and the bottom of the cylinder is zero, while the top has a potential $\Phi = V(\rho, \phi)$. We want to find the potential at any point inside the cylinder. In order that Φ be single valued and vanish at $z = 0$,

$$Q(\phi) = A \sin m\phi + B \cos m\phi \\ Z(z) = \sinh kz \tag{3.105}$$

where $\nu = m$ is an integer and k is a constant to be determined. The radial factor is

$$R(\rho) = CJ_m(k\rho) + DN_m(k\rho) \tag{3.106}$$

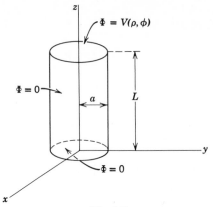

Fig. 3.7

If the potential is finite at $\rho = 0$, $D = 0$. The requirement that the potential vanish at $\rho = a$ means that k can take on only those special values:

$$k_{mn} = \frac{x_{mn}}{a}, \qquad n = 1, 2, 3, \ldots \tag{3.107}$$

where x_{mn} are the roots of $J_m(x_{mn}) = 0$.

Combining all these conditions, we find that the general form of the solution is

$$\Phi(\rho, \phi, z) = \sum_{m=0}^{\infty} \sum_{n=1}^{\infty} J_m(k_{mn}\rho) \sinh(k_{mn}z)[A_{mn} \sin m\phi + B_{mn} \cos m\phi] \tag{3.108}$$

At $z = L$, we are given the potential as $V(\rho, \phi)$. Therefore we have

$$V(\rho, \phi) = \sum_{m,n} \sinh(k_{mn}L)J_m(k_{mn}\rho)[A_{mn} \sin m\phi + B_{mn} \cos m\phi]$$

This is a Fourier series in ϕ and a Bessel-Fourier series in ρ. The coefficients are, from (2.43) and (3.97),

$$\left. \begin{aligned} A_{mn} &= \frac{2 \operatorname{cosech}(k_{mn}L)}{\pi a^2 J_{m+1}^2(k_{mn}a)} \int_0^{2\pi} d\phi \int_0^a d\rho\ \rho V(\rho, \phi)J_m(k_{mn}\rho) \sin m\phi \\ \text{and} \\ B_{mn} &= \frac{2 \operatorname{cosech}(k_{mn}L)}{\pi a^2 J_{m+1}^2(k_{mn}a)} \int_0^{2\pi} d\phi \int_0^a d\rho\ \rho V(\rho, \phi)J_m(k_{mn}\rho) \cos m\phi \end{aligned} \right\} \tag{3.109}$$

with the proviso that, for $m = 0$, we use $\frac{1}{2}B_{0n}$ in the series.

The particular form of expansion (3.108) is indicated by the requirement that the potential vanish at $z = 0$ for arbitrary ρ and at $\rho = a$ for arbitrary z. For different boundary conditions the expansion would take a different form. An example where the potential is zero on the end faces and equal to $V(\phi, z)$ on the side surface is left as Problem 3.6 for the reader.

The Fourier-Bessel series (3.108) is appropriate for a finite interval in $\rho, 0 \leq \rho \leq a$. If $a \to \infty$, the series goes over into an integral in a manner entirely analogous to the transition from a trigonometric Fourier series to a Fourier integral. Thus, for example, if the potential in charge-free space is finite for $z \geq 0$ and vanishes for $z \to \infty$, the general form of the solution for $z \geq 0$ must be

$$\Phi(\rho, \phi, z) = \sum_{m=0}^{\infty} \int_0^{\infty} dk \; e^{-kz} J_m(k\rho)[A_m(k) \sin m\phi + B_m(k) \cos m\phi] \quad (3.110)$$

If the potential is specified over the whole plane $z = 0$ to be $V(\rho, \phi)$ the coefficients are determined by

$$V(\rho, \phi) = \sum_{m=0}^{\infty} \int_0^{\infty} dk J_m(k\rho)[A_m(k) \sin m\phi + B_m(k) \cos m\phi]$$

The variation in ϕ is just a Fourier series. Consequently the coefficients $A_m(k)$ and $B_m(k)$ are separately specified by the integral relations:

$$\frac{1}{\pi}\int_0^{2\pi} V(\rho, \phi)\begin{Bmatrix} \sin m\phi \\ \cos m\phi \end{Bmatrix} d\phi = \int_0^{\infty} J_m(k'\rho)\begin{Bmatrix} A_m(k') \\ B_m(k') \end{Bmatrix} dk' \quad (3.111)$$

These radial integral equations of the first kind can be easily solved, since they are *Hankel transforms*. For our purposes, the integral relation,

$$\int_0^{\infty} x J_m(kx) J_m(k'x) \, dx = \frac{1}{k} \delta(k' - k) \quad (3.112)$$

can be exploited to invert equations (3.111). Multiplying both sides by $\rho J_m(k\rho)$ and integrating over ρ, we find with the help of (3.112) that the coefficients are determined by integrals over the whole area of the plane $z = 0$:

$$\begin{Bmatrix} A_m(k) \\ B_m(k) \end{Bmatrix} = \frac{k}{\pi}\int_0^{\infty} d\rho \, \rho \int_0^{2\pi} d\phi \, V(\rho, \phi) J_m(k\rho)\begin{Bmatrix} \sin m\phi \\ \cos m\phi \end{Bmatrix} \quad (3.113)$$

As usual, for $m = 0$, we must use $\frac{1}{2}B_0(k)$ in series (3.110).

3.8 Expansion of Green's Functions in Spherical Coordinates

In order to handle problems involving distributions of charge as well as boundary values for the potential (i.e., solutions of Poisson's equation) it is necessary to determine the Green's function $G(\mathbf{x}, \mathbf{x}')$ which satisfies the appropriate boundary conditions. Often these boundary conditions are specified on surfaces of some separable coordinate system, e.g., spherical or cylindrical boundaries. Then it is convenient to express the Green's function as a series of products of the functions appropriate to the coordinates in question. We first illustrate the type of expansion involved by considering spherical coordinates.

For the case of no boundary surfaces, except at infinity, we already have the expansion of the Green's function, namely (3.70):

$$\frac{1}{|\mathbf{x} - \mathbf{x}'|} = 4\pi \sum_{l=0}^{\infty} \sum_{m=-l}^{l} \frac{1}{2l+1} \frac{r_<^l}{r_>^{l+1}} Y_{lm}^*(\theta', \phi') Y_{lm}(\theta, \phi)$$

Suppose that we wish to obtain a similar expansion for the Green's function appropriate for the "exterior" problem with a spherical boundary at $r = a$. The result is readily found from the image form of the Green's function (2.22). Using expansion (3.70) for both terms in (2.22), we obtain:

$$G(\mathbf{x}, \mathbf{x}') = 4\pi \sum_{l,m} \frac{1}{2l+1} \left[\frac{r_<^l}{r_>^{l+1}} - \frac{1}{a}\left(\frac{a^2}{rr'}\right)^{l+1} \right] Y_{lm}^*(\theta', \phi') Y_{lm}(\theta, \phi) \quad (3.114)$$

To see clearly the structure of (3.114) and to verify that it satisfies the boundary conditions, we exhibit the radial factors separately for $r < r'$ and for $r > r'$:

$$\left[\frac{r_<^l}{r_>^{l+1}} - \frac{1}{a}\left(\frac{a^2}{rr'}\right)^{l+1} \right] = \begin{cases} \dfrac{1}{r'^{l+1}}\left[r^l - \dfrac{a^{2l+1}}{r^{l+1}} \right], & r < r' \\[3mm] \left[r'^l - \dfrac{a^{2l+1}}{r'^{l+1}} \right]\dfrac{1}{r^{l+1}}, & r > r' \end{cases} \quad (3.115)$$

First of all, we note that for either r or r' equal to a the radial factor vanishes, as required. Similarly, as r or $r' \to \infty$, the radial factor vanishes. It is symmetric in r and r'. Viewed as a function of r, for fixed r', the radial factor is just a linear combination of the solutions r^l and $r^{-(l+1)}$ of the radial part (3.7) of Laplace's equation. It is admittedly a different linear combination for $r < r'$ and for $r > r'$. The reason for this will become apparent below, and is connected with the fact that the Green's function is a solution of Poisson's equation with a delta function inhomogeneity.

Now that we have seen the general structure of the expansion of a Green's function in separable coordinates we turn to the systematic construction of such expansions from first principles. A Green's function for a potential problem satisfies the equation

$$\nabla_x^2 G(\mathbf{x}, \mathbf{x}') = -4\pi \, \delta(\mathbf{x} - \mathbf{x}') \quad (3.116)$$

subject to the boundary conditions $G(\mathbf{x}, \mathbf{x}') = 0$ for either $\mathbf{x}$ or $\mathbf{x}'$ on the boundary surface S. For spherical boundary surfaces we desire an expansion of the general form (3.114). Accordingly we exploit the fact that the delta function can be written*

$$\delta(\mathbf{x} - \mathbf{x}') = \frac{1}{r^2} \delta(r - r') \, \delta(\phi - \phi') \, \delta(\cos \theta - \cos \theta')$$

and that the completeness relation (3.56) can be used to represent the angular delta functions:

$$\delta(\mathbf{x} - \mathbf{x}') = \frac{1}{r^2} \delta(r - r') \sum_{l=0}^{\infty} \sum_{m=-l}^{l} Y_{lm}^*(\theta', \phi') Y_{lm}(\theta, \phi) \qquad (3.117)$$

Then the Green's function, considered as a function of $\mathbf{x}$, can be expanded as

$$G(\mathbf{x}, \mathbf{x}') = \sum_{l=0}^{\infty} \sum_{m=-l}^{l} A_{lm}(\theta', \phi') g_l(r, r') Y_{lm}(\theta, \phi) \qquad (3.118)$$

Substitution of (3.117) and (3.118) into (3.116) leads to the results

$$A_{lm}(\theta', \phi') = Y_{lm}^*(\theta', \phi') \qquad (3.119)$$

and

$$\frac{1}{r}\frac{d^2}{dr^2}(rg_l(r, r')) - \frac{l(l + 1)}{r^2} g_l(r, r') = -\frac{4\pi}{r^2} \delta(r - r') \qquad (3.120)$$

The radial Green's function is seen to satisfy the homogeneous radial equation (3.7) for $r \neq r'$. Thus it can be written as:

$$g_l(r, r') = \begin{cases} Ar^l + Br^{-(l+1)}, & \text{for } r < r' \\ A'r^l + B'r^{-(l+1)}, & \text{for } r > r' \end{cases}$$

The coefficients A, B, A', B' are functions of r' to be determined by the boundary conditions, the requirement implied by $\delta(r - r')$ in (3.120), and the symmetry of $g_l(r, r')$ in r and r'. Suppose that the boundary sufaces are concentric spheres at $r = a$ and $r = b$. The vanishing of $G(\mathbf{x}, \mathbf{x}')$ for $\mathbf{x}$ on

---

* To express $\delta(\mathbf{x} - \mathbf{x}') = \delta(x_1 - x_1') \, \delta(x_2 - x_2') \, \delta(x_3 - x_3')$ in terms of the coordinates (ξ_1, ξ_2, ξ_3), related to (x_1, x_2, x_3) via the Jacobian $J(x_i, \xi_i)$, we note that the meaningful quantity is $\delta(\mathbf{x} - \mathbf{x}') \, d^3x$. Hence

$$\delta(\mathbf{x} - \mathbf{x}') = \frac{1}{|J(x_i, \xi_i)|} \delta(\xi_1 - \xi_1') \, \delta(\xi_2 - \xi_2') \, \delta(\xi_3 - \xi_3')$$

the surface implies the vanishing of $g_l(r, r')$ for $r = a$ and $r = b$. Consequently $g_l(r, r')$ becomes

$$g_l(r, r') = \begin{cases} A\left(r^l - \dfrac{a^{2l+1}}{r^{l+1}}\right), & r < r' \\ B'\left(\dfrac{1}{r^{l+1}} - \dfrac{r^l}{b^{2l+1}}\right), & r > r' \end{cases} \qquad (3.121)$$

The symmetry in r and r' requires that the coefficients $A(r')$ and $B'(r')$ be such that $g_l(r, r')$ can be written

$$g_l(r, r') = C\left(r_<^l - \frac{a^{2l+1}}{r_<^{l+1}}\right)\left(\frac{1}{r_>^{l+1}} - \frac{r_>^l}{b^{2l+1}}\right) \qquad (3.122)$$

where $r_<$ $(r_>)$ is the smaller (larger) of r and r'. To determine the constant C we must consider the effect of the delta function in (3.120). If we multiply both sides by r and integrate over the interval from $r = r' - \epsilon$ to $r = r' + \epsilon$, where ϵ is very small, we obtain

$$\left[\frac{d}{dr}(rg_l(r, r'))\right]_{r'+\epsilon} - \left[\frac{d}{dr}(rg_l(r, r'))\right]_{r'-\epsilon} = -\frac{4\pi}{r'} \qquad (3.123)$$

Thus there is a discontinuity in slope at $r = r'$, as indicated in Fig. 3.8.
For $r = r' + \epsilon$, $r_> = r$, $r_< = r'$. Hence

$$\left[\frac{d}{dr}(rg_l(r, r'))\right]_{r'+\epsilon} = C\left(r'^l - \frac{a^{2l+1}}{r'^{l+1}}\right)\left[\frac{d}{dr}\left(\frac{1}{r^l} - \frac{r^{l+1}}{b^{2l+1}}\right)\right]_{r=r'}$$
$$= -\frac{C}{r'}\left(1 - \left(\frac{a}{r'}\right)^{2l+1}\right)\left(l + (l+1)\left(\frac{r'}{b}\right)^{2l+1}\right)$$

Similarly

$$\left[\frac{d}{dr}(rg_l(r, r'))\right]_{r'-\epsilon} = \frac{C}{r'}\left(l + 1 + l\left(\frac{a}{r'}\right)^{2l+1}\right)\left(1 - \left(\frac{r'}{b}\right)^{2l+1}\right)$$

Substituting these derivatives into (3.123), we find:

$$C = \frac{4\pi}{(2l+1)\left[1 - \left(\dfrac{a}{b}\right)^{2l+1}\right]} \qquad (3.124)$$

Combination of (3.124), (3.122), (3.119), and (3.118) yields the expansion of the Green's function for a spherical shell bounded by $r = a$ and $r = b$:

$$G(\mathbf{x}, \mathbf{x}') =$$
$$4\pi \sum_{l=0}^{\infty} \sum_{m=-l}^{l} \frac{Y_{lm}^*(\theta', \phi')Y_{lm}(\theta, \phi)}{(2l+1)\left[1 - \left(\dfrac{a}{b}\right)^{2l+1}\right]}\left(r_<^l - \frac{a^{2l+1}}{r_<^{l+1}}\right)\left(\frac{1}{r_>^{l+1}} - \frac{r_>^l}{b^{2l+1}}\right) \qquad (3.125)$$

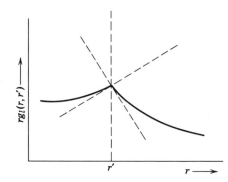

Fig. 3.8 Discontinuity in slope of the radial Green's function.

For the special cases $a \to 0$, $b \to \infty$, and $b \to \infty$, we recover the previous expansions (3.70) and (3.114), respectively. For the "interior" problem with a sphere of radius b we merely let $a \to 0$. Whereas the expansion for a single sphere is most easily obtained from the image solution, the general result (3.125) for a spherical shell is rather difficult to obtain by the method of images, since it involves an infinite set of images.

3.9 Solution of Potential Problems with the Spherical Green's Function Expansion

The general solution to Poisson's equation with specified values of the potential on the boundary surface is (see Section 1.10):

$$\Phi(\mathbf{x}) = \int_V \rho(\mathbf{x}')G(\mathbf{x}, \mathbf{x}')\, d^3x' - \frac{1}{4\pi} \oint_S \Phi(\mathbf{x}') \frac{\partial G}{\partial n'}\, da' \qquad (3.126)$$

For purposes of illustration let us consider the potential *inside* a sphere of radius b. First we will establish the equivalence of the surface integral in (3.126) to the previous method of Section 3.4, equations (3.61) and (3.58). With $a = 0$ in (3.125), the normal derivative, evaluated at $r' = b$, is:

$$\frac{\partial G}{\partial n'} = \frac{\partial G}{\partial r'}\bigg|_{r' = b} = -\frac{4\pi}{b^2} \sum_{l,m} \left(\frac{r}{b}\right)^l Y_{lm}^*(\theta', \phi') Y_{lm}(\theta, \phi) \qquad (3.127)$$

Consequently the solution of Laplace's equation inside $r = b$ with $\Phi = V(\theta', \phi')$ on the surface is, according to (3.126):

$$\Phi(\mathbf{x}) = \sum_{l,m} \left[\int V(\theta', \phi') Y_{lm}^*(\theta', \phi')\, d\Omega' \right] \left(\frac{r}{b}\right)^l Y_{lm}(\theta, \phi) \qquad (3.128)$$

For the case considered, this is the same form of solution as (3.61) with (3.58). There is a *third* form of solution for the sphere, the so-called

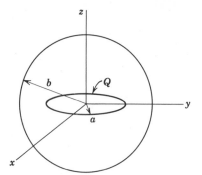

Poisson integral (2.25). The equivalence of this solution to the Green's function expansion solution is implied by the fact that both were derived from the general expression (3.126) and the image Green's function. The explicit demonstration of the equivalence of (2.25) and the series solution (3.61) will be left to the problems.

We now turn to the solution of problems with charge distributed in the volume, so that the volume integral in (3.126) is involved. It is sufficient to consider problems in which the potential vanishes on the boundary surfaces. By linear superposition of a solution of Laplace's equation the general situation can be obtained. The first illustration is that of a hollow grounded sphere of radius b with a concentric ring of charge of radius a and total charge Q. The ring of charge is located in the x-y plane, as shown in Fig. 3.9. The charge density of the ring can be written with the help of delta functions in angle and radius as

$$\rho(\mathbf{x}') = \frac{Q}{2\pi a^2}\,\delta(r' - a)\,\delta(\cos\theta') \tag{3.129}$$

In the volume integral over the Green's function only terms in (3.125) with $m = 0$ will survive because of azimuthal symmetry. Then, using (3.57) and remembering that $a \to 0$ in (3.125), we find

$$\Phi(\mathbf{x}) = \int \rho(\mathbf{x}')G(\mathbf{x}, \mathbf{x}')\,d^3x'$$
$$= Q\sum_{l=0}^{\infty} P_l(0)r_<^l\left(\frac{1}{r_>^{l+1}} - \frac{r_>^l}{b^{2l+1}}\right)P_l(\cos\theta) \tag{3.130}$$

where now $r_<$ ($r_>$) is the smaller (larger) of r and a. Using the fact that $P_{2n+1}(0) = 0$ and $P_{2n}(0) = \dfrac{(-1)^n(2n - 1)!!}{2^n n!}$, (3.130) can be written as:

$$\Phi(\mathbf{x}) = Q\sum_{n=0}^{\infty}\frac{(-1)^n(2n - 1)!!}{2^n n!}\,r_<^{2n}\left(\frac{1}{r_>^{2n+1}} - \frac{r_>^{2n}}{b^{4n+1}}\right)P_{2n}(\cos\theta) \tag{3.131}$$

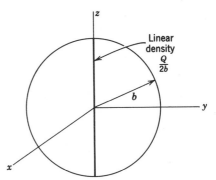

Fig. 3.10 Uniform line charge of length $2b$ and total charge Q inside a grounded, conducting sphere of radius b.

In the limit $b \to \infty$, it will be seen that (3.130) or (3.131) reduces to expression (3.48) for a ring of charge in free space. The present result can be obtained alternatively by using (3.48) and the images for a sphere.

A second example of charge densities, illustrated in Fig. 3.10, is that of a hollow grounded sphere with a uniform line charge of total charge Q located on the z axis between the north and south poles of the sphere. Again with the help of delta functions the volume-charge density can be written:

$$\rho(\mathbf{x}') = \frac{Q}{2b} \frac{1}{2\pi r'^2} [\delta(\cos \theta' - 1) + \delta(\cos \theta' + 1)] \qquad (3.132)$$

The two delta functions in $\cos \theta$ correspond to the two halves of the line charge, above and below the x-y plane. The factor $2\pi r'^2$ in the denominator assures that the charge density has a constant *linear* density $Q/2b$. With this density in (3.126) we obtain

$$\Phi(\mathbf{x}) = \frac{Q}{2b} \sum_{l=0}^{\infty} [P_l(1) + P_l(-1)] P_l(\cos \theta) \int_0^b r_<^l \left(\frac{1}{r_>^{l+1}} - \frac{r_>^l}{b^{2l+1}} \right) dr' \quad (3.133)$$

The integral must be broken up into the intervals $0 \le r' < r$ and $r \le r' \le b$. Then we find

$$\int_0^b = \left(\frac{1}{r^{l+1}} - \frac{r^l}{b^{2l+1}} \right) \int_0^r r'^l \, dr' + r^l \int_r^b \left(\frac{1}{r'^{l+1}} - \frac{r'^l}{b^{2l+1}} \right) dr'$$

$$= \frac{(2l+1)}{l(l+1)} \left(1 - \left(\frac{r}{b} \right)^l \right) \qquad (3.134)$$

For $l = 0$ this result is indeterminant. Applying L'Hospital's rule, we have, for $l = 0$ only,

$$\int_0^b = \lim_{l \to 0} \frac{\dfrac{d}{dl}\left(1 - \left(\dfrac{r}{b} \right)^l \right)}{\dfrac{d}{dl}(l)} = \lim_{l \to 0} \left(-\frac{d}{dl} e^{l \ln (r/b)} \right) = \ln \left(\frac{b}{r} \right) \quad (3.135)$$

This can be verified by direct integration in (3.133) for $l = 0$. Using the
fact that $P_l(-1) = (-1)^l$, the potential (3.133) can be put in the form:

$$\Phi(\mathbf{x}) = \frac{Q}{b}\left\{\ln\left(\frac{b}{r}\right) + \sum_{j=1}^{\infty} \frac{(4j + 1)}{2j(2j + 1)}\left[1 - \left(\frac{r}{b}\right)^{2j}\right]P_{2j}(\cos\theta)\right\} \quad (3.136)$$

The presence of the logarithm for $l = 0$ reminds us that the potential
diverges along the z axis. This is borne out by the series in (3.136), which
diverges for $\cos\theta = \pm 1$, except at $r = b$ exactly.

The surface-charge density on the grounded sphere is readily obtained
from (3.136) by differentiation:

$$\sigma(\theta) = \frac{1}{4\pi}\frac{\partial\Phi}{\partial r}\bigg|_{r=b} = -\frac{Q}{4\pi b^2}\left[1 + \sum_{j=1}^{\infty} \frac{(4j + 1)}{(2j + 1)}P_{2j}(\cos\theta)\right]$$

$$(3.137)$$

The leading term shows that the total charge induced on the sphere is $-Q$,
the other terms integrating to zero over the surface of the sphere.

3.10 Expansion of Green's Functions in Cylindrical Coordinates

The expansion of the potential of a unit point charge in cylindrical
coordinates affords another useful example of Green's function expansions. We will present the initial steps in general enough fashion that the
procedure can be readily adapted to finding Green's functions for potential
problems with cylindrical boundary surfaces. The starting point is the
equation for the Green's function:

$$\nabla_x^2 G(\mathbf{x}, \mathbf{x}') = -\frac{4\pi}{\rho}\delta(\rho - \rho')\,\delta(\phi - \phi')\,\delta(z - z') \quad (3.138)$$

where the delta function has been expressed in cylindrical coordinates.
The ϕ and z delta functions can be written in terms of orthonormal
functions:

$$\left.\begin{aligned}
\delta(z - z') &= \frac{1}{2\pi}\int_{-\infty}^{\infty} dk\, e^{ik(z-z')} = \frac{1}{\pi}\int_0^{\infty} dk\, \cos\left[k(z - z')\right]\\
\delta(\phi - \phi') &= \frac{1}{2\pi}\sum_{m=-\infty}^{\infty} e^{im(\phi-\phi')}
\end{aligned}\right\} \quad (3.139)$$

We expand the Green's function in similar fashion:

$$G(\mathbf{x}, \mathbf{x}') = \frac{1}{2\pi^2}\sum_{m=-\infty}^{\infty}\int_0^{\infty} dk\, e^{im(\phi-\phi')}\cos\left[k(z - z')\right]g_m(\rho, \rho') \quad (3.140)$$

Then substitution into (3.138) leads to an equation for the radial Green's function $g_m(\rho, \rho')$:

$$\frac{1}{\rho}\frac{d}{d\rho}\left(\rho\,\frac{dg_m}{d\rho}\right) - \left(k^2 + \frac{m^2}{\rho^2}\right)g_m = -\frac{4\pi}{\rho}\,\delta(\rho - \rho') \qquad (3.141)$$

For $\rho \neq \rho'$ this is just equation (3.98) for the modified Bessel functions, $I_m(k\rho)$ and $K_m(k\rho)$. Suppose that $\psi_1(k\rho)$ is some linear combination of I_m and K_m which satisfies the correct boundary conditions for $\rho < \rho'$, and that $\psi_2(k\rho)$ is a linearly independent combination which satisfies the proper boundary conditions for $\rho > \rho'$. Then the symmetry of the Green's function in ρ and ρ' requires that

$$g_m(\rho, \rho') = \psi_1(k\rho_<)\psi_2(k\rho_>) \qquad (3.142)$$

The normalization of the product $\psi_1\psi_2$ is determined by the discontinuity in slope implied by the delta function in (3.141):

$$\left.\frac{dg_m}{d\rho}\right|_+ - \left.\frac{dg_m}{d\rho}\right|_- = -\frac{4\pi}{\rho'} \qquad (3.143)$$

where $|_\pm$ means evaluated at $\rho = \rho' \pm \epsilon$. From (3.142) it is evident that

$$\left[\left.\frac{dg_m}{d\rho}\right|_+ - \left.\frac{dg_m}{d\rho}\right|_-\right] = k(\psi_1\psi_2' - \psi_2\psi_1') = kW[\psi_1, \psi_2] \qquad (3.144)$$

where primes mean differentiation with respect to the argument, and $W[\psi_1, \psi_2]$ is the Wronskian of ψ_1 and ψ_2. Equation (3.141) is of the Sturm-Liouville type

$$\frac{d}{dx}\left(p(x)\,\frac{dy}{dx}\right) + g(x)y = 0 \qquad (3.145)$$

and it is well known that the Wronskian of two linearly independent solutions of such an equation is proportional to $[1/p(x)]$. Hence the possibility of satisfying (3.143) for all values of ρ' is assured. Clearly we must demand that the normalization of the product $\psi_1\psi_2$ is such that the Wronskian has the value:

$$W[\psi_1(x), \psi_2(x)] = -\frac{4\pi}{x} \qquad (3.146)$$

If there are no boundary surfaces, the requirement is that $g_m(\rho, \rho')$ be finite at $\rho = 0$ and vanish at $\rho \to \infty$. Consequently $\psi_1(k\rho) = AI_m(k\rho)$ and $\psi_2(k\rho) = K_m(k\rho)$. The constant A is to be determined from the Wronskian condition (3.146). Since the Wronskian is proportional to $(1/x)$ for all values of x, it does not matter where we evaluate it. Using the limiting forms (3.102) and (3.103) for small x [or (3.104) for large x], we find

$$W[I_m(x), K_m(x)] = -\frac{1}{x} \qquad (3.147)$$

so that $A = 4\pi$. The expansion of $1/|\mathbf{x} - \mathbf{x}'|$ therefore becomes:

$$\frac{1}{|\mathbf{x} - \mathbf{x}'|} = \frac{2}{\pi} \sum_{m=-\infty}^{\infty} \int_0^\infty dk \, e^{im(\phi - \phi')} \cos\left[k(z - z')\right] I_m(k\rho_<) K_m(k\rho_>) \quad (3.148)$$

This can also be written entirely in terms of real functions as:

$$\frac{1}{|\mathbf{x} - \mathbf{x}'|} = \frac{4}{\pi} \int_0^\infty dk \cos\left[k(z - z')\right]$$
$$\times \left\{ \tfrac{1}{2} I_0(k\rho_<) K_0(k\rho_>) + \sum_{m=1}^\infty \cos\left[m(\phi - \phi')\right] I_m(k\rho_<) K_m(k\rho_>) \right\}$$
$$(3.149)$$

A number of useful mathematical results can be obtained from this expansion. If we let $\mathbf{x}' \to 0$, only the $m = 0$ term survives, and we obtain the integral representation:

$$\frac{1}{\sqrt{\rho^2 + z^2}} = \frac{2}{\pi} \int_0^\infty \cos kz \, K_0(k\rho) \, dk \qquad (3.150)$$

If we replace ρ^2 in (3.150) by $R^2 = \rho^2 + \rho'^2 - 2\rho\rho' \cos(\phi - \phi')$, then we have on the left-hand side the inverse distance $|\mathbf{x} - \mathbf{x}'|^{-1}$ with $z' = 0$, i.e., just (3.149) with $z' = 0$. Then comparison of the right-hand sides of (3.149) and (3.150) (which must hold for *all* values of z) leads to the identification:

$$K_0(k\sqrt{\rho^2 + \rho'^2 - 2\rho\rho' \cos(\phi - \phi')})$$
$$= I_0(k\rho_<) K_0(k\rho_>) + 2 \sum_{m=1}^\infty \cos\left[m(\phi - \phi')\right] I_m(k\rho_<) K_m(k\rho_>) \quad (3.151)$$

In this last result we can take the limit $k \to 0$ and obtain an expansion for the Green's function for (two-dimensional) polar coordinates:

$$\ln\left(\frac{1}{\sqrt{\rho^2 + \rho'^2 - 2\rho\rho' \cos(\phi - \phi')}}\right)$$
$$= \ln\left(\frac{1}{\rho_>}\right) + \sum_{m=1}^\infty \frac{1}{m}\left(\frac{\rho_<}{\rho_>}\right)^m \cos\left[m(\phi - \phi')\right] \quad (3.152)$$

This representation can be verified by a systematic construction of the two-dimensional Green's function for Poisson's equation along the lines leading to (3.148).

3.11   Eigenfunction Expansions for Green's Functions

Another technique for obtaining expansions of Green's functions is the use of eigenfunctions for some related problem. This approach is intimately connected with the methods of Sections 3.8 and 3.10.

To specify what we mean by eigenfunctions, we consider an elliptic differential equation of the form:

$$\nabla^2 \psi(\mathbf{x}) + [f(\mathbf{x}) + \lambda]\psi(\mathbf{x}) = 0 \qquad (3.153)$$

If the solutions $\psi(\mathbf{x})$ are required to satisfy certain boundary conditions on the surface S of the volume of interest V, then (3.153) will not in general have well-behaved (e.g., finite and continuous) solutions, except for certain values of λ. These values of λ, denoted by λ_n, are called *eigenvalues* (or *characteristic values*) and the solutions $\psi_n(\mathbf{x})$ are called *eigenfunctions.** The eigenvalue differential equation is written:

$$\nabla^2 \psi_n(\mathbf{x}) + [f(\mathbf{x}) + \lambda_n]\psi_n(\mathbf{x}) = 0 \qquad (3.154)$$

By methods similar to those used to prove the orthogonality of the Legendre or Bessel functions it can be shown that the eigenfunctions are orthogonal:

$$\int_V \psi_m{}^*(\mathbf{x})\psi_n(\mathbf{x})\, d^3x = \delta_{mn} \qquad (3.155)$$

where the eigenfunctions are assumed normalized. The spectrum of eigenvalues λ_n may be a discrete set, or a continuum, or both. It will be assumed that the totality of eigenfunctions forms a complete set.

Suppose now that we wish to find the Green's function for the equation:

$$\nabla_x^2 G(\mathbf{x}, \mathbf{x}') + [f(\mathbf{x}) + \lambda]G(\mathbf{x}, \mathbf{x}') = -4\pi\delta(\mathbf{x} - \mathbf{x}') \qquad (3.156)$$

where λ is *not* in general one of the eigenvalues λ_n of (3.154). Furthermore, suppose that the Green's function is to have the same boundary conditions as the eigenfunctions of (3.154). Then the Green's function can be expanded in a series of the eigenfunctions of the form:

$$G(\mathbf{x}, \mathbf{x}') = \sum_n a_n(\mathbf{x}')\psi_n(\mathbf{x}) \qquad (3.157)$$

Substitution into the differential equation for the Green's function leads to the result:

$$\sum_m a_m(\mathbf{x}')(\lambda - \lambda_m)\psi_m(\mathbf{x}) = -4\pi\delta(\mathbf{x} - \mathbf{x}') \qquad (3.158)$$

---

* The reader familiar with wave mechanics will recognize (3.153) as equivalent to the Schrödinger equation for a particle in a potential.

If we multiply both sides by $\psi_n{}^*(\mathbf{x})$ and integrate over the volume V, the orthogonality condition (3.155) reduces the left-hand side to one term, and we find:

$$a_n(\mathbf{x}') = 4\pi \frac{\psi_n{}^*(\mathbf{x}')}{\lambda_n - \lambda} \tag{3.159}$$

Consequently the eigenfunction expansion of the Green's function is:

$$G(\mathbf{x}, \mathbf{x}') = 4\pi \sum_n \frac{\psi_n{}^*(\mathbf{x}')\psi_n(\mathbf{x})}{\lambda_n - \lambda} \tag{3.160}$$

For a continuous spectrum the sum is replaced by an integral.

Specializing the above considerations to Poisson's equation, we place $f(\mathbf{x}) = 0$ and $\lambda = 0$ in (3.156). As a first, essentially trivial, illustration we let (3.154) be the wave equation over all space:

$$(\nabla^2 + k^2)\psi_\mathbf{k}(\mathbf{x}) = 0 \tag{3.161}$$

with the continuum of eigenvalues, k^2, and the eigenfunctions:

$$\psi_\mathbf{k}(\mathbf{x}) = \frac{1}{(2\pi)^{3/2}} e^{i\mathbf{k}\cdot\mathbf{x}} \tag{3.162}$$

These eigenfunctions have delta function normalization:

$$\int \psi_{\mathbf{k}'}{}^*(\mathbf{x})\psi_\mathbf{k}(\mathbf{x}) \, d^3x = \delta(\mathbf{k} - \mathbf{k}') \tag{3.163}$$

Then, according to (3.160), the infinite space Green's function has the expansion:

$$\frac{1}{|\mathbf{x} - \mathbf{x}'|} = \frac{1}{2\pi^2} \int d^3k \, \frac{e^{i\mathbf{k}\cdot(\mathbf{x}-\mathbf{x}')}}{k^2} \tag{3.164}$$

This is just the three-dimensional Fourier integral representation of $1/|\mathbf{x} - \mathbf{x}'|$.

As a second example, consider the Green's function for a Dirichlet problem inside a rectangular box defined by the six planes, $x = y = z = 0$, $x = a$, $y = b$, $z = c$. The expansion is to be made in terms of eigenfunctions of the wave equation:

$$(\nabla^2 + k_{lmn}^2)\psi_{lmn}(x, y, z) = 0 \tag{3.165}$$

where the eigenfunctions which vanish on all the boundary surfaces are

and

$$\left.\begin{array}{l} \psi_{lmn}(x, y, z) = \sqrt{\dfrac{8}{abc}} \sin\left(\dfrac{l\pi x}{a}\right) \sin\left(\dfrac{m\pi y}{b}\right) \sin\left(\dfrac{n\pi z}{c}\right) \\[1.5em] k_{lmn}^2 = \pi^2\left(\dfrac{l^2}{a^2} + \dfrac{m^2}{b^2} + \dfrac{n^2}{c^2}\right) \end{array}\right\} \tag{3.166}$$

The expansion of the Green's function is therefore:

$$G(\mathbf{x}, \mathbf{x}') = \frac{32}{\pi abc} \sum_{l,m,n=1}^{\infty}$$

$$\times \frac{\sin\left(\dfrac{l\pi x}{a}\right) \sin\left(\dfrac{l\pi x'}{a}\right) \sin\left(\dfrac{m\pi y}{b}\right) \sin\left(\dfrac{m\pi y'}{b}\right) \sin\left(\dfrac{n\pi z}{c}\right) \sin\left(\dfrac{n\pi z'}{c}\right)}{\left(\dfrac{l^2}{a^2} + \dfrac{m^2}{b^2} + \dfrac{n^2}{c^2}\right)}$$

$$(3.167)$$

To relate expansion (3.167) to the type of expansions obtained in Sections 3.8 and 3.10, namely, (3.125) for spherical coordinates and (3.148) for cylindrical coordinates, we write down the analogous expansion for the rectangular box. If the x and y coordinates are treated in the manner of (θ, ϕ) or (ϕ, z) in those cases, while the z coordinate is singled out for special treatment, we obtain the Green's function:

$$G(\mathbf{x}, \mathbf{x}') = \frac{16\pi}{ab} \sum_{l,m=1}^{\infty} \sin\left(\frac{l\pi x}{a}\right) \sin\left(\frac{l\pi x'}{a}\right) \sin\left(\frac{m\pi y}{b}\right) \sin\left(\frac{m\pi y'}{b}\right)$$

$$\times \left[\frac{\sinh\left(K_{lm}z_<\right) \sinh\left(K_{lm}(c - z_>)\right)}{K_{lm} \sinh\left(K_{lm}c\right)}\right] \quad (3.168)$$

where $K_{lm} = \pi\left(\dfrac{l^2}{a^2} + \dfrac{m^2}{b^2}\right)^{1/2}$. If (3.167) and (3.168) are to be equal, it must be that the sum over n in (3.167) is just the Fourier series representation on the interval $(0, c)$ of the one-dimensional Green's function in z in (3.168):

$$\frac{\sinh\left(K_{lm}z_<\right) \sinh\left(K_{lm}(c - z_>)\right)}{K_{lm} \sinh\left(K_{lm} c\right)} = \frac{2}{c} \sum_{n=1}^{\infty} \frac{\sin\left(\dfrac{n\pi z'}{c}\right)}{K_{lm}^2 + \left(\dfrac{n\pi}{c}\right)^2} \sin\left(\frac{n\pi z}{c}\right)$$

$$(3.169)$$

The verification that (3.169) is the correct Fourier representation is left as an exercise for the reader.

Further illustrations of this technique will be found in the problems at the end of the chapter.

3.12 Mixed Boundary Conditions; Charged Conducting Disc

The potential problems discussed so far in this chapter have been of the orthodox kind in which the boundary conditions are of one type (usually Dirichlet) over the whole boundary surface. In the uniqueness proof for

Classical Electrodynamics

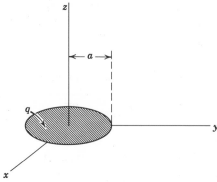

Fig. 3.11

solutions of Laplace's or Poisson's equation (Section 1.9) it was pointed out, however, that *mixed boundary conditions*, where the potential is specified over part of the boundary and its normal derivative is specified over the remainder, also lead to well-defined, unique, boundary-value problems. There is a tendency in existing textbooks to mention the possibility of mixed boundary conditions when making the uniqueness proof and to ignore such problems in subsequent discussion. The reason, as we shall see immediately, is that mixed boundary conditions are much more difficult to handle than the normal type.

To illustrate the difficulties encountered with mixed boundary conditions we consider the apparently simple problem of an isolated, infinitely thin, flat, circular, conducting disc of radius a with a total charge q placed on it, as shown in Fig. 3.11. The charge distributes itself over the disc in such a way as to make its surface an equipotential. We wish to determine the potential everywhere in space and the charge distribution on the disc.

From the geometry of the problem we see that the potential is symmetric about the axis of the disc and with respect to the plane containing the disc. If cylindrical coordinates are chosen with the axis of the disc as the z axis and the origin at the center of the disc, the potential must therefore be of the form [from (3.110)],

$$\Phi(\rho, z) = \int_0^\infty dk\, f(k) e^{-k|z|} J_0(k\rho) \qquad (3.170)$$

The unknown function $f(k)$ must be determined from the boundary conditions at $z = 0$. If the potential were known everywhere over the whole $z = 0$ plane, $f(k)$ could be found by inverting the Hankel transform, as in going from (3.110) to (3.113). Unfortunately the boundary conditions at $z = 0$ are not that simple. For $0 \leq \rho \leq a$ we do know that the potential is constant at an unknown value, $\Phi = V = q/C$, where C is the capacitance of the disc. But for $a < \rho < \infty$, the potential is unknown.

From symmetry, however, we know that the normal derivative of the potential vanishes there. Thus the boundary conditions are mixed:

$$\left.\begin{aligned}
\Phi(\rho, 0) &= V, &&\text{for } 0 \leq \rho \leq a \\
\frac{\partial \Phi}{\partial z}(\rho, 0) &= 0, &&\text{for } a < \rho < \infty
\end{aligned}\right\} \tag{3.171}$$

The connection between the potential of the disc V and the total charge q on it will be established by the fact that at large distances (ρ and/or $z \gg a$) the potential must approach $q/(\rho^2 + z^2)^{1/2}$. From (3.170) and an identity of Problem 3.12c this requirement can be seen to imply

$$\lim_{k \to 0} f(k) = q \tag{3.172}$$

When boundary conditions (3.171) are applied to the general solution (3.170), there results a pair of integral equations of the first kind:

$$\left.\begin{aligned}
\int_0^\infty dk\, f(k) J_0(k\rho) &= V, &&\text{for } 0 \leq \rho \leq a \\
\int_0^\infty dk\, k\, f(k) J_0(k\rho) &= 0, &&\text{for } a < \rho < \infty
\end{aligned}\right\} \tag{3.173}$$

Such pairs of integral equations, with one of the pair holding over one part of the range of the independent variable and the other over the other part of the range, are known as *dual integral equations*. The general theory of dual integral equations is complicated and not highly developed. But the charged disc problem and variations of it have received considerable attention over the years. H. Weber (1873) first solved the present problem by using certain discontinuous integrals involving Bessel functions. Titchmarsh, p. 334, uses Mellin transforms to effect a solution of a somewhat more general pair of dual integral equations. E. T. Copson [*Proc. Edin. Math. Soc.* (2), **8**, 14 (1947)] reduces the disc problem to an integral equation for the surface-charge density of the Abel type. Tranter, p. 50 and Chapter VIII, considers slight generalizations of the pair (3.173). He introduces a systematic technique of finding the most general form satisfying the homogeneous member of the pair and then delimiting that form by substitution into the other equation. The Wiener-Hopf technique can also be used.

For our purposes it is sufficient to observe that the dual integral equations,

$$\left.\begin{aligned}
\int_0^\infty dy\, g(y) J_n(yx) &= x^n, &&\text{for } 0 \leq x < 1 \\
\int_0^\infty dy\, y\, g(y) J_n(yx) &= 0, &&\text{for } 1 < x < \infty
\end{aligned}\right\} \tag{3.174}$$

have the solution,

$$g(y) = \frac{2\Gamma(n+1)}{\sqrt{\pi}\Gamma(n+\frac{1}{2})} j_n(y) = \frac{\Gamma(n+1)}{\Gamma(n+\frac{1}{2})} \left(\frac{2}{y}\right)^{\frac{1}{2}} J_{n+\frac{1}{2}}(y) \qquad (3.175)$$

In this relation $j_n(y)$ is the spherical Bessel function of order n (see Section 16.1). For the set of equations (3.173) the variables are $x = \rho/a$ and $y = ka$, while $n = 0$. Thus the solution is

$$f(k) = \frac{2}{\pi} Va\, j_0(ka) = \frac{2}{\pi} Va\left(\frac{\sin ka}{ka}\right) \qquad (3.176)$$

Remembering the connection (3.172) which determines the potential V in terms of the charge q, we find

$$V = \frac{\pi}{2}\frac{q}{a}$$

This shows that the capacitance of a disc of radius a is

$$C = \frac{2}{\pi} a$$

This value was experimentally established with remarkable precision by Cavendish (*ca.* 1780) by comparing the charges on a disc and a sphere at the same potential.

The potential anywhere in space is found from (3.170) and (3.176) to be

$$\Phi(\rho, z) = q \int_0^\infty dk\, \frac{\sin ka}{ka}\, e^{-k|z|} J_0(k\rho) \qquad (3.177)$$

Values of the potential along the axis and in the plane of the disc can be found readily by putting $\rho = 0$ and $z = 0$ in (3.177). The results are

$$\Phi(0, z) = \frac{q}{a} \tan^{-1}\left(\frac{a}{z}\right)$$

$$\Phi(\rho, 0) = \begin{cases} \dfrac{q}{a} \sin^{-1}\left(\dfrac{a}{\rho}\right), & \text{for } \rho \geq a \\[2ex] \dfrac{\pi}{2}\dfrac{q}{a}, & \text{for } 0 \leq \rho \leq a \end{cases}$$

For arbitrary ρ and z the integral can be transformed into Weber's form of the solution:

$$\Phi(\rho, z) = q \sin^{-1}\left[\frac{2a}{\sqrt{(\rho - a)^2 + z^2} + \sqrt{(\rho + a)^2 + z^2}}\right] \qquad (3.178)$$

The charge density $\sigma(\rho)$ on the surface of the disc is given by

$$\sigma(\rho) = -\frac{1}{2\pi}\frac{\partial \Phi}{\partial z}(\rho, 0) = \frac{q}{2\pi a}\int_0^\infty dk \, \sin ka \, J_0(k\rho)$$

The integral is a well-known discontinuous integral which vanishes for $\rho > a$. For $\rho < a$, the charge density is

$$\sigma(\rho) = \frac{q}{2\pi a}\frac{1}{\sqrt{a^2 - \rho^2}} \tag{3.179}$$

The (integrable) infinity in $\sigma(\rho)$ for $\rho \to a$ is a mathematical singularity which results from the assumption of an infinitely thin disc. In practice the charge is repelled to the outer regions of a thin disc approximately according to (3.179), but near the edge the distribution levels off to a large, but finite, value which depends on the detailed construction of the disc.

We have discussed the charged conducting disc in cylindrical coordinates in order to illustrate the complications of mixed boundary conditions. For this particular example, the mixed boundary conditions can be avoided by separating Laplace's equation in elliptic coordinates. Then the disc can be taken to be the limiting form of an oblate spheroidal surface. See, for example, Smythe, pp. 111, 156, or Jeans, p. 244.

REFERENCES AND SUGGESTED READING

The mathematical apparatus and special functions needed for the solution of potential problems in spherical, cylindrical, spheroidal, and other coordinate systems are discussed in
> Morse and Feshbach, Chapter 10.

A more elementary treatment, with well-chosen examples and problems, can be found in
> Hildebrand, Chapters 4, 5, and 8.

A somewhat old-fashioned source of the theory and practice of Legendre polynomials and spherical harmonics, with many examples and problems, is
> Byerly.

For purely mathematical properties of spherical functions one of the most useful one-volume references is
> Magnus and Oberhettinger.

For more detailed mathematical properties, see
> Watson, for Bessel functions,
> Bateman Manuscript Project books, for all types of special functions.

Electrostatic problems in cylindrical, spherical, and other coordinates are discussed extensively in
> Durand, Chapter XI,
> Jeans, Chapter VIII,
> Smythe, Chapter V,
> Stratton, Chapter III.

PROBLEMS

3.1 The surface of a hollow conducting sphere of inner radius a is divided into an *even number* of equal segments by a set of planes whose common line of intersection is the z axis and which are distributed uniformly in the angle ϕ. (The segments are like the skin on wedges of an apple, or the earth's surface between successive meridians of longitude.) The segments are kept at fixed potentials $\pm V$, alternately.

(*a*) Set up a series representation for the potential inside the sphere for the general case of $2n$ segments, and carry the calculation of the coefficients in the series far enough to determine exactly which coefficients are different from zero. For the nonvanishing terms, exhibit the coefficients as an integral over $\cos \theta$.

(*b*) For the special case of $n = 1$ (two hemispheres) determine explicitly the potential up to and including all terms with $l = 3$. By a coordinate transformation verify that this reduces to result (3.37) of Section 3.3.

3.2 Two concentric spheres have radii a, b ($b > a$) and are divided into two hemispheres by the same horizontal plane. The upper hemisphere of the inner sphere and the lower hemisphere of the outer sphere are maintained at potential V. The other hemispheres are at zero potential.

Determine the potential in the region $a \le r \le b$ as a series in Legendre polynomials. Include terms at least up to $l = 4$. Check your solution against known results in the limiting cases $b \to \infty$, and $a \to 0$.

3.3 A spherical surface of radius R has charge uniformly distributed over its surface with a density $Q/4\pi R^2$, except for a spherical cap at the north pole, defined by the cone $\theta = \alpha$.

(*a*) Show that the potential inside the spherical surface can be expressed as

$$\Phi = \frac{Q}{2} \sum_{l=0}^{\infty} \frac{1}{2l+1} [P_{l+1}(\cos \alpha) - P_{l-1}(\cos \alpha)] \frac{r^l}{R^{l+1}} P_l(\cos \theta)$$

where, for $l = 0$, $P_{l-1}(\cos \alpha) = -1$. What is the potential outside?

(*b*) Find the magnitude and the direction of the electric field at the origin.

(*c*) Discuss the limiting forms of the potential (*a*) and electric field (*b*) as the spherical cap becomes (1) very small, and (2) so large that the area with charge on it becomes a very small cap at the south pole.

3.4 A thin, flat, conducting, circular disc of radius R is located in the x-y plane with its center at the origin, and is maintained at a fixed potential V. With the information that the charge density on a disc at fixed potential is proportional to $(R^2 - \rho^2)^{-\frac{1}{2}}$, where ρ is the distance out from the center of the disc,

(*a*) show that for $r > R$ the potential is

$$\Phi(r, \theta, \phi) = \frac{2V}{\pi} \frac{R}{r} \sum_{l=0}^{\infty} \frac{(-1)^l}{2l+1} \left(\frac{R}{r}\right)^{2l} P_{2l}(\cos \theta)$$

(*b*) find the potential for $r < R$.

3.5 A hollow sphere of inner radius a has the potential specified on its surface to be $\Phi = V(\theta, \phi)$. Prove the equivalence of the two forms of solution for the potential inside the sphere:

(a) $$\Phi(\mathbf{x}) = \frac{a(a^2 - r^2)}{4\pi} \int \frac{V(\theta', \phi')}{(r^2 + a^2 - 2ar \cos \gamma)^{3/2}} \, d\Omega'$$

where $\cos \gamma = \cos \theta \cos \theta' + \sin \theta \sin \theta' \cos (\phi - \phi')$.

(b) $$\Phi(\mathbf{x}) = \sum_{l=0}^{\infty} \sum_{m=-l}^{l} A_{lm} \left(\frac{r}{a}\right)^l Y_{lm}(\theta, \phi)$$

where $A_{lm} = \int d\Omega' \, Y_{lm}^*(\theta', \phi') V(\theta', \phi')$.

3.6 A hollow right circular cylinder of radius b has its axis coincident with the z axis and its ends at $z = 0$ and $z = L$. The potential on the end faces is zero, while the potential on the cylindrical surface is given as $V(\phi, z)$. Using the appropriate separation of variables in cylindrical coordinates, find a series solution for the potential anywhere inside the cylinder.

3.7 For the cylinder in Problem 3.6 the cylindrical surface is made of two equal half-cylinders, one at potential V and the other at potential $-V$, so that

$$V(\phi, z) = \begin{cases} V \text{ for } -\dfrac{\pi}{2} < \phi < \dfrac{\pi}{2} \\[2ex] -V \text{ for } \dfrac{\pi}{2} < \phi < \dfrac{3\pi}{2} \end{cases}$$

(a) Find the potential inside the cylinder.

(b) Assuming $L \gg b$, consider the potential at $z = L/2$ as a function of ρ and ϕ and compare it with two-dimensional Problem 2.8.

3.8 Show that an arbitrary function $f(x)$ can be expanded on the interval $0 \leq x \leq a$ in a modified Fourier-Bessel series

$$f(x) = \sum_{n=1}^{\infty} A_n J_\nu\left(y_{\nu n} \frac{x}{a}\right)$$

where $y_{\nu n}$ is the nth root $\dfrac{dJ_\nu(x)}{dx} = 0$, and the coefficients A_n are given by

$$A_n = \frac{2}{a^2 \left(1 - \dfrac{\nu^2}{y_{\nu n}^2}\right) J_\nu{}^2(y_{\nu n})} \int_0^a f(x) x J_\nu\left(y_{\nu n} \frac{x}{a}\right) dx$$

3.9 An infinite, thin, plane sheet of conducting material has a circular hole of radius a cut in it. A thin, flat disc of the same material and slightly smaller radius lies in the plane, filling the hole, but separated from the sheet by a very narrow insulating ring. The disc is maintained at a fixed potential V, while the infinite sheet is kept at zero potential.

(a) Using appropriate cylindrical coordinates, find an integral expression involving Bessel functions for the potential at any point above the plane.

(b) Show that the potential a perpendicular distance z above the *center* of the disc is

$$\Phi_0(z) = V\left(1 - \frac{z}{\sqrt{a^2 + z^2}}\right)$$

(c) Show that the potential a perpendicular distance z above the *edge* of the disc is

$$\Phi_a(z) = \frac{V}{\pi}[E(k) - (1 - k^2)K(k)]$$

where $k = 2a/(z^2 + 4a^2)^{1/2}$, and $K(k)$, $E(k)$ are the complete elliptic integrals of the first and second kinds.

3.10 Solve for the potential in Problem 3.2, using the appropriate Green's function obtained in the text, and verify that the answer obtained in this way agrees with the direct solution from the differential equation.

3.11 A line charge of length $2d$ with a total charge Q has a linear charge density varying as $(d^2 - z^2)$, where z is the distance from the midpoint. A grounded, conducting, spherical shell of inner radius $b > d$ is centered at the midpoint of the line charge.

(a) Find the potential everywhere inside the spherical shell as an expansion in Legendre polynomials.

(b) Calculate the surface-charge density induced on the shell.

(c) Discuss your answers to (a) and (b) in the limit that $d \ll b$.

3.12 (a) Verify that

$$\frac{1}{\rho}\delta(\rho - \rho') = \int_0^\infty kJ_m(k\rho)J_m(k\rho')\,dk$$

(b) Obtain the following expansion:

$$\frac{1}{|\mathbf{x} - \mathbf{x}'|} = \sum_{m=-\infty}^{\infty}\int_0^\infty dk\, e^{im(\phi-\phi')}J_m(k\rho)J_m(k\rho')e^{-k(z_> - z_<)}$$

(c) By appropriate limiting procedures prove the following expansions:

$$\frac{1}{\sqrt{\rho^2 + z^2}} = \int_0^\infty e^{-k|z|}J_0(k\rho)\,dk$$

$$J_0(k\sqrt{\rho + \rho'^2 - 2\rho\rho'\cos\phi}) = \sum_{m=-\infty}^{\infty}e^{im\phi}J_m(k\rho)J_m(k\rho')$$

$$e^{ik\rho\cos\phi} = \sum_{m=-\infty}^{\infty}i^m e^{im\phi}J_m(k\rho)$$

(d) From the last result obtain an integral representation of the Bessel function:

$$J_m(x) = \frac{1}{2\pi i^m}\int_0^{2\pi}e^{ix\cos\phi - im\phi}d\phi$$

Compare the standard integral representations.

3.13 A unit point charge is located at the point (ρ', ϕ', z') inside a grounded cylindrical box defined by the surfaces $z = 0$, $z = L$, $\rho = a$. Show that the potential inside the box can be expressed in the following alternative forms:

$$\Phi(\mathbf{x}, \mathbf{x}') = \frac{4}{a} \sum_{m=-\infty}^{\infty} \sum_{n=1}^{\infty} \frac{e^{im(\phi-\phi')} J_m\left(\frac{x_{mn}\rho}{a}\right) J_m\left(\frac{x_{mn}\rho'}{a}\right)}{x_{mn}J_{m+1}^2(x_{mn}) \sinh\left(\frac{x_{mn}L}{a}\right)}$$

$$\times \sinh\left[\frac{x_{mn}}{a} z_<\right] \sinh\left[\frac{x_{mn}}{a}(L - z_>)\right]$$

$$\Phi(\mathbf{x}, \mathbf{x}') = \frac{4}{L} \sum_{m=-\infty}^{\infty} \sum_{n=1}^{\infty} e^{im(\phi-\phi')} \sin\left(\frac{n\pi z}{L}\right) \sin\left(\frac{n\pi z'}{L}\right) \frac{I_m\left(\frac{n\pi\rho_<}{L}\right)}{I_m\left(\frac{n\pi a}{L}\right)}$$

$$\times \left[I_m\left(\frac{n\pi a}{L}\right)K_m\left(\frac{n\pi\rho_>}{L}\right) - K_m\left(\frac{n\pi a}{L}\right)I_m\left(\frac{n\pi\rho_>}{L}\right)\right]$$

$$\Phi(\mathbf{x}, \mathbf{x}') = \frac{8}{La^2}$$

$$\times \sum_{m=-\infty}^{\infty} \sum_{k=1}^{\infty} \sum_{n=1}^{\infty} \frac{e^{im(\phi-\phi')} \sin\left(\frac{k\pi z}{L}\right) \sin\left(\frac{k\pi z'}{L}\right) J_m\left(\frac{x_{mn}\rho}{a}\right) J_m\left(\frac{x_{mn}\rho'}{a}\right)}{\left[\left(\frac{x_{mn}}{a}\right)^2 + \left(\frac{k\pi}{L}\right)^2\right] J_{m+1}^2(x_{mn})}$$

Discuss the relation of the last expansion (with its extra summation) to the other two.

3.14 The walls of the conducting cylindrical box of Problem 3.13 are all at zero potential, except for a disc in the upper end, defined by $\rho = b$, at potential V.

(a) Using the various forms of the Green's function obtained in Problem 3.13, find three expansions for the potential inside the cylinder.

(b) For each series, calculate numerically the ratio of the potential at $\rho = 0$, $z = L/2$ to the potential of the disc, assuming $b = L/4 = a/2$. Try to obtain at least two-significant-figure accuracy. Is one series less rapidly convergent than the others? Why?

(Jahnke and Emde have tables of J_0 and J_1 on pp. 156–163, I_0 and I_1 on pp. 226–229, $(2/\pi)K_0$ and $(2/\pi)K_1$ on pp. 236–243. Watson also has numerous tables.)

4

Multipoles, Electrostatics of Macroscopic Media, Dielectrics

This chapter is first concerned with the potential due to localized charge distributions and its expansion in multipoles. The development is made in terms of spherical harmonics, but contact is established with the rectangular components for the first few multipoles. The energy of a multipole in an external field is then discussed. The macroscopic equations of electrostatics are derived by taking into account the response of atoms to an applied field and by suitable averaging procedures. Dielectrics and the appropriate boundary conditions are then described, and some typical boundary-value problems with dielectrics are solved. Simple classical models are used to illustrate the main features of atomic polarizability and susceptibility. Finally the question of electrostatic energy in the presence of dielectrics is discussed.

4.1 Multipole Expansion

A localized distribution of charge is described by the charge density $\rho(\mathbf{x}')$, which is nonvanishing only inside a sphere of radius $R*$ around some origin. The potential outside the sphere can be written as an expansion in spherical harmonics:

$$\Phi(\mathbf{x}) = \sum_{l=0}^{\infty} \sum_{m=-l}^{l} \frac{4\pi}{2l+1} q_{lm} \frac{Y_{lm}(\theta, \phi)}{r^{l+1}} \qquad (4.1)$$

* The sphere of radius R is an arbitrary conceptual device employed merely to divide space into regions with and without charge.

where the particular choice of constant coefficients is made for later convenience. Equation (4.1) is called a multipole expansion; the $l = 0$ term is called the monopole term, $l = 1$ is the dipole term, etc. The reason for these names becomes clear below. The problem to be solved is the determination of the constants q_{lm} in terms of the properties of the charge density $\rho(\mathbf{x}')$. The solution is very easily obtained from the integral (1.17) for the potential:

$$\Phi(\mathbf{x}) = \int \frac{\rho(\mathbf{x}')}{|\mathbf{x} - \mathbf{x}'|} \, d^3x'$$

with expansion (3.70) for $1/|\mathbf{x} - \mathbf{x}'|$. Since we are interested at the moment in the potential outside the charge distribution, $r_< = r'$ and $r_> = r$. Then we find:

$$\Phi(\mathbf{x}) = 4\pi \sum_{l,m} \frac{1}{2l + 1} \left[\int Y_{lm}^*(\theta', \phi') r'^l \rho(\mathbf{x}') \, d^3x' \right] \frac{Y_{lm}(\theta, \phi)}{r^{l+1}} \qquad (4.2)$$

Consequently the coefficients in (4.1) are:

$$q_{lm} = \int Y_{lm}^*(\theta', \phi') r'^l \rho(\mathbf{x}') \, d^3x' \qquad (4.3)$$

These coefficients are called *multipole moments*. To see the physical interpretation of them we exhibit the first few explicitly in terms of cartesian coordinates:

$$q_{00} = \frac{1}{\sqrt{4\pi}} \int \rho(\mathbf{x}') \, d^3x' = \frac{1}{\sqrt{4\pi}} q \qquad (4.4)$$

$$\left. \begin{aligned} q_{11} &= -\sqrt{\frac{3}{8\pi}} \int (x' - iy')\rho(\mathbf{x}') \, d^3x' = -\sqrt{\frac{3}{8\pi}} (p_x - ip_y) \\[2mm] q_{10} &= \sqrt{\frac{3}{4\pi}} \int z' \rho(\mathbf{x}') \, d^3x' = \sqrt{\frac{3}{4\pi}} p_z \end{aligned} \right\} \qquad (4.5)$$

$$\left. \begin{aligned} q_{22} &= \frac{1}{4}\sqrt{\frac{15}{2\pi}} \int (x' - iy')^2 \rho(\mathbf{x}') \, d^3x' = \frac{1}{12}\sqrt{\frac{15}{2\pi}} (Q_{11} - 2iQ_{12} - Q_{22}) \\[2mm] q_{21} &= -\sqrt{\frac{15}{8\pi}} \int z'(x' - iy')\rho(\mathbf{x}') \, d^3x' = -\frac{1}{3}\sqrt{\frac{15}{8\pi}} (Q_{13} - iQ_{23}) \\[2mm] q_{20} &= \frac{1}{2}\sqrt{\frac{5}{4\pi}} \int (3z'^2 - r'^2)\rho(\mathbf{x}') \, d^3x' = \frac{1}{2}\sqrt{\frac{5}{4\pi}} Q_{33} \end{aligned} \right\} \qquad (4.6)$$

Only the moments with $m \geq 0$ have been given, since (3.54) shows that for a real charge density the moments with $m < 0$ are related through

$$q_{l,-m} = (-1)^m q_{lm}^* \qquad (4.7)$$

In equations (4.4)–(4.6), q is the total charge, or monopole moment, $\mathbf{p}$ is the electric dipole moment:

$$\mathbf{p} = \int \mathbf{x}' \rho(\mathbf{x}') \, d^3x' \tag{4.8}$$

and Q_{ij} is the quadrupole moment tensor:

$$Q_{ij} = \int (3x_i' x_j' - r'^2 \delta_{ij}) \rho(\mathbf{x}') \, d^3x' \tag{4.9}$$

We see that the lth multipole coefficients [$(2l + 1)$ in number] are linear combinations of the corresponding multipoles expressed in rectangular coordinates. The expansion of $\Phi(\mathbf{x})$ directly in rectangular coordinates:

$$\Phi(\mathbf{x}) = \frac{q}{r} + \frac{\mathbf{p} \cdot \mathbf{x}}{r^3} + \frac{1}{2} \sum_{i,j} Q_{ij} \frac{x_i x_j}{r^5} + \cdots \tag{4.10}$$

by direct Taylor's series expansion of $1/|\mathbf{x} - \mathbf{x}'|$ will be left as an exercise for the reader. It becomes increasingly cumbersome to continue the expansion in (4.10) beyond the quadrupole terms.

The electric field components for a given multipole can be expressed most easily in terms of spherical coordinates. The negative gradient of a term in (4.1) with definite l, m has spherical components:

$$\left.\begin{aligned}
E_r &= \frac{4\pi(l + 1)}{2l + 1} q_{lm} \frac{Y_{lm}(\theta, \phi)}{r^{l+2}} \\[2mm]
E_\theta &= -\frac{4\pi}{2l + 1} q_{lm} \frac{1}{r^{l+2}} \frac{\partial}{\partial \theta} Y_{lm}(\theta, \phi) \\[2mm]
E_\phi &= -\frac{4\pi}{2l + 1} q_{lm} \frac{1}{r^{l+2}} \frac{im}{\sin \theta} Y_{lm}(\theta, \phi)
\end{aligned}\right\} \tag{4.11}$$

$\partial Y_{lm}/\partial \theta$ and $Y_{lm}/\sin \theta$ can be expressed as linear combinations of other Y_{lm}'s, but the expressions are not particularly illuminating and so will be omitted. The proper way to describe a vector multipole field is by *vector* spherical harmonics, discussed in Chapter 16.

For a dipole $\mathbf{p}$ along the z axis, the fields in (4.11) reduce to the familiar form:

$$\left.\begin{aligned}
E_r &= \frac{2p \cos \theta}{r^3} \\[2mm]
E_\theta &= \frac{p \sin \theta}{r^3} \\[2mm]
E_\phi &= 0
\end{aligned}\right\} \tag{4.12}$$

These dipole fields can be written in vector form by recombining (4.12) or by directly operating with the gradient on the dipole term in (4.10). The result for the field at a point $\mathbf{x}$ due to a dipole $\mathbf{p}$ at the point $\mathbf{x}'$ is:

$$E(\mathbf{x}) = \frac{3\mathbf{n}(\mathbf{p} \cdot \mathbf{n}) - \mathbf{p}}{|\mathbf{x} - \mathbf{x}'|^3} \tag{4.13}$$

where $\mathbf{n}$ is a unit vector directed from $\mathbf{x}'$ to $\mathbf{x}$.

4.2 Multipole Expansion of the Energy of a Charge Distribution in an External Field

If a localized charge distribution described by $\rho(\mathbf{x})$ is placed in an *external* potential $\Phi(\mathbf{x})$, the electrostatic energy of the system is:

$$W = \int \rho(\mathbf{x})\Phi(\mathbf{x}) \, d^3x \tag{4.14}$$

If the potential Φ is slowly varying over the region where $\rho(\mathbf{x})$ is non-negligible, then it can be expanded in a Taylor's series around a suitably chosen origin:

$$\Phi(\mathbf{x}) = \Phi(0) + \mathbf{x} \cdot \nabla\Phi(0) + \frac{1}{2}\sum_i \sum_j x_i x_j \frac{\partial^2 \Phi}{\partial x_i \, \partial x_j}(0) + \cdots \tag{4.15}$$

Utilizing the definition of the electric field $\mathbf{E} = -\nabla\Phi$, the last two terms can be rewritten. Then (4.15) becomes:

$$\Phi(\mathbf{x}) = \Phi(0) - \mathbf{x} \cdot \mathbf{E}(0) - \frac{1}{2}\sum_i \sum_j x_i x_j \frac{\partial E_j}{\partial x_i}(0) + \cdots$$

Since $\nabla \cdot \mathbf{E} = 0$ for the external field, we can subtract

$$\tfrac{1}{6}r^2\nabla \cdot \mathbf{E}(0)$$

from the last term to obtain finally the expansion:

$$\Phi(\mathbf{x}) = \Phi(0) - \mathbf{x} \cdot \mathbf{E}(0) - \frac{1}{6}\sum_i \sum_j (3x_i x_j - r^2\delta_{ij})\frac{\partial E_j}{\partial x_i}(0) + \cdots \tag{4.16}$$

When this is inserted into (4.14) and the definitions of total charge, dipole moment (4.8) and quadrupole moment (4.9), are employed, the energy takes the form:

$$W = q\Phi(0) - \mathbf{p} \cdot \mathbf{E}(0) - \frac{1}{6}\sum_i \sum_j Q_{ij}\frac{\partial E_j}{\partial x_i}(0) + \cdots \tag{4.17}$$

This expansion shows the characteristic way in which the various multipoles interact with an external field—the charge with the potential, the dipole with the electric field, the quadrupole with the field gradient, and so on.

In nuclear physics the quadrupole interaction is of particular interest. Atomic nuclei can possess electric quadrupole moments, and their magnitudes and signs have a bearing on the forces between neutrons and protons, as well as the shapes of the nuclei themselves. The energy levels or states of a nucleus are described by the quantum numbers of total angular momentum J and its projection M along the z axis, as well as others which we will denote by a general index α. A given nuclear state has associated with it a quantum-mechanical charge density* $\rho_{JM\alpha}(\mathbf{x})$, which depends on the quantum numbers (J, M, α), but which is cylindrically symmetric about the z axis. Thus the only nonvanishing quadrupole moment is q_{20} in (4.6), or Q_{33} in (4.9).† The quadrupole moment of a nuclear state is defined as the value of $(1/e)$ Q_{33} with the charge density $\rho_{JM\alpha}(\mathbf{x})$, where e is the protonic charge:

$$Q_{JM\alpha} = \frac{1}{e} \int (3z^2 - r^2)\rho_{JM\alpha}(\mathbf{x}) \, d^3x \tag{4.18}$$

The dimensions of $Q_{JM\alpha}$ are consequently (length)^2. Unless the circumstances are exceptional (e.g., nuclei in atoms with completely closed electronic shells), nuclei are subjected to internal fields which possess field gradients in the neighborhood of the nuclei. Consequently, according to (4.17), the energy of the nuclei will have a contribution from the quadrupole interaction. The states of different M value for the same J will have different quadrupole moments $Q_{JM\alpha}$, and so a degeneracy in M value which may have existed will be removed by the quadrupole coupling to the "external" (crystal lattice, or molecular) electric field. Detection of these small energy differences by radiofrequency techniques allows the determination of the quadrupole moment of the nucleus.‡

The interaction energy between two dipoles $\mathbf{p}_1$ and $\mathbf{p}_2$ can be obtained directly from (4.17) by using the dipole field (4.13). Thus, the mutual potential energy is

$$W_{12} = \frac{\mathbf{p}_1 \cdot \mathbf{p}_2 - 3(\mathbf{n} \cdot \mathbf{p}_1)(\mathbf{n} \cdot \mathbf{p}_2)}{|\mathbf{x}_1 - \mathbf{x}_2|^3} \tag{4.19}$$

where $\mathbf{n}$ is a unit vector in the direction $(\mathbf{x}_1 - \mathbf{x}_2)$. The dipole-dipole interaction is attractive or repulsive, depending on the orientation of the dipoles. For fixed orientation and separation of the dipoles, the value of

---

* See Blatt and Weisskopf, pp. 23 ff., for an elementary discussion of the quantum aspects of the problem.

† Actually Q_{11} and Q_{22} are different from zero, but are not independent of Q_{33}, being given by $Q_{11} = Q_{22} = -\frac{1}{2}Q_{33}$.

‡ "The quadrupole moment of a nucleus," denoted by Q, is defined as the value of $Q_{JM\alpha}$ in the state $M = J$. See Blatt and Weisskopf, *loc. cit.*

the interaction, averaged over the relative positions of the dipoles, is zero. If the moments are generally parallel, attraction (repulsion) occurs when the moments are oriented more or less parallel (perpendicular) to the line joining their centers. For antiparallel moments the reverse is true. The extreme values of the potential energy are equal in magnitude.

4.3 Macroscopic Electrostatics; Effects of Aggregates of Atoms

The equations

$$\left. \begin{array}{l} \nabla \cdot \boldsymbol{\epsilon} = 4\pi\rho' \\[2mm] \nabla \times \boldsymbol{\epsilon} = 0 \end{array} \right\} \tag{4.20}$$

govern electrostatic phenomena of all types, provided the "microscopic" electric field $\boldsymbol{\epsilon}$ is derived from the total "microscopic" charge density ρ'. For problems with a few idealized point charges in the vicinity of mathematically defined boundary surfaces, equations (4.20) are quite acceptable. But there are many physical situations in which a complete specification of the problem in terms of individual charges would be impossible. Any problem involving fields in the presence of matter is a case in point. A macroscopic amount of matter has of the order of $10^{23\pm3}$ charges in it, all of them in motion to a greater or lesser extent because of thermal agitation or zero point vibration.

Setting aside the question of whether electrostatics can be relevant to a situation in which the charges are in incessant motion, let us consider the task of handling macroscopic problems with large numbers of atoms or molecules. Clearly the solution for the electric field:

$$\boldsymbol{\epsilon}(\mathbf{x}) = \int \frac{(\mathbf{x} - \mathbf{x}')}{|\mathbf{x} - \mathbf{x}'|^3} \, \rho'(\mathbf{x}') \, d^3x' \tag{4.21}$$

is not very suitable, since (*a*) it involves a charge density ρ' which must specify the exact positions of very many charges, and (*b*) it fluctuates wildly as the observation point moves by only very small distances (of the order of atomic dimensions). Fortunately, for macroscopic electrostatics we do not want as detailed information as is contained in (4.21). We are content with averages of electric field strengths over regions of the order of 10^{-6} cm³ (i.e., 10^{-2} cm linear dimension) or greater. Since atomic volumes are of the order of 10^{-24} cm³, there are of the order of 10^{18} or more atoms in the volumes of macroscopic interest. This means that the microscopic fluctuations will be entirely averaged out. We will wish to deal with an average $\boldsymbol{\epsilon}(\mathbf{x})$ and $\rho'(\mathbf{x})$. The averages will be over a macroscopically small volume ΔV, large enough, however, to contain very many atoms or molecules:

$$\left.\begin{aligned}
\langle \boldsymbol{\epsilon}(\mathbf{x}) \rangle &= \frac{1}{\Delta V} \int_{\Delta V} \boldsymbol{\epsilon}(\mathbf{x} + \boldsymbol{\xi})\, d^3\xi \\[2mm]
\langle \rho'(\mathbf{x}) \rangle &= \frac{1}{\Delta V} \int_{\Delta V} \rho'(\mathbf{x} + \boldsymbol{\xi})\, d^3\xi
\end{aligned}\right\} \tag{4.22}$$

The averaged quantities are denoted by angle brackets $\langle \;\; \rangle$; the variable $\boldsymbol{\xi}$ ranges over the volume ΔV.

The averaging procedure now allows us to answer the question of whether it is legitimate to talk in static terms when the charges in matter are in thermal motion. At any instant of time the very many charges in the volume ΔV will be in all possible states of motion. An average over them at that instant will yield the same result as an average at some later instant of time. Hence, as far as the averaged quantities are concerned, it is legitimate to talk of static fields and charges.* Furthermore, the averaging can be done as if the atomic charges were fixed in space at the positions they have at some arbitrary instant. Hence the situation can be regarded as electrostatic even at the microscopic level for purposes of calculation.

In the treatment of macroscopic electrostatics it is useful to break up the averaged charge density $\langle \rho'(\mathbf{x}) \rangle$ into two parts, one of which is the averaged charge of the atomic or molecular ions, or excess free charge placed in or on the macroscopic body, and the other of which is the induced or polarization charge. In the absence of external fields, atoms or molecules may or may not have electric dipole moments, but if they do, the moments are randomly oriented. In the presence of a field, the atoms become polarized (or their permanent moments tend to align with the field) and possess on the average a dipole moment These dipole moments can contribute to the averaged charge density $\langle \rho'(\mathbf{x}) \rangle$. Since the induced dipole moments tend to be proportional to the applied field, we will find that the macroscopic version of (4.20) will involve only one constant to characterize the average polarizability of the medium involved.

To see how the induced dipole moments enter the problem we first consider the microscopic field due to one molecule with center of mass at the point $\mathbf{x}_j$ in Fig. 4.1 while the observation point is at $\mathbf{x}$. The molecular charge density is $\rho_j'(\mathbf{x}')$, where $\mathbf{x}'$ is measured from the center of mass of the molecule. It should be noted that ρ_j' in general depends on the position of $\mathbf{x}_j$ of the molecule, since the distortion of the charge cloud depends on the local field present. The microscopic electric field due to the jth

---

* This ignores the very small (at room temperature) induction and radiation fields due to the acceleration of the charges in their thermal motion.

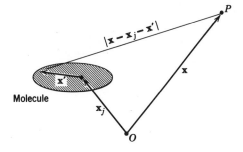

Fig. 4.1   A molecule with center of mass at x_j gives a contribution to the potential at the point P with position x. The internal coordinate x′ is measured from the center of mass.

molecule is

$$\epsilon_j(x) = -\nabla \int_{\mathrm{mol}} \rho_j'(x') \frac{1}{|x - x_j - x'|} \, d^3x' \tag{4.23}$$

For observation points outside the molecule we can expand in multipoles around the center of mass of the molecule. According to (4.10), this leads to

$$\epsilon_j(x) = -\nabla \left[\frac{e_j}{|x - x_j|} + \nabla_j \left(\frac{1}{|x - x_j|} \right) \cdot \mathbf{p}_j + \cdots \right] \tag{4.24}$$

where

$$\left. \begin{array}{l} e_j = \displaystyle\int_{\mathrm{mol}} \rho_j'(x') \, d^3x' \\[3mm] \mathbf{p}_j = \displaystyle\int_{\mathrm{mol}} x' \rho_j'(x') \, d^3x' \end{array} \right\} \tag{4.25}$$

are the molecular charge and the dipole moment, respectively. The quadrupole term in (4.10) could have been retained, but as long as the macroscopic variations of field occur over distances large compared to molecular dimensions it contributes negligibly to the averaged field relative to the dipole term. Both e_j and $\mathbf{p}_j$ are functions of the position of the molecule.

To obtain the microscopic field due to all the molecules we sum over j:

$$\epsilon(x) = -\nabla \sum_j \left[\frac{e_j}{|x - x_j|} + \mathbf{p}_j \cdot \nabla_j \left(\frac{1}{|x - x_j|} \right) \right] \tag{4.26}$$

We now want to average according to (4.22) in order to obtain a macroscopic field. To facilitate this averaging procedure we replace the discrete sum over the molecules by an integral by introducing apparently continuous charge and polarization densities:

$$\left. \begin{array}{l} \rho_{\mathrm{mol}}(x) = \displaystyle\sum_j e_j \delta(x - x_j) \\[3mm] \boldsymbol{\pi}_{\mathrm{mol}}(x) = \displaystyle\sum_j \mathbf{p}_j \delta(x - x_j) \end{array} \right\} \tag{4.27}$$

Then (4.26) can be rewritten formally as:

$$\mathbf{\epsilon}(\mathbf{x}) = -\mathbf{\nabla} \int d^3x'' \left[\frac{\rho_{\text{mol}}(\mathbf{x}'')}{|\mathbf{x} - \mathbf{x}''|} + \mathbf{\pi}_{\text{mol}}(\mathbf{x}'') \cdot \mathbf{\nabla}'' \left(\frac{1}{|\mathbf{x} - \mathbf{x}''|} \right) \right] \quad (4.28)$$

To illustrate the averaging process we consider the first term in (4.28). The averaged value is, by (4.22):

$$\langle \mathbf{\epsilon}_1(\mathbf{x}) \rangle = -\mathbf{\nabla} \left[\frac{1}{\Delta V} \int_{\Delta V} d^3\xi \int d^3x'' \frac{\rho_{\text{mol}}(\mathbf{x}'')}{|\mathbf{x} + \mathbf{\xi} - \mathbf{x}''|} \right] \quad (4.29)$$

where we have used the fact that differentiation and averaging can be interchanged. If the variable of integration $\mathbf{x}''$ is replaced by $\mathbf{x}'' = \mathbf{x}' + \mathbf{\xi}$, then

$$\langle \mathbf{\epsilon}_1(\mathbf{x}) \rangle = -\mathbf{\nabla} \left[\frac{1}{\Delta V} \int_{\Delta V} d^3\xi \int d^3x' \frac{\rho_{\text{mol}}(\mathbf{x}' + \mathbf{\xi})}{|\mathbf{x} - \mathbf{x}'|} \right] \quad (4.30)$$

The equality of (4.29) and (4.30) shows the obvious equivalence of averaging by means of moving the observation point around the volume ΔV centered at $\mathbf{x}$ and averaging by moving the integration point over the molecules in a volume ΔV centered around $\mathbf{x}'$. From definition (4.27) it is clear that the integral of ρ_{mol} over the volume ΔV at $\mathbf{x}'$ just adds up all the molecular charges e_j inside ΔV:

$$\frac{1}{\Delta V} \int_{\Delta V} d^3\xi \rho_{\text{mol}}(\mathbf{x}' + \mathbf{\xi}) = \frac{1}{\Delta V} \sum_{\Delta V} e_j$$

If the macroscopic density of molecules at $\mathbf{x}'$ is $N(\mathbf{x}')$ molecules per unit volume and $\langle e_{\text{mol}}(\mathbf{x}') \rangle$ is the average charge per molecule within the volume ΔV at $\mathbf{x}'$, then

$$\frac{1}{\Delta V} \int_{\Delta V} d^3\xi \rho_{\text{mol}}(\mathbf{x}' + \mathbf{\xi}) = N(\mathbf{x}')\langle e_{\text{mol}}(\mathbf{x}') \rangle \quad (4.31)$$

Now (4.30) can be written

$$\langle \mathbf{\epsilon}_1(\mathbf{x}) \rangle = -\mathbf{\nabla} \int \frac{N(\mathbf{x}')\langle e_{\text{mol}}(\mathbf{x}') \rangle}{|\mathbf{x} - \mathbf{x}'|} d^3x'$$

Exactly similar considerations can be made for the second term in (4.28). With the same definitions of averages we have

$$\frac{1}{\Delta V} \int_{\Delta V} d^3\xi \mathbf{\pi}_{\text{mol}}(\mathbf{x}' + \mathbf{\xi}) = N(\mathbf{x}')\langle \mathbf{p}_{\text{mol}}(\mathbf{x}') \rangle \quad (4.32)$$

Then the averaged form of (4.28) is given by:

$$\langle \mathbf{\epsilon}(\mathbf{x}) \rangle = -\mathbf{\nabla} \int N(\mathbf{x}') \left\{ \frac{\langle e_{\text{mol}}(\mathbf{x}') \rangle}{|\mathbf{x} - \mathbf{x}'|} + \langle \mathbf{p}_{\text{mol}}(\mathbf{x}') \rangle \cdot \mathbf{\nabla}' \left(\frac{1}{|\mathbf{x} - \mathbf{x}'|} \right) \right\} d^3x' \quad (4.33)$$

To obtain the macroscopic equivalent of (4.20) we take the divergence of both sides. Recalling that $\nabla^2(1/|\mathbf{x} - \mathbf{x}'|) = -4\pi\delta(\mathbf{x} - \mathbf{x}')$, we find:

$$\nabla \cdot \langle \boldsymbol{\epsilon}(\mathbf{x}) \rangle = 4\pi \int N(\mathbf{x}')\{\langle e_{\mathrm{mol}}(\mathbf{x}')\rangle\delta(\mathbf{x} - \mathbf{x}') + \langle \mathbf{p}_{\mathrm{mol}}(\mathbf{x}')\rangle \cdot \nabla'\delta(\mathbf{x} - \mathbf{x}')\} \, d^3x'$$

From the properties of the delta function (Section 1.2) it follows that

$$\nabla \cdot \langle \boldsymbol{\epsilon}(\mathbf{x}) \rangle = 4\pi N(\mathbf{x})\langle e_{\mathrm{mol}}(\mathbf{x}) \rangle - 4\pi \nabla \cdot (N(\mathbf{x})\langle \mathbf{p}_{\mathrm{mol}}(\mathbf{x})\rangle) \qquad (4.34)$$

This is of the form of the first equation of (4.20) with the charge density ρ' replaced by two terms, the first being the average charge per unit volume of the molecules and the second being the polarization charge per unit volume. The presence of the divergence in the polarization-charge density seems very natural when one thinks of how this part of the charge density is created. If we consider a small volume in the medium, part of the charge inside that volume may be due to the net charges on the molecules. But there is a contribution arising from the polarization of the charge cloud of the molecules in an external field, since, for example, molecules whose charge once lay totally inside the volume may now have part of their charge cloud outside the volume in question. If the polarization is uniform over the space containing our small volume, then as much charge will be brought in through the surface of the volume as will leave it, and there will be no net effect. But if the polarization is not uniform, there can be a net increase or decrease of charge within the volume, as indicated schematically in Fig. 4.2. This is the physical origin of the polarization-charge density.

In (4.34) the two divergences can be combined so that the equation reads:

$$\nabla \cdot [\langle \boldsymbol{\epsilon} \rangle + 4\pi N \langle \mathbf{p}_{\mathrm{mol}} \rangle] = 4\pi N \langle e_{\mathrm{mol}} \rangle \qquad (4.35)$$

It is customary to introduce certain macroscopic quantities, namely, the electric field $\mathbf{E}$, the polarization $\mathbf{P}$ (electric dipole moment per unit volume),

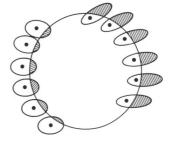

Fig. 4.2   Origin of polarization-charge density. Because of spatial variation of polarization more molecular charge may leave a given small volume than enters it.

the charge density ρ, and the displacement $\mathbf{D}$, defined as follows:

$$\left.\begin{aligned}
\mathbf{E} &= \langle \mathbf{\epsilon} \rangle \\
\mathbf{P} &= N\langle \mathbf{p}_{\text{mol}} \rangle \\
\rho &= N\langle e_{\text{mol}} \rangle \\
\mathbf{D} &= \mathbf{E} + 4\pi\mathbf{P}
\end{aligned}\right\} \tag{4.36}$$

If there are several different kinds of atoms or molecules in the medium and perhaps extra charge is added, these definitions have the obvious generalizations:

$$\left.\begin{aligned}
\mathbf{P} &= \sum_i N_i \langle \mathbf{p}_i \rangle \\
\rho &= \sum_i N_i \langle e_i \rangle + \rho_{\text{ex}}
\end{aligned}\right\} \tag{4.37}$$

where N_i is the number of molecules of type i per unit volume, $\langle e_i \rangle$ is their average charge, and $\langle \mathbf{p}_i \rangle$ is their average dipole moment. ρ_{ex} is the excess (or free) charge density. Usually the molecules are neutral, and the total charge density ρ is just the free charge density.

With the definitions of (4.36) or (4.37), the macroscopic divergence equation becomes:

$$\nabla \cdot \mathbf{D} = 4\pi\rho \tag{4.38}$$

The macroscopic equivalent of the other member of the pair (4.20) can be obtained by taking the curl of (4.33). Obviously the result is

$$\nabla \times \mathbf{E} = 0 \tag{4.39}$$

For macroscopic electrostatic problems in the presence of dielectrics, (4.38) and (4.39) replace the microscopic equations (4.20).

The solution for the electric field (4.33) can be expressed in terms of the macroscopic variables as

$$\mathbf{E}(\mathbf{x}) = -\nabla \int d^3x' \left[\frac{\rho(\mathbf{x}')}{|\mathbf{x} - \mathbf{x}'|} + \mathbf{P}(\mathbf{x}') \cdot \nabla'\left(\frac{1}{|\mathbf{x} - \mathbf{x}'|} \right) \right] \tag{4.40}$$

The second term, describing the dipole field, has already been discussed in Section 1.6.

4.4 Simple Dielectrics and Boundary Conditions

It was mentioned in the previous section that the molecular polarization depends on the local electric field at the molecule. In the absence of a field there is no average polarization.* This means that the polarization

---

* Except for electrets, which have a permanent electric polarization.

P, which is in general a function of **E**, can be expanded as a powers series in the field, at least for small fields. Any component will have an expansion of the form:

$$P_i = \sum_j a_{ij}E_j + \sum_{j,k} b_{ijk}E_jE_k + \cdots$$

A priori it is not clear how important the higher terms will be in practice. Experimentally it is found that the polarization as a function of applied field looks qualitatively as shown in Fig. 4.3. At normal temperatures and for fields attainable in the laboratory the *linear approximation* is *completely adequate*. This is not surprising if it is remembered that interatomic electric fields are of the order of 10^9 volts/cm. Any external field causing polarization is only a small perturbation. For a general anisotropic medium (e.g., certain crystals such as calcite and quartz), there can be six independent elements a_{ij}. But for simple substances, called isotropic, **P** is parallel to **E** with a constant of proportionality χ_e which is independent of the direction of **E**. Then

$$\mathbf{P} = \chi_e\mathbf{E} \qquad (4.41)$$

The constant χ_e is called the electric susceptibility of the medium. We then find the displacement proportional to **E**:

$$\mathbf{D} = \epsilon\mathbf{E} \qquad (4.42)$$

where

$$\epsilon = 1 + 4\pi\chi_e \qquad (4.43)$$

is the dielectric constant.

If the dielectric is not only isotropic, but also uniform, ϵ is independent of position. Then the divergence equation can be written

$$\nabla \cdot \mathbf{E} = \frac{4\pi}{\epsilon}\rho \qquad (4.44)$$

and all problems *in that medium* are reduced to those of previous chapters, except that the electric fields produced by given charges are reduced by a

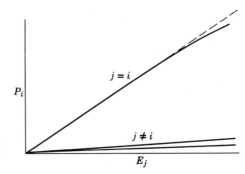

Fig. 4.3 Components of polarization as a function of applied electric field.

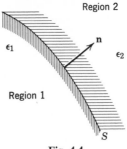

Fig. 4.4

factor $1/\epsilon$. The reduction can be understood in terms of a polarization of the atoms which produce fields in opposition to that of the given charge. One immediate consequence is that the capacitance of a capacitor is increased by a factor of ϵ if the empty space between the electrodes is filled with a dielectric with dielectric constant ϵ (true only to the extent that fringing fields can be neglected).

An important consideration is the boundary conditions on the field quantities **E** and **D** at surfaces where the dielectric properties vary discontinuously. Consider a surface S as shown in Fig. 4.4. The unit vector **n** is normal to the surface and points from region 1 with dielectric constant ϵ_1 to region 2 with dielectric constant ϵ_2. In exactly the same manner as in Section 1.6 we find, by taking a Gaussian pill box with end faces in regions 1 and 2 parallel to the surface S, that

$$(\mathbf{D}_2 - \mathbf{D}_1) \cdot \mathbf{n} = 4\pi\sigma \tag{4.45}$$

where σ is the surface-charge density (*not* including polarization charge). Similarly, by applying Stokes's theorem to $\nabla \times \mathbf{E} = 0$, we find that

$$(\mathbf{E}_1 - \mathbf{E}_2) \times \mathbf{n} = 0 \tag{4.46}$$

These boundary conditions on the normal component of **D** and the tangential component of **E** replace the microscopic conditions (1.22) and below. The macroscopic equivalent of (1.22) can be recovered from (4.45) by extracting the polarization-charge density from the left-hand side.

4.5 Boundary-Value Problems with Dielectrics

The methods of previous chapters for the solution of electrostatic boundary-value problems can readily be extended to handle the presence of dielectrics. In this section we will treat a few examples of the various techniques applied to dielectric media.

To illustrate the method of images for dielectrics we consider a point charge q embedded in a semi-infinite dielectric ϵ_1 a distance d away from a plane interface which separates the first medium from another semi-infinite dielectric ϵ_2. The surface may be taken as the plane $z = 0$, as shown in Fig. 4.5. We must find the appropriate solution to the equations:

$$
\left.
\begin{aligned}
\epsilon_1 \nabla \cdot \mathbf{E} &= 4\pi\rho, & z &> 0 \\
\epsilon_2 \nabla \cdot \mathbf{E} &= 0, & z &< 0 \\
\nabla \times \mathbf{E} &= 0, & &\text{everywhere}
\end{aligned}
\right\}
\qquad (4.47)
$$

and

subject to the boundary conditions at $z = 0$:

$$
\lim_{z \to 0^+}
\begin{bmatrix}
\epsilon_1 E_z \\
E_x \\
E_y
\end{bmatrix}
= \lim_{z \to 0^-}
\begin{bmatrix}
\epsilon_2 E_z \\
E_x \\
E_y
\end{bmatrix}
\qquad (4.48)
$$

Since $\nabla \times \mathbf{E} = 0$ everywhere, $\mathbf{E}$ is derivable in the usual way from a potential Φ. In attempting to use the image method it is natural to locate an image charge q' at the symmetrical position A' shown in Fig. 4.6. Then for $z > 0$ the potential at a point P described by cylindrical coordinates (ρ, ϕ, z) will be

$$
\Phi = \frac{1}{\epsilon_1}\left(\frac{q}{R_1} + \frac{q'}{R_2}\right), \qquad z > 0 \qquad (4.49)
$$

where $R_1 = \sqrt{\rho^2 + (d - z)^2}$, $R_2 = \sqrt{\rho^2 + (d + z)^2}$. So far the procedure is completely analogous to the problem with a conducting material in place of the dielectric ϵ_2 for $z < 0$. But we now must specify the potential for $z < 0$. Since there are no charges in the region $z < 0$, it must be a solution of Laplace's equation without singularities in that region. Clearly the simplest assumption is that for $z < 0$ the potential is equivalent to that of a charge q'' at the position A of the actual charge q:

$$
\Phi = \frac{1}{\epsilon_2}\frac{q''}{R_1}, \qquad z < 0 \qquad (4.50)
$$

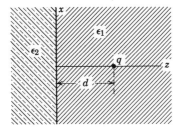

Fig. 4.5

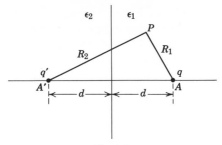

Fig. 4.6

Since
$$\left.\frac{\partial}{\partial z}\left(\frac{1}{R_1}\right)\right|_{z=0} = -\left.\frac{\partial}{\partial z}\left(\frac{1}{R_2}\right)\right|_{z=0} = \frac{d}{(\rho^2 + d^2)^{3/2}}$$

while
$$\left.\frac{\partial}{\partial \rho}\left(\frac{1}{R_1}\right)\right|_{z=0} = \left.\frac{\partial}{\partial \rho}\left(\frac{1}{R_2}\right)\right|_{z=0} = \frac{-\rho}{(\rho^2 + d^2)^{3/2}}$$

the boundary conditions (4.48) lead to the requirements:

$$q - q' = q''$$

$$\frac{1}{\epsilon_1}(q + q') = \frac{1}{\epsilon_2}q''$$

These can be solved to yield the image charges q' and q'':

$$\left. \begin{aligned} q' &= -\left(\frac{\epsilon_2 - \epsilon_1}{\epsilon_2 + \epsilon_1}\right)q \\[2mm] q'' &= \left(\frac{2\epsilon_2}{\epsilon_2 + \epsilon_1}\right)q \end{aligned} \right\} \tag{4.51}$$

For the two cases $\epsilon_2 > \epsilon_1$ and $\epsilon_2 < \epsilon_1$ the lines of force are shown qualitatively in Fig. 4.7.

The polarization-charge density is given by $-\boldsymbol{\nabla} \cdot \mathbf{P}$. Inside either dielectric, $\mathbf{P} = \chi_e \mathbf{E}$, so that $-\boldsymbol{\nabla} \cdot \mathbf{P} = -\chi_e \boldsymbol{\nabla} \cdot \mathbf{E} = 0$, except at the point charge q. At the surface, however, χ_e takes a discontinuous jump, $\Delta\chi_e = (1/4\pi)(\epsilon_1 - \epsilon_2)$ as z passes through $z = 0$. This implies that there is a polarization surface-charge density on the plane $z = 0$:

$$\sigma_{\text{pol}} = -(\mathbf{P}_2 - \mathbf{P}_1) \cdot \mathbf{n} \tag{4.52}$$

where $\mathbf{n}$ is the unit normal from dielectric 1 to dielectric 2, and $\mathbf{P}_i$ is the polarization in the dielectric i at $z = 0$. Since

$$\mathbf{P}_i = \left(\frac{\epsilon_i - 1}{4\pi}\right)\mathbf{E} = -\left(\frac{\epsilon_i - 1}{4\pi}\right)\boldsymbol{\nabla}\Phi$$

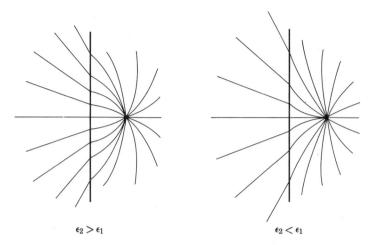

$\epsilon_2 > \epsilon_1$ $\epsilon_2 < \epsilon_1$

Fig. 4.7   Lines of electric force for a point charge embedded in a dielectric ϵ_1 near a semi-infinite slab of dielectric ϵ_2.

it is a simple matter to show that the polarization-charge density is

$$\sigma_{\text{pol}} = -\frac{q}{2\pi}\frac{(\epsilon_2 - \epsilon_1)}{\epsilon_1(\epsilon_2 + \epsilon_1)}\frac{d}{(\rho^2 + d^2)^{3/2}} \tag{4.53}$$

In the limit $\epsilon_2 \gg \epsilon_1$ the dielectric ϵ_2 behaves much like a conductor in that the field inside it becomes very small and the surface-charge density (4.53) approaches the value appropriate to a conducting surface.

The second illustration of electrostatic problems involving dielectrics is that of a dielectric sphere of radius a with dielectric constant ϵ placed in an initially uniform electric field which at large distances from the sphere is directed along the z axis and has magnitude E_0, as indicated in Fig. 4.8. Both inside and outside the sphere there are no free charges. Consequently the problem is one of solving Laplace's equation with the proper boundary conditions at $r = a$. From the axial symmetry of the geometry we can

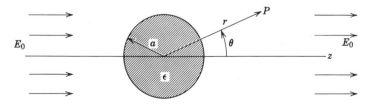

E_0

E_0

Fig. 4.8

take the solution to be of the form:

INSIDE:

$$\Phi_{\text{in}} = \sum_{l=0}^{\infty} A_l r^l P_l(\cos\theta) \tag{4.54}$$

OUTSIDE:

$$\Phi_{\text{out}} = \sum_{l=0}^{\infty} [B_l r^l + C_l r^{-(l+1)}] P_l(\cos\theta) \tag{4.55}$$

From the boundary condition at infinity ($\Phi \to -E_0 z = -E_0 r \cos\theta$) we find that the only nonvanishing B_l is $B_1 = -E_0$. The other coefficients are determined from the boundary conditions at $r = a$:

TANGENTIAL E: $\qquad -\dfrac{1}{a}\dfrac{\partial\Phi_{\text{in}}}{\partial\theta}\bigg|_{r=a} = -\dfrac{1}{a}\dfrac{\partial\Phi_{\text{out}}}{\partial\theta}\bigg|_{r=a}$

NORMAL D: $\qquad -\epsilon\dfrac{\partial\Phi_{\text{in}}}{\partial r}\bigg|_{r=a} = -\dfrac{\partial\Phi_{\text{out}}}{\partial r}\bigg|_{r=a}$
$$\tag{4.56}$$

The first boundary condition leads to the relations:

$$A_1 = -E_0 + \frac{C_1}{a^3}$$
$$A_l = \frac{C_l}{a^{2l+1}}, \qquad \text{for } l \neq 1$$
$$\tag{4.57}$$

while the second gives:

$$\epsilon A_1 = -E_0 - 2\frac{C_1}{a^3}$$
$$\epsilon l A_l = -(l+1)\frac{C_l}{a^{2l+1}}, \qquad \text{for } l \neq 1$$
$$\tag{4.58}$$

The second equations in (4.57) and (4.58) can be satisfied simultaneously only with $A_l = C_l = 0$ for all $l \neq 1$. The remaining coefficients are given in terms of the applied electric field E_0:

$$A_1 = -\left(\frac{3}{2+\epsilon}\right)E_0$$
$$C_1 = \left(\frac{\epsilon-1}{\epsilon+2}\right)a^3 E_0$$
$$\tag{4.59}$$

The potential is therefore

$$\Phi_{\text{in}} = -\left(\frac{3}{\epsilon+2}\right)E_0 r \cos\theta$$
$$\Phi_{\text{out}} = -E_0 r \cos\theta + \left(\frac{\epsilon-1}{\epsilon+2}\right)E_0 \frac{a^3}{r^2}\cos\theta$$
$$\tag{4.60}$$

The potential inside the sphere describes a constant electric field parallel to the applied field with magnitude

$$E_{in} = \frac{3}{\epsilon + 2} E_0 < E_0 \qquad (4.61)$$

Outside the sphere the potential is equivalent to the applied field E_0 plus the field of an electric dipole at the origin with dipole moment:

$$p = \left(\frac{\epsilon - 1}{\epsilon + 2}\right) a^3 E_0 \qquad (4.62)$$

oriented in the direction of the applied field. The dipole moment can be interpreted as the volume integral of the polarization **P**. The polarization is

$$\mathbf{P} = \left(\frac{\epsilon - 1}{4\pi}\right)\mathbf{E} = \frac{3}{4\pi}\left(\frac{\epsilon - 1}{\epsilon + 2}\right)\mathbf{E}_0 \qquad (4.63)$$

It is constant throughout the volume of the sphere and has a volume integral given by (4.62). The polarization surface-charge density is, according to (4.52), $\sigma_{pol} = (\mathbf{P} \cdot \mathbf{r})/r$:

$$\sigma_{pol} = \frac{3}{4\pi}\left(\frac{\epsilon - 1}{\epsilon + 2}\right) E_0 \cos \theta \qquad (4.64)$$

This can be thought of as producing an internal field directed oppositely to the applied field, so reducing the field inside the sphere to its value (4.61), as sketched in Fig. 4.9.

The problem of a spherical cavity of radius a in a dielectric medium with dielectric constant ϵ and with an applied electric field E_0 parallel to the z axis, as shown in Fig. 4.10, can be handled in exactly the same way as the dielectric sphere. In fact, inspection of boundary conditions (4.56) shows that the results for the cavity can be obtained from those of the sphere by the replacement $\epsilon \rightarrow (1/\epsilon)$. Thus, for example, the field inside the cavity

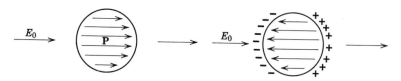

Fig. 4.9   Dielectric sphere in a uniform field E_0, showing the polarization on the left and the polarization charge with its associated, opposing, electric field on the right.

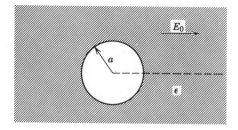

Fig. 4.10. Spherical cavity in a dielectric with a uniform field applied.

is uniform, parallel to $\mathbf{E}_0$, and of magnitude:

$$E_{in} = \frac{3\epsilon}{2\epsilon + 1} E_0 > E_0 \qquad (4.65)$$

Similarly, the field outside is the applied field plus that of a dipole at the origin *oriented oppositely* to the applied field and with dipole moment:

$$p = \left(\frac{\epsilon - 1}{2\epsilon + 1}\right) a^3 E_0 \qquad (4.66)$$

4.6 Molecular Polarizability and Electric Susceptibility

In this section and the next we will consider the relation between molecular properties and the macroscopically defined parameter, the electric susceptibility χ_e. Our discussion will be in terms of simple classical models of the molecular properties, although a proper treatment necessarily would involve quantum-mechanical considerations. Fortunately, the simpler properties of dielectrics are amenable to classical analysis.

Before examining how the detailed properties of the molecules are related to the susceptibility we must make a distinction between the fields acting on the molecules in the medium and the external field. The susceptibility is defined through the relation $\mathbf{P} = \chi_e \mathbf{E}$, where $\mathbf{E}$ is the macroscopic electric field. In rarefied media where molecular separations are large there is little difference between the macroscopic field and that acting on any molecule or group of molecules. But in dense media with closely packed molecules the polarization of neighboring molecules gives rise to an internal field $\mathbf{E}_i$ at any given molecule in addition to the average macroscopic field $\mathbf{E}$, so that the total field at the molecule is $\mathbf{E} + \mathbf{E}_i$. The internal field can be written as

$$\mathbf{E}_i = \left(\frac{4\pi}{3} + s\right)\mathbf{P} \qquad (4.67)$$

where $s\mathbf{P}$ is the contribution of molecules close to the given molecule, and $(4\pi/3)\mathbf{P}$ is the contribution of the more distant molecules. It is customary to consider the two parts separately by imagining a spherical surface of size large microscopically but small macroscopically surrounding a molecule, as shown in Fig. 4.11, and determining the field at the center due to the polarization of the molecules exterior to the sphere and the resulting charge density induced on the surface of the sphere. This charge density is $-\mathbf{P} \cdot \mathbf{n}$, where $\mathbf{n}$ is the outward normal from the spherical surface. The resulting field at the center is obviously parallel to $\mathbf{P}$ and has the magnitude:

$$E_i^{(1)} = \int_{\text{sphere}} r^2 \, d\Omega \, \frac{(-P \cos \theta)(-\cos \theta)}{r^2} = \frac{4\pi}{3} P \qquad (4.68)$$

giving the first term in (4.67).

The field $s\mathbf{P}$ due to the molecules near by is more difficult to determine. Lorentz (p. 138) showed that for atoms in a simple cubic lattice $s = 0$ at any lattice site. The argument depends on the symmetry of the problem, as can be seen as follows. Suppose that inside the sphere we have a cubic array of dipoles such as are shown in Fig. 4.12, with all their moments constant in magnitude and oriented along the same direction (remember that the sphere is macroscopically small). The positions of the dipoles are given by the coordinates $\mathbf{x}_{ijk}$ with the components along the coordinate axes (ia, ja, ka), where a is the lattice spacing, and i, j, k each take on positive and negative integer values. The field at the origin due to all the dipoles is, according to (4.13),

$$\mathbf{E} = \sum_{i,j,k} \frac{3(\mathbf{p} \cdot \mathbf{x}_{ijk})\mathbf{x}_{ijk} - x_{ijk}^2 \mathbf{p}}{x_{ijk}^5} \qquad (4.69)$$

The x component of the field can be written in the form:

$$E_1 = \sum_{ijk} \frac{3(i^2 p_1 + ij p_2 + ik p_3) - (i^2 + j^2 + k^2)p_1}{a^3(i^2 + j^2 + k^2)^{5/2}} \qquad (4.70)$$

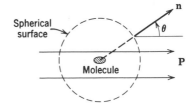

Fig. 4.11 Calculation of the internal field—
contribution from distant molecules.

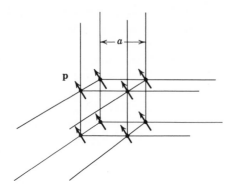

Fig. 4.12 Calculation of the internal field—contribution from nearby molecules in a simple cubic lattice.

Since the indices run equally over positive and negative values, the cross terms involving $(ijp_2 + ikp_3)$ vanish. By symmetry the sums:

$$\sum_{ijk} \frac{i^2}{(i^2 + j^2 + k^2)^{5/2}} = \sum_{ijk} \frac{j^2}{(i^2 + j^2 + k^2)^{5/2}} = \sum_{ijk} \frac{k^2}{(i^2 + j^2 + k^2)^{5/2}}$$

are all equal. Consequently

$$E_1 = \sum_{ijk} \frac{[3i^2 - (i^2 + j^2 + k^2)]p_1}{a^3(i^2 + j^2 + k^2)^{5/2}} = 0 \tag{4.71}$$

Similar arguments show that the y and z components vanish also. Hence $s = 0$ for a simple cubic lattice.

If $s = 0$ for a highly symmetric situation, it seems plausible that $s = 0$ also for completely random situations. Hence we expect amorphous substances like glass to have no internal field due to near-by molecules. Although calculations taking into account the structural details of the substance are necessary to obtain an accurate answer, it is a good working assumption that $s \simeq 0$ for almost all materials.

The polarization vector **P** was defined in (4.36) as

$$\mathbf{P} = N\langle\mathbf{p}_{mol}\rangle$$

where $\langle\mathbf{p}_{mol}\rangle$ is the average dipole moment of the molecules. This dipole moment is approximately proportional to the electric field acting on the molecule. To exhibit this dependence on electric field we define the *molecular polarizability* γ_{mol} as the ratio of the average molecular dipole moment to the applied field at the molecule. Taking account of the internal field (4.67), this gives:

$$\langle\mathbf{p}_{mol}\rangle = \gamma_{mol}(\mathbf{E} + \mathbf{E}_i) \tag{4.72}$$

γ_{mol} is, in principle, a function of the electric field, but for a wide range of field strengths is a constant which characterizes the response of the molecules to an applied field (see Section 4.4). Equation (4.72) can be combined with (4.36) and (4.67) to yield:

$$\mathbf{P} = N\gamma_{mol}\left(\mathbf{E} + \frac{4\pi}{3}\mathbf{P}\right) \tag{4.73}$$

where we have assumed $s = 0$. Solving for $\mathbf{P}$ in terms of $\mathbf{E}$ and using the fact that $\mathbf{P} = \chi_e\mathbf{E}$ defines the electric susceptibility of a substance, we find

$$\chi_e = \frac{N\gamma_{mol}}{1 - \frac{4\pi}{3}N\gamma_{mol}} \tag{4.74}$$

as the relation between susceptibility (the macroscopic parameter) and molecular polarizability (the microscopic parameter). Since the dielectric constant is $\epsilon = 1 + 4\pi\chi_e$, it can be expressed in terms of γ_{mol}, or alternatively the molecular polarizability can be expressed in terms of the dielectric constant:

$$\gamma_{mol} = \frac{3}{4\pi N}\left(\frac{\epsilon - 1}{\epsilon + 2}\right) \tag{4.75}$$

This is called the *Clausius-Mossotti equation*, since Mossotti (in 1850) and Clausius independently (in 1879) established that for any given substance $(\epsilon - 1)/(\epsilon + 2)$ should be proportional to the density of the substance.* The relation holds best for dilute substances such as gases. For liquids and solids, (4.75) is only approximately valid, especially if the dielectric constant is large. The interested reader can refer to the books by Böttcher, Debye, and Fröhlich for further details.

4.7   Models for the Molecular Polarizability

The polarization of a collection of atoms or molecules arises in two ways:

(*a*) the applied field distorts the charge distributions and so produces an induced dipole moment in each molecule;

(*b*) the applied field tends to line up the initially randomly oriented permanent dipole moments of the molecules.

To estimate the induced moments we will consider a simple model of

---

* At optical frequencies, $\epsilon = n^2$, where n is the index of refraction. With n^2 replacing ϵ in (4.75), the equation is sometimes called the *Lorentz-Lorenz equation* (1880).

harmonically bound charges (electrons). Each charge e is bound under the action of a restoring force

$$\mathbf{F} = -m\omega_0^2\mathbf{x} \qquad (4.76)$$

where m is the mass of the charge, and ω_0 the frequency of oscillation about equilibrium. Under the action of an electric field $\mathbf{E}$ the charge is displaced from its equilibrium by an amount $\mathbf{x}$ given by

$$m\omega_0^2\mathbf{x} = e\mathbf{E}$$

Consequently the induced dipole moment is

$$\mathbf{p}_{\text{mol}} = e\mathbf{x} = \frac{e^2}{m\omega_0^2}\mathbf{E} \qquad (4.77)$$

This means that the polarizability is $\gamma = e^2/m\omega_0^2$. If there are Z electrons per molecule, f_j having a restoring force constant $m\omega_j^2$ ($\sum_j f_j = Z$), then the molecular polarizability due to the electrons is:

$$\gamma_{\text{el}} = \frac{e^2}{m}\sum_j \frac{f_j}{\omega_j^2} \qquad (4.78)$$

To get a feeling for the order of magnitude of γ_{el} we can make two different estimates. Since γ has the dimensions of a volume, its magnitude must be of the order of molecular dimensions or less, namely $\gamma_{\text{el}} \lesssim 10^{-23}$ cm^3. Alternatively, we note that the binding frequencies of electrons in atoms must be of the order of light frequencies. Taking a typical wavelength of light as 3000 angstroms, we find $\omega \simeq 6 \times 10^{15}\,\text{sec}^{-1}$. Then $\gamma_{\text{el}} \sim (e^2/m\omega^2) \sim 6 \times 10^{-24}\,\text{cm}^3$, consistent with the molecular volume estimate. For gases at NTP the number of molecules per cubic centimeter is $N = 2.7 \times 10^{19}$, so that their susceptibilities should be of the order of $\chi_e \lesssim 10^{-4}$. This means dielectric constants differing from unity by a few parts in 10^3, or less. Experimentally, typical values of dielectric constant are 1.00054 for air, 1.0072 for ammonia vapor, 1.0057 for methyl alcohol, 1.000068 for helium. For solid or liquid dielectrics, $N \sim 10^{22} - 10^{23}$ molecules/cm^3. Consequently, the susceptibility can be of the order of unity (to within a factor $10^{\pm 1}$) as is observed.*

The possibility that thermal agitation of the molecules could modify the result (4.78) for the induced dipole polarizability needs consideration. In statistical mechanics the probability distribution of particles in phase

---

* See *Handbook of Chemistry and Physics*, Chemical Rubber Publishing Co., or *American Institute of Physics Handbook*, McGraw-Hill, New York, (1957), for tables of dielectric constants of various substances.

space ($\mathbf{p}$, $\mathbf{q}$ space) is proportional to the Boltzmann factor

$$\exp\left(-H/kT\right) \tag{4.79}$$

where H is the Hamiltonian. In the simple problem of a harmonically bound electron with an applied field in the z direction, the Hamiltonian is

$$H = \frac{1}{2m}\,\mathbf{p}^2 + \frac{m}{2}\,\omega_0{}^2\mathbf{x}^2 - eEz \tag{4.80}$$

where here $\mathbf{p}$ is the momentum of the electron. The average value of the dipole moment is

$$\langle p_{\text{mol}}\rangle = \frac{\displaystyle\int d^3p \int d^3x\,(ez)\exp\left(-H/kT\right)}{\displaystyle\int d^3p \int d^3x\,\exp\left(-H/kT\right)} \tag{4.81}$$

The integration over (d^3p) and ($dx\,dy$) can be done immediately to yield

$$\langle p_{\text{mol}}\rangle = \frac{e\displaystyle\int dz\,z\,\exp\left[-\frac{1}{kT}\left(\frac{m\omega_0{}^2}{2}z^2 - eEz\right)\right]}{\displaystyle\int dz\,\exp\left[-\frac{1}{kT}\left(\frac{m\omega_0{}^2}{2}z^2 - eEz\right)\right]}$$

An integration by parts in the numerator yields the result:

$$\langle p_{\text{mol}}\rangle = \frac{e^2}{m\omega_0{}^2}\,E$$

the same as was found in (4.77) by elementary means, ignoring thermal motion. Thus the molecular polarizability (4.78) holds even in the presence of thermal motion.

The second type of polarizability is that caused by the partial orientation of randomly oriented permanent dipole moments. This orientation polarization is important in "polar" substances such as HCl and H_2O and was first discussed by Debye (1912). All molecules are assumed to possess a permanent dipole moment $\mathbf{p}_0$ which can be oriented in any direction in space. In the absence of a field thermal agitation keeps the molecules randomly oriented so that there is no net dipole moment. With an applied field there is a tendency to line up along the field in the configuration of lowest energy. Consequently there will be an average dipole moment. To calculate this we note that the Hamiltonian of the molecule is given by

$$H = H_0 - \mathbf{p}_0 \cdot \mathbf{E} \tag{4.82}$$

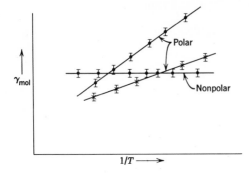

Fig. 4.13. Variation of molecular polarizability γ_{mol} with temperature for polar and nonpolar substances. γ_{mol} is plotted versus T^{-1}.

where H_0 is a function of only the "internal" coordinates of the molecule. Using the Boltzmann factor (4.79), we can write the average dipole moment as:

$$\langle p_{mol} \rangle = \frac{\int d\Omega \, p_0 \cos\theta \exp\left(\frac{p_0 E \cos\theta}{kT}\right)}{\int d\Omega \exp\left(\frac{p_0 E \cos\theta}{kT}\right)} \tag{4.83}$$

where we have chosen **E** along the z axis, integrated out all the irrelevant variables, and noted that only the component of $\mathbf{p}_0$ parallel to the field is different from zero. In general, $(p_0 E/kT)$ is very small compared to unity, except at low temperatures. Hence we can expand the exponentials and obtain the result:

$$\langle p_{mol} \rangle \simeq \frac{1}{3} \frac{p_0^2}{kT} E \tag{4.84}$$

We note that the orientation polarization depends inversely on the temperature, as might be expected of an effect in which the applied field must overcome the opposition of thermal agitation.

In general both types of polarization, induced (electronic) and orientation, are present, and the general form of the molecular polarization is

$$\gamma_{mol} \simeq \gamma_{el} + \frac{1}{3} \frac{p_0^2}{kT} \tag{4.85}$$

This shows a temperature dependence of the form $(a + b/T)$ so that the two types of polarization can be separated experimentally, as indicated in Fig. 4.13. For "polar" molecules, such as HCl and H_2O, the observed permanent dipole moments are of the order of an electronic charge times 10^{-8} cm, in accordance with molecular dimensions.

4.8   Electrostatic Energy in Dielectric Media

In Section 1.11 we discussed the energy of a system of charges in free space. The result

$$W = \tfrac{1}{2} \int \rho(\mathbf{x}) \Phi(\mathbf{x}) \, d^3x \qquad (4.86)$$

for the energy due to a charge density $\rho(\mathbf{x})$ and a potential $\Phi(\mathbf{x})$ cannot in general be taken over as it stands in our macroscopic description of dielectric media. The reason becomes clear when we recall how (4.86) was obtained. We thought of the final configuration of charge as being created by assembling bit by bit the elemental charges, bringing each one in from infinitely far away against the action of the then existing electric field. The total work done was given by (4.86). With dielectric media work is done not only to bring real (macroscopic) charge into position, but also to produce a certain state of polarization in the medium. If ρ and Φ in (4.86) represent macroscopic variables, it is certainly not evident that (4.86) represents the total work, including that done on the dielectric.

In order to be general in our description of dielectrics we will not initially make any assumptions about linearity, uniformity, etc., of the response of a dielectric to an applied field. Rather, let us consider a small change in the energy δW due to some sort of change $\delta \rho$ in the charge density ρ existing in all space. The work done to accomplish this change is

$$\delta W = \int \delta\rho(\mathbf{x}) \Phi(\mathbf{x}) \, d^3x \qquad (4.87)$$

where $\Phi(\mathbf{x})$ is the potential due to the charge density $\rho(\mathbf{x})$ already present. Since $\nabla \cdot \mathbf{D} = 4\pi\rho$, we can relate the change $\delta\rho$ to a change in the displacement of $\delta\mathbf{D}$:

$$\delta\rho = \frac{1}{4\pi} \nabla \cdot (\delta\mathbf{D}) \qquad (4.88)$$

Then the energy change δW can be cast into the form:

$$\delta W = \frac{1}{4\pi} \int \mathbf{E} \cdot \delta\mathbf{D} \, d^3x \qquad (4.89)$$

where we have used $\mathbf{E} = -\nabla\Phi$ and have assumed that $\rho(\mathbf{x})$ was a localized charge distribution. The total electrostatic energy can now be written down formally, at least, by allowing $\mathbf{D}$ to be brought from an initial value $\mathbf{D} = 0$ to its final value $\mathbf{D}$:

$$W = \frac{1}{4\pi} \int d^3x \int_0^D \mathbf{E} \cdot \delta\mathbf{D} \qquad (4.90)$$

If the medium is *linear*, then

$$\mathbf{E} \cdot \delta\mathbf{D} = \tfrac{1}{2}\delta(\mathbf{E} \cdot \mathbf{D}) \qquad (4.91)$$

and the total electrostatic energy is

$$W = \frac{1}{8\pi} \int \mathbf{E} \cdot \mathbf{D} \, d^3x \qquad (4.92)$$

This last result can be transformed into (4.86) by using $\mathbf{E} = -\nabla\Phi$ and $\nabla \cdot \mathbf{D} = 4\pi\rho$, or by going back to (4.87) and assuming that ρ and Φ are connected linearly. Thus we see that (4.86) is valid macroscopically only if the behavior is linear. Otherwise the energy of a final configuration must be calculated from (4.90) and might conceivably depend on the past history of the system (hysteresis effects).

A problem of considerable interest is the change in energy when a dielectric object is placed in an electric field whose sources are fixed. Suppose that initially the electric field $\mathbf{E}_0$ due to a certain distribution of charges $\rho_0(\mathbf{x})$ exists in a medium of dielectric constant ϵ_0 which may be a function of position. The initial electrostatic energy is

$$W_0 = \frac{1}{8\pi} \int \mathbf{E}_0 \cdot \mathbf{D}_0 \, d^3x$$

where $\mathbf{D}_0 = \epsilon_0\mathbf{E}_0$. Then with the sources fixed in position a dielectric object of volume V_1 is introduced into the field, changing the field from $\mathbf{E}_0$ to $\mathbf{E}$. The presence of the object can be described by a dielectric constant $\epsilon(\mathbf{x})$, which has the value ϵ_1 inside V_1 and ϵ_0 outside V_1. To avoid mathematical difficulties we can imagine $\epsilon(\mathbf{x})$ to be a smoothly varying function of position which falls rapidly but continuously from ϵ_1 to ϵ_0 at the edge of the volume V_1. The energy now has the value

$$W_1 = \frac{1}{8\pi} \int \mathbf{E} \cdot \mathbf{D} \, d^3x$$

where $\mathbf{D} = \epsilon\mathbf{E}$. The difference in the energy can be written:

$$W = \frac{1}{8\pi} \int (\mathbf{E} \cdot \mathbf{D} - \mathbf{E}_0 \cdot \mathbf{D}_0) \, d^3x$$

$$= \frac{1}{8\pi} \int (\mathbf{E} \cdot \mathbf{D}_0 - \mathbf{D} \cdot \mathbf{E}_0) \, d^3x + \frac{1}{8\pi} \int (\mathbf{E} + \mathbf{E}_0) \cdot (\mathbf{D} - \mathbf{D}_0) \, d^3x \quad (4.93)$$

The second integral can be shown to vanish by the following argument. Since $\nabla \times (\mathbf{E} + \mathbf{E}_0) = 0$, we can write

$$\mathbf{E} + \mathbf{E}_0 = -\nabla\Phi$$

Then the second integral becomes:

$$I = -\frac{1}{8\pi} \int \nabla\Phi \cdot (\mathbf{D} - \mathbf{D_0}) \, d^3x$$

Integration by parts transforms this into

$$I = \frac{1}{8\pi} \int \Phi\nabla \cdot (\mathbf{D} - \mathbf{D_0}) \, d^3x = 0$$

since $\nabla \cdot (\mathbf{D} - \mathbf{D_0}) = 0$ because the source charge density $\rho_0(\mathbf{x})$ is assumed unaltered by the insertion of the dielectric object. Consequently the energy change is

$$W = \frac{1}{8\pi} \int (\mathbf{E} \cdot \mathbf{D_0} - \mathbf{D} \cdot \mathbf{E_0}) \, d^3x \qquad (4.94)$$

The integration appears to be over all space, but is actually only over the volume V_1 of the object, since, outside V_1, $\mathbf{D} = \epsilon_0\mathbf{E}$. Therefore we can write

$$W = -\frac{1}{8\pi} \int_{V_1} (\epsilon_1 - \epsilon_0)\mathbf{E} \cdot \mathbf{E_0} \, d^3x \qquad (4.95)$$

If the medium surrounding the dielectric body is free space, then $\epsilon_0 = 1$. Using the definition of polarization $\mathbf{P}$, (4.95) can be expressed in the form:

$$W = -\tfrac{1}{2}\int_{V_1} \mathbf{P} \cdot \mathbf{E_0} \, d^3x \qquad (4.96)$$

where $\mathbf{P}$ is the polarization of the dielectric. This shows that the energy density of a dielectric placed in a field $\mathbf{E_0}$ whose sources are fixed is given by

$$w = -\tfrac{1}{2}\mathbf{P} \cdot \mathbf{E_0} \qquad (4.97)$$

This result is analogous to the dipole term in the energy (4.17) of a charge distribution in an external field. The factor $\tfrac{1}{2}$ is due to the fact that (4.97) represents the energy density of a polarizable dielectric in an external field, rather than a permanent dipole. It is the same factor $\tfrac{1}{2}$ which appears in (4.91).

Equations (4.95) and (4.96) show that a dielectric body will tend to move towards regions of increasing field $\mathbf{E_0}$ provided $\epsilon_1 > \epsilon_0$. To calculate the force acting we can imagine a small generalized displacement of the body $\delta\xi$. Then there will be a change in the energy δW. Since the charges are held fixed, there is no external source of energy and the change in field energy must be compensated for by a change in the mechanical energy of the body. This means that there is a force acting on the body:

$$F_\xi = -\left(\frac{\partial W}{\partial \xi}\right)_Q \qquad (4.98)$$

where the subscript Q has been placed on the partial derivative to indicate that the sources of the field are kept fixed.

In practical situations involving the motion of dielectrics the electric fields are often produced by a configuration of electrodes held at *fixed potentials* by connection to an external source such as a battery. As the distribution of dielectric varies, charge will flow to or from the battery to the electrodes in order to maintain the potentials constant. This means that energy is being supplied from the external source, and it is of interest to compare the energy supplied in that way with the energy change found above *for fixed sources* of the field. We will treat only linear media so that (4.86) is valid. It is sufficient to consider small changes in an already existing configuration. From (4.86) it is evident that the change in energy accompanying the changes $\delta\rho(\mathbf{x})$ and $\delta\Phi(\mathbf{x})$ in charge density and potential is

$$\delta W = \tfrac{1}{2}\int \left[\rho\delta\Phi + \Phi\delta\rho\right] d^3x \qquad (4.99)$$

Comparison with (4.87) shows that, if the dielectric properties are not changed, the two terms in (4.99) are equal. If, however, the dielectric properties are altered,

$$\epsilon(\mathbf{x}) \to \epsilon(\mathbf{x}) + \delta\epsilon(\mathbf{x}) \qquad (4.100)$$

the contributions in (4.99) are not necessarily the same. In fact, we have just calculated the change in energy brought about by introducing a dielectric body into an electric field whose sources were fixed ($\delta\rho = 0$). The reason for this difference is the existence of the polarization charge. The change in dielectric properties implied by (4.100) can be thought of as a change in the polarization-charge density. If then (4.99) is interpreted as an integral over both free and polarization-charge densities (i.e., a microscopic equation), the two contributions are always equal. However, it is often convenient to deal with macroscopic quantities. Then the equality holds only if the dielectric properties are unchanged.

The process of altering the dielectric properties in some way (by moving the dielectric bodies, by changing their susceptibilities, etc.) in the presence of electrodes at fixed potentials can be viewed as taking place in two steps. In the first step the electrodes are disconnected from the batteries and the charges on them held fixed ($\delta\rho = 0$). With the change (4.100) in dielectric properties, the energy change is

$$\delta W_1 = \tfrac{1}{2} \int \rho \delta\Phi_1 \, d^3x \qquad (4.101)$$

where $\delta\Phi_1$ is the change in potential produced. This can be shown to yield the result (4.95). In the second step the batteries are connected again to the electrodes to restore their potentials to the original values. There will be a flow of charge $\delta\rho_2$ from the batteries accompanying the change in potential* $\delta\Phi_2 = -\delta\Phi_1$. Therefore the energy change in the second step is

$$\delta W_2 = \tfrac{1}{2} \int (\rho\delta\Phi_2 + \Phi\delta\rho_2) \, d^3x = -2\delta W_1 \qquad (4.102)$$

since the two contributions are equal. In the second step we find the external sources changing the energy in the opposite sense and by twice the amount of the initial step. Consequently the net change is

$$\delta W = -\tfrac{1}{2} \int \rho\delta\Phi_1 \, d^3x \qquad (4.103)$$

Symbolically

$$\delta W_V = -\delta W_Q \qquad (4.104)$$

where the subscript denotes the quantity held fixed. If a dielectric with $\epsilon > 1$ moves into a region of greater field strength, the energy increases instead of decreases. For a generalized displacement $d\xi$ the mechanical force acting is now

$$F_\xi = +\left(\frac{\partial W}{\partial \xi}\right)_V \qquad (4.105)$$

REFERENCES AND SUGGESTED READING

The derivation of the macroscopic equations of electrostatics by averaging over aggregates of atoms is presented by
Rosenfeld, Chapter II,
Mason and Weaver, Chapter I, Part III,
Van Vleck, Chapter 1.
Rosenfeld also treats the classical electron theory of dielectrics. Van Vleck's book is devoted to electric and magnetic susceptibilities. Specific works on electric polarization phenomena are those of
Böttcher,
Debye,
Fröhlich.

* Note that it is necessary merely to know that $\delta\Phi_2 = -\delta\Phi_1$ on the electrodes, since that is the only place where free charge resides.

Boundary-value problems with dielectrics are discussed in all the references on electrostatics in Chapters 2 and 3.

Our treatment of forces and energy with dielectric media is brief. More extensive discussions, including forces on liquid and solid dielectrics, the electric stress tensor, electrostriction, and thermodynamic effects, may be found in

Abraham and Becker, Band 1, Chapter V,

Durand, Chapters VI and VII,

Landau and Lifshitz, *Electrodynamics of Continuous Media*,

Maxwell, Vol. 1, Chapter V,

Panofsky and Phillips, Chapter 6,

Stratton, Chapter II.

PROBLEMS

4.1 Calculate the multipole moments q_{lm} of the charge distributions shown below. Try to obtain results for the nonvanishing moments valid for all l, but in each case find the first *two* sets of nonvanishing moments at the very least.

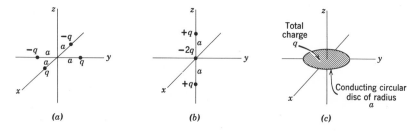

(a)   (b)   (c)

(*d*) For the charge distribution (*b*) write down the multipole expansion for the potential. Keeping only the lowest-order term in the expansion, plot the potential in the *x-y* plane as a function of distance from the origin for distances greater than *a*.

(*e*) Calculate directly from Coulomb's law the exact potential for (*b*) in the *x-y* plane. Plot it as a function of distance and compare with the result found in (*d*).

Divide out the asymptotic form in parts (*d*) and (*e*) in order to see the behavior at large distances more clearly.

4.2 A nucleus with quadrupole moment Q finds itself in a cylindrically symmetric electric field with a gradient $(\partial E_z/\partial z)_0$ along the z axis at the position of the nucleus.

(*a*) Show that the energy of quadrupole interaction is

$$W = -\frac{e}{4}Q\left(\frac{\partial E_z}{\partial z}\right)_0$$

(*b*) If it is known that $Q = 2 \times 10^{-24}$ cm^2 and that W/h is 10 Mc/sec, where h is Planck's constant, calculate $(\partial E_z/\partial z)_0$ in units of e/a_0^3, where $a_0 = \hbar^2/me^2 = 0.529 \times 10^{-8}$ cm is the Bohr radius in hydrogen.

(c) Nuclear-charge distributions can be approximated by a constant charge density throughout a spheroidal volume of semimajor axis a and semiminor axis b. Calculate the quadrupole moment of such a nucleus, assuming that the total charge is Ze. Given that Eu^{153} $(Z = 63)$ has a quadrupole moment $Q = 2.5 \times 10^{-24}$ cm^2 and a mean radius

$$R = (a + b)/2 = 7 \times 10^{-13} \text{ cm},$$

determine the fractional difference in radius $(a - b)/R$.

4.3 A localized distribution of charge has a charge density

$$\rho(\mathbf{r}) = \frac{1}{64\pi} r^2 e^{-r} \sin^2 \theta$$

(a) Make a multipole expansion of the potential due to this charge density and determine all the nonvanishing multipole moments. Write down the potential at large distances as a finite expansion in Legendre polynomials.

(b) Determine the potential explicitly at any point in space, and show that near the origin

$$\Phi(\mathbf{r}) \simeq \frac{1}{4} - \frac{r^2}{120} P_2(\cos \theta)$$

(c) If there exists at the origin a nucleus with a quadrupole moment $Q = 10^{-24}$ cm^2, determine the magnitude of the interaction energy, assuming that the unit of charge in $\rho(\mathbf{r})$ above is the electronic charge and the unit of length is the hydrogen Bohr radius $a_0 = \hbar^2/me^2 = 0.529 \times 10^{-8}$ cm. Express your answer as a frequency by dividing by Planck's constant h.

The charge density in this problem is that for the $m = \pm 1$ states of the $2p$ level in hydrogen, while the quadrupole interaction is of the same order as found in molecules.

4.4 A very long, right circular, cylindrical shell of dielectric constant ϵ and inner and outer radii a and b, respectively, is placed in a previously uniform electric field E_0 with its axis perpendicular to the field. The medium inside and outside the cylinder has a dielectric constant of unity.

(a) Determine the potential and electric field in the three regions, neglecting end effects.

(b) Sketch the lines of force for a typical case of $b \simeq 2a$.

(c) Discuss the limiting forms of your solution appropriate for a solid dielectric cylinder in a uniform field, and a cylindrical cavity in a uniform dielectric.

4.5 A point charge q is located in free space a distance d from the center of a dielectric sphere of radius a $(a < d)$ and dielectric constant ϵ.

(a) Find the potential at all points in space as an expansion in spherical harmonics.

(b) Calculate the rectangular components of the electric field *near* the center of the sphere.

(c) Verify that, in the limit $\epsilon \to \infty$, your result is the same as that for the conducting sphere.

4.6 Two concentric conducting spheres of inner and outer radii a and b, respectively, carry charges $\pm Q$. The empty space between the spheres is half-filled by a hemispherical shell of dielectric (of dielectric constant ϵ), as shown in the figure.

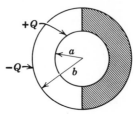

(*a*) Find the electric field everywhere between the spheres.
(*b*) Calculate the surface-charge distribution on the inner sphere.
(*c*) Calculate the polarization-charge density induced on the surface of the dielectric at $r = a$.

4.7 The following data on the variation of dielectric constant with pressure are taken from the *Smithsonian Physical Tables*, 9th ed., p. 424:

Air at 292°K

Pressure (atm)	ϵ	
20	1.0108	Relative density of
40	1.0218	air as a function of
60	1.0333	pressure is given in
80	1.0439	*AIP Handbook*, p.
100	1.0548	4–83.

Pentane (C_5H_{12}) at 303°K

Pressure (atm)	Density (gm/cm³)	ϵ
1	0.613	1.82
10^3	0.701	1.96
4×10^3	0.796	2.12
8×10^3	0.865	2.24
12×10^3	0.907	2.33

Test the Clausius-Mossotti relation between dielectric constant and density for air and pentane in the ranges tabulated. Does it hold exactly? Approximately? If approximately, discuss fractional variations in density and $(\epsilon - 1)$. For pentane, compare the Clausius-Mossotti relation to the cruder relation, $(\epsilon - 1) \propto$ density.

4.8 Water vapor is a polar gas whose dielectric constant exhibits an appreciable temperature dependence. The following table gives experimental data on this effect. Assuming that water vapor obeys the ideal gas law, calculate the molecular polarizability as a function of inverse temperature and plot it. From the slope of the curve, deduce a value for the permanent dipole moment of the H_2O molecule (express the dipole moment in esu—stat-coulomb-centimeters).

T (°K)	Pressure (cm Hg)	$(\epsilon - 1) \times 10^5$
393	56.49	400.2
423	60.93	371.7
453	65.34	348.8
483	69.75	328.7

4.9 Two long, coaxial, cylindrical conducting surfaces of radii a and b are lowered vertically into a liquid dielectric. If the liquid rises a distance h between the electrodes when a potential difference V is established between them, show that the susceptibility of the liquid is

$$\chi_e = \frac{(b^2 - a^2)\rho g h \ln (b/a)}{V^2}$$

where ρ is the density of the liquid, g is the acceleration due to gravity, and the susceptibility of air is neglected.

5

Magnetostatics

5.1 Introduction and Definitions

In the preceding chapters various aspects of electrostatics (i.e., the
fields and interactions of stationary charges and boundaries) have been
studied. We now turn to steady-state magnetic phenomena. From an
historical point of view, magnetic phenomena have been known and
studied for at least as long as electric phenomena. Lodestones were known
in ancient times; the mariner's compass is a very old invention; Gilbert's
researches on the earth as a giant magnet date from before 1600. In
contrast to electrostatics, the basic laws of magnetic fields did not follow
straightforwardly from man's earliest contact with magnetic materials.
The reasons are several, but they all stem from the radical difference
between magnetostatics and electrostatics: *there are no free magnetic
charges.* This means that magnetic phenomena are quite different from
electric phenomena and that for a long time no connection was established
between them. The basic entity in magnetic studies was what we now know
as a magnetic dipole. In the presence of magnetic materials the dipole
tends to align itself in a certain direction. That direction is by definition
the direction of the magnetic-flux density, denoted by $\mathbf{B}$, provided the
dipole is sufficiently small and weak that it does not perturb the existing
field. The magnitude of the flux density can be defined by the mechanical
torque $\mathbf{N}$ exerted on the magnetic dipole:

$$\mathbf{N} = \boldsymbol{\mu} \times \mathbf{B} \qquad (5.1)$$

where $\boldsymbol{\mu}$ is the magnetic moment of the dipole, defined in some suitable
set of units.*

* In analogy with the 100 strokes of cat's fur on an amber rod, we might define our unit
of dipole strength as that of a $\frac{1}{2}$-inch finishing nail which has been stroked slowly 100
times with a certain "standard" lodestone held in a certain standard orientation. With
a little thought we might even think of a more reliable and reproducible standard!

132

Already, in the definition of the magnetic-flux density **B** (sometimes called the *magnetic induction*), we have a more complicated situation than for the electric field. Further quantitative elucidation of magnetic phenomena did not occur until the connection between currents and magnetic fields was established. A current corresponds to charges in motion and is described by a current density **J**, measured in units of positive charge crossing unit area per unit time, the direction of motion of the charges defining the direction of **J**. In electrostatic units, current density is measured in statcoulombs per square centimeter-second, and is sometimes called statamperes per square centimeter, while in mks units it is measured in coulombs per square meter-second or amperes per square meter. If the current density is confined to wires of small cross section, we usually integrate over the cross-sectional area and speak of a current of so many statamperes or amperes flowing along the wire.

Conservation of charge demands that the charge density at any point in space be related to the current density in that neighborhood by a continuity equation:

$$\frac{\partial \rho}{\partial t} + \nabla \cdot \mathbf{J} = 0 \tag{5.2}$$

This expresses the physical fact that a decrease in charge inside a small volume with time must correspond to a flow of charge out through the surface of the small volume, since the total number of charges must be conserved. Steady-state magnetic phenomena are characterized by no change in the net charge density anywhere in space. Consequently in magnetostatics

$$\nabla \cdot \mathbf{J} = 0 \tag{5.3}$$

We now proceed to discuss the experimental connection between current and magnetic-flux density and to establish the basic laws of magnetostatics.

5.2 Biot and Savart Law

In 1819 Oersted observed that wires carrying electric currents produced deflections of permanent magnetic dipoles placed in their neighborhood. Thus the currents were sources of magnetic-flux density. Biot and Savart (1820), first, and Ampère (1820–1825), in much more elaborate and thorough experiments, established the basic experimental laws relating the magnetic induction **B** to the currents and established the law of force between one current and another. Although not in the form in which

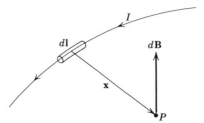

Fig. 5.1   Elemental magnetic induction $d\mathbf{B}$ due to current element $I\,d\mathbf{l}$.

Ampère deduced it, the basic relation is the following. If $d\mathbf{l}$ is an element of length (pointing in the direction of current flow) of a filamentary wire which carries a current I and $\mathbf{x}$ is the coordinate vector from the element of length to an observation point P, as shown in Fig. 5.1, then the elemental flux density $d\mathbf{B}$ at the point P is given in magnitude and direction by

$$d\mathbf{B} = kI\,\frac{(d\mathbf{l} \times \mathbf{x})}{|\mathbf{x}|^3} \tag{5.4}$$

It should be noted that (5.4) is an inverse square law, just as is Coulomb's law of electrostatics. However, the vector character is very different.

If, instead of a current flowing there is a single charge q moving with a velocity $\mathbf{v}$, then the flux density will be*

$$\mathbf{B} = kq\,\frac{\mathbf{v} \times \mathbf{x}}{|\mathbf{x}|^3} = k\mathbf{v} \times \mathbf{E} \tag{5.5}$$

where $\mathbf{E}$ is the electrostatic field of the charge q. (This flux density is, however, time varying. We shall restrict the discussions in the present chapter to steady-state current flow.)

In (5.4) and (5.5) the constant k depends on the system of units used, as discussed in detail in the Appendix. If current is measured in esu, but the flux density is measured in emu, the constant is $k = 1/c$, where c is found experimentally to be equal to the velocity of light *in vacuo* ($c = 2.998 \times 10^{10}$ cm/sec). This system of units is called the Gaussian system. To insert the velocity of light into our equations at this stage seems a little artificial, but it has the advantage of measuring charge and current in a consistent set of units so that the continuity equation (5.2) retains its simple form, without factors of c. We will adopt the Gaussian system here.

Assuming that linear superposition holds, the basic law (5.4) can be integrated to determine the magnetic-flux density due to various config-urations of current-carrying wires. For example, the magnetic induction

---

* True only for particles moving with velocities small compared to that of light.

B of the long straight wire shown in Fig. 5.2 carrying a current I can be seen to be directed along the normal to the plane containing the wire and the observation point, so that the lines of magnetic induction are concentric circles around the wire. The magnitude of **B** is given by

$$|\mathbf{B}| = \frac{IR}{c} \int_{-\infty}^{\infty} \frac{dl}{(R^2 + l^2)^{3/2}} = \frac{2I}{cR} \tag{5.6}$$

where R is the distance from the observation point to the wire. This is the experimental result first found by Biot and Savart and is known as the Biot-Savart law. Note that the magnitude of the induction **B** varies with R in the same way as the electric field due to a long line charge of uniform linear-charge density. This analogy shows that in some circumstances there may be a correspondence between electrostatic and magnetostatic problems, even though the vector character of the fields is different. We shall see more of that in later sections.

Ampère's experiments did not deal directly with the determination of the relation between currents and magnetic induction, but were concerned rather with the force which one current-carrying wire experiences in the presence of another. Since we have already introduced the idea that a current element produces a magnetic induction, we phrase the force law as the force experienced by a current element $I_1 \, dl_1$ in the presence of a magnetic induction **B**. The elemental force is

$$d\mathbf{F} = \frac{I_1}{c} (d\mathbf{l}_1 \times \mathbf{B}) \tag{5.7}$$

I_1 is the current in the element (measured in esu), **B** is the flux density (in emu), and c is the velocity of light. If the external field **B** is due to a closed current loop #2 with current I_2, then the total force which a closed current

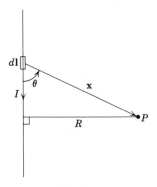

Fig. 5.2

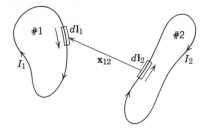

Fig. 5.3 Two Ampèrian current loops.

loop #1 with current I_1 experiences is [from (5.4) and (5.7)]:

$$\mathbf{F}_{12} = \frac{I_1 I_2}{c^2} \oint \oint \frac{d\mathbf{l}_1 \times (d\mathbf{l}_2 \times \mathbf{x}_{12})}{|\mathbf{x}_{12}|^3} \tag{5.8}$$

The line integrals are taken around the two loops; $\mathbf{x}_{12}$ is the vector distance from line element $d\mathbf{l}_2$ to $d\mathbf{l}_1$, as shown in Fig. 5.3. This is the mathematical statement of Ampère's observations about forces between current-carrying loops. By manipulating the integrand it can be put in a form which is symmetric in $d\mathbf{l}_1$ and $d\mathbf{l}_2$ and which explicitly satisfies Newton's third law. Thus

$$\frac{d\mathbf{l}_1 \times (d\mathbf{l}_2 \times \mathbf{x}_{12})}{|\mathbf{x}_{12}|^3} = -(d\mathbf{l}_1 \cdot d\mathbf{l}_2) \frac{\mathbf{x}_{12}}{|\mathbf{x}_{12}|^3} + d\mathbf{l}_2 \left(\frac{d\mathbf{l}_1 \cdot \mathbf{x}_{12}}{|\mathbf{x}_{12}|^3} \right) \tag{5.9}$$

The second term involves a perfect differential in the integral over $d\mathbf{l}_1$. Consequently it gives no contribution to the integral (5.8), provided the paths are closed or extend to infinity. Then Ampère's law of force between current loops becomes

$$\mathbf{F}_{12} = -\frac{I_1 I_2}{c^2} \oint \oint \frac{(d\mathbf{l}_1 \cdot d\mathbf{l}_2)\mathbf{x}_{12}}{|\mathbf{x}_{12}|^3} \tag{5.10}$$

showing symmetry in the integration, apart from the necessary vectorial dependence on $\mathbf{x}_{12}$.

Each of two long, parallel, straight wires a distance d apart, carrying currents I_1 and I_2, experiences a force per unit length directed perpendicularly towards the other wire and of magnitude,

$$F = \frac{2I_1 I_2}{c^2 d} \tag{5.11}$$

The force is attractive (repulsive) if the currents flow in the same (opposite) directions. The forces which exist between current-carrying wires can be used to define magnetic-flux density in a way that is independent of permanent magnetic dipoles.* We will see later that the torque expression (5.1) and the force result (5.7) are intimately related.

If a current density $J(x)$ is in an external magnetic-flux density $B(x)$, the elementary force law implies that the total force on the current distribution is

$$F = \frac{1}{c}\int J(x) \times B(x)\, d^3x \qquad (5.12)$$

Similarly the total torque is

$$N = \frac{1}{c}\int x \times (J \times B)\, d^3x \qquad (5.13)$$

These general results will be applied to localized current distributions in Section 5.6.

5.3 The Differential Equations of Magnetostatics and Ampère's Law

The basic law (5.4) for the magnetic induction can be written down in general form for a current density $J(x)$:

$$B(x) = \frac{1}{c}\int J(x') \times \frac{(x - x')}{|x - x'|^3}\, d^3x' \qquad (5.14)$$

This expression for $B(x)$ is the magnetic analog of electric field in terms of the charge density:

$$E(x) = \int \rho(x') \frac{(x - x')}{|x - x'|^3}\, d^3x' \qquad (5.15)$$

Just as this result for E was not as convenient in some situations as differential equations, so (5.14) is not the most useful form for magnetostatics, even though it contains in principle a description of all the phenomena.

In order to obtain the differential equations equivalent to (5.14) we transform (5.14) into the form:

$$B(x) = \frac{1}{c}\nabla \times \int \frac{J(x')}{|x - x'|}\, d^3x' \qquad (5.16)$$

* In fact, (5.11) is the basis of the internationally accepted standard of current (actually I/c here). See the Appendix.

From (5.16) it follows immediately that the divergence of **B** vanishes:

$$\nabla \cdot \mathbf{B} = 0 \tag{5.17}$$

This is the first equation of magnetostatics and corresponds to $\nabla \times \mathbf{E} = 0$ in electrostatics. By analogy with electrostatics we now calculate the curl of **B**:

$$\nabla \times \mathbf{B} = \frac{1}{c}\nabla \times \nabla \times \int \frac{\mathbf{J}(\mathbf{x}')}{|\mathbf{x} - \mathbf{x}'|}\, d^3x' \tag{5.18}$$

With the identity $\nabla \times (\nabla \times \mathbf{A}) = \nabla(\nabla \cdot \mathbf{A}) - \nabla^2\mathbf{A}$ for an arbitrary vector field **A**, expression (5.18) can be transformed into

$$\nabla \times \mathbf{B} = \frac{1}{c}\nabla \int \mathbf{J}(\mathbf{x}') \cdot \nabla\left(\frac{1}{|\mathbf{x} - \mathbf{x}'|}\right) d^3x' - \frac{1}{c}\int \mathbf{J}(\mathbf{x}')\nabla^2\left(\frac{1}{|\mathbf{x} - \mathbf{x}'|}\right) d^3x' \tag{5.19}$$

Using the fact that

$$\nabla\left(\frac{1}{|\mathbf{x} - \mathbf{x}'|}\right) = -\nabla'\left(\frac{1}{|\mathbf{x} - \mathbf{x}'|}\right)$$

and

$$\nabla^2\left(\frac{1}{|\mathbf{x} - \mathbf{x}'|}\right) = -4\pi\, \delta(\mathbf{x} - \mathbf{x}')$$

the integrals in (5.19) can be written:

$$\nabla \times \mathbf{B} = -\frac{1}{c}\nabla \int \mathbf{J}(\mathbf{x}') \cdot \nabla'\left(\frac{1}{|\mathbf{x} - \mathbf{x}'|}\right) d^3x' + \frac{4\pi}{c}\mathbf{J}(\mathbf{x}) \tag{5.20}$$

Integration by parts yields

$$\nabla \times \mathbf{B} = \frac{4\pi}{c}\mathbf{J} + \frac{1}{c}\nabla \int \frac{\nabla' \cdot \mathbf{J}(\mathbf{x}')}{|\mathbf{x} - \mathbf{x}'|}\, d^3x' \tag{5.21}$$

But for steady-state magnetic phenomena $\nabla \cdot \mathbf{J} = 0$, so that we obtain

$$\nabla \times \mathbf{B} = \frac{4\pi}{c}\mathbf{J} \tag{5.22}$$

This is the second equation of magnetostatics, corresponding to $\nabla \cdot \mathbf{E} = 4\pi\rho$ in electrostatics.

In electrostatics Gauss's law (1.11) is the integral form of the equation $\nabla \cdot \mathbf{E} = 4\pi\rho$. The integral equivalent of (5.22) is called *Ampère's law*. It is obtained by applying Stokes's theorem to the integral of the normal

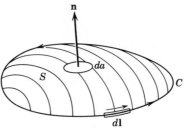

Fig. 5.4

component of (5.22) over an open surface S bounded by a closed curve C, as shown in Fig. 5.4. Thus

$$\int_S \nabla \times \mathbf{B} \cdot \mathbf{n}\, da = \frac{4\pi}{c} \int_S \mathbf{J} \cdot \mathbf{n}\, da \qquad (5.23)$$

is transformed into

$$\oint_C \mathbf{B} \cdot d\mathbf{l} = \frac{4\pi}{c} \int_S \mathbf{J} \cdot \mathbf{n}\, da \qquad (5.24)$$

Since the surface integral of the current density is the total current I passing through the closed curve C, Ampère's law can be written in the form:

$$\oint_C \mathbf{B} \cdot d\mathbf{l} = \frac{4\pi}{c} I \qquad (5.25)$$

Just as Gauss's law can be used for calculation of the electric field in highly symmetric situations, so Ampère's law can be employed in analogous circumstances.

5.4 Vector Potential

The basic differential laws of magnetostatics are given by

$$\left.\begin{array}{c} \nabla \times \mathbf{B} = \dfrac{4\pi}{c} \mathbf{J} \\[2mm] \nabla \cdot \mathbf{B} = 0 \end{array}\right\} \qquad (5.26)$$

The problem is how to solve them. If the current density is zero in the region of interest, $\nabla \times \mathbf{B} = 0$ permits the expression of the vector magnetic induction $\mathbf{B}$ as the gradient of a magnetic scalar potential, $\mathbf{B} = -\nabla \Phi_M$. Then (5.26) reduces to Laplace's equation for Φ_M, and all our techniques for handling electrostatic problems can be brought to bear. There are a large number of problems which fall into this class, but we will defer discussion of them until later in the chapter. The reason is that the boundary conditions are different from those encountered in electrostatics, and the problems usually involve macroscopic media with magnetic properties different from free space with charges and currents.

A general method of attack is to exploit the second equation in (5.26). If $\nabla \cdot \mathbf{B} = 0$ everywhere, $\mathbf{B}$ must be the curl of some vector field $\mathbf{A}(\mathbf{x})$, called the *vector potential,*

$$\mathbf{B}(\mathbf{x}) = \nabla \times \mathbf{A}(\mathbf{x}) \qquad (5.27)$$

We have, in fact, already written $\mathbf{B}$ in this form (5.16). Evidently, from (5.16), the general form of $\mathbf{A}$ is

$$\mathbf{A}(\mathbf{x}) = \frac{1}{c} \int \frac{\mathbf{J}(\mathbf{x}')}{|\mathbf{x} - \mathbf{x}'|} \, d^3x' + \nabla \Psi(\mathbf{x}) \qquad (5.28)$$

The added gradient of an arbitrary scalar function Ψ shows that, for a given magnetic induction $\mathbf{B}$, the vector potential can be freely transformed according to

$$\mathbf{A} \to \mathbf{A} + \nabla \Psi \qquad (5.29)$$

This transformation is called a *gauge transformation.* Such transformations on $\mathbf{A}$ are possible because (5.27) specifies only the curl of $\mathbf{A}$. For a complete specification of a vector field it is necessary to state both its curl and its divergence. The freedom of gauge transformations allows us to make $\nabla \cdot \mathbf{A}$ have any convenient functional form we wish.

If (5.27) is substituted into the first equation in (5.26), we find

$$\left. \begin{array}{c} \nabla \times (\nabla \times \mathbf{A}) = \dfrac{4\pi}{c} \mathbf{J} \\[2em] \nabla(\nabla \cdot \mathbf{A}) - \nabla^2 \mathbf{A} = \dfrac{4\pi}{c} \mathbf{J} \end{array} \right\} \qquad (5.30)$$

or

If we now exploit the freedom implied by (5.29), we can make the convenient choice of gauge,* $\nabla \cdot \mathbf{A} = 0$. Then each rectangular component of the vector potential satisfies Poisson's equation,

$$\nabla^2 \mathbf{A} = -\frac{4\pi}{c} \mathbf{J} \qquad (5.31)$$

---

* The choice is called the *Coulomb gauge,* for a reason which will become apparent only in Section 6.5.

From our discussions of electrostatics it is clear that the solution for $\mathbf{A}$ in unbounded space is (5.28) with $\Psi = 0$:

$$A(x) = \frac{1}{c} \int \frac{J(x')}{|x - x'|} \, d^3x' \qquad (5.32)$$

The condition $\Psi = 0$ can be understood as follows. Our choice of gauge, $\nabla \cdot \mathbf{A} = 0$, reduces to $\nabla^2\Psi = 0$, since the first term in (5.28) has zero divergence because of $\nabla' \cdot \mathbf{J} = 0$. If $\nabla^2\Psi = 0$ holds in all space, Ψ must vanish identically.

5.5 Vector Potential and Magnetic Induction for a Circular Current Loop

As an illustration of the calculation of magnetic fields from given current distributions we consider the problem of a circular loop of radius a, lying in the x-y plane, centered at the origin, and carrying a current I, as shown in Fig. 5.5. The current density $\mathbf{J}$ has only a component in the ϕ direction,

$$J_\phi = I\delta(\cos \theta') \frac{\delta(r' - a)}{a} \qquad (5.33)$$

The delta functions restrict current flow to a ring of radius a. Only a ϕ component of $\mathbf{J}$ means that $\mathbf{A}$ will have only a ϕ component also. *But* this component A_ϕ *cannot* be calculated by merely substituting J_ϕ into (5.32). Equation (5.32) holds only for rectangular components of $\mathbf{A}$.* Thus we write rectangular components of $\mathbf{J}$:

$$\left.\begin{array}{l} J_x = -J_\phi \sin \phi' \\ J_y = J_\phi \cos \phi' \end{array}\right\} \qquad (5.34)$$

Since the geometry is cylindrically symmetric, we may choose the observation point in the x-z plane ($\phi = 0$) for purposes of calculation. Then it is clear that the x component of the vector potential vanishes, leaving only

---

* The reason is that the vector Poisson's equation (5.31) can be treated as three uncoupled scalar equations, $\nabla^2 A_i = (-4\pi/c)J_i$, only if the components A_i, J_i are rectangular components. If $\mathbf{A}$ is resolved into orthogonal components with unit vectors which are functions of position, the differential operation involved in (5.31) mixes the components together, giving coupled equations. See Morse and Feshbach, pp. 51 and 116–117.

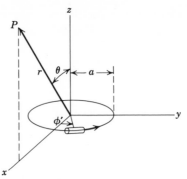

Fig. 5.5

the y component, which is A_ϕ. Thus

$$A_\phi(r, \theta) = \frac{I}{ca} \int r'^2 \, dr' \, d\Omega' \, \frac{\cos \phi' \, \delta(\cos \theta') \, \delta(r' - a)}{|\mathbf{x} - \mathbf{x}'|} \qquad (5.35)$$

where $|\mathbf{x} - \mathbf{x}'| = [r^2 + r'^2 - 2rr'(\cos \theta \cos \theta' + \sin \theta \sin \theta' \cos \phi')]^{\frac{1}{2}}$.
We first consider the straightforward evaluation of (5.35). Integration over the delta functions leaves the result

$$A_\phi(r, \theta) = \frac{Ia}{c} \int_0^{2\pi} \frac{\cos \phi' \, d\phi'}{(a^2 + r^2 - 2ar \sin \theta \cos \phi')^{\frac{1}{2}}} \qquad (5.36)$$

This integral can be expressed in terms of the complete elliptic integrals K and E:

$$A_\phi(r, \theta) = \frac{4Ia}{c\sqrt{a^2 + r^2 + 2ar \sin \theta}} \left[\frac{(2 - k^2)K(k) - 2E(k)}{k^2} \right] \qquad (5.37)$$

where the argument of the elliptic integrals is

$$k^2 = \frac{4ar \sin \theta}{a^2 + r^2 + 2ar \sin \theta}$$

The components of magnetic induction,

$$\left. \begin{aligned} B_r &= \frac{1}{r \sin \theta} \frac{\partial}{\partial \theta} (\sin \theta A_\phi) \\[2mm] B_\theta &= -\frac{1}{r} \frac{\partial}{\partial r} (r A_\phi) \\[2mm] B_\phi &= 0 \end{aligned} \right\} \qquad (5.38)$$

can also be expressed in terms of elliptic integrals. But the results are not particularly illuminating (useful, however, for computation).

For small k^2, corresponding to $a \gg r$, $a \ll r$, or $\theta \ll 1$, the square bracket in (5.37) reduces to $(\pi k^2/16)$. Then the vector potential becomes approximately

$$A_\phi(r, \theta) = \frac{I\pi a^2}{c} \frac{r \sin \theta}{(a^2 + r^2 + 2ar \sin \theta)^{3/2}} \tag{5.39}$$

The corresponding fields are

$$
\left.
\begin{aligned}
B_r &\simeq \frac{I\pi a^2}{c} \cos \theta \frac{(2a^2 + 2r^2 + ar \sin \theta)}{(a^2 + r^2 + 2ar \sin \theta)^{5/2}} \\[2mm]
B_\theta &\simeq -\frac{I\pi a^2}{c} \sin \theta \frac{(2a^2 - r^2 + ar \sin \theta)}{(a^2 + r^2 + 2ar \sin \theta)^{5/2}}
\end{aligned}
\right\} \tag{5.40}
$$

These can easily be specialized to the three regions, near the axis ($\theta \ll 1$), near the center of the loop ($r \ll a$), and far from the loop ($r \gg a$).

Of particular interest are the fields far from the loop:

$$
\left.
\begin{aligned}
B_r &= 2\left(\frac{I\pi a^2}{c}\right) \frac{\cos \theta}{r^3} \\[2mm]
B_\theta &= \left(\frac{I\pi a^2}{c}\right) \frac{\sin \theta}{r^3}
\end{aligned}
\right\} \tag{5.41}
$$

Comparison with the electrostatic dipole fields (4.12) shows that the magnetic fields far away from a circular current loop are dipole in character. By analogy with electrostatics we define the magnetic dipole moment of the loop to be

$$m = \frac{\pi I a^2}{c} \tag{5.42}$$

We will see in the next section that this is a special case of a general result—localized current distributions give dipole fields at large distances; the magnetic moment of a plane current loop is the product of the area of the loop times I/c.

Although we have obtained a complete solution to the problem in terms of elliptic integrals, we will illustrate the use of a spherical harmonic expansion to point out similarities and differences between the magnetostatic and electrostatic problems. Thus we return to (5.35) and substitute the spherical expansion (3.70) for $|\mathbf{x} - \mathbf{x}'|^{-1}$:

$$A_\phi = \frac{4\pi I}{ca} \, \text{Re} \sum_{l,m} \frac{Y_{lm}(\theta, \phi)}{2l + 1}$$

$$\times \int r'^2 \, dr' \, d\Omega' \, \delta(\cos \theta') \, \delta(r' - a) e^{i\phi'} \frac{r_<^l}{r_>^{l+1}} Y_{lm}^*(\theta', \phi') \quad (5.43)$$

The presence of $e^{i\phi'}$ means that only $m = +1$ will contribute to the sum. Hence

$$A_\phi = \frac{8\pi^2 Ia}{c} \, \text{Re} \sum_{l=1}^{\infty} \frac{Y_{l1}(\theta, \phi)}{2l + 1} \frac{r_<^l}{r_>^{l+1}} \left[Y_{l,1}^* \left(\theta' = \frac{\pi}{2}, \phi' \right) e^{i\phi'} \right] \quad (5.44)$$

where now $r_<$ ($r_>$) is the smaller (larger) of a and r. The square-bracketed quantity is a number depending on l:

$$\left[\quad \right] = \sqrt{\frac{2l + 1}{4\pi l(l + 1)}} \, P_l^1(0) = \begin{cases} 0, & \text{for } l \text{ even} \\ \sqrt{\frac{2l + 1}{4\pi l(l + 1)}} \left[\frac{(-1)^{n+1} \Gamma(n + \frac{3}{2})}{\Gamma(n + 1) \Gamma(\frac{3}{2})} \right], & \\ & \text{for } l = 2n + 1 \end{cases} \quad (5.45)$$

Then A_ϕ can be written

$$A_\phi = -\frac{\pi Ia}{c} \sum_{n=0}^{\infty} \frac{(-1)^n (2n - 1)!!}{2^n (n + 1)!} \frac{r_<^{2n+1}}{r_>^{2n+2}} P_{2n+1}^1(\cos \theta) \quad (5.46)$$

where $(2n - 1)!! = (2n - 1)(2n - 3)(\cdots) \times 5 \times 3 \times 1$, and the $n = 0$ coefficient in the sum is unity by definition. To evaluate the radial component of $\mathbf{B}$ from (5.38) we need

$$\frac{d}{dx} \left(\sqrt{1 - x^2} P_l^1(x) \right) = l(l + 1) P_l(x) \quad (5.47)$$

Then we find

$$B_r = \frac{2\pi Ia}{cr} \sum_{n=0}^{\infty} \frac{(-1)^n (2n + 1)!!}{2^n n!} \frac{r_<^{2n+1}}{r_>^{2n+2}} P_{2n+1}(\cos \theta) \quad (5.48)$$

The θ component of $\mathbf{B}$ is similarly

$$B_\theta = -\frac{\pi Ia^2}{c} \sum_{n=0}^{\infty} \frac{(-1)^n (2n + 1)!!}{2^n (n + 1)!} \begin{Bmatrix} -\left(\frac{2n + 2}{2n + 1} \right) \frac{1}{a^3} \left(\frac{r}{a} \right)^{2n} \\ \frac{1}{r^3} \left(\frac{a}{r} \right)^{2n} \end{Bmatrix} P_{2n+1}^1(\cos \theta) \quad (5.49)$$

The upper line holds for $r < a$, and the lower line for $r > a$. For $r \gg a$, only the $n = 0$ term in the series is important. Then, since $P_1^1(\cos \theta) = -\sin \theta$, (5.48) and (5.49) reduce to (5.41). For $r \ll a$, the leading term is again $n = 0$. The fields are then equivalent to a magnetic induction $2\pi I/ac$ in the z direction, a result that can be found by elementary means.

We note a characteristic difference between this problem and a corresponding cylindrically symmetric electrostatic problem. Associated Legendre polynomials appear, as well as ordinary Legendre polynomials. This can be traced to the vector character of the current and vector potential, as opposed to the scalar properties of charge and electrostatic potential.

Another mode of attack on the problem of the loop is to employ an expansion in cylindrical waves. Instead of (3.70) as a representation of $|\mathbf{x} - \mathbf{x}'|^{-1}$ we may use the cylindrical form (3.148) or (3.149). The application of this technique to the circular loop will be left to the problems. It is generally useful for any current distribution which involves current flowing only in the ϕ direction.

5.6 Magnetic Fields of a Localized Current Distribution; Magnetic Moment

We now consider the properties of a general current distribution which is localized in a small region of space, "small" being relative to the scale of length of interest to the observer. The proper treatment of this problem, in analogy with the electrostatic multipole expansion, demands a discussion of *vector* spherical harmonics. These are presented in Chapter 16 in connection with multipole radiation. We will be content here with only the lowest order of approximation. Starting with (5.32), we expand the denominator in powers of $\mathbf{x}'$ measured relative to a suitable origin in the localized current distribution, shown schematically in Fig. 5.6:

$$\frac{1}{|\mathbf{x} - \mathbf{x}'|} = \frac{1}{|\mathbf{x}|} + \frac{\mathbf{x} \cdot \mathbf{x}'}{|\mathbf{x}|^3} + \cdots \qquad (5.50)$$

Then a given component of the vector potential will have the expansion,

$$A_i(\mathbf{x}) = \frac{1}{c|\mathbf{x}|} \int J_i(\mathbf{x}') \, d^3x' + \frac{\mathbf{x}}{c|\mathbf{x}|^3} \cdot \int J_i(\mathbf{x}') \mathbf{x}' \, d^3x' + \cdots \qquad (5.51)$$

For a localized steady-state current distribution the volume integral of $\mathbf{J}$ vanishes because $\nabla \cdot \mathbf{J} = 0$. Consequently the first term, corresponding to the monopole term in an electrostatic expansion, vanishes.

The integrand of the second term can be manipulated into a more convenient form by using the triple vector product. Thus

$$(\mathbf{x} \cdot \mathbf{x}')\mathbf{J} = (\mathbf{x} \cdot \mathbf{J})\mathbf{x}' - \mathbf{x} \times (\mathbf{x}' \times \mathbf{J}) \tag{5.52}$$

The volume integral of the first term on the right can be shown to be the negative of the integral of the left-hand side of (5.52). Thus we consider the integral,

$$\int J_j x_i' \, d^3x' = \int \nabla' \cdot (x_j' \mathbf{J}) x_i' \, d^3x' = -\int x_j' (\mathbf{J} \cdot \nabla') x_i' \, d^3x'$$

$$= -\int x_j' J_i \, d^3x' \tag{5.53}$$

The step from the first integral to the second depends on $\nabla \cdot \mathbf{J} = 0$; the following step involves an integration by parts. With this identity (5.52) can be written in integrated form as

$$\int (\mathbf{x} \cdot \mathbf{x}')\mathbf{J}(\mathbf{x}') \, d^3x' = -\tfrac{1}{2}\mathbf{x} \times \int [\mathbf{x}' \times \mathbf{J}(\mathbf{x}')] \, d^3x' \tag{5.54}$$

We now *define* the magnetic moment of the current distribution $\mathbf{J}$ as

$$\mathbf{m} = \frac{1}{2c} \int \mathbf{x}' \times \mathbf{J}(\mathbf{x}') \, d^3x' \tag{5.55}$$

Note that it is sometimes useful to consider the integrand in (5.55) as a magnetic-moment density or magnetization. We denote the magnetization due to the current density $\mathbf{J}$ by

$$\mathcal{M} = \frac{1}{2c} (\mathbf{x} \times \mathbf{J}) \tag{5.56}$$

The vector potential (5.51) can be expressed in terms of $\mathbf{m}$ as

$$\mathbf{A}(\mathbf{x}) = \frac{\mathbf{m} \times \mathbf{x}}{|\mathbf{x}|^3} \tag{5.57}$$

This is the lowest nonvanishing term in the expansion of $\mathbf{A}$ for a localized steady-state current distribution. The magnetic induction $\mathbf{B}$ can be

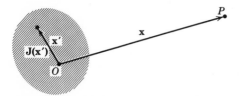

Fig. 5.6 Localized current density $\mathbf{J}(\mathbf{x}')$ gives rise to a magnetic induction at the point P with coordinate $\mathbf{x}$.

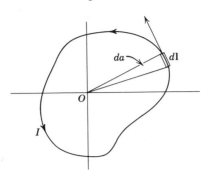

Fig. 5.7

calculated directly by evaluating the curl of (5.57):

$$\mathbf{B(x)} = \frac{3\mathbf{n(n \cdot m)} - \mathbf{m}}{|\mathbf{x}|^3} + \frac{8\pi}{3}\mathbf{m}\,\delta(\mathbf{x}) \qquad (5.58)$$

Here $\mathbf{n}$ is a unit vector in the direction $\mathbf{x}$. Since (5.57) and (5.58) have meaning only outside the current distribution, we drop the delta function term. The magnetic induction (5.58) has exactly the form (4.13) of the field of a dipole. This is the generalization of the result found for the circular loop in the last section. Far away from any localized current distribution the magnetic induction is that of a magnetic dipole of dipole moment given by (5.55).

If the current is confined to a plane, but otherwise arbitrary, loop, the magnetic moment can be expressed in a simple form. If the current I flows in a closed circuit whose line element is $d\mathbf{l}$, (5.55) becomes

$$\mathbf{m} = \frac{I}{2c}\oint \mathbf{x} \times d\mathbf{l} \qquad (5.59)$$

For a plane loop such as that in Fig. 5.7, the magnetic moment is perpendicular to the plane of the loop. Since $\frac{1}{2}(\mathbf{x} \times d\mathbf{l}) = da$, where da is the triangular element of the area defined by the two ends of $d\mathbf{l}$ and the origin, the loop integral in (5.59) gives the total area of the loop. Hence the magnetic moment has magnitude,

$$|\mathbf{m}| = \frac{I}{c} \times (\text{Area}) \qquad (5.60)$$

regardless of the shape of the circuit.

If the current distribution is provided by a number of charged particles with charges q_i and masses M_i in motion with velocities $\mathbf{v}_i$, the magnetic moment can be expressed in terms of the orbital angular momentum of the particles. The current density is

$$\mathbf{J} = \sum q_i\mathbf{v}_i\,\delta(\mathbf{x} - \mathbf{x}_i) \qquad (5.61)$$

where $\mathbf{x}_i$ is the position of the ith particle. Then the magnetic moment (5.55) becomes

$$\mathbf{m} = \frac{1}{2c} \sum_i q_i (\mathbf{x}_i \times \mathbf{v}_i) \tag{5.62}$$

The vector product $(\mathbf{x}_i \times \mathbf{v}_i)$ is proportional to the ith particle's orbital angular momentum, $\mathbf{L}_i = M_i(\mathbf{x}_i \times \mathbf{v}_i)$. Thus (5.62) becomes

$$\mathbf{m} = \sum_i \frac{q_i}{2M_i c} \, \mathbf{L}_i \tag{5.63}$$

If all the particles in motion have the same charge to mass ratio ($q_i/M_i = e/M$), the magnetic moment can be written in terms of the *total* orbital angular momentum $\mathbf{L}$:

$$\mathbf{m} = \frac{e}{2Mc} \sum_i \mathbf{L}_i = \frac{e}{2Mc} \, \mathbf{L} \tag{5.64}$$

This is the well-known classical connection between angular momentum and magnetic moment which holds for orbital motion even on the atomic scale. But this classical connection fails for the intrinsic moment of electrons and other elementary particles. For electrons, the intrinsic moment is slightly more than twice as large as implied by (5.64), with the spin angular momentum $\mathbf{S}$ replacing $\mathbf{L}$. Thus we speak of the electron having a g factor of 2(1.00117). The departure of the magnetic moment from its classical value has its origins in relativistic and quantum-mechanical effects which we cannot consider here.

5.7 Force and Torque on a Localized Current Distribution in an External Magnetic Induction

If a localized distribution of current is placed in an external magnetic induction $\mathbf{B}(\mathbf{x})$, it experiences forces and torques according to Ampère's laws. The general expressions for the total force and torque are given by (5.12) and (5.13). If the external magnetic induction varies slowly over the region of current, a Taylor's series expansion can be utilized to find the dominant terms in the force and torque. A component of $\mathbf{B}$ can be expanded around a suitable origin,

$$B_i(\mathbf{x}) = B_i(0) + \mathbf{x} \cdot \nabla B_i(0) + \cdots \tag{5.65}$$

The force (5.12) then becomes

$$\mathbf{F} = -\frac{1}{c} \mathbf{B}(0) \times \int \mathbf{J}(\mathbf{x}') \, d^3x' + \frac{1}{c} \int \mathbf{J}(\mathbf{x}') \times [(\mathbf{x}' \cdot \nabla)\mathbf{B}(0)] \, d^3x' + \cdots \tag{5.66}$$

Since the volume integral of **J** vanishes for steady-state currents, the lowest-order term is the one involving the gradient of **B**. Because the integrand involves **J** and **x**, in addition to ∇B, we expect that the integral can be somehow transformed into the magnetic moment (5.55). To accomplish this we use

$$\mathbf{J} \times [(\mathbf{x}' \cdot \nabla)\mathbf{B}] = \mathbf{J} \times \nabla(\mathbf{x}' \cdot \mathbf{B}) = -\nabla \times [\mathbf{J}(\mathbf{x}' \cdot \mathbf{B})] \quad (5.67)$$

The first step depends on the fact that $\nabla \times \mathbf{B} = 0$ for the external field, and that the gradient operator operates only on **B**. Then the force can be written

$$\mathbf{F} = -\frac{1}{c}\nabla \times \int \mathbf{J}(\mathbf{x}' \cdot \mathbf{B})\, d^3x' + \cdots \quad (5.68)$$

Use can now be made of identity (5.54) with the fixed vector **x** replaced by **B**. Then we obtain

$$\mathbf{F} = \nabla \times (\mathbf{B} \times \mathbf{m}) = (\mathbf{m} \cdot \nabla)\mathbf{B} = \nabla(\mathbf{m} \cdot \mathbf{B}) \quad (5.69)$$

where **m** is the magnetic moment (5.55). The second form in (5.69) follows from $\nabla \cdot \mathbf{B} = 0$, while the third depends on $\nabla \times \mathbf{B} = 0$.

A localized current distribution in a nonuniform magnetic induction experiences a force proportional to its magnetic moment **m** and given by (5.69). One simple application of this result is the time-average force on a charged particle spiraling in a nonuniform magnetic field. As is well known, a charged particle in a *uniform* magnetic induction moves in a circle at right angles to the field and with constant velocity parallel to the field, tracing out a helical path. The circular motion is, on the time average, equivalent to a circular loop of current which will have a magnetic moment given by (5.60). If the field is not uniform but has a small gradient (so that in one turn around the helix the particle does not feel significantly different field strengths), then the motion of the particle can be discussed in terms of the force on the equivalent magnetic moment. Consideration of the signs of the moment and the force shows that charged particles tend to be repelled by regions of high flux density, independent of the sign of their charge. This is the basis of the so-called "magnetic mirrors" discussed in Section 12.10 from another point of view.

The total torque on the localized current distribution is found in a similar way by inserting expansion (5.65) into (5.13). Here the zeroth-order term in the expansion contributes. Keeping only this leading term, we have

$$\mathbf{N} = \frac{1}{c}\int \mathbf{x}' \times [\mathbf{J} \times \mathbf{B}(0)]\, d^3x' \quad (5.70)$$

Writing out the triple vector product, we get

$$N = \frac{1}{c} \int \left[(x' \cdot B)J - (x' \cdot J)B \right] d^3x' \tag{5.71}$$

The first integral is the same one considered in (5.68). Hence we can write down its value immediately. The second integral vanishes for a localized steady-state current distribution, as can be seen from the identity, $\nabla \cdot (x^2 J) = 2(x \cdot J) + x^2 \nabla \cdot J$. The leading term in the torque is therefore

$$N = m \times B(0) \tag{5.72}$$

This is the familiar expression for the torque on a dipole, discussed in Section 5.1 as one of the ways of defining the magnitude and direction of the magnetic induction.

The potential energy of a permanent magnetic moment (or dipole) in an external magnetic field can be obtained from either the force (5.69) or the torque (5.72). If we interpret the force as the negative gradient of a potential energy U, we find

$$U = -m \cdot B \tag{5.73}$$

For a magnetic moment in a uniform field the torque (5.72) can be interpreted as the negative derivative of U with respect to the angle between B and m. This well-known result for the potential energy of a dipole shows that the dipole tends to orient itself parallel to the field in the position of lowest potential energy.

We remark in passing that (5.73) is *not* the total energy of the magnetic moment in the external field. In bringing the dipole m into its final position in the field, work must be done to keep the current J which produces m constant. Even though the final situation is a steady-state, there is a transient period initially in which the relevant fields are time dependent. This lies outside our present considerations. Consequently we will leave the discussion of the energy of magnetic fields to Section 6.2, after having treated Faraday's law of induction.

5.8 Macroscopic Equations

So far we have dealt with the basic laws (5.17) and (5.22) of steady-state magnetic fields as microscopic equations in the sense of Chapter 4. We have assumed that the current density J was a completely known function of position. In macroscopic problems this is often not true. The atoms in matter have electrons which give rise to effective atomic currents the current density of which is a rapidly fluctuating quantity. Only its average over a macroscopic volume is known or pertinent. Furthermore, the atomic electrons possess intrinsic magnetic moments which cannot be expressed in terms of a current density. These moments can give rise to dipole fields which vary appreciably on the atomic scale of dimensions.

To treat these atomic contributions we proceed similarly to Section 4.3. The derivation of the macroscopic equations will only be sketched here. A somewhat more complete discussion will be given in Section 6.10. The reason is that for time-varying fields there is a contribution to the atomic current from the time derivative of the polarization **P**. Hence all the contributions to the current appear only in the general, time-dependent problem.

The total current density can be divided into:

(*a*) conduction-current density **J**, representing the actual transport of charge;

(*b*) atomic-current density $\mathbf{J}_a$, representing the circulating currents inside atoms or molecules.

The total vector potential due to all currents is

$$\mathbf{a} = \frac{1}{c} \int \frac{\mathbf{J}(\mathbf{x}') \, d^3x'}{|\mathbf{x} - \mathbf{x}'|} + \frac{1}{c} \int \frac{\mathbf{J}_a(\mathbf{x}') \, d^3x'}{|\mathbf{x} - \mathbf{x}'|} \tag{5.74}$$

We use a small **a** for the microscopic vector potential, just as we used **ε** for the microscopic electric field in Chapter 4. For the atomic contribution we first consider a single molecule, and then average over molecules. The discussion proceeds exactly as in Section 5.6 for a localized current distribution. For a molecule with center at $\mathbf{x}_j$ the vector potential at **x** is given approximately by

$$\mathbf{a}_{\text{mol}}(\mathbf{x}) = \frac{\mathbf{m}_{\text{mol}} \times (\mathbf{x} - \mathbf{x}_j)}{|\mathbf{x} - \mathbf{x}_j|^3} \tag{5.75}$$

To take into account the intrinsic magnetic moments of the electrons, as well as the orbital contribution, we interpret $\mathbf{m}_{\text{mol}}$ as the *total* molecular magnetic moment. If we now sum up over all molecules, averaging as in Section 4.3, the macroscopic vector potential can be written

$$\mathbf{A}(\mathbf{x}) = \frac{1}{c} \int \frac{\mathbf{J}(\mathbf{x}')}{|\mathbf{x} - \mathbf{x}'|} \, d^3x' + \int \frac{\mathbf{M}(\mathbf{x}') \times (\mathbf{x} - \mathbf{x}')}{|\mathbf{x} - \mathbf{x}'|^3} \, d^3x' \tag{5.76}$$

where **M(x)** is the macroscopic magnetization (magnetic moment per unit volume) defined by

$$\mathbf{M} = N \langle \mathbf{m}_{\text{mol}} \rangle \tag{5.77}$$

where N is the number of molecules per unit volume.

The magnetization contribution to $\mathbf{A}$ in (5.76) can be rewritten in a more useful form:

$$\int \mathbf{M}(\mathbf{x}') \times \frac{(\mathbf{x} - \mathbf{x}')}{|\mathbf{x} - \mathbf{x}'|^3} \, d^3x' = \int \mathbf{M}(\mathbf{x}') \times \boldsymbol{\nabla}' \frac{1}{|\mathbf{x} - \mathbf{x}'|} \, d^3x' \qquad (5.78)$$

Then the identity, $\boldsymbol{\nabla} \times (\phi \mathbf{M}) = \boldsymbol{\nabla}\phi \times \mathbf{M} + \phi \boldsymbol{\nabla} \times \mathbf{M}$, can be used to obtain

$$\int \mathbf{M}(\mathbf{x}') \times \frac{(\mathbf{x} - \mathbf{x}')}{|\mathbf{x} - \mathbf{x}'|^3} \, d^3x' = \int \frac{\boldsymbol{\nabla}' \times \mathbf{M}}{|\mathbf{x} - \mathbf{x}'|} \, d^3x' - \int \boldsymbol{\nabla}' \times \left(\frac{\mathbf{M}}{|\mathbf{x} - \mathbf{x}'|} \right) d^3x' \quad (5.79)$$

The last integral can be converted to a surface integral of $\dfrac{\mathbf{n} \times \mathbf{M}}{|\mathbf{x} - \mathbf{x}'|}$, and so vanishes if $\mathbf{M}$ is assumed to be mathematically well behaved and localized within a finite volume. Combining the first term in (5.79) with the conduction-current term in (5.76), we can write the vector potential as

$$\mathbf{A}(\mathbf{x}) = \frac{1}{c} \int \frac{\mathbf{J}(\mathbf{x}') + c\boldsymbol{\nabla}' \times \mathbf{M}(\mathbf{x}')}{|\mathbf{x} - \mathbf{x}'|} \, d^3x' \qquad (5.80)$$

We see that the magnetization contributes to the vector potential as an effective current density $\mathbf{J}_M$:

$$\mathbf{J}_M = c(\boldsymbol{\nabla} \times \mathbf{M}) \qquad (5.81)$$

There is one questionable step in the derivation of (5.80). That is the use of the dipole vector potential (5.75) for all molecules, even those near the point $\mathbf{x}$. If a molecule lies within a sphere of radius a few molecular diameters d of $\mathbf{x}$, its vector potential will differ appreciably from the dipole form (5.75), being much less singular. Thus in (5.80) the contribution from that sphere around $\mathbf{x}$ is in error. To estimate its importance we note that the magnitude of the vector potential per unit volume near $\mathbf{x}$ is $|\boldsymbol{\nabla} \times \mathbf{M}|/R$, while the volume within a distance R to $(R + dR)$ of the point $\mathbf{x}$ is $4\pi R^2 \, dR$. Hence the contribution to $\mathbf{A}$ from the immediate neighborhood of $\mathbf{x}$ is in error at most by an amount of the order of $d^2 |\boldsymbol{\nabla} \times \mathbf{M}| \sim (d^2/L) \langle M \rangle$, where L is a macroscopic dimension measuring the spatial variation of $\mathbf{M}$. Since the whole vector potential is of the order of $\langle M \rangle L$, the relative error made in using the dipole approximation everywhere is of the order of d^2/L^2. This is completely negligible unless the macroscopic length L becomes microscopic; then the whole development fails.

To obtain the macroscopic equivalent of the curl equation (5.22) we calculate $\mathbf{B}$ from (5.80) or, what is the same thing, write down (5.22) with the total current $(\mathbf{J} + \mathbf{J}_M)$ replacing $\mathbf{J}$:

$$\boldsymbol{\nabla} \times \mathbf{B} = \frac{4\pi}{c} \mathbf{J} + 4\pi \boldsymbol{\nabla} \times \mathbf{M} \qquad (5.82)$$

The $\nabla \times \mathbf{M}$ term can be combined with $\mathbf{B}$ to define a new macroscopic field $\mathbf{H}$, called the *magnetic field*,

$$\mathbf{H} = \mathbf{B} - 4\pi\mathbf{M} \tag{5.83}$$

Then the macroscopic equations, replacing (5.26), are

$$\left.\begin{aligned} \nabla \times \mathbf{H} &= \frac{4\pi}{c}\,\mathbf{J} \\[2mm] \nabla \cdot \mathbf{B} &= 0 \end{aligned}\right\} \tag{5.84}$$

The introduction of $\mathbf{H}$ as a macroscopic field is completely analogous to the introduction of $\mathbf{D}$ for the electrostatic field. The macroscopic equations (5.84) have their electrostatic counterparts,

$$\left.\begin{aligned} \nabla \cdot \mathbf{D} &= 4\pi\rho \\[2mm] \nabla \times \mathbf{E} &= 0 \end{aligned}\right\} \tag{5.85}$$

We emphasize that the fundamental fields are $\mathbf{E}$ and $\mathbf{B}$. They satisfy the homogeneous equations in (5.84) and (5.85). The derived fields, $\mathbf{D}$ and $\mathbf{H}$, are introduced as a matter of convenience in order to take into account in an average way the contributions to ρ and $\mathbf{J}$ of the atomic charges and currents.

In analogy with dielectric media we expect that the properties of magnetic media can be described by a small number of constants characteristic of the material. Thus in the simplest case we would expect that $\mathbf{B}$ and $\mathbf{H}$ are proportional:

$$\mathbf{B} = \mu\mathbf{H} \tag{5.86}$$

where μ is a constant characteristic of the material called the *permeability*.* This simple result does hold for materials other than the ferromagnetic substances. But for these nonmagnetic materials μ generally differs from unity by only a few parts in 10^5 ($\mu > 1$ for paramagnetic substances, $\mu < 1$ for diamagnetic substances). For the ferromagnetic substances, (5.86) must be replaced by a nonlinear functional relationship,

$$\mathbf{B} = \mathbf{F}(\mathbf{H}) \tag{5.87}$$

The phenomenon of hysteresis, shown schematically in Fig. 5.8, implies that $\mathbf{B}$ is not a single-valued function of $\mathbf{H}$. In fact, the function $\mathbf{F}(\mathbf{H})$ depends on the history of preparation of the material. The incremental permeability of $\mu(\mathbf{H})$ is defined as the derivative of $\mathbf{B}$ with respect to $\mathbf{H}$,

---

* To be consistent with the electrostatic relation $\mathbf{D} = \epsilon\mathbf{E}$, expressing the derived quantity $\mathbf{D}$ as a factor times $\mathbf{E}$, we should write $\mathbf{H} = \mu'\mathbf{B}$. But traditional usage is that of (5.86). It makes most substances have $\mu > 1$. Perhaps that is more comforting than $\mu' < 1$.

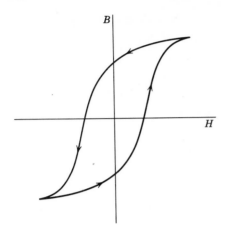

Fig. 5.8 Hysteresis loop giving **B** in a ferromagnetic material as a function of **H**.

assuming that **B** and **H** are parallel. For high-permeability substances, $\mu(\mathbf{H})$ can be as high as 10^6. Most untreated ferromagnetic materials have a linear relation (5.86) between **B** and **H** for very small fields. Typical values of initial permeability range from 10 to 10^4.

The complicated relationship between **B** and **H** in ferromagnetic materials makes analysis of magnetic boundary-value problems inherently more difficult than that of similar electrostatic problems. But the very large values of permeability sometimes allow simplifying assumptions on the boundary conditions. We will see that explicitly in the next section.

5.9 Boundary Conditions on B and H

Before we can solve magnetic boundary-value problems, we must establish the boundary conditions satisfied by **B** and **H** at the interface between two media of different magnetic properties. If a small Gaussian

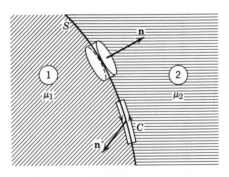

Fig. 5.9

pillbox is oriented so that its faces are in regions 1 and 2 and parallel to the surface boundary, S, as shown in Fig. 5.9, Gauss's theorem can be applied to $\nabla \cdot \mathbf{B} = 0$ to yield

$$(\mathbf{B_2} - \mathbf{B_1}) \cdot \mathbf{n} = 0 \qquad (5.88)$$

where $\mathbf{n}$ is the unit normal to the surface directed from region 1 into region 2, and the subscripts refer to values at the surface in the two media.

If we now consider a small, narrow circuit C, as shown in Fig. 5.9, with normal $\mathbf{n}'$ parallel to the interface and surface S, Stokes's theorem can be applied to the curl equation in (5.84) to give

$$\oint_C \mathbf{H} \cdot d\mathbf{l} = \frac{4\pi}{c} \int_S \mathbf{J} \cdot \mathbf{n}' \, da \qquad (5.89)$$

The contributions to the line integral are the tangential values of $\mathbf{H}$ in the two regions, while the surface integral is proportional to the surface-current density $\mathbf{K}$ (charge/length × time) in the limit of vanishing width to the loop. Thus (5.89) becomes

$$(\mathbf{H_2} - \mathbf{H_1}) \cdot (\mathbf{n}' \times \mathbf{n}) = \frac{4\pi}{c} \mathbf{n}' \cdot \mathbf{K}$$

or

$$\mathbf{n} \times (\mathbf{H_2} - \mathbf{H_1}) = \frac{4\pi}{c} \mathbf{K}$$

$$(5.90)$$

We express these boundary conditions in terms of the magnetic field $\mathbf{H}$ and the permeability μ. For simplicity assuming no surface currents, we have

$$\mathbf{H_2} \cdot \mathbf{n} = \left(\frac{\mu_1}{\mu_2}\right) \mathbf{H_1} \cdot \mathbf{n}$$

$$\mathbf{H_2} \times \mathbf{n} = \mathbf{H_1} \times \mathbf{n}$$

$$(5.91)$$

If $\mu_1 \gg \mu_2$, the normal component of $\mathbf{H_2}$ is much larger than the normal component of $\mathbf{H_1}$, as shown in Fig. 5.10. In the limit $(\mu_1/\mu_2) \to \infty$, the

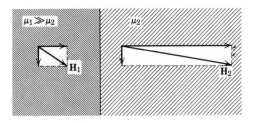

Fig. 5.10

magnetic field H_2 is normal to the boundary surface, independent of the direction of H_1 (barring the exceptional case of H_1 exactly parallel to the interface). The boundary condition on H at the surface of a very high-permeability material is thus the same as for the electric field at the surface of a conductor. We may therefore use electrostatic potential theory for the magnetic field. The surfaces of the high-permeability material are approximately "equipotentials," and the lines of H are normal to these equipotentials. This analogy is exploited in many magnet-design problems. The type of field is decided upon, and the pole faces are shaped to be equipotential surfaces.

5.10   Uniformly Magnetized Sphere

To illustrate the different methods possible for the solution of a boundary-value problem in magnetostatics, we consider in Fig. 5.11 the simple problem of a sphere of radius a, with a uniform permanent magnetization M of magnitude M_0 and parallel to the z axis, embedded in a nonpermeable medium. Outside the sphere, $\nabla \cdot B = \nabla \times B = 0$. Consequently, for $r > a$, $B = H$ can be written as the negative gradient of a magnetic scalar potential which satisfies Laplace's equation,

$$\left. \begin{array}{l} B_{out} = -\nabla \Phi_M \\ \\ \nabla^2 \Phi_M = 0 \end{array} \right\} \qquad (5.92)$$

With the boundary condition that $B \to 0$ for $r \to \infty$, the general solution for the potential is

$$\Phi_M(r, \theta) = \sum_{l=0}^{\infty} \alpha_l \frac{P_l(\cos \theta)}{r^{l+1}} \qquad (5.93)$$

Past experience tells us that only the lowest few terms in this expansion will appear, probably just $l = 1$.

Inside a magnetized object we cannot in general use equations (5.92) because $\nabla \times B \neq 0$. This causes no difficulty in the present simple situation because (5.83) implies that B, H, and M are all parallel in the absence of applied fields.

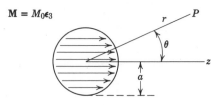

$M = M_0 \epsilon_3$

Fig. 5.11

Hence we assume that

$$\left.\begin{array}{l} \mathbf{B}_{in} = B_0\boldsymbol{\epsilon}_3 \\ \mathbf{H}_{in} = (B_0 - 4\pi M_0)\boldsymbol{\epsilon}_3 \end{array}\right\} \tag{5.94}$$

The boundary conditions at the surface of the sphere are that B_r and H_θ be continuous. Thus, from (5.92), (5.93), and (5.94), we obtain

$$\left.\begin{array}{l} B_0 \cos\theta = \sum_{l=0}^{\infty} \frac{(l+1)\alpha_l P_l(\cos\theta)}{a^{l+2}} \\[2mm] -(B_0 - 4\pi M_0)\sin\theta = -\sum_{l=0}^{\infty} \frac{\alpha_l}{a^{l+2}} \frac{dP_l(\cos\theta)}{d\theta} \end{array}\right\} \tag{5.95}$$

Evidently only the $l = 1$ term survives in the expansion. We find the unknown constants α_1 and B_0 to be

$$\left.\begin{array}{l} \alpha_1 = \dfrac{4\pi}{3} M_0 a^3 \\[4mm] B_0 = \dfrac{8\pi}{3} M_0 \end{array}\right\} \tag{5.96}$$

The fields outside the sphere are those of a dipole (5.41) of dipole moment,

$$\mathbf{m} = \frac{4\pi}{3} a^3\mathbf{M} \tag{5.97}$$

The fields inside are

$$\left.\begin{array}{l} \mathbf{B}_{in} = \dfrac{8\pi}{3} \mathbf{M} \\[4mm] \mathbf{H}_{in} = -\dfrac{4\pi}{3} \mathbf{M} \end{array}\right\} \tag{5.98}$$

We note that $\mathbf{B}_{in}$ is parallel to $\mathbf{M}$, while $\mathbf{H}_{in}$ is antiparallel. The lines of $\mathbf{B}$ and $\mathbf{H}$ are shown in Fig. 5.12. The lines of $\mathbf{B}$ are continuous closed paths, but those of $\mathbf{H}$ terminate on the surface. The surface appears to have a "magnetic-charge" density on it. This fictitious charge is related to the divergence of the magnetization (see below).

The solution both inside and outside the sphere could have been obtained from electrostatic potential theory if we had chosen to discuss $\mathbf{H}$ rather than $\mathbf{B}$. We can treat the equations,

$$\left.\begin{array}{l} \nabla \times \mathbf{H} = 0 \\ \nabla \cdot \mathbf{H} = -4\pi\nabla \cdot \mathbf{M} \end{array}\right\} \tag{5.99}$$

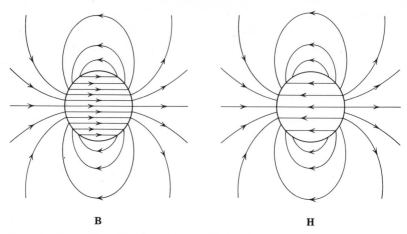

B   **H**

Fig. 5.12 Lines of **B** and lines of **H** for a uniformly magnetized sphere. The lines of **B** are closed curves, but the lines of **H** originate on the surface of the sphere where the magnetic "charge," $-\nabla \cdot \mathbf{M}$, resides.

These equations show that **H** is derivable from a potential, and that $-\nabla \cdot \mathbf{M}$ acts as a magnetic-charge density. Thus, with $\mathbf{H} = -\nabla \Phi_M$, we find

$$\nabla^2 \Phi_M = 4\pi \nabla \cdot \mathbf{M} \qquad (5.100)$$

Since **M** is constant in magnitude and direction, its divergence is zero inside the sphere. But there is a contribution because **M** vanishes outside the sphere. We write the solution for Φ_M inside and outside the sphere as

$$\Phi_M(\mathbf{x}) = -\int \frac{\nabla' \cdot \mathbf{M}(\mathbf{x}')}{|\mathbf{x} - \mathbf{x}'|} \, d^3x' \qquad (5.101)$$

Then we use the vector identity $\nabla \cdot (\phi \mathbf{M}) = \mathbf{M} \cdot \nabla \phi + \phi \nabla \cdot \mathbf{M}$ to obtain

$$\Phi_M(\mathbf{x}) = -\int \nabla' \cdot \frac{\mathbf{M}(\mathbf{x}')}{|\mathbf{x} - \mathbf{x}'|} \, d^3x' + \int \mathbf{M}(\mathbf{x}') \cdot \nabla' \left(\frac{1}{|\mathbf{x} - \mathbf{x}'|} \right) d^3x' \quad (5.102)$$

The first integral vanishes on integration over any volume containing the sphere. If we convert the derivative with respect to $\mathbf{x}'$ into one with respect to $\mathbf{x}$ according to the rule $\nabla' \rightarrow -\nabla$ when operating on any function of $|\mathbf{x} - \mathbf{x}'|$, the potential can be written

$$\Phi_M(\mathbf{x}) = -\nabla \cdot \int \frac{\mathbf{M}(\mathbf{x}')}{|\mathbf{x} - \mathbf{x}'|} \, d^3x' = -\nabla \cdot \left[M_0 \boldsymbol{\epsilon}_3 \int_0^a r'^2 \, dr' \int d\Omega' \frac{1}{|\mathbf{x} - \mathbf{x}'|} \right]$$

$$(5.103)$$

Only the $l = 0$ part of $|\mathbf{x} - \mathbf{x}'|^{-1}$ contributes to the integral. Therefore

$$\Phi_M(\mathbf{x}) = -4\pi M_0 \nabla \cdot \left[\boldsymbol{\epsilon}_3 \int_0^a \frac{r'^2\, dr'}{r_>} \right] \tag{5.104}$$

The integral yields different values, depending on whether r lies inside or outside the sphere. We find easily

$$\Phi_M(\mathbf{x}) = \frac{4\pi M_0 a^2}{3} \left(\frac{r_<}{r_>^2} \right) \cos \theta \tag{5.105}$$

where $r_<$ ($r_>$) is the smaller (larger) of r and a. This potential yields a dipole field outside with magnetic moment (5.97) and the constant value $\mathbf{H}_{\text{in}}$ (5.98) inside, in agreement with the first method of solution.*

Finally we solve the problem using the generally applicable vector potential. Referring to (5.80), we see that the vector potential is given by

$$\mathbf{A}(\mathbf{x}) = \int \frac{\nabla' \times \mathbf{M}(\mathbf{x}')}{|\mathbf{x} - \mathbf{x}'|}\, d^3x' \tag{5.106}$$

Since $\mathbf{M}$ is constant inside the sphere, the curl vanishes there. But because of the discontinuity of $\mathbf{M}$ at the surface, there is a surface integral contribution to $\mathbf{A}$. If we consider (5.79), the required surface integral can be recovered:

$$\mathbf{A}(\mathbf{x}) = -\int \nabla' \times \left(\frac{\mathbf{M}(\mathbf{x}')}{|\mathbf{x} - \mathbf{x}'|} \right) d^3x' = \oint \frac{\mathbf{M}(\mathbf{x}') \times \mathbf{n}}{|\mathbf{x} - \mathbf{x}'|}\, da' \tag{5.107}$$

The quantity $c(\mathbf{M} \times \mathbf{n})$ can be considered as a surface-current density. The equivalence of a uniform magnetization throughout a certain volume to a surface-current density $c(\mathbf{M} \times \mathbf{n})$ over its surface is a general result for arbitrarily shaped volumes. This equivalence is often useful in treating fields due to permanent magnets.

For the sphere with $\mathbf{M}$ in the z direction, $(\mathbf{M} \times \mathbf{n})$ has only an azimuthal component,

$$(\mathbf{M} \times \mathbf{n})_\phi = M_0 \sin \theta' \tag{5.108}$$

To determine $\mathbf{A}$ we choose our observation point in the x-z plane for calculational convenience, just as in Sections 5.5. Then only the y component of $-(\mathbf{n} \times \mathbf{M})$ enters. The azimuthal component of the vector potential is then

$$A_\phi(\mathbf{x}) = M_0 a^2 \int d\Omega' \frac{\sin \theta' \cos \phi'}{|\mathbf{x} - \mathbf{x}'|} \tag{5.109}$$

---

* The development from (5.101) to (5.105) is unnecessarily complicated for the simple calculation at hand. For the uniformly magnetized sphere it is easy to show that $\nabla \cdot \mathbf{M} = -M_0 \cos \theta\, \delta(r - a)$. Substitution into (5.101) and use of (3.70) yields (5.105) directly. Equation (5.103) is still useful, of course, for more complicated distributions of magnetization.

where $\mathbf{x}'$ has coordinates (a, θ', ϕ'). The angular factor can be written

$$\sin \theta' \cos \phi' = -\sqrt{\frac{8\pi}{3}} \operatorname{Re}[Y_{1,1}(\theta', \phi')] \qquad (5.110)$$

Thus with expansion (3.70) for $|\mathbf{x} - \mathbf{x}'|$ only the $l = 1$, $m = 1$ term will survive. Consequently

$$A_\phi(\mathbf{x}) = \frac{4\pi}{3} M_0 a^2 \left(\frac{r_<}{r_>^2}\right) \sin \theta \qquad (5.111)$$

where $r_<$ $(r_>)$ is the smaller (larger) of r and a. With only a ϕ component of $\mathbf{A}$, the components of the magnetic induction $\mathbf{B}$ are given by (5.38). Equation (5.111) evidently gives the uniform $\mathbf{B}$ inside and the dipole field outside, as found before.

The different techniques used here illustrate the variety of ways of solving steady-state magnetic problems, in this case with a specified distribution of magnetization. The scalar potential method is applicable provided no currents are present. But for the general problem with currents we must use the vector potential (apart from special techniques for particularly simple geometries).

5.11 Magnetized Sphere in an External Field; Permanent Magnets

In Section 5.10 we discussed the fields due to a uniformly magnetized sphere. Because of the linearity of the field equations we can superpose a uniform magnetic induction $\mathbf{B}_0 = \mathbf{H}_0$ throughout all space. Then we have the problem of a uniformly magnetized sphere in an external field. From (5.98) we find that the magnetic induction and field inside the sphere are now

$$\left.\begin{aligned}
\mathbf{B}_{\text{in}} &= \mathbf{B}_0 + \frac{8\pi}{3} \mathbf{M} \\[2mm]
\mathbf{H}_{\text{in}} &= \mathbf{B}_0 - \frac{4\pi}{3} \mathbf{M}
\end{aligned}\right\} \qquad (5.112)$$

We now imagine that the sphere is not a permanently magnetized object, but rather a paramagnetic or diamagnetic substance of permeability μ. Then the magnetization $\mathbf{M}$ is a result of the application of the external field. To find the magnitude of $\mathbf{M}$ we use (5.86):

$$\mathbf{B}_{\text{in}} = \mu \mathbf{H}_{\text{in}} \qquad (5.113)$$

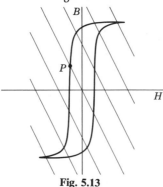

Fig. 5.13

Thus

$$B_0 + \frac{8\pi}{3} M = \mu\left(B_0 - \frac{4\pi}{3} M\right) \tag{5.114}$$

This gives a magnetization,

$$M = \frac{3}{4\pi}\left(\frac{\mu - 1}{\mu + 2}\right) B_0 \tag{5.115}$$

We note that this is completely analogous to the polarization P of a dielectric sphere in a uniform electric field (4.63).

For a ferromagnetic substance the arguments of the last paragraph fail. Equation (5.115) implies that the magnetization vanishes when the external field vanishes. The existence of permanent magnets contradicts this result. The nonlinear relation (5.87) and the phenomenon of hysteresis allow the creation of permanent magnets. We can solve equations (5.112) for one relation between H_{in} and B_{in} by eliminating M:

$$B_{in} + 2H_{in} = 3B_0 \tag{5.116}$$

The hysteresis curve provides the other relation between B_{in} and H_{in}, so that specific values can be found for any external field. Equation (5.116) corresponds to lines with slope -2 on the hysteresis diagram with intercepts $3B_0$ on the y axis, as in Fig. 5.13. Suppose, for example, that the external field is increased until the ferromagnetic sphere becomes saturated and decreased to zero. The internal B and H will then be given by the point marked P in Fig. 5.13. The magnetization can be found from (5.112) with $B_0 = 0$.

The relation (5.116) between B_{in} and H_{in} is specific to the sphere. For other geometries other relations pertain. The problem of the ellipsoid can be solved exactly and shows that the slope of the lines (5.116) range from zero for a flat disc to $-\infty$ for a long needle-like object. Thus a larger internal magnetic induction can be obtained with a rod geometry than with spherical or oblate spheroidal shapes.

5.12 Magnetic Shielding; Spherical Shell of Permeable Material in a Uniform Field

Suppose that a certain magnetic induction $\mathbf{B}_0$ exists in a region of empty space initially. A permeable body is now placed in the region. The lines of magnetic induction are modified. From our remarks at the end of Section 5.9 concerning media of very high permeability we would expect that the field lines would tend to be normal to the surface of the body. Carrying the analogy with conductors further, if the body is hollow, we would expect that the field in the cavity would be smaller than the external field, vanishing in the limit $\mu \to \infty$. Such a reduction in field is said to be due to the *magnetic shielding* provided by the permeable material. It is of considerable practical importance, since essentially field-free regions are often necessary or desirable for experimental purposes or for the reliable working of electronic devices.

As an example of the phenomenon of magnetic shielding we consider a spherical shell of inner (outer) radius a (b), made of material of permeability μ, and placed in a formerly uniform constant magnetic induction $\mathbf{B}_0$, as shown in Fig. 5.14. We wish to find the fields $\mathbf{B}$ and $\mathbf{H}$ everywhere in space, but most particularly in the cavity ($r < a$), as functions of μ. Since there are no currents present, the magnetic field $\mathbf{H}$ is derivable from a scalar potential, $\mathbf{H} = -\nabla \Phi_M$. Furthermore, since $\mathbf{B} = \mu \mathbf{H}$, the divergence equation $\nabla \cdot \mathbf{B} = 0$ becomes $\nabla \cdot \mathbf{H} = 0$ in the various regions. Thus the potential Φ_M satisfies Laplace's equation everywhere. The problem reduces to finding the proper solutions in the different regions to satisfy the boundary conditions (5.88) and (5.90) at $r = a$ and $r = b$.

For $r > b$, the potential must be of the form,

$$\Phi_M = -B_0 r \cos\theta + \sum_{l=0}^{\infty} \frac{\alpha_l}{r^{l+1}} P_l(\cos\theta) \tag{5.117}$$

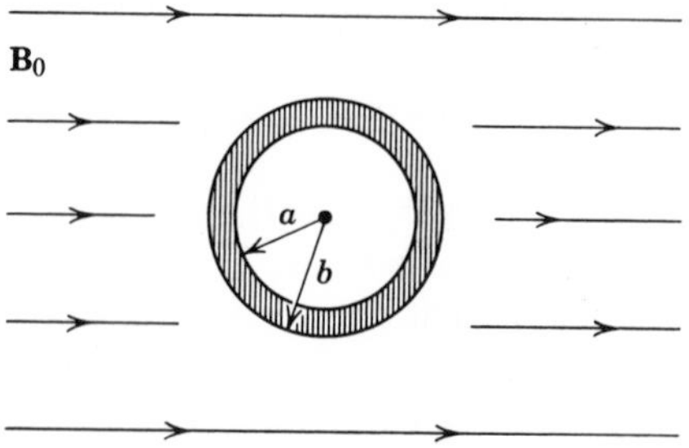

Fig. 5.14

in order to give the uniform field, $\mathbf{H} = \mathbf{B} = \mathbf{B}_0$, at large distances. For the inner regions, the potential must be

$$a < r < b \qquad \Phi_M = \sum_{l=0}^{\infty}\left(\beta_l r^l + \gamma_l \frac{1}{r^{l+1}}\right)P_l(\cos\theta)$$

$$r < a \qquad \Phi_M = \sum_{l=0}^{\infty}\delta_l r^l P_l(\cos\theta) \qquad\qquad (5.118)$$

The boundary conditions at $r = a$ and $r = b$ are that H_θ and B_r be continuous. In terms of the potential Φ_M these conditions become

$$\frac{\partial\Phi_M}{\partial\theta}(b_+) = \frac{\partial\Phi_M}{\partial\theta}(b_-) \qquad \frac{\partial\Phi_M}{\partial\theta}(a_+) = \frac{\partial\Phi_M}{\partial\theta}(a_-)$$

$$\frac{\partial\Phi_M}{\partial r}(b_+) = \mu\frac{\partial\Phi_M}{\partial r}(b_-) \qquad \mu\frac{\partial\Phi_M}{\partial r}(a_+) = \frac{\partial\Phi_M}{\partial r}(a_-) \qquad (5.119)$$

The notation $b_\pm$ means the limit $r \to b$ approached from $r \gtrless b$, and similarly for $a_\pm$. These four conditions, which hold for all angles θ, are sufficient to determine the unknown constants in (5.117) and (5.118). All coefficients with $l \neq 1$ vanish. The $l = 1$ coefficients satisfy the four simultaneous equations

$$\alpha_1 - b^3\beta_1 - \gamma_1 = b^3 B_0$$

$$2\alpha_1 + \mu b^3\beta_1 - 2\mu\gamma_1 = -b^3 B_0$$

$$a^3\beta_1 + \gamma_1 - a^3\delta_1 = 0 \qquad (5.120)$$

$$\mu a^3\beta_1 - 2\mu\gamma_1 - a^3\delta_1 = 0.$$

The solutions for α_1 and δ_1 are

$$\alpha_1 = \left[\frac{(2\mu+1)(\mu-1)}{(2\mu+1)(\mu+2) - 2\dfrac{a^3}{b^3}(\mu-1)^2}\right](b^3 - a^3)B_0$$

$$\delta_1 = -\left[\frac{9\mu}{(2\mu+1)(\mu+2) - 2\dfrac{a^3}{b^3}(\mu-1)^2}\right]B_0 \qquad (5.121)$$

The potential outside the spherical shell corresponds to a uniform field $\mathbf{B}_0$ plus a dipole field (5.41) with dipole moment α_1 oriented parallel to $\mathbf{B}_0$. Inside the cavity, there is a uniform magnetic field parallel to $\mathbf{B}_0$ and equal in magnitude to $-\delta_1$. For $\mu \gg 1$, the dipole moment α_1 and the

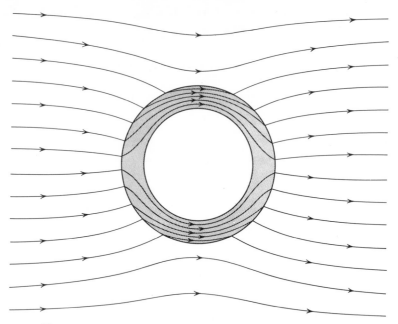

Fig. 5.15   Shielding effect of a shell of highly permeable material.

inner field $-\delta_1$ become

$$\left.\begin{array}{l} \alpha_1 \to b^3 B_0 \\[4mm] -\delta_1 \to \dfrac{9}{2\mu\left(1 - \dfrac{a^3}{b^3}\right)}\, B_0 \end{array}\right\} \qquad (5.122)$$

We see that the inner field is proportional to μ^{-1}. Consequently a shield made of high-permeability material with $\mu \sim 10^3$ to 10^6 causes a great reduction in the field inside it, even with a relatively thin shell. Figure 5.15 shows the behavior of the lines of **B**. The lines tend to pass through the permeable medium if possible.

REFERENCES AND SUGGESTED READING

Problems in steady-state current flow in an extended resistive medium are analogous to electrostatic potential problems, with the current density replacing the displacement and the conductivity replacing the dielectric constant. But the boundary conditons are generally different. Steady-state current flow is treated in
Jeans, Chapters IX and X,
Smythe, Chapter VI.

Magnetic fields due to specified current distributions and boundary-value problems in magnetostatics are discussed, with numerous examples, by
  Durand, Chapters XIV and XV,
  Smythe, Chapters VII and XII.
The atomic theory of magnetic properties rightly falls in the domain of quantum mechanics. Semiclassical discussions are given by
  Abraham and Becker, Band II, Sections 29–34,
  Durand, pp. 551–573, and Chapter XVII,
  Landau and Lifshitz, *Electrodynamics of Continuous Media,*
  Rosenfeld, Chapter IV.
Quantum-mechanical treatments appear in books devoted entirely to the electrical and magnetic properties of matter, such as
  Van Vleck.

PROBLEMS

5.1 Starting with the differential expression

$$d\mathbf{B} = \frac{I\,d\mathbf{l} \times \mathbf{x}}{c}\frac{}{x^3}$$

for the magnetic induction produced by an increment $I\,d\mathbf{l}$ of current, show explicitly that for a closed loop carrying a current I the magnetic induction at an observation point P is

$$\mathbf{B} = -\frac{I}{c}\,\nabla\Omega$$

where Ω is the solid angle subtended by the loop at the point P. This is an alternative form of Ampère's law for current loops.

5.2 (*a*) For a solenoid wound with N turns per unit length and carrying a current I, show that the magnetic-flux density on the axis is given approximately by

$$B_z = \frac{2\pi NI}{c}\,(\cos\theta_1 + \cos\theta_2)$$

where the angles are defined in the figure.

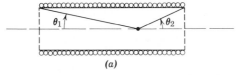

(a)

(*b*) For a long solenoid of length L and radius a show that *near the axis* and *near the center of the solenoid* the magnetic induction is mainly parallel to the axis, but has a small radial component

$$B_\rho \simeq \frac{96\pi NI}{c}\left(\frac{a^2 z\rho}{L^4}\right)$$

correct to order a^2/L^2, and for $z \ll L$, $\rho \ll a$. The coordinate z is measured from the center point of the axis.

(*c*) Show that at the end of a long solenoid the magnetic induction near the axis has components

$$B_z \simeq \frac{2\pi NI}{c}, \quad B_\rho \simeq \frac{\pi NI}{c}\left(\frac{\rho}{a}\right)$$

5.3 A cylindrical conductor of radius *a* has a hole of radius *b* bored parallel to, and centered a distance *d* from, the cylinder axis ($d + b < a$). The current density is uniform throughout the remaining metal of the cylinder and is parallel to the axis. Use Ampère's law and the principle of linear superposition to find the magnitude and the direction of the magnetic-flux density in the hole.

5.4 A circular current loop of radius *a* carrying a current *I* lies in the *x-y* plane with its center at the origin.

(*a*) Show that the only nonvanishing component of the vector potential is

$$A_\phi(\rho, z) = \frac{4Ia}{c}\int_0^\infty dk \cos kz\, I_1(k\rho_<)K_1(k\rho_>)$$

where $\rho_<$ ($\rho_>$) is the smaller (larger) of *a* and ρ.

(*b*) Show that an alternative expression for A_ϕ is

$$A_\phi(\rho, z) = \frac{2\pi Ia}{c}\int_0^\infty dk\, e^{-k|z|} J_1(ka) J_1(k\rho)$$

(*c*) Write down integral expressions for the components of magnetic induction, using the expressions of (*a*) and (*b*). Evaluate explicitly the components of **B** on the axis by performing the necessary integrations.

5.5 Two concentric circular loops of radii *a*, *b* and currents *I*, *I'*, respectively ($b < a$), have an angle α between their planes. Show that the torque on one of the loops is about the line of intersection of the two planes containing the loops and has the magnitude:

$$N = \frac{2\pi^2 II'b^2}{ac^2}\sum_{n=0}^\infty \frac{(n+1)}{(2n+1)}\left[\frac{\Gamma(n+\frac{3}{2})}{\Gamma(n+2)\,\Gamma(\frac{3}{2})}\right]^2 \left(\frac{b}{a}\right)^{2n} P_{2n+1}^1(\cos\alpha)$$

where $P_l^1(\cos\alpha)$ is an associated Legendre polynomial. Determine the sense of the torque for α an acute angle and the currents in the same (opposite) directions.

5.6 A sphere of radius *a* carries a uniform charge distribution on its surface. The sphere is rotated about a diameter with constant angular velocity ω. Find the vector potential and magnetic-flux density both inside and outside the sphere.

5.7 A long, hollow, right circular cylinder of inner (outer) radius *a* (*b*), and of relative permeability μ, is placed in a region of initially uniform magnetic-flux density $\mathbf{B}_0$ at right angles to the field. Find the flux density at all points in space, and sketch the logarithm of the ratio of the magnitudes of **B** on the cylinder axis to $\mathbf{B}_0$ as a function of $\log_{10}\mu$ for $a^2/b^2 = 0.5, 0.1$. Neglect end effects.

5.8 A current distribution $\mathbf{J}(\mathbf{x})$ exists in a medium of unit permeability adjacent to a semi-infinite slab of material having permeability μ and filling the half-space, $z < 0$.

(a) Show that for $z > 0$ the magnetic induction can be calculated by replacing the medium of permeability μ by an image current distribution, $\mathbf{J}^*$, with components,

$$\left(\frac{\mu - 1}{\mu + 1}\right) J_x(x, y, -z), \quad \left(\frac{\mu - 1}{\mu + 1}\right) J_y(x, y, -z), \quad -\left(\frac{\mu - 1}{\mu + 1}\right) J_z(x, y, -z)$$

(b) Show that for $z < 0$ the magnetic induction appears to be due to a current distribution $\left(\dfrac{2\mu}{\mu + 1}\right) \mathbf{J}$ in a medium of unit permeability.

5.9 A circular loop of wire having a radius a and carrying a current I is located in vacuum with its center a distance d away from a semi-infinite slab of permeability μ. Find the force acting on the loop when

(a) the plane of the loop is parallel to the face of the slab,

(b) the plane of the loop is perpendicular to the face of the slab.

(c) Determine the limiting form of your answers to (a) and (b) when $d \gg a$. Can you obtain these limiting values in some simple and direct way?

5.10 A magnetically "hard" material is in the shape of a right circular cylinder of length L and radius a. The cylinder has a permanent magnetization M_0, uniform throughout its volume and parallel to its axis.

(a) Determine the magnetic field $\mathbf{H}$ and magnetic induction $\mathbf{B}$ at all points on the axis of the cylinder, both inside and outside.

(b) Plot the ratios $\mathbf{B}/4\pi M_0$ and $\mathbf{H}/4\pi M_0$ on the axis as functions of z for $L/a = 5$.

5.11 (a) Starting from the force equation (5.12) and the fact that a magnetization $\mathbf{M}$ is equivalent to a current density $\mathbf{J}_M = c(\boldsymbol{\nabla} \times \mathbf{M})$, show that, in the absence of macroscopic currents, the total magnetic force on a body with magnetization $\mathbf{M}$ can be written

$$\mathbf{F} = -\int (\boldsymbol{\nabla} \cdot \mathbf{M}) \mathbf{B}_e \, d^3x$$

where $\mathbf{B}_e$ is the magnetic induction due to all other except the one in question.

(b) Show that an alternative expression for the total force is

$$\mathbf{F} = -\int (\boldsymbol{\nabla} \cdot \mathbf{M}) \mathbf{H} \, d^3x$$

where $\mathbf{H}$ is the *total* magnetic field, including the field of the magnetized body.

Hint: The results of (a) and (b) differ by a self-force term which can be omitted (why?).

5.12 A magnetostatic field is due entirely to a localized distribution of permanent magnetization.

(a) Show that

$$\int \mathbf{B} \cdot \mathbf{H} \, d^3x = 0$$

provided the integral is taken over all space.

(*b*) From the potential energy (5.73) of a dipole in an external field show that for a continuous distribution of permanent magnetization the magneto-static energy can be written

$$W = \frac{1}{8\pi} \int \mathbf{H} \cdot \mathbf{H}\, d^3x = -\frac{1}{2} \int \mathbf{M} \cdot \mathbf{H}\, d^3x$$

apart from an additive constant which is independent of the orientation or position of the various constituent magnetized bodies.

5.13 Show that in general a long, straight bar of uniform cross-sectional area A with uniform lengthwise magnetization M, when placed with its flat end against an infinitely permeable flat surface, adheres with a force given approximately by

$$F \simeq 2\pi AM^2$$

5.14 A right circular cylinder of length L and radius a has a uniform lengthwise magnetization M.

(*a*) Show that, when it is placed with its flat end against an infinitely permeable plane surface, it adheres with a force

$$F = 8\pi aLM^2 \left[\frac{K(k) - E(k)}{k} - \frac{K(k_1) - E(k_1)}{k_1} \right]$$

where

$$k = \frac{2a}{\sqrt{4a^2 + L^2}}, \quad k_1 = \frac{a}{\sqrt{a^2 + L^2}}$$

(*b*) Find the limiting form for the force if $L \gg a$.

6

Time-Varying Fields, Maxwell's Equations, Conservation Laws

In the previous chapters we have dealt with steady-state problems in electricity and in magnetism. Similar mathematical techniques were employed, but electric and magnetic phenomena were treated as independent. The only link between them was the fact that currents which produce magnetic fields are basically electrical in character, being charges in motion. The almost independent nature of electric and magnetic phenomena disappears when we consider time-dependent problems. Time-varying magnetic fields give rise to electric fields and vice-versa. We then must speak of *electromagnetic fields*, rather than electric or magnetic fields. The full import of the interconnection between electric and magnetic fields and their essential sameness becomes clear only within the framework of special relativity (Chapter 11). For the present we will content ourselves with examining the basic phenomena and deducing the set of equations known as *Maxwell's equations*, which describe the behavior of electromagnetic fields. General properties of these equations will be established so that the basic groundwork of electrodynamics will have been laid. Subsequent chapters will then explore the many ramifications.

In our desire to proceed to other things, we will leave out a number of topics which, while of interest in themselves, can be studied elsewhere. Some of these are quasi-stationary fields, circuit theory, inductance calculations, eddy currents, and induction heating. None of these subjects involves new concepts beyond what are developed in this chapter and previous ones. The interested reader will find references at the end of the chapter.

6.1 Faraday's Law of Induction

The first quantitative observations relating time-dependent electric and magnetic fields were made by Faraday (1831) in experiments on the behavior of currents in circuits placed in time-varying magnetic fields. It was observed by Faraday that a transient current is induced in a circuit if (a) the steady current flowing in an adjacent circuit is turned on or off, (b) the adjacent circuit with a steady current flowing is moved relative to the first circuit, (c) a permanent magnet is thrust into or out of the circuit. No current flows unless either the adjacent current changes or there is relative motion. Faraday interpreted the transient current flow as being due to a changing magnetic flux linked by the circuit. The changing flux induces an electric field around the circuit, the line integral of which is called the *electromotive force, $\mathscr{E}$*. The electromotive force causes a current flow, according to Ohm's law.

We now express Faraday's observations in quantitative mathematical terms. Let the circuit C be bounded by an open surface S with unit normal **n**, as in Fig. 6.1. The magnetic induction in the neighborhood of the circuit is **B**. The magnetic flux linking the circuit is defined by

$$F = \int_S \mathbf{B} \cdot \mathbf{n}\, da \tag{6.1}$$

The electromotive force around the circuit is

$$\mathscr{E} = \oint_C \mathbf{E}' \cdot d\mathbf{l} \tag{6.2}$$

where $\mathbf{E}'$ is the electric field at the element $d\mathbf{l}$ of the circuit C. Faraday's observations are summed up in the mathematical law,

$$\mathscr{E} = -k\frac{dF}{dt} \tag{6.3}$$

The induced electromotive force around the circuit is proportional to the time rate of change of magnetic flux linking the circuit. The sign is specified by Lenz's law, which states that the induced current (and accompanying magnetic flux) is in such a direction as to oppose the change of flux through the circuit.

The constant of proportionality k depends on the choice of units for the electric and magnetic field quantities. It is not, as might at first be supposed, an independent empirical constant to be determined from experiment. As we will see immediately, once the units and dimensions in

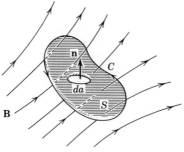

Fig. 6.1

Ampère's law have been chosen, the magnitude and dimensions of k follow from the assumption of Galilean invariance for Faraday's law. For Gaussian units, $k = c^{-1}$, where c is the velocity of light.

Before the development of special relativity (and even afterwards, when dealing with relative speeds small compared with the velocity of light), it was understood, although not often explicitly stated, by all physicists that physical laws should be invariant under Galilean transformations. That is, physical phenomena are the same when viewed by two observers moving with a constant velocity $\mathbf{v}$ relative to one another, provided the coordinates in space and time are related by the Galilean transformation, $\mathbf{x}' = \mathbf{x} + \mathbf{v}t$, $t' = t$. In particular, consider Faraday's observations. It is obvious (i.e., experimentally verified) that the same current is induced in a circuit whether *it* is moved while the circuit through which current is flowing is stationary or it is held fixed while the current-carrying circuit is moved in the same relative manner.

Let us now consider Faraday's law for a moving circuit and see the consequences of Galilean invariance. Expressing (6.3) in terms of the integrals over $\mathbf{E}'$ and $\mathbf{B}$, we have

$$\oint_C \mathbf{E}' \cdot d\mathbf{l} = -k \frac{d}{dt} \int_S \mathbf{B} \cdot \mathbf{n} \, da \qquad (6.4)$$

The induced electromotive force is proportional to the *total* time derivative of the flux—the flux can be changed by changing the magnetic induction or by changing the shape or orientation or position of the circuit. In form (6.4) we have a far-reaching generalization of Faraday's law. The circuit C can be thought of as any closed geometrical path in space, not necessarily coincident with an electric circuit. Then (6.4) becomes a relation between the fields themselves. It is important to note, however, that the electric field, $\mathbf{E}'$ is the electric field at $d\mathbf{l}$ in the coordinate system in which $d\mathbf{l}$ is at rest, since it is that field which causes current to flow if a circuit is actually present.

If the circuit C is moving with a velocity $\mathbf{v}$ in some direction, as shown in Fig. 6.2, the total time derivative in (6.4) must take into account this motion. The flux through the circuit may change because (a) the flux changes with time at a point, or (b) the translation of the circuit changes the location of the boundary. It is easy to show that the result for the total time derivative of flux through the moving circuit is*

$$\frac{d}{dt}\int_{S} \mathbf{B} \cdot \mathbf{n}\, da = \int_{S} \frac{\partial \mathbf{B}}{\partial t} \cdot \mathbf{n}\, da + \oint_{C} (\mathbf{B} \times \mathbf{v}) \cdot d\mathbf{l} \qquad (6.5)$$

Equation (6.4) can now be written in the form,

$$\oint_{C} [\mathbf{E}' - k(\mathbf{v} \times \mathbf{B})] \cdot d\mathbf{l} = -k \int_{S} \frac{\partial \mathbf{B}}{\partial t} \cdot \mathbf{n}\, da \qquad (6.6)$$

This is an equivalent statement of Faraday's law applied to the moving circuit C. But we can choose to interpret it differently. We can think of the circuit C and surface S as instantaneously at a certain position in space in the laboratory. Applying Faraday's law (6.4) to that fixed circuit, we find

$$\oint_{C} \mathbf{E} \cdot d\mathbf{l} = -k \int_{S} \frac{\partial \mathbf{B}}{\partial t} \cdot \mathbf{n}\, da \qquad (6.7)$$

where $\mathbf{E}$ is now the electric field in the laboratory. The assumption of Galilean invariance implies that the left-hand sides of (6.6) and (6.7) must be equal. This means that the electric field $\mathbf{E}'$ in the moving coordinate system of the circuit is

$$\mathbf{E}' = \mathbf{E} + k(\mathbf{v} \times \mathbf{B}) \qquad (6.8)$$

To determine the constant k we merely observe the significance of $\mathbf{E}'$. A charged particle (e.g., one of the conduction electrons) in a moving circuit

Fig. 6.2

* For a general vector field there is an added term, $\displaystyle\int_{S} (\boldsymbol{\nabla} \cdot \mathbf{B})\mathbf{v} \cdot \mathbf{n}\, da$, which gives the contribution of the sources of the vector field swept over by the moving circuit. The general result follows most easily from the use of the convective derivative,

$$\frac{d}{dt} = \frac{\partial}{\partial t} + \mathbf{v} \cdot \boldsymbol{\nabla}$$

will experience a force $q\mathbf{E}'$. When viewed from the laboratory, the charge represents a current $\mathbf{J} = q\mathbf{v}\, \delta(\mathbf{x} - \mathbf{x}_0)$. From the magnetic force law (5.7) or (5.12) it is evident that this current experiences a force in agreement with (6.8) provided the constant k is equal to c^{-1}.

We have thus reached the conclusion that Faraday's law takes the form

$$\oint_C \mathbf{E}' \cdot d\mathbf{l} = -\frac{1}{c}\frac{d}{dt}\int_S \mathbf{B}\cdot \mathbf{n}\, da \qquad (6.9)$$

where $\mathbf{E}'$ is the electric field at $d\mathbf{l}$ in its rest frame of coordinates. The time derivative on the right is a *total* time derivative. If the circuit C is moving with a velocity $\mathbf{v}$, the electric field in the moving frame is

$$\mathbf{E}' = \mathbf{E} + \frac{1}{c}(\mathbf{v}\times\mathbf{B}) \qquad (6.10)$$

These considerations are valid only for nonrelativistic velocities. Galilean invariance is not rigorously valid, but holds only for relative velocities small compared to the velocity of light. Expression (6.10) is correct to *first* order in v/c, but in error by terms of order v^2/c^2 (see Section 11.10). Evidently, for laboratory experiments with macroscopic circuits, (6.9) and (6.10) are completely adequate.

Faraday's law (6.9) can be put in differential form by use of Stokes's theorem, provided the circuit is held fixed in the chosen reference frame (in order to have $\mathbf{E}$ and $\mathbf{B}$ defined in the *same* frame). The transformation of the electromotive force integral into a surface integral leads to

$$\int_S \left(\nabla\times\mathbf{E} + \frac{1}{c}\frac{\partial\mathbf{B}}{\partial t}\right)\cdot\mathbf{n}\, da = 0$$

Since the circuit C and bounding surface S are arbitrary, the integrand must vanish at all points in space.

Thus the differential form of Faraday's law is

$$\nabla\times\mathbf{E} + \frac{1}{c}\frac{\partial\mathbf{B}}{\partial t} = 0 \qquad (6.11)$$

We note that this is the time-dependent generalization of the statement, $\nabla\times\mathbf{E} = 0$, for electrostatic fields.

6.2   Energy in the Magnetic Field

In discussing steady-state magnetic fields in Chapter 5 we avoided the question of field energy and energy density. The reason was that the creation of a steady-state configuration of currents and associated magnetic fields involves an initial transient period during which the currents and fields are brought from zero to their final values. For such time-varying fields there are induced electromotive forces which cause the sources of current to do work. Since the energy in the field is by definition the total work done to establish it, we must consider these contributions.

Suppose for a moment that we have only a single circuit with a constant current I flowing in it. If the flux through the circuit changes, an electromotive force $\mathscr{E}$ is induced around it. In order to keep the current constant, the sources of current must do work at the rate,

$$\frac{dW}{dt} = -I\mathscr{E} = \frac{1}{c}I\frac{dF}{dt}$$

This is in addition to ohmic losses in the circuit which are not to be included in the magnetic-energy content. Thus, if the flux change through a circuit carrying a current I is δF, the work done by the sources is

$$\delta W = \frac{1}{c}I\,\delta F$$

Now we consider the problem of the work done in establishing a general steady-state distribution of currents and fields. We can imagine that the build-up process occurs at an infinitesimal rate so that $\nabla \cdot \mathbf{J} = 0$ holds to any desired degree of accuracy. Then the current distribution can be broken up into a network of elementary current loops, the typical one of which is an elemental tube of current of cross-sectional area $\Delta\sigma$ following a closed path C and spanned by a surface S with normal $\mathbf{n}$, as shown in Fig. 6.3.

We can express the increment of work done against the induced emf in terms of the change in magnetic induction through the loop:

$$\Delta(\delta W) = \frac{J\,\Delta\sigma}{c}\int_{S}\mathbf{n}\cdot\delta\mathbf{B}\,da$$

where the extra Δ comes from the fact that we are considering only one elemental circuit. If we express $\mathbf{B}$ in terms of the vector potential $\mathbf{A}$, then we have

$$\Delta(\delta W) = \frac{J\,\Delta\sigma}{c}\int_{S}(\nabla\times\delta\mathbf{A})\cdot\mathbf{n}\,da$$

With application of Stokes's theorem this can be written

$$\Delta(\delta W) = \frac{J\,\Delta\sigma}{c}\oint_{C}\delta\mathbf{A}\cdot d\mathbf{l}$$

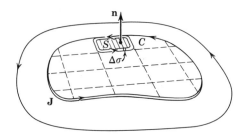

Fig. 6.3 Distribution of current density broken up into elemental current loops.

But $J \, \Delta\sigma \, d\mathbf{l}$ is equal to $\mathbf{J} \, d^3x$, by definition, since $d\mathbf{l}$ is parallel to $\mathbf{J}$. Evidently the sum over all such elemental loops will be the volume integral. Hence the total increment of work done by the external sources due to a change $\delta\mathbf{A}(\mathbf{x})$ in the vector potential is

$$\delta W = \frac{1}{c} \int \delta\mathbf{A} \cdot \mathbf{J} \, d^3x \qquad (6.12)$$

An expression involving the magnetic fields rather than $\mathbf{J}$ and $\delta\mathbf{A}$ can be obtained by using Ampère's law:

$$\nabla \times \mathbf{H} = \frac{4\pi}{c} \mathbf{J}$$

Then

$$\delta W = \frac{1}{4\pi} \int \delta\mathbf{A} \cdot (\nabla \times \mathbf{H}) \, d^3x \qquad (6.13)$$

The vector identity,

$$\nabla \cdot (\mathbf{P} \times \mathbf{Q}) = \mathbf{Q} \cdot (\nabla \times \mathbf{P}) - \mathbf{P} \cdot (\nabla \times \mathbf{Q})$$

can be used to transform (6.13):

$$\delta W = \frac{1}{4\pi} \int [\mathbf{H} \cdot (\nabla \times \delta\mathbf{A}) + \nabla \cdot (\mathbf{H} \times \delta\mathbf{A})] \, d^3x \qquad (6.14)$$

If the field distribution is assumed to be localized, the second integral vanishes. With the definition of $\mathbf{B}$ in terms of $\mathbf{A}$, the energy increment can be written:

$$\delta W = \frac{1}{4\pi} \int \mathbf{H} \cdot \delta\mathbf{B} \, d^3x \qquad (6.15)$$

This relation is the magnetic equivalent of the electrostatic equation (4.89). In its present form it is applicable to all magnetic media, including ferromagnetic substances. If we assume that the medium is para- or diamagnetic, so that a linear relation exists between $\mathbf{H}$ and $\mathbf{B}$, then

$$\mathbf{H} \cdot \delta\mathbf{B} = \tfrac{1}{2} \, \delta(\mathbf{H} \cdot \mathbf{B})$$

If we now bring the fields up from zero to their final values, the total magnetic energy will be

$$W = \frac{1}{8\pi} \int \mathbf{H} \cdot \mathbf{B} \, d^3x \qquad (6.16)$$

This is the magnetic analog of (4.92).

The magnetic equivalent of (4.86) where the electrostatic energy is expressed in terms of charge density and potential, can be obtained from (6.12) by assuming a linear relation between $\mathbf{J}$ and $\mathbf{A}$. Then we find the magnetic energy to be

$$W = \frac{1}{2c} \int \mathbf{J} \cdot \mathbf{A} \, d^3x \qquad (6.17)$$

The magnetic problem of the change in energy when an object of permeability μ_1 is placed in a magnetic field whose current sources are fixed can be treated in close analogy with the electrostatic problem of Section 4.8. The role of $\mathbf{E}$ is taken by $\mathbf{B}$, that of $\mathbf{D}$ by $\mathbf{H}$. The original medium has permeability μ_0 and existing magnetic induction $\mathbf{B}_0$. After the object is in place the fields are $\mathbf{B}$ and $\mathbf{H}$. It is left as an exercise for the reader to verify that for fixed sources of the field the change in energy is

$$W = \frac{1}{8\pi} \int_{V_1} (\mathbf{B} \cdot \mathbf{H}_0 - \mathbf{H} \cdot \mathbf{B}_0) \, d^3x \qquad (6.18)$$

where the integration is over the volume of the object. This can be written in the alternative forms:

$$W = \frac{1}{8\pi} \int_{V_1} (\mu_1 - \mu_0)\mathbf{H} \cdot \mathbf{H}_0 \, d^3x = \frac{1}{8\pi} \int_{V_1} \left(\frac{1}{\mu_0} - \frac{1}{\mu_1}\right) \mathbf{B} \cdot \mathbf{B}_0 \, d^3x \qquad (6.19)$$

Both μ_1 and μ_0 can be functions of position, but they are assumed independent of field strength.

If the object is in otherwise free space ($\mu_0 = 1$), the change in energy can be expressed in terms of the magnetization as

$$W = \tfrac{1}{2} \int_{V_1} \mathbf{M} \cdot \mathbf{B}_0 \, d^3x \qquad (6.20)$$

It should be noted that (6.20) is equivalent to the electrostatic result (4.96), except for sign. This sign change arises because the energy W consists of the total energy change occurring when the permeable body is introduced in the field, including the work done by the sources against the induced electromotive forces. In this respect the magnetic problem with fixed currents is analogous to the electrostatic problem with fixed potentials on the surfaces which determine the fields. By an analysis equivalent to that at the end of Section 4.8 we can show that for a small displacement the work done against the induced emf's is twice as large as, and of the opposite sign to, the potential-energy change of the body. Thus, to find the force acting on the body under a generalized displacement ξ, we calculate the *positive* derivative of W with respect to the displacement:

$$F_\xi = \left(\frac{\partial W}{\partial \xi}\right)_J \tag{6.21}$$

The subscript J implies fixed source currents.

The difference between (6.20) and the potential energy (5.73) for a permanent magnetic moment in an external field (apart from the factor $\frac{1}{2}$, which is traced to the linear relation assumed between **M** and **B**) comes from the fact that (6.20) is the total energy required to produce the configuration, whereas (5.73) includes only the work done in establishing the permanent magnetic moment in the field, not the work done in creating the magnetic moment and keeping it permanent.

6.3   Maxwell's Displacement Current;   Maxwell's Equations

The basic laws of electricity and magnetism which we have discussed so far can be summarized in differential form by these four equations:

Coulomb's law: $\qquad\qquad\qquad \nabla \cdot \mathbf{D} = 4\pi\rho$

Ampère's law: $\qquad\qquad\qquad \nabla \times \mathbf{H} = \dfrac{4\pi}{c}\,\mathbf{J}$

Faraday's law: $\qquad\qquad \nabla \times \mathbf{E} + \dfrac{1}{c}\dfrac{\partial \mathbf{B}}{\partial t} = 0$

Absence of free magnetic poles: $\qquad \nabla \cdot \mathbf{B} = 0$

$$\tag{6.22}$$

These equations are written in macroscopic form and in Gaussian units. Let us recall that all but Faraday's law were derived from steady-state observations. Consequently, from a logical point of view there is no *a priori* reason to expect that the static equations hold unchanged for time-dependent fields. In fact, the equations in set (6.22) are inconsistent as they stand.

It required the genius of J. C. Maxwell, spurred on by Faraday's observations, to see the inconsistency in equations (6.22) and to modify them into a consistent set which implied new physical phenomena, at that time unknown but subsequently verified in all details by experiment. For this brilliant stroke in 1865, the modified set of equations is justly known as Maxwell's equations.

The faulty equation is Ampère's law. It was derived for steady-state current phenomena with $\nabla \cdot \mathbf{J} = 0$. This requirement on the divergence of $\mathbf{J}$ is contained right in Ampère's law, as can be seen by taking the divergence of both sides:

$$\frac{4\pi}{c} \nabla \cdot \mathbf{J} = \nabla \cdot (\nabla \times \mathbf{H}) \equiv 0 \tag{6.23}$$

While $\nabla \cdot \mathbf{J} = 0$ is valid for steady-state problems, the complete relation is given by the continuity equation for charge and current:

$$\nabla \cdot \mathbf{J} + \frac{\partial \rho}{\partial t} = 0 \tag{6.24}$$

What Maxwell saw was that the continuity equation could be converted into a vanishing divergence by using Coulomb's law (6.22). Thus

$$\nabla \cdot \mathbf{J} + \frac{\partial \rho}{\partial t} = \nabla \cdot \left(\mathbf{J} + \frac{1}{4\pi} \frac{\partial \mathbf{D}}{\partial t} \right) = 0 \tag{6.25}$$

Then Maxwell replaced $\mathbf{J}$ in Ampère's law by its generalization,

$$\mathbf{J} \to \mathbf{J} + \frac{1}{4\pi} \frac{\partial \mathbf{D}}{\partial t} \tag{6.26}$$

for time-dependent fields. Thus Ampère's law became

$$\nabla \times \mathbf{H} = \frac{4\pi}{c} \mathbf{J} + \frac{1}{c} \frac{\partial \mathbf{D}}{\partial t} \tag{6.27}$$

still the same, experimentally verified, law for steady-state phenomena, but now mathematically consistent with the continuity equation (6.24) for time-dependent fields. Maxwell called the added term in (6.26) the *displacement current*. This necessary addition to Ampère's law is of crucial importance for rapidly fluctuating fields. Without it there would be no electromagnetic radiation, and the greatest part of the remainder of this book would have to be omitted. It was Maxwell's prediction that light was an electromagnetic wave phenomenon, and that electromagnetic waves of all frequencies could be produced, which drew the attention of all physicists and stimulated so much theoretical and experimental research into electromagnetism during the last part of the nineteenth century.

The set of four equations,

$$\left. \begin{array}{ll} \nabla \cdot \mathbf{D} = 4\pi\rho & \nabla \times \mathbf{H} = \dfrac{4\pi}{c} \mathbf{J} + \dfrac{1}{c} \dfrac{\partial \mathbf{D}}{\partial t} \\[3mm] \nabla \cdot \mathbf{B} = 0 & \nabla \times \mathbf{E} + \dfrac{1}{c} \dfrac{\partial \mathbf{B}}{\partial t} = 0 \end{array} \right\} \tag{6.28}$$

known as Maxwell's equations, forms the basis of all electromagnetic phenomena. When combined with the Lorentz force equation and Newton's second law of motion, these equations provide a complete description of the classical dynamics of interacting charged particles and electromagnetic fields (see Section 6.9 and Chapters 10 and 12). For macroscopic media the dynamical response of the aggregates of atoms is summarized in the *constitutive relations* which connect **D** and **J** with **E**, and **H** with **B** (e.g., $\mathbf{D} = \epsilon\mathbf{E}$, $\mathbf{J} = \sigma\mathbf{E}$, $\mathbf{B} = \mu\mathbf{H}$ for an isotropic, permeable, conducting dielectric).

The units employed in writing Maxwell's equations (6.28) are those of the previous chapters, namely, Gaussian. For the reader more at home in other units, such as mks, Table 2 of the Appendix summarizes essential equations in the commoner systems. Table 3 of the Appendix allows the conversion of any equation from Gaussian to mks units, while Table 4 gives the corresponding conversions for given amounts of any variable.

6.4   Vector and Scalar Potentials

Maxwell's equations consist of a set of coupled first-order partial differential equations relating the various components of electric and magnetic fields. They can be solved as they stand in simple situations. But it is often convenient to introduce potentials, obtaining a smaller number of second-order equations, while satisfying some of Maxwell's equations identically. We are already familiar with this concept in electrostatics and magnetostatics, where we used the scalar potential Φ and the vector potential **A**.

Since $\nabla \cdot \mathbf{B} = 0$ still holds, we can define **B** in terms of a vector potential:

$$\mathbf{B} = \nabla \times \mathbf{A} \tag{6.29}$$

Then the other homogeneous equation in (6.28), Faraday's law, can be written

$$\nabla \times \left(\mathbf{E} + \frac{1}{c}\frac{\partial \mathbf{A}}{\partial t} \right) = 0 \tag{6.30}$$

This means that the quantity with vanishing curl in (6.30) can be written as the gradient of some scalar function, namely, a scalar potential Φ:

or

$$\left.\begin{aligned} \mathbf{E} + \frac{1}{c}\frac{\partial \mathbf{A}}{\partial t} &= -\nabla\Phi \\[2mm] \mathbf{E} &= -\nabla\Phi - \frac{1}{c}\frac{\partial \mathbf{A}}{\partial t} \end{aligned}\right\} \tag{6.31}$$

The definition of **B** and **E** in terms of the potentials **A** and Φ according to (6.29) and (6.31) satisfies identically the two homogeneous Maxwell's equations. The dynamic behavior of **A** and Φ will be determined by the two inhomogeneous equations in (6.28).

At this stage it is convenient to restrict our considerations to the microscopic form of Maxwell's equations. Then the inhomogeneous equations in (6.28) can be written in terms of the potentials as

$$\nabla^2\Phi + \frac{1}{c}\frac{\partial}{\partial t}(\nabla \cdot \mathbf{A}) = -4\pi\rho \qquad (6.32)$$

$$\nabla^2\mathbf{A} - \frac{1}{c^2}\frac{\partial^2\mathbf{A}}{\partial t^2} - \nabla\left(\nabla \cdot \mathbf{A} + \frac{1}{c}\frac{\partial\Phi}{\partial t}\right) = -\frac{4\pi}{c}\mathbf{J} \qquad (6.33)$$

We have now reduced the set of four Maxwell's equations to two equations. But they are still coupled equations. The uncoupling can be accomplished by exploiting the arbitrariness involved in the definition of the potentials. Since **B** is defined through (6.29) in terms of **A**, the vector potential is arbitrary to the extent that the gradient of some scalar function Λ can be added. Thus **B** is left unchanged by the transformation,

$$\mathbf{A} \to \mathbf{A}' = \mathbf{A} + \nabla\Lambda \qquad (6.34)$$

In order that the electric field (6.31) be unchanged as well, the scalar potential must be simultaneously transformed,

$$\Phi \to \Phi' = \Phi - \frac{1}{c}\frac{\partial\Lambda}{\partial t} \qquad (6.35)$$

The freedom implied by (6.34) and (6.35) means that we can choose a set of potentials (**A**, Φ) such that

$$\nabla \cdot \mathbf{A} + \frac{1}{c}\frac{\partial\Phi}{\partial t} = 0 \qquad (6.36)$$

This will uncouple the pair of equations (6.32) and (6.33) and leave two inhomogeneous wave equations, one for Φ and one for **A**:

$$\nabla^2\Phi - \frac{1}{c^2}\frac{\partial^2\Phi}{\partial t^2} = -4\pi\rho \qquad (6.37)$$

$$\nabla^2\mathbf{A} - \frac{1}{c^2}\frac{\partial^2\mathbf{A}}{\partial t^2} = -\frac{4\pi}{c}\mathbf{J} \qquad (6.38)$$

Equations (6.37) and (6.38), plus (6.36), form a set of equations equivalent in all respects to Maxwell's equations.

6.5 Gauge Transformations; Lorentz Gauge; Coulomb Gauge

The transformation (6.34) and (6.35) is called a *gauge transformation*, and the invariance of the fields under such transformations is called *gauge invariance*. The relation (6.36) between **A** and Φ is called the *Lorentz condition*. To see that potentials can always be found to satisfy the Lorentz condition, suppose that the potentials **A**, Φ which satisfy (6.32) and (6.33) do not satisfy (6.36). Then let us make a gauge transformation to potentials **A**′, Φ′ and demand that **A**′, Φ′ satisfy the Lorentz condition:

$$\nabla \cdot \mathbf{A}' + \frac{1}{c}\frac{\partial \Phi'}{\partial t} = 0 = \nabla \cdot \mathbf{A} + \frac{1}{c}\frac{\partial \Phi}{\partial t} + \nabla^2 \Lambda - \frac{1}{c^2}\frac{\partial^2 \Lambda}{\partial t^2} \tag{6.39}$$

Thus, provided a gauge function Λ can be found to satisfy

$$\nabla^2 \Lambda - \frac{1}{c^2}\frac{\partial^2 \Lambda}{\partial t^2} = -\left(\nabla \cdot \mathbf{A} + \frac{1}{c}\frac{\partial \Phi}{\partial t}\right) \tag{6.40}$$

the new potentials **A**′, Φ′ will satisfy the Lorentz condition and the wave equations (6.37) and (6.38).

Even for potentials which satisfy the Lorentz condition (6.36) there is arbitrariness. Evidently the *restricted gauge transformation*,

$$\left.\begin{array}{l} \mathbf{A} \rightarrow \mathbf{A} + \nabla\Lambda \\[2mm] \Phi \rightarrow \Phi - \dfrac{1}{c}\dfrac{\partial \Lambda}{\partial t} \end{array}\right\} \tag{6.41}$$

where

$$\nabla^2 \Lambda - \frac{1}{c^2}\frac{\partial^2 \Lambda}{\partial t^2} = 0 \tag{6.42}$$

preserves the Lorentz condition, provided **A**, Φ satisfy it initially. All potentials in this restricted class are said to belong to the *Lorentz gauge*. The Lorentz gauge is commonly used, first because it leads to the wave equations (6.37) and (6.38) which treat Φ and **A** on equivalent footings, and second because it is a concept which is independent of the coordinate system chosen and so fits naturally into the considerations of special relativity (see Section 11.9).

Another useful gauge for the potentials is the so-called *Coulomb* or *transverse gauge*. This is the gauge in which

$$\nabla \cdot \mathbf{A} = 0 \tag{6.43}$$

From (6.32) we see that the scalar potential satisfies Poisson's equation,

$$\nabla^2 \Phi = -4\pi\rho \tag{6.44}$$

with solution,

$$\Phi(\mathbf{x}, t) = \int \frac{\rho(\mathbf{x}', t)}{|\mathbf{x} - \mathbf{x}'|} \, d^3x' \tag{6.45}$$

The scalar potential is just the *instantaneous* Coulomb potential due to the charge density $\rho(\mathbf{x}, t)$. This is the origin of the name "Coulomb gauge."

The vector potential satisfies the inhomogeneous wave equation,

$$\nabla^2 \mathbf{A} - \frac{1}{c^2} \frac{\partial^2 \mathbf{A}}{\partial t^2} = -\frac{4\pi}{c} \mathbf{J} + \frac{1}{c} \nabla \frac{\partial \Phi}{\partial t} \tag{6.46}$$

The "current" term involving the potential can, in principle, be calculated from (6.45). Formally, we use the continuity equation to write

$$\nabla \frac{\partial \Phi}{\partial t} = -\nabla \int \frac{\nabla' \cdot \mathbf{J}(\mathbf{x}', t)}{|\mathbf{x} - \mathbf{x}'|} \, d^3x' \tag{6.47}$$

If the current is written as the sum of a longitudinal and transverse part,

$$\mathbf{J} = \mathbf{J}_l + \mathbf{J}_t \tag{6.48}$$

where $\nabla \times \mathbf{J}_l = 0$ and $\nabla \cdot \mathbf{J}_t = 0$, then the parts can be written

$$\mathbf{J}_l = -\frac{1}{4\pi} \nabla \int \frac{\nabla' \cdot \mathbf{J}}{|\mathbf{x} - \mathbf{x}'|} \, d^3x' \tag{6.49}$$

$$\mathbf{J}_t = \frac{1}{4\pi} \nabla \times \nabla \times \int \frac{\mathbf{J}}{|\mathbf{x} - \mathbf{x}'|} \, d^3x' \tag{6.50}$$

This can be proved by using the vector identity, $\nabla \times (\nabla \times \mathbf{J}) = \nabla(\nabla \cdot \mathbf{J}) - \nabla^2 \mathbf{J}$, together with $\nabla^2(1/|\mathbf{x} - \mathbf{x}'|) = -4\pi \, \delta(\mathbf{x} - \mathbf{x}')$. Comparison of (6.47) with (6.49) shows that

$$\nabla \frac{\partial \Phi}{\partial t} = 4\pi \mathbf{J}_l \tag{6.51}$$

Therefore the source for the wave equation for $\mathbf{A}$ can be expressed entirely in terms of the *transverse* current (6.50):

$$\nabla^2 \mathbf{A} - \frac{1}{c^2} \frac{\partial^2 \mathbf{A}}{\partial t^2} = -\frac{4\pi}{c} \mathbf{J}_t \tag{6.52}$$

This is, of course, the origin of the name "transverse gauge."

The Coulomb or transverse gauge is often used when no sources are present. Then $\Phi = 0$, and $\mathbf{A}$ satisfies the homogeneous wave equation. The fields are given by

$$
\left.
\begin{aligned}
\mathbf{E} &= -\frac{1}{c}\frac{\partial \mathbf{A}}{\partial t} \\[2mm]
\mathbf{B} &= \nabla \times \mathbf{A}
\end{aligned}
\right\}
\tag{6.53}
$$

6.6   Green's Function for the Time-Dependent Wave Equation

The wave equations (6.37), (6.38), and (6.52) all have the basic structure,

$$
\nabla^2 \psi - \frac{1}{c^2}\frac{\partial^2 \psi}{\partial t^2} = -4\pi f(\mathbf{x}, t)
\tag{6.54}
$$

where $f(\mathbf{x}, t)$ is a known source distribution. The factor c is the velocity of propagation in the medium.

To solve (6.54) it is useful to find a Green's function for the equation, just as in electrostatics. Since the time is involved, the Green's function will depend on the variables $(\mathbf{x}, \mathbf{x}', t, t\,)$, and will satisfy the equation,

$$
\left(\nabla_x^2 - \frac{1}{c^2}\frac{\partial^2}{\partial t^2}\right)G(\mathbf{x}, t; \mathbf{x}', t') = -4\pi\,\delta(\mathbf{x} - \mathbf{x}')\,\delta(t - t')
\tag{6.55}
$$

Then in infinite space with no boundary surfaces the solution of (6.54) will be

$$
\psi(\mathbf{x}, t) = \int G(\mathbf{x}, t; \mathbf{x}', t')f(\mathbf{x}', t')\, d^3x'\, dt'
\tag{6.56}
$$

Of course, the Green's function will have to satisfy certain boundary conditions demanded by physical considerations.

The basic Green's function satisfying (6.55) is a function only of the differences in coordinates $(\mathbf{x} - \mathbf{x}')$ and times $(t - t')$. To find G we consider the Fourier transform of both sides of (6.55). The delta functions on the right have the representation,

$$
\delta(\mathbf{x} - \mathbf{x}')\,\delta(t - t') = \frac{1}{(2\pi)^4}\int d^3k \int d\omega\, e^{i\mathbf{k}\cdot(\mathbf{x}-\mathbf{x}')} e^{-i\omega(t-t')}
\tag{6.57}
$$

We therefore write the representation of G as

$$
G(\mathbf{x}, t; \mathbf{x}', t') = \int d^3k \int d\omega\, g(\mathbf{k}, \omega) e^{i\mathbf{k}\cdot(\mathbf{x}-\mathbf{x}')} e^{-i\omega(t-t')}
\tag{6.58}
$$

Classical Electrodynamics

The Fourier transform $g(\mathbf{k}, \omega)$ is to be determined. When (6.57) and (6.58) are substituted into the defining equation (6.55), it turns out that $g(\mathbf{k}, \omega)$ is

$$g(\mathbf{k}, \omega) = \frac{1}{4\pi^3} \frac{1}{\left(k^2 - \dfrac{\omega^2}{c^2}\right)} \qquad (6.59)$$

When $g(\mathbf{k}, \omega)$ is substituted into (6.58) and the integrations over $\mathbf{k}$ and ω are begun, there appears a singularity in the integrand at $k^2 = \omega^2/c^2$. Consequently solution (6.59) is meaningless without some rule as to how to handle the singularities. The rule cannot come from the mathematics. It must come from physical considerations. The Green's function satisfying (6.55) represents the wave disturbance caused by a point source at $\mathbf{x}'$ which is turned on only for an infinitesimal time interval at $t' = t$. We know that such a wave disturbance propagates outwards as a spherically diverging wave with a velocity c. Hence we demand that our solution for G have the following properties:

(a) $G = 0$ everywhere for $t < t'$.

(b) G represent outgoing waves for $t > t'$.

If we think of the ω integration in (6.58), the singularities in $g(\mathbf{k}, \omega)$ occur at $\omega = \pm ck$. We can do the ω integration as a Cauchy integral in the complex ω plane. For $t > t'$ the integral along the real axis in (6.58) is equivalent to the contour integral around a path C closed in the lower half-plane, since the contribution on the semicircle at infinity vanishes exponentially. On the other hand, for $t' > t$, the contour must be closed in the upper half-plane, as shown in Fig. 6.4 by path C'.

In order to make G vanish for $t < t'$ we must imagine that the poles at $\omega = \pm ck$ are displaced below the real axis, as in Fig. 6.4. Then the integral over C for $t > t'$ will give a nonvanishing contribution, while the integral

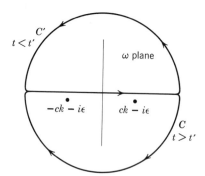

Fig. 6.4 Complex ω plane with contour C for $t > t'$ and contour C' for $t < t'$.

over C' for $t < t'$ will vanish. The displacement of the poles can be accomplished mathematically by writing $(\omega + i\epsilon)$ in place of ω in (6.59). Then the Green's function is given by

$$G(\mathbf{x}, t; \mathbf{x}', t') = \frac{1}{4\pi^3} \int d^3k \int d\omega \, \frac{e^{i\mathbf{k}\cdot\mathbf{R} - i\omega\tau}}{k^2 - \frac{1}{c^2}(\omega + i\epsilon)^2} \qquad (6.60)$$

where $\mathbf{R} = \mathbf{x} - \mathbf{x}'$, $\tau = t - t'$, and ϵ is a positive infinitesimal.

The integration over ω for $\tau > 0$ can be done with Cauchy's theorem applied around the contour C of Fig. 6.4, giving

$$G = \frac{c}{2\pi^2} \int d^3k \, e^{i\mathbf{k}\cdot\mathbf{R}} \frac{\sin(c\tau k)}{k} \qquad (6.61)$$

The integration over d^3k can be accomplished by first integrating over angles. Then

$$G = \frac{2c}{\pi R} \int_0^\infty dk \, \sin(kR) \sin(c\tau k) \qquad (6.62)$$

Since the integrand is even in k, the integral can be written over the whole interval, $-\infty < k < \infty$. With a change of variable $x = ck$, (6.62) can be written

$$G = \frac{1}{2\pi R} \int_{-\infty}^\infty dx (e^{i[\tau - (R/c)]x} - e^{i[\tau + (R/c)]x}) \qquad (6.63)$$

From (2.52) we see that the integrals are just Dirac delta functions. The argument of the second one never vanishes (remember, $\tau > 0$). Hence only the first integral contributes, and the Green's function is

$$G = \frac{1}{R} \delta\left(\tau - \frac{R}{c}\right)$$

or, more explicitly,

$$G(\mathbf{x}, t; \mathbf{x}', t') = \frac{\delta\left(t' + \frac{|\mathbf{x} - \mathbf{x}'|}{c} - t\right)}{|\mathbf{x} - \mathbf{x}'|} \qquad (6.64)$$

This Green's function is sometimes called the *retarded Green's function* because it exhibits the causal behavior associated with a wave disturbance. The effect observed at the point $\mathbf{x}$ at time t is due to a disturbance which originated at an earlier or *retarded* time $t' = t - \dfrac{|\mathbf{x} - \mathbf{x}'|}{c}$ at the point $\mathbf{x}'$.

The solution for the wave equation (6.54) in the absence of boundaries is

$$\psi(\mathbf{x}, t) = \int \frac{\delta\left(t' + \dfrac{|\mathbf{x} - \mathbf{x}'|}{c} - t\right)}{|\mathbf{x} - \mathbf{x}'|} f(\mathbf{x}', t')\, d^3x'\, dt' \qquad (6.65)$$

The integration over dt' can be performed to yield the so-called "retarded solution,"

$$\psi(\mathbf{x}, t) = \int \frac{[f(\mathbf{x}', t')]_{\text{ret}}}{|\mathbf{x} - \mathbf{x}'|}\, d^3x' \qquad (6.66)$$

The square bracket $[\ ]_{\text{ret}}$ means that the time t' is to be evaluated at the retarded time, $t' = t - \dfrac{|\mathbf{x} - \mathbf{x}'|}{c}$.

6.7 Initial-Value Problem; Kirchhoff's Surface-Integral Representation

Solution (6.66) is a particular integral of the inhomogeneous wave equation (6.54). To it can be added any solution of the homogeneous wave equation necessary to satisfy the boundary conditions. From the table at the end of Section 1.9 we see that the proper boundary conditions are Cauchy boundary conditions (ψ and $\partial\psi/\partial n$ given) on an "open surface." For the three-dimensional wave equation an open surface is defined as a three-dimensional volume specified by one functional relationship between the four coordinates (x, y, z, t). The customary open surface is ordinary three-dimensional space at a fixed time, $t = t_0$. Then the problem is an initial-value problem with $\psi(\mathbf{x}, t_0)$ and $\partial\psi(\mathbf{x}, t_0)/\partial t$ given for all $\mathbf{x}$. We wish to determine $\psi(\mathbf{x}, t)$ for all times $t > t_0$.

To discuss the initial-value problem and also an integral representation of Kirchhoff for closed bounding surfaces, we use Green's theorem (1.35), integrated over time from $t' = t_0$ to $t' = t_1 > t$:

$$\int_{t_0}^{t_1} dt' \int_V d^3x'(\phi \nabla'^2 \psi - \psi \nabla'^2 \phi) = \int_{t_0}^{t_1} dt' \oint_S \left(\phi \frac{\partial\psi}{\partial n'} - \psi \frac{\partial\phi}{\partial n'}\right) da' \qquad (6.67)$$

We now choose $\psi = \psi$ and $\phi = G$. With wave equations (6.54) and (6.55) the left-hand side can be written

$$\text{L-HS} = \int_{t_0}^{t_1} dt' \int_V d^3x' \left[4\pi\psi(\mathbf{x}', t')\, \delta(\mathbf{x}' - \mathbf{x})\, \delta(t' - t) \right.$$

$$\left. - 4\pi f(\mathbf{x}', t')G + \frac{1}{c^2}\left(G \frac{\partial^2\psi}{\partial t'^2} - \psi \frac{\partial^2 G}{\partial t'^2}\right) \right] \qquad (6.68)$$

The first two terms in (6.68) will evidently give the particular integral (6.66). The last two terms can be integrated by parts with respect to the time to give

$$
\text{L-HS} = 4\pi\psi(\mathbf{x}, t) - 4\pi\int_{t_0}^{t_1} dt' \int_V d^3x' f(\mathbf{x}', t')G
$$

$$
+ \frac{1}{c^2}\int_V d^3x' \left(G\frac{\partial\psi}{\partial t'} - \psi\frac{\partial G}{\partial t'}\right)\bigg|_{t'=t_0}^{t'=t_1} \quad (6.69)
$$

Since $G = 0$ at $t' = t_1 > t$, the upper limit vanishes. We can thus combine (6.69) and (6.67) to give the integral representation for $\psi(\mathbf{x}, t)$ inside the volume V, bounded by the surface S, at times $t > t_0$:

$$
\psi(\mathbf{x}, t) = \int_V d^3x' \frac{[f(\mathbf{x}', t')]_{\text{ret}}}{|\mathbf{x} - \mathbf{x}'|} + \frac{1}{4\pi c^2}\int_V d^3x' \left(G\frac{\partial\psi}{\partial t'} - \psi\frac{\partial G}{\partial t'}\right)\bigg|_{t'=t_0}
$$

$$
+ \frac{1}{4\pi}\int_{t_0}^{t_1} dt' \oint_S da' \left(G\frac{\partial\psi}{\partial n'} - \psi\frac{\partial G}{\partial n'}\right) \quad (6.70)
$$

We have written the first term in (6.70) in the usual form (6.66) by using the explicit result (6.64) for G. We now do the same for the other terms.

For simplicity, first we consider the infinite-domain initial-value problem with ψ and $\partial\psi/\partial t$ as given functions of space at $t = t_0 = 0$:

$$
\psi(\mathbf{x}, 0) = F(\mathbf{x}), \quad \frac{\partial\psi}{\partial t}(\mathbf{x}, 0) = D(\mathbf{x}) \quad (6.71)
$$

Then the surface integral in (6.70) can be omitted. To simplify the notation we take the observation point at the origin and use spherical coordinates in the integrals. Then we have

$$
\psi(0, t) = \int d\Omega' \int_0^\infty r' \, dr' f\left(r', \Omega', t' = t - \frac{r'}{c}\right) + \frac{1}{4\pi c^2}\int d\Omega' \int_0^\infty r' \, dr'
$$

$$
\times \left[D(r', \Omega')\delta\left(t - \frac{r'}{c}\right) - F(r', \Omega)\frac{\partial}{\partial t'}\delta\left(t' + \frac{r'}{c} - t\right)\bigg|_{t'=0}\right] \quad (6.72)
$$

The derivative of the delta function can be written

$$
\frac{\partial}{\partial t'}\delta\left(t' + \frac{r'}{c} - t\right)\bigg|_{t'=0} = c^2\delta'(r' - ct)
$$

Then, with the properties of the delta function summarized in Section 1.2, (6.72) becomes

$$
\psi(0, t) = \int d\Omega' \int_0^\infty r' \, dr' f\left(r', \Omega', t' = t - \frac{r'}{c}\right)
$$

$$
+ \frac{1}{4\pi}\int d\Omega' \left[tD(ct, \Omega') + \frac{\partial}{\partial t}(tF(ct, \Omega'))\right] \quad (6.73)
$$

This is called *Poisson's solution* to the initial-value problem. With no sources present ($f = 0$), only values of the initial field at distances ct from the origin contribute at time t.

The initial-value problem for the wave equation has been extensively studied in one, two, three, and more dimensions. The reader is referred to Morse and Feshbach, pp. 843–847, and to the more mathematical treatment of Hadamard.

The other result which we wish to obtain from (6.70) is the so-called Kirchhoff representation of the field inside the volume V in terms of the values of ψ and its derivatives on the boundary surface S. We thus assume that there are no sources within V and that the initial values of ψ and $\partial\psi/\partial t$ vanish. (Alternatively, we can assume that the initial time is in the remote past so that there are no more contributions from the initial-value solution (6.73) within the volume V.) Then the field inside V is given by

$$\psi(\mathbf{x}, t) = \frac{1}{4\pi}\int_{t_0}^{t_1} dt' \oint_S da' \left(G \frac{\partial \psi}{\partial n'} - \psi \frac{\partial G}{\partial n'}\right) \tag{6.74}$$

With G given by (6.64) we can calculate $\partial G/\partial n'$:

$$\mathbf{\nabla}'G = \frac{\partial G}{\partial R}\mathbf{\nabla}'R = -\frac{\mathbf{R}}{R}\frac{\partial}{\partial R}\left[\frac{\delta\left(t' + \dfrac{R}{c} - t\right)}{R}\right]$$

$$= -\frac{\mathbf{R}}{R}\left\{-\frac{\delta\left(t' + \dfrac{R}{c} - t\right)}{R^2} + \frac{\delta'\left(t' + \dfrac{R}{c} - t\right)}{cR}\right\} \tag{6.75}$$

The term involving the derivative of the delta function can be integrated by parts with respect to the time t'. Then the Kirchhoff integral representation is

$$\psi(\mathbf{x}, t) = \frac{1}{4\pi}\oint_S \mathbf{n}\cdot\left[\frac{\mathbf{\nabla}'\psi(\mathbf{x}', t')}{R} - \frac{\mathbf{R}}{R^3}\psi(\mathbf{x}', t') - \frac{\mathbf{R}}{cR^2}\frac{\partial\psi(\mathbf{x}', t')}{dt'}\right]_{\text{ret}} da' \tag{6.76}$$

where $\mathbf{R} = \mathbf{x} - \mathbf{x}'$, and $\mathbf{n}$ is a unit normal to the surface S. We emphasize that (6.76) is *not* a solution for the field ψ, but only an integral representation in terms of its value and the values of its space and time derivatives on the surface. These cannot be specified arbitrarily; they are known only when the appropriate Cauchy boundary-value problem has been solved.

The Kirchhoff integral (6.76) is a mathematical statement of Huygens' principle and is used as the starting point in discussing optical-diffraction problems. Diffraction is discussed in detail in Chapter 9, Section 9.5 and below.

6.8   Poynting's Theorem

The forms of the laws of conservation of energy and momentum are important results to establish for the electromagnetic field. We begin by considering conservation of energy, often called *Poynting's theorem* (1884). For a single charge q the rate of doing work by external electromagnetic fields **E** and **B** is $q\mathbf{v} \cdot \mathbf{E}$, where **v** is the velocity of the charge. The magnetic field does no work, since the magnetic force is perpendicular to the velocity. If there exists a continuous distribution of charge and current, the total rate of doing work by the fields in a finite volume V is

$$\int_V \mathbf{J} \cdot \mathbf{E} \, d^3x \tag{6.77}$$

This power represents a conversion of electromagnetic energy into mechanical or thermal energy. It must be balanced by a corresponding rate of decrease of energy in the electromagnetic field within the volume V. In order to exhibit this conservation law explicitly, we will use Maxwell's equations to express (6.77) in other terms. Thus we use the Ampère-Maxwell law to eliminate **J**:

$$\int_V \mathbf{J} \cdot \mathbf{E} \, d^3x = \frac{1}{4\pi} \int_V \left[c\mathbf{E} \cdot (\nabla \times \mathbf{H}) - \mathbf{E} \cdot \frac{\partial \mathbf{D}}{\partial t} \right] d^3x \tag{6.78}$$

If we now employ the vector identity,

$$\nabla \cdot (\mathbf{E} \times \mathbf{H}) = \mathbf{H} \cdot (\nabla \times \mathbf{E}) - \mathbf{E} \cdot (\nabla \times \mathbf{H})$$

and use Faraday's law, the right side of (6.78) becomes

$$\int_V \mathbf{J} \cdot \mathbf{E} \, d^3x = \frac{-1}{4\pi} \int_V \left[c\nabla \cdot (\mathbf{E} \times \mathbf{H}) + \mathbf{E} \cdot \frac{\partial \mathbf{D}}{\partial t} + \mathbf{H} \cdot \frac{\partial \mathbf{B}}{\partial t} \right] d^3x \tag{6.79}$$

To proceed further we must make two assumptions. The first one is not fundamental, and is made for simplicity only. We assume that the macroscopic medium involved is *linear* in its electric and magnetic properties. Then the two time derivatives in (6.79) can be interpreted, according to equations (4.92) and (6.16), as the time derivatives of the electrostatic and magnetic energy densities. We now make our second assumption, namely, that the sum of (4.92) and (6.16) represents the total electromagnetic energy, even for time-varying fields. Then if the total energy density is denoted by

$$u = \frac{1}{8\pi} (\mathbf{E} \cdot \mathbf{D} + \mathbf{B} \cdot \mathbf{H}) \tag{6.80}$$

190 Classical Electrodynamics

(6.79) can be written

$$-\int_V \mathbf{J} \cdot \mathbf{E} \, d^3x = \int_V \left[\frac{\partial u}{\partial t} + \frac{c}{4\pi} \nabla \cdot (\mathbf{E} \times \mathbf{H}) \right] d^3x \qquad (6.81)$$

Since the volume V is arbitrary, this can be cast into the form of a differential continuity equation or conservation law,

$$\frac{\partial u}{\partial t} + \nabla \cdot \mathbf{S} = -\mathbf{J} \cdot \mathbf{E} \qquad (6.82)$$

The vector $\mathbf{S}$, representing energy flow, is called *Poynting's vector*. It is given by

$$\mathbf{S} = \frac{c}{4\pi} (\mathbf{E} \times \mathbf{H}) \qquad (6.83)$$

and has the dimensions of (energy/area $\times$ time). Since only its divergence appears in the conservation law, Poynting's vector is arbitrary to the extent that the curl of any vector field can be added to it. Such an added term can, however, have no physical consequences. Hence it is customary to make the specific choice (6.83).

The physical meaning of the integral or differential form (6.81) or (6.82) is that the time rate of change of electromagnetic energy within a certain volume, plus the energy flowing out through the boundary surfaces of the volume per unit time, is equal to the negative of the total work done by the fields on the sources within the volume. This is the statement of conservation of energy. If nonlinear effects, such as hysteresis in ferromagnetic materials, are envisioned, the simple law (6.82) is no longer valid, but must be supplemented by terms giving the hysteresis power loss.

6.9 Conservation Laws for a System of Charged Particles and Electromagnetic Fields

The statements (6.81) and (6.82) of Poynting's theorem have emphasized the energy of the electromagnetic fields. The work done per unit time per unit volume by the fields ($\mathbf{J} \cdot \mathbf{E}$) is a conversion of electromagnetic into mechanical or heat energy. Since matter is ultimately composed of charged particles (electrons and atomic nuclei), we can think of this rate of conversion as a rate of increase of energy of the charged particles per unit volume. Then we can interpret Poynting's theorem for the microscopic fields as a statement of conservation of energy of the combined system of particles and fields. If we denote the total energy of the particles within the volume V as E_{mech} and assume that no particles move out of the volume, we have

$$\frac{dE_{\text{mech}}}{dt} = \int_V \mathbf{J} \cdot \mathbf{E} \, d^3x \qquad (6.84)$$

Then Poynting's theorem expresses the conservation of energy for the combined system as

$$\frac{dE}{dt} = \frac{d}{dt} (E_{\text{mech}} + E_{\text{field}}) = -\oint_S \mathbf{n} \cdot \mathbf{S} \, da \qquad (6.85)$$

where the total field energy within V is

$$E_{\text{field}} = \int_V u \, d^3x = \frac{1}{8\pi} \int_V (\mathbf{E}^2 + \mathbf{B}^2) \, d^3x \qquad (6.86)$$

The conservation of linear momentum can be similarly considered. We have seen that the force on a charge q in an external field $\mathbf{E}$ is $q\mathbf{E}$. From the basic law (5.12) for forces on currents we can deduce that the magnetic force on a charge q moving with velocity $\mathbf{v}$ in an external magnetic induction $\mathbf{B}$ is $(q/c)\,\mathbf{v} \times \mathbf{B}$. Thus the total electromagnetic force on a charged particle is

$$\mathbf{F} = q\left(\mathbf{E} + \frac{\mathbf{v}}{c} \times \mathbf{B} \right) \qquad (6.87)$$

This is called the *Lorentz force*. Although we have deduced it within the framework of steady-state phenomena, it is well verified for all charged particles with arbitrarily large velocities.

From Newton's second law we can write the rate of change of the particle's momentum as

$$\frac{d\mathbf{p}}{dt} = q\left(\mathbf{E} + \frac{\mathbf{v}}{c} \times \mathbf{B} \right) \qquad (6.88)$$

If the sum of all the momenta of all the particles in the volume V is denoted by $\mathbf{P}_{\text{mech}}$, we can write

$$\frac{d\mathbf{P}_{\text{mech}}}{dt} = \int_V \left(\rho\mathbf{E} + \frac{1}{c} \mathbf{J} \times \mathbf{B} \right) d^3x \qquad (6.89)$$

We have converted the sum over particles to an integral over charge and current densities for convenience in manipulation. The particulate nature can be recovered at any stage by making use of delta functions, as in Section 1.2. In the same manner as for Poynting's theorem, we use Maxwell's equations to eliminate ρ and $\mathbf{J}$ from (6.89):

$$\rho = \frac{1}{4\pi} \nabla \cdot \mathbf{E}, \quad \mathbf{J} = \frac{c}{4\pi} \left(\nabla \times \mathbf{B} - \frac{1}{c} \frac{\partial \mathbf{E}}{\partial t} \right) \qquad (6.90)$$

Note that we have written only **E** and **B** in (6.90), and not **H** or **D**. The reason is, as mentioned earlier, that we are imagining all the charges as treated in the mechanical part of the system and so use the microscopic equations which involve only **E** and **B**. Some remarks will be made in the next section on the differences which arise when some of the particles, namely, the bound atoms, are included in the "field" energy and momentum through the dielectric constant and permeability. (See also Problem 6.8.)

With (6.90) substituted into (6.89) the integrand becomes

$$\rho\mathbf{E} + \frac{1}{c}\mathbf{J} \times \mathbf{B} = \frac{1}{4\pi}\left[\mathbf{E}(\nabla \cdot \mathbf{E}) + \frac{1}{c}\mathbf{B} \times \frac{\partial\mathbf{E}}{\partial t} - \mathbf{B} \times (\nabla \times \mathbf{B})\right] \quad (6.91)$$

Then writing

$$\mathbf{B} \times \frac{\partial\mathbf{E}}{\partial t} = -\frac{\partial}{\partial t}(\mathbf{E} \times \mathbf{B}) + \mathbf{E} \times \frac{\partial\mathbf{B}}{\partial t}$$

and adding $\mathbf{B}(\nabla \cdot \mathbf{B}) = 0$ to the square bracket, we obtain

$$\rho E + \frac{1}{c}\mathbf{J} \times \mathbf{B} = \frac{1}{4\pi}[\mathbf{E}(\nabla \cdot \mathbf{E}) + \mathbf{B}(\nabla \cdot \mathbf{B}) - \mathbf{E} \times (\nabla \times \mathbf{E})$$

$$- \mathbf{B} \times (\nabla \times \mathbf{B})] - \frac{1}{4\pi c}\frac{\partial}{\partial t}(\mathbf{E} \times \mathbf{B}) \quad (6.92)$$

The rate of change of mechanical momentum (6.89) can now be written

$$\frac{d\mathbf{P}_{\text{mech}}}{dt} + \frac{d}{dt}\int_V \frac{1}{4\pi c}(\mathbf{E} \times \mathbf{B})\,d^3x = \frac{1}{4\pi}\int_V [\mathbf{E}(\nabla \cdot \mathbf{E}) - \mathbf{E} \times (\nabla \times \mathbf{E})$$

$$+ \mathbf{B}(\nabla \cdot \mathbf{B}) - \mathbf{B} \times (\nabla \times \mathbf{B})]\,d^3x \quad (6.93)$$

We may tentatively identify the volume integral on the left as the total electromagnetic momentum $\mathbf{P}_{\text{field}}$ in the volume V:

$$\mathbf{P}_{\text{field}} = \frac{1}{4\pi c}\int_V (\mathbf{E} \times \mathbf{B})\,d^3x \quad (6.94)$$

The integrand can be interpreted as a density of electromagnetic momentum. We note that this momentum density is proportional to the energy-flux density **S**, with proportionality constant c^{-2}.

To complete the identification of the volume integral of $\frac{1}{4\pi c}(\mathbf{E} \times \mathbf{B})$ as electromagnetic momentum, and to establish (6.93) as the conservation law for momentum, we must convert the volume integral on the right into a surface integral of the normal component of something which can be identified as momentum flow.

Evidently the terms in the volume integral (6.93) transform like vectors. Consequently, if they are to be combined into the divergence of some quantity, that quantity must be a tensor of the second rank. While it is possible to deal with rectangular components of momentum, instead of the vectorial form (6.93), the tensor can be handled within the framework of vector operations by introducing a corresponding *dyadic*. If a tensor in three dimensions is denoted by T_{ij} ($i, j = 1, 2, 3$), and $\boldsymbol{\epsilon}_i$ are the unit base vectors of the coordinate axes, the dyadic corresponding to the tensor T_{ij} is defined to be

$$\overleftrightarrow{\mathbf{T}} = \sum_{i=1}^{3} \sum_{j=1}^{3} \boldsymbol{\epsilon}_i T_{ij} \boldsymbol{\epsilon}_j \tag{6.95}$$

The unit vector on the left can form scalar or vector products from the left, and correspondingly for the unit vector on the right. Given the dyadic, we can determine the tensor elements by taking the appropriate scalar products:

$$T_{ij} = \boldsymbol{\epsilon}_i \cdot \overleftrightarrow{\mathbf{T}} \cdot \boldsymbol{\epsilon}_j \tag{6.96}$$

A special dyadic is the identity $\overleftrightarrow{\mathbf{I}}$ formed with the unit second-rank tensor:

$$\overleftrightarrow{\mathbf{I}} = \boldsymbol{\epsilon}_1 \boldsymbol{\epsilon}_1 + \boldsymbol{\epsilon}_2 \boldsymbol{\epsilon}_2 + \boldsymbol{\epsilon}_3 \boldsymbol{\epsilon}_3 \tag{6.97}$$

The scalar product of any vector or vector operation with $\overleftrightarrow{\mathbf{I}}$ from either the left or right merely gives the original vector quantity.

With these sketchy remarks about dyadics, we now proceed with the vector manipulations needed to convert the volume integral on the right side of (6.93) into a surface integral. Using the vector identity,

$$\tfrac{1}{2} \boldsymbol{\nabla}(\mathbf{B} \cdot \mathbf{B}) = (\mathbf{B} \cdot \boldsymbol{\nabla})\mathbf{B} + \mathbf{B} \times (\boldsymbol{\nabla} \times \mathbf{B})$$

the terms involving $\mathbf{B}$ in (6.93) can be written

$$\mathbf{B}(\boldsymbol{\nabla} \cdot \mathbf{B}) - \mathbf{B} \times (\boldsymbol{\nabla} \times \mathbf{B}) = \mathbf{B}(\boldsymbol{\nabla} \cdot \mathbf{B}) + (\mathbf{B} \cdot \boldsymbol{\nabla})\mathbf{B} - \tfrac{1}{2}\boldsymbol{\nabla}B^2 \tag{6.98}$$

This can be identified as the divergence of a dyadic:

$$\mathbf{B}(\boldsymbol{\nabla} \cdot \mathbf{B}) + (\mathbf{B} \cdot \boldsymbol{\nabla})\mathbf{B} - \tfrac{1}{2}\boldsymbol{\nabla}B^2 = \boldsymbol{\nabla} \cdot (\mathbf{B}\mathbf{B} - \tfrac{1}{2}\overleftrightarrow{\mathbf{I}}B^2) \tag{6.99}$$

The electric field term in (6.93) can be put in this same form. Consequently the conservation of linear momentum becomes

$$\frac{d}{dt}(\mathbf{P}_{\text{mech}} + \mathbf{P}_{\text{field}}) = \int_V \boldsymbol{\nabla} \cdot \overleftrightarrow{\mathbf{T}} \, d^3x = \oint_S \mathbf{n} \cdot \overleftrightarrow{\mathbf{T}} \, da \tag{6.100}$$

The tensor-dyadic $\overleftrightarrow{\mathbf{T}}$, called *Maxwell's stress tensor*, is

$$\overleftrightarrow{\mathbf{T}} = \frac{1}{4\pi} [\mathbf{E}\mathbf{E} + \mathbf{B}\mathbf{B} - \tfrac{1}{2}\overleftrightarrow{\mathbf{I}}(E^2 + B^2)] \tag{6.101}$$

The elements of the tensor are

$$T_{ij} = \frac{1}{4\pi} [E_i E_j + B_i B_j - \tfrac{1}{2}\delta_{ij}(E^2 + B^2)] \tag{6.102}$$

Evidently $(-\mathbf{n} \cdot \overset{\leftrightarrow}{\mathbf{T}})$ in (6.100) represents the normal flow of momentum per unit area out of the volume V through the surface S. Or, in other words, $(-\mathbf{n} \cdot \overset{\leftrightarrow}{\mathbf{T}})$ is the force per unit area transmitted across surface S. This can be used to calculate the forces acting on material objects in electromagnetic fields by enclosing the objects with a boundary surface S and adding up the total electromagnetic force according to the right-hand side of (6.100).

The conservation of angular momentum of the combined system of particles and fields can be treated in the same way as we have handled energy and linear momentum. This is left as a problem for the student (Problem 6.9).

6.10 Macroscopic Equations

Although the equations of electrodynamics have been written in macroscopic form for the most part in this chapter, the reader will be aware that the derivation of the macroscopic equations from the microscopic ones was done separately for electro*statics* and magneto*statics* in Sections 4.3 and 5.8. Thus there arises the question of whether the derivation still holds good for time-dependent fields. It is intuitively obvious that it must, since Maxwell's addition of the displacement current was done at the macroscopic level. Nevertheless, it is useful to examine briefly the derivation to see in particular how the time variation of the polarization $\mathbf{P}$ gives rise to a current contribution and so converts the microscopic displacement current $\partial \mathbf{E}/\partial t$ into the macroscopic displacement current $\partial \mathbf{D}/\partial t$.

The basic assumption inherent in our previous discussions was that the macroscopic fields $\mathbf{E}$ and $\mathbf{B}$ which satisfy the two homogeneous Maxwell's equations (6.28) are the averages of the corresponding microscopic fields $\boldsymbol{\epsilon}$ and $\boldsymbol{\beta}$:

$$\mathbf{E}(\mathbf{x}, t) = \langle \boldsymbol{\epsilon} \rangle, \quad \mathbf{B}(\mathbf{x}, t) = \langle \boldsymbol{\beta} \rangle \tag{6.103}$$

The averages now involve a temporal and a spatial average, e.g.,

$$\langle \boldsymbol{\epsilon} \rangle = \frac{1}{\Delta V \Delta T} \int d^3\xi \int d\tau \boldsymbol{\epsilon}(\mathbf{x} + \boldsymbol{\xi}, t + \tau) \tag{6.104}$$

where the volume ΔV and the time interval ΔT are small compared to macroscopic quantities.

Relations (6.103) imply that the macroscopic potentials Φ and $\mathbf{A}$ are the averages of their microscopic counterparts,

$$\Phi(\mathbf{x}, t) = \langle \phi \rangle, \quad \mathbf{A}(\mathbf{x}, t) = \langle \mathbf{a} \rangle \tag{6.105}$$

since the fields $\mathbf{E}$ and $\mathbf{B}$ (or $\boldsymbol{\epsilon}$ and $\boldsymbol{\beta}$) are derived by differentiation according to (6.29) and (6.31).

The derivation of the averaged potentials in terms of the molecular properties proceeds exactly as in Sections 4.3 and 5.8, with two modifications. The first is that, according to our discussion of the solution of the wave equation in Section 6.6, we must have a "retarded" solution. Thus the same steps that led to (4.33) now lead to the averaged scalar potential:

$$\langle \phi \rangle = \int N(\mathbf{x}') \left[\frac{\langle e_{\mathrm{mol}}(\mathbf{x}', t') \rangle}{|\mathbf{x} - \mathbf{x}'|} + \langle \mathbf{p}_{\mathrm{mol}}(\mathbf{x}', t') \rangle \cdot \boldsymbol{\nabla}' \left(\frac{1}{|\mathbf{x} - \mathbf{x}'|} \right) \right]_{\mathrm{ret}} d^3 x' \tag{6.106}$$

To this must be added the retarded contribution of the excess free charge, ρ_{ex}. The second change comes in the vector potential. In the steady-state situation the molecular contribution to the vector potential was the sum of terms like (5.75), representing magnetic dipole contributions. The leading term in the expansion vanished because of the condition $\boldsymbol{\nabla} \cdot \mathbf{J} = 0$. With time-dependent fields this is no longer true. If we retrace our steps to equation (5.51), we see that the leading term in the expansion of $\mathbf{a}_{\mathrm{mol}}$ is

$$\mathbf{a}_{\mathrm{mol}} = \frac{1}{c|\mathbf{x} - \mathbf{x}_j|} \int \mathbf{J}_{\mathrm{mol}}(\mathbf{x}', t') \, d^3 x' + \binom{\text{magnetic dipole}}{\text{term (5.75)}} + \cdots \tag{6.107}$$

For simplicity we omit the retarded symbols temporarily. Using the identity $\boldsymbol{\nabla}' \cdot (x_i' \mathbf{J}) = J_i + x_i' \boldsymbol{\nabla}' \cdot \mathbf{J}$, the first term can be transformed into

$$\mathbf{a}_{\mathrm{mol}} = \frac{-1}{c|\mathbf{x} - \mathbf{x}_j|} \int \mathbf{x}' (\boldsymbol{\nabla}' \cdot \mathbf{J}_{\mathrm{mol}}) \, d^3 x' + (5.75) + \cdots \tag{6.108}$$

The continuity equation can now be used to write $\boldsymbol{\nabla}' \cdot \mathbf{J}_{\mathrm{mol}} = -\partial \rho_{\mathrm{mol}}/\partial t'$, and the definition of molecular electric dipole moment (4.25) can be employed to cast (6.108) in the form:

$$\mathbf{a}_{\mathrm{mol}}(\mathbf{x}, t) \simeq \left[\frac{1}{c} \frac{\partial \mathbf{p}_j(t')}{\partial t'} \frac{1}{|\mathbf{x} - \mathbf{x}_j|} + \frac{\mathbf{m}_j(t') \times (\mathbf{x} - \mathbf{x}_j)}{|\mathbf{x} - \mathbf{x}_j|^3} \right]_{\mathrm{ret}} \tag{6.109}$$

With time-dependent fields we have a leading term proportional to the time rate of change of the electric dipole moment.* Summing over all

---

* Actually, for time-varying fields not only does the leading term appear, but also from the second term in (5.51) there arises a term which is $\dfrac{\partial \overset{\leftrightarrow}{\mathbf{Q}}}{c \, \partial t} \cdot \boldsymbol{\nabla}' \left(\dfrac{1}{|\mathbf{x} - \mathbf{x}'|} \right)$, where $\overset{\leftrightarrow}{\mathbf{Q}}$ is the quadrupole dyadic of the molecule. Since we kept only electric dipole terms in (6.106), we drop this quadrupole term here.

molecules and averaging according to (6.104) leads to the averaged microscopic vector potential:

$$\langle \mathbf{a} \rangle = \frac{1}{c} \int N(\mathbf{x}') \left[\frac{1}{|\mathbf{x} - \mathbf{x}'|} \frac{\partial}{\partial t'} \langle \mathbf{p}_{\text{mol}}(\mathbf{x}', t') \rangle \right.$$
$$\left. + c \langle \mathbf{m}_{\text{mol}}(\mathbf{x}', t') \rangle \times \mathbf{\nabla}' \left(\frac{1}{|\mathbf{x} - \mathbf{x}'|} \right) \right]_{\text{ret}} d^3x' \quad (6.110)$$

To this must be added the standard contribution from the macroscopic conduction-current density $\mathbf{J}(\mathbf{x}, t)$.

Solutions (6.106) and (6.110), augmented by the free-charge and conduction-current contributions, can be written as

$$\left. \begin{array}{l} \langle \phi \rangle = \int \left[\dfrac{\langle \rho \rangle}{|\mathbf{x} - \mathbf{x}'|} \right]_{\text{ret}} d^3x' \\[3mm] \langle \mathbf{a} \rangle = \dfrac{1}{c} \int \left[\dfrac{\langle \mathbf{J} \rangle}{|\mathbf{x} - \mathbf{x}'|} \right]_{\text{ret}} d^3x' \end{array} \right\} \quad (6.111)$$

With definitions (4.36) and (5.77) of macroscopic polarization $\mathbf{P}$ and magnetization $\mathbf{M}$, the averaged charge and current densities in (6.111) can be expressed as

$$\langle \rho \rangle = \rho - \mathbf{\nabla} \cdot \mathbf{P}$$
$$\langle \mathbf{J} \rangle = \mathbf{J} + c(\mathbf{\nabla} \times \mathbf{M}) + \frac{\partial \mathbf{P}}{\partial t} \quad (6.112)$$

where ρ and $\mathbf{J}$ are the macroscopic charge and current densities.

We are now in a position to verify explicitly the deduction of the macroscopic Maxwell's equations from the microscopic ones. The homogeneous ones follow directly from identification (6.103). Of the inhomogeneous ones, consider the microscopic form of Ampère's law:

$$\mathbf{\nabla} \times \mathbf{\beta} = \frac{4\pi}{c} \mathbf{J} + \frac{1}{c} \frac{\partial \mathbf{\epsilon}}{\partial t} \quad (6.113)$$

Averaging both sides and using (6.112) for $\langle \mathbf{J} \rangle$, we get

$$\mathbf{\nabla} \times \mathbf{B} = \frac{4\pi}{c} \left(\mathbf{J} + c(\mathbf{\nabla} \times \mathbf{M}) + \frac{\partial \mathbf{P}}{\partial t} \right) + \frac{1}{c} \frac{\partial \mathbf{E}}{\partial t} \quad (6.114)$$

With the definitions $\mathbf{H} = \mathbf{B} - 4\pi \mathbf{M}$ and $\mathbf{D} = \mathbf{E} + 4\pi \mathbf{P}$, this becomes

$$\mathbf{\nabla} \times \mathbf{H} = \frac{4\pi}{c} \mathbf{J} + \frac{1}{c} \frac{\partial \mathbf{D}}{\partial t} \quad (6.115)$$

as required. The other equation, $\mathbf{\nabla} \cdot \mathbf{D} = 4\pi\rho$, follows in an even simpler way from (6.112).

As a final remark concerning the macroscopic field equations we discuss the differences between the microscopic and macroscopic forms of Poynting's theorem. We derived the conservation of energy in Section 6.8 in the macroscopic form (6.81). Written out explicitly in terms of all the fields, it is

$$\frac{c}{4\pi}\int_S (\mathbf{E} \times \mathbf{H}) \cdot \mathbf{n} \, da + \frac{1}{4\pi}\int_V \left(\mathbf{E} \cdot \frac{\partial \mathbf{D}}{\partial t} + \mathbf{H} \cdot \frac{\partial \mathbf{B}}{\partial t}\right) d^3x = -\int_V \mathbf{E} \cdot \mathbf{J} \, d^3x$$

$$(6.116)$$

The different fields **E**, **D**, **B**, **H** enter in characteristic ways which can be understood if we establish contact with the microscopic form of Poynting's theorem. We can do this most easily by merely expressing the left side of (6.116) in terms of the basic fields **E** and **B**. Then (6.116) can easily be shown to be

$$\frac{c}{4\pi}\int_S (\mathbf{E} \times \mathbf{B}) \cdot \mathbf{n} \, da + \frac{1}{4\pi}\int_V \left(\mathbf{E} \cdot \frac{\partial \mathbf{E}}{\partial t} + \mathbf{B} \cdot \frac{\partial \mathbf{B}}{\partial t}\right) d^3x$$

$$= -\int_V \mathbf{E} \cdot \left(\mathbf{J} + c\nabla \times \mathbf{M} + \frac{\partial \mathbf{P}}{\partial t}\right) d^3x \quad (6.117)$$

From (6.112) we see that (6.117) looks like the statement of Poynting's theorem for the microscopic fields, except that each quantity is replaced by its average. This is *not* the average of Poynting's theorem for microscopic fields, but differs from it by a set of terms which are the statement of energy conservation for the fluctuating fields measuring the instantaneous departure of $\boldsymbol{\epsilon}$ and $\boldsymbol{\beta}$ from **E** and **B**. Apart from these fluctuating fields, (6.117) can be interpreted as follows.

If we include in the sources of charge and current the electronic motion within the molecules as well as the conduction current, then Poynting's theorem appears in terms of the basic fields **E** and **B** and involves the work done per unit time by the electric field on *all* currents. If we choose to treat the work done on the effective molecular current $\left[\dfrac{\partial \mathbf{P}}{\partial t} + c(\nabla \times \mathbf{M})\right]$ as energy stored or propagated in the medium, that term can be taken over to the left-hand side and included in the energy-density and energy-flow terms characteristic of the medium. Then we return to the macroscopic Poynting's theorem (6.116) with only the work done per unit time by the electric field on the conduction current shown explicitly. It is natural to absorb the energy associated with the effective molecular current into the energy stored in the field, since it is a property of the medium and is in general *stored* energy (i.e., reactive power) which involves no time-average dissipation (not true for magnetic media with hysteresis effects). The power associated with the conduction current is, on the other hand, dissipative, since it involves a conversion of electrical energy into mechanical.

REFERENCES AND SUGGESTED READING

The conservation laws for the energy and momentum of electromagnetic fields are discussed in almost all textbooks. A good treatment of the energy of quasi-stationary currents and forces acting on circuits carrying currents, different from ours, is given by
   Panofsky and Phillips, Chapter 10.
   Stratton, Chapter II,
discusses the Maxwell stress tensor in some detail in considering forces in fluids and solids. The general topics of conservation laws, as well as quasi-stationary circuit theory, inductance calculations, and forces, are dealt with in a lucid manner by
   Abraham and Becker, Band I, Chapters VIII and IX.
Inductance calculations and circuit theory are also presented by
   Smythe, Chapters VIII, IX, and X,
and many engineering textbooks.
The here-neglected subject of eddy currents and induction heating is discussed with many examples by
   Smythe, Chapter XI.
The mathematical topics in this chapter center around the wave equation. The initial-value problem in one, two, three, and more dimenisons is discussed by
   Morse and Feshbach, pp. 843–847,
and, in more mathematical detail, by
   Hadamard.

PROBLEMS

6.1   (a) Show that for a system of current-carrying elements in empty space the total energy in the magnetic field is

$$W = \frac{1}{2c^2}\int d^3x \int d^3x' \frac{\mathbf{J}(\mathbf{x}) \cdot \mathbf{J}(\mathbf{x}')}{|\mathbf{x} - \mathbf{x}'|}$$

where $\mathbf{J}(\mathbf{x})$ is the current density.
   (b) If the current configuration consists of n circuits carrying currents $I_1, I_2, \ldots, I_n$, show that the energy can be expressed as

$$W = \tfrac{1}{2}\sum_{i=1}^{n} L_i I_i^2 + \sum_{i=1}^{n} \sum_{j>i}^{n} M_{ij}I_i I_j$$

Exhibit integral expressions for the self-inductances (L_i) and the mutual inductances (M_{ij}).

6.2   A two-wire transmission line consists of a pair of nonpermeable parallel wires of radii a and b separated by a distance $d > a + b$. A current flows down one wire and back the other. It is uniformly distributed over the cross section of each wire. Show that the self-inductance per unit length is

$$c^2 L = 1 + 2 \ln \left(\frac{d^2}{ab} \right)$$

6.3 A circuit consists of a thin conducting shell of radius a and a parallel return wire of radius b inside. If the current is assumed distributed uniformly throughout the cross section of the wire, calculate the self-inductance per unit length. What is the self-inductance if the inner conductor is a thin hollow tube?

6.4 Show that the mutual inductance of two circular coaxial loops in a homogeneous medium of permeability μ is

$$M_{12} = \frac{4\pi\mu}{c^2} \sqrt{ab} \left[\left(\frac{2}{k} - k \right) K(k) - \frac{2}{k} E(k) \right]$$

where

$$k^2 = \frac{4ab}{(a + b)^2 + d^2}$$

and a, b are the radii of the loops, d is the distance between their centers, and K and E are the complete elliptic integrals.

Find the limiting value when $d \ll a$, b and $a \simeq b$.

6.5 A transmission line consists of two, parallel perfect conductors of arbitrary, but constant, cross section. Current flows down one conductor and returns via the other.

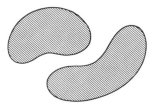

Show that the product of the inductance per unit length L and the capacitance per unit length C is

$$LC = \frac{\mu\epsilon}{c^2}$$

where μ and ϵ are the permeability and the dielectric constant of the medium surrounding the conductors, while c is the velocity of light *in vacuo*.

6.6 Prove that any vector field $\mathbf{F}$ can be decomposed into transverse and longitudinal parts,

$$\mathbf{F} = \mathbf{F}_l + \mathbf{F}_t$$

with $\nabla \cdot \mathbf{F}_t = 0$ and $\nabla \times \mathbf{F}_l = 0$, where $\mathbf{F}_l$ and $\mathbf{F}_t$ are given by (6.49) and (6.50).

6.7   (a) Show that the one-dimensional wave equation,

$$\frac{\partial^2 \psi}{\partial x^2} - \frac{1}{c^2}\frac{\partial^2 \psi}{\partial t^2} = 0$$

has the general solution,

$$\psi(x, t) = \tfrac{1}{2}f\left(t - \frac{x}{c}\right) + \tfrac{1}{2}f\left(t + \frac{x}{c}\right) + \frac{c}{2}\int_{t-(x/c)}^{t+(x/c)} F(t')\,dt'$$

where the boundary conditions are specified by the values of ψ and $\partial\psi/\partial x$ at $x = 0$ for all time:

$$\psi(0, t) = f(t), \quad \frac{\partial \psi}{\partial x}(0, t) = F(t)$$

(b) What is the corresponding solution if the boundary conditions are that, at $t = 0$,

$$\psi(x, 0) = f(x), \quad \frac{\partial \psi}{\partial t}(x, 0) = g(x)$$

6.8   Discuss the conservation of energy and linear momentum for a macroscopic system of sources and electromagnetic fields in a medium described by a dielectric constant ϵ and a permeability μ. Show that the energy density, Poynting's vector, field-momentum density, and Maxwell stress tensor are given by

$$u = \frac{1}{8\pi}(\epsilon E^2 + \mu H^2)$$

$$\mathbf{S} = \frac{c}{4\pi}(\mathbf{E} \times \mathbf{H})$$

$$\mathbf{g} = \frac{\mu\epsilon}{4\pi c}(\mathbf{E} \times \mathbf{H})$$

$$T_{ij} = \frac{1}{4\pi}[\epsilon E_i E_j + \mu H_i H_j - \tfrac{1}{2}\delta_{ij}(\epsilon E^2 + \mu H^2)]$$

What modifications arise if ϵ and μ are functions of position?

6.9   With the same assumptions as in Problem 6.8 discuss the conservation of angular momentum. Show that the differential and integral forms of the conservation law are

$$\frac{\partial}{\partial t}(\mathscr{L}_{\text{mech}} + \mathscr{L}_{\text{field}}) + \nabla \cdot \overset{\leftrightarrow}{\mathbf{M}} = 0$$

and

$$\frac{d}{dt}\int_V (\mathscr{L}_{\text{mech}} + \mathscr{L}_{\text{field}})\,d^3x + \int_S \mathbf{n} \cdot \overset{\leftrightarrow}{\mathbf{M}}\,da = 0$$

where the field angular-momentum density is

$$\mathscr{L}_{\text{field}} = \mathbf{x} \times \mathbf{g} = \frac{\mu\epsilon}{4\pi c}\mathbf{x} \times (\mathbf{E} \times \mathbf{H})$$

and the flux of angular momentum is described by the tensor

$$\overset{\leftrightarrow}{\mathbf{M}} = \overset{\leftrightarrow}{\mathbf{T}} \times \mathbf{x}$$

Note: $\overset{\leftrightarrow}{\mathbf{M}}$ can be written as a third-rank tensor, $M_{ijk} = T_{ij}x_k - T_{ik}x_j$. But in the indices j and k it is antisymmetric and so has only three independent elements. Including the index i, M_{ijk} therefore has nine components and can be written as a pseudo tensor of the second rank, as above.

6.10 A plane wave is incident normally on a perfectly absorbing flat screen.

(*a*) From the law of conservation of linear momentum show that the pressure (called radiation pressure) exerted on the screen is equal to the field energy per unit volume in the wave.

(*b*) In the neighborhood of the earth the flux of electromagnetic energy from the sun is approximately 0.14 watt/cm². If an interplanetary "sailplane" had a sail of mass 10^{-4} gm/cm² of area and negligible other weight, what would be its maximum acceleration in centimeters per square second due to the solar radiation pressure? How does this compare with the acceleration due to the solar "wind" (corpuscular radiation)?

6.11 A circularly polarized plane wave moving in the z direction has a finite extent in the x and y directions. Assuming that the amplitude modulation is slowly varying (the wave is many wavelengths broad), show that the electric and magnetic fields are given approximately by

$$\mathbf{E}(x, y, z, t) \simeq \left[E_0(x, y)(\mathbf{e}_1 \pm i\mathbf{e}_2) + \frac{i}{k}\left(\frac{\partial E_0}{\partial x} \pm i\,\frac{\partial E_0}{\partial y} \right)\mathbf{e}_3 \right] e^{ikz - i\omega t}$$

$$\mathbf{B} \simeq \mp i\sqrt{\mu\epsilon}\,\mathbf{E}$$

where $\mathbf{e}_1$, $\mathbf{e}_2$, $\mathbf{e}_3$ are unit vectors in the x, y, z directions.

6.12 For the circularly polarized wave of Problem 6.11 calculate the time-averaged component of angular momentum parallel to the direction of propagation. Show that the ratio of this component of angular momentum to the energy of the wave is

$$\frac{L_3}{U} = \pm \omega^{-1}$$

Interpret this result in terms of quanta of radiation (photons). Show that for a cylindrically symmetric, finite plane wave the transverse components of angular momentum vanish.

7

Plane Electromagnetic Waves

This chapter is concerned with plane waves in unbounded, or perhaps semi-infinite, media. The basic properties of plane waves in nonconducting media—their transverse nature, the various states of polarization—are treated first. Then the behavior of one-dimensional wave packets is discussed; group velocity is introduced; dispersive effects are considered. Reflection and refraction of radiation at a plane interface between dielectrics are presented. Then plane waves in a conducting medium are described, and a simple model of electrical conductivity is discussed. Finally the conductivity model is modified to apply to a tenuous plasma, or electron gas, and the propagation of transverse waves in a plasma in the presence of an external static magnetic field is considered.

7.1 Plane Waves in a Nonconducting Medium

A basic feature of Maxwell's equations for the electromagnetic field is the existence of traveling wave solutions which represent the transport of energy from one point to another. The simplest and most fundamental electromagnetic waves are transverse, plane waves. We proceed to see how such solutions can be obtained in simple nonconducting media described by spatially constant permeability and susceptibility. In the absence of sources, Maxwell's equations in an infinite medium are:

$$\left. \begin{array}{ll} \nabla \cdot \mathbf{E} = 0 & \nabla \times \mathbf{E} + \dfrac{1}{c}\dfrac{\partial \mathbf{B}}{\partial t} = 0 \\[3mm] \nabla \cdot \mathbf{B} = 0 & \nabla \times \mathbf{B} - \dfrac{\mu\epsilon}{c}\dfrac{\partial \mathbf{E}}{\partial t} = 0 \end{array} \right\} \tag{7.1}$$

where the medium is characterized by the parameters μ, ϵ. By combining the two curl equations and making use of the vanishing divergences, we find easily that each cartesian component of **E** and **B** satisfies the wave equation:

$$\nabla^2 u - \frac{1}{v^2}\frac{\partial^2 u}{\partial t^2} = 0 \tag{7.2}$$

where

$$v = \frac{c}{\sqrt{\mu\epsilon}} \tag{7.3}$$

is a constant of the dimensions of velocity characteristic of the medium. The wave equation (7.2) has the well-known plane-wave solutions:

$$u = e^{i\mathbf{k}\cdot\mathbf{x}-i\omega t} \tag{7.4}$$

where the frequency ω and the magnitude of the wave vector **k** are related by

$$k = \frac{\omega}{v} = \sqrt{\mu\epsilon}\,\frac{\omega}{c} \tag{7.5}$$

If we consider waves propagating in only one direction, say, the x direction, the fundamental solution is

$$u(x, t) = Ae^{ikx-i\omega t} + Be^{-ikx-i\omega t} \tag{7.6}$$

Using (7.5), this can be written

$$u_k(x, t) = Ae^{ik(x-vt)} + Be^{-ik(x+vt)} \tag{7.7}$$

If v is not a function of k (i.e., a nondispersive medium, with $\mu\epsilon$ independent of frequency), we know by the Fourier integral theorem (2.50) and (2.51) that by linear superposition we can construct from $u_k(x, t)$ a general solution of the form:

$$u(x, t) = f(x - vt) + g(x + vt) \tag{7.8}$$

where $f(z)$ and $g(z)$ are arbitrary functions. It is easy to verify directly that this is a solution of the wave equation (7.2). Equation (7.8) represents waves traveling to the right and to the left with velocities of propagation equal to v, which is called the phase velocity of the wave. If v is a function of k, the situation is not as simple—the initial waves $f(x)$ and $g(x)$ are not propagated without distortion at velocity v (see Section 7.3). For each frequency component, however, v given by (7.3) is still the phase velocity.

The basic plane wave (7.4) and (7.5) satisfies the scalar-wave equation (7.2). But we still must consider the vector nature of the electromagnetic fields and the requirement of satisfying Maxwell's equations. With the convention that the physical electric and magnetic fields are obtained by taking the real parts of complex quantities, we assume that the plane-wave fields are of the form:

$$\left.\begin{array}{l} \mathbf{E}(\mathbf{x}, t) = \boldsymbol{\epsilon}_1 E_0 e^{i\mathbf{k} \cdot \mathbf{x} - i\omega t} \\ \mathbf{B}(\mathbf{x}, t) = \boldsymbol{\epsilon}_2 B_0 e^{i\mathbf{k} \cdot \mathbf{x} - i\omega t} \end{array}\right\} \tag{7.9}$$

where $\boldsymbol{\epsilon}_1$, $\boldsymbol{\epsilon}_2$ are two constant unit vectors, and E_0, B_0 are complex amplitudes which are constant in space and time. The requirements $\nabla \cdot \mathbf{E} = 0$ and $\nabla \cdot \mathbf{B} = 0$ demand that

$$\boldsymbol{\epsilon}_1 \cdot \mathbf{k} = 0, \qquad \boldsymbol{\epsilon}_2 \cdot \mathbf{k} = 0 \tag{7.10}$$

This means that $\mathbf{E}$ and $\mathbf{B}$ are both perpendicular to the direction of propagation $\mathbf{k}$. Such a wave is called a *transverse wave*. The curl equations provide further restrictions. Substitution of (7.9) into the first curl equation in (7.1) leads to the relation:

$$i\left[(\mathbf{k} \times \boldsymbol{\epsilon}_1)E_0 - \frac{\omega}{c} \boldsymbol{\epsilon}_2 B_0\right] e^{i\mathbf{k} \cdot \mathbf{x} - i\omega t} = 0 \tag{7.11}$$

Equation (7.11) (really several equations) has the solution:

$$\boldsymbol{\epsilon}_2 = \frac{\mathbf{k} \times \boldsymbol{\epsilon}_1}{k} \tag{7.12}$$

and

$$B_0 = \sqrt{\mu\epsilon}\, E_0 \tag{7.13}$$

This shows that $(\boldsymbol{\epsilon}_1, \boldsymbol{\epsilon}_2, \mathbf{k})$ form a set of orthogonal vectors and that $\mathbf{E}$ and $\mathbf{B}$ are in phase and in constant ratio, as indicated in Fig. 7.1. The wave described by (7.9), (7.12), and (7.13) is a transverse wave propagating in the direction $\mathbf{k}$. It represents a time-averaged flux of energy given by the

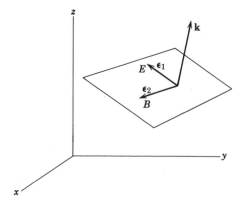

Fig. 7.1   Propagation vector $\mathbf{k}$ and two orthogonal polarization vectors $\boldsymbol{\epsilon}_1$ and $\boldsymbol{\epsilon}_2$.

real part of the complex Poynting's vector:

$$\mathbf{S} = \frac{1}{2}\frac{c}{4\pi}\,\mathbf{E}\times\mathbf{H}^* \qquad (7.14)$$

The energy flow (energy per unit area per unit time) is

$$\mathbf{S} = \frac{c}{8\pi}\sqrt{\frac{\epsilon}{\mu}}\,|E_0|^2\boldsymbol{\epsilon}_3 \qquad (7.15)$$

where $\boldsymbol{\epsilon}_3$ is a unit vector in the direction of $\mathbf{k}$. The time-averaged density u is correspondingly

$$u = \frac{1}{16\pi}\left(\epsilon\mathbf{E}\cdot\mathbf{E}^* + \frac{1}{\mu}\mathbf{B}\cdot\mathbf{B}^*\right) \qquad (7.16)$$

This gives

$$u = \frac{\epsilon}{8\pi}|E_0|^2 \qquad (7.17)$$

The ratio of the magnitude of (7.15) to (7.17) shows that the velocity of energy flow is $v = c/\sqrt{\mu\epsilon}$, as expected from (7.8).

7.2 Linear and Circular Polarization

The plane wave (7.9) is a wave with its electric field vector always in the direction $\boldsymbol{\epsilon}_1$. Such a wave is said to be linearly polarized with polarization vector $\boldsymbol{\epsilon}_1$. To describe a general state of polarization we need another linearly polarized wave which is independent of the first. Clearly the two waves

$$\left.\begin{array}{l} \mathbf{E}_1 = \boldsymbol{\epsilon}_1 E_1 e^{i\mathbf{k}\cdot\mathbf{x}-i\omega t} \\[4pt] \mathbf{E}_2 = \boldsymbol{\epsilon}_2 E_2 e^{i\mathbf{k}\cdot\mathbf{x}-i\omega t} \\[4pt] \mathbf{B}_j = \sqrt{\mu\epsilon}\,\dfrac{\mathbf{k}\times\mathbf{E}_j}{k}, \qquad j=1,2 \end{array}\right\} \qquad (7.18)$$

with

represent two such linearly independent solutions. The amplitudes E_1 and E_2 are complex numbers to allow the possibility of a phase difference between the waves. A general solution for a plane wave propagating in the direction $\mathbf{k}$ is given by a linear combination of $\mathbf{E}_1$ and $\mathbf{E}_2$:

$$\mathbf{E}(\mathbf{x},t) = (\boldsymbol{\epsilon}_1 E_1 + \boldsymbol{\epsilon}_2 E_2)e^{i\mathbf{k}\cdot\mathbf{x}-i\omega t} \qquad (7.19)$$

If E_1 and E_2 have the *same phase*, (7.19) represents a *linearly polarized* wave, with its polarization vector making an angle $\theta = \tan^{-1}(E_2/E_1)$ with $\boldsymbol{\epsilon}_1$ and a magnitude $E = \sqrt{E_1^2 + E_2^2}$, as shown in Fig. 7.2.

If E_1 and E_2 have *different phases*, the wave (7.19) is *elliptically polarized*.

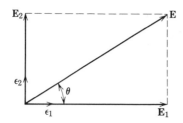

Fig. 7.2 Electric field of a linearly polarized wave.

To understand what this means let us consider the simplest case, *circular polarization*. Then E_1 and E_2 have the same magnitude, but differ in phase by 90°. The wave (7.19) becomes:

$$\mathbf{E}(\mathbf{x}, t) = E_0(\boldsymbol{\epsilon}_1 \pm i\boldsymbol{\epsilon}_2)e^{i\mathbf{k} \cdot \mathbf{x} - i\omega t} \qquad (7.20)$$

with E_0 the common real amplitude. We imagine axes chosen so that the wave is propagating in the positive z direction, while $\boldsymbol{\epsilon}_1$ and $\boldsymbol{\epsilon}_2$ are in the x and y directions, respectively. Then the components of the actual electric field, obtained by taking the real part of (7.20), are

$$\left.\begin{aligned} E_x(\mathbf{x}, t) &= E_0 \cos{(kz - \omega t)} \\ E_y(\mathbf{x}, t) &= \mp E_0 \sin{(kz - \omega t)} \end{aligned}\right\} \qquad (7.21)$$

At a *fixed point in space*, the fields (7.21) are such that the electric vector is constant in magnitude, but sweeps around in a circle at a frequency ω, as shown in Fig. 7.3. For the upper sign ($\boldsymbol{\epsilon}_1 + i\boldsymbol{\epsilon}_2$), the rotation is counter-clockwise when the observer is facing into the oncoming wave. This wave is called *left circularly polarized* in optics. In the terminology of modern physics, however, such a wave is said to have *positive helicity*. The latter description seems more appropriate because such a wave has a positive projection of angular momentum on the z axis (see Problem 6.12). For the lower sign ($\boldsymbol{\epsilon}_1 - i\boldsymbol{\epsilon}_2$), the rotation of $\mathbf{E}$ is clockwise when looking into

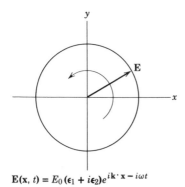

$$\mathbf{E}(\mathbf{x}, t) = E_0(\boldsymbol{\epsilon}_1 + i\boldsymbol{\epsilon}_2)e^{i\mathbf{k} \cdot \mathbf{x} - i\omega t}$$

Fig. 7.3 Electric field of a circularly polarized wave.

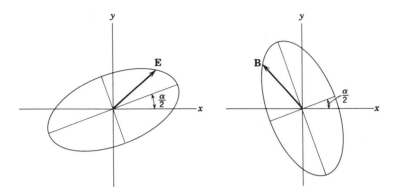

Fig. 7.4   Electric field and magnetic induction for an elliptically polarized wave.

the wave; the wave is right circularly polarized (optics); it has *negative helicity*.

The two circularly polarized waves (7.20) form an equally acceptable set of basic fields for description of a general state of polarization. We introduce the complex orthogonal unit vectors:

$$\boldsymbol{\epsilon}_{\pm} = \frac{1}{\sqrt{2}} (\boldsymbol{\epsilon}_1 \pm i\boldsymbol{\epsilon}_2) \tag{7.22}$$

with properties

$$\left. \begin{aligned} \boldsymbol{\epsilon}_{\pm}{}^* \cdot \boldsymbol{\epsilon}_{\mp} &= 0 \\ \boldsymbol{\epsilon}_{\pm}{}^* \cdot \boldsymbol{\epsilon}_3 &= 0 \\ \boldsymbol{\epsilon}_{\pm}{}^* \cdot \boldsymbol{\epsilon}_{\pm} &= 1 \end{aligned} \right\} \tag{7.23}$$

Then a general representation, equivalent to (7.19), is

$$\mathbf{E}(\mathbf{x}, t) = (E_+\boldsymbol{\epsilon}_+ + E_-\boldsymbol{\epsilon}_-)e^{i\mathbf{k}\cdot\mathbf{x}-i\omega t} \tag{7.24}$$

where E_+ and E_- are complex amplitudes. If E_+ and E_- have different magnitudes, but the same phase, (7.24) represents an elliptically polarized wave with principal axes of the ellipse in the directions of $\boldsymbol{\epsilon}_1$ and $\boldsymbol{\epsilon}_2$. The ratio of semimajor to semiminor axis is $(1 + r)/(1 - r)$, where $E_-/E_+ = r$. If the amplitudes have a phase difference between them, $E_-/E_+ = re^{i\alpha}$, then it is easy to show that the ellipse traced out by the E vector has its axes rotated by an angle $(\alpha/2)$. Figure 7.4 shows the general case of elliptical polarization and the ellipses traced out by both **E** and **B** at a given point in space.

For $r = \pm 1$ we get back a linearly polarized wave.

7.3 Superposition of Waves in One Dimension; Group Velocity

In the previous sections plane-wave solutions to Maxwell's equations were found and their properties discussed. Only monochromatic waves, those with a definite frequency and wave number, were treated. In actual circumstances such idealized solutions do not arise. Even in the most monochromatic light source or the most sharply tuned radio transmitter or receiver, one deals with a finite (although perhaps small) spread of frequencies or wavelengths. This spread may originate in the finite duration of a pulse, in inherent broadening in the source, or in many other ways. Since the basic equations are linear, it is in principle an elementary matter to make the appropriate linear superposition of solutions with different frequencies. In general, however, there are several new features which arise.

1. If the medium is dispersive (i.e., the dielectric constant is a function of the frequency of the fields), the phase velocity is not the same for each frequency component of the wave. Consequently different components of the wave travel with different speeds and tend to change phase with respect to one another. This leads to a change in the shape of a pulse, for example, as it travels along.

2. In a dispersive medium the velocity of energy flow may differ greatly from the phase velocity, or may even lack precise meaning.

3. In a dissipative medium, a pulse of radiation will be attenuated as it travels with or without distortion, depending on whether the dissipative effects are or are not sensitive functions of frequency.

The essentials of these dispersive and dissipative effects are implicit in the ideas of Fourier series and integrals (Section 2.9). For simplicity, we consider scalar waves in only one dimension. The scalar amplitude $u(x, t)$ can be thought of as one of the components of the electromagnetic field. The basic solution to the wave equation (7.2) has been exhibited in (7.6). The relationship between frequency ω and wave number k is given by (7.5) for the electromagnetic field. Either ω or k can be viewed as the independent variable when one considers making a linear superposition. Initially we will find it most convenient to use k as an independent variable. To allow for the possibility of dispersion we will consider ω as a general function of k:

$$\omega = \omega(k) \tag{7.25}$$

Since the dispersive properties cannot depend on whether the wave travels to the left or to the right, ω must be an even function of k, $\omega(-k) = \omega(k)$. For most wavelengths ω is a smoothly varying function of k. But at certain frequencies there are regions of "anomalous dispersion" where ω varies rapidly over a narrow interval of wavelengths. With the general form (7.25), our subsequent discussion can apply equally well to electromagnetic waves, sound waves, de Broglie matter waves, etc. For the present we assume that k and $\omega(k)$ are real, and so exclude dissipative effects.

From the basic solutions (7.6) we can build up a general solution of the form

$$u(x, t) = \frac{1}{\sqrt{2\pi}} \int_{-\infty}^{\infty} A(k)e^{ikx - i\omega(k)t} \, dk \qquad (7.26)$$

The factor $1/\sqrt{2\pi}$ has been inserted to conform with the Fourier integral notation of (2.50) and (2.51). The amplitude $A(k)$ describes the properties of the linear superposition of the different waves. It is given by the transform of the spatial amplitude $u(x, t)$, evaluated at $t = 0$*:

$$A(k) = \frac{1}{\sqrt{2\pi}} \int_{-\infty}^{\infty} u(x, 0)e^{-ikx} \, dx \qquad (7.27)$$

If $u(x, 0)$ represents a harmonic wave e^{ik_0x} for all x, the orthogonality relation (2.52) shows that $A(k) = \sqrt{2\pi}\delta(k - k_0)$, corresponding to a monochromatic traveling wave $u(x, t) = e^{ik_0x - i\omega(k_0)t}$, as required. If, however, at $t = 0$, $u(x, 0)$ represents a finite wave train with a length of order Δx, as shown in Fig. 7.5, then the amplitude $A(k)$ is not a delta function. Rather, it is a peaked function with a breadth of the order of Δk, centered around a wave number k_0 which is the dominant wave number in the modulated wave $u(x, 0)$. If Δx and Δk are defined as the rms deviations from the average values of x and k [defined in terms of the *intensities* $|u(x, 0)|^2$ and $|A(k)|^2$], it is possible to draw the general conclusion:

$$\Delta x \, \Delta k \geq \tfrac{1}{2} \qquad (7.28)$$

The reader may readily verify that, for most reasonable pulses or wave packets which do not cut off too violently, Δx times Δk lies near the lower limiting value in (7.28). This means that short wave trains with only a few wavelengths present have a very wide distribution of wave numbers of monochromatic waves, and conversely that long sinusoidal wave trains are almost monochromatic. Relation (7.28) applies equally well to distributions in time and in frequency.

The next question is the behavior of a pulse or finite wave train in time.

* The following discussion slights somewhat the initial-value problem. For a second-order differential equation we must specify not only $u(x, 0)$ but also $\partial u(x, 0)/\partial t$. This omission is of no consequence for the rest of the material in this section. It is remedied in the following section.

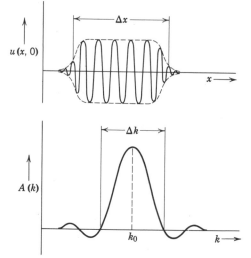

Fig. 7.5 A harmonic wave train of finite extent and its Fourier spectrum in wave number.

The pulse shown at $t = 0$ in Fig. 7.5 begins to move as time goes on. The different frequency or wave-number components in it move at different phase velocities. Consequently there is a tendency for the original coherence to be lost and for the pulse to become distorted in shape. At the very least, we might expect it to propagate with a rather different velocity from, say, the average phase velocity of its component waves. The general case of a highly dispersive medium or a very sharp pulse with a great spread of wave numbers present is difficult to treat. But the propagation of a pulse which is not too broad in its wave-number spectrum, or a pulse in a medium for which the frequency depends weakly on wave number, can be handled in the following approximate way. The wave at time t is given by (7.26). If the distribution $A(k)$ is fairly sharply peaked around some value k_0, then the frequency $\omega(k)$ can be expanded around that value of k:

$$\omega(k) = \omega_0 + \frac{d\omega}{dk}\bigg|_0 (k - k_0) + \cdots \tag{7.29}$$

and the integral performed. Thus

$$u(x, t) \simeq \frac{e^{i[k_0(d\omega/dk)|_0 - \omega_0]t}}{\sqrt{2\pi}} \int_{-\infty}^{\infty} A(k)e^{i[x - (d\omega/dk)|_0 t]k}\, dk \tag{7.30}$$

From (7.27) and its inverse it is apparent that the integral in (7.30) is just $u(x', 0)$, where $x' = x - (d\omega/dk)|_0 t$:

$$u(x, t) \simeq u\left(x - \frac{d\omega}{dk}\bigg|_0 t, 0\right) e^{i[k_0(d\omega/dk)|_0 - \omega_0]t} \tag{7.31}$$

This shows that, apart from an overall phase factor, the pulse travels along undistorted in shape with a velocity, called the *group velocity*:

$$v_g = \frac{d\omega}{dk}\Big|_0 \tag{7.32}$$

If an energy density is associated with the magnitude of the wave (or its absolute square), it is clear that in this approximation the transport of energy occurs with the group velocity, since that is the rate at which the pulse travels along.

For light waves the relation between ω and k is given by

$$\omega(k) = \frac{ck}{n(k)} \tag{7.33}$$

where c is the velocity of light in vacuum, and $n(k)$ is the index of refraction expressed as a function of k. The phase velocity is

$$v_p = \frac{\omega(k)}{k} = \frac{c}{n(k)} \tag{7.34}$$

and is greater or smaller than c depending on whether $n(k)$ is smaller or larger than unity. For most optical wavelengths $n(k)$ is greater than unity in almost all substances. The group velocity (7.32) is

$$v_g = \frac{c}{[n(\omega) + \omega(dn/d\omega)]} \tag{7.35}$$

In this equation it is more convenient to think of n as a function of ω than of k. For normal dispersion $(dn/d\omega) > 0$, and also $n > 1$; then the velocity of energy flow is less than the phase velocity and also less than c. In regions of anomalous dispersion, however, $dn/d\omega$ can become large and negative. Then the group velocity differs greatly from the phase velocity, often becoming larger than c.* The behavior of group and phase velocities as a function of frequency in the neighborhood of a region of anomalous dispersion is shown in Fig. 7.6.

---

* There is no cause for alarm that our ideas of special relativity are here violated; group velocity is no longer a meaningful concept. A large value of $dn/d\omega$ is equivalent to a rapid variation of ω as a function of k. Consequently the approximations made in (7.29) ff. are no longer valid. The behavior of the pulse is much more involved.

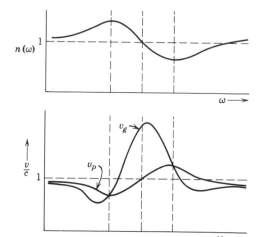

Fig. 7.6 Index of refraction $n(\omega)$ as a function of frequency ω at a region of anomalous dispersion; phase velocity v_p and group velocity v_g as functions of ω.

7.4 Illustration of Propagation of a Pulse in a Dispersive Medium

To illustrate the ideas of the previous section and to show the validity of the concept of group velocity we will now consider a specific model for the dependence of frequency on wave number and will calculate without approximations the propagation of a pulse in this model medium. Before specifying the particular model it is necessary to state the initial-value problem in more detail than was done in (7.26) and (7.27). As noted there, the proper specification of an initial-value problem for the wave equation demands the initial values of both function $u(x, 0)$ and time derivative $\partial u(x, 0)/\partial t$. If we agree to take the real part of (7.26) to obtain $u(x, t)$,

$$u(x, t) = \frac{1}{2}\frac{1}{\sqrt{2\pi}}\int_{-\infty}^{\infty} A(k)e^{ikx - i\omega(k)t}\, dk + \text{c. c.} \qquad (7.36)$$

then it is easy to show that $A(k)$ is given in terms of the initial values by:

$$A(k) = \frac{1}{\sqrt{2\pi}}\int_{-\infty}^{\infty} e^{-ikx}\left[u(x, 0) + \frac{i}{\omega(k)}\frac{\partial u}{\partial t}(x, 0)\right] dx \qquad (7.37)$$

We will take a Gaussian modulated oscillation

$$u(x, 0) = e^{-x^2/2L^2}\cos k_0 x \qquad (7.38)$$

as the initial shape of the pulse. For simplicity, we will assume that

$$\frac{\partial u}{\partial t}(x, 0) = 0 \qquad (7.39)$$

This means that at times immediately before $t = 0$ the wave consisted of two pulses, both moving towards the origin, such that at $t = 0$ they coalesced into the shape given by (7.38). Clearly at later times we expect each pulse to re-emerge on the other side of the origin. Consequently the initial distribution (7.38) may be expected to split into two identical packets, one moving to the left and one to the right. The Fourier amplitude $A(k)$ for the pulse described by (7.38) and (7.39) is:

$$A(k) = \frac{1}{\sqrt{2\pi}} \int_{-\infty}^{\infty} e^{-ikx} e^{-x^2/2L^2} \cos k_0 x \, dx$$

$$= \frac{L}{2} \left[e^{-(L^2/2)(k-k_0)^2} + e^{-(L^2/2)(k+k_0)^2} \right] \tag{7.40}$$

The symmetry $A(-k) = A(k)$ is a reflection of the presence of two pulses traveling away from the origin, as will be seen below.

In order to calculate the wave form at later times we must specify $\omega = \omega(k)$. As a model allowing exact calculation and showing the essential dispersive effects, we assume

$$\omega(k) = \nu \left(1 + \frac{a^2 k^2}{2} \right) \tag{7.41}$$

where ν is a constant frequency, and a is a constant length which is a typical wavelength where dispersive effects become important. Since the pulse (7.38) is a modulated wave of wave number $k = k_0$, the approximate arguments of the preceding section imply that the two pulses will travel with the group velocity

$$v_g = \frac{d\omega}{dk}(k_0) = \nu a^2 k_0 \tag{7.42}$$

and will be essentially unaltered in shape provided the pulse is not too narrow in space.

The exact behavior of the wave as a function of time is given by (7.36), with (7.40) for $A(k)$:

$$u(x, t) = \frac{L}{2\sqrt{2\pi}} \, \mathrm{Re} \int_{-\infty}^{\infty} \left[e^{-(L^2/2)(k-k_0)^2} + e^{-(L^2/2)(k+k_0)^2} \right] e^{ikx - i\nu t[1 + (a^2 k^2/2)]} \, dk$$

$$\tag{7.43}$$

The integrals can be performed by appropriately completing the squares in the exponents. The result is

$$u(x, t) = \tfrac{1}{2} \, \mathrm{Re} \left\{ \frac{\exp\left[-\dfrac{(x - va^2 k_0 t)^2}{2 L^2 \left(1 + \dfrac{ia^2 vt}{L^2}\right)} \right]}{\left(1 + \dfrac{ia^2 vt}{L^2}\right)^{\!\frac{1}{2}}} \right. $$

$$\left. \times \exp\left[ik_0 x - iv\left(1 + \frac{a^2 k_0^2}{2}\right)t \right] + (k_0 \rightarrow -k_0) \right\} $$

(7.44)

Equation (7.44) represents two pulses traveling in opposite directions. The peak amplitude of each pulse travels with the group velocity (7.42), while the modulation envelope remains Gaussian in shape. The width of the Gaussian is not constant, however, but increases with time. The width of the envelope is

$$L(t) = \left[L^2 + \left(\frac{a^2 vt}{L}\right)^2 \right]^{\frac{1}{2}} \tag{7.45}$$

Thus the dispersive effects on the pulse are greater (for a given elapsed time), the sharper the envelope. The criterion for a small change in shape is that $L \gg a$. Of course, at long times the width of the Gaussian increases linearly with time

$$L(t) \rightarrow \frac{a^2 vt}{L} \tag{7.46}$$

but the time of attainment of this asymptotic form depends on the ratio (L/a). A measure of how rapidly the pulse spreads is provided by a comparison of $L(t)$ given by (7.45), with $v_g t = va^2 k_0 t$. Figure 7.7 shows two examples of curves of the position of peak amplitude $(v_g t)$ and the positions $v_g t \pm L(t)$, which indicate the spread of the pulse, as functions of time. On the left the pulse is not too narrow compared to the wavelength k_0^{-1} and so does not spread too rapidly. The pulse on the right, however, is so narrow initially that it is very rapidly spread out and scarcely represents a pulse after a short time.

Although the above results have been derived for a special choice (7.38) of initial pulse shape and dispersion relation (7.41), their implications are of a more general nature. We have seen in Section 7.3 that the average velocity of a pulse is the group velocity $v_g = d\omega/dk = \omega'$. The spreading of the pulse can be accounted for by noting that a pulse with an initial

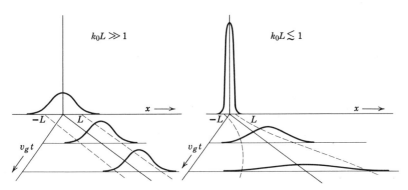

Fig. 7.7 Change in shape of a wave packet as it travels along. The broad packet, containing many wavelengths ($k_0 L \gg 1$), is distorted comparatively little, while the narrow packet ($k_0 L < 1$) broadens and diffuses out rapidly.

spatial width Δx_0 must have inherent in it a spread of wave numbers $\Delta k \sim (1/\Delta x_0)$. This means that the group velocity, when evaluated for various k values within the pulse, has a spread in it of the order

$$\Delta v_g \sim \omega'' \, \Delta k \sim \frac{\omega''}{\Delta x_0} \qquad (7.47)$$

At a time t this implies a spread in position of the order of $\Delta v_g t$. If we combine the uncertainties in position by taking the square root of the sum of squares, we obtain the width $\Delta x(t)$ at time t:

$$\Delta x(t) \simeq \sqrt{(\Delta x_0)^2 + \left(\frac{\omega'' t}{\Delta x_0}\right)^2} \qquad (7.48)$$

We note that (7.48) agrees exactly with (7.45) if we put $\Delta x_0 = L$. The expression (7.48) for $\Delta x(t)$ shows the general result that, if $\omega'' \neq 0$, a narrow pulse spreads rapidly because of its broad spectrum of wave numbers, and vice versa. All these ideas carry over immediately into wave mechanics. They form the basis of the Heisenberg uncertainty principle. In wave mechanics, the frequency is identified with energy divided by Planck's constant, while wave number is momentum divided by Planck's constant.

The problem of wave packets in a dissipative, as well as dispersive, medium is rather complicated. Certain aspects can be discussed analytically, but the analytical expressions are not readily interpreted physically. Wave packets are attenuated and distorted appreciably as they propagate. The reader may refer to Stratton, pp. 301–309, for a discussion of the problem, including numerical examples.

7.5   Reflection and Refraction of Electromagnetic Waves at a Plane Interface between Dielectrics

The reflection and refraction of light at a plane surface between two media of different dielectric properties are familiar phenomena. The various aspects of the phenomena divide themselves into two classes.

(1) Kinematic properties:

(a) Angle of reflection equals angle of incidence.

(b) Snell's law: $\dfrac{\sin i}{\sin r} = \dfrac{n'}{n}$, where i, r are the angles of incidence and refraction, while n, n' are the corresponding indices of refraction.

(2) Dynamic properties:

(a) Intensities of reflected and refracted radiation.

(b) Phase changes and polarization.

The kinematic properties follow immediately from the wave nature of the phenomena and the fact that there are boundary conditions to be satisfied. But they do not depend on the nature of the waves or the boundary conditions. On the other hand, the dynamic properties depend entirely on the specific nature of electromagnetic fields and their boundary conditions.

The coordinate system and symbols appropriate to the problem are shown in Fig. 7.8. The media below and above the plane $z = 0$ have permeabilities and dielectric constants μ, ϵ and μ', ϵ', respectively. A plane wave with wave vector $\mathbf{k}$ and frequency ω is incident from medium μ, ϵ. The refracted and reflected waves have wave vectors $\mathbf{k}'$ and $\mathbf{k}''$, respectively, and $\mathbf{n}$ is a unit normal directed from medium μ, ϵ into medium μ', ϵ'.

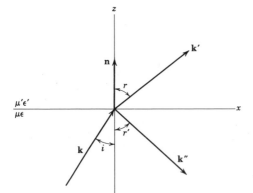

Fig. 7.8 Incident wave $\mathbf{k}$ strikes plane interface between different media, giving rise to a reflected wave $\mathbf{k}''$ and a refracted wave $\mathbf{k}'$.

According to (7.18), the three waves are

INCIDENT

$$\left.\begin{array}{l} \mathbf{E} = \mathbf{E}_0 e^{i\mathbf{k}\cdot\mathbf{x}-i\omega t} \\[2mm] \mathbf{B} = \sqrt{\mu\epsilon}\,\dfrac{\mathbf{k}\times\mathbf{E}}{k} \end{array}\right\} \qquad (7.49)$$

REFRACTED

$$\left.\begin{array}{l} \mathbf{E}' = \mathbf{E}_0' e^{i\mathbf{k}'\cdot\mathbf{x}-i\omega t} \\[2mm] \mathbf{B}' = \sqrt{\mu'\epsilon'}\,\dfrac{\mathbf{k}'\times\mathbf{E}'}{k'} \end{array}\right\} \qquad (7.50)$$

REFLECTED

$$\left.\begin{array}{l} \mathbf{E}'' = \mathbf{E}_0'' e^{i\mathbf{k}''\cdot\mathbf{x}-i\omega t} \\[2mm] \mathbf{B}'' = \sqrt{\mu\epsilon}\,\dfrac{\mathbf{k}''\times\mathbf{E}''}{k} \end{array}\right\} \qquad (7.51)$$

The wave numbers have the magnitudes

$$\left.\begin{array}{l} |\mathbf{k}| = |\mathbf{k}''| = k = \dfrac{\omega}{c}\sqrt{\mu\epsilon} \\[3mm] |\mathbf{k}'| = k' = \dfrac{\omega}{c}\sqrt{\mu'\epsilon'} \end{array}\right\} \qquad (7.52)$$

The existence of boundary conditions at $z = 0$, which boundary conditions must be satisfied at all points on the plane at all times, implies that the spatial (and time) variation of all fields must be the same at $z = 0$. Consequently, we must have the phase factors all equal at $z = 0$,

$$(\mathbf{k}\cdot\mathbf{x})_{z=0} = (\mathbf{k}'\cdot\mathbf{x})_{z=0} = (\mathbf{k}''\cdot\mathbf{x})_{z=0} \qquad (7.53)$$

independent of the nature of the boundary conditions. Equation (7.53) contains the kinematic aspects of reflection and refraction. We see immediately that all three wave vectors must lie in a plane. Furthermore, in the notation of Fig. 7.8,

$$k \sin i = k' \sin r = k'' \sin r' \qquad (7.54)$$

Since $k'' = k$, we find $i = r'$; the angle of incidence equals the angle of reflection. Snell's law is

$$\frac{\sin i}{\sin r} = \frac{k'}{k} = \sqrt{\frac{\mu'\epsilon'}{\mu\epsilon}} = \frac{n'}{n} \qquad (7.55)$$

The dynamic properties are contained in the boundary conditions—normal components of **D** and **B** are continuous; tangential components of

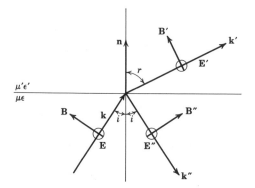

Fig. 7.9 Reflection and refraction
with polarization perpendicular to
the plane of incidence.

E and **H** are continuous. In terms of fields (7.49)–(7.51) these boundary
conditions at $z = 0$ are:

$$\left.\begin{array}{c}
[\epsilon(\mathbf{E}_0 + \mathbf{E}_0'') - \epsilon'\mathbf{E}_0'] \cdot \mathbf{n} = 0 \\[4pt]
[\mathbf{k} \times \mathbf{E}_0 + \mathbf{k}'' \times \mathbf{E}_0'' - \mathbf{k}' \times \mathbf{E}_0'] \cdot \mathbf{n} = 0 \\[4pt]
(\mathbf{E}_0 + \mathbf{E}_0'' - \mathbf{E}_0') \times \mathbf{n} = 0 \\[4pt]
\left[\dfrac{1}{\mu}(\mathbf{k} \times \mathbf{E}_0 + \mathbf{k}'' \times \mathbf{E}_0'') - \dfrac{1}{\mu'}(\mathbf{k}' \times \mathbf{E}_0')\right] \times \mathbf{n} = 0
\end{array}\right\} \quad (7.56)$$

In applying these boundary conditions it is convenient to consider two
separate situations, one in which the incident plane wave is linearly
polarized with its polarization vector perpendicular to the plane of
incidence (the plane defined by **k** and **n**), and the other in which the
polarization vector is parallel to the plane of incidence. The general case
of arbitrary elliptic polarization can be obtained by appropriate linear
combinations of the two results, following the methods of Section 7.2.

We first consider the electric field perpendicular to the plane of incidence,
as shown in Fig. 7.9. All the electric fields are shown directed away from
the viewer. The orientations of the **B** vectors are chosen to give a positive
flow of energy in the direction of the wave vectors. Since the electric
fields are all parallel to the surface, the first boundary condition in (7.56)
yields nothing. The third and fourth equations in (7.56) give

$$\left.\begin{array}{c}
E_0 + E_0'' - E_0' = 0 \\[6pt]
\sqrt{\dfrac{\epsilon}{\mu}}\,(E_0 - E_0'')\cos i - \sqrt{\dfrac{\epsilon'}{\mu'}}\,E_0'\cos r = 0
\end{array}\right\} \quad (7.57)$$

while the second, using Snell's law, duplicates the third. The relative amplitudes of the refracted and reflected waves can be found from (7.57). These are:

E PERPENDICULAR TO PLANE OF INCIDENCE

$$\frac{E_0{}'}{E_0} = \frac{2}{1 + \dfrac{\mu \tan i}{\mu' \tan r}} \rightarrow \frac{2 \cos i \sin r}{\sin (i + r)}$$

$$\frac{E_0{}''}{E_0} = \frac{1 - \dfrac{\mu \tan i}{\mu' \tan r}}{1 + \dfrac{\mu \tan i}{\mu' \tan r}} \rightarrow - \frac{\sin (i - r)}{\sin (i + r)}$$

(7.58)

The expression on the right in each case is the result appropriate for $\mu' = \mu$, as is generally true for optical frequencies.

If the electric field is parallel to the plane of incidence, as shown in Fig. 7.10, the boundary conditions involved are normal D, tangential E, and tangential H [the first, third, and fourth equations in (7.56)]. The tangential E and H continuous demand that

$$\cos i \, (E_0 - E_0{}'') - \cos r \, E_0{}' = 0$$

$$\sqrt{\frac{\epsilon}{\mu}} (E_0 + E_0{}'') - \sqrt{\frac{\epsilon'}{\mu'}} E_0{}' = 0$$

(7.59)

Normal D continuous, plus Snell's law, merely duplicates the second of these equations. The relative amplitudes of refracted and reflected fields are therefore

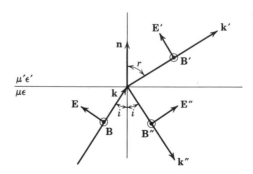

Fig. 7.10 Reflection and refraction with polarization parallel to the plane of incidence.

E PARALLEL TO PLANE OF INCIDENCE

$$\left.\begin{aligned}
\frac{E_0{}'}{E_0} &= 2\sqrt{\frac{\mu\epsilon}{\mu'\epsilon'}} \; \frac{\sin 2i}{\sin 2r + \dfrac{\mu}{\mu'}\sin 2i} \rightarrow \frac{2\cos i \sin r}{\sin(i+r)\cos(i-r)} \\[2em]
\frac{E_0{}''}{E_0} &= \frac{\dfrac{\mu}{\mu'}\sin 2i - \sin 2r}{\sin 2r + \dfrac{\mu}{\mu'}\sin 2i} \rightarrow \frac{\tan(i-r)}{\tan(i+r)}
\end{aligned}\right\} \qquad (7.60)$$

Again the results on the right apply for $\mu' = \mu$.
For normal incidence ($i = 0$), both (7.58) and (7.60) reduce to

$$\left.\begin{aligned}
\frac{E_0{}'}{E_0} &= \frac{2}{\sqrt{\dfrac{\mu\epsilon'}{\mu'\epsilon}} + 1} \rightarrow \frac{2n}{n'+n} \\[2em]
\frac{E_0{}''}{E_0} &= \frac{\sqrt{\dfrac{\mu\epsilon'}{\mu'\epsilon}} - 1}{\sqrt{\dfrac{\mu\epsilon'}{\mu'\epsilon}} + 1} \rightarrow \frac{n'-n}{n'+n}
\end{aligned}\right\} \qquad (7.61)$$

For the reflected wave the sign convention is that for polarization parallel to the plane of incidence. This means that if $n' > n$ there is a phase reversal for the reflected wave.

7.6 Polarization by Reflection and Total Internal Reflection

Two aspects of the dynamical relations on reflection and refraction are worthy of mention. The first is that for polarization parallel to the plane of incidence there is an angle of incidence, called *Brewster's angle*, for which there is no reflected wave. Putting $\mu' = \mu$ for simplicity, we see from (7.60) that there will be no reflected wave when $i + r = \pi/2$. From Snell's law (7.55) we find that this specifies Brewster's angle to be

$$i_B = \tan^{-1}\left(\frac{n'}{n}\right) \qquad (7.62)$$

For a typical ratio $(n'/n) = 1.5$, $i_B \simeq 56°$. If a plane wave of mixed polarization is incident on a plane interface at the Brewster angle, the reflected radiation is *completely* plane *polarized* with polarization vector *perpendicular* to the plane of incidence. This behavior can be utilized to produce beams of plane-polarized light, but is not as efficient as other means employing anisotropic properties of some dielectric media. Even if the unpolarized wave is reflected at angles other than the Brewster angle, there is a tendency for the reflected wave to be predominantly polarized perpendicular to the plane of incidence. The success of dark glasses which selectively transmit only one direction of polarization depends on this fact. In the domain of radiofrequencies, receiving antennas can be so oriented as to discriminate against surface-reflected waves (and also waves reflected from the ionosphere) in favor of the directly transmitted wave.

The second phenomenon is called *total internal reflection*. The word internal implies that the incident and reflected waves are in a medium of larger index of refraction than the refracted wave ($n > n'$). Snell's law (7.55) shows that, if $n > n'$, then $r > i$. Consequently, $r = \pi/2$ when $i = i_0$, where

$$i_0 = \sin^{-1}\left(\frac{n'}{n}\right) \tag{7.63}$$

For waves incident at $i = i_0$, the refracted wave is propagated parallel to the surface. There can be no energy flow across the surface. Hence at that angle of incidence there must be total reflection. What happens if $i > i_0$? To answer this we first note that, for $i > i_0$, $\sin r > 1$. This means that r is a complex angle with a purely imaginary cosine.

$$\cos r = i\sqrt{\left(\frac{\sin i}{\sin i_0}\right)^2 - 1} \tag{7.64}$$

The meaning of these complex quantities becomes clear when we consider the propagation factor for the refracted wave:

$$e^{i\mathbf{k'}\cdot\mathbf{x}} = e^{ik'(x\sin r + z\cos r)} = e^{-k'[(\sin i/\sin i_0)^2 - 1]^{\frac{1}{2}}z}\, e^{ik'(\sin i/\sin i_0)x} \tag{7.65}$$

This shows that, for $i > i_0$, the refracted wave is propagated only parallel to the surface and is attenuated exponentially beyond the interface. The attenuation occurs within a very few wavelengths of the boundary, except for $i \simeq i_0$.

Even though fields exist on the other side of the surface it is clear that there is no energy flow through the surface. Hence total internal reflection occurs for $i \geq i_0$. The lack of energy flow can be verified by calculating the time-averaged normal component of the Poynting's vector just inside the surface:

$$\mathbf{S}\cdot\mathbf{n} = \frac{c}{8\pi}\,\mathrm{Re}\,[\mathbf{n}\cdot(\mathbf{E'}\times\mathbf{H'}^*)] \tag{7.66}$$

with $\mathbf{H}' = \dfrac{c}{\mu'\omega}\,(\mathbf{k}' \times \mathbf{E}')$, we find

$$\mathbf{S} \cdot \mathbf{n} = \frac{c^2}{8\pi\omega\mu'}\,\mathrm{Re}\,[(\mathbf{n} \cdot \mathbf{k}')\,|\mathbf{E}_0'|^2] \qquad (7.67)$$

But $\mathbf{n} \cdot \mathbf{k}' = k' \cos r$ is purely imaginary, so that $\mathbf{S} \cdot \mathbf{n} = 0$.

The phenomenon of total internal reflection is exploited in many applications where it is required to transmit light without loss in intensity. In nuclear physics Lucite or other plastic "light pipes" are used to carry light emitted from a scintillation crystal because of the passage of an ionizing particle to a photomultiplier tube, where it is converted into a useful electric signal. The photomultiplier must often be some distance away from the scintillation crystal because of space limitations or magnetic fields which disturb its performance. If the light pipe is large in cross section compared to a wavelength of the radiation involved, the considerations presented here for a plane interface have approximate validity. When the dielectric medium has cross-sectional dimensions of the order of a wavelength, however, the precise geometry must be taken into account. Then the propagation is that of a dielectric wave guide (see Section 8.8).

7.7 Waves in a Conducting Medium

If the medium in which waves are propagating is a conductor, there are characteristic differences in the propagation, when compared with non-conducting media. If the medium is characterized by a conductivity σ, as well as a dielectric constant ϵ and permeability μ, Maxwell's equations are supplemented by Ohm's law:

$$\mathbf{J} = \sigma\mathbf{E} \qquad (7.68)$$

Hence they take the form:

$$\left.\begin{array}{ll} \nabla \cdot \mu\mathbf{H} = 0 & \nabla \times \mathbf{E} + \dfrac{\mu}{c}\dfrac{\partial \mathbf{H}}{\partial t} = 0 \\[3mm] \nabla \cdot \epsilon\mathbf{E} = 0 & \nabla \times \mathbf{H} - \dfrac{\epsilon}{c}\dfrac{\partial \mathbf{E}}{\partial t} - \dfrac{4\pi\sigma}{c}\mathbf{E} = 0 \end{array}\right\} \qquad (7.69)$$

In the insulating dielectric we found that the *time-varying* fields were *transverse*, i.e., the field vectors $\mathbf{E}$ and $\mathbf{H}$ were perpendicular to the direction in which the spatial variation occurred. In the limit of zero frequency we know from our study of electro- and magnetostatics that the *static* fields in a dielectric are *longitudinal*, in the sense that the fields are derivable from scalar potentials and so point in the direction of the spatial variation.

If the conductivity is not zero, modifications arise. For simplicity, consider fields which vary in only one spatial variable, ξ. We decompose the fields into longitudinal and transverse parts:

$$\left.\begin{array}{l} \mathbf{E}(\xi, t) = \mathbf{E}_{\text{long}}(\xi, t) + \mathbf{E}_{\text{tr}}(\xi, t) \\[2mm] \mathbf{H}(\xi, t) = \mathbf{H}_{\text{long}}(\xi, t) + \mathbf{H}_{\text{tr}}(\xi, t) \end{array}\right\} \tag{7.70}$$

Then, because of the properties of curl operation, we find that the transverse parts of $\mathbf{E}$ and $\mathbf{H}$ satisfy the two curl equations in (7.69), leading to transverse waves (see below), while the longitudinal parts satisfy the equations:

$$\left.\begin{array}{ll} \dfrac{\partial H_{\text{long}}}{\partial \xi} = 0 & \dfrac{\partial H_{\text{long}}}{\partial t} = 0 \\[4mm] \dfrac{\partial E_{\text{long}}}{\partial \xi} = 0 & \left(\dfrac{\partial}{\partial t} + \dfrac{4\pi\sigma}{\epsilon}\right) E_{\text{long}} = 0 \end{array}\right\} \tag{7.71}$$

The first pair of equations shows that the only longitudinal magnetic field possible is a static uniform field. This is the same situation as in an insulator. But the second pair in (7.71) shows that the longitudinal electric field is uniform in space, while having the time variation:

$$E_{\text{long}}(\xi, t) = E_0 e^{-4\pi\sigma t/\epsilon} \tag{7.72}$$

Consequently, no static longitudinal fields can exist in a conducting medium in the absence of an applied current density. For good conductors like copper, $\sigma \sim 10^{17} \text{ sec}^{-1}$ so that disturbances are damped out in an extremely short time.

We now consider the transverse fields in the conducting medium. Assuming that the fields vary as $\exp(i\mathbf{k} \cdot \mathbf{x} - i\omega t)$, the first curl equation of (7.69) yields:

$$\mathbf{H} = \frac{c}{\mu\omega}(\mathbf{k} \times \mathbf{E}) \tag{7.73}$$

while the second gives

$$i(\mathbf{k} \times \mathbf{H}) + i\epsilon\frac{\omega}{c}\mathbf{E} - \frac{4\pi\sigma}{c}\mathbf{E} = 0 \tag{7.74}$$

Elimination of either $\mathbf{H}$ or $\mathbf{E}$ from this last equation with (7.73) yields

$$\left[k^2 - \left(\mu\epsilon\frac{\omega^2}{c^2} + 4\pi i\frac{\mu\omega\sigma}{c^2}\right)\right]\left\{\begin{array}{c}\mathbf{E}\\\mathbf{H}\end{array}\right\} = 0 \tag{7.75}$$

This means that the propagation vector k is complex:

$$k^2 = \mu\epsilon\frac{\omega^2}{c^2}\left(1 + i\frac{4\pi\sigma}{\omega\epsilon}\right) \tag{7.76}$$

The first term corresponds to the displacement-current and the second to the conduction-current contribution. In taking the square root to find k the branch is chosen to give the familiar results when $\sigma = 0$. Then one finds, assuming that σ is real,

$$k = \alpha + i\beta$$

where

$$\left.\begin{array}{c}\alpha\\\beta\end{array}\right\} = \sqrt{\mu\epsilon}\,\frac{\omega}{c}\left[\frac{\sqrt{1 + \left(\dfrac{4\pi\sigma}{\omega\epsilon}\right)^2} \pm 1}{2}\right]^{\frac{1}{2}} \tag{7.77}$$

For a poor conductor $\left(\dfrac{4\pi\sigma}{\omega\epsilon} \ll 1\right)$ we find approximately

$$k = \alpha + i\beta \simeq \sqrt{\mu\epsilon}\,\frac{\omega}{c} + i\,\frac{2\pi}{c}\sqrt{\frac{\mu}{\epsilon}}\,\sigma \tag{7.78}$$

correct to first order in $(\sigma/\omega\epsilon)$. In this limit Re $k \gg$ Im k and the attenuation of the wave (Im k) is independent of frequency, aside from the possible frequency variation of the conductivity. For a good conductor $\left(\dfrac{4\pi\sigma}{\omega\epsilon} \gg 1\right)$, on the other hand, α and β are approximately equal:

$$k \simeq (1 + i)\,\frac{\sqrt{2\pi\omega\mu\sigma}}{c} \tag{7.79}$$

where only the lowest-order terms in $(\omega\epsilon/\sigma)$ have been kept.

The waves propagating as $\exp{(i\mathbf{k} \cdot \mathbf{x} - i\omega t)}$ are damped, transverse waves. The fields can be written as

$$\left.\begin{array}{l}\mathbf{E} = \mathbf{E}_0 e^{-\beta\mathbf{n}\cdot\mathbf{x}}\,e^{i\alpha\mathbf{n}\cdot\mathbf{x} - i\omega t}\\[2mm]\mathbf{H} = \mathbf{H}_0 e^{-\beta\mathbf{n}\cdot\mathbf{x}}\,e^{i\alpha\mathbf{n}\cdot\mathbf{x} - i\omega t}\end{array}\right\} \tag{7.80}$$

where $\mathbf{n}$ is a unit vector in the direction of $\mathbf{k}$. The divergence equation for $\mathbf{E}$ shows that $\mathbf{E}_0 \cdot \mathbf{n} = 0$, while the relation between $\mathbf{H}$ and $\mathbf{E}$ (7.73) gives

$$\mathbf{H}_0 = \frac{c}{\mu\omega}(\alpha + i\beta)\mathbf{n} \times \mathbf{E}_0 \tag{7.81}$$

This shows that $\mathbf{H}$ and $\mathbf{E}$ are out of phase in a conductor. Defining the magnitude and phase of k:

$$\left.\begin{array}{l}|k| = \sqrt{\alpha^2 + \beta^2} = \sqrt{\mu\epsilon}\,\dfrac{\omega}{c}\left[1 + \left(\dfrac{4\pi\sigma}{\omega\epsilon}\right)^2\right]^{\frac{1}{4}}\\[4mm]\phi = \tan^{-1}\dfrac{\beta}{\alpha} = \frac{1}{2}\tan^{-1}\left(\dfrac{4\pi\sigma}{\omega\epsilon}\right)\end{array}\right\} \tag{7.82}$$

equation (7.81) can be written in the form:

$$\mathbf{H}_0 = \sqrt{\frac{\epsilon}{\mu}} \left[1 + \left(\frac{4\pi\sigma}{\omega\epsilon} \right)^2 \right]^{1/4} e^{i\phi} \mathbf{n} \times \mathbf{E}_0 \qquad (7.83)$$

The interpretation of (7.83) is that $\mathbf{H}$ lags $\mathbf{E}$ in time by the phase angle ϕ and has a relative amplitude:

$$\frac{|\mathbf{H}_0|}{|\mathbf{E}_0|} = \sqrt{\frac{\epsilon}{\mu}} \left[1 + \left(\frac{4\pi\sigma}{\omega\epsilon} \right)^2 \right]^{1/4} \qquad (7.84)$$

In very good conductors we see that the magnetic field is very large compared to the electric field and lags in phase by almost 45°. The field energy is almost entirely magnetic in nature.

The waves given by (7.80) show an exponential damping with distance. This means that an electromagnetic wave entering a conductor is damped to $1/e = 0.369$ of its initial amplitude in a distance:

$$\delta = \frac{1}{\beta} \simeq \frac{c}{\sqrt{2\pi\mu\omega\sigma}} \qquad (7.85)$$

the last form being the approximation for good conductors. The distance δ is called the *skin depth* or the *penetration depth*.* For a conductor like copper, $\delta \simeq 0.85$ cm for frequencies of 60 cps, and $\delta \simeq 0.71 \times 10^{-3}$ cm for 100 Mc/sec. This rapid attenuation of waves means that in high-frequency circuits current flows only on the surface of the conductors. One simple consequence is that the high-frequency inductance of circuit elements is somewhat smaller than the low-frequency inductance because of the expulsion of flux from the interior of the conductors.

The problem of reflection and refraction at an interface between conducting media is rather complicated and will not be treated here. The interested reader may refer to Stratton, pp. 500 ff., for a discussion of this point. See, however, Section 8.1 for a treatment of fields at the interface between a dielectric and a good conductor.

7.8   Simple Model for Conductivity

The simplest model of conduction, due originally to Drude (1900), is that in a metal there are a certain number n_0 of electrons per unit volume free to move under the action of applied electric fields, but subject to

---

* For reference, the skin depth (7.85) appears in mks units as $\delta = (2/\mu\omega\sigma)^{1/2}$.

damping force due to collisions. Thus the equation of motion of such an electron is

$$m \frac{d\mathbf{v}}{dt} + mg\mathbf{v} = e\mathbf{E}(\mathbf{x}, t) \qquad (7.86)$$

where g is the damping constant.* For rapidly oscillating fields the displacement of the electron is small compared to a wavelength so that approximately

$$m \frac{d\mathbf{v}}{dt} + mg\mathbf{v} = e\mathbf{E}_0 e^{-i\omega t} \qquad (7.87)$$

where $\mathbf{E}_0$ is the electric field at the average position of the electron. The steady-state solution for the velocity of the electron is:

$$\mathbf{v} = \frac{e}{m(g - i\omega)} \mathbf{E}_0 e^{-i\omega t} \qquad (7.88)$$

so that the conductivity is given by

$$\sigma = \frac{n_0 e^2}{m(g - i\omega)} \qquad (7.89)$$

Assuming one free electron per atom, a metal such as copper ($n_0 \simeq 8 \times 10^{22}$ electrons/cm³, $\sigma \simeq 5 \times 10^{17}$ sec⁻¹) has an empirical damping constant $g \simeq 3 \times 10^{13}$ sec⁻¹. This shows that for frequencies of the order of, or smaller than, microwave frequencies ($\sim 10^{10}$ sec⁻¹) metallic conductivities are essentially real (i.e., current in phase with the field) and independent of frequency. At higher frequencies (in the infrared and beyond), however, the conductivity is complex and depends markedly on frequency in a manner qualitatively described by the simple result (7.89).

7.9   Transverse Waves in a Tenuous Plasma

In certain situations, such as the ionosphere or a tenuous plasma, the damping of the motion of the free electrons due to collisions becomes negligible. Then the "conductivity" becomes purely imaginary:

$$\sigma_{\text{plasma}} \simeq i \frac{n_0 e^2}{m\omega} \qquad (7.90)$$

---

* The damping constant g is some sort of average rate of collisions involving appreciable momentum transfer. Collisions occur between electrons and lattice vibrations, lattice imperfections, and impurities. The proper calculation of g involves quantum-mechanical considerations, including the effects of the Pauli exclusion principle. See A. H. Wilson, *Theory of Metals*, 2nd ed., Cambridge University Press (1953).

Quotation marks are placed on "conductivity" because there is no resistive loss of energy if the current and electric field are out of phase. The propagation of transverse electromagnetic waves in a tenuous plasma is governed by equation (7.76) of Section 7.7, with σ_{plasma} (7.90) inserted for σ:*

$$k^2 \simeq \frac{\omega^2}{c^2}\left(1 - \frac{\omega_p^2}{\omega^2}\right) \tag{7.91}$$

where

$$\omega_p^2 = \frac{4\pi n_0 e^2}{m} \tag{7.92}$$

is called the *plasma frequency*. Since the wave number can be written as $k = n\omega/c$, where n is the index of refraction, we see that the index of refraction of a plasma is given by

$$n^2 \simeq 1 - \frac{\omega_p^2}{\omega^2} \tag{7.93}$$

For high-frequency radiation ($\omega > \omega_p$) the index of refraction is real and the waves propagate freely. For frequencies lower than the plasma frequency ω_p, n is purely imaginary. Consequently such electromagnetic waves incident on a plasma will be reflected from the surface. Within the plasma the fields will fall off exponentially with distance from the surface. The penetration depth δ_{plasma} is given by

$$\delta_{\text{plasma}} = \frac{c}{\sqrt{\omega_p^2 - \omega^2}} \simeq \frac{c}{\omega_p} \tag{7.94}$$

the last value being valid for $\omega \ll \omega_p$. On the laboratory scale, plasma densities are in the range $n_0 \simeq 10^{12}\text{--}10^{16}$ electrons/cm^3. This means $\omega_p \simeq 6 \times 10^{10}\text{--}6 \times 10^{12}$ sec^{-1}, so that typical penetration depths are of the order of 0.5 cm–5 $\times$ 10^{-3} cm for static or low-frequency fields. The expulsion of fields from within a plasma is a well-known effect in controlled thermonuclear processes and is exploited in attempts at confinement of hot plasmas (see Sections 10.5 and 10.6).

The simple result (7.93) for the index of refraction of a plasma is modified by the presence of an external static magnetic induction. This circumstance arises not only in the laboratory, but also in the ionosphere, where the earth's dipole field provides the external magnetic induction. To illustrate the influence of the external field we consider the simple

---

* Sometimes this equation is solved for ω^2 as a function of k:

$$\omega^2 \simeq \omega_p^2 + c^2 k^2$$

Then it is called a dispersion relation for $\omega = \omega(k)$.

problem of a tenuous electronic plasma of uniform density with a strong, static, uniform, magnetic induction $\mathbf{B}_0$ and transverse waves propagating parallel to the direction of $\mathbf{B}_0$. If the amplitude of the electronic motion is small and collisions are neglected, the equation of motion is approximately:

$$m \frac{d\mathbf{v}}{dt} \simeq e\mathbf{E}e^{-i\omega t} + e \frac{\mathbf{v}}{c} \times \mathbf{B}_0 \tag{7.95}$$

where the influence of the $\mathbf{B}$ field of the transverse wave has been neglected compared to the static induction $\mathbf{B}_0$. It is convenient to consider the transverse waves as circularly polarized. Then

$$\mathbf{E} = E(\boldsymbol{\epsilon}_1 \pm i\boldsymbol{\epsilon}_2) \tag{7.96}$$

while $\mathbf{B}_0$ is in the direction of $\boldsymbol{\epsilon}_3$. Since we are looking for a steady-state solution, we will assume that the velocity of the electron is of the form:

$$\mathbf{v}(t) = v(\boldsymbol{\epsilon}_1 \pm i\boldsymbol{\epsilon}_2)e^{-i\omega t} \tag{7.97}$$

Then from (7.95), using (7.96), we find immediately

$$v = \frac{ie}{m(\omega \pm \omega_B)} E \tag{7.98}$$

where ω_B is the frequency of precession of a charged particle in a magnetic field,

$$\omega_B = \frac{eB_0}{mc} \tag{7.99}$$

Result (7.98) can be understood by noting that, in a coordinate system precessing with frequency ω_B, the electron is driven by a rotating electric field of effective frequency $\omega \pm \omega_B$, depending on the sign of the circular polarization.

The current density in the plasma due to electronic motion is

$$\mathbf{J} = en_0\mathbf{v} = \frac{in_0e^2}{m(\omega \pm \omega_B)} \mathbf{E} \tag{7.100}$$

When this current density is added to the displacement current, Maxwell's generalization of Ampère's law becomes:

$$\nabla \times \mathbf{H} = -i \frac{\omega}{c} \left[1 - \frac{\omega_p{}^2}{\omega(\omega \pm \omega_B)} \right] \mathbf{E} \tag{7.101}$$

The factor in square brackets can be interpreted as the dielectric constant or square of the index of refraction:

$$n_{\pm}{}^2 = 1 - \frac{\omega_p{}^2}{\omega(\omega \pm \omega_B)} \tag{7.102}$$

This is the extension of (7.93) to include a static magnetic induction. It is not completely general, since it applies only to waves propagating along the static field direction. But even in this simple example we see the essential characteristic that waves of right-handed and left-handed circular polarizations propagate differently. The ionosphere is birefringent. For propagation in directions other than parallel to the static field B_0 it is straightforward to show that, if terms of the order of ω_B^2 are neglected compared to ω^2 and $\omega\omega_B$, the index of refraction is still given by (7.102). But the precession frequency (7.99) is now to be interpreted as that due to only the component of B_0 parallel to the direction of propagation. This means that ω_B in (7.102) is a function of angle—the medium is not only birefringent, but also anisotropic.

For the ionosphere a typical maximum density of free electrons is $n_0 \simeq 10^4$–10^6 electrons/cm^3, corresponding to a plasma frequency of the order of $\omega_p \simeq 6 \times 10^6$–$6 \times 10^7$ sec^{-1}. If we take a value of 0.3 gauss as representative of the earth's magnetic field, the precession frequency is $\omega_B \simeq 6 \times 10^6$ sec^{-1}.

Figure 7.11 shows $n_{\pm}^2$ as a function of frequency for two values of the ratio of (ω_p/ω_B). In both examples there are wide intervals of frequency where one of n_+^2 or n_-^2 is positive while the other is negative. At such frequencies one state of circular polarization cannot propagate in the plasma. Consequently a wave of that polarization incident on the plasma will be totally reflected. The other state of polarization will be partially transmitted. Thus, when a linearly polarized wave is incident on a plasma, the reflected wave will be elliptically polarized, with its major axis generally rotated away from the direction of the polarization of the incident wave.

The behavior of radio waves reflected from the ionosphere is explicable in terms of these ideas, but the presence of several layers of plasma with densities and relative positions varying with height and time makes the problem considerably more complicated than our simple example. The electron densities at various heights can be inferred by studying the reflection of pulses of radiation transmitted vertically upwards. The number of free electrons per unit volume increases slowly with height in a given layer of the ionosphere, as shown in Fig. 7.12, reaches a maximum, and then falls abruptly with further increase in height. A pulse of a given frequency ω_1 enters the layer without reflection because of the slow change in n_0. When the density n_0 is large enough, however, $\omega_p(h_1) \simeq \omega_1$. Then the indices of refraction (7.102) vanish and the pulse is reflected. The actual density n_0 where the reflection occurs is given by the roots of the right-hand side of (7.102). By observing the time interval between the initial transmission and reception of the reflected signal the height h_1

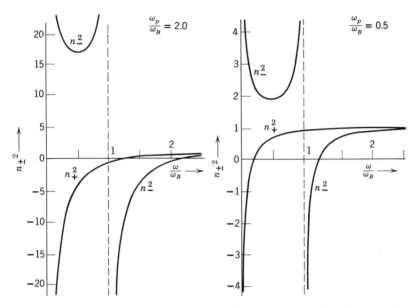

Fig. 7.11 Indices of refraction as a function of frequency for model of the ionosphere (tenuous electronic plasma in a static, uniform magnetic induction). $n_{\pm}(\omega)$ apply to right and left circularly polarized waves propagating parallel to the magnetic field. ω_B is the gyration frequency; ω_p is the plasma frequency. The two sets of curves correspond to $\omega_p/\omega_B = 2.0, 0.5$.

corresponding to that density can be found. By varying the frequency ω_1 and studying the change in time intervals the electron density as a function of height can be determined. If the frequency ω_1 is too high, the index of refraction does not vanish and very little reflection occurs. The frequency above which reflections disappear determines the maximum electron density in a given layer.

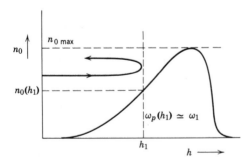

Fig. 7.12 Electron density as a function of height in a layer of the ionosphere (schematic).

REFERENCES AND SUGGESTED READING

The whole subject of optics as an electromagnetic phenomenon is treated authoritatively by
   Born and Wolf.
Their first chapter covers plane waves, polarization, and reflection and refraction, among other topics. A very complete discussion of plane waves incident on boundaries of dielectrics and conductors is given by
   Stratton, Chapter IX.
Another good treatment of electromagnetic waves in both isotropic and anisotropic media is that of
   Landau and Lifshitz, *Electrodynamics of Continuous Media*, Chapters X and XI.
A more elementary, but clear and thorough, approach to plane waves and their properties appears in
   Adler, Chu, and Fano, Chapters 7 and 8.
The propagation of waves in dispersive media is discussed in detail in the book by
   Brillouin.
The distortion and attenuation of pulses in dissipative materials are covered by
   Stratton, pp. 301–309.

PROBLEMS

7.1 An approximately monochromatic plane wave packet in one dimension has the instantaneous form, $u(x, 0) = f(x)e^{ik_0x}$, with $f(x)$ the modulation envelope. For each of the forms $f(x)$ below, calculate the wave-number spectrum $|A(k)|^2$ of the packet, sketch $|u(x, 0)|^2$ and $|A(k)|^2$, evaluate explicitly the rms deviations from the means, Δx and Δk (defined in terms of the intensities $|u(x, 0)|^2$ and $|A(k)|^2$), and test inequality (7.28).

   (a) $f(x) = Ne^{-\alpha|x|/2}$

   (b) $f(x) = Ne^{-\alpha^2x^2/4}$

   (c) $f(x) = \begin{cases} N(1 - \alpha|x|) & \text{for } \alpha|x| < 1 \\ 0 & \text{for } \alpha|x| > 1 \end{cases}$

   (d) $f(x) = \begin{cases} N & \text{for } |x| < a \\ 0 & \text{for } |x| > a \end{cases}$

7.2 A plane wave is incident on a layered interface as shown in the figure (p. 232). The indices of refraction of the three nonpermeable media are n_1, n_2, n_3. The thickness of the intermediate layer is d.

   (a) Calculate the transmission and reflection coefficients (ratios of transmitted and reflected Poynting's flux to the incident flux), and sketch their behavior as a function of frequency for $n_1 = 1, n_2 = 2, n_3 = 3$; $n_1 = 3, n_2 = 2, n_3 = 1$; and $n_1 = 2, n_2 = 4, n_3 = 1$.

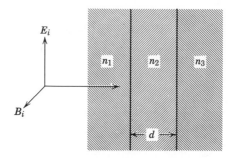

(*b*) The medium n_1 is part of an optical system (e.g., a lens); medium n_3 is air ($n_3 = 1$). It is desired to put an optical coating (medium n_2) on the surface so that there is no reflected wave for a frequency ω_0. What thickness d and index of refraction n_2 are necessary?

7.3 Two plane semi-infinite slabs of the same uniform, isotropic, nonpermeable, lossless dielectric with index of refraction n are parallel and separated by an air gap ($n = 1$) of width d. A plane electromagnetic wave of frequency ω is incident on the gap from one of the slabs with angle of incidence i. For linear polarization *both* parallel to *and* perpendicular to the plane of incidence,

(*a*) calculate the ratio of power transmitted into the second slab to the incident power and the ratio of reflected to incident power;

(*b*) for i greater than the critical angle for total internal reflection, sketch the ratio of transmitted power to incident power as a function of d measured in units of wavelength in the gap.

7.4 A plane polarized electromagnetic wave of frequency ω in free space is incident normally on the flat surface of a nonpermeable medium of conductivity σ and dielectric constant ϵ.

(*a*) Calculate the amplitude and phase of the reflected wave relative to the incident wave for arbitrary σ and ϵ.

(*b*) Discuss the limiting cases of a very poor and a very good conductor, and show that for a good conductor the reflection coefficient (ratio of reflected to incident intensity) is approximately

$$R \simeq 1 - 2\frac{\omega}{c}\delta$$

where δ is the skin depth.

7.5 A plane polarized electromagnetic wave $\mathbf{E} = \mathbf{E}_i e^{i\mathbf{k}\cdot\mathbf{x} - i\omega t}$ is incident normally on a flat uniform sheet of an *excellent* conductor ($\sigma \gg \omega$) having a thickness t. Assuming that in space and in the conducting sheet $\mu = \epsilon = 1$, discuss the reflection and transmission of the incident wave.

(*a*) Show that the amplitudes of the reflected and transmitted waves, correct to the first order in $(\omega/\sigma)^{\frac{1}{2}}$, are:

$$\frac{E_r}{E_i} = \frac{-(1 - \beta)(1 - e^{-2\lambda})}{(1 - e^{-2\lambda}) + \beta(1 + 3e^{-2\lambda})}$$

$$\frac{E_t}{E_i} = \frac{4\beta e^{-\lambda}}{(1 - e^{-2\lambda}) + \beta(1 + 3e^{-2\lambda})}$$

where

$$\beta = \sqrt{\frac{\omega}{8\pi\sigma}}\,(1 - i) = \frac{\omega\delta}{2c}\,(1 - i)$$

$$\lambda = (1 - i)t/\delta$$

and $\delta = c/\sqrt{2\pi\omega\sigma}$ is the penetration depth.

(b) Verify that for zero thickness and infinite thickness you obtain the proper limiting results.

(c) Show that, except for sheets of very small thickness, the transmission coefficient is

$$T = \frac{32(\mathrm{Re}\ \beta)^2 e^{-2t/\delta}}{1 - 2e^{-2t/\delta}\cos(2t/\delta) + e^{-4t/\delta}}$$

Sketch $\log T$ as a function of (t/δ), assuming $\mathrm{Re}\ \beta = 10^{-2}$.
Define "very small thickness."

7.6 Plane waves propagate in a homogeneous, nonpermeable, but *anisotropic* dielectric. The dielectric is characterized by a tensor ϵ_{ij}, but if coordinate axes are chosen as the principal axes the components of displacement along these axes are related to the electric-field components by $D_i = \epsilon_i E_i$ ($i = 1, 2, 3$), where ϵ_i are the eigenvalues of the matrix ϵ_{ij}.

(a) Show that plane waves with frequency ω and wave vector $\mathbf{k}$ must satisfy

$$\mathbf{k} \times (\mathbf{k} \times \mathbf{E}) + \frac{\omega^2}{c^2}\mathbf{D} = 0$$

(b) Show that for a given wave vector $\mathbf{k} = k\mathbf{n}$ there are two distinct modes of propagation with different phase velocities $v = \omega/k$ which satisfy the Fresnel equation,

$$\sum_{i=1}^{3} \frac{n_i^2}{v^2 - v_i^2} = 0$$

where $v_i = c/\sqrt{\epsilon_i}$ is called a principal velocity, and n_i is the component of $\mathbf{n}$ along the ith principal axis.

(c) Show that $\mathbf{D}_a \cdot \mathbf{D}_b = 0$, where $\mathbf{D}_a$, $\mathbf{D}_b$ are the displacements associated with the two modes of propagation.

7.7 A homogeneous, isotropic, nonpermeable dielectric is characterized by an index of refraction $n(\omega)$ which is in general complex in order to describe absorptive processes.

(a) Show that the general solution for plane waves in one dimension can be written

$$u(x, t) = \frac{1}{\sqrt{2\pi}}\int_{-\infty}^{\infty} d\omega\, e^{-i\omega t}[A(\omega)e^{i(\omega/c)n(\omega)x} + B(\omega)e^{-i(\omega/c)n(\omega)x}]$$

where $u(x, t)$ is a component of $\mathbf{E}$ or $\mathbf{B}$.

(b) If $u(x, t)$ is real, show that $n(-\omega) = n^*(\omega)$.

(c) Show that, if $u(0, t)$ and $\partial u(0, t)/\partial x$ are the boundary values of u and its derivative at $x = 0$, the coefficients $A(\omega)$ and $B(\omega)$ are

$$\begin{Bmatrix} A(\omega) \\ B(\omega) \end{Bmatrix} = \frac{1}{2}\frac{1}{\sqrt{2\pi}}\int_{-\infty}^{\infty} dt\, e^{i\omega t}\left[u(0, t) \mp \frac{ic}{\omega n(\omega)}\frac{\partial u}{\partial x}(0, t)\right]$$

7.8 A very long plane-wave train of frequency ω_0 with a sharp front edge is incident normally from vacuum on a semi-infinite dielectric described by an index of refraction $n(\omega)$ and occupying the half-space $x > 0$. Just outside the dielectric (at $x = 0$) the *incident* electric field is

$$E_0(0, t) = \theta(t)e^{-\epsilon t} \sin \omega_0 t$$

where $\theta(t)$ is the step function ($\theta(t) = 0$ for $t < 0$, $\theta(t) = 1$ for $t > 0$). The exponential decay constant ϵ is a positive infinitesimal.

(a) Using the results of Section 7.5 determine the transmitted field $E_0'(x, t)$ at any point in the dielectric as an integral over real frequencies.

(b) Prove that a sufficient condition for causality (that no signal propagate faster than the speed of light in vacuum) in this problem is that the index of refraction as a function of *complex* ω be an analytic function, regular in the upper half ω plane with nonvanishing imaginary part there, and approaching unity for $|\omega| \to \infty$.

(c) Generalize the argument of (b) to apply to any incident wave train.

7.9 (a) Show that, if the index of refraction $n(\omega)$ is analytic in the upper half complex ω plane and approaches unity for large $|\omega|$, its real and imaginary parts are related for real frequencies by the *dispersion relation*,

$$\operatorname{Re} n(\omega) = 1 + \frac{2}{\pi} P \int_0^\infty \frac{\omega'}{\omega'^2 - \omega^2} \operatorname{Im} n(\omega') \, d\omega'$$

where P stands for Cauchy principal value. Write the other dispersion relation, expressing the imaginary part as an integral over the real.

(b) Show by direct calculation with the dispersion relation that in a frequency range where resonant absorption occurs there is necessarily anomalous dispersion.

(c) The elementary classical model for an index of refraction is based on a collection of damped electronic oscillators and gives an index of refraction,

$$n(\omega) \simeq 1 + \frac{2\pi N e^2}{m} \sum_k \frac{f_k}{\omega_k^2 - \omega^2 - i\nu_k \omega}$$

where ω_k is the resonant frequency of the kth type of oscillator, ν_k its damping constant, and f_k the number of such oscillators per atom. Verify that this index of refraction has the appropriate properties to satisfy the dispersion relation of (a).

8

Wave Guides
and Resonant Cavities*

Electromagnetic fields in the presence of metallic boundaries form a practical aspect of the subject of considerable importance. At high frequencies where the wavelengths are of the order of meters or less the only practical way of generating and transmitting electromagnetic radiation involves metallic structures with dimensions comparable to the wavelengths involved. In this chapter we consider first the fields in the neighborhood of a conductor and discuss their penetration into the surface and the accompanying resistive losses. Then the problems of waves guided in hollow metal pipes and of resonant cavities are treated from a fairly general viewpoint, with specific illustrations included along the way. Finally, dielectric wave guides are briefly described as an alternative method of transmission.

---

* In this chapter certain formulas, denoted by an asterisk on the equation number, are written so that they can be read as formulas in mks units provided the first factor in square brackets is omitted. For example, (8.12) is

$$\frac{dP_{\text{loss}}}{da} = \left[\frac{1}{4\pi}\right]\frac{\mu\omega\delta}{4}|\mathbf{H}_{\parallel}|^2$$

The corresponding equation in mks form is

$$\frac{dP_{\text{loss}}}{da} = \frac{\mu\omega\delta}{4}|\mathbf{H}_{\parallel}|^2$$

where all symbols are to be interpreted as mks symbols, perhaps with entirely different magnitudes and dimensions from those of the corresponding Gaussian symbols.

If an asterisk appears and there is no square bracket, the formula can be interpreted equally in Gaussian or mks symbols.

General rules for conversion of any equation into its corresponding mks form are given in Table 3 of the Appendix.

8.1 Fields at the Surface of and within a Conductor

As was mentioned at the end of Section 7.7, the problem of reflection and refraction of waves at an interface of two conducting media is somewhat complicated. The most important and useful features of the phenomenon can, however, be obtained with an approximate treatment valid if one medium is a good conductor. Furthermore, the method, within its range of validity, is applicable to situations more general than plane waves incident.

First consider a surface with unit normal $\mathbf{n}$ directed outward from a *perfect* conductor on one side into a nonconducting medium on the other side. Then, just as in the static case, there is no electric field inside the conductors. The charges inside a perfect conductor are assumed to be so mobile that they move instantly in response to changes in the fields, no matter how rapid, and always produce the correct surface-charge density Σ (capital Σ is used to avoid confusion with the conductivity σ):

$$\mathbf{n} \cdot \mathbf{D} = [4\pi]\Sigma \qquad (8.1)^*$$

in order to give zero electric field inside the perfect conductor. Similarly, for time-varying magnetic fields, the surface charges move in response to the tangential magnetic field to produce always the correct surface current $\mathbf{K}$:

$$\mathbf{n} \times \mathbf{H} = \left[\frac{4\pi}{c}\right]\mathbf{K} \qquad (8.2)^*$$

in order to have zero magnetic field inside the perfect conductor. The other two boundary conditions are on normal $\mathbf{B}$ and tangential $\mathbf{E}$:

$$\left.\begin{array}{r} \mathbf{n} \cdot (\mathbf{B} - \mathbf{B}_c) = 0 \\ \mathbf{n} \times (\mathbf{E} - \mathbf{E}_c) = 0 \end{array}\right\} \qquad (8.3)^*$$

where the subscript c refers to the conductor. From these boundary conditions we see that just outside the surface of a perfect conductor only *normal* $\mathbf{E}$ and *tangential* $\mathbf{H}$ fields can exist, and that the fields drop abruptly to zero inside the perfect conductor. This behavior is indicated schematically in Fig. 8.1.

For a good, but not perfect, conductor the fields in the neighborhood of its surface must behave approximately the same as for a perfect conductor. In Section 7.7 we have seen that inside a conductor the fields are attenuated exponentially in a characteristic length δ, called the *skin depth*. For good conductors and moderate frequencies, δ is a small fraction of a centimeter. Consequently, boundary conditions (8.1) and (8.2) are

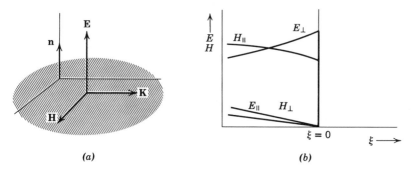

Fig. 8.1 Fields near the surface of a perfect conductor.

approximately true for a good conductor, aside from a thin transitional layer at the surface.

If we wish to examine that thin transitional region, however, care must be taken. First of all, Ohm's law (7.68) shows that with a finite conductivity there cannot actually be a surface layer of current, as implied in (8.2). Instead, the boundary condition on the magnetic field is

$$\mathbf{n} \times (\mathbf{H} - \mathbf{H}_c) = 0 \qquad (8.4)^*$$

To explore the changes produced by a finite, rather than an infinite, conductivity we employ a successive approximation scheme. First we assume that just outside the conductor there exists only a normal electric field $\mathbf{E}_\perp$ and a tangential magnetic field $\mathbf{H}_{\parallel}$, as for a perfect conductor. The values of these fields are assumed to have been obtained from the solution of an appropriate boundary-value problem. Then we use the boundary conditions and Maxwell's equations in the conductor to find the fields within the transition layer and small corrections to the fields outside. In solving Maxwell's equations within the conductor we make use of the fact that the spatial variation of the fields normal to the surface is much more rapid than the variations parallel to the surface. This means that we can safely neglect all derivatives with respect to coordinates parallel to the surface compared to the normal derivative.

If there exists a tangential $\mathbf{H}_{\parallel}$ outside the surface, boundary condition (8.4) implies the same $\mathbf{H}_{\parallel}$ inside the surface. With the neglect of the displacement current in the conductor, the curl equations in (7.69) become

$$\left. \begin{aligned} \mathbf{E}_c &\simeq \frac{c}{4\pi\sigma} \nabla \times \mathbf{H}_c \\[2mm] \mathbf{H}_c &= -\frac{ic}{\mu\omega} \nabla \times \mathbf{E}_c \end{aligned} \right\} \qquad (8.5)$$

where a harmonic variation $e^{-i\omega t}$ has been assumed. If $\mathbf{n}$ is the unit normal *outward* from the conductor and ξ is the normal coordinate *inward* into the conductor, then the gradient operator can be written

$$\nabla \simeq -\mathbf{n}\,\frac{\partial}{\partial \xi}$$

neglecting the other derivatives when operating on the fields within the conductor. With this approximation (8.5) become:

$$\left.\begin{aligned}
\mathbf{E}_c &\simeq -\frac{c}{4\pi\sigma}\,\mathbf{n}\,\times\,\frac{\partial \mathbf{H}_c}{\partial \xi}\\[2mm]
\mathbf{H}_c &\simeq \frac{ic}{\mu\omega}\,\mathbf{n}\,\times\,\frac{\partial \mathbf{E}_c}{\partial \xi}
\end{aligned}\right\} \qquad (8.6)$$

These can be combined to yield

$$\frac{\partial^2}{\partial \xi^2}(\mathbf{n}\,\times\,\mathbf{H}_c) + \frac{2i}{\delta^2}(\mathbf{n}\,\times\,\mathbf{H}_c) \simeq 0 \qquad (8.7)$$

and

$$\mathbf{n}\cdot\mathbf{H}_c \simeq 0 \qquad (8.8)$$

where δ is the skin depth defined by (7.85). The second equation shows that inside the conductor $\mathbf{H}$ is parallel to the surface, consistent with our boundary conditions. The solution for $\mathbf{H}_c$ is:

$$\mathbf{H}_c = \mathbf{H}_\| e^{-\xi/\delta}\, e^{i\xi/\delta} \qquad (8.9)$$

where $\mathbf{H}_\|$ is the tangential magnetic field outside the surface. From (8.6) the electric field in the conductor is approximately:

$$\mathbf{E}_c \simeq \sqrt{\frac{\mu\omega}{8\pi\sigma}}\,(1 - i)(\mathbf{n}\,\times\,\mathbf{H}_\|)e^{-\xi/\delta}\, e^{i\xi/\delta} \qquad (8.10)$$

These solutions for $\mathbf{H}$ and $\mathbf{E}$ inside the conductor exhibit the properties discussed in Section 7.7: (*a*) rapid exponential decay, (*b*) phase difference, (*c*) magnetic field much larger than the electric field. Furthermore, they show that, for a good conductor, the fields in the conductor are parallel to the surface* and propagate normal to it, with magnitudes which depend only on the tangential magnetic field $\mathbf{H}_\|$ which exists just outside the surface.

---

* From the continuity of the tangential component of $\mathbf{H}$ and the equation connecting $\mathbf{E}$ to $\nabla \times \mathbf{H}$ on either side of the surface, one can show that there exists in the conductor a small normal component of electric field, $\mathbf{E}_c \cdot \mathbf{n} \simeq (i\omega\epsilon/4\pi\sigma)E_\perp$, but this is of the next order in small quantities compared with (8.10).

From the boundary condition on tangential **E** (8.3) we find that *just outside* the surface there exists a small tangential electric field given by (8.10), evaluated at $\xi = 0$:

$$\mathbf{E}_{\|} \simeq \sqrt{\frac{\mu\omega}{8\pi\sigma}} \, (1 - i)(\mathbf{n} \times \mathbf{H}_{\|}) \qquad (8.11)$$

In this approximation there is also a small normal component of **B** just outside the surface. This can be obtained from Faraday's law of induction and gives $\mathbf{B}_{\perp}$ of the same order of magnitude as $\mathbf{E}_{\|}$. The amplitudes of the fields both inside and outside the conductor are indicated schematically in Fig. 8.2.

The existence of a small tangential component of **E** outside the surface, in addition to the normal **E** and tangential **H**, means that there is a power flow into the conductor. The time-average power absorbed per unit area is

$$\frac{dP_{\text{loss}}}{da} = -\frac{c}{8\pi} \operatorname{Re}\left[\mathbf{n} \cdot \mathbf{E} \times \mathbf{H}^*\right] = \left[\frac{1}{4\pi}\right] \frac{\mu\omega\delta}{4} \, |\mathbf{H}_{\|}|^2 \qquad (8.12)^*$$

This result can be given a simple interpretation as ohmic losses in the body

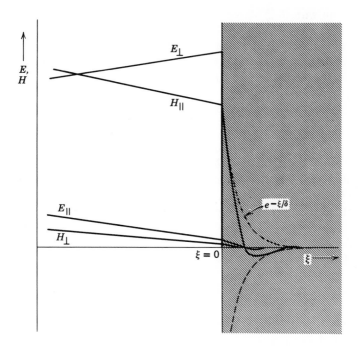

Fig. 8.2   Fields near the surface of a good, but not perfect, conductor.

of the conductor. According to Ohm's law, there exists a current density $\mathbf{J}$ near the surface of the conductor:

$$\mathbf{J} = \sigma \mathbf{E}_c = \sqrt{\frac{\mu\omega\sigma}{8\pi}}\,(1 - i)(\mathbf{n} \times \mathbf{H}_{\parallel})e^{-\xi(1-i)/\delta} \qquad (8.13)$$

The time-average rate of dissipation of energy per unit volume in ohmic losses is $\tfrac{1}{2}\mathbf{J} \cdot \mathbf{E}^* = (1/2\sigma)\,|\mathbf{J}|^2$, so that the total rate of energy dissipation in the conductor for the volume lying beneath an area element ΔA is

$$\frac{1}{2\sigma}\,\Delta A \int_0^\infty d\xi\, \mathbf{J} \cdot \mathbf{J}^* = \Delta A\,\frac{\mu\omega}{8\pi}\,|\mathbf{H}_{\parallel}|^2 \int_0^\infty e^{-2\xi/\delta}\, d\xi = \Delta A\,\frac{\mu\omega\delta}{16\pi}\,|\mathbf{H}_{\parallel}|^2$$

This is the same rate of energy dissipation as given by the Poynting's vector result (8.12).

The current density $\mathbf{J}$ is confined to such a small thickness just below the surface of the conductor that it is equivalent to an effective surface current $\mathbf{K}_{\text{eff}}$:

$$\mathbf{K}_{\text{eff}} = \int_0^\infty \mathbf{J}\, d\xi = \left[\frac{c}{4\pi}\right]\mathbf{n} \times \mathbf{H}_{\parallel} \qquad (8.14)*$$

Comparison with (8.2) shows that a good conductor behaves effectively like a perfect conductor, with the idealized surface current replaced by an equivalent surface current which is actually distributed throughout a very small, but finite, thickness at the surface. The power loss can be written in terms of the effective surface current:

$$\frac{dP_{\text{loss}}}{da} = \frac{1}{2\sigma\delta}\,|\mathbf{K}_{\text{eff}}|^2 \qquad (8.15)*$$

This shows that $1/\sigma\delta$ plays the role of a surface resistance of the conductor. Equation (8.15), with $\mathbf{K}_{\text{eff}}$ given by (8.14), or (8.12) will allow us to calculate approximately the resistive losses for practical cavities, transmission lines, and wave guides, provided we have solved for the fields in the idealized problem of infinite conductivity.

8.2 Cylindrical Cavities and Wave Guides

A practical situation of great importance is the propagation or excitation of electromagnetic waves in hollow metallic cylinders. If the cylinder has end surfaces, it is called a cavity; otherwise, a wave guide. In our discussion of this problem the boundary surfaces will be assumed to be perfect conductors. The losses occurring in practice can be accounted for adequately by the methods of Section 8.1. A cylindrical surface S of general cross-sectional contour is shown in Fig. 8.3. For simplicity, the cross-sectional size and shape are assumed constant along the cylinder axis. With a sinusoidal time dependence $e^{-i\omega t}$ for the fields inside the cylinder, Maxwell's equations take the form:

$$\left.\begin{aligned}
\mathbf{\nabla} \times \mathbf{E} = i\,\frac{\omega}{c}\,\mathbf{B} \qquad \mathbf{\nabla} \cdot \mathbf{B} = 0 \\[2mm]
\mathbf{\nabla} \times \mathbf{B} = -i\mu\epsilon\,\frac{\omega}{c}\,\mathbf{E} \qquad \mathbf{\nabla} \cdot \mathbf{E} = 0
\end{aligned}\right\} \tag{8.16}$$

where it is assumed that the cylinder is filled with a uniform nondissipative medium having dielectric constant ϵ and permeability μ. If follows that both $\mathbf{E}$ and $\mathbf{B}$ satisfy

$$\left(\nabla^2 + \mu\epsilon\,\frac{\omega^2}{c^2}\right)\begin{Bmatrix} \mathbf{E} \\ \mathbf{B} \end{Bmatrix} = 0 \tag{8.17}$$

Because of the cylindrical geometry it is useful to single out the spatial variation of the fields in the z direction and to assume

$$\begin{aligned}
\mathbf{E}(x, y, z, t) \\
\mathbf{B}(x, y, z, t)
\end{aligned}\Bigg\} = \begin{cases} \mathbf{E}(x, y)e^{\pm ikz - i\omega t} \\ \mathbf{B}(x, y)e^{\pm ikz - i\omega t} \end{cases} \tag{8.18}$$

Appropriate linear combinations can be formed to give traveling or standing waves in the z direction. The wave number k is, at present, an unknown parameter which may be real or complex. With this assumed z dependence of the fields the wave equation reduces to the two-dimensional form:

$$\left[\nabla_t^{\,2} + \left(\mu\epsilon\,\frac{\omega^2}{c^2} - k^2\right)\right]\begin{Bmatrix} \mathbf{E} \\ \mathbf{B} \end{Bmatrix} = 0 \tag{8.19}$$

where $\nabla_t^{\,2}$ is the transverse part of the Laplacian operator:

$$\nabla_t^{\,2} = \nabla^2 - \frac{\partial^2}{\partial z^2} \tag{8.20}$$

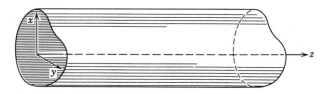

Fig. 8.3 Hollow, cylindrical wave guide of arbitrary cross-sectional shape.

It is also useful to separate the fields into components parallel to and transverse to the z axis:

$$\mathbf{E} = \mathbf{E}_z + \mathbf{E}_t \tag{8.21}$$

where the parallel field is

$$\mathbf{E}_z = (\mathbf{e}_3 \cdot \mathbf{E})\mathbf{e}_3 \tag{8.22}$$

and the transverse field is

$$\mathbf{E}_t = (\mathbf{e}_3 \times \mathbf{E}) \times \mathbf{e}_3 \tag{8.23}$$

and $\mathbf{e}_3$ is a unit vector in the z direction. Similar definitions hold for the magnetic-flux density $\mathbf{B}$. Manipulation of the curl equations in (8.16) and use of the explicit z dependence (8.18) lead to the determination of the transverse fields in terms of the axial components:

$$
\left.
\begin{aligned}
\mathbf{B}_t &= \frac{1}{\left(\mu\epsilon \dfrac{\omega^2}{c^2} - k^2\right)} \left[\nabla_t\!\left(\frac{\partial B_z}{\partial z}\right) + i\mu\epsilon\,\frac{\omega}{c}\,\mathbf{e}_3 \times \nabla_t E_z \right] \\
\mathbf{E}_t &= \frac{1}{\left(\mu\epsilon \dfrac{\omega^2}{c^2} - k^2\right)} \left[\nabla_t\!\left(\frac{\partial E_z}{\partial z}\right) - i\,\frac{\omega}{c}\,\mathbf{e}_3 \times \nabla_t B_z \right]
\end{aligned}
\right\} \tag{8.24}
$$

These relations show that it is sufficient to determine E_z and B_z as the appropriate solutions of the two-dimensional wave equation (8.19). The other components can then be calculated from (8.24).

The boundary values on the surface of the cylinder will be taken as those for a perfect conductor:

$$\mathbf{n} \cdot \mathbf{B} = 0, \qquad \mathbf{n} \times \mathbf{E} = 0 \tag{8.25}$$

where $\mathbf{n}$ is a unit normal at the surface. Since Maxwell's equations and the boundary conditions are internally consistent, it is sufficient to note that the vanishing of tangential $\mathbf{E}$ at the surface requires

$$E_z \big|_S = 0 \tag{8.26}$$

For the normal components of $\mathbf{B}$, using the expression for $\mathbf{B}_t$ (8.24), we find that $\mathbf{n} \cdot \mathbf{B} = 0$ implies

$$\frac{\partial B_z}{\partial n}\bigg|_S = 0 \tag{8.27}$$

where $\partial/\partial n$ is the normal derivative at a point on the surface.

The two-dimensional wave equations (8.19) for E_z and B_z, together with the boundary conditions on E_z and B_z at the surface of the cylinder, specify eigenvalue problems of the usual sort. For a given frequency ω, only certain values of the axial wave number k will be consistent with the differential equation and the boundary conditions (typical wave-guide situation); or, for a given k, only certain frequencies ω will be allowed (typical resonant-cavity situation). Because the boundary conditions on E_z and B_z are different, they cannot generally be satisfied simultaneously. Consequently the fields divide themselves into two distinct categories:

TRANSVERSE MAGNETIC (TM)

$$B_z = 0 \text{ everywhere}$$

The boundary condition is

$$E_z \big|_S = 0$$

TRANSVERSE ELECTRIC (TE)

$$E_z = 0 \text{ everywhere}$$

The boundary condition is

$$\frac{\partial B_z}{\partial n}\bigg|_S = 0$$

The designations "Electric (or E) Waves" and "Magnetic (or H) Waves" are sometimes used instead of Transverse Magnetic and Transverse Electric, respectively, corresponding to specification of the axial component of the field. In addition to these two types of fields there is a degenerate mode, called the *Transverse Electromagnetic* (TEM) mode, in which both E_z and B_z vanish. From (8.24) we see that, in order to have nonvanishing transverse components when both E_z and B_z vanish, the axial wave number must satisfy the condition:

$$k = \sqrt{\mu\epsilon}\,\frac{\omega}{c} \tag{8.28}$$

Thus TEM waves travel as if they were in an infinite medium without boundary surfaces. From the two-dimensional wave equation (8.19) we now find

$$\nabla_t^2 \begin{Bmatrix} \mathbf{E}_{\mathrm{TEM}} \\ \mathbf{B}_{\mathrm{TEM}} \end{Bmatrix} = 0 \tag{8.29}$$

showing that each component of the transverse fields satisfies Laplace's equation of electrostatics in two dimensions. It is easy to show (a) that both $\mathbf{E}_{\mathrm{TEM}}$ and $\mathbf{B}_{\mathrm{TEM}}$ are derivable from scalar potentials satisfying Laplace's equation and (b) that $\mathbf{B}_{\mathrm{TEM}}$ is everywhere perpendicular to $\mathbf{E}_{\mathrm{TEM}}$. In fact, from Faraday's law of induction we find

$$\mathbf{B}_{\mathrm{TEM}} = \frac{c}{i\omega}\frac{\partial}{\partial z}(\mathbf{e}_3 \times \mathbf{E}_{\mathrm{TEM}}) \tag{8.30}$$

With z-dependence $e^{i\sqrt{\mu\epsilon}\omega z/c}$, we have

$$\mathbf{B}_{\text{TEM}} = \sqrt{\mu\epsilon}\, \mathbf{e}_3 \times \mathbf{E}_{\text{TEM}} \qquad (8.31)^*$$

which is just the relation for plane waves in an infinite medium.

An immediate consequence of (8.29) is that the TEM mode cannot exist inside a single hollow, cylindrical conductor of infinite conductivity. The surface is an equipotential; hence the electric field vanishes inside. It is necessary to have two or more cylindrical surfaces in order to support the TEM mode. The familiar coaxial cable and the parallel-wire transmission line are structures for which this is the dominant mode. (See Problems 8.1 and 8.2.)

8.3 Wave Guides

We now consider the propagation of electromagnetic waves along a hollow wave guide of uniform cross section. With the z-dependence e^{ikz}, the transverse components of the fields for the two types of waves are related, according to (8.24), as follows:

$$
\left.
\begin{aligned}
\text{TM WAVES:} \quad & \mathbf{B}_t = \frac{\mu\epsilon\omega}{ck}\, \mathbf{e}_3 \times \mathbf{E}_t \\[2mm]
\text{TE WAVES:} \quad & \mathbf{E}_t = -\frac{\omega}{ck}\, \mathbf{e}_3 \times \mathbf{B}_t
\end{aligned}
\right\} \qquad (8.32)
$$

The transverse fields are in turn determined by the longitudinal fields:

$$
\begin{aligned}
\text{TM WAVES:} \quad & \mathbf{E}_t = \frac{ik}{\gamma^2}\, \boldsymbol{\nabla}_t \psi \\[2mm]
\text{TE WAVES:} \quad & \mathbf{B}_t = \frac{ik}{\gamma^2}\, \boldsymbol{\nabla}_t \psi
\end{aligned}
\qquad (8.33)
$$

where ψ is E_z (B_z) for TM (TE) waves. The scalar function ψ satisfies the two-dimensional wave equation (8.19):

$$(\nabla_t^2 + \gamma^2)\psi = 0 \qquad (8.34)$$

where

$$\gamma^2 = \mu\epsilon\,\frac{\omega^2}{c^2} - k^2 \qquad (8.35)$$

subject to the boundary condition:

$$\psi\,|_S = 0, \quad \text{or} \quad \frac{\partial\psi}{\partial n}\bigg|_S = 0 \qquad (8.36)$$

for TM (TE) waves.

Equation (8.34) for ψ, together with boundary condition (8.36), specifies an eigenvalue problem. It is easy to see that the constant γ^2 must be non-negative. Roughly speaking, it is because ψ must be oscillatory in order to satisfy boundary condition (8.36) on opposite sides of the cylinder. There will be a spectrum of eigenvalues γ_λ^2 and corresponding solutions ψ_λ, $\lambda = 1, 2, 3, \ldots$, which form an orthogonal set. These different solutions are called the *modes of the guide*. For a given frequency ω, the wave number k is determined for each value of λ:

$$k_\lambda^2 = \mu\epsilon \frac{\omega^2}{c^2} - \gamma_\lambda^2 \tag{8.37}$$

If we define a *cutoff frequency* ω_λ,

$$\omega_\lambda = [c] \frac{\gamma_\lambda}{\sqrt{\mu\epsilon}} \tag{8.38}*$$

then the wave number can be written:

$$k_\lambda = \left[\frac{1}{c}\right] \sqrt{\mu\epsilon} \sqrt{\omega^2 - \omega_\lambda^2} \tag{8.39}*$$

We note that, for $\omega > \omega_\lambda$, the wave number k_λ is real; waves of the λ mode can propagate in the guide. For frequencies less than the cutoff frequency, k_λ is imaginary; such modes cannot propagate and are called *cutoff modes*. The behavior of the axial wave number as a function of frequency is shown qualitatively in Fig. 8.4. We see that at any given frequency only a finite number of modes can propagate. It is often convenient to choose the dimensions of the guide so that at the operating frequency only the lowest mode can occur. This is shown by the vertical arrow on the figure.

Since the wave number k_λ is always less than the free-space value $\sqrt{\mu\epsilon}\omega/c$, the wavelength in the guide is always greater than the free-space

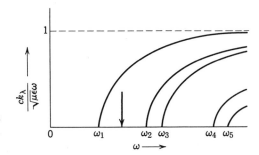

Fig. 8.4 Wave number k_λ versus frequency ω for various modes λ. ω_λ is the cutoff frequency.

wavelength. In turn, the phase velocity v_p is larger than the infinite space value:

$$v_p = \frac{\omega}{k_\lambda} = \frac{c}{\sqrt{\mu\epsilon}} \frac{1}{\sqrt{1 - \left(\frac{\omega_\lambda}{\omega}\right)^2}} > \frac{c}{\sqrt{\mu\epsilon}} \tag{8.40}$$

The phase velocity becomes infinite exactly at cutoff.

8.4   Modes in a Rectangular Wave Guide

As an important illustration of the general features described in Section 8.3 we consider the propagation of TE waves in a rectangular wave guide with inner dimensions a, b, as shown in Fig. 8.5. The wave equation for $\psi = B_z$ is

$$\left(\frac{\partial^2}{\partial x^2} + \frac{\partial^2}{\partial y^2} + \gamma^2\right)\psi = 0 \tag{8.41}$$

with boundary conditions $\partial\psi/\partial n = 0$ at $x = 0$, a and $y = 0$, b. The solution for ψ is consequently

$$\psi_{mn}(x, y) = B_0 \cos\left(\frac{m\pi x}{a}\right) \cos\left(\frac{n\pi y}{b}\right) \tag{8.42}$$

where

$$\gamma_{mn}{}^2 = \pi^2\left(\frac{m^2}{a^2} + \frac{n^2}{b^2}\right) \tag{8.43}$$

The single index λ specifying the modes previously is now replaced by the two positive integers m, n. In order that there be nontrivial solutions, m and n cannot both be zero. The cutoff frequency ω_{mn} is given by

$$\omega_{mn} = [c]\frac{\pi}{\sqrt{\mu\epsilon}}\left(\frac{m^2}{a^2} + \frac{n^2}{b^2}\right)^{\frac{1}{2}} \tag{8.44}*$$

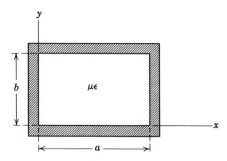

Fig. 8.5

If $a > b$, the lowest cutoff frequency, that of the dominant TE mode, occurs for $m = 1$, $n = 0$:

$$\omega_{10} = \frac{\pi c}{\sqrt{\mu\epsilon}\, a} \tag{8.45}$$

This corresponds to one-half of a free-space wavelength across the guide. The explicit fields for this mode, denoted by $TE_{1,0}$, are:

$$\begin{aligned} B_z &= B_0 \cos\left(\frac{\pi x}{a}\right) e^{ikz - i\omega t} \\ B_x &= -\frac{ika}{\pi} B_0 \sin\left(\frac{\pi x}{a}\right) e^{ikz - i\omega t} \\ E_y &= i\frac{\omega a}{\pi c} B_0 \sin\left(\frac{\pi x}{a}\right) e^{ikz - i\omega t} \end{aligned} \tag{8.46}$$

The presence of a factor i in B_x (and E_y) means that there is a spatial (or temporal) phase difference of 90° between B_x (and E_y) and B_z in the propagation direction. It happens that the $TE_{1,0}$ mode has the lowest cutoff frequency of both TE and TM modes,* and so is the one used in most practical situations. For a typical choice $a = 2b$ the ratio of cutoff frequencies ω_{mn} for the next few modes to ω_{10} are as follows:

$m \downarrow$ \ $n \rightarrow$	0	1	2	3
0		2.00	4.00	6.00
1	1.00	2.24	4.13	
2	2.00	2.84	4.48	
3	3.00	3.61	5.00	
4	4.00	4.48	5.66	
5	5.00	5.39		
6	6.00			

There is a frequency range from cutoff to twice cutoff where the $TE_{1,0}$ mode is the only propagating mode. Beyond that frequency other modes rapidly begin to enter. The field configurations of the $TE_{1,0}$ mode and other modes are shown in many books, e.g., *American Institute of Physics Handbook*, McGraw-Hill, New York (1957), p. 5–61.

* This is evident if we note that for the TM modes E_z is of the form

$$E_z = E_0 \sin\left(\frac{m\pi x}{a}\right) \sin\left(\frac{n\pi y}{b}\right)$$

while γ^2 is still given by (8.43). The lowest mode has $m = n = 1$. Its cutoff frequency is greater than that of the $TE_{1,0}$ mode by the factor $\left(1 + \frac{a^2}{b^2}\right)^{1/2}$.

8.5 Energy Flow and Attenuation in Wave Guides

The general discussion of Section 8.3 for a cylindrical wave guide of arbitrary cross-sectional shape can be extended to include the flow of energy along the guide and the attenuation of the waves due to losses in the walls having finite conductivity. The treatment will be restricted to one mode at a time; degenerate modes will be mentioned only briefly. The flow of energy is described by the complex Poynting's vector:

$$S = \left[\frac{c}{4\pi}\right]\frac{1}{2}(\mathbf{E} \times \mathbf{H}^*) \qquad (8.47)^*$$

whose real part gives the time-averaged flux of energy. For the two types of field we find, using (8.24):

$$S = \frac{\omega k}{8\pi\gamma^4} \begin{Bmatrix} \epsilon\left[\mathbf{e}_3 |\mathbf{\nabla}_t\psi|^2 + i\frac{\gamma^2}{k}\,\psi\mathbf{\nabla}_t\psi^*\right] \\ \frac{1}{\mu}\left[\mathbf{e}_3 |\mathbf{\nabla}_t\psi|^2 - i\frac{\gamma^2}{k}\,\psi^*\mathbf{\nabla}_t\psi\right] \end{Bmatrix} \qquad (8.48)$$

where the upper (lower) line is for TM (TE) modes. Since ψ is generally real,* we see that the transverse component of S represents reactive energy flow and does not contribute to the time-average flux of energy. On the other hand, the axial component of S gives the time-averaged flow of energy along the guide. To evaluate the total power flow P we integrate the axial component of S over the cross-sectional area A:

$$P = \int_A S \cdot \mathbf{e}_3 \, da = \frac{\omega k}{8\pi\gamma^4} \begin{Bmatrix} \epsilon \\ \frac{1}{\mu} \end{Bmatrix} \int_A (\mathbf{\nabla}_t\psi)^* \cdot (\mathbf{\nabla}_t\psi) \, da \qquad (8.49)$$

By means of Green's first identity (1.34) applied to two dimensions, (8.49) can be written:

$$P = \frac{\omega k}{8\pi\gamma^4} \begin{Bmatrix} \epsilon \\ \frac{1}{\mu} \end{Bmatrix} \left[\oint_C \psi^* \frac{\partial\psi}{\partial n}\, dl - \int_A \psi^*\nabla_t^2\psi \, da\right] \qquad (8.50)$$

where the first integral is around the curve C which defines the boundary surface of the cylinder. This integral vanishes for both types of fields

---

* It is possible to excite a guide in such a manner that a given mode or linear combination of modes has a complex ψ. Then a time-averaged transverse energy flow can occur. Since it is a circulatory flow, however, it really only represents stored energy and is not of great practical importance.

because of boundary conditions (8.36). By means of the wave equation
(8.34) the second integral may be reduced to the normalization integral for
ψ. Consequently the transmitted power is

$$P = \left[\frac{c}{4\pi}\right]\frac{1}{2\sqrt{\mu\epsilon}}\left(\frac{\omega}{\omega_\lambda}\right)^2\left(1 - \frac{\omega_\lambda^2}{\omega^2}\right)^{\!\frac{1}{2}}\!\begin{Bmatrix}\epsilon \\ \frac{1}{\mu}\end{Bmatrix}\int_A \psi^*\psi\,da \qquad (8.51)*$$

where the upper (lower) line is for TM (TE) modes, and we have exhibited
all the frequency dependence explicitly.

It is straightforward to calculate the field energy per unit length of the
guide in the same way as the power flow. The result is

$$U = \left[\frac{1}{4\pi}\right]\frac{1}{2}\left(\frac{\omega}{\omega_\lambda}\right)^2\!\begin{Bmatrix}\epsilon \\ \frac{1}{\mu}\end{Bmatrix}\int_A \psi^*\psi\,da \qquad (8.52)*$$

Comparison with the power flow P shows that P and U are proportional.
The constant of proportionality has the dimensions of velocity (velocity
of energy flow) and is just the group velocity:

$$\frac{P}{U} = \frac{k}{\omega}\frac{c^2}{\mu\epsilon} = \frac{c}{\sqrt{\mu\epsilon}}\sqrt{1 - \frac{\omega_\lambda^2}{\omega^2}} = v_g \qquad (8.53)$$

as can be verified by a direct calculation of $v_g = d\omega/dk$ from (8.39),
assuming that the dielectric filling the guide is nondispersive. We note
that v_g is always less than the velocity of waves in an infinite medium and
falls to zero at cutoff. The product of phase velocity (8.40) and group
velocity is constant:

$$v_p v_g = \frac{c^2}{\mu\epsilon} \qquad (8.54)$$

an immediate consequence of the fact that $\omega\,\Delta\omega \propto k\,\Delta k$.

Our considerations so far have applied to wave guides with perfectly
conducting walls. The axial wave number k_λ was either real or purely
imaginary. If the walls have a finite conductivity, there will be ohmic
losses and the power flow along the guide will be attenuated. For walls
with large conductivity the wave number will have a small imaginary part:

$$k_\lambda \simeq k_\lambda^{(0)} + i\beta_\lambda \qquad (8.55)*$$

where $k_\lambda^{(0)}$ is the value for perfectly conducting walls. The attenuation
constant β_λ can be found either by solving the boundary-value problem
over again with boundary conditions appropriate for finite conductivity,
or by calculating the ohmic losses by the methods of Section 8.1 and

using conservation of energy. We will use the latter technique. The power flow along the guide will be given by

$$P(z) = P_0 e^{-2\beta_\lambda z} \tag{8.56)*}$$

Thus the attenuation constant is given by

$$\beta_\lambda = -\frac{1}{2P}\frac{dP}{dz} \tag{8.57)*}$$

where $-dP/dz$ is the power dissipated in ohmic losses per unit length of the guide. According to the results of Section 8.1, this power loss is

$$-\frac{dP}{dz} = \left[\frac{c^2}{16\pi^2}\right]\frac{1}{2\sigma\delta\mu^2}\oint_C |\mathbf{n} \times \mathbf{B}|^2\, dl \tag{8.58)*}$$

where the integral is around the boundary of the guide. With fields (8.32) and (8.33) it is easy to show that for a given mode:

$$-\frac{dP}{dz} = \frac{c^2}{32\pi^2\sigma\delta\mu^2}\left(\frac{\omega}{\omega_\lambda}\right)^2\oint_C \left\{ \begin{array}{c} \dfrac{c^2}{\omega_\lambda{}^2}\left|\dfrac{\partial\psi}{\partial n}\right|^2 \\[2mm] \dfrac{c^2}{\mu\epsilon\omega_\lambda{}^2}\left(1-\dfrac{\omega_\lambda{}^2}{\omega^2}\right)|\mathbf{n}\times\boldsymbol{\nabla}_t\psi|^2 + \dfrac{\omega_\lambda{}^2}{\omega^2}|\psi|^2 \end{array}\right\} dl$$
$$\tag{8.59}$$

where again the upper (lower) line applies to TM (TE) modes.

Since the transverse derivatives of ψ are determined entirely by the size and shape of the wave guide, the frequency dependence of the power loss is explicitly exhibited in (8.59). In fact, the integrals in (8.59) may be simply estimated from the fact that for each mode:

$$\left(\nabla_t{}^2 + \frac{\mu\epsilon\omega_\lambda{}^2}{c^2}\right)\psi = 0 \tag{8.60}$$

This means that, in some average sense, and barring exceptional circumstances, the transverse derivatives of ψ must be of the order of magnitude of $\sqrt{\mu\epsilon}(\omega_\lambda/c)\psi$:

$$\left\langle\left|\frac{\partial\psi}{\partial n}\right|^2\right\rangle \sim \langle|\mathbf{n}\times\boldsymbol{\nabla}_t\psi|^2\rangle \sim \mu\epsilon\frac{\omega_\lambda{}^2}{c^2}\langle|\psi|^2\rangle \tag{8.61}$$

Consequently the line integrals in (8.59) can be related to the normalization integral of $|\psi|^2$ over the area. For example,

$$\oint_C \frac{c^2}{\omega_\lambda{}^2}\left|\frac{\partial\psi}{\partial n}\right|^2 dl = \xi_\lambda\mu\epsilon\frac{C}{A}\int_A |\psi|^2\, da \tag{8.62}$$

where C is the circumference and A is the area of cross section, while ξ_λ is a dimensionless number of the order of unity. Without further knowledge of the shape of the guide we can obtain the order of magnitude of the attenuation constant β_λ and exhibit completely its frequency dependence. Thus, using (8.59) with (8.62) and (8.51), plus the frequency dependence of the skin depth (7.85), we find

$$\beta_\lambda = \left[\frac{c}{4\pi}\right]\sqrt{\frac{\epsilon}{\mu}}\frac{1}{\sigma\delta_\lambda}\left(\frac{C}{2A}\right)\frac{\left(\dfrac{\omega}{\omega_\lambda}\right)^{\frac{1}{2}}}{\left(1-\dfrac{\omega_\lambda^2}{\omega^2}\right)^{\frac{1}{2}}}\left[\xi_\lambda+\eta_\lambda\left(\frac{\omega_\lambda}{\omega}\right)^2\right] \qquad (8.63)*$$

where σ is the conductivity (assumed independent of frequency), δ_λ is the skin depth at the cutoff frequency, and ξ_λ, η_λ are dimensionless numbers of the order of unity. For TM modes, $\eta_\lambda = 0$.

For a given cross-sectional geometry it is a straightforward matter to calculate the dimensionless parameters ξ_λ and η_λ in (8.63). For the TE modes with $n = 0$ in a rectangular guide, the values are $\xi_{m,0} = a/(a + b)$ and $\eta_{m,0} = 2b/(a + b)$. For reasonable relative dimensions, these parameters are of order unity, as expected.

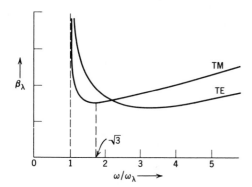

Fig. 8.6 Attenuation constant β_λ as a function of frequency for typical TE and TM modes. For TM modes the minimum attenuation occurs at $\omega/\omega_\lambda = \sqrt{3}$, regardless of cross-sectional shape.

The general behavior of β_λ as a function of frequency is shown in Fig. 8.6. Minimum attenuation occurs at a frequency well above cutoff. For TE modes the relative magnitudes of ξ_λ and η_λ depend on the shape of the guide and on λ. Consequently no general statement can be made about the exact frequency for minimum attenuation. But for TM modes the minimum always occurs at $\omega_{min} = \sqrt{3}\omega_\lambda$. At high frequencies the attenuation increases as $\omega^{\frac{1}{2}}$. In the microwave region typical attenuation constants for copper guides are of the order $\beta_\lambda \sim 10^{-4}\omega_\lambda/c$, giving $1/e$ distances of 200–400 meters.

The approximations employed in obtaining (8.63) break down close to cutoff. Evidence for this is the physically impossible, infinite value of (8.63) at $\omega = \omega_\lambda$. A treatment of the problem by perturbation theory with the boundary condition (8.11) yields the more accurate result,

$$k_\lambda^2 = k_\lambda^{(0)2} + 2(1 + i)k_\lambda^{(0)}\beta_\lambda \tag{8.64}$$

where β_λ is still given by (8.63) For $k_\lambda^{(0)} \gg \beta_\lambda$ this reduces to our previous expression (8.55). But at cutoff ($k_\lambda^{(0)} = 0$) the wave number is now finite with real and imaginary parts of the order of the geometrical mean of ω_λ/c and a typical value of β_λ, say at $\omega \sim 2\omega_\lambda$.

In the discussion so far we have considered only one mode at a time. This procedure fails whenever a TE and a TM mode have the same cutoff frequency, as occurs in the rectangular guide, for example, with $n \neq 0$, $m \neq 0$. The reason for the failure is that the boundary condition (8.11) for finite conductivity couples the degenerate modes. The calculation of the attenuation then involves so-called degenerate state perturbation theory, and the expression for β takes the form,

$$\beta = \tfrac{1}{2}(\beta_{\mathrm{TM}} + \beta_{\mathrm{TE}}) \pm \sqrt{(\beta_{\mathrm{TM}} - \beta_{\mathrm{TE}})^2 + |K|^2} \tag{8.65}$$

where β_{TM} and β_{TE} are the values found above, while K is a coupling parameter. The two values of β in (8.65) give the attenuation for the two orthogonal, mixed modes which satisfy the perturbed boundary conditions.*

8.6 Resonant Cavities

Although an electromagnetic cavity resonator can be of any shape whatsoever, an important class of cavities is produced by placing end faces on a length of cylindrical wave guide. We will assume that the end surfaces are plane and perpendicular to the axis of the cylinder. As usual, the walls of the cavity are taken to have infinite conductivity, while the cavity is filled with a lossless dielectric with constants μ, ϵ. Because of reflections at the end surfaces the z dependence of the fields will be that appropriate to standing waves:

$$A \sin kz + B \cos kz$$

If the plane boundary surfaces are at $z = 0$ and $z = d$, the boundary conditions can be satisfied at each surface only if

$$k = p\frac{\pi}{d}, \qquad p = 0, 1, 2, \ldots \tag{8.66}$$

---

* For the theory of perturbation of boundary conditions in guides and cavities, see G. Goubau, *Electromagnetic Waveguides and Cavities*, Pergamon Press, New York, 1961; Sect. 25. Attenuation for degenerate modes in guides is treated by R. Müller, Z. Naturforsch., *4a*, 218 (1949), and for the rectangular cavity by the same author in Sect. 37 of the book by Goubau.

For TM fields the vanishing of E_t at $z = 0$ and $z = d$ requires

$$E_z = \psi(x, y) \cos\left(\frac{p\pi z}{d}\right), \qquad p = 0, 1, 2, \ldots \qquad (8.67)$$

Similarly for TE fields, the vanishing of B_z at $z = 0$ and $z = d$ requires

$$B_z = \psi(x, y) \sin\left(\frac{p\pi z}{d}\right), \qquad p = 1, 2, 3, \ldots \qquad (8.68)$$

Then from (8.24) we find the transverse fields:

TM FIELDS

$$\mathbf{E}_t = -\frac{p\pi}{d\gamma^2} \sin\left(\frac{p\pi z}{d}\right) \nabla_t \psi$$

$$\mathbf{B}_t = \frac{i\mu\epsilon\omega}{c\gamma^2} \cos\left(\frac{p\pi z}{d}\right) \mathbf{e}_3 \times \nabla_t \psi \qquad (8.69)$$

TE FIELDS

$$\mathbf{E}_t = -\frac{i\omega}{c\gamma^2} \sin\left(\frac{p\pi z}{d}\right) \mathbf{e}_3 \times \nabla_t \psi$$

$$\mathbf{B}_t = \frac{p\pi}{d\gamma^2} \cos\left(\frac{p\pi z}{d}\right) \nabla_t \psi \qquad (8.70)$$

The boundary conditions at the ends of the cavity are now explicitly satisfied. There remains the eigenvalue problem (8.34)–(8.36), as before. But now the constant γ^2 is:

$$\gamma^2 = \mu\epsilon \frac{\omega^2}{c^2} - \left(\frac{p\pi}{d}\right)^2 \qquad (8.71)$$

For each value of p the eigenvalue γ_λ^2 determines an eigenfrequency of resonance frequency $\omega_{\lambda p}$:

$$\omega_{\lambda p}^2 = \frac{[c^2]}{\mu\epsilon}\left[\gamma_\lambda^2 + \left(\frac{p\pi}{d}\right)^2\right] \qquad (8.72)*$$

and the corresponding fields of that resonant mode. The resonance frequencies form a discrete set which can be determined graphically on the figure of axial wave number k versus frequency in a wave guide (see p. 245) by demanding that $k = p\pi/d$. It is usually expedient to choose the various dimensions of the cavity so that the resonant frequency of operation lies well separated from other resonant frequencies. Then the cavity will be relatively stable in operation and insensitive to perturbing effects associated with frequency drifts, changes in loading, etc.

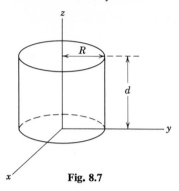

Fig. 8.7

An important practical resonant cavity is the right circular cylinder, perhaps with a piston to allow tuning by varying the height. The cylinder is shown in Fig. 8.7, with inner radius R and length d. For a TM mode the transverse wave equation for $\psi = E_z$, subject to the boundary condition $E_z = 0$ at $\rho = R$, has the solution:

$$\psi(\rho, \phi) = J_m(\gamma_{mn}\rho)e^{\pm im\phi} \tag{8.73}$$

where

$$\gamma_{mn} = \frac{x_{mn}}{R} \tag{8.74}$$

x_{mn} is the nth root of the equation, $J_m(x) = 0$. These roots are given on page 72, below equation (3.92). The integers m and n take on the values $m = 0, 1, 2, \ldots$, and $n = 1, 2, 3, \ldots$. The resonance frequencies are given by

$$\omega_{mnp} = \frac{[c]}{\sqrt{\mu\epsilon}} \sqrt{\frac{x_{mn}^2}{R^2} + \frac{p^2\pi^2}{d^2}} \tag{8.75}*$$

The lowest TM mode has $m = 0$, $n = 1$, $p = 0$, and so is designated $TM_{0,1,0}$. Its resonance frequency is

$$\omega_{010} = \frac{2.405}{\sqrt{\mu\epsilon}} \frac{c}{R} \tag{8.76}$$

The explicit expressions for the fields are

$$\left. \begin{aligned} E_z &= E_0 J_0\left(\frac{2.405\rho}{R}\right)e^{-i\omega t} \\ B_\phi &= -i\sqrt{\mu\epsilon}E_0 J_1\left(\frac{2.405\rho}{R}\right)e^{-i\omega t} \end{aligned} \right\} \tag{8.77}$$

The resonant frequency for this mode is independent of d. Consequently simple tuning is impossible.

For TE modes, the basic solution (8.73) still applies, but the boundary condition on B_z $\left(\dfrac{\partial \psi}{\partial \rho}\bigg|_R = 0\right)$ makes

$$\gamma_{mn} = \frac{x'_{mn}}{R} \tag{8.78}$$

where x'_{mn} is the nth root of $J_m'(x) = 0$. These roots, for a few values of m and n, are tabulated below:

Roots of $J_m'(x) = 0$

$m = 0$: $x'_{0n} = 3.832, 7.016, 10.174, \ldots$

$m = 1$: $x'_{1n} = 1.841, 5.331, \; 8.536, \ldots$

$m = 2$: $x'_{2n} = 3.054, 6.706, \; 9.970, \ldots$

$m = 3$: $x'_{3n} = 4.201, 8.015, 11.336, \ldots$

The resonance frequencies are given by

$$\omega_{mnp} = \frac{[c]}{\sqrt{\mu\epsilon}}\left(\frac{x'^2_{mn}}{R^2} + \frac{p^2\pi^2}{d^2}\right)^{\!\frac{1}{2}} \tag{8.79}*$$

where $m = 0, 1, 2, \ldots$, but $n, p = 1, 2, 3, \ldots$. The lowest TE mode has $m = n = p = 1$, and is denoted $TE_{1,1,1}$. Its resonance frequency is

$$\omega_{111} = \frac{1.841}{\sqrt{\mu\epsilon}}\frac{c}{R}\left(1 + 2.912\frac{R^2}{d^2}\right)^{\!\frac{1}{2}} \tag{8.80}$$

while the fields are derivable from

$$B_z = B_0 J_1\!\left(\frac{1.841\rho}{R}\right)\cos\phi \sin\left(\frac{\pi z}{d}\right)e^{-i\omega t} \tag{8.81}$$

by means of (8.70). For d large enough ($d > 2.03R$), the resonance frequency ω_{111} is smaller than that for the lowest TM mode (8.76). Then the $TE_{1,1,1}$ mode is the fundamental oscillation of the cavity. Because the frequency depends on the ratio d/R it is possible to provide easy tuning by making the separation of the end faces adjustable.

8.7 Power Losses in a Cavity; Q of a Cavity

In the preceding section it was found that resonant cavities had discrete frequencies of oscillation with a definite field configuration for each resonance frequency. This implies that, if one were attempting to excite a particular mode of oscillation in a cavity by some means, no fields of the right sort could be built up unless the exciting frequency were exactly equal to the chosen resonance frequency. In actual fact there will not be a delta function singularity, but rather a narrow band of frequencies around the eigenfrequency over which appreciable excitation can occur. An important source of this smearing out of the sharp frequency of oscillation is the dissipation of energy in the cavity walls and perhaps in the dielectric filling the cavity. A measure of the sharpness of response of the cavity to external excitation is the Q of the cavity, defined as 2π times the ratio of the time-averaged energy stored in the cavity to the energy loss per cycle:

$$Q = \omega_0 \frac{\text{Stored energy}}{\text{Power loss}} \qquad (8.82)^*$$

Here ω_0 is the resonance frequency, assuming no losses. By conservation of energy the power dissipated in ohmic losses is the negative of the time rate of change of stored energy U. Thus from (8.82) we can write an equation for the behavior of U as a function of time:

$$\frac{dU}{dt} = -\frac{\omega_0}{Q} U$$

with solution

$$U(t) = U_0 e^{-\omega_0 t/Q} \qquad (8.83)$$

If an initial amount of energy U_0 is stored in the cavity, it decays away exponentially with a decay constant inversely proportional to Q. The time dependence in (8.83) implies that the oscillations of the fields in the cavity are damped as follows:

$$E(t) = E_0 e^{-\omega_0 t/2Q} e^{-i\omega_0 t} \qquad (8.84)$$

A damped oscillation such as this has not a pure frequency, but a superposition of frequencies around $\omega = \omega_0$. Thus,

$$E(t) = \frac{1}{\sqrt{2\pi}} \int_{-\infty}^{\infty} E(\omega) e^{-i\omega t}\, d\omega$$

where

$$E(\omega) = \frac{1}{\sqrt{2\pi}} \int_0^{\infty} E_0 e^{-\omega_0 t/2Q} e^{i(\omega-\omega_0)t}\, dt \qquad (8.85)$$

The integral in (8.85) is elementary and leads to a frequency distribution for the energy in the cavity having a Lorentz line shape:

$$|E(\omega)|^2 \propto \frac{1}{(\omega - \omega_0)^2 + (\omega_0/2Q)^2} \qquad (8.86)$$

The resonance shape (8.86), shown in Fig. 8.8, has a full width at half-maximum (confusingly called the half-width) equal to ω_0/Q. For a constant input voltage, the energy of oscillation in the cavity as a function of frequency will follow the resonance curve in the neighborhood of a particular resonant frequency. Thus, if $\Delta\omega$ is the frequency separation between half-power points, the Q of the cavity is

$$Q = \frac{\omega_0}{\Delta\omega} \tag{8.87}$$

Q values of several hundreds or thousands are common for microwave cavities.

To determine the Q of a cavity we must calculate the time-averaged energy stored in it and then determine the power loss in the walls. The computations are very similiar to those done in Section 8.5 for attenuation in wave guides. We will consider here only the cylindrical cavities of Section 8.6, assuming no degeneracies (see the footnote on p. 252). The energy stored in the cavity for the mode λ, p is, according to (8.67)–(8.70):

$$U = \left[\frac{1}{4\pi}\right]\frac{d}{4}\begin{Bmatrix}\epsilon \\ \frac{1}{\mu}\end{Bmatrix}\left[1 + \left(\frac{p\pi}{\gamma_\lambda d}\right)^2\right]\int\!\!\int_A |\psi|^2\, da \tag{8.88}*$$

where the upper (lower) line applies to TM (TE) modes. For the TM modes with $p = 0$ the result must be multiplied by 2.

The power loss can be calculated by a modification of (8.58):

$$P_{\text{loss}} = \left[\frac{c^2}{16\pi^2}\right]\frac{1}{2\sigma\delta\mu^2}\left[\oint_C dl \int_0^d dz |\mathbf{n} \times \mathbf{B}|^2_{\text{sides}} + 2\int_A da |\mathbf{n} \times \mathbf{B}|^2_{\text{ends}}\right] \tag{8.89}*$$

For TM modes with $p \neq 0$ it is easy to show that

$$P_{\text{loss}} = \left[\frac{c^2}{16\pi^2}\right]\frac{\epsilon}{\sigma\delta\mu}\left[1 + \left(\frac{p\pi}{\gamma_\lambda d}\right)^2\right]\left(1 + \xi_\lambda \frac{Cd}{4A}\right)\int_A |\psi|^2\, da \tag{8.90}*$$

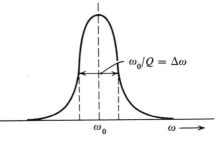

$\omega_0/Q = \Delta\omega$

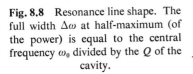

Fig. 8.8 Resonance line shape. The full width $\Delta\omega$ at half-maximum (of the power) is equal to the central frequency ω_0 divided by the Q of the cavity.

where the dimensionless number ξ_λ is the same one that appears in (8.62), C is the circumference of the cavity, and A is its cross-sectional area. For $p = 0$, ξ_λ must be replaced by $2\xi_\lambda$. Combining (8.88) and (8.89) according to (8.82), and using definition (7.85) for the skin depth δ, we find the Q of the cavity:

$$Q = \frac{\mu}{\mu_c}\frac{d}{\delta}\frac{1}{2\left(1 + \xi_\lambda \dfrac{Cd}{4A}\right)} \tag{8.91}*$$

where μ_c is the permeability of the metal walls of the cavity. For $p = 0$ modes, (8.91) must be multiplied by 2 and ξ_λ replaced by $2\xi_\lambda$. This expression for Q has an intuitive physical interpretation when written in the form:

$$Q = \frac{\mu}{\mu_c}\left(\frac{V}{S\delta}\right) \times \text{(Geometrical factor)} \tag{8.92}*$$

where V is the volume of the cavity, and S its total surface area. The Q of a cavity is evidently, apart from a geometrical factor, the ratio of the volume occupied by the fields to the volume of the conductor into which the fields penetrate because of the finite conductivity. For TM modes in cylindrical cavities the geometrical factor is

$$\frac{\left(1 + \dfrac{Cd}{2A}\right)}{\left(1 + \xi_\lambda \dfrac{Cd}{4A}\right)} \tag{8.93}$$

for $p \neq 0$, and is

$$\frac{2\left(1 + \dfrac{Cd}{2A}\right)}{\left(1 + \xi_\lambda \dfrac{Cd}{2A}\right)} \tag{8.94}$$

for $p = 0$ modes. For TE modes in the cylindrical cavity the geometrical factor is somewhat more complicated, but of the same order of magnitude. For the $TM_{0,1,0}$ mode in a circular cylindrical cavity with fields (8.77), $\xi_\lambda = 1$ (true for all TM modes), so that the geometrical factor is 2 and Q is:

$$Q = 2\left(\frac{\mu}{\mu_c}\right)\frac{1}{\left(1 + \dfrac{d}{R}\right)}\frac{d}{\delta} \tag{8.95}*$$

For the $TE_{1,1,1}$ mode calculation yields a geometrical factor*

$$\left(1 + \frac{d}{R}\right) \frac{\left(1 + 0.344\frac{d^2}{R^2}\right)}{\left(1 + 0.209\frac{d}{R} + 0.242\frac{d^3}{R^3}\right)} \tag{8.96}$$

and a Q:

$$Q = \frac{\mu}{2\mu_c}\left(\frac{d}{\delta}\right)\frac{\left(1 + 0.344\frac{d^2}{R^2}\right)}{\left(1 + 0.209\frac{d}{R} + 0.242\frac{d^3}{R^3}\right)} \tag{8.97}*$$

Expression (8.92) for Q applies not only to cylindrical cavities but also to cavities of arbitrary shape, with an appropriate geometrical factor of the order of unity.

8.8 Dielectric Wave Guides

In Sections 8.2–8.5 we considered wave guides made of hollow metal cylinders with fields only inside the hollow. Other guiding structures are possible. The parallel-wire transmission line is an example. The general requirement for a guide of electromagnetic waves is that there be a flow of energy only along the guiding structure and not perpendicular to it. This means that the fields will be appreciable only in the immediate neighborhood of the guiding structure. For hollow wave guides these requirements are satisfied in a trivial way. But for an open structure like the parallel-wire line the fields extend somewhat away from the conductors, falling off like ρ^{-2} for the TEM mode, and exponentially for higher modes.

A dielectric cylinder, such as shown in Fig. 8.9, can serve as a wave guide, with some properties very similar to those of a hollow metal guide if its dielectric constant is large enough. There are, however, characteristic differences which arise because of the very different boundary conditions to be satisfied at the surface of the cylinder. The general considerations of Section 8.2 still apply, except that the transverse behavior of the fields is governed by two equations like (8.19), one for inside the cylinder and one for outside:

INSIDE

$$\left[\nabla_t^2 + \left(\mu_1\epsilon_1\frac{\omega^2}{c^2} - k^2\right)\right]\begin{Bmatrix}\mathbf{E}\\\mathbf{B}\end{Bmatrix} = 0 \tag{8.98}$$

* Note that this factor varies by only 30 per cent as the cylinder geometry is changed from $d/R \gg 1$ to $d/R \ll 1$.

OUTSIDE

$$\left[\nabla_t^2 + \left(\mu_0\epsilon_0\,\frac{\omega^2}{c^2} - k^2\right)\right]\!\begin{Bmatrix}\mathbf{E}\\\mathbf{B}\end{Bmatrix} = 0 \qquad (8.99)$$

Both dielectric (μ_1, ϵ_1) and surrounding medium (μ_0, ϵ_0) are assumed to be uniform and isotropic in their properties. The axial propagation constant k must be the same inside and outside the cylinder in order to satisfy boundary conditions at all points on the surface at all times.

In the usual way, inside the dielectric cylinder the transverse Laplacian of the fields must be negative so that the constant

$$\gamma^2 = \mu_1\epsilon_1\,\frac{\omega^2}{c^2} - k^2 \qquad (8.100)$$

is positive. Outside the cylinder, however, the requirement of no transverse flow of energy demands that the fields fall off exponentially. (There is no TEM mode for a dielectric guide.) Consequently, the quantity in (8.99) equivalent to γ^2 must be negative. Therefore we define a quantity β^2:

$$\beta^2 = k^2 - \mu_0\epsilon_0\,\frac{\omega^2}{c^2} \qquad (8.101)$$

and demand that acceptable wave guide solutions have β^2 positive (β real).

The oscillatory solutions (inside) must be matched to the exponential solutions (outside) at the boundary of the dielectric cylinder. The boundary conditions are continuity of normal **B** and **D** and tangential **E** and **H**, rather than the vanishing of normal **B** and tangential **E** (8.25) appropriate for hollow conductors. Because of the more involved boundary conditions the types of fields do not separate into TE and TM modes, except in special circumstances such as azimuthal symmetry in

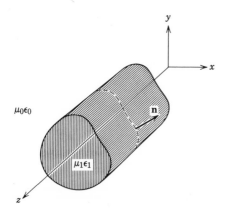

Fig. 8.9   Section of dielectric wave guide.

circular cylinders, to be discussed below. In general, axial components of both **E** and **B** exist. Such waves are sometimes designated as HE modes.

To illustrate some of the features of the dielectric wave guide we consider a circular cylinder of radius a consisting of nonpermeable dielectric with dielectric constant ϵ_1 in an external nonpermeable medium with dielectric constant ϵ_0. As a simplifying assumption we take the fields to have no azimuthal variation. Then in cylindrical coordinates the radial equations for E_z or B_z are Bessel's equations:

$$\left. \begin{array}{l} \left(\dfrac{d^2}{d\rho^2} + \dfrac{1}{\rho} \dfrac{d}{d\rho} + \gamma^2 \right)\psi = 0, \qquad \rho \leq a \\[4mm] \left(\dfrac{d^2}{d\rho^2} + \dfrac{1}{\rho} \dfrac{d}{d\rho} - \beta^2 \right)\psi = 0, \qquad \rho > a \end{array} \right\} \tag{8.102}$$

The solution, satisfying the requirements of finiteness at the origin and at infinity, is found from Section 3.6 to be:

$$\psi = \left\{ \begin{array}{ll} J_0(\gamma\rho), & \rho \leq a \\[2mm] AK_0(\beta\rho), & \rho > a \end{array} \right\} \tag{8.103}$$

The other components of **E** and **B** can be found from (8.24) when the relative amounts of E_z and B_z are known. With no ϕ dependence to the fields, (8.24) reduces to

INSIDE

$$\left. \begin{array}{ll} B_\rho = \dfrac{ik}{\gamma^2} \dfrac{\partial B_z}{\partial \rho}, & B_\phi = \dfrac{i\epsilon_1\omega}{\gamma^2 c} \dfrac{\partial E_z}{\partial \rho} \\[4mm] E_\phi = -\dfrac{\omega}{ck} B_\rho, & E_\rho = \dfrac{ck}{\epsilon_1\omega} B_\phi \end{array} \right\} \tag{8.104}$$

and similar expressions for $\rho > a$. The fact that the fields arrange themselves in two groups, (B_ρ, E_ϕ) depending on B_z, and (B_ϕ, E_ρ) depending on E_z, suggests that we attempt to obtain solutions of the TE or TM type, as for the metal wave guides. For the TE modes, the fields are explicitly

$$\left. \begin{array}{l} B_z = J_0(\gamma\rho) \\[3mm] B_\rho = -\dfrac{ik}{\gamma} J_1(\gamma\rho) \\[3mm] E_\phi = \dfrac{i\omega}{c\gamma} J_1(\gamma\rho) \end{array} \right\} \qquad \rho \leq a \tag{8.105}$$

and

$$B_z = AK_0(\beta\rho)$$

$$\left.\begin{array}{l} B_\rho = \dfrac{ikA}{\beta} K_1(\beta\rho) \\[2mm] E_\phi = -\dfrac{i\omega A}{c\beta} K_1(\beta\rho) \end{array}\right\} \quad \rho > a \qquad (8.106)$$

These fields must satisfy the standard boundary conditions at $\rho = a$. This leads to the two conditions,

$$\left.\begin{array}{l} AK_0(\beta a) = J_0(\gamma a) \\[2mm] -\dfrac{A}{\beta} K_1(\beta a) = \dfrac{J_1(\gamma a)}{\gamma} \end{array}\right\} \qquad (8.107)$$

Upon elimination of the constant A we obtain the determining equations for γ, β, and therefore k:

$$\left.\begin{array}{l} \dfrac{J_1(\gamma a)}{\gamma J_0(\gamma a)} + \dfrac{K_1(\beta a)}{\beta K_0(\beta a)} = 0 \\[4mm] \gamma^2 + \beta^2 = (\epsilon_1 - \epsilon_0)\dfrac{\omega^2}{c^2} \end{array}\right\} \qquad (8.108)$$

and, from (8.100) and (8.101),

The general behavior of the two parts of the first equation in (8.108) is shown in Fig. 8.10*a*. Figure 8.10*b* shows the two curves superposed

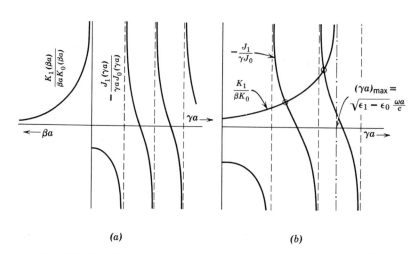

(a)   (b)

Fig. 8.10   Graphical determination of the axial propagation constant for a dielectric wave guide.

according to the second equation in (8.108). The frequency is assumed to be high enough that two modes, marked by the circles at the intersections of the two curves, exist. The vertical asymptotes are given by the roots of $J_0(x) = 0$. If the maximum value of γa is smaller than the first root ($x_{01} = 2.405$), there can be no intersection of the two curves for real β. Hence the lowest "cutoff" frequency for $TE_{0,n}$ waves is given by

$$\omega_{01} = \frac{2.405c}{\sqrt{\epsilon_1 - \epsilon_0}\, a} \qquad (8.109)$$

At this frequency $\beta^2 = 0$, but the axial wave number k is still real and equal to its free-space value $\sqrt{\epsilon_0}\,\omega/c$. Immediately below this "cutoff" frequency, the system no longer acts as a guide but as an antenna, with energy being radiated radially. For frequencies well above cutoff, β and k are of the same order of magnitude and are large compared to γ provided ϵ_1 and ϵ_0 are not nearly equal.

For TM modes, the first equation in (8.108) is replaced by

$$\frac{J_1(\gamma a)}{\gamma J_0(\gamma a)} + \frac{\epsilon_0}{\epsilon_1} \frac{K_1(\beta a)}{\beta K_0(\beta a)} = 0 \qquad (8.110)$$

It is evident that all the qualitative features shown in Fig. 8.10 are retained for the TM waves. The lowest "cutoff" frequency for $TM_{0,n}$ waves is clearly the same as for $TE_{0,n}$ waves. For $\epsilon_1 \gg \epsilon_0$, provided the maximum value of γa does not fall very near one of the roots of $J_0(x) = 0$, (8.110) shows that the propagation constants are determined by $J_1(\gamma a) \simeq 0$. This is just the determining equation for *TE waves* in a metallic wave guide. The reason for the equivalence of the TM modes in a dielectric guide and the TE modes in a hollow metallic guide can be traced to the symmetry of Maxwell's equations under the interchange of $\mathbf{E}$ and $\mathbf{B}$ (with appropriate sign changes and factors of $\sqrt{\mu\epsilon}$), plus the correspondence between the vanishing of normal $\mathbf{B}$ at the metallic surface and the almost vanishing of normal $\mathbf{E}$ at the dielectric surface (due to continuity of normal $\mathbf{D}$ with $\epsilon_1 \gg \epsilon_0$).

If $\epsilon_1 \gg \epsilon_0$, then from (8.100) and (8.101) it is clear that the outside decay constant β is much larger than γ, except near cutoff. This means that the fields do not extend appreciably outside the dielectric cylinder. Figure 8.11 shows qualitatively the behavior of the fields for the $TE_{0,1}$ mode. The other modes behave similarly. As mentioned earlier, modes with azimuthal dependence to the fields have longitudinal components of both $\mathbf{E}$ and $\mathbf{B}$. Although the mathematics is somewhat more involved (see Problem 8.6), the qualitative features of propagation—short wavelength along the cylinder, rapid decrease of fields outside the cylinder, etc.—are the same as for the circularly symmetric modes.

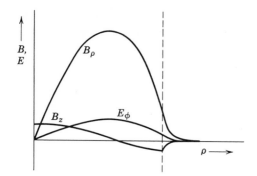

Fig. 8.11 Radial variation of fields of $TE_{0,1}$ mode in dielectric guide. For $\epsilon_1 \gg \epsilon_0$, the fields are confined mostly inside the dielectric.

Dielectric wave guides have not been used for microwave propagation, except for special applications. One reason is that it is difficult to obtain suitable dielectrics with sufficiently low losses at microwave frequencies. In a recent application at optical frequencies very fine dielectric filaments, each coated with a thin layer of material of much lower index of refraction, are closely bundled together to form image-transfer devices.* The filaments are sufficiently small in diameter ($\sim$ 10 microns) that wave-guide concepts are useful, even though the propagation is usually a mixture of several modes.

REFERENCES AND SUGGESTED READING

Wave guides and resonant cavities are dealt with in numerous electrical and communi-cations engineering books. Among the physics textbooks which discuss guides, trans-mission lines, and cavities are
   Panofsky and Phillips, Chapter 12,
   Slater,
   Sommerfeld, *Electrodynamics*, Sections 22–25,
   Stratton, Sections 9.18–9.22.
The mathematical tools for the discussion of these boundary-value problems are presented by
   Morse and Feshbach, especially Chapter 13.
Information on special functions may be found in the ever-reliable
   Magnus and Oberhettinger.
Numerical values of Bessel functions are given by
   Jahnke and Emde,
   Watson.

PROBLEMS

8.1 A transmission line consisting of two concentric circular cylinders of metal with conductivity σ and skin depth δ, as shown on p. 265, is filled with a

* B. O'Brien, *Physics Today*, **13**, 52 (1960).

uniform lossless dielectric (μ, ϵ). A TEM mode is propagated along this line.
(a) Show that the time-averaged power flow along the line is

$$P = \left[\frac{c}{4\pi}\right]\sqrt{\frac{\mu}{\epsilon}}\,\pi a^2\,|H_0|^2 \ln\left(\frac{b}{a}\right)$$

where H_0 is the peak value of the azimuthal magnetic field at the surface of
the inner conductor.

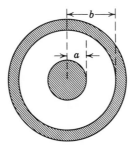

(b) Show that the transmitted power is attenuated along the line as

$$P(z) = P_0 e^{-2\gamma z}$$

where

$$\gamma = \left[\frac{c}{4\pi}\right]\frac{1}{2\sigma\delta}\sqrt{\frac{\epsilon}{\mu}}\,\frac{\left(\dfrac{1}{a}+\dfrac{1}{b}\right)}{\ln\left(\dfrac{b}{a}\right)}$$

(c) The characteristic impedance Z_0 of the line is defined as the ratio of
the voltage between the cylinders to the axial current flowing in one of them
at any position z. Show that for this line

$$Z_0 = \left[\frac{4\pi}{c}\right]\frac{1}{2\pi}\sqrt{\frac{\mu}{\epsilon}}\ln\left(\frac{b}{a}\right)$$

(d) Show that the series resistance and inductance per unit length of the
line are

$$R = \frac{1}{2\pi\sigma\delta}\left(\frac{1}{a}+\frac{1}{b}\right)$$

$$L = \left[\frac{4\pi}{c^2}\right]\left\{\frac{\mu}{2\pi}\ln\left(\frac{b}{a}\right) + \frac{\mu_c\delta}{4\pi}\left(\frac{1}{a}+\frac{1}{b}\right)\right\}$$

where μ_c is the permeability of the conductor. The correction to the
inductance comes from the penetration of the flux into the conductors by a
distance of order δ.

8.2 A transmission line consists of two identical thin strips of metal, shown in
cross section on p. 266. Assuming that $b \gg a$, discuss the propagation

of a TEM mode on this line, repeating the derivations of Problem 8.1.
Show that

$$P = \left[\frac{c}{4\pi}\right]\frac{ab}{2}\sqrt{\frac{\mu}{\epsilon}}\,|H_0|^2$$

$$\gamma = \left[\frac{c}{4\pi}\right]\frac{1}{a\sigma\delta}\sqrt{\frac{\epsilon}{\mu}}$$

$$Z_0 = \left[\frac{4\pi}{c}\right]\sqrt{\frac{\mu}{\epsilon}}\left(\frac{a}{b}\right)$$

$$R = \frac{1}{2\sigma\delta b}$$

$$L = \left[\frac{4\pi}{c^2}\right]\left(\frac{\mu a + \mu_c\delta}{b}\right)$$

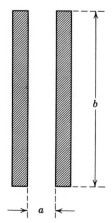

where the symbols have the same meanings as in Problem 8.1.

8.3 Transverse electric and magnetic waves are propagated along a hollow, right circular cylinder of brass with inner radius R.

(*a*) Find the cutoff frequencies of the various TE and TM modes. Determine numerically the lowest cutoff frequency (the dominant mode) in terms of the tube radius and the ratio of cutoff frequencies of the next four higher modes to that of the dominant mode.

(*b*) Calculate the attenuation constant of the wave guide as a function of frequency for the lowest two modes and plot it as a function of frequency.

8.4 A wave guide is constructed so that the cross section of the guide forms a right triangle with sides of length a, a, $\sqrt{2a}$, as shown on p. 267. The medium inside has $\mu = \epsilon = 1$.

(*a*) Assuming infinite conductivity for the walls, determine the possible modes of propagation and their cutoff frequencies.

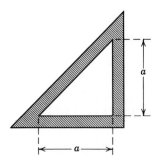

(*b*) For the lowest modes of each type calculate the attenuation constant, assuming that the walls have large, but finite, conductivity. Compare the result with that for a square guide of side *a* made from the same material.

8.5 A resonant cavity of copper consists of a hollow, right circular cylinder of inner radius R and length L, with flat end faces.

(*a*) Determine the resonant frequencies of the cavity for all types of waves. With $(c/\sqrt{\mu\epsilon}\,R)$ as a unit of frequency, plot the lowest four resonant frequencies of each type as a function of R/L for $0 < R/L < 2$. Does the same mode have the lowest frequency for all R/L?

(*b*) If $R = 2$ cm, $L = 3$ cm, and the cavity is made of pure copper, what is the numerical value of Q for the lowest resonant mode?

8.6 A right circular cylinder of nonpermeable dielectric with dielectric constant ϵ and radius a serves as a dielectric wave guide in vacuum.

(*a*) Discuss the propagation of waves along such a guide, assuming that the azimuthal variation of the fields is $e^{im\phi}$.

(*b*) For $m = \pm 1$, determine the mode with the lowest cutoff frequency and discuss the properties of its fields (cutoff frequency, spatial variation, etc.), assuming that $\epsilon \gg 1$.

9

Simple Radiating Systems and Diffraction

In Chapters 7 and 8 we have discussed the properties of electromagnetic waves and their propagation in both bounded and unbounded geometries. But nothing has been said about how to produce these waves. In the present chapter we remedy this omission to some extent by presenting a discussion of radiation by a localized oscillating system of charges and currents. The treatment is straightforward, with little in the way of elegant formalism. It is by its nature restricted to rather simple radiating systems. A more systematic approach to radiation by localized distributions of charge and current is left to Chapter 16, where multipole fields are discussed.

The second half of the chapter is devoted to the subject of diffraction. Since the customary scalar Kirchhoff theory is discussed in many books, the emphasis has been placed on the vector properties of the electromagnetic field in diffraction.

9.1 Fields and Radiation of a Localized Oscillating Source

For a system of charges and currents varying in time we can make a Fourier analysis of the time dependence and handle each Fourier component separately. We therefore lose no generality by considering the potentials, fields, and radiation from a localized system of charges and currents which vary sinusoidally in time:

$$\left.\begin{array}{l} \rho(\mathbf{x}, t) = \rho(\mathbf{x})e^{-i\omega t} \\ \mathbf{J}(\mathbf{x}, t) = \mathbf{J}(\mathbf{x})e^{-i\omega t} \end{array}\right\} \tag{9.1}$$

As usual, the real part of such expressions is to be taken to obtain physical quantities. The electromagnetic potentials and fields are assumed to have the same time dependence.

It was shown in Chapter 6 that the solution for the vector potential $\mathbf{A}(\mathbf{x}, t)$ in the Lorentz gauge is

$$\mathbf{A}(\mathbf{x}, t) = \frac{1}{c}\int d^3x' \int dt' \, \frac{\mathbf{J}(\mathbf{x}', t')}{|\mathbf{x} - \mathbf{x}'|} \, \delta\left(t' + \frac{|\mathbf{x} - \mathbf{x}'|}{c} - t\right) \qquad (9.2)$$

provided no boundary surfaces are present. The Dirac delta function assures the causal behavior of the fields. With the sinusoidal time dependence (9.1), the solution for $\mathbf{A}$ becomes

$$\mathbf{A}(\mathbf{x}) = \frac{1}{c}\int \mathbf{J}(\mathbf{x}') \, \frac{e^{ik|\mathbf{x}-\mathbf{x}'|}}{|\mathbf{x} - \mathbf{x}'|} \, d^3x' \qquad (9.3)$$

where $k = \omega/c$ is the wave number, and a sinusoidal time dependence is understood. The magnetic induction is given by

$$\mathbf{B} = \nabla \times \mathbf{A} \qquad (9.4)$$

while, outside the source, the electric field is

$$\mathbf{E} = \frac{i}{k} \nabla \times \mathbf{B} \qquad (9.5)$$

Given a current distribution $\mathbf{J}(\mathbf{x}')$, the fields can, in principle at least, be determined by calculating the integral in (9.3). We will consider one or two examples of direct integration of the source integral in Section 9.4. But at present we wish to establish certain simple, but general, properties of the fields in the limit that the source of current is confined to a small region, very small in fact compared to a wavelength. If the source dimensions are of order d and the wavelength is $\lambda = 2\pi c/\omega$, and if $d \ll \lambda$, then there are three spatial regions of interest:

The near (static) zone: $d \ll r \ll \lambda$

The intermediate (induction) zone: $d \ll r \sim \lambda$

The far (radiation) zone: $d \ll \lambda \ll r$

We will see that the fields have very different properties in the different zones. In the near zone the fields have the character of static fields with radial components and variation with distance which depends in detail on the properties of the source. In the far zone, on the other hand, the fields are transverse to the radius vector and fall off as r^{-1}, typical of radiation fields.

For the near zone where $r \ll \lambda$ (or $kr \ll 1$) the exponential in (9.3) can be replaced by unity. Then the vector potential is of the form already considered in Chapter 5. The inverse distance can be expanded using (3.70), with the result,

$$\lim_{kr \to 0} \mathbf{A}(\mathbf{x}) = \frac{1}{c} \sum_{l,m} \frac{4\pi}{2l+1} \frac{Y_{lm}(\theta, \phi)}{r^{l+1}} \int \mathbf{J}(\mathbf{x}') r'^{l} Y_{lm}^{*}(\theta', \phi') \, d^{3}x' \qquad (9.6)$$

This shows that the near fields are quasi-stationary, oscillating harmonically as $e^{-i\omega t}$, but otherwise static in character.

In the far zone ($kr \gg 1$) the exponential in (9.3) oscillates rapidly and determines the behavior of the vector potential. In this region it is sufficient to approximate

$$|\mathbf{x} - \mathbf{x}'| \simeq r - \mathbf{n} \cdot \mathbf{x}' \qquad (9.7)$$

where $\mathbf{n}$ is a unit vector in the direction of $\mathbf{x}$. Furthermore, if only the leading term in kr is desired, the inverse distance in (9.3) can be replaced by r. Then the vector potential is

$$\lim_{kr \to \infty} \mathbf{A}(\mathbf{x}) = \frac{e^{ikr}}{cr} \int \mathbf{J}(\mathbf{x}') e^{-ik\mathbf{n} \cdot \mathbf{x}'} \, d^{3}x'. \qquad (9.8)$$

This demonstrates that in the far zone the vector potential behaves as an outgoing spherical wave. It is easy to show that the fields calculated from (9.4) and (9.5) are transverse to the radius vector and fall off as r^{-1}. They thus correspond to radiation fields. If the source dimensions are small compared to a wavelength it is appropriate to expand the integral in (9.8) in powers of k:

$$\lim_{kr \to \infty} \mathbf{A}(\mathbf{x}) = \frac{e^{ikr}}{cr} \sum_{n} \frac{(-ik)^{n}}{n!} \int \mathbf{J}(\mathbf{x}')(\mathbf{n} \cdot \mathbf{x}')^{n} \, d^{3}x' \qquad (9.9)$$

The magnitude of the nth term is given by

$$\frac{1}{n!} \int \mathbf{J}(\mathbf{x}')(k\mathbf{n} \cdot \mathbf{x}')^{n} \, d^{3}x' \qquad (9.10)$$

Since the order of magnitude of $\mathbf{x}'$ is d and kd is small compared to unity by assumption, the successive terms in the expansion of $\mathbf{A}$ evidently fall off rapidly with n. Consequently the radiation emitted from the source will come mainly from the first nonvanishing term in the expansion (9.9). We will examine the first few of these in the following sections.

In the intermediate or induction zone the two alternative approximations leading to (9.6) and (9.8) cannot be made; all powers of kr must be retained. Without marshalling the full apparatus of vector multipole fields, described in Chapter 16, we can abstract enough for our immediate purpose. The key result is the exact expansion (16.22) for the Green's function appearing in (9.3). For points outside the source (9.3) then becomes

$$\mathbf{A}(x) = \frac{4\pi ik}{c} \sum_{l,m} h_{l}^{(1)}(kr) Y_{lm}(\theta, \phi) \int \mathbf{J}(\mathbf{x}') j_{l}(kr') Y_{lm}^{*}(\theta', \phi') \, d^{3}x' \qquad (9.11)$$

If the source dimensions are small compared to a wavelength, $j_l(kr')$ can be approximated by (16.12). Then the result for the vector potential is of the form of (9.6), but with the replacement,

$$\frac{1}{r^{l+1}} \to \frac{e^{ikr}}{r^{l+1}} (1 + a_1(ikr) + a_2(ikr)^2 + \cdots + a_l(ikr)^l) \quad (9.12)$$

The numerical coefficients a_i come from the explicit expressions for the spherical Hankel functions. The right hand side of (9.12) shows the transition from the static-zone result (9.6) for $kr \ll 1$ to the radiation-zone form (9.9) for $kr \gg 1$.

9.2 Electric Dipole Fields and Radiation

If only the first term in (9.9) is kept, the vector potential is

$$\mathbf{A}(\mathbf{x}) = \frac{e^{ikr}}{cr} \int \mathbf{J}(\mathbf{x}') \, d^3x' \quad (9.13)$$

Examination of (9.11) and (9.12) shows that (9.13) is the $l = 0$ part of the series and that it is valid everywhere outside the source, not just in the far zone. The integral can be put in more familiar terms by an integration by parts:

$$\int \mathbf{J} \, d^3x' = -\int \mathbf{x}'(\nabla' \cdot \mathbf{J}) \, d^3x' = -i\omega \int \mathbf{x}'\rho(\mathbf{x}') \, d^3x' \quad (9.14)$$

from the continuity equation,

$$i\omega\rho = \nabla \cdot \mathbf{J} \quad (9.15)$$

Thus the vector potential is

$$\mathbf{A}(\mathbf{x}) = -ik\mathbf{p} \frac{e^{ikr}}{r} \quad (9.16)$$

where

$$\mathbf{p} = \int \mathbf{x}'\rho(\mathbf{x}') \, d^3x' \quad (9.17)$$

is the electric dipole moment, as defined in electrostatics by (4.8).

The electric dipole fields from (9.4) and (9.5) are

$$\mathbf{B} = k^2(\mathbf{n} \times \mathbf{p}) \frac{e^{ikr}}{r} \left(1 - \frac{1}{ikr}\right)$$

$$\mathbf{E} = k^2(\mathbf{n} \times \mathbf{p}) \times \mathbf{n} \frac{e^{ikr}}{r} + [3\mathbf{n}(\mathbf{n} \cdot \mathbf{p}) - \mathbf{p}]\left(\frac{1}{r^3} - \frac{ik}{r^2}\right)e^{ikr} \quad (9.18)$$

We note that the magnetic induction is transverse to the radius vector at all distances, but that the electric field has components parallel and perpendicular to $\mathbf{n}$.

In the radiation zone the fields take on the limiting forms,

$$\mathbf{B} = k^2(\mathbf{n} \times \mathbf{p}) \frac{e^{ikr}}{r}$$

$$\mathbf{E} = \mathbf{B} \times \mathbf{n} \quad (9.19)$$

showing the typical behavior of radiation fields.

In the near zone, on the other hand, the fields approach

$$\left.\begin{array}{l} \mathbf{B} = ik(\mathbf{n} \times \mathbf{p})\,\dfrac{1}{r^2} \\[4mm] \mathbf{E} = [3\mathbf{n}(\mathbf{n} \cdot \mathbf{p}) - \mathbf{p}]\,\dfrac{1}{r^3} \end{array}\right\} \qquad (9.20)$$

The electric field, apart from its oscillations in time, is just the static electric dipole field (4.13). The magnetic induction is a factor (kr) smaller than the electric field in the region where $kr \ll 1$. Thus the fields in the near zone are dominantly electric in nature. The magnetic induction vanishes, of course, in the static limit $k \to 0$. Then the near zone extends to infinity.

The time-averaged power radiated per unit solid angle by the oscillating dipole moment $\mathbf{p}$ is

$$\frac{dP}{d\Omega} = \frac{c}{8\pi}\,\mathrm{Re}\,[r^2\mathbf{n} \cdot \mathbf{E} \times \mathbf{B}^*] \qquad (9.21)$$

where $\mathbf{E}$ and $\mathbf{B}$ are given by (9.19). Thus we find

$$\frac{dP}{d\Omega} = \frac{c}{8\pi}\,k^4|\mathbf{n} \times (\mathbf{n} \times \mathbf{p})|^2 \qquad (9.22)$$

The state of polarization of the radiation is given by the vector inside the absolute value signs. If the components of $\mathbf{p}$ all have the same phase, the angular distribution is a typical dipole pattern,

$$\frac{dP}{d\Omega} = \frac{c}{8\pi}\,k^4|\mathbf{p}|^2 \sin^2 \theta \qquad (9.23)$$

where the angle θ is measured from the direction of $\mathbf{p}$. The total power radiated is

$$P = \frac{ck^4}{3}\,|\mathbf{p}|^2 \qquad (9.24)$$

A simple example of an electric dipole radiator is a centerfed, linear antenna whose length d is small compared to a wavelength. The antenna is assumed to be oriented along the z axis, extending from $z = (d/2)$ to $z = -(d/2)$ with a narrow gap at the center for purposes of excitation, as shown in Fig. 9.1. The current is in the same direction in each half of the antenna, having a value I_0 at the gap and falling approximately linearly to zero at the ends:

$$I(z)e^{-i\omega t} = I_0\left(1 - \frac{2|z|}{d}\right)e^{-i\omega t} \qquad (9.25)$$

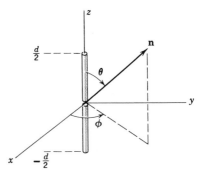

Fig. 9.1 Short, center-fed, linear antenna.

From the continuity equation (9.15) the linear-charge density ρ' (charge per unit length) is constant along each arm of the antenna, with the value,

$$\rho'(z) = \pm \frac{2iI_0}{\omega d} \tag{9.26}$$

the upper (lower) sign being appropriate for positive (negative) values of z. The dipole moment (9.17) is parallel to the z axis and has the magnitude

$$p = \int_{-(d/2)}^{(d/2)} z\rho'(z)\,dz = \frac{iI_0 d}{2\omega} \tag{9.27}$$

The angular distribution of radiated power is

$$\frac{dP}{d\Omega} = \frac{I_0^{\,2}}{32\pi c}(kd)^2 \sin^2\theta \tag{9.28}$$

while the total power radiated is

$$P = \frac{I_0^{\,2}(kd)^2}{12c} \tag{9.29}$$

We see that for a fixed input current the power radiated increases as the square of the frequency, at least in the long-wavelength domain where $kd \ll 1$.

9.3 Magnetic Dipole and Electric Quadrupole Fields

The next term in expansion (9.9) leads to a vector potential,

$$\mathbf{A}(\mathbf{x}) = \frac{e^{ikr}}{cr}\left(\frac{1}{r} - ik\right)\int \mathbf{J}(\mathbf{x}')(\mathbf{n}\cdot\mathbf{x}')\,d^3x' \tag{9.30}$$

where we have included the correct terms from (9.12) in order to make (9.30) valid everywhere outside the source. This vector potential can be written as the sum of two terms, one of which gives a transverse magnetic induction and the other of which gives a transverse electric field. These physically distinct contributions can be separated by writing the integrand in (9.30) as the sum of a part symmetric in $\mathbf{J}$ and $\mathbf{x}'$ and a part that is antisymmetric. Thus

$$\frac{1}{c}(\mathbf{n} \cdot \mathbf{x}')\mathbf{J} = \frac{1}{2c}\left[(\mathbf{n} \cdot \mathbf{x}')\mathbf{J} + (\mathbf{n} \cdot \mathbf{J})\mathbf{x}'\right] + \frac{1}{2c}(\mathbf{x}' \times \mathbf{J}) \times \mathbf{n} \qquad (9.31)$$

The second, antisymmetric part is recognizable as the magnetization due to the current $\mathbf{J}$:

$$\mathscr{M} = \frac{1}{2c}(\mathbf{x} \times \mathbf{J}) \qquad (9.32)$$

The first, symmetric term will be shown to be related to the electric quadrupole moment density.

Considering only the magnetization term, we have the vector potential,

$$\mathbf{A}(\mathbf{x}) = ik(\mathbf{n} \times \mathbf{m})\frac{e^{ikr}}{r}\left(1 - \frac{1}{ikr}\right) \qquad (9.33)$$

where $\mathbf{m}$ is the magnetic dipole moment,

$$\mathbf{m} = \int\mathscr{M}\, d^3x = \frac{1}{2c}\int(\mathbf{x} \times \mathbf{J})\, d^3x \qquad (9.34)$$

The fields can be determined by noting that the vector potential (9.33) is proportional to the magnetic induction (9.18) for an electric dipole. This means that the magnetic induction for the present magnetic dipole source will be equal to the electric field for the electric dipole, with the substitution $\mathbf{p} \rightarrow \mathbf{m}$. Thus we find

$$\mathbf{B} = k^2(\mathbf{n} \times \mathbf{m}) \times \mathbf{n}\frac{e^{ikr}}{r} + [3\mathbf{n}(\mathbf{n} \cdot \mathbf{m}) - \mathbf{m}]\left(\frac{1}{r^3} - \frac{ik}{r^2}\right)e^{ikr} \qquad (9.35)$$

Similarly, the electric field for a magnetic dipole source is the negative of the magnetic field for an electric dipole:

$$\mathbf{E} = -k^2(\mathbf{n} \times \mathbf{m})\frac{e^{ikr}}{r}\left(1 - \frac{1}{ikr}\right) \qquad (9.36)$$

All the arguments concerning the behavior of the fields in the near and far zones are the same as for the electric dipole source, with the interchanges $\mathbf{E} \rightarrow \mathbf{B}$, $\mathbf{B} \rightarrow -\mathbf{E}$, $\mathbf{p} \rightarrow \mathbf{m}$. Similarly the radiation pattern and total power radiated are the same for the two kinds of dipole. The only difference in the radiation fields is in the polarization. For an electric dipole the electric vector lies in the plane defined by $\mathbf{n}$ and $\mathbf{p}$, while for a magnetic dipole it is perpendicular to the plane defined by $\mathbf{n}$ and $\mathbf{m}$.

The integral of the symmetric term in (9.31) can be transformed by an integration by parts and some rearrangement:

$$\frac{1}{2c} \int [(\mathbf{n} \cdot \mathbf{x}')\mathbf{J} + (\mathbf{n} \cdot \mathbf{J})\mathbf{x}'] \, d^3x' = -\frac{ik}{2} \int \mathbf{x}'(\mathbf{n} \cdot \mathbf{x}')\rho(\mathbf{x}') \, d^3x' \quad (9.37)$$

The continuity equation (9.15) has been used to replace $\nabla \cdot \mathbf{J}$ by $i\omega\rho$. Since the integral involves second moments of the charge density, this symmetric part corresponds to an electric quadrupole source. The vector potential is

$$\mathbf{A}(\mathbf{x}) = -\frac{k^2}{2} \frac{e^{ikr}}{r} \left(1 - \frac{1}{ikr} \right) \int \mathbf{x}'(\mathbf{n} \cdot \mathbf{x}')\rho(\mathbf{x}') \, d^3x' \quad (9.38)$$

The complete fields are somewhat complicated to write down. We will content ourselves with the fields in the radiation zone. Then it is easy to see that

$$\left. \begin{array}{l} \mathbf{B} = ik\mathbf{n} \times \mathbf{A} \\[2mm] \mathbf{E} = ik(\mathbf{n} \times \mathbf{A}) \times \mathbf{n} \end{array} \right\} \quad (9.39)$$

Consequently the magnetic induction is

$$\mathbf{B} = -\frac{ik^3}{2} \frac{e^{ikr}}{r} \int (\mathbf{n} \times \mathbf{x}')(\mathbf{n} \cdot \mathbf{x}')\rho(\mathbf{x}') \, d^3x' \quad (9.40)$$

With definition (4.9) for the quadrupole moment tensor,

$$Q_{\alpha\beta} = \int (3x_\alpha x_\beta - r^2\delta_{\alpha\beta})\rho(\mathbf{x}) \, d^3x \quad (9.41)$$

the integral in (9.40) can be written

$$\mathbf{n} \times \int \mathbf{x}'(\mathbf{n} \cdot \mathbf{x}')\rho(\mathbf{x}') \, d^3x' = \tfrac{1}{3}\mathbf{n} \times \mathbf{Q}(\mathbf{n}) \quad (9.42)$$

The vector $\mathbf{Q}(\mathbf{n})$ is defined as having components,

$$Q_\alpha = \sum_\beta Q_{\alpha\beta} n_\beta \quad (9.43)$$

We note that it depends in magnitude and direction on the direction of observation as well as on the properties of the source. With these definitions we have the magnetic induction,

$$\mathbf{B} = -\frac{ik^3}{6} \frac{e^{ikr}}{r} \mathbf{n} \times \mathbf{Q}(\mathbf{n}) \quad (9.44)$$

and the time-averaged power radiated per unit solid angle,

$$\frac{dP}{d\Omega} = \frac{c}{288\pi} k^6 |\mathbf{n} \times \mathbf{Q}(\mathbf{n})|^2 \quad (9.45)$$

The general angular distribution is complicated. But the total power radiated can be calculated in a straightforward way. With the definition of $\mathbf{Q(n)}$ we can write the angular dependence as

$$|\mathbf{n} \times \mathbf{Q(n)}|^2 = \mathbf{Q}^* \cdot \mathbf{Q} - |\mathbf{n} \cdot \mathbf{Q}|^2$$

$$= \sum_{\alpha,\beta,\gamma} Q_{\alpha\beta}^* Q_{\alpha\gamma} n_\beta n_\gamma - \sum_{\alpha,\beta,\gamma,\delta} Q_{\alpha\beta}^* Q_{\gamma\delta} n_\alpha n_\beta n_\gamma n_\delta \qquad (9.46)$$

The necessary angular integrals over products of the rectangular components of $\mathbf{n}$ are readily found to be

$$\left.\begin{aligned}
\int n_\beta n_\gamma \, d\Omega &= \frac{4\pi}{3}\,\delta_{\beta\gamma} \\[2mm]
\int n_\alpha n_\beta n_\gamma n_\delta \, d\Omega &= \frac{4\pi}{15}\,(\delta_{\alpha\beta}\delta_{\gamma\delta} + \delta_{\alpha\gamma}\delta_{\beta\delta} + \delta_{\alpha\delta}\delta_{\beta\gamma})
\end{aligned}\right\} \qquad (9.47)$$

Then we find

$$\int |\mathbf{n} \times \mathbf{Q(n)}|^2 \, d\Omega = 4\pi\left\{\frac{1}{3}\sum_{\alpha,\beta}|Q_{\alpha\beta}|^2 - \frac{1}{15}\left[\sum_\alpha Q_{\alpha\alpha}^* \sum_\gamma Q_{\gamma\gamma} + 2\sum_{\alpha,\beta}|Q_{\alpha\beta}|^2\right]\right\} \qquad (9.48)$$

Since $Q_{\alpha\beta}$ is a tensor whose main diagonal sum is zero, the first term in the square brackets vanishes identically. Thus we obtain the final result for the total power radiated by a quadrupole source:

$$P = \frac{ck^6}{360}\sum_{\alpha,\beta}|Q_{\alpha\beta}|^2 \qquad (9.49)$$

The radiated power varies as the sixth power of the frequency for fixed quadrupole moments, compared to the fourth power for dipole radiation.

A simple example of a radiating quadrupole source is an oscillating spheroidal distribution of charge. The off-diagonal elements of $Q_{\alpha\beta}$ vanish. The diagonal elements may be written

$$Q_{33} = Q_0, \qquad Q_{11} = Q_{22} = -\tfrac{1}{2}Q_0 \qquad (9.50)$$

Then the angular distribution of radiated power is

$$\frac{dP}{d\Omega} = \frac{ck^6}{128\pi}\,Q_0^2 \sin^2\theta \cos^2\theta \qquad (9.51)$$

This is a four-lobed pattern, as shown in Fig. 9.2, with maxima at $\theta = \pi/4$ and $3\pi/4$. The total power radiated by this quadrupole is

$$P = \frac{ck^6 Q_0^2}{240} \qquad (9.52)$$

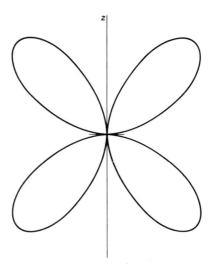

Fig. 9.2 Quadrupole radiation pattern.

The labor involved in manipulating higher terms in expansion (9.9) of the vector potential (9.8) becomes increasingly prohibitive as the expansion is extended beyond the electric quadrupole terms. Another disadvantage of the present approach is that physically distinct fields such as those of the magnetic dipole and the electric quadrupole must be disentangled from the separate terms in (9.9). Finally, the present technique is useful only in the long-wavelength limit. A systematic development of multipole radiation is given in Chapter 16. It involves a fairly elaborate mathematical apparatus, but the price paid is worth while. The treatment allows all multipole orders to be handled in the same way; the results are valid for all wavelengths; the physically different electric and magnetic multipoles are clearly separated from the beginning.

9.4   Center-fed Linear Antenna

For certain radiating systems the geometry of current flow is sufficiently simple that integral (9.3) for the vector potential can be found in relatively simple, closed form. As an example of such a system we consider a thin, linear antenna of length d which is excited across a small gap at its midpoint. The antenna is assumed to be oriented along the z axis with its gap at the origin, as indicated in Fig. 9.3. If damping due to the emission of radiation is neglected, the current along the antenna can be taken as sinusoidal in time and space with wave number $k = \omega/c$, and is symmetric on the two arms of the antenna. The current vanishes at the ends of the antenna. Hence the current density can be written

$$\mathbf{J}(\mathbf{x}) = I \sin \left(\frac{kd}{2} - k\,|z| \right) \delta(x)\,\delta(y)\boldsymbol{\epsilon}_3 \qquad (9.53)$$

for $|z| < (d/2)$. The delta functions assure that the current flows only along the z axis. I is the peak value of the current if $kd \geq \pi$. The current at the gap is $I_0 = I \sin (kd/2)$.

With the current density (9.53) the vector potential is in the z direction and in the radiation zone has the form [from (9.7)]:

$$\mathbf{A}(\mathbf{x}) = \boldsymbol{\epsilon}_3 \frac{Ie^{ikr}}{cr} \int_{-(d/2)}^{(d/2)} \sin \left(\frac{kd}{2} - k\,|z| \right) e^{-ikz \cos \theta}\, dz \qquad (9.54)$$

The result of straightforward integration is

$$\mathbf{A}(\mathbf{x}) = \boldsymbol{\epsilon}_3 \frac{2Ie^{ikr}}{ckr} \left[\frac{\cos \left(\dfrac{kd}{2} \cos \theta \right) - \cos \left(\dfrac{kd}{2} \right)}{\sin^2 \theta} \right] \qquad (9.55)$$

Since the magnetic induction in the radiation zone is given by $\mathbf{B} = ik\mathbf{n} \times \mathbf{A}$, its magnitude is $|\mathbf{B}| = k \sin \theta\,|A_3|$. Thus the time-averaged power radiated per unit solid angle is

$$\frac{dP}{d\Omega} = \frac{I^2}{2\pi c} \left| \frac{\cos \left(\dfrac{kd}{2} \cos \theta \right) - \cos \left(\dfrac{kd}{2} \right)}{\sin \theta} \right|^2 \qquad (9.56)$$

The electric vector is in the direction of the component of $\mathbf{A}$ perpendicular to $\mathbf{n}$. Consequently the polarization of the radiation lies in the plane containing the antenna and the radius vector to the observation point.

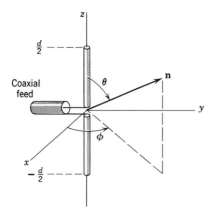

Fig. 9.3   Center-fed, linear antenna.

The angular distribution (9.56) depends on the value of kd. In the long-wavelength limit ($kd \ll 1$) it is easy to show that it reduces to the dipole result (9.28). For the special values $kd = \pi$ (2π), corresponding to a half (two halves) of a wavelength of current oscillation along the antenna, the angular distributions are

$$\frac{dP}{d\Omega} = \frac{I^2}{2\pi c} \begin{cases} \dfrac{\cos^2\left(\dfrac{\pi}{2}\cos\theta\right)}{\sin^2\theta}, & kd = \pi \\[4ex] \dfrac{4\cos^4\left(\dfrac{\pi}{2}\cos\theta\right)}{\sin^2\theta}, & kd = 2\pi \end{cases} \tag{9.57}$$

These angular distributions are shown in Chapter 16 in Fig. 16.4, where they are compared to multipole expansions. The half-wave antenna distribution is seen to be quite similar to a simple dipole pattern, but the full-wave antenna has a considerably sharper distribution.

The full-wave antenna distribution can be thought of as due to the coherent superposition of the fields of two half-wave antennas, one above the other, excited in phase. The intensity at $\theta = \pi/2$, where the waves add algebraically, is 4 times that of a half-wave antenna. At angles away from $\theta = \pi/2$ the amplitudes tend to interfere, giving the narrower pattern. By suitable arrangement of a set of basic antennas, such as the half-wave antenna, with the phasing of the currents appropriately chosen, arbitrary radiation patterns can be formed by coherent superposition. The interested reader should refer to the electrical engineering literature for detailed treatments of antenna arrays.

For the half-wave and full-wave antennas the angular distributions can be integrated over angles to give

$$P = \frac{I^2}{c} \begin{cases} \dfrac{1}{2}\displaystyle\int_0^{2\pi}\left(\dfrac{1-\cos t}{t}\right)dt, & kd = \pi \\[3ex] 2\displaystyle\int_0^{2\pi}\left(\dfrac{1-\cos t}{t}\right)dt - \dfrac{1}{2}\displaystyle\int_0^{4\pi}\left(\dfrac{1-\cos t}{t}\right)dt, & kd = 2\pi \end{cases} \tag{9.58}$$

The integrals in (9.58) can be expressed in terms of the cosine integral,

$$Ci(x) = -\int_x^\infty \frac{\cos t}{t}\, dt \tag{9.59}$$

as follows:

$$\int_0^x\left(\frac{1-\cos t}{t}\right)dt = \ln(\gamma x) - Ci(x) \tag{9.60}$$

where $\gamma = 1.781 \ldots$ is Euler's constant. Tables of the cosine integral are given by Jahnke and Emde, pp. 6–9. The numerical results for the power radiated are

$$P = \frac{I^2}{2c} \begin{cases} 2.44, & kd = \pi \\ 6.70, & kd = 2\pi \end{cases} \tag{9.61}$$

For a given peak current I the full-wave, center-fed antenna radiates nearly 3 times as much power as the half-wave antenna. The coefficient of $I^2/2$ has the dimensions of a resistance and is called the *radiation resistance* R_{rad} of the antenna. The value in ohms is obtained from (9.61) by multiplying the numbers by 30 (actually the multiplier is the numerical value of the velocity of light divided by appropriate powers of 10). Thus the half- and full-wave center-fed antennas have radiation resistances of 73.2 ohms and 201 ohms, respectively.

The reader should be warned that the idealized problem of an infinitely thin, linear antenna with a sinusoidal current distribution is a somewhat simplified version of what occurs in practice. Finite lateral dimensions, ohmic and radiative losses, nonsinusoidal current distributions, finite gaps for excitation, etc., all introduce complications. These problems are important in practical applications and are treated in detail in an extensive literature on antenna design, to which the interested reader may refer.

9.5 Kirchhoff's Integral for Diffraction

The general problem of diffraction involves a wave incident on one or more obstacles or apertures in absorbing or conducting surfaces. The wave is scattered and perhaps absorbed, leading to radiation propagating in directions other than the incident direction. The calculation of the radiation emerging from a diffracting system is the aim of all diffraction theories. The earliest systematic attempt was that of G. Kirchhoff (1882), based on the ideas of superposition of elemental wavelets due to Huygens. In this section we will discuss Kirchhoff's method and point out some of its deficiencies, and in the next section derive vector theorems which correspond to the basic scalar theorem of Kirchhoff.

The customary geometrical situation in diffraction is two spatial regions I and II separated by a boundary surface S, as shown in Fig. 9.4. For example, S may be an infinite metallic sheet with certain apertures in it. The incident wave, generated by sources in region I, approaches S from one side and is diffracted at the boundary surface, giving rise to scattered waves, one transmitted and one reflected. It is usual to consider only the transmitted wave and call its distribution in angle the *diffraction pattern*

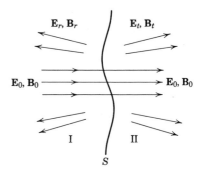

Fig. 9.4 Diffracting system. The surface S, with certain apertures in it, gives rise to reflected and transmitted fields in regions I and II in addition to the fields which would be present in the absence of the surface.

of the system. If the incident wave is described by the fields E_0, B_0, the reflected wave by the fields E_r, B_r, and the transmitted wave by E_t, B_t, then the total fields in regions I and II are $E = E_0 + E_s$, $B = B_0 + B_s$, where s stands for r or t. The basic problem is to determine (E_t, B_t) and (E_r, B_r) from the incident fields (E_0, B_0) and the properties of the boundary surface S. To connect the fields in region I with those in region II boundary conditions for E and B must be satisfied on S, the form of these boundary conditions depending on the properties of S.

The method of attack used in solving such problems is the Green's theorem technique, as applied to the wave equation in Chapter 6. Consider a scalar field $\psi(x, t)$ defined on and inside a *closed* surface S and satisfying the source-free wave equation in that region. The field $\psi(x, t)$ can be thought of as a rectangular component of E or B. We proved in Chapter 6 that the value of ψ inside S could be written in terms of the value of ψ and its normal derivative on the surface as

$$\psi(x, t) = \frac{1}{4\pi} \oint_S \frac{1}{R} \left[\nabla'\psi(x', t') - \frac{R}{R^2} \psi(x', t') - \frac{R}{cR} \frac{\partial \psi(x', t')}{\partial t'} \right]_{\text{ret}} \cdot n \, da' \tag{9.62}$$

where $R = x - x'$, n is the outwardly directed normal to the surface, and ret means evaluated at a time $t' = t - (R/c)$. If a harmonic time dependence $e^{-i\omega t}$ is assumed, this integral form for $\psi(x, t)$ can be written:

$$\psi(x) = \frac{1}{4\pi} \oint_S \frac{e^{ikR}}{R} n \cdot \left[\nabla'\psi + ik\left(1 + \frac{i}{kR}\right)\frac{R}{R} \psi \right] da' \tag{9.63}$$

To adapt (9.63) to diffraction problems we consider the closed surface S to be made up of two surfaces S_1 and S_2. Surface S_1 will be chosen as a convenient one for the particular problem to be solved (e.g., the conducting screen with apertures in it), while surface S_2 will be taken as a sphere or hemisphere of very large radius (tending to infinity) in region II, as shown in Fig. 9.5. Since the fields in region II are the transmitted fields which originate from the diffracting region, they will be outgoing waves in the neighborhood of S_2. This means that the fields, and therefore $\psi(\mathbf{x})$, will satisfy the *radiation condition*,

$$\psi \to f(\theta, \phi)\frac{e^{ikr}}{r}, \qquad \frac{1}{\psi}\frac{\partial \psi}{\partial r} \to \left(ik - \frac{1}{r}\right) \qquad (9.64)$$

With this condition on ψ it can readily be seen that the integral in (9.63) over the hemisphere S_2 vanishes inversely as the hemisphere radius as that radius goes to infinity. Then we obtain the Kirchhoff integral for $\psi(\mathbf{x})$ in region II:

$$\psi(\mathbf{x}) = -\frac{1}{4\pi}\int_{S_1}\frac{e^{ikR}}{R}\,\mathbf{n}\cdot\left[\nabla'\psi + ik\left(1 + \frac{i}{kR}\right)\frac{\mathbf{R}}{R}\,\psi\right]da' \qquad (9.65)$$

where $\mathbf{n}$ is now a unit vector normal to S_1 and pointing *into* region II.

In order to apply the Kirchhoff formula (9.65) to a diffraction problem it is necessary to know the values of ψ and $\partial\psi/\partial n$ on the surface S_1. Unless we have already solved the problem exactly, these values are not known. If, for example, S_1 is a plane, perfectly conducting screen with an opening in it and ψ represents the component of electric field parallel to S_1, then we know that ψ vanishes everywhere on S_1, except in the opening. But the value of ψ in the opening is undetermined. Without additional knowledge, only approximate solutions can be found by making some assumption about ψ and $\partial\psi/\partial n$ on S_1. The *Kirchhoff approximation* consists of the assumptions:

1. ψ and $\partial\psi/\partial n$ vanish everywhere on S_1 except in the openings.
2. The values of ψ and $\partial\psi/\partial n$ in the openings are equal to the values of the incident wave in the absence of any screens or obstacles.

The standard diffraction calculations of classical optics are all based on the Kirchhoff approximation. It should be obvious that the recipe can have only very approximate validity. There is a basic mathematical inconsistency in the assumptions. It was shown for Laplace's equation (and equally well for the Helmholtz wave equation) in Section 1.9 that the

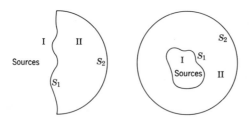

Fig. 9.5 Possible diffraction geometries. Region I contains the sources of radiation. Region II is the diffraction region, where the fields satisfy the radiation condition.

solution inside a closed volume is determined uniquely by specifying ψ (Dirichlet boundary condition) *or* $\partial\psi/\partial n$ (Neumann boundary condition) on the surface. Both ψ and $\partial\psi/\partial n$ cannot be given on the surface. The Kirchhoff approximation works best in the short-wavelength limit in which the diffracting openings have dimensions large compared to a wavelength. Being a scalar theory, even there it cannot account for details of the polarization of the diffracted radiation. In the intermediate- and long-wavelength limit, the scalar approximation fails badly, aside from the drastic approximations inherent in the basic assumptions listed above.

Since the diffraction of electromagnetic radiation is a boundary-value problem in vector fields, we expect that a considerable improvement can be made by developing vector equivalents to the Kirchhoff integral (9.65).

9.6 Vector Equivalents of Kirchhoff Integral

To obtain vector equivalents to the Kirchhoff integral (9.63) we first note that with the definition,

$$G(\mathbf{x}, \mathbf{x}') = \frac{1}{4\pi} \frac{e^{ikR}}{R} \tag{9.66}$$

the scalar form (9.63) can be written

$$\psi(\mathbf{x}) = \oint_S [G\mathbf{n} \cdot \nabla'\psi - \psi\mathbf{n} \cdot \nabla'G]\, da' \tag{9.67}$$

By writing down the result (9.67) for each rectangular component of the electric or magnetic field and combining them vectorially, we can obtain the vector theorem,

$$\mathbf{E}(\mathbf{x}) = \oint_S [G(\mathbf{n} \cdot \nabla')\mathbf{E} - \mathbf{E}(\mathbf{n} \cdot \nabla'G)]\, da' \tag{9.68}$$

with a corresponding relation for **B**. This result is not a particularly convenient one for calculations. It can be transformed into a more useful form by a succession of vector manipulations. First the integrand in (9.68) can be written

$$[\,] = (\mathbf{n} \cdot \nabla')(G\mathbf{E}) - 2\mathbf{E}(\mathbf{n} \cdot \nabla'G) \tag{9.69}$$

Then the vector identities,

$$\left. \begin{array}{l} \mathbf{n} \times (\mathbf{E} \times \nabla'G) = \mathbf{E}(\mathbf{n} \cdot \nabla'G) - (\mathbf{n} \cdot \mathbf{E})\nabla'G \\[2mm] \nabla'G \times (\mathbf{n} \times \mathbf{E}) = \mathbf{n}(\mathbf{E} \cdot \nabla'G) - \mathbf{E}(\mathbf{n} \cdot \nabla'G) \end{array} \right\} \tag{9.70}$$

can be combined to eliminate the last term in (9.69):

$$[\,] = (\mathbf{n} \cdot \nabla')(G\mathbf{E}) - \mathbf{n} \times (\mathbf{E} \times \nabla'G) - \mathbf{n}(\mathbf{E} \cdot \nabla'G)$$
$$-(\mathbf{n} \cdot \mathbf{E})\nabla'G - (\mathbf{n} \times \mathbf{E}) \times \nabla'G \tag{9.71}$$

Now the curl of the product of a vector and a scalar is used to transform the second term in (9.71), while the fact that $\nabla' \cdot E = 0$ is used to re-express the third term. The result is

$$[\ ] = (n \cdot \nabla')(GE) + n \times \nabla' \times (GE) - n\nabla' \cdot (GE)$$
$$-(n \cdot E)\nabla'G - (n \times E) \times \nabla'G - Gn \times (\nabla' \times E)$$
$$(9.72)$$

While it may not appear very fruitful to transform the two terms in (9.68) into six terms, we will now show that the surface integral of the first three terms in (9.72), involving the product (GE), vanishes identically. To do this we make use of the following easily proved identities connecting surface integrals over a *closed* surface S to volume integrals over the interior of S:

$$
\left.
\begin{aligned}
\oint_S A \cdot n \, da &= \int_V \nabla \cdot A \, d^3x \\
\oint_S (n \times A) \, da &= \int_V \nabla \times A \, d^3x \\
\int_S \phi n \, da &= \int_V \nabla \phi \, d^3x
\end{aligned}
\right\}
\qquad (9.73)
$$

where A and ϕ are any well-behaved vector and scalar functions. With these identities the surface integral of the first three terms in (9.72) can be written

$$\oint_S [(n \cdot \nabla')(GE) + n \times \nabla' \times (GE) - n\nabla' \cdot (GE)] \, da'$$
$$= \int_V [\nabla'^2(GE) + \nabla' \times \nabla' \times (GE) - \nabla'(\nabla' \cdot (GE))] \, d^3x' \qquad (9.74)$$

From the expansion, $\nabla \times \nabla \times A = \nabla(\nabla \cdot A) - \nabla^2 A$, it is evident that the volume integral vanishes identically.*

With the surface integral of the first three terms in (9.72) identically zero, the remaining three terms give an alternative form for the vector Kirchhoff relation (9.68). From Maxwell's equations we have $\nabla \times E = ik B$, so that the final result for the electric field anywhere inside the volume

---

* The reader may well be concerned that theorems (9.73) do not apply, since the vector function (GE) is singular at the point $x' = x$. But if the singularity is excluded by taking the surface S as an outer surface S' and a small sphere S'' around $x' = x$, the contribution of the integral over S'' can be shown to vanish in the limit that the radius of S'' goes to zero. Hence result (9.74) is valid, even though G is singular inside the volume of interest.

bounded by the surface S is

$$\mathbf{E}(\mathbf{x}) = - \oint_S [ik(\mathbf{n} \times \mathbf{B})G + (\mathbf{n} \times \mathbf{E}) \times \nabla'G + (\mathbf{n} \cdot \mathbf{E})\nabla'G] \, da' \quad (9.75)$$

The analogous expression for the magnetic induction is

$$\mathbf{B}(\mathbf{x}) = - \oint_S [-ik(\mathbf{n} \times \mathbf{E})G + (\mathbf{n} \times \mathbf{B}) \times \nabla'G + (\mathbf{n} \cdot \mathbf{B})\nabla'G] \, da' \quad (9.76)$$

In (9.75) and (9.76) the unit vector $\mathbf{n}$ is the usual outwardly directed normal. These integrals have an obvious interpretation in terms of equivalent sources of charge and current. The normal component of $\mathbf{E}$ in (9.75) is evidently an effective surface-charge density. Similarly, according to (8.14), the tangential component of magnetic induction $(\mathbf{n} \times \mathbf{B})$ acts as an effective surface current. The other terms $(\mathbf{n} \cdot \mathbf{B})$ and $(\mathbf{n} \times \mathbf{E})$ are effective magnetic surface charge and current densities, respectively.

Vector formulas (9.75) and (9.76) serve as vector equivalents to the Huygens-Kirchhoff scalar integral (9.63). If the fields $\mathbf{E}$ and $\mathbf{B}$ are assumed to obey the radiation condition (9.64) with the added vectorial relationship, $\mathbf{E} = \mathbf{B} \times (\mathbf{r}/r)$, it is easy to show that the surface integral at infinity vanishes. Then, in the notation of Fig. 9.5, the electric field (9.75) is

$$\mathbf{E}(\mathbf{x}) = \int_{S_1} [(\mathbf{n} \times \mathbf{E}) \times \nabla'G + (\mathbf{n} \cdot \mathbf{E})\nabla'G + ik(\mathbf{n} \times \mathbf{B})G] \, da' \quad (9.77)$$

where S_1 is the surface appropriate to the diffracting system, and $\mathbf{n}$ is now directed *into* the region of interest.

The vector theorem (9.77) is a considerable improvement over the scalar expression (9.65) in that the vector nature of the electromagnetic fields is fully included. But to calculate the diffracted fields it is still necessary to know the values of $\mathbf{E}$ and $\mathbf{B}$ on the surface S_1. The Kirchhoff approximations of the previous section can be applied in the short-wave-length limit. But the sudden discontinuity of $\mathbf{E}$ and $\mathbf{B}$ from the unperturbed values in the "illuminated" region to zero in the "shadow" region on the back side of the diffracting system must be compensated for mathematically by line currents around the boundaries of the openings.*

A very convenient formula can be obtained from (9.77) for the special case of plane boundary surface S_1. We imagine that the surface S_1 containing the sources in the right-hand side of Fig. 9.5 is changed in shape into a large, flat pancake, as shown in Fig. 9.6. The region II of "transmitted" fields now becomes two regions, II and II', connected together only by an annular opening at infinity. We denote the two sides

* For a discussion of these line currents, see Stratton, pp. 468–470, and Silver, Chapter 5.

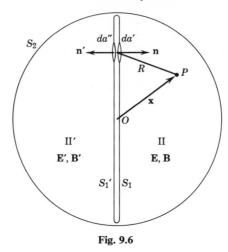

Fig. 9.6

of the disc by S_1 and S_1'. The unit vectors $\mathbf{n}$ and $\mathbf{n}' = -\mathbf{n}$ are directed into regions II and II', respectively. Our aim is to obtain an integral form for the fields in region II in terms of the fields specified on the *right-hand* surface S_1. This is analogous to the geometrical situation shown in the left side of Fig. 9.5. We do not care about the values of the fields in region II'. In fact, the hypothetical sources inside the disc will be imagined to be such that the fields in region II' give a contribution to the surface integral (9.77) which makes the final expression for the diffracted fields in region II especially useful. Once we have obtained the desired result [equation (9.82) below] for the fields in region II as an integral over the surface S_1, we will forget about the manner of derivation and ignore the whole left-hand side of Fig. 9.6. Our interest is in the diffracted fields in region II caused by apertures or obstacles located on the plane surface S_1.

If the fields in regions II and II' are $\mathbf{E}$, $\mathbf{B}$ and $\mathbf{E}'$, $\mathbf{B}'$, respectively, then from the figure it is evident that when the thickness of the disc becomes vanishingly small, integral (9.77) may be written

$$\mathbf{E}(\mathbf{x}) = \int_{S_1} [(\mathbf{n} \times (\mathbf{E} - \mathbf{E}')) \times \nabla'G + \mathbf{n} \cdot (\mathbf{E} - \mathbf{E}')\nabla'G$$
$$+ ik\mathbf{n} \times (\mathbf{B} - \mathbf{B}')G]\, da' \quad (9.78)$$

The field $\mathbf{E}(\mathbf{x})$ on the left side is either $\mathbf{E}$ or $\mathbf{E}'$, depending on where the point $\mathbf{x}$ lies. But the integral is over the right-hand surface S_1 only.

One of the most common applications is to conducting surfaces with apertures in them. The boundary conditions at a perfectly conducting surface are $\mathbf{n} \times \mathbf{E} = 0$, $\mathbf{n} \cdot \mathbf{B} = 0$, but $\mathbf{n} \cdot \mathbf{E} \neq 0$, $\mathbf{n} \times \mathbf{B} \neq 0$. In calculating the surface integral in (9.78) it would be desirable to integrate only over the apertures in the surface rather than over all of it. The first term in (9.78) exists only in the apertures if the screen is perfectly con-ducting. Consequently we try to choose the fields in region II′ so that the other terms vanish everywhere on S_1. Evidently we must choose

$$\left.\begin{array}{l} (\mathbf{n} \cdot \mathbf{E}')_{S_1'} = (\mathbf{n} \cdot \mathbf{E})_{S_1} \\ (\mathbf{n} \times \mathbf{B}')_{S_1'} = (\mathbf{n} \times \mathbf{B})_{S_1} \end{array}\right\} \tag{9.79}$$

Of course, the fields $\mathbf{E}'$, $\mathbf{B}'$ must satisfy Maxwell's equations and the radiation condition in region II′ if $\mathbf{E}$, $\mathbf{B}$ satisfy them in region II. It is easy to show that the required relationship, giving (9.79) on the surfaces, is

$$\begin{aligned} \mathbf{n} \times \mathbf{E}'(\mathbf{x}') &= -\mathbf{n} \times \mathbf{E}(\mathbf{x}) \\ \mathbf{n} \cdot \mathbf{E}'(\mathbf{x}') &= \mathbf{n} \cdot \mathbf{E}(\mathbf{x}) \\ \mathbf{n} \times \mathbf{B}'(\mathbf{x}') &= \mathbf{n} \times \mathbf{B}(\mathbf{x}) \\ \mathbf{n} \cdot \mathbf{B}'(\mathbf{x}') &= -\mathbf{n} \cdot \mathbf{B}(\mathbf{x}) \end{aligned} \tag{9.80}$$

where the point $\mathbf{x}'$ is the mirror image of $\mathbf{x}$ in the plane S_1. The fields at mirror-image points have the opposite (same) values of tangential and outwardly directed normal components of electric field (magnetic induction).

With conditions (9.80) in (9.78) we obtain the simple result for the field $\mathbf{E}(\mathbf{x})$ in terms of an integral over the plane surface S_1 bounding region II,*

$$\mathbf{E}(\mathbf{x}) = 2 \int_{S_1} (\mathbf{n} \times \mathbf{E}) \times \nabla'G \, da' \tag{9.81}$$

where $(\mathbf{n} \times \mathbf{E})$ is the tangential electric field on S_1, $\mathbf{n}$ is a unit normal directed into region II, and G is the Green's function (9.66). Since $\nabla' = -\nabla$ when operating on G, (9.81) can be put in the alternate form,

$$\mathbf{E}(\mathbf{x}) = 2\nabla \times \int_{S_1} \mathbf{n} \times \mathbf{E}(\mathbf{x}')G(\mathbf{x}, \mathbf{x}') \, da' \tag{9.82}$$

For a diffraction system consisting of apertures in a perfectly conducting plane screen the integral over S_1 may be confined to the apertures only. Result (9.81) or (9.82) is exact if the correct tangential component of $\mathbf{E}$ over the apertures is inserted. In practice, we must make some approxi-mation as to the form of the aperture field. But, for plane conducting screens at least, only the tangential electric field need be approximated and the boundary conditions on the screen are correctly satisfied [as can be verified explicitly from (9.82)].

---

* This form for plane screens was first obtained by W. R. Smythe, *Phys. Rev.*, **72**, 1066 (1947), using an argument based on the fields due to a double current sheet filling the apertures, rather than the present Green's-theorem technique.

9.7 Babinet's Principle of Complementary Screens

Before discussing examples of diffraction we wish to establish a useful relation called *Babinet's principle*. Babinet's principle relates the diffraction fields of one diffracting screen to those of the complementary screen. We first discuss the principle in the scalar Kirchhoff approximation. The diffracting screen is assumed to lie in some surface S which divides space into regions I and II in the sense of Section 9.5. The screen occupies all of the surface S except for certain apertures. The complementary screen is that diffracting screen which is obtained by replacing the apertures by screen and the screen by apertures. If the surface of the original screen is S_a and that of the complementary screen is S_b, then $S_a + S_b = S$, as shown schematically in Fig. 9.7.

If there are sources inside S (in region I) which give rise to a field $\psi(\mathbf{x})$, then in the absence of either screen the field $\psi(\mathbf{x})$ in region II is given by the Kirchhoff integral (9.65) where the surface integral is over the entire surface S. With the screen S_a in position, the field $\psi_a(\mathbf{x})$ in region II is given in the Kirchhoff approximation by (9.65) with the source field ψ in the integrand and the surface integral only over S_b (the apertures). Similarly, for the complementary screen S_b, the field $\psi_b(\mathbf{x})$ is given in the same approximation by a surface integral over S_a. Evidently, then, we have the following relation between the diffraction fields ψ_a and ψ_b:

$$\psi_a + \psi_b = \psi \qquad (9.83)$$

This is Babinet's principle as usually formulated in optics. If ψ represents an incident plane wave, for example, Babinet's principle says that the

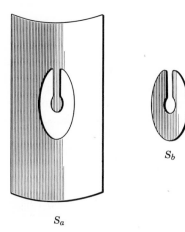

S_b

S_a

Fig. 9.7 A diffraction screen S_a and its complementary diffraction screen S_b.

diffraction pattern away from the incident direction is the same for the original screen and its complement.

The above formulation of Babinet's principle is unsatisfactory in two aspects: it is a statement about scalar fields, and it is based on the Kirchhoff approximation. The second deficiency can be remedied by defining the complementary problem as not only involving complementary screens but also involving complementary boundary conditions (Dirichlet versus Neumann) for the scalar fields. But since we are interested in the electromagnetic field, we will not pursue the scalar problem further.

A rigorous statement of Babinet's principle for electromagnetic fields can be made for a thin, plane, perfectly conducting screen and its complement. We start by considering certain fields E_0, B_0 incident on the screen with metallic surface S_a (see Fig. 9.7) in otherwise empty space. The presence of the screen gives rise to transmitted and reflected fields, as shown in Fig. 9.4. These transmitted and reflected fields will be denoted collectively as scattered fields, E_s, B_s, unless we need to be more specific. For a perfectly conducting screen, the surface current K induced by the incident fields must be such that at all points on the screen's surface S_a, $n \times E_s = -n \times E_0$. For a thin, plane surface, the symmetry of the problem implies that the tangential components of scattered magnetic field at the surface must be equal and opposite, being given from (5.90) by

$$n \times H_t = \frac{2\pi}{c} K = -n \times H_r \qquad (9.84)$$

where n points into the transmitted region II. As a matter of fact, by the same arguments that led from (9.79) to (9.80), it can be established that at any point x in region II and its mirror-image point x' in region I, the scattered fields satisfy the symmetry conditions,

$$\left.\begin{array}{r} n \times E_r(x') = n \times E_t(x) \\ n \cdot E_r(x') = -n \cdot E_t(x) \\ n \times B_r(x') = -n \times B_t(x) \\ n \cdot B_r(x') = n \cdot B_t(x) \end{array}\right\} \qquad (9.85)$$

It will be noted that these relations differ from those in (9.80) by having the signs of $E_r(x')$ and $B_r(x')$ reversed. As we see from the work of Smythe (*op. cit.*, Section 9.6), the fields of (9.80) correspond to a double layer of current. The present fields have the symmetries (9.85) appropriate to a single, plane, current sheet radiating in both directions.

An integral expression for the scattered magnetic induction can now be written down in terms of the surface current K. Since B is the curl of the vector potential, we have

$$\mathbf{B}_s = \mathbf{\nabla} \times \frac{4\pi}{c} \int_{S_a} \mathbf{K} G \, da' \qquad (9.86)$$

where G is the Green's function (9.66), and the integration goes over the metallic surface S_a of the screen. If we substitute for $\mathbf{K}$ from (9.84), we can write the magnetic induction in region II as

$$\mathbf{B}_t(\mathbf{x}) = 2\mathbf{\nabla} \times \int_{S_a} \mathbf{n} \times \mathbf{B}_t(\mathbf{x}')G(\mathbf{x}, \mathbf{x}') \, da \qquad (9.87)$$

This result is identical with (9.82) except that

(1) the roles of $\mathbf{E}$ and $\mathbf{B}$ have been interchanged,
(2) the present integration is only over the body of the screen, whereas that in (9.82) is only over the apertures,
(3) the total electric field appears in (9.82), whereas only the scattered fields occur in (9.87).

The comparison of (9.87) with (9.82) forms the basis of Babinet's principle. If we write down the result (9.82) *for the complement* of the screen with metallic surface S_a, we have

$$\mathbf{E}'(\mathbf{x}) = 2\mathbf{\nabla} \times \int_{S_a} \mathbf{n} \times \mathbf{E}'(\mathbf{x}')G(\mathbf{x}, \mathbf{x}') \, da' \qquad (9.88)$$

The integration is only over S_a, since that is the aperture in the complementary screen. The field $\mathbf{E}'$ in region II is the sum,

$$\mathbf{E}' = \mathbf{E}_0' + \mathbf{E}_t' \qquad (9.89)$$

where $\mathbf{E}_0'$ is the incident electric field of the complementary diffraction problem, and $\mathbf{E}_t'$ the corresponding transmitted or diffracted field. Evidently the two expressions (9.87) and (9.88) turn into one another under the transformation,

$$\mathbf{B}_t \rightarrow \pm(\mathbf{E}_0' + \mathbf{E}_t') \qquad (9.90)$$

It is easy to show that the other fields transform at the same time according to

$$\mathbf{E}_t \rightarrow \mp(\mathbf{B}_0' + \mathbf{B}_t') \qquad (9.91)$$

the sign difference arising from the fact that the fields must represent outgoing radiation in both cases. Since we could have started with the complementary screen initially, it is clear that (9.90) and (9.91) must hold equally with the primed and unprimed quantities interchanged. Comparison of the two sets of expressions shows that the incident fields of the original and complementary diffraction problems must be related according to

$$\mathbf{E}_0' = -\mathbf{B}_0, \qquad \mathbf{B}_0' = \mathbf{E}_0 \qquad (9.92)$$

The complementary problem involves not only the complementary screen,

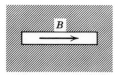

Fig. 9.8   Equivalent radiators according to Babinet's principle.

but also a complementary set of incident fields with the roles of **E** and **B** interchanged.

The statement of Babinet's principle is therefore as follows: a diffracting system consists of a source producing fields $\mathbf{E}_0$, $\mathbf{B}_0$ incident on a thin, plane, perfectly conducting screen with certain apertures in it. The complementary diffracting system consists of a source producing fields $\mathbf{E}_0' = -\mathbf{B}_0$, $\mathbf{B}_0' = \mathbf{E}_0$ incident on the complementary screen. If the transmitted (diffraction) fields on the opposite side of the screens from the source are $\mathbf{E}_t$, $\mathbf{B}_t$ and $\mathbf{E}_t'$, $\mathbf{B}_t'$ for the diffracting system and its complement, respectively, then they are related by

$$\left.\begin{array}{c} \mathbf{E}_t + \mathbf{B}_t' = -\mathbf{E}_0 = -\mathbf{B}_0' \\ \mathbf{B}_t - \mathbf{E}_t' = -\mathbf{B}_0 = +\mathbf{E}_0' \end{array}\right\} \tag{9.93}$$

These are the vector analogs of the scalar relation (9.83).

If a plane wave is incident on the diffracting screen, Babinet's principle states that, in directions other than the incident direction, the intensity of the diffraction pattern of the screen and its complement will be the same, the fields being related by

$$\left.\begin{array}{c} \mathbf{E}_t = -\mathbf{B}_t' \\ \mathbf{B}_t = \mathbf{E}_t' \end{array}\right\} \tag{9.94}$$

The polarization of the wave incident on the complementary screen must, of course, be rotated according to (9.92).

The rigorous vector formulation of Babinet's principle is very useful in microwave problems. For example, consider a narrow slot cut in an infinite, plane, conducting sheet and illuminated with fields that have the magnetic induction along the slot and the electric field perpendicular to it, as shown in Fig. 9.8. The radiation pattern from the slot will be the same as that of a thin linear antenna with its driving electric field along the antenna, as considered in Sections 9.2 and 9.4. The polarization of the radiation will be opposite for the two systems. Elaboration of these ideas makes it possible to design antenna arrays by cutting suitable slots in the sides of wave guides.*

* See, for example, Silver, Chapter 9.

9.8 Diffraction by a Circular Aperture

The subject of diffraction has been extensively studied since Kirchhoff's original work, both in optics, where the scalar theory based on (9.65) generally suffices, and in microwave generation and transmission, where more accurate solutions are needed. There exist specialized treatises devoted entirely to the subject of diffraction and scattering. We will content ourselves with a few examples to illustrate the use of the scalar and vector theorems (9.65) and (9.82) and to compare the accuracy of the approximation schemes.

Historically, diffraction patterns were classed as Fresnel diffraction and Fraunhofer diffraction, depending on the distance of the observation point from the diffracting system. Generally the diffracting system (e.g., an aperture in an opaque screen) has dimensions comparable to, or large compared to, a wavelength. Then the observation point may be in the near zone, less than a wavelength away from the diffracting system. The near-zone fields are complicated in structure and of little interest. Points many wavelengths away from the diffracting system, but still near the system in terms of its own dimensions, are said to lie in the Fresnel zone. Further away, at distances large compared to both the dimensions of the diffracting system and the wavelength, lies the Fraunhofer zone. The Fraunhofer zone corresponds to the radiation zone of Section 9.1. The diffraction patterns in the Fresnel and Fraunhofer zones show characteristic differences which come from the fact that for Fresnel diffraction the region of the diffracting system nearest the observation point is of greatest importance, whereas for Fraunhofer diffraction the whole diffracting system contributes. We will consider only Fraunhofer diffraction, leaving examples of Fresnel diffraction to the problems at the end of the chapter.

If the observation point is far from the diffracting system, expansion (9.7) can be used for $R = |\mathbf{x} - \mathbf{x}'|$. Keeping only lowest-order terms in $(1/kr)$, the scalar Kirchhoff expression (9.65) becomes

$$\psi(\mathbf{x}) = -\frac{e^{ikr}}{4\pi r} \int_{S_1} e^{-i\mathbf{k}\cdot\mathbf{x}'} \left[\mathbf{n} \cdot \nabla'\psi(\mathbf{x}') + i\mathbf{k} \cdot \mathbf{n}\psi(\mathbf{x}') \right] da' \qquad (9.95)$$

where $\mathbf{x}'$ is the coordinate of the element of surface area da', r is the length of the vector $\mathbf{x}$ from the origin O to the observation point P, and $\mathbf{k} = k(\mathbf{x}/r)$ is the wave vector in the direction of observation, as indicated in Fig. 9.9. For a plane surface the vector expression (9.82) reduces in this limit to

$$\mathbf{E}(\mathbf{x}) = \frac{ie^{ikr}}{2\pi r} \mathbf{k} \times \int_{S_1} \mathbf{n} \times \mathbf{E}(\mathbf{x}')e^{-i\mathbf{k}\cdot\mathbf{x}'} da' \qquad (9.96)$$

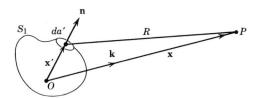

Fig. 9.9

As an example of diffraction we consider a plane wave incident at an angle α on a thin, perfectly conducting screen with a circular hole of radius a in it. The polarization vector of the incident wave lies in the plane of incidence. Figure 9.10 shows an appropriate system of coordinates. The screen lies in the x-y plane with the opening centered at the origin. The wave is incident from below, so that the domain $z > 0$ is the region of diffraction fields. The plane of incidence is taken to be the x-z plane. The incident wave's electric field, written out explicitly in rectangular components, is

$$\mathbf{E}_i = E_0(\boldsymbol{\epsilon}_1 \cos \alpha - \boldsymbol{\epsilon}_3 \sin \alpha)e^{ik(\cos \alpha\, z + \sin \alpha\, x)} \tag{9.97}$$

In calculating the diffraction field with (9.95) or (9.96) we will make the customary approximation that the exact field in the surface integral may be replaced by the incident field. For the vector relation (9.96) we need

$$(\mathbf{n} \times \mathbf{E}_i)_{z=0} = E_0\boldsymbol{\epsilon}_2 \cos \alpha e^{ik \sin \alpha\, x'} \tag{9.98}$$

Then, introducing plane polar coordinates for the integration over the opening, we have

$$\mathbf{E}(\mathbf{x}) = \frac{ie^{ikr}E_0 \cos \alpha}{2\pi r} (\mathbf{k} \times \boldsymbol{\epsilon}_2) \int_0^a \rho\, d\rho \int_0^{2\pi} d\beta\, e^{ik\rho[\sin \alpha \cos \beta - \sin \theta \cos(\phi - \beta)]} \tag{9.99}$$

where θ, ϕ are the spherical angles of $\mathbf{k}$. If we define the angular function,

$$\xi = (\sin^2 \theta + \sin^2 \alpha - 2 \sin \theta \sin \alpha \cos \phi)^{1/2} \tag{9.100}$$

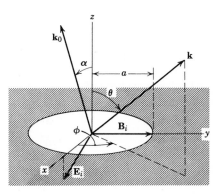

Fig. 9.10 Diffraction by a circular hole of radius a.

the angular integral can be transformed into

$$\frac{1}{2\pi}\int_0^{2\pi} d\beta = \frac{1}{2\pi}\int_0^{2\pi} d\beta' e^{-ik\rho\xi\cos\beta'} = J_0(k\rho\xi) \tag{9.101}$$

Then the radial integral in (9.99) can be done directly. The resulting electric field in the vector Kirchhoff approximation is

$$\mathbf{E(x)} = \frac{ie^{ikr}}{r}\, a^2 E_0 \cos\alpha (\mathbf{k}\times\boldsymbol{\epsilon}_2)\frac{J_1(ka\xi)}{ka\xi} \tag{9.102}$$

The time-averaged diffracted power per unit solid angle is

$$\frac{dP}{d\Omega} = P_i \cos\alpha\, \frac{(ka)^2}{4\pi}\,(\cos^2\phi + \cos^2\theta\sin^2\phi)\left|\frac{2J_1(ka\xi)}{ka\xi}\right|^2 \tag{9.103}$$

where

$$P_i = \left(\frac{cE_0^{\,2}}{8\pi}\right)\pi a^2 \cos\alpha \tag{9.104}$$

is the total power *normally incident* on the aperture. If the opening is large compared to a wavelength ($ka \gg 1$), the factor $[2J_1(ka\xi)/ka\xi]^2$ peaks sharply to a value of unity at $\xi = 0$ and falls rapidly to zero (with small secondary maxima) within a region $\Delta\xi \sim (1/ka)$ on either side of $\xi = 0$. This means that the main part of the wave passes through the opening in the manner of geometrical optics; only slight diffraction effects occur. For $ka \sim 1$ the Bessel-function varies comparatively slowly in angle; the transmitted wave is distributed in directions very different from the incident direction. For $ka \ll 1$, the angular distribution is entirely determined by the factor $(\mathbf{k}\times\boldsymbol{\epsilon}_2)$ in (9.102). But in this limit the assumption of an unperturbed field in the aperture breaks down badly.

The total transmitted power can be obtained by integrating (9.103) over all angles in the forward hemisphere. The ratio of transmitted power to incident power is called the *transmission coefficient T*:

$$T = \frac{\cos\alpha}{\pi}\int_0^{2\pi} d\phi \int_0^{\pi/2} (\cos^2\phi + \cos^2\theta\sin^2\phi)\left|\frac{J_1(ka\xi)}{\xi}\right|^2 \sin\theta\, d\theta \tag{9.105}$$

In the two extreme limits $ka \gg 1$ and $ka \ll 1$, the transmission coefficient approaches the values,

$$T \to \begin{cases} \cos\alpha, & ka \gg 1 \\[2mm] \tfrac{1}{3}(ka)^2\cos\alpha, & ka \ll 1 \end{cases} \tag{9.106}$$

The long-wavelength limit ($ka \ll 1$) is suspect because of our approximations, but it shows that the transmission is small for very small holes.

For normal incidence ($\alpha = 0$) the transmission coefficient (9.105) can be written

$$T = \int_0^{\pi/2} J_1^2(ka \sin \theta)\left(\frac{2}{\sin \theta} - \sin \theta\right) d\theta \qquad (9.107)$$

With the help of the integral relations,

$$\left.\begin{array}{l} \displaystyle\int_0^{\pi/2} J_n^2(z \sin \theta) \frac{d\theta}{\sin \theta} = \int_0^{2z} \frac{J_{2n}(t)}{t} dt \\[1em] \displaystyle\int_0^{\pi/2} J_n^2(z \sin \theta)\sin \theta \, d\theta = \frac{1}{2z}\int_0^{2z} J_{2n}(t) \, dt \end{array}\right\} \qquad (9.108)$$

and the recurrence formulas (3.87) and (3.88), the transmission coefficient can be put in the alternative forms,

$$T = \left\{\begin{array}{l} \displaystyle 1 - \frac{1}{ka} \sum_{m=0}^{\infty} J_{2m+1}(2ka) \\[1.5em] \displaystyle 1 - \frac{1}{2ka} \int_0^{2ka} J_0(t) \, dt \end{array}\right. \qquad (9.109)$$

The transmission coefficient increases more or less monotonically as ka increases, with small oscillations superposed. For $ka \gg 1$, the second form in (9.109) can be used to obtain an asymptotic expression,

$$T \simeq 1 - \frac{1}{2ka} - \frac{1}{2\sqrt{\pi}(ka)^{3/2}} \sin\left(2ka - \frac{\pi}{4}\right) + \cdots \qquad (9.110)$$

which exhibits the small oscillations explicitly. These approximate expressions (9.109) and (9.110) for T give the general behavior as a function of ka, but are not very accurate. Exact calculations, as well as more accurate approximate ones, have been made for the circular opening. These are compared with each other in the book by King and Wu (Fig. 41, p.126). The correct asymptotic expression does not contain the $1/2ka$ term in (9.110), and the coefficient of the term in $(ka)^{-3/2}$ is twice as large.

We now wish to compare our results of the *vector* Kirchhoff approximation with the usual scalar theory based on (9.95). For a wave not normally incident the question immediately arises as to what to choose for the scalar function $\psi(\mathbf{x})$. Perhaps the most consistent assumption is to take the magnitude of the electric or magnetic field. Then the diffracted intensity is treated consistently as proportional to the absolute square of (9.95). If a component of $\mathbf{E}$ or $\mathbf{B}$ is chosen for ψ, we must then decide whether to keep or throw away radial components of the diffracted field in calculating the diffracted power. Choosing the magnitude of $\mathbf{E}$ for ψ, we have, by straightforward calculation with (9.95),

$$\psi(\mathbf{x}) = -ik\,\frac{e^{ikr}}{r}\,a^2 E_0 \left(\frac{\cos\alpha + \cos\theta}{2}\right) \frac{J_1(ka\xi)}{ka\xi} \qquad (9.111)$$

as the scalar equivalent of (9.102). The power radiated per unit solid angle in the *scalar* Kirchhoff approximation is

$$\frac{dP}{d\Omega} \simeq P_i\,\frac{(ka)^2}{4\pi}\cos\alpha\left(\frac{\cos\alpha + \cos\theta}{2\cos\alpha}\right)^2 \left|\frac{2J_1(ka\xi)}{ka\xi}\right|^2 \qquad (9.112)$$

where P_i is given by (9.104).

If we compare the vector Kirchhoff result (9.103) with (9.112), we see similarities and differences. Both formulas contain the same "diffraction" distribution factor $[J_1(ka\xi)/ka\xi]^2$ and the same dependence on wave number. But the scalar result has no azimuthal dependence (apart from that contained in ξ), whereas the vector expression does. The azimuthal variation comes from the polarization properties of the field, and must be absent in a scalar approximation. For normal incidence ($\alpha = 0$) and $ka \gg 1$ the polarization dependence is unimportant. The diffraction is

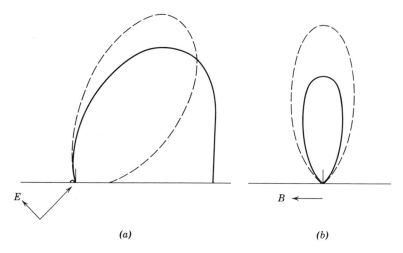

(a)   (b)

Fig. 9.11 Fraunhofer diffraction pattern for a circular opening one wavelength in diameter in a thin, plane, conducting sheet. The plane wave is incident on the screen at 45°. The solid curves are the vector Kirchhoff approximation, while the dotted curves are the scalar approximation. (a) The intensity distribution in the plane of incidence (*E* plane). (b) The intensity distribution (enlarged 2.5 times) perpendicular to the plane of incidence (*H* plane).

confined to very small angles in the forward direction. Then both scalar and vector approximations reduce to the common expression,

$$\frac{dP}{d\Omega} \simeq P_i \frac{(ka)^2}{\pi} \left| \frac{J_1(ka \sin \theta)}{ka \sin \theta} \right|^2 \tag{9.113}$$

The vector and scalar Kirchhoff approximations are compared in Fig. 9.11 for the angle of incidence equal to 45° and for an aperture one wavelength in diameter ($ka = \pi$). The angular distribution is shown in the plane of incidence (containing the electric field vector of the incident wave) and a plane perpendicular to it. The solid (dotted) curve gives the vector (scalar) approximation in each case. We see that for $ka = \pi$ there is a considerable disagreement between the two approximations. There is reason to believe that the vector Kirchhoff result is close to the correct one, even though the approximation breaks down seriously for $ka \lesssim 1$. The vector approximation and exact calculations for a rectangular opening yield results in surprisingly good agreement, even down to $ka \sim 1$.*

9.9 Diffraction by Small Apertures

In the large-aperture or short-wavelength limit we have seen that a reasonably good description of the diffracted fields is obtained by approximating the tangential electric field in the aperture by its unperturbed incident value. For longer wavelengths this approximation begins to fail. When the apertures have dimensions small compared to a wavelength, an entirely different approach is necessary. We will consider a thin, flat, perfectly conducting sheet with a small hole in it. The dimensions of the hole are assumed to be very small compared to a wavelength of the electromagnetic fields which are assumed to exist on one side of the sheet. The problem is to calculate the diffracted fields on the other side of the sheet. Since the sheet is assumed flat, the simple vector theorem (9.82) is appropriate. Evidently the problem is solved if we can determine the electric field in the plane of the hole.

As pointed out by Bethe (1942), the fields in the neighborhood of the aperture can be treated by static or quasi-static methods. In the absence of the aperture the electromagnetic fields near the conducting plane consist of a normal electric field $\mathbf{E}_0$ and a tangential magnetic induction $\mathbf{B}_0$ on one side, and no fields on the other. By "near the conducting plane," we mean at distances small compared to a wavelength. If a small hole is

---

* See J. A. Stratton and L. J. Chu, *Phys. Rev.*, **56**, 99 (1939), for a series of figures comparing the vector Kirchhoff approximation with exact calculations by P. M. Morse and P. J. Rubenstein, *Phys. Rev.*, **54**, 895 (1938).

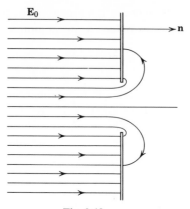

Fig. 9.12

now cut in the plane, the fields will be altered and will penetrate through the hole to the other side. But far away from the hole (in terms of its dimensions), although still "near the conducting plane," the fields will be the same as if the hole were not there, namely, normal $\mathbf{E}_0$ and tangential $\mathbf{B}_0$. The electric field lines might appear as shown in Fig. 9.12. Since the departures of the fields $\mathbf{E}$ and $\mathbf{B}$ from their unperturbed values $\mathbf{E}_0$ and $\mathbf{B}_0$ occur only in a region with dimensions small compared to a wavelength, the task of determining $\mathbf{E}$ or $\mathbf{B}$ near the aperture becomes a problem in electrostatics or magnetostatics, apart from the overall sinusoidal time dependence $e^{-i\omega t}$. For the electric field, it is a standard potential problem of knowing the "asymptotic" values of $\mathbf{E}$ on either side of the perfectly conducting sheet which is an equipotential surface. Similarly for the magnetic induction, $\mathbf{B}$ must be found to yield $\mathbf{B}_0$ and zero "asymptotically" on either side of the sheet, with no normal component on the surface. Then the electric field due to the time variation of $\mathbf{B}$ can be calculated and combined with the "electrostatic" electric field to give the total electric field near the opening.

For a circular opening of radius a small compared to a wavelength, for example, the tangential electric field in the plane of the opening can be shown to be

$$\mathbf{E}_{\text{tan}} = E_0 \frac{\boldsymbol{\rho}}{\pi\sqrt{a^2 - \rho^2}} + \frac{2ik}{\pi}(\mathbf{n} \times \mathbf{B}_0)\sqrt{a^2 - \rho^2} \qquad (9.114)$$

where $E_0 = \mathbf{E}_0 \cdot \mathbf{n}$ is the magnitude of the normal electric field in the absence of the hole, $\mathbf{B}_0$ is the tangential magnetic induction in the absence of the hole, $\mathbf{n}$ is the unit vector normal to the surface and directed *into* the diffraction region [as in (9.82)], and $\boldsymbol{\rho}$ is the radius vector in the plane measured from the center of the opening. With this tangential field determined in the static limit it is a straightforward matter to determine the diffracted fields and power from (9.82). The calculations for the circular opening will be left to the problems at the end of the chapter (Problems 9.10 and 9.11).

9.10 Scattering by a Conducting Sphere in the Short-Wavelength Limit

Another type of problem which is essentially diffraction is the scattering of waves by an obstacle. We will consider the scattering of a plane electromagnetic wave by a perfectly conducting obstacle whose dimensions are large compared to a wavelength. For a thin, flat obstacle, the techniques of Section 9.8, perhaps with Babinet's principle, can be used. But for other obstacles we base the calculation on vector theorem (9.77) for the scattered fields. If we consider only the fields in the radiation zone ($kr \gg 1$), the integral (9.77) for the scattered field $\mathbf{E}_s$ becomes

$$\mathbf{E}_s \to \frac{e^{ikr}}{4\pi ir} \int_{S_1} [(\mathbf{n} \times \mathbf{E}_s) \times \mathbf{k} + (\mathbf{n} \cdot \mathbf{E}_s)\mathbf{k} - k(\mathbf{n} \times \mathbf{B}_s)]e^{-i\mathbf{k}\cdot\mathbf{x}'} \, da'$$

$$(9.115)$$

where $\mathbf{k}$ is the wave vector of the scattered wave, and S_1 is the surface of the obstacle. It will be somewhat easier to calculate with the magnetic induction $\mathbf{B}_s = (\mathbf{k} \times \mathbf{E}_s)/k$:

$$\mathbf{B}_s \to \frac{e^{ikr}}{4\pi ir} \mathbf{k} \times \int_{S_1} \left[(\mathbf{n} \times \mathbf{E}_s) \times \frac{\mathbf{k}}{k} - \mathbf{n} \times \mathbf{B}_s \right] e^{-i\mathbf{k}\cdot\mathbf{x}'} da' \quad (9.116)$$

In the absence of knowledge about the correct fields $\mathbf{E}_s$ and $\mathbf{B}_s$ on the surface of the obstacle, we must make some approximations. If the wavelength is short compared to the dimensions of the obstacle, the surface can be divided approximately into an *illuminated* region and a *shadow* region.* The boundary between these regions is sharp only in the limit of geometrical optics. The transition region can be shown to have a width of the order of $(2/kR)^{1/3}R$, where R is a typical radius of curvature of the surface. Since R is of the order of magnitude of the dimensions of the obstacle, the short-wavelength limit will approximately satisfy the geometrical condition. In the shadow region the scattered fields on the surface must be very nearly equal and opposite to the incident fields. In the illuminated region, the scattered tangential electric field and normal magnetic induction must be equal and opposite to the corresponding incident fields in order to satisfy the boundary conditions on the surface

---

* For a very similar treatment of the scattering of a *scalar* wave by a sphere, see Morse and Feshbach, pp. 1551–1555.

of the perfectly conducting obstacle. On the other hand, the tangential $\mathbf{B}_s$ and normal $\mathbf{E}_s$ in the illuminated region will be approximately equal to the incident values, just as for an infinite, flat, conducting sheet, to the extent that the wavelength is small compared to the radius of curvature. Thus we obtain the following approximate values for the scattered fields on the surface of the obstacle:

Shadow Region   *Illuminated Region*

$\mathbf{E}_s \simeq - \mathbf{E}_i$ $\mathbf{n} \times \mathbf{E}_s = -\mathbf{n} \times \mathbf{E}_i$

$\mathbf{B}_s \simeq -\mathbf{B}_i$ $\mathbf{n} \cdot \mathbf{B}_s = -\mathbf{n} \cdot \mathbf{B}_i$

$\mathbf{n} \times \mathbf{B}_s \simeq \mathbf{n} \times \mathbf{B}_i$

$\mathbf{n} \cdot \mathbf{E}_s \simeq \mathbf{n} \cdot \mathbf{E}_i$

where $\mathbf{E}_i$, $\mathbf{B}_i$ are the fields of the incident wave. With these boundary values the scattered magnetic induction (9.116) can be written as

$$\mathbf{B}_s \simeq \frac{e^{ikr}}{4\pi i r} \mathbf{k} \times (\mathbf{F}_{\text{sh}} + \mathbf{F}_{\text{ill}}) \tag{9.117}$$

where

$$\mathbf{F}_{\text{sh}} = \int_{\text{sh}} \left[\frac{\mathbf{k}}{k} \times (\mathbf{n} \times \mathbf{E}_i) + \mathbf{n} \times \mathbf{B}_i \right] e^{-i\mathbf{k}\cdot\mathbf{x}'} \, da' \tag{9.118}$$

is the integral over the shadow region, and

$$\mathbf{F}_{\text{ill}} = \int_{\text{ill}} \left[\frac{\mathbf{k}}{k} \times (\mathbf{n} \times \mathbf{E}_i) - \mathbf{n} \times \mathbf{B}_i \right] e^{-i\mathbf{k}\cdot\mathbf{x}'} \, da' \tag{9.119}$$

is the integral over the illuminated region.

If the incident wave is a plane wave with wave vector $\mathbf{k}_0$,

$$\left. \begin{aligned} \mathbf{E}_i(\mathbf{x}) &= \mathbf{E}_0 e^{i\mathbf{k}_0 \cdot \mathbf{x}} \\ \mathbf{B}_i(\mathbf{x}) &= \frac{\mathbf{k}_0}{k} \times \mathbf{E}_i(\mathbf{x}) \end{aligned} \right\} \tag{9.120}$$

the integrals over the shadow and illuminated regions of the obstacle's surface are

$$\mathbf{F}_{\text{sh}} = \frac{1}{k} \int_{\text{sh}} [(\mathbf{k} + \mathbf{k}_0) \times (\mathbf{n} \times \mathbf{E}_0) + (\mathbf{n} \cdot \mathbf{E}_0)\mathbf{k}_0] e^{i(\mathbf{k}_0 - \mathbf{k})\cdot\mathbf{x}'} \, da'$$

$$\mathbf{F}_{\text{ill}} = \frac{1}{k} \int_{\text{ill}} [(\mathbf{k} - \mathbf{k}_0) \times (\mathbf{n} \times \mathbf{E}_0) - (\mathbf{n} \cdot \mathbf{E}_0)\mathbf{k}_0] e^{i(\mathbf{k}_0 - \mathbf{k})\cdot\mathbf{x}'} \, da' \tag{9.121}$$

These integrals behave very differently as functions of the scattering angle. In the short-wavelength limit the magnitudes of $\mathbf{k} \cdot \mathbf{x}'$ and $\mathbf{k}_0 \cdot \mathbf{x}'$ are large compared to unity. Thus the exponential factors in (9.121) will oscillate rapidly and cause the integrands to have very small average values except in the forward direction where $\mathbf{k} \simeq \mathbf{k}_0$. In that direction the second term in both $\mathbf{F}_{sh}$ and $\mathbf{F}_{ill}$ is unimportant, since the scattered field (9.117) is proportional to $\mathbf{k} \times \mathbf{F}$. The behavior of the two contributions is thus governed by the first terms in (9.121), at least in the forward direction. We see that $\mathbf{F}_{sh}$ and $\mathbf{F}_{ill}$ are proportional to $(\mathbf{k} \pm \mathbf{k}_0)$, respectively; the shadow integral will be large and the integral from the illuminated region will go to zero. As the scattering angle departs from the forward direction the shadow integral will vanish rapidly, both the exponential and the vector factor in the integrand having the same tendency. On the other hand, the integral from the illuminated region will be small in the forward direction and can be expected to be small at all angles, the exponential and the vector factor in the integrand having opposite tendencies. The shadow integral is evidently the diffraction contribution, while the integral from the illuminated region is the reflected wave.

To proceed much further we must specify the shape of the obstacle. We will assume that it is a perfectly conducting sphere of radius a. Since the shadow integral is large only in the forward direction, we will evaluate it approximately by placing $\mathbf{k} = \mathbf{k}_0$ everywhere except in the exponential. Then, omitting the second term in (9.121) and using spherical coordinates on the surface of the sphere, we obtain

$$\mathbf{F}_{sh} \simeq -2\mathbf{E}_0 a^2 \int_0^{\pi/2} \sin\alpha \, d\alpha \cos\alpha \, e^{ika(1-\cos\theta)\cos\alpha} \int_0^{2\pi} d\beta \, e^{-ika\sin\theta\sin\alpha\cos(\beta-\phi)} \tag{9.122}$$

The angles θ, ϕ and α, β are those of $\mathbf{k}$ and $\mathbf{n}$ relative to $\mathbf{k}_0$. The exponential factor involving $(1-\cos\theta)$ can be set equal to unity, since at small angles its exponent is a factor $\theta/2$ smaller than the other exponent. The integral over β is $2\pi J_0(ka\sin\theta\sin\alpha)$. Hence

$$\mathbf{F}_{sh} \simeq -4\pi a^2 \mathbf{E}_0 \int_0^{\pi/2} J_0(ka\theta\sin\alpha)\cos\alpha\sin\alpha \, d\alpha \tag{9.123}$$

where we have approximated $\sin\theta \simeq \theta$. The integral over α can be written as the integral $\int_0^{ka\theta} x J_0(x) \, dx = ka\theta \, J_1(ka\theta)$. Therefore the shadow scattering integral is

$$\mathbf{F}_{sh} \simeq -4\pi^2 a \mathbf{E}_0 \frac{J_1(ka\theta)}{ka\theta} \tag{9.124}$$

We see that this is essentially the diffraction field of a circular aperture (9.102).

The integral over the illuminated region, giving the reflected or back-scattered wave, is somewhat harder to evaluate. We must consider arbitrary scattering angles, since there is no enhancement in the forward direction. Then the integral consists of a relatively slowly varying vector function of angles times a rapidly varying exponential. As is well known, the dominant contribution to such an integral comes from the region of integration where the phase of the exponential is stationary. The phase factor is

$$f(\alpha, \beta) = (\mathbf{k}_0 - \mathbf{k}) \cdot \mathbf{x}' = ka[(1 - \cos \theta) \cos \alpha - \sin \theta \sin \alpha \cos (\beta - \phi)]$$

(9.125)

The stationary point is easily shown to be at angles α_0, β_0, where

$$\left.\begin{aligned} \alpha_0 &= \frac{\pi}{2} + \frac{\theta}{2} \\ \beta_0 &= \phi \end{aligned}\right\}$$

(9.126)

These angles are evidently just those appropriate for reflection from the sphere according to geometrical optics. At this point the unit vector $\mathbf{n}$ points in the direction of $(\mathbf{k} - \mathbf{k}_0)$. If we expand the phase factor around $\alpha = \alpha_0$, $\beta = \beta_0$, we obtain

$$f(\alpha, \beta) = -2ka \sin \frac{\theta}{2}\left[1 - \frac{1}{2}\left(x^2 + \cos^2 \frac{\theta}{2} y^2\right) + \cdots\right] \quad (9.127)$$

where $x = \alpha - \alpha_0$, $y = \beta - \beta_0$. Then integral (9.121) can be approximated by evaluating the square bracket at $\alpha = \alpha_0$, $\beta = \beta_0$:

$$\mathbf{F}_{ill} \simeq a^2 \sin \theta [2(\mathbf{n}_0 \cdot \mathbf{E}_0) \, \mathbf{n}_0 - \mathbf{E}_0] e^{-2ika \sin (\theta/2)}$$

$$\times \int dx \, e^{i[ka \sin (\theta/2)]x^2} \int dy \, e^{i[ka \sin (\theta/2) \cos^2 (\theta/2)]y^2} \quad (9.128)$$

where $\mathbf{n}_0$ is a unit vector in the direction $(\mathbf{k} - \mathbf{k}_0)$. Provided θ is not too small, the phase factors oscillate rapidly for large x or y. Hence the integration can be extended to $\pm\infty$ in each integral without error. Using the result,

$$\int_{-\infty}^{\infty} e^{i\lambda x^2} \, dx = \left(\frac{\pi}{\lambda}\right)^{1/2} e^{i\pi/4}$$

(9.129)

we obtain

$$\mathbf{F}_{ill} \simeq i \frac{2\pi a}{k} e^{-2ika \sin (\theta/2)} [2(\mathbf{n}_0 \cdot \mathbf{E}_0)\mathbf{n}_0 - \mathbf{E}_0]$$

(9.130)

After some vector algebra the contribution to the scattered field from the illuminated part of the sphere can be written

$$\mathbf{E}_s^{(ill)} \simeq -\frac{a}{2} E_0 \frac{e^{ikr}}{r} e^{-2ika \sin (\theta/2)} \boldsymbol{\epsilon}_{ill}$$

(9.131)

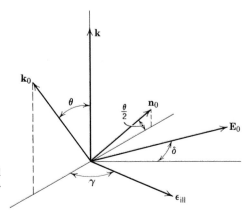

Fig. 9.13 Polarization of reflected wave relative to the incident polarization.

where the polarization vector ϵ_{ill} has a direction defined in Fig. 9.13. If the polarization vector of the incident wave E_0 makes an angle δ with the normal to the plane containing the wave vectors k and k_0, the azimuthal angle γ of ϵ_{ill}, measured from the plane containing k and k_0, is given by $\gamma = (\pi/2) - \delta$. We note that the reflected field (9.131) is constant in magnitude as a function of angle, although it has a rapidly varying phase.

The scattered electric field due to the shadow region is, from (9.124) and (9.117),

$$E_s^{(\text{sh})} \simeq ika^2 \, \frac{J_1(ka\theta)}{ka\theta} \, \frac{e^{ikr}}{r} \, \frac{(k \times E_0) \times k}{k^2} \tag{9.132}$$

Comparison of the two contributions to the scattered wave shows that in the forward direction the shadow field is larger by a factor $ka \gg 1$. But for angles much larger than $\theta \sim (1/ka)$ the shadow field becomes very small and the isotropic reflected field dominates. The power scattered per unit solid angle can be expressed in the form:

$$\frac{dP_s}{d\Omega} \simeq P_i \begin{cases} \dfrac{(ka)^2}{4\pi} \left| \dfrac{2J_1(ka\theta)}{ka\theta} \right|^2, & \theta \lesssim \dfrac{1}{ka} \\[3mm] \dfrac{1}{4\pi}, & \theta \gg \dfrac{1}{ka} \end{cases} \tag{9.133}$$

where $P_i = (cE_0^2 a^2/8)$ is the incident power per unit area times the projected area (πa^2) of the sphere. At small angles the scattering is a typical diffraction pattern [see (9.113)]. At large angles the scattering is isotropic. At intermediate angles the two amplitudes interfere, causing the scattered power to have sharp minimum values considerably smaller than the

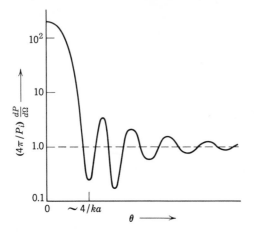

Fig. 9.14 Diffraction pattern for a conducting sphere, showing the forward peak due to shadow scattering, the isotropic reflected contribution, and the interference maxima and minima.

isotropic value at certain angles, as shown in Fig. 9.14. The amount of interference depends on the orientation of the incident polarization vector relative to the plane of observation containing $\mathbf{k}$ and $\mathbf{k}_0$. For $\mathbf{E}_0$ in this plane the interference is much greater than for $\mathbf{E}_0$ perpendicular to it.*

The total power scattered is obtained by integrating over all angles. Neglecting the interference terms, the total scattered power is the sum of the integrals of the diffraction peak and the isotropic reflected part. The integrals are easily shown to be equal in magnitude. Hence

$$P_s = P_i + P_i = 2P_i \qquad (9.134)$$

We sometimes rephase this result by saying that the effective area of the sphere for scattering (its scattering cross section) is $2\pi a^2$. One factor of πa^2 comes from the direct reflection; the other comes from the diffraction scattering which must accompany the formation of a shadow behind the obstacle.

Scattering of electromagnetic waves by a conducting sphere is treated by another method, especially in the long-wavelength limit, in Section 16.9.

REFERENCES AND SUGGESTED READING

The simple theory of radiation from a localized source distribution is discussed in all modern textbooks. Treatments analogous to that given here may be found in
   Panofsky and Phillips, Chapter 13,
   Stratton, Chapter VIII.

* See King and Wu, Appendix, for numerous graphs of scattering by spheres as a function of ka.

More complete discussions of antennas and antenna arrays are given in engineering works, such as

Jordan,
Kraus,
Schelkunoff,
Silver.

The subject of diffraction has a very extensive literature. A comprehensive treatment of both the scalar Kirchhoff and the vector theory, with many examples and excellent figures, is given by

Born and Wolf, Chapters VIII, IX, and XI.

A more elementary discussion of the scalar theory is found in

Slater and Frank, Chapters XIII and XIV.

Mathematical techniques for diffraction problems are discussed by

Baker and Copson,
Morse and Feshbach, Chapter 11.

The vector theorems used in Sections 9.6–9.10 are presented by

Morse and Feshbach, Chapter 13,
Silver, Chapter 5,
Stratton, Sections 8.14 and 8.15.

A specialized monograph on the scattering of electromagnetic waves by obstacles, with an emphasis on useful numerical results, is the book by

King and Wu.

PROBLEMS

9.1 Discuss the power flow and energy content of the complete electric dipole fields (9.18) in terms of the complex Poynting's vector $\mathbf{S} = (c/8\pi)(\mathbf{E} \times \mathbf{B}^*)$ and the time-averaged energy density $u = (1/16\pi)(\mathbf{E} \cdot \mathbf{E}^* + \mathbf{B} \cdot \mathbf{B}^*)$. The real part of $\mathbf{S}$ gives the true, resistive power flow, while the imaginary part represents circulating, reactive power.

(*a*) Show that the real part of $\mathbf{S}$ is in the radial direction and is given by r^{-2} times equation (9.23).

(*b*) Show that the imaginary part of $\mathbf{S}$ has components in the r and θ directions given by

$$\operatorname{Im} S_r = \frac{ck}{8\pi r^5} |\mathbf{p}|^2 \sin^2 \theta$$

$$\operatorname{Im} S_\theta = -\frac{ck |\mathbf{p}|^2}{4\pi r^5} (1 + k^2 r^2) \sin \theta \cos \theta$$

Make a sketch to show the direction of circulating power flow by suitably oriented arrows, the length of each arrow being proportional to the magnitude of $\operatorname{Im} S$ at that point.

(*c*) Calculate the time-averaged energy density:

$$u = \frac{1}{16\pi} \frac{|3\mathbf{n}(\mathbf{n} \cdot \mathbf{p}) - \mathbf{p}|^2}{r^6} + \frac{k^2 |\mathbf{n} \cdot \mathbf{p}|^2}{4\pi r^4} + \frac{k^4 |\mathbf{n} \times \mathbf{p}|^2}{8\pi r^2}$$

(*d*) Derive Poynting's theorem for the complex Poynting's vector. To what is $\operatorname{Im}(\boldsymbol{\nabla} \cdot \mathbf{S})$ equal? Verify that this holds true for the results of (*b*) and (*c*).

9.2 A radiating quadrupole consists of a square of side a with charges $\pm q$ at alternate corners. The square rotates with angular velocity ω about an axis normal to the plane of the square and through its center. Calculate the quadrupole moments, the radiation fields, the angular distribution of radiation, and the total radiated power in the long-wavelength approximation.

9.3 Two halves of a spherical metallic shell of radius R and infinite conductivity are separated by a very small insulating gap. An alternating potential is applied between the two halves of the sphere so that the potentials are $\pm V \cos \omega t$. In the long-wavelength limit, find the radiation fields, the angular distribution of radiated power, and the total radiated power from the sphere.

9.4 A thin linear antenna of length d is excited in such a way that the sinusoidal current makes a full wavelength of oscillation as shown in the figure.

(*a*) Calculate exactly the power radiated per unit solid angle and plot the angular distribution of radiation.

(*b*) Determine the total power radiated and find a numerical value for the radiation resistance.

9.5 Treat the linear antenna of Problem 9.4 by the long-wavelength multipole expansion method.

(*a*) Calculate the multipole moments (electric dipole, magnetic dipole, and electric quadrupole).

(*b*) Compare the angular distribution for the lowest nonvanishing multipole with the exact distribution of Problem 9.4.

(*c*) Determine the total power radiated for the lowest multipole and the corresponding radiation resistance.

9.6 A perfectly conducting flat screen occupies one-half of the x-y plane (i.e., $x < 0$). A plane wave of intensity I_0 and wave number k is incident along the z axis from the region $z < 0$. Discuss the values of the diffracted fields in the plane parallel to the x-y plane defined by $z = Z > 0$. Let the coordinates of the observation point be $(X, 0, Z)$.

(*a*) Show that, for the usual scalar Kirchhoff approximation and in the limit $Z \gg X$, the diffracted field is

$$\psi(X, 0, Z) \simeq I_0^{1/2} e^{ikZ - i\omega t} \left(\frac{1+i}{2i} \right) \sqrt{\frac{2}{\pi}} \int_{-\xi}^{\infty} e^{it^2} \, dt$$

where $\xi = (k/2Z)^{1/2} X$.

(*b*) Show that the intensity can be written

$$I = |\psi|^2 = \frac{I_0}{2}[(C(\xi) + \tfrac{1}{2})^2 + (S(\xi) + \tfrac{1}{2})^2]$$

where $C(\xi)$ and $S(\xi)$ are the so-called Fresnel integrals. Determine the asymptotic behavior of I for ξ large and positive (illuminated region) and ξ

large and negative (shadow region). What is the value of I at $X = 0$? Make a sketch of I as a function of X for fixed Z.

(c) Use the vector formula (9.82) to obtain a result equivalent to that of part (a). Compare the two expressions.

9.7 A linearly polarized plane wave of amplitude E_0 and wave number k is incident on a circular opening of radius a in an otherwise perfectly conducting flat screen. The incident wave vector makes an angle α with the normal to the screen. The polarization vector is perpendicular to the plane of incidence.

(a) Calculate the diffracted fields and the power per unit solid angle transmitted through the opening, using the vector Kirchhoff formula (9.82) with the assumption that the tangential electric field in the opening is the unperturbed incident field.

(b) Compare your result in part (a) with the standard scalar Kirchhoff approximation and with the result in Section 9.8 for the polarization vector in the plane of incidence.

9.8 A rectangular opening with sides of length a and $b \geq a$ defined by $x = \pm(a/2)$, $y = \pm(b/2)$ exists in a flat, perfectly conducting plane sheet filling the x-y plane. A plane wave is normally incident with its polarization vector, making an angle β with the long edges of the opening.

(a) Calculate the diffracted fields and power per unit solid angle with the vector Kirchhoff relation (9.82), assuming that the tangential electric field in the opening is the incident unperturbed field.

(b) Calculate the corresponding result of the scalar Kirchhoff approximation.

(c) For $b = a$, $\beta = 45°$, $ka = 4\pi$, compute the vector and scalar approximations to the diffracted power per unit solid angle as a function of the angle θ for $\phi = 0$. Plot a graph showing a comparison between the two results.

9.9 A cylindrical coaxial transmission line of inner radius a and outer radius b has its axis along the negative z axis. Both inner and outer conductors end at $z = 0$, and the outer one is connected to an infinite plane flange occupying the whole x-y plane (except for the annulus of radius b around the origin). The transmission line is excited at frequency ω in its dominant TEM mode, with the peak voltage between the cylinders being V. Use the vector Kirchhoff approximation to discuss the radiated fields, the angular distribution of radiation, and the total power radiated.

9.10 Discuss the diffraction due to a small, circular hole of radius a in a flat, perfectly conducting sheet, assuming that $ka \ll 1$.

(a) If the fields near the screen on the incident side are normal $\mathbf{E}_0 e^{-i\omega t}$ and tangential $\mathbf{B}_0 e^{-i\omega t}$, show that the diffracted electric field in the Fraunhofer zone is

$$\mathbf{E} = \frac{e^{ikr - i\omega t}}{3\pi r} k^2 a^3 \left[2\frac{\mathbf{k}}{k} \times \mathbf{B}_0 + \frac{\mathbf{k}}{k} \times \left(\mathbf{E}_0 \times \frac{\mathbf{k}}{k} \right) \right]$$

where $\mathbf{k}$ is the wave vector in the direction of observation.

(b) Determine the angular distribution of the diffracted radiation and show that the total power transmitted through the hole is

$$P = \frac{c}{54\pi^2} k^4 a^6 (4B_0^2 + E_0^2)$$

9.11 Specialize the discussion of Problem 9.10 to the diffraction of a plane wave by the small, circular hole. Treat the general case of oblique incidence at an angle α to the normal, with polarization in and perpendicular to the plane of incidence.

(*a*) Calculate the angular distributions of the diffracted radiation and compare them to the vector Kirchhoff approximation results of Section 9.8 and Problem 9.7 in the limit $ka \ll 1$.

(*b*) Show that the transmission coefficients [defined above (9.105)] for the two states of polarization are

$$T_{\shortparallel} = \frac{64}{27\pi^2} (ka)^4 \left(\frac{4 + \sin^2 \alpha}{4 \cos \alpha} \right)$$

$$T_{\perp} = \frac{64}{27\pi^2} (ka)^4 \cos \alpha$$

Note that these transmission coefficients are a factor $(ka)^2$ smaller than those given by the vector Kirchhoff approximation in the same limit.

10

Magnetohydrodynamics
and Plasma Physics

10.1 Introduction and Definitions

Magnetohydrodynamics and plasma physics both deal with the behavior of the combined system of electromagnetic fields and a conducting liquid or gas. Conduction occurs when there are free or quasi-free electrons which can move under the action of applied fields. In a solid conductor, the electrons are actually bound, but can move considerable distances on the atomic scale within the crystal lattice before making collisions. Dynamical effects such as conduction and Hall effect are observed when fields are applied to the solid conductor, but mass motion does not in general occur. The effects of the applied fields on the atoms themselves are taken up as stresses in the lattice structure. For a fluid, on the other hand, the fields act on both electrons and ionized atoms to produce dynamical effects, including bulk motion of the medium itself. This mass motion in turn produces modifications in the electromagnetic fields. Consequently we must deal with a complicated coupled system of matter and fields.

The distinction between magnetohydrodynamics and the physics of plasmas is not a sharp one. Nevertheless there are clearly separated domains in which the ideas and concepts of only one or the other are applicable. One way of seeing the distinction is to look at the way in which the relation $\mathbf{J} = \sigma\mathbf{E}$ is established for a conducting substance. In the simple model of Section 7.8 the electrons are imagined to be accelerated by the applied fields, but to be altered in direction by collisions, so that their motion in the direction of the field is opposed by an effective frictional force $\nu m \mathbf{v}$, where ν is the collision frequency. Ohm's law just represents a balance between the applied force and the frictional drag. When the frequency of the applied fields is comparable to ν, the electrons have time to accelerate and decelerate between collisions. Then inertial effects enter and the conductivity becomes complex. Unfortunately at these same frequencies the description of collisions in terms of a frictional force tends to lose its validity. The whole process becomes more complicated. At frequencies well above the collision frequency another thing happens. The electrons and ions are accelerated in opposite directions by electric fields and tend to separate. Strong electrostatic restoring forces are set up by this charge separation. Oscillations occur in the charge density. These high-frequency oscillations are called *plasma oscillations* and are to be distinguished from lower-frequency oscillations which involve motion of the fluid, but no charge separation. These low-frequency oscillations are called *magnetohydrodynamic waves*.

In conducting liquids or dense ionized gases the collision frequency is sufficiently high even for very good conductors that there is a wide frequency range where Ohm's law in its simple form is valid. Under the action of applied fields the electrons and ions move in such a way that, apart from a high-frequency jitter, there is no separation of charge. Electric fields arise from motion of the fluid which causes a current flow, or as a result of time-varying magnetic fields or charge distributions external to the fluid. The mechanical motion of the system can then be described in terms of a single conducting fluid with the usual hydrodynamic variables of density, velocity, and pressure. At low frequencies it is customary to neglect the displacement current in Ampère's law. This is then the approximation which is called *magnetohydrodynamics*.

In less dense ionized gases the collision frequency is smaller. There may still be a low-frequency domain where the magnetohydrodynamic equations are applicable to quasi-stationary processes. Frequently astrophysical applications fall in this category. At higher frequencies, however, the neglect of charge separation and of the displacement current is not allowable. The separate inertial effects of the electrons and ions must be included in the description of the motion. This is the domain which we call *plasma physics*. There is here a range of physical conditions where a two-fluid model of electrons and ions gives an approximately correct description of various phenomena. But for high temperatures and low densities, the finite velocity spreads of the particles about their mean values must be included. Then the description is made in terms of the Boltzmann equation, with or without short-range correlations. We will not attempt to go into such details here. At still higher temperatures and lower densities, the electrostatic restoring forces become so weak that the length scale of charge separation becomes large compared to the size of the volume being considered. Then the collective behavior implicit in a fluid model is gone completely. We have left a few rapidly moving charged particles interacting via Coulomb collisions. A plasma is, by definition, an ionized gas in which the length which divides the small-scale individual-particle behavior from the large-scale collective behavior is small compared to the characteristic lengths of interest. This length, called the *Debye screening radius*, will be discussed in Section 10.10. It is numerically equal to 7.91 $(T/n)^{\frac{1}{2}}$ cm, where T is the absolute temperature in degrees Kelvin and n is the number of electrons per cubic centimeter. For all but the hottest or most tenuous plasmas it is small compared to 1 cm.

10.2 Magnetohydrodynamic Equations

We first consider the behavior of an electrically neutral, conducting fluid in electromagnetic fields. For simplicity, we assume the fluid to be nonpermeable. It is described by a matter density $\rho(\mathbf{x}, t)$, a velocity $\mathbf{v}(\mathbf{x}, t)$, a pressure $p(\mathbf{x}, t)$ (taken to be a scalar), and a real conductivity σ. The hydrodynamic equations are the continuity equation

$$\frac{\partial \rho}{\partial t} + \boldsymbol{\nabla} \cdot (\rho \mathbf{v}) = 0 \tag{10.1}$$

and the force equation:

$$\rho \frac{d\mathbf{v}}{dt} = -\boldsymbol{\nabla} p + \frac{1}{c} (\mathbf{J} \times \mathbf{B}) + \mathbf{F}_v + \rho \mathbf{g} \tag{10.2}$$

In addition to the pressure and magnetic-force terms we have included viscous and gravitational forces. For an incompressible fluid the viscous force can be written

$$\mathbf{F}_v = \eta \nabla^2 \mathbf{v} \tag{10.3}$$

where η is the coefficient of viscosity. It should be emphasized that the time derivative of the velocity on the left side of (10.2) is the *convective derivative*,

$$\frac{d}{dt} = \frac{\partial}{\partial t} + \mathbf{v} \cdot \boldsymbol{\nabla} \tag{10.4}$$

which gives the total time rate of change of a quantity moving instantaneously with the velocity $\mathbf{v}$.

With the neglect of the displacement current, the electromagnetic fields in the fluid are described by

$$\left. \begin{aligned} \mathbf{\nabla} \times \mathbf{E} + \frac{1}{c}\frac{\partial \mathbf{B}}{\partial t} &= 0 \\[2mm] \mathbf{\nabla} \times \mathbf{B} &= \frac{4\pi}{c}\,\mathbf{J} \end{aligned} \right\} \tag{10.5}$$

The condition $\mathbf{\nabla} \cdot \mathbf{J} = 0$, equivalent to the neglect of displacement currents, follows from the second equation in (10.5). The two divergence equations have been omitted in (10.5). It follows from Faraday's law that $(\partial/\partial t)\,\mathbf{\nabla} \cdot \mathbf{B} = 0$, and the requirement $\mathbf{\nabla} \cdot \mathbf{B} = 0$ can be imposed as an initial condition. With the neglect of the displacement current, it is appropriate to ignore Coulomb's law as well. The reason is that the electric field is completely determined by the curl equations and Ohm's law (see below). If the displacement current is retained in Ampère's law and $\mathbf{\nabla} \cdot \mathbf{E} = 4\pi\rho_e$ is taken into account, corrections of only the order of (v^2/c^2) result. For normal magnetohydrodynamic problems these are completely negligible.

To complete the specification of dynamical equations we must specify the relation between the current density $\mathbf{J}$ and the fields $\mathbf{E}$ and $\mathbf{B}$. For a simple conducting medium of conductivity σ, Ohm's law applies, and the current density is

$$\mathbf{J}' = \sigma\mathbf{E}' \tag{10.6}$$

where $\mathbf{J}'$ and $\mathbf{E}'$ are measured in the rest frame of the medium. For a medium moving with velocity $\mathbf{v}$ relative to the laboratory, we must transform both the current density and the electric field appropriately. The transformation of the field is given by equation (6.10). The current density in the laboratory is evidently

$$\mathbf{J} = \mathbf{J}' + \rho_e\mathbf{v} \tag{10.7}$$

where ρ_e is the electrical charge density. For a one-component conducting fluid, $\rho_e = 0$. Consequently, Ohm's law assumes the form,

$$\mathbf{J} = \sigma\left(\mathbf{E} + \frac{\mathbf{v}}{c} \times \mathbf{B}\right) \tag{10.8}$$

Sometimes it is possible to assume that the conductivity of the fluid is effectively infinite. Then under the action of fields $\mathbf{E}$ and $\mathbf{B}$ the fluid flows in such a way that

$$\mathbf{E} + \frac{1}{c}(\mathbf{v} \times \mathbf{B}) = 0 \tag{10.9}$$

is satisfied.

Equations (10.1), (10.2), (10.5), and (10.8), supplemented by an equation of state for the fluid, form the equations of magnetohydrodynamics. In the next section we will consider some of the simpler aspects of them and will elaborate the basic concepts involved.

10.3 Magnetic Diffusion, Viscosity, and Pressure

The behavior of a fluid in the presence of electromagnetic fields is governed to a large extent by the magnitude of the conductivity. The effects are both electromagnetic and mechanical. We first consider the electromagnetic effects. We will see that, depending on the conductivity, quite different behaviors of the fields occur. The time dependence of the magnetic field can be written, using (10.8) to eliminate $\mathbf{E}$, in the form:

$$\frac{\partial \mathbf{B}}{\partial t} = \nabla \times (\mathbf{v} \times \mathbf{B}) + \frac{c^2}{4\pi\sigma} \nabla^2 \mathbf{B} \qquad (10.10)$$

Here it is assumed that σ is constant in space. For a fluid at rest (10.10) reduces to the diffusion equation

$$\frac{\partial \mathbf{B}}{\partial t} = \frac{c^2}{4\pi\sigma} \nabla^2 \mathbf{B} \qquad (10.11)$$

This means that an initial configuration of magnetic field will decay away in a diffusion time

$$\tau = \frac{4\pi\sigma L^2}{c^2} \qquad (10.12)$$

where L is a length characteristic of the spatial variation of $\mathbf{B}$. The time τ is of the order of 1 sec for a copper sphere of 1 cm radius, of the order of 10^4 years for the molten core of the earth, and of the order of 10^{10} years for a typical magnetic field in the sun.

For times short compared to the diffusion time τ (or, in other words, when the conductivity is so large that the second term in (10.10) can be neglected) the temporal behavior of the magnetic field is given by

$$\frac{\partial \mathbf{B}}{\partial t} = \nabla \times (\mathbf{v} \times \mathbf{B}) \qquad (10.13)$$

From (6.5) it can be shown that this is equivalent to the statement that the magnetic flux through any loop moving with the local fluid velocity is constant in time. We say that the lines of force are frozen into the fluid and are carried along with it. Since the conductivity is effectively infinite, the velocity **w** of the lines of force (defined to be perpendicular to **B**) is given by (10.9):

$$\mathbf{w} = c\,\frac{(\mathbf{E}\,\times\,\mathbf{B})}{B^2} \qquad (10.14)$$

This so-called "$E \times B$ drift" of both fluid and lines of force can be understood in terms of individual particle orbits of the electrons and ions in crossed electric and magnetic fields (see Section 12.8).

A useful parameter to distinguish between situations in which diffusion of the field lines relative to the fluid occurs and those in which the lines of force are frozen in is the *magnetic Reynolds number* R_M. If V is a velocity typical of the problem and L is a corresponding length, then the magnetic Reynolds number is defined as

$$R_M = \frac{V\tau}{L} \qquad (10.15)$$

where τ is the diffusion time (10.12). Transport of the lines of force with the fluid dominates over diffusion if $R_M \gg 1$. For liquids like mercury or sodium in the laboratory $R_M < 1$, except for very high velocities. But in geophysical and astrophysical applications R_M can be very large compared to unity.

The mechanical behavior of the system can be studied with the force equation (10.2). Substituting for **J** from (10.8), we find

$$\rho\,\frac{d\mathbf{v}}{dt} = \mathbf{F} - \frac{\sigma B^2}{c^2}\,(\mathbf{v}_\perp - \mathbf{w}) \qquad (10.16)$$

where **F** is the sum of all the nonelectromagnetic forces, and $\mathbf{v}_\perp$ is the component of velocity perpendicular to **B**. From (10.16) it is apparent that flow parallel to **B** is governed by the nonelectromagnetic forces alone. The velocity of flow of the fluid perpendicular to **B**, on the other hand, decays from some initially arbitrary value in a time of the order of

$$\tau' = \frac{\rho c^2}{\sigma B^2} \qquad (10.17)$$

to a value

$$\mathbf{v}_\perp = \mathbf{w} + \frac{c^2}{\sigma B^2}\,\mathbf{F}_\perp \qquad (10.18)$$

In the limit of infinite conductivity this result reduces to that of (10.14), as expected. The term proportional to B^2 in (10.16) is an effective viscous or frictional force which tends to prevent flow of the fluid perpendicular to the lines of magnetic force. Sometimes it is described as a magnetic viscosity. If ordinary viscosity, here lumped into **F**, is comparable to the magnetic viscosity, then the decay time τ' is shortened by an obvious factor involving the ratio of the two viscosities.

The above considerations have shown that if the conductivity is large the lines of force are frozen into the fluid and move along with it. Any departure from that state decays rapidly away. In considering the mechanical or electromagnetic effects we treated the opposite quantities as given, but the equations are, of course, coupled. In the limit of very large conductivity it is convenient to relate the current density **J** in the force equation to the magnetic induction **B** via Ampère's law and to use the infinite conductivity expression (10.9) to eliminate **E** from Faraday's law to yield (10.13). The magnetic force term in (10.2) can now be written

$$\frac{1}{c}(\mathbf{J} \times \mathbf{B}) = -\frac{1}{4\pi}\mathbf{B} \times (\nabla \times \mathbf{B}) \tag{10.19}$$

With the vector identity

$$\tfrac{1}{2}\nabla(\mathbf{B} \cdot \mathbf{B}) = (\mathbf{B} \cdot \nabla)\mathbf{B} + \mathbf{B} \times (\nabla \times \mathbf{B}) \tag{10.20}$$

Equation (10.19) can be transformed into

$$\frac{1}{c}(\mathbf{J} \times \mathbf{B}) = -\nabla\left(\frac{B^2}{8\pi}\right) + \frac{1}{4\pi}(\mathbf{B} \cdot \nabla)\mathbf{B} \tag{10.21}$$

This equation shows that the magnetic force is equivalent to a *magnetic hydrostatic pressure,*

$$p_M = \frac{B^2}{8\pi} \tag{10.22}$$

plus a term which can be thought of as an additional tension along the lines of force. The result (10.21) can also be derived from the Maxwell stress tensor (see Section 6.9).

If we neglect viscous effects and assume that the gravitational force is derivable from a potential $\mathbf{g} = -\nabla\psi$, the force equation (10.2) takes the form

$$\rho\frac{d\mathbf{v}}{dt} = -\nabla(p + p_M + \rho\psi) + \frac{1}{4\pi}(\mathbf{B} \cdot \nabla)\mathbf{B} \tag{10.23}$$

In some simple geometrical situations, such as **B** having only one component, the additional tension vanishes. Then the static properties of the fluid are described by

$$p + p_M + \rho\psi = \text{constant} \tag{10.24}$$

This shows that, apart from gravitational effects, any change in mechanical pressure must be balanced by an opposite change in magnetic pressure. If the fluid is to be confined within a certain region so that p falls rapidly to zero outside that region, the magnetic pressure must rise equally rapidly in order to confine the fluid. This is the principle of the pinch effect discussed in Section 10.5.

10.4   Magnetohydrodynamic Flow between Boundaries with Crossed Electric and Magnetic Fields

To illustrate the competition between freezing in of lines of force and diffusion through them and between the $E \times B$ drift and behavior imposed by boundary conditions, we consider the simple example of an incompressible, but viscous, conducting fluid flowing in the x direction between two nonconducting boundary surfaces at $z = 0$ and $z = a$, as shown in Fig. 10.1. The surfaces move with velocities V_1 and V_2, respectively, in the x direction. A uniform magnetic field B_0 acts in the z-direction. The system is infinite in the x and y directions. We will look for a steady-state solution for flow in the x direction in which the various quantities depend only upon z.

If the fields do not vary in time, it is clear from Maxwell's equations (10.5) that any electric field present must be an electrostatic field derivable from a potential and determined solely by the boundary conditions, i.e. an arbitrary external field. Expression (10.14) for the velocity of the lines of force when σ is infinite implies that there is an electric field in the y direction. If we assume that to be the only component of $\mathbf{E}$, then it must be a constant, E_0. Because the moving fluid will tend to carry the lines of force with it, we expect an x component $B_x(z)$ of magnetic induction, as well as the z component B_0.

The continuity equation (10.1) reduces to $\nabla \cdot \mathbf{v} = 0$ for an incompressible fluid. This is satisfied identically by a velocity in the x direction which depends only on z. The force equation, neglecting gravity, has the steady-state form:

$$\nabla p = \frac{1}{c} (\mathbf{J} \times \mathbf{B}) + \eta \nabla^2 \mathbf{v} \tag{10.25}$$

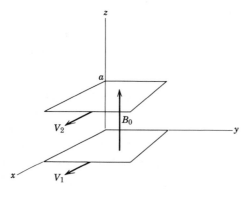

Fig. 10.1   Flow of viscous conducting fluid in a magnetic field between two plane surfaces moving with different velocities.

The only component of $\mathbf{J}$ that is nonvanishing is $J_y(z)$:

$$J_y(z) = \sigma\left[E_0 - \frac{1}{c} B_0 v(z)\right] \tag{10.26}$$

where v is the x component of velocity. When we write out the three component equations in (10.25), we find

$$\left.\begin{aligned}
\frac{\partial p}{\partial x} &= \frac{\sigma B_0}{c}\left(E_0 - \frac{B_0}{c} v\right) + \eta \frac{\partial^2 v}{\partial z^2} \\[2mm]
\frac{\partial p}{\partial y} &= 0 \\[2mm]
\frac{\partial p}{\partial z} &= -\frac{\sigma B_x}{c}\left(E_0 - \frac{B_0}{c} v\right)
\end{aligned}\right\} \tag{10.27}$$

The magnetic force in the z direction is just balanced by the pressure gradient. If we assume no pressure gradient in the x direction, the first of these equations can be written:

$$\frac{\partial^2 v}{\partial z^2} - \left(\frac{M}{a}\right)^2 v = -\left(\frac{M}{a}\right)^2 \frac{cE_0}{B_0} \tag{10.28}$$

where

$$M = \left(\frac{\sigma B_0{}^2 a^2}{\eta c^2}\right)^{\!1/2} \tag{10.29}$$

is called the *Hartmann number*. From (10.17) M^2 can be seen to be the ratio of magnetic to normal viscosity. The solution to (10.28), subject to the boundary conditions $v(0) = V_1$ and $v(a) = V_2$, is readily found to be

$$v(z) = \frac{V_1}{\sinh M} \sinh\left[M\left(\frac{a-z}{a}\right)\right] + \frac{V_2}{\sinh M} \sinh\left(\frac{Mz}{a}\right)$$

$$+ \frac{cE_0}{B_0}\left\{1 - \frac{\sinh\left[M\left(\frac{a-z}{a}\right)\right] + \sinh\left(\frac{Mz}{a}\right)}{\sinh M}\right\} \tag{10.30}$$

In the limit $B_0 \to 0$, $M \to 0$, we obtain the standard laminar-flow result

$$v(z) = V_1 + \frac{z}{a}(V_2 - V_1) \tag{10.31}$$

In the other limit of $M \gg 1$ we expect the magnetic viscosity to dominate and the flow to be determined almost entirely by the $E \times B$ drift. If we approximate $v(z)$ for $z \ll a$ and $M \gg 1$, we obtain

$$v(z) \simeq \frac{cE_0}{B_0} + \left(V_1 - \frac{cE_0}{B_0}\right)e^{-Mz/a} \tag{10.32}$$

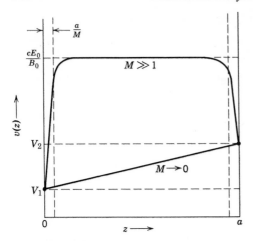

Fig. 10.2 Velocity profiles for large and small Hartmann numbers M. For $M \to 0$, laminar flow occurs. For $M \gg 1$, the flow is given by the **E** × **B** drift velocity, except in the immediate neighborhood of the boundaries.

This shows that, while $v(z) = V_1$ exactly at the surface, there is a rapid transition in a distance of order (a/M) to the $E \times B$ drift value (cE_0/B_0). Near $z = a$, (10.32) is changed by replacing V_1 by V_2 and z by $(a - z)$. The velocity profile in the two limits (10.31) and (10.32) is shown in Fig. 10.2.

The magnetic field $B_x(z)$ is determined by the equation

$$\frac{\partial B_x}{\partial z} = \frac{4\pi}{c} J_y = \frac{4\pi\sigma}{c} \left(E_0 - \frac{1}{c} B_0 v \right) \tag{10.33}$$

The boundary conditions on B_x at $z = 0$ and $z = a$ are indeterminate unless we know the detailed history of how the steady state was created or can use some symmetry argument. All we know is that the difference in B_x is related to the total current flowing in the y direction per unit length in the x direction:

$$B_x(a) - B_x(0) = \frac{4\pi}{c} \int_0^a J_y(z) \, dz \tag{10.34}$$

This indeterminacy stems from the one-dimensional nature of the problem. For simplicity we will calculate the magnetic field only for the case when the total current in the y direction is zero.* Then we can assume that B_x vanishes at $z = 0$ and $z = a$. Using (10.30) for the velocity in (10.33), it is easy to show that then

$$B_x(z) = B_0 \left[\left(\frac{4\pi\sigma a^2}{c^2} \right) \left(\frac{V_2 - V_1}{2a} \right) \right] \left\{ \frac{\cosh \dfrac{M}{2} - \cosh \left(\dfrac{M}{2} - \dfrac{Mz}{a} \right)}{M \sinh \dfrac{M}{2}} \right\} \tag{10.35}$$

---

* This requirement means that $cE_0/B_0 = \frac{1}{2}(V_1 + V_2)$.

The dimensionless coefficient in square brackets in (10.35) may be identified as the magnetic Reynolds number (10.15), since $(V_2 - V_1)/2$ is a typical velocity in the problem and a is a typical length. In the two limits $M \ll 1$ and $M \gg 1$, (10.35) reduces to

$$B_x(z) \simeq R_M B_0 \begin{cases} \dfrac{z}{a}\left(1 - \dfrac{z}{a}\right), & \text{for } M \ll 1 \\[2mm] \dfrac{1}{M}\left[1 - \left(e^{-\frac{Mz}{a}} + e^{-M\left(\frac{a-z}{a}\right)}\right)\right], & \text{for } M \gg 1 \end{cases} \tag{10.36}$$

Figure 10.3 shows the behavior of the lines of force in the two limiting cases. Only for large R_M is there appreciable transport of the lines of force. And for a given R_M, the transport is less the larger the Hartmann number.

For liquid mercury at room temperature the relevant physical constants are

$$\sigma = 9.4 \times 10^{15} \text{ sec}^{-1}$$
$$\eta = 1.5 \times 10^{-2} \text{ poise}$$
$$\rho = 13.5 \text{ gm/cm}^3$$

The diffusion time (10.12) is $\tau = 1.31 \times 10^{-4} [L \text{ (cm)}]^2$ sec. The Hartmann number (10.29) is $M = 2.64 \times 10^{-2} B_0$ (gauss) a (cm). With $L \simeq a \simeq 1$ cm, this gives a magnetic Reynolds number $R_M \sim 10^{-4} V$. Consequently unless the flow velocity is very large, there is no significant transport of lines of force for laboratory experiments with mercury. On the other hand, if the magnetic induction B_0 is of the order of 10^4 gauss, then $M \sim 250$ and the velocity flow is almost completely specified by the $E \times B$ drift (10.14).

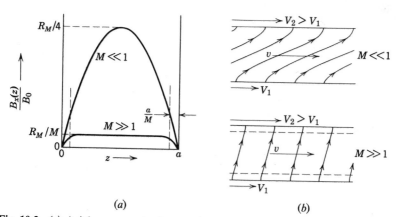

(a)   (b)

Fig. 10.3 (a) Axial component of magnetic induction between the boundary surfaces for large and small Hartmann numbers. (b) Transport of lines of magnetic induction in direction of flow.

In geomagnetic problems with the earth's core and in astrophysical problems the parameters (e.g., the length scale) are such that $R_M \gg 1$ occurs often and transport of the lines of force becomes very important.

10.5 Pinch Effect

The confinement of a plasma or conducting fluid by self-magnetic fields is of considerable interest in thermonuclear research, as well as in other applications. To illustrate the principles we consider an infinite cylinder of conducting fluid with an axial current density $J_z = J(r)$ and a resulting azimuthal magnetic induction $B_\phi = B(r)$. For simplicity, the current density, magnetic field, pressure, etc., are assumed to depend only on the distance r from the cylinder axis, and viscous and gravitational effects are neglected. We first ask whether a steady-state condition can exist in which the material is mainly confined within a certain radius $r = R$ by the action of its own magnetic induction. For a steady state with $\mathbf{v} = 0$ the equation of motion (10.23) of the fluid reduces to

$$ 0 = -\frac{dp}{dr} - \frac{d}{dr}\left(\frac{B^2}{8\pi}\right) - \frac{B^2}{4\pi r} \tag{10.37} $$

Ampère's law in integral form relates $B(r)$ to the current enclosed:

$$ B(r) = \frac{4\pi}{cr} \int_0^r r J(r)\, dr \tag{10.38} $$

A number of results can be obtained without specifying the form of $J(r)$, aside from physical limitations of finiteness, etc. From Ampère's law it is evident that, if the fluid lies almost entirely inside $r = R$, then the magnetic induction outside the fluid is

$$ B(r) = \frac{2I}{cr} \tag{10.39} $$

where

$$ I = \int_0^R 2\pi r J(r)\, dr $$

is the total current flowing in the cylinder. Equation (10.37) can be written as

$$ \frac{dp}{dr} = -\frac{1}{8\pi r^2}\frac{d}{dr}(r^2 B^2) \tag{10.40} $$

with the solution:

$$ p(r) = p_0 - \frac{1}{8\pi} \int_0^r \frac{1}{r^2}\frac{d}{dr}(r^2 B^2)\, dr \tag{10.41} $$

Here p_0 is the pressure of the fluid at $r = 0$. If the matter is confined to $r \leq R$, the pressure drops to zero at $r = R$. Consequently the axial pressure p_0 is given by

$$p_0 = \frac{1}{8\pi} \int_0^R \frac{1}{r^2} \frac{d}{dr} (r^2 B^2) \, dr \tag{10.42}$$

The upper limit of integration can be replaced by infinity, since the integrand vanishes for $r \geq R$, as can be seen from (10.39). With this expression (10.42) for p_0, (10.41) can be written as

$$p(r) = \frac{1}{8\pi} \int_r^R \frac{1}{r^2} \frac{d}{dr} (r^2 B^2) \, dr \tag{10.43}$$

The average pressure inside the cylinder can be related to the total current I and radius R without specifying the detailed radial behavior. Thus

$$\langle p \rangle = \frac{2\pi}{\pi R^2} \int_0^R r p(r) \, dr \tag{10.44}$$

Integration by parts and use of (10.40) gives

$$\langle p \rangle = \frac{I^2}{2\pi R^2 c^2} \tag{10.45}$$

as the relation between average pressure, total current, and radius of the cylinder of fluid or plasma confined by its own magnetic field. Note that the average pressure of the matter is equal to the magnetic pressure $(B^2/8\pi)$ at the surface of the cylinder. In thermonuclear work, hot plasmas with temperatures of the order of $10^{8}\,°K$ ($kT \sim 10$ kev) and densities of the order of 10^{15} particles/cm^3 are envisioned. These conditions correspond to a pressure of approximately $10^{15} \times 10^{8}k \simeq 1.4 \times 10^7$ dynes/cm^2, or 14 atmospheres. A magnetic induction of approximately 19 kilogauss at the surface, corresponding to a current of $9 \times 10^4 R$ (cm) amperes, is necessary for confinement. This shows that extremely high currents are needed to confine very hot plasmas.

So far the radial behavior of the system has not been discussed. Two simple examples will serve to illustrate the possibilities. One is that the current density $J(r)$ is constant for $r < R$. Then $B(r) = (2Ir/cR^2)$ for $r < R$. Equation (10.43) then yields a parabolic dependence for pressure versus radius:

$$p(r) = \frac{I^2}{\pi c^2 R^2} \left(1 - \frac{r^2}{R^2} \right) \tag{10.46}$$

The axial pressure p_0 is then twice the average pressure $\langle p \rangle$. The radial dependences of the various quantities are sketched in Fig. 10.4.

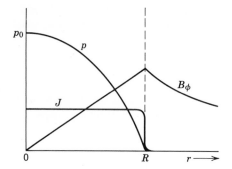

Fig. 10.4 Variation of azimuthal magnetic induction and pressure with radius in a cylindrical plasma column with a uniform current density *J*.

The other model has the current density confined to a very thin layer on the surface, as is appropriate for a highly conducting fluid or plasma. The magnetic induction is given by (10.39) for $r > R$, but vanishes inside the cylinder. Then the pressure p is constant inside the cylinder and equal to the value (10.45). This is sketched in Fig. 10.5.

10.6 Dynamic Model of the Pinch Effect

The simple considerations of the previous section are valid for a static or quasi-static situation. In actual practice with plasmas, such circumstances do not arise. Generally, at some time early in the history of current flow down the plasma the pressure p is much too small to resist the magnetic pressure outside. Consequently the radius of the cylinder of plasma is forced inwards; the plasma column is pinched. This has the desirable consequence that the plasma is pulled away from its confining walls. If the pinched configuration were stable for a sufficiently long time, it would be possible to heat the plasma to very high temperatures without burning up the walls of the confining vessel.

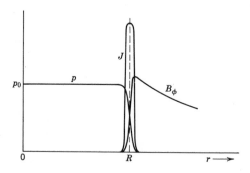

Fig. 10.5. Variation of azimuthal magnetic induction and pressure in a cylindrical plasma column with a surface-current density.

A simple model, first discussed by M. Rosenbluth, exhibits the essential dynamical features. Suppose that a plasma is created in a hollow conducting cylinder of radius R_0 and length L. A voltage difference V is applied between the ends of the cylinder so that a current I flows in the plasma. This produces an azimuthal magnetic induction B_ϕ which causes the plasma to pinch inwards. The radius of the plasma column at time $t > 0$ is $R(t)$. The conductivity of the plasma is taken to be virtually infinite. Then the current all flows on the surface, and the magnetic induction

$$B_\phi = \frac{2I}{cr} \tag{10.47}$$

exists only between $r = R(t)$ and $r = R_0$. Because of the assumption of infinite conductivity the electric field at the plasma surface, in the moving frame of reference in which the interface is at rest, vanishes:

$$\mathbf{E}' = \mathbf{E} + \frac{\mathbf{v}}{c} \times \mathbf{B} = 0 \tag{10.48}$$

If we now apply Faraday's law of induction to the dotted loop shown in Fig. 10.6, the inner arm of which is moving inwards with the interface, we find that the only contribution to the line integral of $\mathbf{E}$ comes from the side of the loop in the conducting wall. Thus

$$-\frac{V}{L} = -\frac{1}{c}\frac{d}{dt} \int_{R(t)}^{R_0} B_\phi \, dr = -\frac{2}{c^2}\frac{d}{dt}\left(I \ln \frac{R_0}{R}\right) \tag{10.49}$$

This is the standard inductive relation between current, voltage, and dimensions (inductance). The integral of this equation is

$$I \ln \left(\frac{R_0}{R}\right) = \frac{c^2}{2} E_0 \int_0^t f(t) \, dt \tag{10.50}$$

where $E_0 f(t) = V/L$ is the applied electric field. The function $f(t)$ is assumed known and is normalized so that E_0 is the peak value of applied

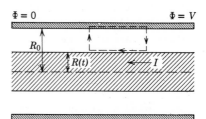

Fig. 10.6   Plasma column inside a hollow, cylindrical conductor.

field. In order to proceed further we must relate the current I in a dynamic way to the behavior of the plasma radius R.

The desired dynamical connection between I and R is essentially the momentum-balance equation, or Newton's second law. Some assumption about the plasma must be made. If the mean free path for collisions is short compared to the radius, the dynamic behavior is characteristic of hydrodynamic shock waves. But for a hot, tenuous plasma the mean free path is comparable to, or larger than, the radius. Then a model with particles moving freely inside the plasma is more appropriate. If the velocity $\dot{R}$ of the plasma surface is large compared to thermal speeds, each particle approaches the interface with a velocity $\dot{R}$ in the frame of reference in which the interface is at rest. As the particle penetrates into the outer region, it starts feeling the magnetic induction, is turned around, and leaves the surface with velocity $\dot{R}$. Consequently each particle colliding with the plasma surface receives a momentum transfer $2M\dot{R}$. The number colliding with unit area of the surface per unit time is $N\dot{R}$, where N is the initial number of particles per unit volume. Therefore the rate of transfer of momentum (i.e., pressure) is

$$p = 2NM\dot{R}^2 = 2\rho\dot{R}^2 \qquad (10.51)$$

where ρ is the initial mass density. At the surface of the plasma there is a magnetic pressure $(B^2/8\pi)$ due to the discontinuity in magnetic induction from zero inside to the value B just outside. These pressures must balance. Consequently, using (10.47), we find that the current is related to the velocity by:

$$I^2 = 4\pi\rho c^2 R^2 \left(\frac{dR}{dt}\right)^2 \qquad (10.52)$$

Equation (10.52) depends on a rather simplified model of the mechanical-momentum transfer rate in which each particle collides only once with the interface. In fact, the velocity of the interface increases with time so that the surface catches up with particles which were reflected earlier and hits them again and again. This effect can be approximated by the "snow-plow" model in which the interface is imagined to carry along with it all the material which it hits as it moves in. Then the magnetic pressure and rate of change of momentum are related by

$$\frac{d}{dt}\left[M(R)\dot{R}\right] = -2\pi R \frac{B^2}{8\pi} \qquad (10.53)$$

where $M(R)$ is the mass carried along by the snowplow:

$$M(R) = \pi\rho(R_0{}^2 - R^2) \qquad (10.54)$$

This leads to the relation

$$I^2 = -\pi \rho c^2 R \frac{d}{dt}\left[(R_0{}^2 - R^2)\frac{dR}{dt}\right] \qquad (10.55)$$

between current and radius. In the initial stages when $R \lesssim R_0$ the snow-plow model and free-particle model give the same relation between current and radius to within a factor of 2*, and do not differ by an order of magnitude even at later times.

The equation of motion for $R(t)$ is obtained by substituting I^2 from either (10.52) or (10.55) into the inductive relation (10.50). Choosing the free-particle model as an illustration, we obtain

$$2R \ln\left(\frac{R_0}{R}\right)\frac{dR}{dt} = -\frac{cE_0}{\sqrt{4\pi\rho}}\int_0^t f(t')\,dt' \qquad (10.56)$$

where the signs of the square root have been taken to give $\dot{R} < 0$. Without knowledge of $E_0(t)$ we cannot solve this equation. Nevertheless, some idea of the solution can be obtained by introducing the dimensionless variables:

$$\left.\begin{aligned} \tau &= \left(\frac{c^2 E_0{}^2}{4\pi\rho}\right)^{\!1/4}\frac{t}{R_0} \\[2mm] x &= \frac{R}{R_0} \end{aligned}\right\} \qquad (10.57)$$

Then (10.56) becomes

$$2x \ln x \frac{dx}{d\tau} = \int_0^\tau f(\tau')\,d\tau' \qquad (10.58)$$

For the snowplow model the equivalent equation is

$$\frac{d}{d\tau}\left[(1 - x^2)\frac{dx}{d\tau}\right] = -\frac{\left[\int_0^\tau f(\tau')\,d\tau'\right]^2}{x(\ln x)^2} \qquad (10.59)$$

Without solving these equations it is evident that x changes significantly in times such that $\tau \sim 1$. This means that the scaling law for the radial velocity of the pinch is

$$|\dot{R}| \sim v_0 \equiv \left(\frac{c^2 E_0{}^2}{4\pi\rho}\right)^{\!1/4} \qquad (10.60)$$

This result emerges whatever dynamic model is used, including a hydro-dynamic one. Typical experimental conditions for a fast pinch in small-scale hydrogen or deuterium plasmas involve applied electric fields of the

---

* The factor of 2 comes from the fact that in the one case the particles are elastically reflected and suffer a velocity change of $2\dot{R}$, while in the other the particles collide inelastically with the interface and receive a velocity change of $\dot{R}$.

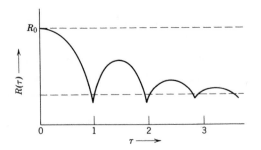

Fig. 10.7 Radius of plasma column as a function of time after initiation of current flow. The characteristic velocity of pinching is given by (10.60).

order of 10^3 volts/cm and initial densities of the order of 10^{-8} gm/cm^3 ($\sim 3 \times 10^{15}$ deuterons/cm^3). Then v_0 is of the order of 10^7 cm/sec. The current flowing is, according to (10.52) or (10.55),

$$I = \frac{c^2 R_0 E_0}{v_0} F\left(x, \frac{dx}{d\tau}\right) \tag{10.61}$$

where F is a dimensionless function of the order of unity. For a tube radius of 10 cm and the conditions described, the current I is measured in units of 10^5 or 10^6 amperes.

The discussion of the pinching action presented so far is obviously valid only for short times after the initiation of current flow. The simplified models indicate that in a time of the order of R_0/v_0 the radius of the plasma column goes to zero. It is clear, however, that before that will occur (even approximately) the behavior will be modified. In the hydrodynamic limit, the radial shock waves caused by the pinch will be reflected off the axis and move outwards, striking the interface and retarding its inward motion or even reversing it. This phenomenon is known as bouncing. It is evidently present also in the free-particle model. Consequently the general behavior of radius R as a function of time is expected to be as shown in Fig. 10.7. Although no proper analysis has been made of the subsequent bounces, it is conjectured that there is an approach to a steady state at some radius less than R_0.

10.7 Instabilities in a Pinched-Plasma Column

In the laboratory long-lived pinched plasmas are extremely difficult to produce. The dynamic behavior of the previous section is found to be followed at least qualitatively for times up to around the first bounce. But then the plasma column is observed to break up rapidly. The reason for the disintegration of the column is the growth of instabilities. The

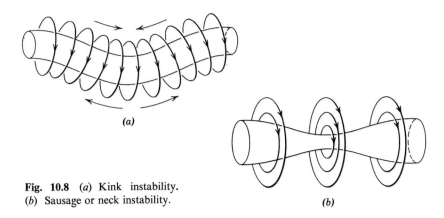

(a)

Fig. 10.8 (*a*) Kink instability.
(*b*) Sausage or neck instability.

(b)

column is unstable against various departures from cylindrical geometry. Small distortions are amplified rapidly and destroy the column in a very short time. The detailed analysis of instabilities is sufficiently complex that we will attempt only qualitative arguments. Two of the simpler unstable distortions will be described.

The first is the *kink instability*, shown in Fig. 10.8*a*. The lines of azimuthal magnetic induction near the column are bunched together above, and separated below, the column by the distortion downwards. Thus the magnetic pressure changes are in such a direction as to increase the distortion. The distortion is unstable.

The second type of distortion is called a *sausage* or *neck instability*, shown in Fig. 10.8*b*. In the neighborhood of the constriction the azimuthal induction increases, causing a greater inwards pressure at the neck than elsewhere. This serves to enhance the existing distortion.

Both types of instability are hindered by axial magnetic fields within the plasma column. For the sausage distortion the lines of axial induction are compressed by the constriction, causing an increased pressure inside to oppose the increased pressure of the azimuthal field, as indicated schematically in Fig. 10.9. It is easy to see that the fractional changes in

Fig. 10.9 Hindering neck instability with outward pressure of trapped axial magnetic fields.

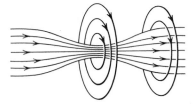

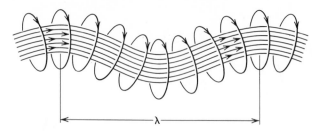

Fig. 10.10   Hindering kink instability with tension of trapped axial fields.

the two magnetic pressures, assuming a sharp boundary to the plasma, are

$$\frac{\Delta p_\phi}{p_\phi} = \frac{2x}{R}, \qquad \frac{\Delta p_z}{p_z} = \frac{4x}{R} \qquad (10.62)$$

where x is the small inwards displacement. Consequently, if

$$B_z^{\ 2} > \tfrac{1}{2}B_\phi^{\ 2} \qquad (10.63)$$

the column is stable against sausage distortions.

For kinks the axial magnetic field lines are stretched, rather than compressed laterally together. But the result is the same; namely the increased tension in the field lines inside opposes the external forces and tends to stabilize the column. It is evident from Fig. 10.10 that a short-wavelength kink of a given lateral displacement will cause the lines of force to stretch relatively more than a long-wavelength kink. Consequently, for a given ratio of internal axial field to external azimuthal field, there will be a tendency to stabilize short-wavelength kinks, but not very long-wavelength ones. If the fields are approximately equal, analysis shows that if the wavelength of the kink $\lambda < 14\ R$ the disturbance is stabilized.

For longer-wavelength kinks stabilization can be achieved by the action of the outer conductor, provided the plasma radius is not too small compared to the radius of the conductor. The azimuthal field lines are trapped between the conductor and the plasma boundary, as shown in Fig. 10.11. If the plasma column moves too close to the walls, the lines of force are crowded together between it and the walls, causing an increased magnetic pressure and restoring force.

Fig. 10.11 Stabilization of long-wavelength kinks with outer conductor.

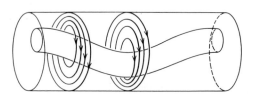

It is clear qualitatively that it must be possible, by a combination of trapped axial field and conducting walls, to create a stable configuration, at least in the approximation of a highly conducting plasma with a sharp boundary. Detailed analysis* confirms this qualitative conclusion and sets limits on the quantities involved. It is important to have as little axial field outside the plasma as possible and to keep the plasma radius of the order of one-half or one-third of the cylinder radius. If the axial field outside the plasma is too large, the combined B_z and B_ϕ cause helical instabilities that are troublesome in toroidal geometries. If, however, the axial field outside the plasma is made very large, the pitch of the helix becomes so great that there is much less than one turn of the helix in a plasma column of finite length. Then it turns out that there is the possibility of stability again. Stabilization by means of a strong axial field produced by currents external to the plasma is the basis of some fusion devices, e.g., the Stellarator.

The idealized situation of a sharp plasma boundary is difficult to create experimentally, and even when created is destroyed by diffusion of the plasma through the lines of force in times of the order of $4\pi\sigma R^2/c^2$ (see Section 10.3). For a hydrogen plasma of 1 ev energy per particle this time is of the order of 10^{-4} sec for $R \sim 10$ cm, while for a 10 kev plasma it is of the order of 10^2 sec. Clearly the thermonuclear experimenter must try to create initially as hot a plasma as possible in order to make the initial diffusion time long enough to allow further heating.

10.8 Magnetohydrodynamic Waves

In ordinary hydrodynamics the only type of small-amplitude wave motion possible is that of longitudinal, compressional (sound) waves. These propagate with a velocity s related to the derivative of pressure with respect to density at constant entropy:

$$s^2 = \left(\frac{\partial p}{\partial \rho}\right)_0 \tag{10.64}$$

If the adiabatic law $p = K\rho^\gamma$ is assumed, $s^2 = \gamma p_0/\rho_0$, where γ is the ratio of specific heats. In magnetohydrodynamics another type of wave motion is possible. It is associated with the transverse motion of lines of magnetic force. The tension in the lines of force tends to restore them to straight-line form, thereby causing a transverse oscillation. By analogy with

* V. D. Shafranov, *Atomnaya Energ.* I, **5**, 38 (1956); R. J. Tayler, *Proc. Phys. Soc.* (*London*), **B70**, 1049 (1957); M. Rosenbluth, *Los Alamos Report LA*-2030 (1956). See also *Proceedings of the Second International Conference on Peaceful Uses of Atomic Energy*, Vol. 31 (1958), papers by Braginsky and Shafranov (p. 43) and Tayler (p. 160).

Classical Electrodynamics

ordinary sound waves whose velocity squared is of the order of the hydro-static pressure divided by the density, we expect that these magnetohydro-dynamic waves, called *Alfvén waves*, will have a velocity

$$v_A \sim \left(\frac{B_0^2}{8\pi\rho_0}\right)^{\frac{1}{2}} \tag{10.65}$$

where $B_0^2/8\pi$ is the magnetic pressure.

To examine the wave motion of a conducting fluid in the presence of a magnetic field, we consider a compressible, nonviscous, perfectly con-ducting fluid in a magnetic field in the absence of gravitational forces. The appropriate equations governing its behavior are:

$$\left. \begin{aligned} &\frac{\partial \rho}{\partial t} + \nabla \cdot (\rho \mathbf{v}) = 0 \\[2mm] &\rho \frac{\partial \mathbf{v}}{\partial t} + \rho (\mathbf{v} \cdot \nabla)\mathbf{v} = -\nabla p - \frac{1}{4\pi} \mathbf{B} \times (\nabla \times \mathbf{B}) \\[2mm] &\frac{\partial \mathbf{B}}{\partial t} = \nabla \times (\mathbf{v} \times \mathbf{B}) \end{aligned} \right\} \tag{10.66}$$

These must be supplemented by an equation of state relating the pressure to the density. We assume that the equilibrium velocity is zero, but that there exists a spatially uniform, static, magnetic induction $\mathbf{B}_0$ throughout the uniform fluid of constant density ρ_0. Then we imagine small-amplitude departures from the equilibrium values:

$$\left. \begin{aligned} \mathbf{B} &= \mathbf{B}_0 + \mathbf{B}_1(\mathbf{x}, t) \\ \rho &= \rho_0 + \rho_1(\mathbf{x}, t) \\ \mathbf{v} &= \mathbf{v}_1(\mathbf{x}, t) \end{aligned} \right\} \tag{10.67}$$

If equations (10.66) are linearized in the small quantities, then they become:

$$\left. \begin{aligned} &\frac{\partial \rho_1}{\partial t} + \rho_0 \nabla \cdot \mathbf{v}_1 = 0 \\[2mm] &\rho_0 \frac{\partial \mathbf{v}_1}{\partial t} + s^2 \nabla \rho_1 + \frac{\mathbf{B}_0}{4\pi} \times (\nabla \times \mathbf{B}_1) = 0 \\[2mm] &\frac{\partial \mathbf{B}_1}{\partial t} - \nabla \times (\mathbf{v}_1 \times \mathbf{B}_0) = 0 \end{aligned} \right\} \tag{10.68}$$

where s^2 is the square of the sound velocity (10.64). These equations can be combined to yield an equation for $\mathbf{v}_1$ alone:

$$\frac{\partial^2 \mathbf{v}_1}{\partial t^2} - s^2 \nabla(\nabla \cdot \mathbf{v}_1) + \mathbf{v}_A \times \nabla \times [\nabla \times (\mathbf{v}_1 \times \mathbf{v}_A)] = 0 \tag{10.69}$$

where we have introduced a vectorial Alfvén velocity:

$$\mathbf{v}_A = \frac{\mathbf{B}_0}{\sqrt{4\pi\rho}} \qquad (10.70)$$

The wave equation (10.69) for $\mathbf{v}_1$ is somewhat involved, but it allows simple solutions for waves propagating parallel or perpendicular to the magnetic field direction.* With $\mathbf{v}_1(\mathbf{x}, t)$ a plane wave with wave vector $\mathbf{k}$ and frequency ω:

$$\mathbf{v}_1(\mathbf{x}, t) = \mathbf{v}_1 e^{i\mathbf{k}\cdot\mathbf{x} - i\omega t} \qquad (10.71)$$

equation (10.69) becomes:

$$-\omega^2 \mathbf{v}_1 + (s^2 + v_A^2)(\mathbf{k}\cdot\mathbf{v}_1)\mathbf{k}$$
$$+ \mathbf{v}_A\cdot\mathbf{k}[(\mathbf{v}_A\cdot\mathbf{k})\mathbf{v}_1 - (\mathbf{v}_A\cdot\mathbf{v}_1)\mathbf{k} - (\mathbf{k}\cdot\mathbf{v}_1)\mathbf{v}_A] = 0 \quad (10.72)$$

If $\mathbf{k}$ *is perpendicular* to $\mathbf{v}_A$ the last term vanishes. Then the solution for $\mathbf{v}_1$ is a *longitudinal* magnetosonic wave with a phase velocity:

$$u_{\text{long}} = \sqrt{s^2 + v_A^2} \qquad (10.73)$$

Note that this wave propagates with a velocity which depends on the sum of hydrostatic and magnetic pressures, apart from factors of the order of unity. If $\mathbf{k}$ *is parallel* to $\mathbf{v}_A$, (10.72) reduces to

$$(k^2 v_A^2 - \omega^2)\mathbf{v}_1 + \left(\frac{s^2}{v_A^2} - 1\right) k^2(\mathbf{v}_A\cdot\mathbf{v}_1)\mathbf{v}_A = 0 \qquad (10.74)$$

There are two types of wave motion possible in this case. There is an ordinary longitudinal wave ($\mathbf{v}_1$ parallel to $\mathbf{k}$ and $\mathbf{v}_A$) with phase velocity equal to the sound velocity s. But there is also a *transverse* wave ($\mathbf{v}_1\cdot\mathbf{v}_A = 0$) with a phase velocity equal to the Alfvén velocity v_A. This Alfvén wave is a purely magnetohydrodynamic phenomenon which depends only on the magnetic field (tension) and the density (inertia).

For mercury at room temperature the Alfvén velocity is [B_0 (gauss)/13.1]) cm/sec, compared with sound speed of 1.45×10^5 cm/sec. At all laboratory field strengths the Alfvén velocity is much less than the speed of sound. In astrophysical problems, on the other hand, the Alfvén velocity can become very large because of the much smaller densities. In the sun's photosphere, for example, the density is of the order of 10^{-7} gm/cm³ ($\sim 6 \times 10^{16}$ hydrogen atoms/cm³) so that $v_A \simeq 10^3\, B_0$ cm/sec. Solar magnetic fields appear to be of the order of 1 or 2 gauss at the surface, with much larger values around sunspots. For comparison, the velocity of sound is of the order of 10^6 cm/sec in both the photosphere and the chromosphere.

---

* The determination of the characteristics of the waves for arbitrary direction of propagation is left to Problem 10.3.

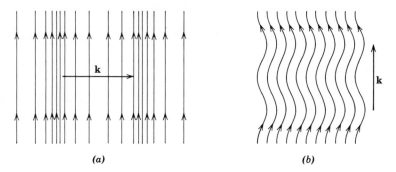

Fig. 10.12   Magnetohydrodynamic waves.

The magnetic fields of these different waves can be found from the third equation in (10.68):

$$\mathbf{B_1} = \begin{cases} \dfrac{k}{\omega} v_1 \mathbf{B_0} & \text{for } \mathbf{k} \perp \mathbf{B_0} \\[2mm] 0 & \text{for the longitudinal } \mathbf{k} \parallel \mathbf{B_0} \\[2mm] -\dfrac{k}{\omega} B_0 \mathbf{v_1} & \text{for the transverse } \mathbf{k} \parallel \mathbf{B_0} \end{cases} \qquad (10.75)$$

The magnetosonic wave moving perpendicular to $\mathbf{B_0}$ causes compressions and rarefactions in the lines of force without changing their direction, as indicated in Fig. 10.12a. The Alfvén wave parallel to $\mathbf{B_0}$ causes the lines of force to oscillate back and forth laterally (Fig. 10.12b). In either case the lines of force are "frozen in" and move with the fluid.

If the conductivity of the fluid is not infinite or viscous effects are present, we anticipate dissipative losses and a consequent damping of oscillations. The second and third equations in (10.68) are modified by additional terms:

$$\left. \begin{aligned} \rho_0 \frac{\partial \mathbf{v_1}}{\partial t} &= -s^2 \, \nabla \rho_1 - \frac{\mathbf{B_0}}{4\pi} \times (\nabla \times \mathbf{B_1}) + \eta \, \nabla^2 \mathbf{v_1} \\[2mm] \frac{\partial \mathbf{B_1}}{\partial t} &= \nabla \times (\mathbf{v_1} \times \mathbf{B_0}) + \frac{c^2}{4\pi\sigma} \nabla^2 \mathbf{B_1} \end{aligned} \right\} \qquad (10.76)$$

where η is the viscosity* and σ is the conductivity. Since both additions cause dispersion in the phase velocity, their effects are most easily seen when a plane wave solution is being sought. For plane waves it is evident

---

* Use of the simple viscous force (10.3) is not really allowed for a compressible fluid. But it can be expected to give the correct qualitative behavior.

that these equations are equivalent to

$$
\left.\begin{aligned}
\rho_0 \frac{\partial \mathbf{v}_1}{\partial t} &= \frac{1}{\left(1 + i\,\dfrac{\eta k^2}{\rho_0 \omega}\right)} \left[-s^2 \boldsymbol{\nabla}\rho_1 - \frac{\mathbf{B}_0}{4\pi} \times (\boldsymbol{\nabla} \times \mathbf{B}_1) \right] \\[2ex]
\frac{\partial \mathbf{B}_1}{\partial t} &= \frac{1}{\left(1 + i\,\dfrac{c^2 k^2}{4\pi\sigma\omega}\right)} \boldsymbol{\nabla} \times (\mathbf{v}_1 \times \mathbf{B}_0)
\end{aligned}\right\}
\qquad (10.77)
$$

Consequently equation (10.72) relating k and ω is modified by (*a*) multi-plying s^2 and ω^2 by the factor $\left(1 + i\,\dfrac{c^2 k^2}{4\pi\sigma\omega}\right)$, and (*b*) multiplying ω^2 by the factor $\left(1 + i\,\dfrac{\eta k^2}{\rho_0 \omega}\right)$.

For the important case of the Alfvén wave parallel to the field, the relation between ω and k becomes

$$
k^2 v_A{}^2 = \omega^2\left(1 + i\,\frac{c^2 k^2}{4\pi\sigma\omega}\right)\left(1 + i\,\frac{\eta k^2}{\rho_0 \omega}\right)
\qquad (10.78)
$$

If the resistive and viscous correction terms are small, the wave number is approximately

$$
k \simeq \frac{\omega}{v_A} + i\,\frac{\omega^2}{2v_A{}^3}\left(\frac{c^2}{4\pi\sigma} + \frac{\eta}{\rho_0}\right)
\qquad (10.79)
$$

This shows that the attenuation increases rapidly with frequency (or wave number), but decreases with increasing magnetic field strength. In terms of the diffusion time τ of Section 10.3, the imaginary part of the wave number shows that, apart from viscosity effects, the wave travels for a time τ before falling to $1/e$ of its original intensity, where the length parameter in τ (10.12) is the wavelength of oscillation. For the opposite extreme in which the resistive and/or viscous terms dominate, the wave number is given by the vanishing of the two factors on the right-hand side of (10.78). Thus k has equal real and imaginary parts and the wave is damped out rapidly, independent of the magnitude of the magnetic field.

The considerations of magnetohydrodynamic waves given above are valid only at comparatively low frequencies, since the displacement current was ignored in Ampère's law. It is evident that, if the frequency is high enough, the behavior of the fields must go over into the "ionospheric" behavior described in Section 7.9, where charge-separation effects play an important role. But even when charge-separation effects are neglected in the magnetohydrodynamic description, the displacement current modifies the propagation of the Alfvén and magnetosonic waves. The form of

Ampère's law, including the displacement current, is:

$$\mathbf{\nabla} \times \mathbf{B} = \frac{4\pi}{c} \mathbf{J} - \frac{1}{c^2} \frac{\partial}{\partial t} (\mathbf{v} \times \mathbf{B}) \tag{10.80}$$

where we have used the infinite conductivity approximation (10.9) in eliminating the electric field $\mathbf{E}$. Thus the current to be inserted into the force equation for fluid motion is now

$$\mathbf{J} = \frac{c}{4\pi} \left[\mathbf{\nabla} \times \mathbf{B} + \frac{1}{c^2} \frac{\partial}{\partial t} (\mathbf{v} \times \mathbf{B}) \right] \tag{10.81}$$

In the linearized set of equations (10.68) the second one is then generalized to read:

$$\rho_0 \left[\frac{\partial \mathbf{v}_1}{\partial t} + \frac{1}{c^2} \mathbf{v}_A \times \left(\frac{\partial \mathbf{v}_1}{\partial t} \times \mathbf{v}_A \right) \right] = -s^2 \mathbf{\nabla} \rho_1 - \frac{\mathbf{B}_0}{4\pi} \times (\mathbf{\nabla} \times \mathbf{B}_1) \tag{10.82}$$

This means that the wave equation for $\mathbf{v}_1$ is altered to the form:

$$\frac{\partial^2}{\partial t^2} \left[\mathbf{v}_1 \left(1 + \frac{v_A{}^2}{c^2} \right) - \frac{\mathbf{v}_A}{c^2} (\mathbf{v}_A \cdot \mathbf{v}_1) \right] - s^2 \mathbf{\nabla}(\mathbf{\nabla} \cdot \mathbf{v}_1)$$
$$+ \mathbf{v}_A \times \mathbf{\nabla} \times \mathbf{\nabla} \times (\mathbf{v}_1 \times \mathbf{v}_A) = 0 \tag{10.83}$$

Inspection shows that for $\mathbf{v}_1$ parallel to $\mathbf{v}_A$ (i.e., $\mathbf{B}_0$) there is no change from before. But for transverse $\mathbf{v}_1$ (either magnetosonic with $\mathbf{k}$ perpendicular to $\mathbf{B}_0$, or Alfvén waves with $\mathbf{k}$ parallel to $\mathbf{B}_0$) the square of the frequency is multiplied by a factor $[1 + (v_A{}^2/c^2)]$. Thus the phase velocity of Alfvén waves becomes

$$u_A = \frac{c v_A}{\sqrt{c^2 + v_A{}^2}} \tag{10.84}$$

In the usual limit where $v_A \ll c$, the velocity is approximately equal to v_A; the displacement current is unimportant. But, if $v_A \gg c$, then the phase velocity is equal to the velocity of light. From the point of view of electromagnetic waves, the transverse Alfvén wave can be thought of as a wave in a medium with an index of refraction given by

$$u_A = \frac{c}{n} \tag{10.85}$$

Thus

$$n^2 = 1 + \frac{c^2}{v_A{}^2} = 1 + \frac{4\pi \rho_0 c^2}{B_0{}^2} \tag{10.86}$$

Caution must be urged in using this index of refraction for the propagation of electromagnetic waves in a plasma. It is valid only at frequencies where charge-separation effects are unimportant.

10.9   High-Frequency Plasma Oscillations

The magnetohydrodynamic approximation considered in the previous sections is based on the concept of a single-component, electrically neutral fluid with a scalar conductivity σ to describe its interaction with the electromagnetic field. As discussed in the introduction to this chapter a conducting fluid or plasma is, however, a multicomponent fluid with electrons and one or more types of ions present. At low frequencies or long wavelengths the description in terms of a single fluid is valid because the collision frequency ν is large enough (and the mean free path short enough) that the electrons and ions always maintain local electrical neutrality, while on the average drifting in opposite directions according to Ohm's law under the action of electric fields. At higher frequencies the single-fluid model breaks down. The electrons and ions tend to move independently, and charge separations occur. These charge separations produce strong restoring forces. Consequently oscillations of an electrostatic nature are set up. If a magnetic field is present, other effects occur. The electrons and ions tend to move in circular or helical orbits in the magnetic field with orbital frequencies given by

$$\omega_B = \frac{eB}{mc} \tag{10.87}$$

When the fields are strong enough or the densities low enough that the orbital frequencies are comparable to the collision frequency, the concept of a scalar conductivity breaks down and the current flow exhibits a marked directional dependence relative to the magnetic field (see Problem 10.5). At still higher frequencies the greater inertia of the ions implies that they will be unable to follow the rapid fluctuation of the fields. Only the electrons partake in the motion. The ions merely provide a uniform background of positive charge to give electrical neutrality on the average. The idea of a uniform background of charge, and indeed the concept of an electron fluid, is valid only when we are considering a scale of length which is at least large compared to interparticle spacings ($l \gg n_0^{-\frac{1}{3}}$). In fact, there is another limit, the Debye screening length, which for plasmas at reasonable temperatures is greater than $n_0^{-\frac{1}{3}}$, and which forms the actual dividing line between small-scale individual-particle motion and collective fluid motion (see the following section).

To avoid undue complications we consider only the high-frequency behavior of a plasma, ignoring the dynamical effects of the ions. We also ignore the effects of collisions. The electrons of charge e and mass m are

Classical Electrodynamics

described by a density $n(\mathbf{x}, t)$ and an average velocity $\mathbf{v}(\mathbf{x}, t)$. The equilibrium-charge density of ions and electrons is $\mp e n_0$. The dynamical equations for the electron fluid are

$$
\left.
\begin{aligned}
&\frac{\partial n}{\partial t} + \mathbf{\nabla} \cdot (n\mathbf{v}) = 0 \\
&\frac{\partial \mathbf{v}}{\partial t} + (\mathbf{v} \cdot \mathbf{\nabla})\mathbf{v} = \frac{e}{m}\left(\mathbf{E} + \frac{\mathbf{v}}{c} \times \mathbf{B}\right) - \frac{1}{mn}\mathbf{\nabla}p
\end{aligned}
\right\}
\tag{10.88}
$$

where the effects of the thermal kinetic energy of the electrons are described by the electron pressure p (here assumed a scalar). The charge and current densities are:

$$
\left.
\begin{aligned}
\rho_e &= e(n - n_0) \\
\mathbf{J} &= en\mathbf{v}
\end{aligned}
\right\}
\tag{10.89}
$$

Thus Maxwell's equations can be written

$$
\left.
\begin{aligned}
&\mathbf{\nabla} \cdot \mathbf{E} = 4\pi e(n - n_0) \\
&\mathbf{\nabla} \cdot \mathbf{B} = 0 \\
&\mathbf{\nabla} \times \mathbf{E} + \frac{1}{c}\frac{\partial \mathbf{B}}{\partial t} = 0 \\
&\mathbf{\nabla} \times \mathbf{B} - \frac{1}{c}\frac{\partial \mathbf{E}}{\partial t} = \frac{4\pi en}{c}\mathbf{v}
\end{aligned}
\right\}
\tag{10.90}
$$

We now assume that the static situation is the electron fluid at rest with $n = n_0$ and no fields present, and consider small departures from that state due to some initial disturbance. The linearized equations of motion are

$$
\left.
\begin{aligned}
&\frac{\partial n}{\partial t} + n_0 \mathbf{\nabla} \cdot \mathbf{v} = 0 \\
&\frac{\partial \mathbf{v}}{\partial t} - \frac{e}{m}\mathbf{E} + \frac{1}{mn_0}\left(\frac{\partial p}{\partial n}\right)_0 \mathbf{\nabla}n = 0 \\
&\mathbf{\nabla} \cdot \mathbf{E} - 4\pi en = 0 \\
&\mathbf{\nabla} \times \mathbf{B} - \frac{1}{c}\frac{\partial \mathbf{E}}{\partial t} - \frac{4\pi en_0}{c}\mathbf{v} = 0
\end{aligned}
\right\}
\tag{10.91}
$$

plus the two homogeneous Maxwell's equations. Here $n(\mathbf{x}, t)$ and $\mathbf{v}(\mathbf{x}, t)$ represent departures from equilibrium. If an external magnetic field $\mathbf{B}_0$ is present a $[(\mathbf{v}/c) \times \mathbf{B}_0]$ term must be kept in the force equation (see Problem 10.7), but the fluctuation field $\mathbf{B}$ is of first order in small quantities so that $(\mathbf{v} \times \mathbf{B})$ is second order. The continuity equation is actually not an independent equation, but may be derived by combining the last two equations in (10.91).

Since the force equation in (10.91) is independent of magnetic field, we suspect that there exist solutions of a purely electrostatic nature, with $\mathbf{B} = 0$. The continuity and force equations can be combined to yield a wave equation for the density fluctuations:

$$\frac{\partial^2 n}{\partial t^2} + \left(\frac{4\pi e^2 n_0}{m}\right) n - \frac{1}{m}\left(\frac{\partial p}{\partial n}\right)_0 \nabla^2 n = 0 \tag{10.92}$$

On the other hand, the time derivative of Ampère's law and the force equation can be combined to give an equation for the fields:

$$\frac{\partial^2 \mathbf{E}}{\partial t^2} + \left(\frac{4\pi e^2 n_0}{m}\right)\mathbf{E} - \frac{1}{m}\left(\frac{\partial p}{\partial n}\right)_0 \nabla(\nabla \cdot \mathbf{E}) = c\nabla \times \frac{\partial \mathbf{B}}{\partial t} \tag{10.93}$$

The structures of the left-hand sides of these two equations are essentially identical. Consequently no inconsistency arises if we put $\partial \mathbf{B}/\partial t = 0$. Having excluded static fields already, we conclude that $\mathbf{B} = 0$ is a possibility. If $\partial \mathbf{B}/\partial t = 0$, then Faraday's law implies $\nabla \times \mathbf{E} = 0$. Hence $\mathbf{E}$ is a longitudinal field derivable from a scalar potential. It is immediately evident that each component of $\mathbf{E}$ satisfies the same equation (10.92) as the density fluctuations. If the pressure term in (10.92) is neglected, we find that the density, velocity, and electric field all oscillate with the plasma frequency ω_p:

$$\omega_p{}^2 = \frac{4\pi n_0 e^2}{m} \tag{10.94}$$

If the pressure term is included, we obtain a dispersion relation for the frequency:

$$\omega^2 = \omega_p{}^2 + \frac{1}{m}\left(\frac{\partial p}{\partial n}\right)_0 k^2 \tag{10.95}$$

The determination of the coefficient of k^2 takes some care. The adiabatic law $p = p_0(n/n_0)^\gamma$ can be assumed, but the customary acoustical value $\gamma = \frac{5}{3}$ for a gas of particles with 3 external, but no internal, degrees of freedom is not valid. The reason is that the frequency of the present density oscillations is much higher than the collision frequency, contrary to the acoustical limit. Consequently the one-dimensional nature of the density oscillations is maintained. A value of γ appropriate to 1 translational degree of freedom must be used. Since $\gamma = (m + 2)/m$, where m is the number of degrees of freedom, we have in this case $\gamma = 3$. Then

$$\frac{1}{m}\left(\frac{\partial p}{\partial n}\right)_0 = 3\frac{p_0}{mn_0} \tag{10.96}$$

If we use $p_0 = n_0 KT$ and define the rms velocity component in one direction (parallel to the electric field),

$$m\langle u^2 \rangle = KT \qquad (10.97)$$

then the dispersion equation can be written

$$\omega^2 = \omega_p{}^2 + 3\langle u^2 \rangle k^2 \qquad (10.98)$$

This relation is an approximate one, valid for long wavelengths, and is actually just the first two terms in an expansion involving higher and higher moments of the velocity distribution of the electrons (see Problem 10.6). In form (10.98) the dispersion equation has a validity beyond the ideal gas law which was used in the derivation. For example, it applies to plasma oscillations in a degenerate Fermi gas of electrons in which all cells in velocity space are filled inside a sphere of radius equal to the Fermi velocity V_F. Then the average value of the square of a component of velocity is

$$\langle u^2 \rangle = \tfrac{1}{5} V_F{}^2 \qquad (10.99)$$

Quantum effects appear explicitly in the dispersion equation only in higher-order terms in the expansion in powers of k^2.

The oscillations described above are longitudinal electrostatic oscillations in which the oscillating magnetic field vanishes identically. This means that they cannot give rise to radiation in an unbounded plasma. There are, however, modes of oscillation in a plasma which are transverse electromagnetic waves. To see the various possibilities of plasma oscillations we assume that all variables vary as $\exp{(i\mathbf{k} \cdot \mathbf{x} - i\omega t)}$ and look for a defining relationship between ω and $\mathbf{k}$, as we did for the magnetohydrodynamic waves in Section 10.8. With this assumption the linearized equations (10.91) and the two homogeneous Maxwell's equations can be written:

$$\left.\begin{array}{l} n = \dfrac{\mathbf{k} \cdot \mathbf{v}}{\omega}\, n_0 \\[2mm] \mathbf{v} = \dfrac{ie\mathbf{E}}{m\omega} + \dfrac{3\langle u^2 \rangle}{\omega}\dfrac{n}{n_0}\mathbf{k} \\[2mm] \mathbf{k} \cdot \mathbf{E} = -i4\pi en \\[2mm] \mathbf{k} \cdot \mathbf{B} = 0 \\[2mm] \mathbf{k} \times \mathbf{B} = -\dfrac{\omega}{c}\mathbf{E} - i\dfrac{4\pi e n_0}{c}\mathbf{v} \\[2mm] \mathbf{k} \times \mathbf{E} = \dfrac{\omega}{c}\mathbf{B} \end{array}\right\} \qquad (10.100)$$

Maxwell's equations can be solved for **v** in terms of **k** and **E**:

$$\mathbf{v} = \left(\frac{ie}{m\omega}\right) \frac{1}{\omega_p^{\,2}} [(\omega^2 - c^2 k^2)\mathbf{E} + c^2(\mathbf{k} \cdot \mathbf{E})\mathbf{k}] \qquad (10.101)$$

Then the force equation and the divergence of **E** can be used to eliminate **v** in order to obtain an equation for **E** alone:

$$(\omega^2 - \omega_p^{\,2} - c^2 k^2)\mathbf{E} + (c^2 - 3\langle u^2 \rangle)(\mathbf{k} \cdot \mathbf{E})\mathbf{k} = 0 \qquad (10.102)$$

If we write **E** in terms of components parallel and perpendicular to **k**:

$$\left.\begin{array}{l} \mathbf{E} = \mathbf{E}_{\parallel} + \mathbf{E}_{\perp} \\[2mm] \text{where} \qquad \mathbf{E}_{\parallel} = \left(\dfrac{\mathbf{k} \cdot \mathbf{E}}{k^2}\right)\mathbf{k} \end{array}\right\} \qquad (10.103)$$

then (10.102) can be written as two equations:

$$\left.\begin{array}{l} (\omega^2 - \omega_p^{\,2} - 3\langle u^2 \rangle k^2)\mathbf{E}_{\parallel} = 0 \\[2mm] (\omega^2 - \omega_p^{\,2} - c^2 k^2)\mathbf{E}_{\perp} = 0 \end{array}\right\} \qquad (10.104)$$

The first of these results shows that the longitudinal waves satisfy the dispersion relation (10.98) already discussed, while the second shows that there are two transverse waves (two states of polarization) which have the dispersion relation:

$$\omega^2 = \omega_p^{\,2} + c^2 k^2 \qquad (10.105)$$

Equation (10.105) is just the dispersion equation for the transverse electromagnetic waves described in Section 7.9 from another point of view. In the absence of external fields the electrostatic oscillations and the transverse electromagnetic oscillations are not coupled together. But in the presence of an external magnetic induction, for example, the force equation has an added term involving the magnetic field and the oscillations are coupled (see Problem 10.7).

10.10   Short-Wavelength Limit for Plasma Oscillations and the Debye Screening Distance

In the discussion of plasma oscillations so far no mention has been made of the range of wave numbers over which the description in terms of collective oscillations applies. Certainly $n_0^{1/3}$ is one upper bound on the wave-number scale. A clue to a more relevant upper bound can be obtained by examining the dispersion relation (10.98) for the longitudinal oscillations. For long wavelengths the frequency of oscillation is very closely $\omega = \omega_p$. It is only for wave numbers comparable to the *Debye wave number k_D,*

$$k_D{}^2 = \frac{\omega_p{}^2}{\langle u^2 \rangle}$$ (10.106)

that appreciable departures of the frequency from ω_p occur.

For wave numbers $k \ll k_D$, the phase and group velocities of the longitudinal plasma oscillations are:

$$\left. \begin{aligned} v_p &\simeq \frac{\omega_p}{k} \\[2mm] v_g &\simeq \frac{3\langle u^2 \rangle}{v_p} \end{aligned} \right\}$$ (10.107)

From the definition of k_D we see that for such wave numbers the phase velocity is much larger than, and the group velocity much smaller than, the rms thermal velocity $\langle u^2 \rangle^{1/2}$. As the wave number increases towards k_D, the phase velocity decreases from large values down towards $\langle u^2 \rangle^{1/2}$. Consequently for wave numbers of the order of k_D the wave travels with a small enough velocity that there are appreciable numbers of electrons traveling somewhat faster than, or slower than, or at about the same speed as, the wave. The phase velocity lies in the tail of the thermal distribution. The circumstance that the wave's velocity is comparable with the electronic thermal velocities is the source of an energy-transfer mechanism which causes the destruction of the oscillation. The mechanism is the trapping of particles by the moving wave with a resultant transfer of energy out of the wave motion into the particles. The consequent damping of the wave is called *Landau damping*.

A detailed calculation of Landau damping is out of place here. But we can describe qualitatively the physical mechanism. Fig. 10.13 shows a distribution of electron velocities with a certain rms spread and a Maxwellian tail out to higher velocities. For small k the phase velocity

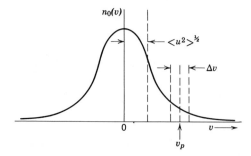

Fig. 10.13 Thermal velocity distribution of electrons.

lies far out on the tail and negligible damping occurs. But as $k \to k_D$ the phase velocity lies within the tail, as shown in Fig. 10.13, with a significant number of electrons having thermal speeds comparable to v_p. There is then a velocity band Δv around $v = v_p$ where electrons are moving sufficiently slowly relative to the wave that they can be trapped in the potential troughs and carried along at velocity v_p by the wave. If there are more particles in Δv moving initially slower than v_p than there are particles moving faster (as shown in the figure), the trapping process will cause a net increase in the energy of the particles at the expense of the wave. This is the mechanism of Landau damping. Detailed calculations show that the damping can be expressed in terms of an imaginary part of the frequency given by

$$\text{Im } \omega \simeq -\omega_p \sqrt{\frac{\pi}{8}} \left(\frac{k_D}{k}\right)^3 e^{-(k_D{}^2/2k^2)} \tag{10.108}$$

provided $k \ll k_D$. To obtain (10.108) a Maxwellian distribution of velocities was assumed. For $k \geqslant k_D$ the damping constant is larger than given by (10.108) and rapidly becomes much larger than the real part of the frequency, as given by (10.98).

The Landau formula (10.108) shows that for $k \ll k_D$ the longitudinal plasma oscillations are virtually undamped. But the damping becomes important as soon as $k \sim k_D$ (even for $k = 0.5k_D$, $\text{Im } \omega \simeq -0.7\omega_p$). For wave numbers larger than the Debye wave number the damping is so great that it is meaningless to speak of organized oscillations.

Another, rather different consideration leads to the same limiting Debye wave number as the boundary of collective oscillatory effects. We know that an electronic plasma is a collection of electrons with a uniform background of positive charge. On a very small scale of length we must describe the behavior in terms of a succession of very many two-body Coulomb collisions. But on a larger scale the electrons tend to cooperate. If a local surplus of positive charge appears anywhere, the electrons rush over to neutralize it. This collective response to charge fluctuations is what gives rise to large-scale plasma oscillations. But in addition to, or, better, because of, the collective oscillations the cooperative response of the electrons also tends to reduce the long-range nature of the Coulomb interaction between particles. An individual electron is, after all, a local fluctuation in the charge density. The surrounding electrons are repelled in such a way that they tend to screen out the Coulomb field of the chosen electron, turning it into a short-range interaction. That something like this must occur is obvious when one realizes that the only source of electrostatic interaction is the Coulomb force between the particles. If some of it is effectively taken away to cause long-wavelength collective plasma oscillations, the residue must be a sum of short-range interactions between particles.

A nonrigorous derivation of the screening effect described above was first given by Debye and Hückel in their theory of electrolytes. The basic argument is as follows. Suppose that we have a plasma with a distribution of electrons in thermal equilibrium in an electrostatic potential Φ. Then they are distributed according to the Boltzmann factor $e^{-H/KT}$ where H is the electronic Hamiltonian. The spatial density of electrons is therefore

$$n(\mathbf{x}) = n_0\, e^{-(e\Phi/KT)} \qquad (10.109)$$

Now we imagine a test charge Ze placed at the origin in this distribution of electrons with its uniform background of positive ions (charge density $-en_0$). The resulting potential Φ will be determined by Poisson's equation

$$\nabla^2\Phi = -4\pi Ze\,\delta(\mathbf{x}) - 4\pi en_0[e^{-(e\Phi/KT)} - 1] \qquad (10.110)$$

If $(e\Phi/KT)$ is assumed small, the equation can be linearized:

$$\nabla^2\Phi - k^2{}_D\Phi \simeq -4\pi Ze\,\delta(\mathbf{x}) \qquad (10.111)$$

where

$$k_D{}^2 = \frac{4\pi n_0 e^2}{KT} \qquad (10.112)$$

is an alternative way of writing (10.106). Equation (10.111) has the spherically symmetric solution:

$$\Phi(r) = Ze\,\frac{e^{-k_D r}}{r} \qquad (10.113)$$

showing that the electrons move in such a way as to screen out the Coulomb field of a test charge in a distance of the order of $k_D{}^{-1}$. The balance between thermal kinetic energy and electrostatic energy determines the magnitude of the screening radius. Numerically

$$k_D{}^{-1} = 6.91\left(\frac{T}{n_0}\right)^{1/2} \text{cm} \qquad (10.114)$$

where T is in degrees Kelvin, and n_0 is the number of electrons per cubic centimeter. For a typical hot plasma with $T = 10^{6\circ} K$ and $n_0 = 10^{15}$ cm^{-3}, we find $k_D{}^{-1} \simeq 2.2 \times 10^{-4}$ cm.

For the degenerate electron gas at low temperatures the Debye wave number k_D is replaced by a Fermi wave number k_F:

$$k_F \sim \frac{\omega_p}{V_F} \qquad (10.115)$$

where V_F is the velocity at the surface of the Fermi sphere. This magnitude of screening radius can be deduced from a Fermi-Thomas generalization of the Debye-Hückel approach. It fits in naturally with the dispersion relation (10.98) and the mean square velocity (10.99).

The Debye-Hückel screening distance provides a natural dividing line between the small-scale collisions of pairs of particles and the large-scale collective effects such as plasma oscillations. It is a happy and not fortuitous happening that plasma oscillations of shorter wavelengths can independently be shown not to exist because of severe damping.

REFERENCES AND SUGGESTED READING

The subject of magnetohydrodynamics and plasma physics has a rapidly growing literature. Many of the available books are collections of papers presented at conferences and symposia. Although these are useful to someone who has some knowledge of the field, they are not suited for beginning study. Two works on magnetohydrodynamics which are coherent presentations of the subject are
> Alfvén,
> Cowling.
A short discussion of magnetohydrodynamics appears in
> Landau and Lifshitz, *Electrodynamics of Continuous Media*, Chapter VIII.
Corresponding books devoted mainly to the physics of plasmas are
> Chandrasekhar,
> Linhart,
> Simon,
> Spitzer.
The subject of controlled thermonuclear reactions, with much material on the fundamental physics of plasmas, is treated thoroughly by
> Glasstone and Lovberg,
> Rose and Clark.

PROBLEMS

10.1 An infinitely long, solid, right circular, metallic cylinder has radius $(R/2)$ and conductivity σ. It is tightly surrounded by, but insulated from, a hollow cylinder of the same material of inner radius $(R/2)$ and outer radius R. Equal and opposite total currents, distributed uniformly over the cross-sectional areas, flow in the inner cylinder and in the hollow outer one. At $t = 0$ the applied voltages are short-circuited.

(*a*) Find the distribution of magnetic induction inside the cylinders before $t = 0$.

(*b*) Find the distribution as a function of time after $t = 0$, neglecting the displacement current.

(*c*) What is the behavior of the magnetic induction as a function of time for long times? Define what you mean by long times.

10.2 A comparatively stable self-pinched column of plasma can be produced by trapping an axial magnetic induction inside the plasma before the pinch begins. Suppose that the plasma column initially fills a conducting tube of radius R_0 and that a uniform axial magnetic induction B_{z0} is present in the tube. Then a voltage is applied along the tube so that axial currents flow and an azimuthal magnetic induction is built up.

(a) Show that, if quasi-equilibrium conditions apply, the pressure-balance relation can be written:

$$\left[p(r) + \frac{B_z^2}{8\pi} + \frac{B_\phi^2}{8\pi} \right]_{r_1}^{r_2} + \frac{1}{4\pi} \int_{r_1}^{r_2} \frac{B_\phi^2}{r}\, dr = 0$$

(b) If the plasma has a sharp boundary and such a large conductivity that currents flow only in a thin layer on the surface, show that for a quasi-static situation the radius $R(t)$ of the plasma column is given by the equation

$$\frac{R_0}{R} \ln\left(\frac{R_0}{R} \right) = \frac{1}{t_0} \int_0^t f(t)\, dt$$

where

$$t_0 = \frac{B_{z0} R_0}{c E_0}$$

and $E_0 f(t)$ is the applied electric field.

(c) If the initial axial field is 100 gauss, and the applied electric field has an initial value of 1 volt/cm and falls almost linearly to zero in 1 millisecond, determine the final radius if the initial radius is 50 cm. These conditions are of the same order of magnitude as those appropriate for the British toroidal apparatus (Zeta), but external inductive effects limit the pinching effect to less than the value found here. See E. P. Butt *et al.*, *Proceedings of the Second International Conference on Peaceful Uses of Atomic Energy*, Vol. 32, p. 42 (1958).

10.3 Magnetohydrodynamic waves can occur in a compressible, nonviscous, perfectly conducting fluid in a uniform static magnetic induction $\mathbf{B}_0$. If the propagation direction is not parallel or perpendicular to $\mathbf{B}_0$, the waves are not separated into purely longitudinal (magnetosonic) or transverse (Alfvén) waves. Let the angle between the propagation direction $\mathbf{k}$ and the field $\mathbf{B}_0$ be θ.

(a) Show that there are three different waves with phase velocities given by

$$u_1^2 = (v_A \cos \theta)^2$$
$$u_{2,3}^2 = \tfrac{1}{2}(s^2 + v_A^2) \pm \tfrac{1}{2}[(s^2 + v_A^2)^2 - 4s^2 v_A^2 \cos^2 \theta]^{\frac{1}{2}}$$

where s is the sound velocity in the fluid, and $v_A = (B_0^2/4\pi\rho_0)^{\frac{1}{2}}$ is the Alfvén velocity.

(b) Find the velocity eigenvectors for the three different waves, and prove that the first (Alfvén) wave is always transverse, while the other two are neither longitudinal nor transverse.

(c) Evaluate the phase velocities and eigenvectors of the mixed waves in the approximation that $v_A \gg s$. Show that for one wave the only appreciable component of velocity is parallel to the magnetic field, while for the other the only component is perpendicular to the field and in the plane containing $\mathbf{k}$ and $\mathbf{B}_0$.

10.4 An incompressible, nonviscous, perfectly conducting fluid with constant density ρ_0 is acted upon by a gravitational potential ψ and a uniform, static, magnetic induction $\mathbf{B}_0$.

(a) Show that magnetohydrodynamic waves of arbitrary amplitude and form $B_1(x, t)$, $v(x, t)$ can exist, described by the equations

$$(\mathbf{B}_0 \cdot \nabla)\mathbf{B}_1 = \pm \sqrt{4\pi\rho_0}\, \frac{\partial \mathbf{B}_1}{\partial t}$$

$$\mathbf{B}_1 = \pm \sqrt{4\pi\rho_0}\, \mathbf{v}$$

$$p + \rho_0 \psi + \frac{(\mathbf{B}_0 + \mathbf{B}_1)^2}{8\pi} = \text{constant}$$

(b) Suppose that at $t = 0$ a certain disturbance $\mathbf{B}_1(\mathbf{x}, 0)$ exists in the fluid such that it satisfies the above equations with the upper sign. What is the behavior of the disturbance at later times?

10.5 The force equation for an electronic plasma, including a phenomenological collision term, but neglecting the hydrostatic pressure (zero temperature approximation) is

$$\frac{\partial \mathbf{v}}{\partial t} + (\mathbf{v} \cdot \nabla)\mathbf{v} = \frac{e}{m}\left(\mathbf{E} + \frac{\mathbf{v}}{c} \times \mathbf{B}\right) - \nu\, \mathbf{v}$$

where ν is the collision frequency.

(a) Show that in the presence of static, uniform, external, electric, and magnetic fields, the linearized steady-state expression for Ohm's law becomes

$$J_i = \sum_j \sigma_{ij} E_j$$

where the conductivity tensor is

$$\sigma_{ij} = \frac{\omega_p^2}{4\pi\nu\left(1 + \dfrac{\omega_B^2}{\nu^2}\right)}\begin{pmatrix} 1 & \dfrac{\omega_B}{\nu} & 0 \\ -\dfrac{\omega_B}{\nu} & 1 & 0 \\ 0 & 0 & \left(1 + \dfrac{\omega_B^2}{\nu^2}\right) \end{pmatrix}$$

and $\omega_p(\omega_B)$ is the electronic plasma (precession) frequency. The direction of $\mathbf{B}$ is chosen as the z axis.

(b) Suppose that at $t = 0$ an external electric field $\mathbf{E}$ is suddenly applied in the x direction, there being a magnetic induction $\mathbf{B}$ in the z direction. The current is zero at $t = 0$. Find expressions for the components of the current at all times, including the transient behavior.

10.6 The effects of finite temperature on a plasma can be described approximately by means of the correlationless Boltzmann (Vlasov) equation. Let $f(\mathbf{x}, \mathbf{v}, t)$ be the distribution function for electrons of charge e and mass m in a one-component plasma. The Vlasov equation is

$$\frac{df}{dt} \equiv \frac{\partial f}{\partial t} + \mathbf{v} \cdot \nabla_x f + \mathbf{a} \cdot \nabla_v f = 0$$

where ∇_x and ∇_v are gradients with respect to coordinate and velocity, and $\mathbf{a}$ is the acceleration of a particle. For electrostatic oscillations of the plasma $\mathbf{a} = e\mathbf{E}/m$, where $\mathbf{E}$ is the macroscopic electric field satisfying

$$\nabla \cdot \mathbf{E} = 4\pi e \left[\int f(\mathbf{x}, \mathbf{v}, t)\, d^3v - n_0 \right]$$

If $f_0(\mathbf{v})$ is the normalized equilibrium distribution of electrons

$$\left[n_0 \int f_0(\mathbf{v})\, d^3v = n_0 \right]$$

(*a*) show that the dispersion relation for small-amplitude longitudinal plasma oscillations is

$$\frac{k^2}{\omega_p{}^2} = \int \frac{\mathbf{k} \cdot \nabla_v f_0}{\mathbf{k} \cdot \mathbf{v} - \omega}\, d^3v$$

(*b*) assuming that the phase velocity of the wave is large compared to thermal velocities, show that the dispersion relation gives

$$\frac{\omega^2}{\omega_p{}^2} \simeq 1 + 2\frac{\langle \mathbf{k} \cdot \mathbf{v} \rangle}{\omega} + 3\frac{\langle (\mathbf{k} \cdot \mathbf{v})^2 \rangle}{\omega^2} + \cdots$$

where $\langle\ \rangle$ means averaged over the equilibrium distribution $f_0(\mathbf{v})$. Relate this result to that obtained in the text with the electronic fluid model.

(*c*) What is the meaning of the singularity in the dispersion relation when $\mathbf{k} \cdot \mathbf{v} = \omega$?

10.7 Consider the problem of waves in an electronic plasma when an external magnetic field $\mathbf{B}_0$ is present. Use the fluid model, neglecting the pressure term as well as collisions.

(*a*) Write down the linearized equations of motion and Maxwell's equations, assuming all variables vary as $\exp(i\mathbf{k} \cdot \mathbf{x} - i\omega t)$.

(*b*) Show that the dispersion relation for the frequencies of the different modes in terms of the wave number can be written

$$\omega^2(\omega^2 - \omega_p{}^2)(\omega^2 - \omega_p{}^2 - k^2c^2)^2$$
$$= \omega_B{}^2(\omega^2 - k^2c^2)[\omega^2(\omega^2 - \omega_p{}^2 - k^2c^2) + \omega_p{}^2c^2(\mathbf{k} \cdot \mathbf{b})^2]$$

where $\mathbf{b}$ is a unit vector in the direction of $\mathbf{B}_0$; ω_p and ω_B are the plasma and precession frequencies, respectively.

(*c*) Assuming $\omega_B \ll \omega_p$, solve approximately for the various roots for the cases (i) $\mathbf{k}$ parallel to $\mathbf{b}$, (ii) $\mathbf{k}$ perpendicular to $\mathbf{b}$. Sketch your results for ω^2 versus k^2 in the two cases.

11

Special Theory of Relativity

The special theory of relativity has been treated extensively in many books. Its history is interwoven with the history of electromagnetism. In fact, one can say that the development of Maxwell's equations with the unification of electricity and magnetism and optics forced special relativity on us. Lorentz laid the groundwork in his studies of electrodynamics, while Einstein contributed crucial concepts and placed the theory on a consistent and general basis. Even though special relativity had its origin in electromagnetism and optics, it is now believed to apply to all types of interactions except, of course, large-scale gravitational phenomena. In modern physics the theory serves as a touchstone for possible forms for the interactions between elementary particles. Only theories consistent with special relativity need to be considered. This often severely limits the possibilities. Since the experimental basis and the development of the theory are described in detail in many places, we will content ourselves with a summary of the key points.

11.1 Historical Background and Key Experiments

In the forty years before 1900 electromagnetism and optics were correlated and explained in triumphal fashion by the wave theory based on Maxwell's equations. Since previous experience with wave motion had always involved a medium for the propagation of waves, it was natural for physicists to assume that light needed a medium through which to propagate. In view of the known facts about light it was necessary to assume that this medium, called the *ether*, permeated all space, was of negligible density, and had negligible interaction with matter. It existed solely as a vehicle for the propagation of electromagnetic waves. The hypothesis of

347

an ether set electromagnetic phenomena apart from the rest of physics, For a long time it had been known that the laws of mechanics were the same in different coordinate systems moving uniformly relative to one another—the laws of mechanics are invariant under Galilean transformations. The existence of an ether implied that the laws of electromagnetism were not invariant under Galilean coordinate transformations. There was a preferred coordinate system in which the ether was at rest. There the velocity of light in vacuum was equal to c. In other coordinate frames the velocity of light was presumably not c.

To avoid setting electromagnetism apart from the rest of physics by a failure of Galilean relativity there are several avenues open. Some of these are:

1. Assume that the velocity of light is equal to c with respect to a coordinate system in which the source is at rest.

2. Assume that the preferred reference frame for light is the coordinate system in which the medium through which the light is propagating is at rest.

3. Assume that, although the ether has a very small interaction with matter, the interaction is enough that it can be carried along with astronomical bodies such as the earth.

A large number of experiments led to the abandonment of all these hypotheses and the birth of the special theory of relativity. Three basic experiments are:

(1) Observation of the aberration of star positions during the year,

(2) Fizeau's experiment on the velocity of light in moving fluids (1859),

(3) Michelson-Morley experiment to detect motion through the ether (1887).

The aberration of star light (the small shift in apparent position of distant stars during the year) is an ancient phenomenon which finds a simple explanation in the motion of our earth in its orbit around the sun at a velocity of the order of 3×10^6 cm/sec. Suppose that the star light is incident normal to the earth's surface while the velocity of the earth in orbit is parallel to the surface. Figure 11.1 shows that a telescope must be inclined at an angle α, where

$$\alpha \simeq \frac{v}{c} \sim 10^{-4} \text{ radian} \tag{11.1}$$

in order that the light pass down it to the observer as the telescope moves along. Six months later the velocity vector $\mathbf{v}$ will be in the opposite direction. The star will then appear at an angle α on the other side of the vertical. The apparent positions of stars trace out elliptical paths on the celestial sphere during the year with angular spreads of the order of (11.1).

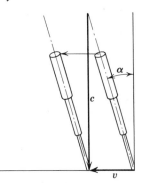

Fig. 11.1   Aberration of star positions.

This simple explanation of aberration contradicts the hypothesis that the velocity of light is determined by the transmitting medium (our atmosphere in this case) or that the ether is dragged along by the earth.   In neither case would aberration occur.

Fizeau's experiment involved measuring, by means of an interferometer, the velocity of light in liquids flowing in a pipe, both in the direction of and opposed to the propagation of the light.   If the index of refraction of the liquid is n, then depending on which of the various hypotheses one chooses, he expects the velocity to be

$$u = \frac{c}{n}, \qquad \frac{c}{n} \pm v \tag{11.2}$$

where v is the velocity of flow.   The actual result found by Fizeau was, within experimental error,

$$u = \frac{c}{n} \pm v\left(1 - \frac{1}{n^2}\right) \tag{11.3}$$

We note that this result can be made consistent with the ether being dragged along by the earth only by assuming that smaller bodies are partially successful in carrying the ether with them.   Even then the assumption is rather artificial in that their effectiveness at carrying the ether involves their indices of refraction.*

The Michelson-Morley experiment was designed to detect a motion of the earth relative to a preferred reference frame (the ether at rest) in which the velocity of light is c.   The basic apparatus is shown schematically in Fig. 11.2.   A laboratory light source S is focused on a thinly silvered glass plate P which divides the light into two beams at right angles to each other, one of which goes to mirror M_1 and is reflected back through the plate

---

* Actually formula (11.3) is a theoretical one proposed in 1818 by Fresnel on the basis of a model in which the density of the elastic ether in matter is proportional to n^2.

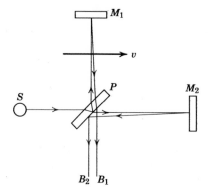

Fig. 11.2   Michelson-Morley experiment.

to B_1 and the other of which goes to mirror M_2, back to the plate, and is reflected to B_2. Conditions are such that the two beams travel almost the same path length. Small differences in path length or in the times taken to traverse the paths are detected by observing shifts in interference fringes produced by the two beams. The whole apparatus was attached to a stone slab floating in mercury so that it could be rotated. Suppose that velocity v of the earth through the ether is parallel to the light path from P to M_2. Then the velocity of light relative to the apparatus on the path from P to M_2 and return is $c \pm v$. If the path distance from P to M_2 is d_2, the time taken by the light to go from P to M_2 and return is

$$t_2 = d_2\left(\frac{1}{c-v} + \frac{1}{c+v}\right) = \frac{2d_2}{c}\frac{1}{\left(1 - \dfrac{v^2}{c^2}\right)} \qquad (11.4)$$

For the path from P to M_1 and return it is convenient to view things from the preferred coordinate frame. Then it is evident that the path length traversed is greater than d_1, the distance from P to M_1, because the mirror is moving with velocity v through the ether.

Figure 11.3 shows the geometrical relations. Evidently $\sin \alpha = v/c$, so that the effective path length is

$$2d_1 \sec \alpha = 2d_1 \frac{1}{\sqrt{1 - \dfrac{v^2}{c^2}}}$$

and the time taken is

$$t_1 = \frac{2d_1}{c}\frac{1}{\sqrt{1 - \dfrac{v^2}{c^2}}} \qquad (11.5)$$

The difference between the two transit times is

$$\Delta t = t_2 - t_1 = \frac{2}{c}\left[\frac{d_2}{1 - \dfrac{v^2}{c^2}} - \frac{d_1}{\sqrt{1 - \dfrac{v^2}{c^2}}}\right] \qquad (11.6)$$

If we assume that $v \ll c$, we can expand the denominators, obtaining

$$\Delta t \simeq \frac{2}{c}\left[(d_2 - d_1) + \frac{v^2}{c^2}\left(d_2 - \frac{d_1}{2}\right)\right] \qquad (11.7)$$

If the apparatus is now rotated through 90°, the transit times become

$$t_2' = \frac{2d_2}{c}\frac{1}{\sqrt{1 - \dfrac{v^2}{c^2}}}$$

$$\qquad (11.8)$$

$$t_1' = \frac{2d_1}{c}\frac{1}{1 - \dfrac{v^2}{c^2}}$$

and the difference, to lowest order, is

$$\Delta t' \simeq \frac{2}{c}\left[(d_2 - d_1) + \frac{v^2}{c^2}\left(\frac{d_2}{2} - d_1\right)\right] \qquad (11.7')$$

Since Δt and $\Delta t'$ are not the same, we expect a shift in the interference fringes upon rotation of the apparatus, the shift being proportional to

$$\Delta t' - \Delta t = -\frac{1}{c}(d_1 + d_2)\frac{v^2}{c^2} \qquad (11.9)$$

Since the orbital velocity of the earth is about 3×10^6 cm/sec, we expect $v^2/c^2 \sim 10^{-8}$, at least at some time of the year. With $(d_1 + d_2) \sim 3 \times 10^2$ cm, the time difference (11.9) is 10^{-16} sec. This means that the relevant length (to be compared to a wavelength of light) is $c\,|\Delta t' - \Delta t| \sim 3 \times 10^{-6}$ cm $= 300\,\text{Å}$. Since visible light has wavelengths of the order of

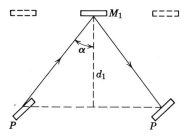

Fig. 11.3

3000 Å, the expected effect is a fringe shift of about one-tenth of a fringe. The accuracy of the Michelson-Morley experiment was such that a relative velocity of 10^6 cm/sec would have been seen (i.e., one-third of the above estimate). No fringe shift was found. Since the original work of Michelson the experiment has been repeated many times with modifications such as very unequal path lengths. No evidence for relative motion through the ether has been found. A summary of all the available evidence has been given by Shankland *et al.*, *Revs. Modern Phys.*, **27**, 167 (1955).

The negative result of the Michelson-Morley experiment can be explained on the ether-drag hypothesis. But that hypothesis is inconsistent with the aberration of star light. Only the so-called emission theories, where the velocity of light is constant relative to the source, are consistent with all three of the experiments cited. And we will see in the next section that other experiments exclude such theories. On the positive side the Michelson-Morley experiment can be thought of as restoring electromagnetism to the rest of physics in the matter of relativity. No observable effects were found which depended on the motion of the apparatus relative to some conjectured absolute reference frame. It should be mentioned, however, that FitzGerald and Lorentz (1892) explained the null result while still retaining the ether concept by postulating that all material objects are contracted in their direction of motion as they move through the ether. The rule of contraction is

$$L(v) = L_0 \sqrt{1 - \frac{v^2}{c^2}} \qquad (11.10)$$

It is clear from (11.4) or (11.7) that this hypothesis leads to a zero result for $(\Delta t' - \Delta t)$ in place of (11.9). The FitzGerald-Lorentz contraction hypothesis was perhaps the last gasp of the ether advocates, and it contains the germ of the special theory of relativity. The contraction phenomenon is present in special relativity, but in a more general way applying to all systems in relative motion with one another. Going along with it is the phenomenon of time dilatation (not postulated by FitzGerald or Lorentz), an experimentally well-founded effect. These are discussed in Section 11.3.

11.2   The Postulates of Special Relativity and the Lorentz Transformation

In 1904 Lorentz showed that Maxwell's equations in vacuum were invariant under a transformation of coordinates given by (11.19) below, and now called a *Lorentz transformation*, provided the field strengths were suitably transformed. By supposing that all matter was essentially electromagnetic in origin and so transformed in the same way as Maxwell's equations, Lorentz was able to deduce the contraction law (11.10). Then Poincaré showed that the transformation of charge and current densities could be made in such a way that all the equations of electrodynamics are invariant in form under Lorentz transformations. In 1905, almost at the same time as Poincaré and without knowledge of Lorentz's paper, Einstein formulated special relativity in a general and complete way, obtaining the results of Lorentz and Poincaré, but showing that the ideas were of much wider applicability. Instead of basing his discussion on electrodynamics, Einstein showed that just two postulates were necessary, one of them involving only a very general property of light.

The two postulates of Einstein were:

1. POSTULATE OF RELATIVITY

   The laws of nature and the results of all experiments performed in a given frame of reference are independent of the translational motion of the system as a whole. Thus there exists a triply infinite set of equivalent reference frames moving with constant velocities in rectilinear paths relative to one another in which all physical phenomena occur in an identical manner.

For brevity these equivalent coordinate systems are called *Galilean reference frames*. The postulate of relativity is consistent with all our experience with mechanics where only relative motion between bodies is relevant. It is also consistent with the Michelson-Morley experiment and makes meaningless the question of detecting motion relative to the ether.

2. POSTULATE OF THE CONSTANCY OF THE VELOCITY OF LIGHT

   The velocity of light is independent of the motion of the source.

This hypothesis, untested when Einstein proposed it, is necessary and decisive in obtaining the Lorentz transformation and all its consequences (see below). Because our classical concept of time as a variable independent of the spatial coordinates is destroyed by this postulate, its acceptance was resisted vehemently for a number of years. Many ingenious attempts were made to invent theories which would explain all the observed facts without this assumption. One notable one was Ritz's version of electrodynamics, which kept the two homogeneous Maxwell's equations intact but modified the equations involving the sources in such a way that the velocity of light was equal to c only when measured relative to the source. Experiments have proved all such theories wrong and have established the constancy of the velocity of light independent of the motion of the source. One such experiment is the Michelson-Morley interferometer experiment performed

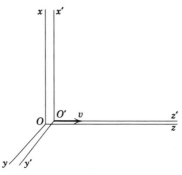

Fig. 11.4

with starlight, rather than a terrestrial light source. No effect was observed
which could be attributed to a change in the velocity of light due to the
relative motion of the star and the earth. Another experiment on the light
from rotating binary stars showed that the velocity of light depends
negligibly (if $c' = c + kv$, then $k < 0.002$) on the motion of the stars
toward or away from us.

The constancy of the velocity of light, independent of the motion of the
source, allows us to deduce the connection between space-time coordinates
in different Galilean reference frames. To see how this is possible we
consider two coordinate systems K and K'. System K' has its axes parallel
to those of K, but it is moving with a velocity v in the positive z direction
relative to the system K, as shown in Fig. 11.4. Points in space and time
in the two systems are specified by (x, y, z, t) and (x', y', z', t'), respectively.
For convenience we suppose that a common origin of time $t = t' = 0$ is
chosen at the instant when the two sets of coordinate axes exactly overlap.
Now we imagine an observer in each reference frame equipped with the
necessary apparatus (e.g., a network of correlated clocks and photocells
at known distances from the origin) to detect the arrival time of a light
signal from the origin at various points in space. If there is a light source
at rest in the system K (and so moving with velocity v in the negative z
direction in system K') which is flashed on and off rapidly at $t = t' = 0$,
then Einstein's second postulate implies that each observer will see his
photocell network respond to a spherical shell of radiation moving outward from his origin of coordinates with velocity c. Consequently the
arrival time t of the pulse at a detector located at (x, y, z) in system K will
satisfy the equation:

$$x^2 + y^2 + z^2 - c^2t^2 = 0 \qquad (11.11)$$

Similarly, in system K' the wave front is described by

$$x'^2 + y'^2 + z'^2 - c^2t'^2 = 0 \qquad (11.12)$$

Relations (11.11) and (11.12) seem to violate the first postulate of relativity. If two observers in different coordinate systems both see spherical pulses centered on a fixed origin in each system, the spheres must be different! This apparent contradiction is resolved when we allow the possibility that events which are simultaneous in one coordinate system are not necessarily simultaneous in another coordinate system moving relative to the first. We can now anticipate that time is no longer an absolute quantity independent of spatial variables and of relative motion.

To obtain a connection between the coordinates (x', y', z', t') of system K' and (x, y, z, t) of system K it is only necessary to assume that the transformation is linear. This seems very plausible and is equivalent to the assumption that space-time is homogeneous and isotropic. If the transformation is linear, the only possible connection between the quadratic forms (11.11) and (11.12) is

$$x'^2 + y'^2 + z'^2 - c^2 t'^2 = \lambda^2(x^2 + y^2 + z^2 - c^2 t^2) \qquad (11.13)$$

where $\lambda = \lambda(v)$ with $\lambda(0) = 1$. The presence of λ allows for the possibility of an overall scale change in going from K to K'. But shells of radiation are spheres in both systems. From the hypothesis that K' is moving parallel to the z axis of K, it is evident that the transformation of x', y' must be

$$x' = \lambda x, \qquad y' = \lambda y \qquad (11.14)$$

independent of the time, because motion parallel to the z axis in K' must remain so in K. Then the most general linear connection between z', t' and z, t is

$$z' = \lambda(a_1 z + a_2 t), \qquad t' = \lambda(b_1 t + b_2 z) \qquad (11.15)$$

A factor λ has been extracted for convenience. The coefficients a_1, a_2, b_1, b_2 are functions of v with the following limiting values as $v \to 0$:

$$\lim_{v \to 0} \begin{Bmatrix} a_1 \\ a_2 \\ b_1 \\ b_2 \end{Bmatrix} = \begin{Bmatrix} 1 \\ 0 \\ 1 \\ 0 \end{Bmatrix} \qquad (11.16)$$

The origin of K' moves with a velocity v in the system K. Consequently its position is specified by $z = vt$. This means that $a_2 = -va_1$ in (11.15). If equations (11.15) are now substituted into (11.13), three algebraic relations between $a_1, b_1,$ and b_2 are obtained. These are easily solved to give the following values, with signs chosen to agree with (11.16):

$$a_1 = b_1 = \frac{1}{\sqrt{1 - \dfrac{v^2}{c^2}}}$$

$$b_2 = -\frac{v}{c^2}\, a_1 \tag{11.17}$$

There remains the problem of the determination of $\lambda(v)$. If a third reference frame K'' is considered to be moving with a velocity $-v$ parallel to the z axis relative to K', the coordinates (x'', y'', z'', t'') can be obtained in terms of (x', y', z', t') from the above results merely by the change $v \to -v$. But the system K'' is just the original system K, so that $x'' = x$, $y'' = y$, $z'' = z$, $t'' = t$. This leads to the requirement that

$$\lambda(v)\,\lambda(-v) = 1 \tag{11.18}$$

But $\lambda(v)$ must be independent of the sign of v, since it represents a scale change in the transverse direction. Consequently $\lambda(v) = 1$. Then we can write down the *Lorentz transformation*, connecting coordinates in K' to those in K:

$$\left. \begin{aligned} z' &= \frac{z - vt}{\sqrt{1 - \dfrac{v^2}{c^2}}}, \qquad x' = x, \; y' = y \\[2em] t' &= \frac{t - \dfrac{v}{c^2}z}{\sqrt{1 - \dfrac{v^2}{c^2}}} \end{aligned} \right\} \tag{11.19}$$

Transformation (11.19) represents a special case in which the relative motion of K and K' is parallel to the z axis. It is a straight-forward matter to write down the result for an arbitrary velocity $\mathbf{v}$ of translation of K' relative to K, as shown in Fig. 11.5. Equation (11.19) clearly applies to parallel and perpendicular components of the coordinates relative to $\mathbf{v}$:

$$\left. \begin{aligned} \mathbf{x}_\parallel' &= \frac{1}{\sqrt{1 - \dfrac{v^2}{c^2}}}(\mathbf{x}_\parallel - \mathbf{v}t), \qquad \mathbf{x}_\perp' = \mathbf{x}_\perp \\[2em] t' &= \frac{1}{\sqrt{1 - \dfrac{v^2}{c^2}}}\left(t - \frac{\mathbf{v}\cdot\mathbf{x}}{c^2}\right) \end{aligned} \right\} \tag{11.20}$$

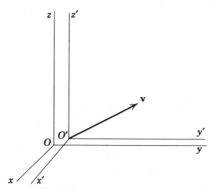

Fig. 11.5

With the definition that $\mathbf{x}_{\parallel} = [(\mathbf{v} \cdot \mathbf{x})\mathbf{v}]/v^2$ and $\mathbf{x}_{\perp} = \mathbf{x} - \mathbf{x}_{\parallel}$, equations (11.20) can be combined to yield the *general Lorentz transformation*:*

$$\left.\begin{array}{l} \mathbf{x}' = \mathbf{x} + \left(\dfrac{1}{\sqrt{1 - \dfrac{v^2}{c^2}}} - 1 \right) \dfrac{\mathbf{x} \cdot \mathbf{v}}{v^2} \mathbf{v} - \dfrac{1}{\sqrt{1 - \dfrac{v^2}{c^2}}} \mathbf{v}t \\[3em] t' = \dfrac{1}{\sqrt{1 - \dfrac{v^2}{c^2}}} \left(t - \dfrac{\mathbf{x} \cdot \mathbf{v}}{c^2} \right) \end{array}\right\} \qquad (11.21)$$

It should be noted that (11.21) represents a *single* Lorentz transformation to a reference frame K' moving with velocity $\mathbf{v}$ relative to the system K. Successive Lorentz transformations do not in general commute. It is easy to show that they commute only if the successive velocities are parallel. Consequently three successive transformations corresponding to the components of the velocity $\mathbf{v}$ in three mutually perpendicular directions yield different results, depending on the order in which the transformations are applied, and none agrees with (11.21). (See Problem 11.2.)

11.3 FitzGerald-Lorentz Contraction and Time Dilatation

As has already been mentioned, FitzGerald and Lorentz proposed the contraction rule (11.10) for the dimensions of an object parallel to its

* The word "general" is not really applicable to transformation (11.21). The connotation here is that the direction of the velocity $\mathbf{v}$ is arbitrary. But a more general transformation would allow the axes in K' to be rotated relative to those in K. Even this Lorentz transformation is not the most general, since it is still *homogeneous* in the coordinates. The general *inhomogeneous* Lorentz transformation allows translation of the origin in space-time as well. See Møller, Section 18.

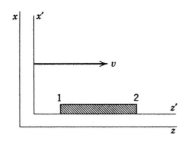

Fig. 11.6 Moving rod. FitzGerald-Lorentz contraction.

motion at velocity v through the ether; Lorentz was able to give the rule an electrodynamic basis from the properties of Maxwell's equations under Lorentz transformations. We now show that the contraction of lengths in the direction of motion is a more general phenomenon which applies to all relative motion. Consider a rod of length L_0 at rest parallel to the z' axis in the system K' of the previous section, as indicated schematically in Fig. 11.6. By definition $L_0 = z_2' - z_1'$, where z_1' and z_2' are the coordinates of the end points of the rod in K'. In the system K the length L of the rod is, again by definition, $L = z_2 - z_1$, where z_2 and z_1 are the instantaneous coordinates of the end points of the rod, observed at the same time t. From (11.19) the length in K' is

$$L_0 = z_2' - z_1' = \frac{z_2 - z_1}{\sqrt{1 - \dfrac{v^2}{c^2}}} = \frac{L}{\sqrt{1 - \dfrac{v^2}{c^2}}} \qquad (11.22)$$

which is just the FitzGerald-Lorentz result (11.10). Note that in the system K the length is defined at equal times t. The fact that this is not at equal times in the system K' is not relevant for the definition of length in the system K. This again illustrates that simultaneity is only a relative concept.

Another consequence of the special theory of relativity is time dilatation. A clock moving relative to an observer is found to run more slowly than one at rest relative to him. The most fundamental "clocks" which we have available are the unstable elementary particles. Each type of particle decays at rest with a well-defined lifetime (mean life) which is unaltered by external fields, apart from nuclear or atomic force fields which cause transformations that are well understood.* These particles can serve therefore as "clocks" which can be examined at rest and in motion. Suppose that we consider a meson of lifetime τ_0 at rest in the system K'

---

* For example, negative mu mesons can become bound in hydrogen-like orbits around nuclei with binding energies that are not negligible compared to the energy liberated in their decay. Since the rate of decay depends sensitively on the energy release (closely as the fifth power of ΔE), tightly bound negative mu mesons exhibit a considerably slower rate of decay than unbound ones.

which is moving with uniform velocity v relative to the system K. We assume that the meson is created at the origin of K' at time $t' = t = 0$. As seen from the system K the position of the meson is given by $z = vt$. If it lives a time τ_0 in K', then at its instant of decay, we find

$$t' = \tau_0 = \frac{t - \frac{v}{c^2} z}{\sqrt{1 - \frac{v^2}{c^2}}} = t \sqrt{1 - \frac{v^2}{c^2}} \tag{11.23}$$

The time t is the meson's lifetime τ as observed in the system K. Consequently

$$\tau = \frac{\tau_0}{\sqrt{1 - \frac{v^2}{c^2}}} \tag{11.24}$$

When viewed from K the moving meson lives longer than a meson at rest in K. The "clock" in motion is observed to run more slowly than an identical one at rest.

Time dilatation has been observed in cosmic rays with high-energy mu mesons. These mesons are produced as secondary particles at a height of the order of 10 or 20 km, and a large fraction of them reach the earth's surface. Since the mean lifetime of a mu meson is $\tau_0 = 2.2 \times 10^{-6}$ sec, it could travel no more than $c\tau_0 = 0.66$ km on the average before decaying if no time dilatation occurred. Clearly dilatation factors of the order of 10 or more are involved, consistent with the high energies (velocities approaching the velocity of light) of these particles.

A laboratory experiment exhibiting time dilatation with pi mesons is not difficult to perform. Charged pi mesons have a mean lifetime $\tau_0 = 2.56 \times 10^{-8}$ sec. An experiment studying the numbers of charged pi mesons decaying in flight per unit length as a function of distance from the point of production was done at Columbia University.* The mesons had a velocity $v \simeq 0.75c$. The numbers of mesons decaying per unit distance should follow an exponential law $N(x) = N_0 e^{-x/\lambda}$, where λ is the mean free path in the laboratory and x is the distance from the source (corrected for finite solid angles, etc.). Figure 11.7 shows schematically the results of the experiment. The mean free path is $\lambda = 8.5 \pm 0.6$ meters. Since $\lambda = v\tau$, the laboratory lifetime is $\tau = 3.8 \pm 0.3 \times 10^{-8}$ sec. Consequently

$$\frac{\tau}{\tau_0} = \frac{3.8 \pm 0.3}{2.56} = 1.5 \pm 0.1$$

* Durbin, Loar, and Havens, *Phys. Rev.*, **88**, 179 (1952). This experiment was performed to measure the lifetime of the pi meson. The validity of time dilatation was assumed. But with independent knowledge of the lifetime, the argument can be inverted as we do here.

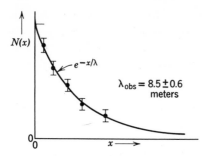

$\lambda_{obs} = 8.5 \pm 0.6$
meters

Fig. 11.7 Number of pi mesons decaying per unit distance as a function of distance from the point of production (schematic).

This value is to be compared with 1.51 calculated from (11.24) with $v = 0.75c$. The laboratory experiment on time dilatation is not so dramatic as the cosmic rays but has the great virtue of being performed under controlled conditions in a comparatively small space.

11.4 Addition of Velocities: Aberration and the Fizeau Experiment; Doppler Shift

The Lorentz transformation (11.19) can be used to obtain the addition law for velocities. Suppose that there is a velocity vector $\mathbf{u}'$ in the system K' which makes polar angles θ', ϕ' with the z' axis as shown in Fig. 11.8. The system K' is moving relative to the system K with a velocity $\mathbf{v}$ in the z direction. We want to know the components of velocity $\mathbf{u}$ as seen in the system K. From (11.19), or rather the inverse transformation, the differential expressions for dx, dy, dz, dt can be obtained:

$$dx = dx', \qquad dy = dy'$$

$$\left. dz = \frac{1}{\sqrt{1 - \dfrac{v^2}{c^2}}} (dz' + v \, dt'), \qquad dt = \frac{1}{\sqrt{1 - \dfrac{v^2}{c^2}}} \left(dt' + \frac{v}{c^2} dz' \right) \right\} \quad (11.25)$$

This means that the components of velocity are

$$\left. u_x = \frac{\sqrt{1 - \dfrac{v^2}{c^2}} \, u_x{}'}{\left(1 + \dfrac{v u_z{}'}{c^2}\right)}, \qquad u_y = \frac{\sqrt{1 - \dfrac{v^2}{c^2}} \, u_y{}'}{\left(1 + \dfrac{v u_z{}'}{c^2}\right)} \right\}$$

$$u_z = \frac{u_z{}' + v}{\left(1 + \dfrac{v u_z{}'}{c^2}\right)} \qquad\qquad\qquad (11.26)$$

The magnitude of **u** and the angles θ, ϕ of **u** in the system K are easily found. Since $u_x{}'/u_y{}' = u_x/u_y$, the azimuthal angles are equal, $\phi = \phi'$. Similarly

$$\tan \theta = \sqrt{1 - \frac{v^2}{c^2}} \; \frac{u' \sin \theta'}{u' \cos \theta' + v}$$

and

$$u = \frac{\sqrt{u'^2 + v^2 + 2u'v \cos \theta' - \left[\frac{u'v \sin \theta'}{c} \right]^2}}{\left(1 + \frac{u'v}{c^2} \cos \theta'\right)} \qquad (11.27)$$

The inverse results for (u', θ') in terms of (u, θ) can be obtained by interchanging $u \leftrightarrow u'$, $\theta \leftrightarrow \theta'$, and also changing the sign of v.

Study of (11.26) or (11.27) shows that for velocities u' and v both small compared to c the addition law is just that of Galilean relativity, $\mathbf{u} = \mathbf{u}' + \mathbf{v}$. But for either velocity approaching that of light, modifications appear. It is impossible to obtain a velocity greater than that of light by adding two velocities, even if each is very close to c. For the simple case of parallel velocities the addition law is

$$u = \frac{u' + v}{1 + \frac{u'v}{c^2}} \qquad (11.28)$$

If $u' = c$, then (11.28) shows that $u = c$ also. This is merely an explicit statement of Einstein's second postulate.

The laws of addition of velocities are in accord with both the aberration of starlight and the Fizeau experiment. For the aberration, the velocity u'

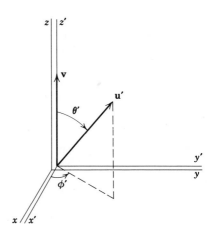

Fig. 11.8   Addition of velocities.

is that of light in the system K', namely, $u' = c$, and $\mathbf{v}$ is the orbital velocity of the earth. Then the angle θ is related to θ' by

$$\tan \theta = \sqrt{1 - \frac{v^2}{c^2}} \; \frac{\sin \theta'}{\cos \theta' + \dfrac{v}{c}} \tag{11.29}$$

For starlight incident normally on the earth $\theta' = \pi/2$. Then

$$\tan \theta = \frac{c}{v} \sqrt{1 - \frac{v^2}{c^2}} \tag{11.30}$$

The angle θ is the complement of the angle α in (11.1). Thus

$$\tan \alpha = \frac{v/c}{\sqrt{1 - \dfrac{v^2}{c^2}}} \tag{11.31}$$

completely consistent with observation. (Since $v/c \sim 10^{-4}$, the departure of the radical from unity is far beyond the realm of observation.)

In the simplest version of the Fizeau experiment the liquid flows with velocity v parallel or antiparallel to the path of light. If the liquid has an index of refraction n we may assume that light propagates with a velocity $u' = c/n$ relative to the liquid. From (11.28) the velocity of light observed in the laboratory is

$$u = \frac{\dfrac{c}{n} \pm v}{1 \pm \dfrac{v}{nc}} \simeq \frac{c}{n} \pm v\left(1 - \frac{1}{n^2}\right) \tag{11.32}$$

The latter expression is the expansion to lowest order of the exact result. This is in agreement with the Fresnel formula (11.3). Actually there is an added term in (11.32) if the index of refraction is a function of wavelength. It comes about because of the Doppler shift in wavelength in the moving liquid. The increase $\Delta \lambda$ in wavelength in the moving medium is

$$\Delta \lambda = \pm \lambda n \frac{v}{c} \tag{11.33}$$

correct to lowest order in v/c for the parallel and antiparallel velocities, respectively. Consequently the appropriate velocity of light in the liquid is

$$\frac{c}{n(\lambda + \Delta\lambda)} \simeq \frac{c}{n(\lambda)} - \frac{c}{n^2} \frac{dn}{d\lambda} \Delta\lambda$$

Then the corrected expression to replace (11.32) is

$$u = \frac{c}{n} \pm v\left(1 - \frac{1}{n^2} - \frac{\lambda}{n}\frac{dn}{d\lambda}\right) \tag{11.34}$$

The correction due to dispersive effects has been observed.

The relativistic Doppler shift formulas can be obtained from the fact that the phase of a light wave is an invariant quantity. Actually, the phase of any plane wave is invariant under a Lorentz transformation, the reason being that the phase can be associated with mere counting which is independent of coordinate frame. Consider a plane wave of frequency ω and wave vector $\mathbf{k}$ in the reference frame K. An observer at the point P with coordinate $\mathbf{x}$ is equipped to record the number of wave crests which reach him in a certain time. If the wave crest passing the origin at $t = 0$ is the first one which he records (when it reaches him), then at time t he will have counted

$$\frac{1}{2\pi}(\mathbf{k}\cdot\mathbf{x} - \omega t)$$

wave crests. Now imagine another reference frame K' which moves relative to the frame K with a velocity $\mathbf{v}$ parallel to the z axis, and has its origin coincident with that of K at $t = 0$. An observer in K' at the point P' with coordinate $\mathbf{x}'$ is equipped similarly to the one in K. He begins counting when the wave crest passing the origin reaches him, and continues counting until time t'. If the point P' is such that at the end of the counting period it coincides with the point P, then both observers must have counted the same number of wave crests. But the observer in K' has counted

$$\frac{1}{2\pi}(\mathbf{k}'\cdot\mathbf{x}' - \omega't')$$

wave crests, where $\mathbf{k}'$ and ω' are the wave vector and frequency of the plane wave in K'. Thus the phase of the wave is invariant. Consequently we have

$$\mathbf{k}'\cdot\mathbf{x}' - \omega't' = \mathbf{k}\cdot\mathbf{x} - \omega t \tag{11.35}$$

Using the transformation formulas (11.19), we find

$$\left.\begin{aligned}
&k_x' = k_x, \qquad k_y' = k_y \\
&k_z' = \frac{1}{\sqrt{1 - \dfrac{v^2}{c^2}}}\left(k_z - \frac{v}{c^2}\omega\right), \qquad \omega' = \frac{1}{\sqrt{1 - \dfrac{v^2}{c^2}}}(\omega - vk_z)
\end{aligned}\right\} \tag{11.36}$$

For light waves, $|\mathbf{k}| = \omega/c$ and $|\mathbf{k}'| = \omega'/c$. Hence these results can be expressed in the form:

$$\left.\begin{aligned}
\omega' &= \frac{\omega}{\sqrt{1 - \dfrac{v^2}{c^2}}}\left(1 - \frac{v}{c}\cos\theta\right) \\[2em]
\tan\theta' &= \frac{\sqrt{1 - \dfrac{v^2}{c^2}}\sin\theta}{\cos\theta - \dfrac{v}{c}}
\end{aligned}\right\} \tag{11.37}$$

where θ and θ' are the angles of $\mathbf{k}$ and $\mathbf{k}'$ relative to the direction of $\mathbf{v}$. This last equation is just the inverse of (11.29).

It is sometimes useful to have the frequency ω' expressed in terms of the angle θ' of the wave in the frame K'. From the inverse of the first equation in (11.37), it is evident that the desired expression is

$$\omega' = \frac{\sqrt{1 - \dfrac{v^2}{c^2}}\,\omega}{\left(1 + \dfrac{v}{c}\cos\theta'\right)} \tag{11.38}$$

The first equation in (11.37) is the customary Doppler shift, modified by the radical in the denominator. The presence of the square root shows that there is a transverse Doppler shift, even when $\theta = \pi/2$. This relativistic transverse Doppler shift has been observed spectroscopically with atoms in motion (Ives-Stilwell experiment, 1938). It also has been observed using a precise resonance-absorption technique, with a nuclear gamma-ray source on the axis of a rapidly rotating cylinder and the absorber attached to the circumference of the cylinder.

11.5 Thomas Precession

In 1926, Uhlenbeck and Goudsmit introduced the idea of electron spin and showed that, if the electron had a g factor of 2, the anomalous Zeeman effect could be explained, as well as the existence of multiplet splittings. There was a difficulty, however, in that the observed fine-structure intervals were only one-half the theoretically expected values. If a g factor of unity were chosen, the fine-structure intervals were given correctly, but the Zeeman effect was then the normal one. The complete explanation of spin, including correctly the g factor and the proper fine-structure interaction, came only with the relativistic electron theory of Dirac. But within the framework of an empirical spin angular momentum and a g factor of 2, Thomas showed that the origin of the discrepancy was a relativistic kinematic effect which, when included properly, gave both the anomalous Zeeman effect and the correct fine-structure splittings. The *Thomas precession*, as it is called, also gives a qualitative explanation for a spin-orbit interaction in atomic nuclei and shows why the doublets are "inverted" in nuclei.

The Uhlenbeck-Goudsmit hypothesis was that an electron possessed a spin angular momentum **S** (which could take on quantized values of $\pm\hbar/2$ along any axis) and a magnetic moment μ related to **S** by

$$\mu = \frac{e}{mc}\mathbf{S} \tag{11.39}$$

The customary relation between magnetic moment and angular momentum is given by (5.64). Equation (11.39) shows that the electron has a g factor of 2. Suppose that an electron moves with a velocity **v** in external fields **E** and **B**. Then the equation of motion for its angular momentum in its rest frame is

$$\frac{d\mathbf{S}}{dt} = \mu \times \mathbf{B}' \tag{11.40}$$

where **B**′ is the magnetic induction in that frame. We will show in Section 11.10 that in a coordinate system moving with the electron the magnetic induction is

$$\mathbf{B}' \simeq \left(\mathbf{B} - \frac{\mathbf{v}}{c} \times \mathbf{E}\right) \tag{11.41}$$

where we have neglected terms of the order of (v^2/c^2). Then (11.40) becomes

$$\frac{d\mathbf{S}}{dt} = \mu \times \left(\mathbf{B} - \frac{\mathbf{v}}{c} \times \mathbf{E}\right) \tag{11.42}$$

Equation (11.42) is equivalent to an energy of interaction of the electron spin:

$$U' = -\mu \cdot \left(\mathbf{B} - \frac{\mathbf{v}}{c} \times \mathbf{E}\right) \tag{11.43}$$

In an atom the electric force $e\mathbf{E}$ can be approximated as the negative gradient of a spherically symmetric average potential energy $V(r)$. For one-electron atoms this is, of course, exact. Thus

$$e\mathbf{E} = -\frac{\mathbf{r}}{r}\frac{dV}{dr} \tag{11.44}$$

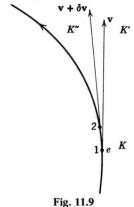

Fig. 11.9

Then the spin-interaction energy can be written

$$U' = - \frac{e}{mc} \mathbf{S} \cdot \mathbf{B} + \frac{1}{m^2 c^2} (\mathbf{S} \cdot \mathbf{L}) \frac{1}{r} \frac{dV}{dr} \qquad (11.45)$$

where $\mathbf{L} = m(\mathbf{r} \times \mathbf{v})$ is the electron's orbital angular momentum. This interaction energy gives the anomalous Zeeman effect correctly, but has a spin-orbit interaction which is twice too large.

The error in (11.45) can be traced to the incorrectness of (11.40) as an equation of motion for the electron spin. The left-hand side of (11.40) gives the rate of change of spin in the rest frame of the electron. This is equal to the applied torque ($\boldsymbol{\mu} \times \mathbf{B}'$) only if the electron's rest frame is not a rotating coordinate system. If, as Thomas first pointed out, that co-ordinate system rotates, then the time rate of change of any vector $\mathbf{G}$ in that system is*

$$\frac{d\mathbf{G}}{dt} = \left(\frac{d\mathbf{G}}{dt} \right)_{\text{nonrot}} - \boldsymbol{\omega}_T \times \mathbf{G} \qquad (11.46)$$

where $\boldsymbol{\omega}_T$ is the angular velocity of rotation found by Thomas. When applied to the electron spin, (11.46) gives an equation of motion:

$$\frac{d\mathbf{S}}{dt} = \mathbf{S} \times \left(\frac{e\mathbf{B}'}{mc} + \boldsymbol{\omega}_T \right) \qquad (11.47)$$

The corresponding energy of interaction is

$$U = U' - \mathbf{S} \cdot \boldsymbol{\omega}_T \qquad (11.48)$$

where U' is the electromagnetic spin interaction (11.45).

The origin of the Thomas precessional frequency $\boldsymbol{\omega}_T$ is the acceleration experienced by the electron as it moves in its atomic orbit. Figure 11.9 shows the electron at position 1 at a time t with instantaneous velocity

---

* See, for example, Goldstein, p. 133.

vector **v**, and at position 2 an infinitesimal time later $(t + \delta t)$ with velocity $\mathbf{v} + \delta\mathbf{v}$. The increment in velocity is related to the electron's acceleration **a** by $\delta\mathbf{v} = \mathbf{a}\,\delta t$. At time t the electron's rest frame K' and the laboratory frame K are related by a Lorentz transformation with velocity **v**. At time $t + \delta t$ the electron's rest frame has now changed to K'', related to K by a Lorentz transformation with velocity $\mathbf{v} + \delta\mathbf{v}$. The question now arises, "How are the coordinate frames K'' and K' related? That is, how do the axes in the electron's rest frame behave in time?" As viewed from the laboratory, in a time δt the electron undergoes an infinitesimal change in velocity $\delta\mathbf{v}$. Consequently we might anticipate that K'' and K' would be connected by a simple infinitesimal Lorentz transformation. If so, (11.45) would be correct as it stands. To see that the connection is more than a mere Lorentz transformation we note that the transformation from K' to K'' is equivalent to two successive Lorentz transformations, one with velocity $-\mathbf{v}$, and the other with velocity $\mathbf{v} + \delta\mathbf{v}$:

$$K' \boxed{-\mathbf{v}} \to K \boxed{\mathbf{v} + \delta\mathbf{v}} \to K'' \tag{11.49}$$

Now it is generally true that two successive Lorentz transformations are equivalent to a single Lorentz transformation plus a rotation. Using the general formula (11.21) twice, it is a straightforward matter to show that the time variables in K'' and K' are related by

$$t'' = t' - \frac{\mathbf{x}'}{c^2} \cdot \frac{1}{\sqrt{1 - \frac{v^2}{c^2}}}\left[\delta\mathbf{v} + \left(\frac{1}{\sqrt{1 - \frac{v^2}{c^2}}} - 1\right)\frac{\mathbf{v}\cdot\delta\mathbf{v}}{v^2}\mathbf{v}\right] \tag{11.50}$$

correct to first order in $\delta\mathbf{v}$. This shows that the direct transformation from K' to K'' involves an infinitesimal Lorentz transformation with a velocity

$$\Delta\mathbf{v} = \frac{1}{\sqrt{1 - \frac{v^2}{c^2}}}\left[\delta\mathbf{v} + \left(\frac{1}{\sqrt{1 - \frac{v^2}{c^2}}} - 1\right)\frac{\mathbf{v}\cdot\delta\mathbf{v}}{v^2}\mathbf{v}\right] \tag{11.51}$$

The corresponding transformation of the coordinates is

$$\mathbf{x}'' = \mathbf{x}' + \left(\frac{1}{\sqrt{1 - \frac{v^2}{c^2}}} - 1\right)\mathbf{x}' \times \left(\frac{\mathbf{v}\times\delta\mathbf{v}}{v^2}\right) - \Delta\mathbf{v}\,t' \tag{11.52}$$

Comparison with $\mathbf{x}'' = \mathbf{x}' + \mathbf{x}' \times \Delta\mathbf{\Omega}$ for a rotation of axes by an infinitesimal angle $\Delta\mathbf{\Omega}$ shows that the coordinate axes in K'' are rotated relative to those in K' by an angle

$$\Delta\mathbf{\Omega} = \left(\frac{1}{\sqrt{1 - \frac{v^2}{c^2}}} - 1\right)\frac{\mathbf{v}\times\delta\mathbf{v}}{v^2} \tag{11.53}$$

This shows that the coordinate axes in the electron's rest frame precess with an angular velocity

$$\boldsymbol{\omega}_T = \left(\frac{1}{\sqrt{1 - \dfrac{v^2}{c^2}}} - 1\right)\frac{\mathbf{v} \times \mathbf{a}}{v^2} \simeq \frac{1}{2c^2}\,\mathbf{v} \times \mathbf{a} \qquad (11.54)$$

where the result on the right is an approximation valid if $v \ll c$. We emphasize the purely kinematic origin of the Thomas precession by noting that nothing has been said about the cause of the acceleration. If a component of acceleration exists perpendicular to $\mathbf{v}$, then there is a Thomas precession, independent of other effects such as precession of the magnetic moment in a magnetic field.

For electrons in atoms the acceleration is caused by the screened Coulomb field (11.44). Thus the Thomas angular velocity is

$$\boldsymbol{\omega}_T \simeq \frac{1}{2c^2}\frac{\mathbf{r} \times \mathbf{v}}{m}\frac{1}{r}\frac{dV}{dr} = \frac{1}{2m^2c^2}\mathbf{L}\frac{1}{r}\frac{dV}{dr} \qquad (11.55)$$

It is evident from (11.48) and (11.45) that the extra contribution to the energy from the Thomas precession just reduces the spin-orbit coupling by a factor of $\frac{1}{2}$ (sometimes called the *Thomas factor*), yielding

$$U = -\frac{e}{mc}\,\mathbf{S} \cdot \mathbf{B} + \frac{1}{2m^2c^2}\,\mathbf{S} \cdot \mathbf{L}\frac{1}{r}\frac{dV}{dr} \qquad (11.56)$$

as the correct spin-orbit interaction energy for an atomic electron.

In atomic nuclei the nucleons experience strong accelerations due to the specifically nuclear forces. The electromagnetic forces are comparatively weak. In an approximate way one can treat the nucleons as moving separately in a short-range, spherically symmetric, attractive, potential well, $V_N(r)$. Then each nucleon will experience in addition a spin-orbit interaction given by (11.48) with the negligible electromagnetic contribution U' omitted:

$$U_N \simeq -\mathbf{S} \cdot \boldsymbol{\omega}_T \qquad (11.57)$$

where the acceleration in $\boldsymbol{\omega}_T$ is determined by $V_N(r)$. The form of $\boldsymbol{\omega}_T$ is the same as (11.55) with V replaced by V_N. Thus the nuclear spin-orbit interaction is approximately

$$U_N \simeq -\frac{1}{2M^2c^2}\,\mathbf{S} \cdot \mathbf{L}\frac{1}{r}\frac{dV_N}{dr} \qquad (11.58)$$

In comparing (11.58) with atomic formula (11.56) we note that both V and V_N are attractive (although V_N is much larger), so that the signs of the spin-orbit energies are opposite. This means that in nuclei the single particle levels form "inverted" doublets. With a reasonable form for V_N,

(11.58) is in qualitative agreement with the observed spin-orbit splittings in nuclei.

11.6 Proper Time and the Light Cone

In the previous sections we have explored some of the physical consequences of the special theory of relativity and Lorentz transformations. In the next two sections we want now to discuss some of the more formal aspects and to introduce some notation and concepts which are very useful in a systematic discussion of physical theories within the framework of special relativity.

In Galilean relativity space and time coordinates are unconnected. Consequently under Galilean transformations the infinitesimal elements of distance and time are separately invariant. Thus

$$\left. \begin{aligned} ds^2 &= dx^2 + dy^2 + dz^2 = ds'^2 \\ dt^2 &= dt'^2 \end{aligned} \right\} \tag{11.59}$$

For Lorentz transformations, on the other hand, the time and space coordinates are interrelated. From (11.21) it is easy to show that the invariant "length" element is

$$ds^2 = dx^2 + dy^2 + dz^2 - c^2\, dt^2 \tag{11.60}$$

This leads immediately to the concept of a Lorentz invariant *proper time*. Consider a system, which for definiteness we will think of as a particle, moving with an instantaneous velocity $\mathbf{v}(t)$ relative to some coordinate system K. In the coordinate system K' where the particle is instantaneously at rest the space-time increments are $dx' = dy' = dz' = 0$, $dt' = d\tau$. Then the invariant length (11.60) is

$$-c^2\, d\tau^2 = dx^2 + dy^2 + dz^2 - c^2\, dt^2 \tag{11.61}$$

In terms of the particle velocity $\mathbf{v}(t)$ this can be written

$$d\tau = dt\sqrt{1 - \frac{v^2}{c^2}} \tag{11.62}$$

Equation (11.62) shows the time-dilatation effect already discussed. But much more important, by the manner of its derivation (11.62) shows that the time τ, called the *proper time of the particle*, is a Lorentz invariant quantity. This is of considerable importance later on when we wish to discuss various quantities and their time derivatives. If a quantity behaves in a certain way under Lorentz transformations, then its proper time

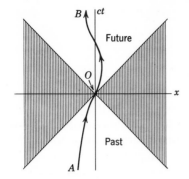

Fig. 11.10 World line of a system and the light cone. The unshaded interior of the cone represents the past and the future, while the shaded region outside the cone is called "elsewhere." A point inside (outside) the light cone is said to have a time-like (space-like) separation from the origin.

derivative will behave in the same way because of the invariance of $d\tau$. But its ordinary time derivative will not have the same transformation properties. From (11.62) we see that a certain proper time interval $(\tau_2 - \tau_1)$ will be seen in the system K as a time interval

$$t_2 - t_1 = \int_{\tau_1}^{\tau_2} \frac{d\tau}{\sqrt{1 - \dfrac{v^2(\tau)}{c^2}}} \qquad (11.63)$$

where t_1 and t_2 are the corresponding times in K.

Another fruitful concept in special relativity is the idea of the light cone and "space-like" and "time-like" separations between two events. Consider Fig. 11.10, in which the time axis (actually ct) is vertical and the space axes are perpendicular to it. For simplicity only one space dimension is shown. At $t = 0$ a physical system, say a particle, is at the origin. Because the velocity of light is an upper bound on all velocities, the space-time domain can be divided into three regions by a "cone," called the *light cone*, whose surface is specified by $x^2 + y^2 + z^2 = c^2 t^2$. Light signals emitted at $t = 0$ from the origin would travel out the $45°$ lines in the figure. But any material system has a velocity less than c. Consequently as time goes on it would trace out a path, called its *world line*, inside the upper half-cone, e.g., the curve OB. Since the path of the system lies inside the *upper half-cone* for times $t > 0$, that region is called the *future*. Similarly the *lower half-cone* is called the *past*. The system may have reached O by a path such as AO lying inside the lower half-cone. The shaded region outside the light cone is called *elsewhere*. A system at O can never reach or come from a point in space-time in elsewhere.

The division of space-time into the past-future region and the elsewhere region can be emphasized by considering the invariant separation between two events $P_1(x_1, y_1, z_1, t_1)$ and $P_2(x_2, y_2, z_2, t_2)$ in space-time:

$$s_{12}^2 = (x_1 - x_2)^2 + (y_1 - y_2)^2 + (z_1 - z_2)^2 - c^2(t_1 - t_2)^2 \quad (11.64)$$

For any two events P_1 and P_2 there are two possibilities: (1) $s_{12}^2 > 0$, (2) $s_{12}^2 < 0$. If $s_{12}^2 > 0$, the events are said to have a space-like separation, because it is always possible to find a Lorentz transformation to a new coordinate system K' where $(t_1' - t_2') = 0$ and

$$s_{12}^2 = (x_1' - x_2')^2 + (y_1' - y_2')^2 + (z_1' - z_2')^2 > 0 \qquad (11.65)$$

That is, the two events are at different space points at the same instant of time. In terms of Fig. 11.10, one of the events is at the origin and the other lies in elsewhere. If $s_{12}^2 < 0$, the events are said to have a time-like separation. Then a Lorentz transformation can be found which will make $x_1' = x_2'$, $y_1' = y_2'$, $z_1' = z_2'$, and

$$s_{12}^2 = -c^2(t_1' - t_2')^2 < 0 \qquad (11.66)$$

In the coordinate system K' the two events are at the same space point, but are separated in time. In Fig. 11.10, one point is at the origin and the other is in the past or future.

The division of the separation of two events in space-time into two classes—space-like separations or time-like separations—is a Lorentz invariant one. Two events with a space-like separation in one coordinate system have a space-like separation in all coordinate systems. This means that two such events cannot be causally connected. Since physical interactions propagate from one point to another with velocities no greater than that of light, only events with time-like separations can be causally related. An event at the origin in Fig. 11.10 can be influenced causally only by the events which occur in the past region of the light cone.

11.7 Lorentz Transformations as Orthogonal Transformations in Four Dimensions

The Lorentz transformation (11.19) and the more general form (11.21) are linear relations between the space-time coordinates (x, y, z, t) and (x', y', z', t'), subject to the constraint,

$$x^2 + y^2 + z^2 - c^2t^2 = x'^2 + y'^2 + z'^2 - c^2t'^2 \qquad (11.67)$$

This constraint is very reminiscent of the constraint involved in the rotation of coordinate axes in three space dimensions. In fact, if we introduce the four space-time coordinates,

$$x_1 = x, \quad x_2 = y, \quad x_3 = z, \quad x_4 = ict \qquad (11.68)$$

then the constraint becomes

$$R^2 = x_1^2 + x_2^2 + x_3^2 + x_4^2 \qquad (11.69)$$

is an invariant under Lorentz transformations. This is then exactly the requirement that Lorentz transformations are rotations in a four-dimensional Euclidean space or, more correctly, are orthogonal transformations in four dimensions. The Lorentz transformation (11.21) can be written in the general form:

$$x_\mu' = \sum_{v=1}^{4} a_{\mu v} x_v, \qquad \mu = 1, 2, 3, 4 \tag{11.70}$$

where the coefficients $a_{\mu v}$ are constants characteristic of the particular transformation. The invariance of R^2 (11.69) forces the transformation coefficients $a_{\mu v}$ to satisfy the orthogonality condition:

$$\sum_{\mu=1}^{4} a_{\mu v} a_{\mu \lambda} = \delta_{v\lambda} \tag{11.71}$$

With (11.71) it is easy to show that the inverse transformation is

$$x_\mu = \sum_{v=1}^{4} x_v' a_{v\mu} \tag{11.72}$$

and that

$$\sum_{\mu=1}^{4} a_{v\mu} a_{\lambda \mu} = \delta_{v\lambda} \tag{11.73}$$

Furthermore, if we solve the four equations (11.70) for x_μ in terms of x_v' and compare the solution to (11.72), we find that the determinant of the coefficients has the value unity:

$$\det |a_{\mu v}| = 1 \tag{11.74}$$

In general the determinant can be ± 1, but the choice of the minus sign implies an inversion followed by a rotation.

To give some substance to the above formalities we exhibit explicitly the transformation coefficients $a_{\mu v}$ for a Lorentz transformation from system K to a system K' moving with a velocity $\mathbf{v}$ parallel to the z axis

$$(a_{\mu v}) = \begin{pmatrix} 1 & 0 & 0 & 0 \\ 0 & 1 & 0 & 0 \\ 0 & 0 & \gamma & i\gamma\beta \\ 0 & 0 & -i\gamma\beta & \gamma \end{pmatrix} \tag{11.75}$$

We have introduced the convenient abbreviations:

$$\left. \begin{aligned} \beta &= \frac{v}{c} \\ \gamma &= \frac{1}{\sqrt{1 - \dfrac{v^2}{c^2}}} \end{aligned} \right\} \tag{11.76}$$

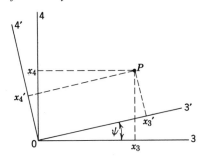

Fig. 11.11 Lorentz transformation as rotation of axes.

With definitions (11.68) and (11.70) it is elementary to show that (11.75) yields exactly the Lorentz transformation (11.19).

The formal representation of transformation (11.75) as a rotation of axes in the x_3, x_4 plane (with x_4 drawn as if it were real) can be accomplished simply. Figure 11.11 shows a rotation of the axes through an angle ψ. The coordinates of the point P relative to the two sets of axes are related by

$$\left.\begin{array}{l} x_3{}' = \cos\psi\,x_3 + \sin\psi\,x_4 \\[4pt] x_4{}' = -\sin\psi\,x_3 + \cos\psi\,x_4 \end{array}\right\} \tag{11.77}$$

Comparison of the coefficients in (11.77) with the transformation coefficients in (11.75) shows that the angle ψ is a complex angle whose tangent is

$$\tan\psi = i\beta \tag{11.78}$$

This result can be obtained directly from (11.77) without reference to (11.75) by noting that the origin $x_3{}' = 0$ moves with a velocity v in the system K. That the angle ψ is complex is emphasized by the fact that its cosine is greater than unity ($\cos\psi = \gamma \geq 1$). Consequently the graphical representation of a Lorentz transformation as a rotation is merely a formal device.

In spite of the formal nature of the x_3, x_4 rotation diagram the phenomena of FitzGerald-Lorentz contraction and time dilatation can be displayed graphically. Figure 11.12 shows the length contraction on the right and time dilatation on the left. The distance L_0 in the frame K' is observed in the frame K as L, represented by the horizontal line at constant time in K. Because of the complex nature of the angle ψ, L appears on the figure as larger than L_0, but mathematically the two lengths are related by

or

$$\left.\begin{array}{l} L\cos\psi = L_0 \\[6pt] L = \dfrac{L_0}{\gamma} \end{array}\right\} \tag{11.79}$$

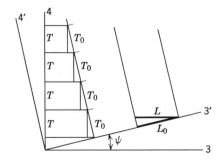

Fig. 11.12 Time dilatation and Fitz-Gerald-Lorentz contraction in terms of a rotation of space-time axes.

in agreement with (11.22). Similarly the time intervals T_0 in the frame K' are seen in the frame K as intervals T, where

$$T = T_0 \cos \psi = \gamma T_0 \qquad (11.80)$$

in accord with (11.24).

Sometimes a graphical display of Lorentz transformations is made using a real time variable $x_0 = ct$, rather than x_4. This is called a *Minkowski diagram* and has the virtue of dealing with real quantities. It has the major disadvantage that the coordinate grids in the two frames K and K' must be scaled according to a rectangular hyperboloid law, as can be seen from (11.67). The interested reader may refer to Minkowski's paper in the collection, *The Principle of Relativity*, by Einstein *et al.*, for a discussion of these diagrams.

11.8 4-Vectors and Tensors; Covariance of the Equations of Physics

The transformation law (11.70) for the coordinates x_μ defines the transformation properties of vectors in the four-dimensional space-time (11.68). Any set of four quantities A_μ which transform in the same way as x_μ is called a 4-*vector*. Under the Lorentz transformation $(a_{\mu\nu})$ A_μ is transformed into $A_\mu{}'$, where

$$A_\mu{}' = \sum_{\nu=1}^{4} a_{\mu\nu} A_\nu \qquad (11.81)$$

If a quantity ϕ is unchanged under a Lorentz transformation, it is called a *scalar* or a *Lorentz scalar*. The four quantities formed by differentiation of a Lorentz scalar with respect to x_μ transform as a 4-vector. This can be shown as follows. Consider

$$\frac{\partial \phi}{\partial x_\mu{}'} = \sum_{\nu=1}^{4} \frac{\partial \phi}{\partial x_\nu} \frac{\partial x_\nu}{\partial x_\mu{}'} \qquad (11.82)$$

From (11.72) it is evident that

$$\frac{\partial x_\nu}{\partial x_\mu{}'} = a_{\mu\nu} \tag{11.83}$$

Consequently

$$\frac{\partial \phi}{\partial x_\mu{}'} = \sum_{\nu=1}^{4} a_{\mu\nu} \frac{\partial \phi}{\partial x_\nu} \tag{11.84}$$

as required for the transformation of a 4-vector. By similar means it is elementary to show that the 4-divergence of a 4-vector is Lorentz invariant:

$$\sum_{\nu=1}^{4} \frac{\partial A_\nu{}'}{\partial x_\nu{}'} = \sum_{\mu=1}^{4} \frac{\partial A_\mu}{\partial x_\mu} \tag{11.85}$$

With $A_\mu = \partial \phi / \partial x_\mu$ in this expression, we find that the four-dimensional Laplacian operator is a Lorentz invariant operator:

$$\Box'^2 \phi \equiv \sum_{\nu=1}^{4} \frac{\partial^2 \phi}{\partial x_\nu{}'^2} = \sum_{\mu=1}^{4} \frac{\partial^2 \phi}{\partial x_\mu{}^2} \equiv \Box^2 \phi \tag{11.86}$$

If $\Box^2$ operates on some other function, such as a 4-vector A_μ, the resulting quantity retains the transformation properties of the function operated on. The scalar product of two 4-vectors A_μ and B_μ is readily proved to be invariant:

$$(A' \cdot B') \equiv \sum_{\mu=1}^{4} A_\mu{}' B_\mu{}' = (A \cdot B) \tag{11.87}$$

Lorentz 4-vectors are tensors of the first rank in a four-dimensional space. Higher-ranks tensors are defined in an analogous way. A second-rank tensor $T_{\mu\nu}$ is a set of sixteen quantities which transforms according to the law:

$$T'_{\mu\nu} = \sum_{\lambda,\sigma=1}^{4} a_{\mu\lambda} a_{\nu\sigma} T_{\lambda\sigma} \tag{11.88}$$

Higher-rank tensors are formed by the inclusion of more and more factors $a_{\mu\nu}$. A tensor of the nth rank is a set of 4^n quantities which have a transformation law involving a product of n coefficients $a_{\mu\nu}$, in obvious generalization of (11.88). Just as the scalar product of two 4-vectors has rank one less than the original quantities, so certain contracted quantities can be formed from higher-rank tensors. For example, the scalar product of a tensor of the second rank and a 4-vector transforms as a 4-vector:

$$B_\mu{}' = \sum_{\nu=1}^{4} T'_{\mu\nu} A'_\nu = \sum_{\lambda=1}^{4} a_{\mu\lambda} \left(\sum_{\nu=1}^{4} T_{\lambda\nu} A_\nu \right) \tag{11.89}$$

This and similar relations can be proved using the orthogonality conditions (11.71) and (11.73).

The volume element in the four-dimensional space-time (11.68) will be defined as the real quantity

$$d^4x \equiv dx_1\, dx_2\, dx_3\, dx_0 \tag{11.90}$$

where $dx_0 = (1/i)\, dx_4 = d(ct)$. The transformation law of the volume element is

$$d^4x' = \frac{\partial(x_1', x_2', x_3', x_4')}{\partial(x_1, x_2, x_3, x_4)}\, d^4x \tag{11.91}$$

But the Jacobian in (11.91) is just the determinant of the $a_{\mu\nu}$ (11.74). Consequently the 4-volume element d^4x is a Lorentz invariant quantity.

The first principle of Einstein is that the laws of physics have the same form in different Lorentz frames. This means that the equations which we write down to describe the physical laws must be *covariant* in form. By covariant we mean that the equation can be written so that both sides have the same, well-defined, transformation properties under Lorentz transformations. Thus physical equations must be relations between 4-vectors, or Lorentz scalars, or in general 4-tensors of the same rank. This is necessary in order that a relation valid in one coordinate frame will also hold in the same form in another. Consider, for example, the inhomogeneous pair of Maxwell's equations. It will be shown in the next section that these can be written in the relativistic form

$$\sum_{\nu=1}^{4} \frac{\partial F_{\mu\nu}}{\partial x_\nu} = \frac{4\pi}{c} J_\mu, \qquad \mu = 1, 2, 3, 4 \tag{11.92}$$

where J_μ is a suitable current 4-vector, and $F_{\mu\nu}$ is the field strength 4-tensor. Since the 4-divergence of a 4-tensor is a 4-vector, (11.92) is a relation between two 4-vectors. In another reference frame K', we expect the same physical laws to take the same form,

$$\sum_{\sigma=1}^{4} \frac{\partial F_{\lambda\sigma}'}{\partial x_\sigma'} = \frac{4\pi}{c} J_\lambda' \tag{11.93}$$

Using transformation (11.81), we find that (11.93) can be expressed in terms of quantities in the original coordinate frame as

$$\sum_{\mu=1}^{4} a_{\lambda\mu}\left(\sum_{\nu=1}^{4} \frac{\partial F_{\mu\nu}}{\partial x_\nu} - \frac{4\pi}{c} J_\mu \right) = 0 \tag{11.94}$$

This shows that, if (11.92) holds in the original frame of reference, then it holds in all equivalent Lorentz frames. If the two sides of (11.92) had not had the same Lorentz transformation properties, this would obviously not be true.

To conclude these formal considerations we introduce some simplifying notation. In what follows:

1. Greek indices will be summed from 1 to 4.
2. Roman indices will represent spatial directions and will be summed from 1 to 3.
3. 4-vectors will be denoted by A_μ with (A_1, A_2, A_3) the components of a space vector $\mathbf{A}$ and $A_4 = iA_0$. This correspondence will sometimes be written

$$A_\mu = (\mathbf{A}, iA_0) \qquad (11.95)$$

Sometimes the subscript on the 4-vector will be omitted, e.g. $f(x)$ means $f(\mathbf{x}, t)$.

4. Scalar products of 4-vectors will be denoted by

$$(A \cdot B) = \mathbf{A} \cdot \mathbf{B} - A_0 B_0 \qquad (11.96)$$

where $\mathbf{A} \cdot \mathbf{B}$ is the ordinary 3-space scalar product.

5. The summation convention will be used. That is, repeated indices are understood to be summed over, even though the summation sign is not written. If the repeated index is roman, the sum is from 1 to 3; if it is Greek, the sum is from 1 to 4. Thus, for example, (11.85) will be written

$$\frac{\partial A_\nu'}{\partial x_\nu'} = \frac{\partial A_\mu}{\partial x_\mu}$$

and (11.89) will be written

$$T'_{\mu\nu}A_\nu' = a_{\mu\lambda}T_{\lambda\nu}A_\nu$$

11.9 Covariance of Electrodynamics

The invariance in form of the equations of electrodynamics under Lorentz transformations was shown by Lorentz and Poincaré before Einstein formulated the special theory of relativity. We will now discuss this covariance and consider its consequences. There are two points of view possible. One is to take some experimentally proven fact such as the invariance of electric charge and try to deduce that the equations must be covariant. The other is to demand that the equations be covariant in form and to show that the transformation properties of the various physical quantities, such as field strengths and charge and current, can be satisfactorily chosen to accomplish this. Although the first view is to some the most satisfying, we will adopt the second course. Classical electrodynamics *is* correct, and it can be cast in covariant form. For simplicity we will consider the microscopic equations, without the derived quantities $\mathbf{D}$ and $\mathbf{H}$.

We begin with the continuity equation for charge and current densities:

$$\frac{\partial \rho}{\partial t} = -\nabla \cdot \mathbf{J} \qquad (11.97)$$

This can be cast in covariant form by introducing the charge-current 4-vector J_μ defined by

$$J_\mu = (\mathbf{J}, ic\rho) \qquad (11.98)$$

Then (11.97) takes on the obviously covariant form:

$$\frac{\partial J_\mu}{\partial x_\mu} = 0 \qquad (11.99)$$

That J_μ is a legitimate 4-vector can be established from the experimentally known invariance of electric charge. This invariance implies that $(\rho\, dx_1\, dx_2\, dx_3)$ is a Lorentz invariant. Since $i\, d^4x = (dx_1\, dx_2\, dx_3\, dx_4)$ is a Lorentz invariant, it follows that ρ transforms like the fourth component of a 4-vector. The transformation properties of $\mathbf{J}$ follow similarly.

The wave equations for the vector potential $\mathbf{A}$ and the scalar potential Φ are

$$\left.\begin{aligned}
\nabla^2 \mathbf{A} - \frac{1}{c^2}\frac{\partial^2 \mathbf{A}}{\partial t^2} &= -\frac{4\pi}{c}\mathbf{J} \\[2ex]
\nabla^2 \Phi - \frac{1}{c^2}\frac{\partial^2 \Phi}{\partial t^2} &= -4\pi\rho
\end{aligned}\right\} \qquad (11.100)$$

with the Lorentz condition

$$\nabla \cdot \mathbf{A} + \frac{1}{c}\frac{\partial \Phi}{\partial t} = 0 \qquad (11.101)$$

The differential operator on the left-hand sides of the wave equations can be recognized as the Lorentz invariant four-dimensional Laplacian (11.86). The right-hand sides of these equations are the components of a 4-vector. Consequently, the requirement of covariance means that the vector and scalar potentials are the space and time parts of a 4-vector potential A_μ:

$$A_\mu = (\mathbf{A}, i\Phi) \qquad (11.102)$$

Then the wave equations can be written

$$\Box^2 A_\mu = -\frac{4\pi}{c}J_\mu, \qquad \mu = 1, 2, 3, 4 \qquad (11.103)$$

while the Lorentz condition becomes

$$\frac{\partial A_\mu}{\partial x_\mu} = 0 \qquad (11.104)$$

We are now ready to consider the field strengths **E** and **B**. They are defined in terms of the potentials by

$$\left. \begin{aligned} \mathbf{E} &= -\nabla\Phi - \frac{1}{c}\frac{\partial \mathbf{A}}{\partial t} \\ \mathbf{B} &= \nabla \times \mathbf{A} \end{aligned} \right\} \tag{11.105}$$

By writing out the components explicitly, for example,

$$\left. \begin{aligned} iE_1 &= \frac{\partial A_1}{\partial x_4} - \frac{\partial A_4}{\partial x_1} \\ B_1 &= \frac{\partial A_3}{\partial x_2} - \frac{\partial A_2}{\partial x_3} \end{aligned} \right\} \tag{11.106}$$

it is evident that the electric field and the magnetic induction are elements of the second-rank, antisymmetric, field-strength tensor $F_{\mu\nu}$:

$$F_{\mu\nu} = \frac{\partial A_\nu}{\partial x_\mu} - \frac{\partial A_\mu}{\partial x_\nu} \tag{11.107}$$

Explicitly, the field-strength tensor is

$$(F_{\mu\nu}) = \begin{pmatrix} 0 & B_3 & -B_2 & -iE_1 \\ -B_3 & 0 & B_1 & -iE_2 \\ B_2 & -B_1 & 0 & -iE_3 \\ iE_1 & iE_2 & iE_3 & 0 \end{pmatrix} \tag{11.108}$$

To complete the demonstration of the covariance of electrodynamics we must consider Maxwell's equations. The inhomogeneous pair are

$$\left. \begin{aligned} \nabla \cdot \mathbf{E} &= 4\pi\rho \\ \nabla \times \mathbf{B} - \frac{1}{c}\frac{\partial \mathbf{E}}{\partial t} &= \frac{4\pi}{c}\mathbf{J} \end{aligned} \right\} \tag{11.109}$$

Since the right-hand sides form the components of a 4-vector, so must the left-hand sides. With definition (11.108) of the field-strength tensor it is easy to show that the left-hand sides in (11.109) are the divergence of the field-strength tensor. Thus (11.109) takes the covariant form

$$\frac{\partial F_{\mu\nu}}{\partial x_\nu} = \frac{4\pi}{c}J_\mu \tag{11.110}$$

Similarly the two homogeneous Maxwell's equations,

$$\nabla \cdot \mathbf{B} = 0, \qquad \nabla \times \mathbf{E} + \frac{1}{c}\frac{\partial \mathbf{B}}{\partial t} = 0 \tag{11.111}$$

can be shown to reduce to the four equations:

$$\frac{\partial F_{\mu\nu}}{\partial x_\lambda} + \frac{\partial F_{\lambda\mu}}{\partial x_\nu} + \frac{\partial F_{\nu\lambda}}{\partial x_\mu} = 0 \qquad (11.112)$$

where λ, μ, ν are any three of the integers 1, 2, 3, 4. Each term in (11.112) transforms like a 4-tensor of the third rank so that the equation is covariant in form, as required.

11.10 Transformation of the Electromagnetic Fields

Since the fields **E** and **B** are elements of the field-strength tensor $F_{\mu\nu}$, their transformation properties can be found from

$$F'_{\mu\nu} = a_{\mu\lambda} a_{\nu\sigma} F_{\lambda\sigma} \qquad (11.113)$$

With transformation (11.75) from a system K to K' moving with velocity v along the x_3 axis, (11.113) gives the transformed fields:

$$\left.\begin{aligned}
E_1' &= \gamma(E_1 - \beta B_2) & B_1' &= \gamma(B_1 + \beta E_2) \\
E_2' &= \gamma(E_2 + \beta B_1) & B_2' &= \gamma(B_2 - \beta E_1) \\
E_3' &= E_3 & B_3' &= B_3
\end{aligned}\right\} \qquad (11.114)$$

The inverse transformation can be obtained from (11.114) by the interchange of primed and unprimed quantities and $\beta \to -\beta$. For a general Lorentz transformation from K to a system K' moving with velocity **v** relative to K, the transformation of the fields is evidently

$$\left.\begin{aligned}
\mathbf{E}_\parallel' &= \mathbf{E}_\parallel & \mathbf{B}_\parallel' &= \mathbf{B}_\parallel \\
\mathbf{E}_\perp' &= \gamma\left(\mathbf{E}_\perp + \frac{\mathbf{v}}{c} \times \mathbf{B}\right) & \mathbf{B}_\perp' &= \gamma\left(\mathbf{B}_\perp - \frac{\mathbf{v}}{c} \times \mathbf{E}\right)
\end{aligned}\right\} \qquad (11.115)$$

Here $\parallel$ and $\perp$ mean parallel and perpendicular to the velocity **v**. Transformation (11.115) shows that **E** and **B** have no independent existence. A purely electric or magnetic field in one coordinate system will appear as a mixture of electric and magnetic fields in another coordinate frame. Of course certain restrictions apply (see Problem 11.10) so that, for example, a purely electrostatic field in one coordinate system cannot be transformed into a purely magnetostatic field in another. But the fields are completely interrelated, and one should properly speak of the electromagnetic field $F_{\mu\nu}$, rather than **E** or **B** separately.

As an example of the transformation of the electromagnetic fields we consider the fields seen by an observer in the system K when a point charge q moves by in a straight-line path with a velocity **v**. The charge is at rest in the system K', and the transformation of the fields is given by the inverse

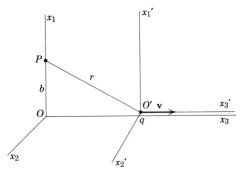

Fig. 11.13 Particle of charge q
moving at constant velocity v
passes an observation point P
at impact parameter b.

of (11.114) or (11.115). We suppose that the charge moves in the positive
x_3 direction and that its closest distance of approach to the observer is b.
Figure 11.13 shows a suitably chosen set of axes. The observer is at the
point P. At $t = t' = 0$ the origins of the two coordinate systems coincide
and the charge q is at its closest distance to the observer. In the frame K'
the observer's point P, where the fields are to be evaluated, has coordinates
$x_1' = b$, $x_2' = 0$, $x_3' = -vt'$, and is a distance $r' = \sqrt{b^2 + (vt')^2}$ away
from q. We will need to express r' in terms of the coordinates of K. The
only coordinate needing transformation is the time $t' = \gamma[t - (v/c^2)x_3] =
\gamma t$, since $x_3 = 0$ for the point P in the frame K. In the rest frame K' of the
charge the electric and magnetic fields are

$$E_1' = \frac{qb}{r'^3}, \qquad E_2' = 0, \qquad E_3' = -\frac{qvt'}{r'^3}$$
$$B_1' = 0, \qquad B_2' = 0, \qquad B_3' = 0 \tag{11.116}$$

In terms of the coordinates of K the nonvanishing field components are

$$E_1' = \frac{qb}{(b^2 + \gamma^2 v^2 t^2)^{3/2}}, \qquad E_3' = -\frac{q\gamma vt}{(b^2 + \gamma^2 v^2 t^2)^{3/2}} \tag{11.117}$$

Then, using the inverse of (11.114), we find the transformed fields in the
system K:

$$E_1 = \gamma E_1' = \frac{\gamma qb}{(b^2 + \gamma^2 v^2 t^2)^{3/2}}$$
$$E_3 = E_3' = -\frac{q\gamma vt}{(b^2 + \gamma^2 v^2 t^2)^{3/2}} \tag{11.118}$$
$$B_2 = \gamma \beta E_1' = \beta E_1$$

with the other components vanishing.

Fields (11.118) exhibit interesting behavior when the velocity of the
charge approaches that of light. First of all there is observed a magnetic

induction in the x_2 direction. This magnetic field becomes almost equal to the transverse electric field E_1 as $\beta \to 1$. Even at nonrelativistic velocities where $\gamma \simeq 1$, this magnetic induction is equivalent to

$$\mathbf{B} \simeq \frac{q}{c} \frac{\mathbf{v} \times \mathbf{r}}{r^3} \qquad (11.119)$$

which is just the Ampère-Biot-Savart expression for the magnetic field of a moving charge. This can obviously be obtained directly from the inverse of (11.115). At high speeds when $\gamma \gg 1$ we see that the peak transverse electric field E_1 ($t = 0$) becomes equal to γ times its nonrelativistic value. In the same limit, however, the duration of appreciable field strengths at the point P is decreased. A measure of the time interval over which the fields are appreciable is evidently

$$\Delta t \sim \frac{b}{\gamma v} \qquad (11.120)$$

As γ increases, the peak fields increase in proportion, but their duration goes in inverse proportion. The time integral of the fields times v is independent of velocity. Figure 11.14 shows this behavior of the transverse electric and magnetic fields and the longitudinal electric field. For $\beta \to 1$ the observer at P sees nearly equal transverse and mutually perpendicular electric and magnetic fields. These are indistinguishable from the fields of a pulse of plane polarized radiation propagating in the x_3 direction.

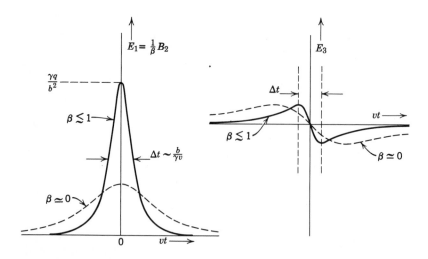

Fig. 11.14 Fields due to a uniformly moving, charged particle as a function of time.

The extra longitudinal electric field varies rapidly from positive to negative and has zero time integral. If the observer's detecting apparatus has any significant inertia, it will not respond to this longitudinal field. Consequently for practical purposes he will see only the transverse fields. This equivalence of the fields of a relativistic charged particle and those of a pulse of electromagnetic radiation will be exploited later in Chapter 15.

That a plane electromagnetic wave in one coordinate frame K will also appear as a plane wave in another coordinate frame K' moving with constant velocity relative to K follows from the invariant form of the wave equation under Lorentz transformations. Thus in the frame K a plane wave is represented by

$$F_{\mu\nu}(\mathbf{x}, t) = f_{\mu\nu} e^{i\mathbf{k} \cdot \mathbf{x} - i\omega t} \tag{11.121}$$

where $f_{\mu\nu}$ are appropriate constant coefficients, and $\mathbf{k}$ and ω are the wave vector and frequency of the wave. In the coordinates system K' the plane will be

$$F'_{\mu\nu}(\mathbf{x}', t') = f'_{\mu\nu} e^{i\mathbf{k}' \cdot \mathbf{x}' - i\omega' t'} \tag{11.122}$$

where the $f'_{\mu\nu}$ are again constant coefficients, and $\mathbf{k}'$ and ω' are the wave vector and frequency as seen in K'. According to (11.113), the two sets of fields are related by

$$f'_{\mu\nu} e^{i\mathbf{k}' \cdot \mathbf{x}' - i\omega' t'} = a_{\mu\lambda} a_{\nu\sigma} f_{\lambda\sigma} e^{i\mathbf{k} \cdot \mathbf{x} - i\omega t} \tag{11.123}$$

In order that (11.123) be true at all points in space-time the phase factors on both sides must be equal:

$$\mathbf{k}' \cdot \mathbf{x}' - \omega' t' = \mathbf{k} \cdot \mathbf{x} - \omega t \tag{11.124}$$

This invariance of the phase means that $\mathbf{k}$ and ω must form the space and time parts of a 4-vector k_μ:

$$k_\mu = \left(\mathbf{k}, \frac{i\omega}{c} \right) \tag{11.125}$$

Then the invariance of phase becomes the obvious invariance of a scalar product $(k \cdot x)$ of two 4-vectors. The relativistic formulas for the Doppler shift follow immediately from (11.125), as was shown in Section 11.4.

11.11 Covariance of the Force Equation and the Conservation Laws

In Section 11.9 the covariance of electrodynamics was discussed from the point of view of charge and current densities and the resulting fields and potentials. We know that the sources of charge and current are ultimately charged particles which can move under the action of fields.

Consequently to complete our discussion we must consider the covariant formulation of the Lorentz force equation and the conservation laws of momentum and energy.

The Lorentz force equation can be written as a force per unit volume (representing the rate of change of mechanical momentum of the sources per unit volume):

$$\mathbf{f} = \rho \mathbf{E} + \frac{1}{c} \mathbf{J} \times \mathbf{B} \qquad (11.126)$$

where $\mathbf{J}$ and ρ are the current and charge densities. Writing out a single component of $\mathbf{f}$, we find

$$f_1 = \rho E_1 + \frac{1}{c}(J_2 B_3 - J_3 B_2) = \frac{1}{c}(F_{12}J_2 + F_{13}J_3 + F_{14}J_4) \quad (11.127)$$

where we have used definitions (11.98) and (11.108). The other components of $\mathbf{f}$ yield similar results, showing that (11.126) can be written as

$$f_k = \frac{1}{c} F_{k\nu} J_\nu, \qquad k = 1, 2, 3 \qquad (11.128)$$

The right-hand side of (11.128) is evidently the space components of a 4-vector. Hence $\mathbf{f}$ must be the space part of a 4-vector $f_\mu = \left(\mathbf{f}, i\dfrac{f_0}{c}\right)$, where:

$$f_\mu = \frac{1}{c} F_{\mu\nu} J_\nu \qquad (11.129)$$

To see the meaning of the fourth component of the force-density 4-vector we write out

$$f_0 = \frac{c}{i} f_4 = \frac{1}{i}(F_{41}J_1 + F_{42}J_2 + F_{43}J_3) = \mathbf{E} \cdot \mathbf{J} \qquad (11.130)$$

But $(\mathbf{E} \cdot \mathbf{J})$ is just the rate at which the field does work on the sources per unit volume, or the rate of change of mechanical energy of the sources per unit volume. Thus we see that the covariant form (11.129) of the Lorentz force equation gives the rate of change of mechanical momentum per unit volume as its space part, and the rate of change of mechanical energy per unit volume as its time part. Alternatively, it may be viewed as giving the space and time derivatives of something of the dimensions of work per unit volume.

The conservation laws for mechanical plus electromagnetic energy and momentum derived in Chapter 6 can be presented in covariant form as the space and time components of a single 4-vector equation. If the inhomogeneous Maxwell's equations (11.110) are used to eliminate J_ν in

(11.129), the force density becomes

$$f_\mu = \frac{1}{4\pi} F_{\mu\nu} \frac{\partial F_{\nu\lambda}}{\partial x_\lambda} \qquad (11.131)$$

The right-hand side of (11.131) can be written as the divergence of a tensor of the second rank. We define the symmetric tensor $T_{\mu\nu}$, called the *electromagnetic stress-energy-momentum tensor*,

$$T_{\mu\nu} = \frac{1}{4\pi}\left[F_{\mu\lambda}F_{\lambda\nu} + \tfrac{1}{4}\delta_{\mu\nu}F_{\lambda\sigma}F_{\lambda\sigma} \right] \qquad (11.132)$$

It will be left to the problems (Problem 11.12) to show that by means of the homogeneous Maxwell's equations and (11.132) force equation (11.131) can be written in the form:

$$f_\mu = \frac{\partial T_{\mu\nu}}{\partial x_\nu} \qquad (11.133)$$

The tensor $T_{\mu\nu}$ can be written out explicitly in terms of the fields using (11.132):

$$(T_{\mu\nu}) = \begin{pmatrix} T_{11} & T_{12} & T_{13} & -icg_1 \\ T_{21} & T_{22} & T_{23} & -icg_2 \\ T_{31} & T_{32} & T_{33} & -icg_3 \\ -icg_1 & -icg_2 & -icg_3 & u \end{pmatrix} \qquad (11.134)$$

where T_{ik} is the symmetric Maxwell's stress tensor defined on page 194, **g** is the electromagnetic momentum density,

$$\mathbf{g} = \frac{1}{4\pi c}\,\mathbf{E} \times \mathbf{B} = \frac{1}{c^2}\,\mathbf{S} \;\Bigg]$$

and u is the energy density,

$$u = \frac{1}{8\pi}(E^2 + B^2) \;\Bigg\} \qquad (11.135)$$

From definition (6.102) of the spatial parts of $T_{\mu\nu}$ [or from (11.132)], we see that the stress-energy-momentum tensor has a vanishing trace:

$$\sum_\mu T_{\mu\mu} = 0 \qquad (11.136)$$

The conservation laws of momentum and energy are merely the three-dimensional integrals of the force equation (11.133). To see this we write out a typical spatial-component equation:

$$f_k = \frac{\partial T_{k\nu}}{\partial x_\nu} = \frac{\partial T_{kj}}{\partial x_j} + \frac{\partial T_{k4}}{\partial x_4} = (\nabla \cdot \overleftrightarrow{\mathbf{T}})_k - \frac{\partial g_k}{\partial t} \qquad (11.137)$$

If we identify the spatial integral of f_k as the rate of change of the kth component of mechanical momentum P_k, then the integral of (11.137) can be written

$$\frac{d}{dt}(\mathbf{P} + \mathbf{G}) = \int_V \mathbf{\nabla} \cdot \overset{\leftrightarrow}{\mathbf{T}} d^3x = \oint_S \mathbf{n} \cdot \overset{\leftrightarrow}{\mathbf{T}} da \qquad (11.138)$$

where G_k is the kth component of total electromagnetic momentum. This is the momentum-conservation law already obtained in Chapter 6. Similarly the fourth component of (11.133) can be written

$$f_0 = \frac{c}{i}f_4 = \mathbf{E} \cdot \mathbf{J} = \frac{c}{i}\frac{\partial T_{4j}}{\partial x_j} + \frac{c}{i}\frac{\partial T_{44}}{\partial x_4} = -\mathbf{\nabla} \cdot \mathbf{S} - \frac{\partial u}{\partial t} \qquad (11.139)$$

With the volume integral of f_0 identified as the rate of change of total mechanical energy T, the conservation of energy law is

$$\frac{d}{dt}(T + U) = -\int_V \mathbf{\nabla} \cdot \mathbf{S} \, d^3x = -\oint_S \mathbf{n} \cdot \mathbf{S} \, da \qquad (11.140)$$

where U is the total electromagnetic energy in the volume V.

REFERENCES AND SUGGESTED READING

The theory of relativity has an extensive literature all its own. To my mind the most lucid, though concise, presentation of special and general relativity is the famous 1921 article (recently brought up to date) by
    Pauli.
In addition, there are a number of textbooks devoted to special relativity at the graduate level, some of which are
    Aharoni,
    Bergmann, Chapters I–IX,
    Møller, Chapters I–VII.
Møller's book is perhaps the most authoritative.
    The flavor of the original theoretical developments can be obtained by consulting the collected papers of
    Einstein, Lorentz, Minkowski, and Weyl.
    The main experiments are summarized briefly, but clearly, in
    Møller, Chapter I,
    Panofsky and Phillips, Chapter 14.
A fuller description of the experimental basis of special relativity, with many references, is presented in
    Condon and Odishaw, Part 6, Chapter 8 by E. L. Hill.
    Thomas precession is discussed by
    Møller, Sections 22 and 47,
in a manner akin to ours. A different approach to the problem is given by
    Corben and Stehle, Section 92.

PROBLEMS

11.1 A possible clock is shown in the figure. It consists of a flashtube F and a photocell P shielded so that each views only the mirror M, located a distance d away, and mounted rigidly with respect to the flashtube-photocell assembly. The electronic innards of the box are such that, when the photocell responds to a light flash from the mirror, the flashtube is triggered with a negligible delay and emits a short flash towards the mirror. The clock thus "ticks" once every $(2d/c)$ seconds when at rest.

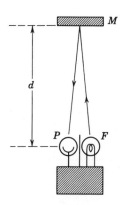

(*a*) Suppose that the clock moves with a uniform velocity v, perpendicular to the line from PF to M, relative to an observer. Using the second postulate of relativity, show by explicit geometrical or algebraic construction that the observer sees the relativistic time dilatation as the clock moves by.

(*b*) Suppose that the clock moves with a velocity v parallel to the line from PF to M. Verify that here, too, the clock is observed to tick more slowly, by the same time dilatation factor.

11.2 (*a*) Show explicitly that two successive Lorentz transformations in the same direction commute and that they are equivalent to a single Lorentz transformation with a velocity

$$v = \frac{v_1 + v_2}{1 + (v_1 v_2/c^2)}$$

This is an alternative way to derive the parallel-velocity addition law.

(*b*) Show explicitly that two successive Lorentz transformations at right angles (v_1 in the x direction, v_2 in the y direction) do not commute. Show further that in whatever order they are applied the result is not the same as a single transformation with $\mathbf{v} = \mathbf{i}v_1 + \mathbf{j}v_2$. Give one or more simple reasons why this result is necessary within the framework of special relativity.

11.3 (*a*) Find the form of the wave equation in system K if it has its standard form in system K' and the two coordinate systems are related by the Galilean transformation $x' = x - vt$, $t' = t$.

(b) Show explicitly that the form of the wave equation is the same in system K as in K' if the coordinates are related by the Lorentz transformation $x' = \gamma(x - vt)$, $t' = \gamma[t - (vx/c^2)]$.

11.4 A coordinate system K' moves with a velocity $\mathbf{v}$ relative to another system K. In K' a particle has a velocity $\mathbf{u}'$ and an acceleration $\mathbf{a}'$. Find the Lorentz transformation law for accelerations, and show that in the system K the components of acceleration parallel and perpendicular to $\mathbf{v}$ are

$$\mathbf{a}_{\parallel} = \frac{\left(1 - \dfrac{v^2}{c^2}\right)^{3/2}}{\left(1 + \dfrac{\mathbf{v} \cdot \mathbf{u}'}{c^2}\right)^3} \mathbf{a}_{\parallel}'$$

$$\mathbf{a}_{\perp} = \frac{\left(1 - \dfrac{v^2}{c^2}\right)}{\left(1 + \dfrac{\mathbf{v} \cdot \mathbf{u}'}{c^2}\right)^3}\left(\mathbf{a}_{\perp}' + \frac{\mathbf{v}}{c^2} \times (\mathbf{a}' \times \mathbf{u}')\right)$$

11.5 Assume that a rocket ship leaves the earth in the year 2000. One of a set of twins born in 1980 remains on earth; the other rides in the rocket. The rocket ship is so constructed that it has an acceleration g in its own rest frame (this makes the occupants feel at home). It accelerates in a straight-line path for 5 years (by its own clocks), decelerates at the same rate for 5 more years, turns around, accelerates for 5 years, decelerates for 5 years, and lands on earth. The twin in the rocket is 40 years old.

(a) What year is it on earth?

(b) How far away from the earth did the rocket ship travel?

11.6 In the reference frame K two very evenly matched sprinters are lined up a distance d apart on the y axis for a race parallel to the x axis. Two starters, one beside each man, will fire their starting pistols at slightly different times, giving a handicap to the better of the two runners. The time difference in K is T.

(a) For what range of time differences will there be a reference frame K' in which there is no handicap, and for what range of time differences is there a frame K' in which there is a true (not apparent) handicap?

(b) Determine explicitly the Lorentz transformation to the frame K' appropriate for each of the two possibilities in (a), finding the velocity of K' relative to K and the space-time positions of each sprinter in K'.

11.7 Using the four-dimensional form of Green's theorem, solve the inhomogeneous wave equations

$$\Box^2 A_{\mu} = \frac{-4\pi}{c} J_{\mu}$$

(a) Show that for a localized charge-current distribution the 4-vector potential is

$$A_{\mu}(x) = \frac{1}{\pi c} \int \frac{J_{\mu}(\xi)}{R^2} d^4\xi$$

where $R^2 = (x - \xi) \cdot (x - \xi)$, x means (x_1, x_2, x_3, x_4), and $d^4\xi = d\xi_1 \, d\xi_2 \, d\xi_3 \, d\xi_4$.
(b) From the definitions of the field strengths $F_{\mu\nu}$ show that

$$F_{\mu\nu} = \frac{2}{\pi c} \int \frac{(J \times R)_{\mu\nu}}{R^4} \, d^4\xi$$

where $(J \times R)_{\mu\nu} = J_\mu R_\nu - J_\nu R_\mu$.

11.8 The three-dimensional formulation of the radiation problem leads to the retarded solution

$$A_\mu(\mathbf{x}, t) = \frac{1}{c} \int \frac{J_\mu(\xi, t')}{r} \bigg|_{t' = t - (r/c)} d^3\xi$$

where $r = |\mathbf{x} - \xi|$. Show the connection between this retarded solution and the solution of Problem 11.7 by explicitly performing the integration over $d\xi_4$.

11.9 A classical point magnetic moment μ at rest has a vector potential

$$\mathbf{A} = \frac{\mu \times \mathbf{r}}{r^3}$$

and no scalar potential. Show that, if the magnetic moment moves with a velocity $\mathbf{v}(v \ll c)$, there is an electric dipole moment $\mathbf{p}$ associated with the magnetic moment, where

$$\mathbf{p} = \frac{\mathbf{v}}{c} \times \mu$$

What can you say if $\mathbf{v}$ is not small in magnitude compared to c? Show that the interaction energy between the moving dipole and fields $\mathbf{E}$ and $\mathbf{B}$ is the same as would be obtained by calculating the magnetic field in the rest frame of the magnetic moment.

11.10 (a) Show that $(B^2 - E^2)$ is an invariant quantity under Lorentz transformations. What is its form in four-dimensional notation?
(b) The symbol $\epsilon_{\lambda\mu\nu\sigma}$ is defined to have the properties

$$\epsilon_{\lambda\mu\nu\sigma} = \begin{cases} 0 \text{ if any two indices are equal} \\ \pm 1 \text{ for an even (odd) permutation of indices} \end{cases}$$

$\epsilon_{\lambda\mu\nu\sigma}$ is a completely antisymmetric unit tensor of the fourth rank (actually a pseudotensor under spatial inversion). Prove that $\epsilon_{\lambda\mu\nu\sigma} F_{\lambda\mu} F_{\nu\sigma}$ (summation convention implied) is a Lorentz invariant, and find its form in terms of $\mathbf{E}$ and $\mathbf{B}$.

11.11 In a certain reference frame a static, uniform, electric field E_0 is parallel to the x axis, and a static, uniform, magnetic induction $B_0 = 2E_0$ lies in the x-y plane, making an angle θ with the x axis. Determine the relative velocity of a reference frame in which the electric and magnetic fields are parallel. What are the fields in that frame for $\theta \ll 1$ and $\theta \to (\pi/2)$?

11.12 Show that the force equation $f_\mu = (1/c)F_{\mu\nu}J_\nu$ can be written as

$$f_\mu = \frac{\partial T_{\mu\nu}}{\partial x_\nu}$$

where

$$T_{\mu\nu} = \frac{1}{4\pi}[F_{\mu\lambda}F_{\lambda\nu} + \tfrac{1}{4}\delta_{\mu\nu}F_{\lambda\sigma}F_{\lambda\sigma}]$$

11.13 A pulse of electromagnetic radiation of finite spatial extent exists in charge- and current-free space.

(a) By means of the divergence theorem in four dimensions, prove that the *total* electromagnetic momentum and energy transform like a 4-vector.

(b) Show that for a plane wave this 4-vector has zero "length," but that for other possible field configurations (e.g., spherically diverging wave) this is not true.

12

Relativistic-Particle
Kinematics and Dynamics

In Chapter 11 the special theory of relativity was developed with particular emphasis on the electromagnetic fields and the covariance of the equations of electrodynamics. Only in Section 11.11 was there a mention of the mechanical origin of the sources of charge and current density. The emphasis on electromagnetic fields is fully justified in the presentation of the first aspects of relativity, since it was the behavior of light which provided the puzzling phenomena that were understood in terms of the special theory of relativity. Furthermore, a large class of problems can be handled without inquiry into the detailed mechanical behavior of the sources of charge and current. Nevertheless, problems which emphasize the fields rather than the sources form only a part of electrodynamic phenomena. There is the converse type of problem in which we are interested in the behavior of charged particles under the action of applied electromagnetic fields. The particles represent charge and current densities, of course, and so act as sources of new fields. But for most applications these fields can be neglected or taken into account in an approximate way. In the present chapter we wish to explore the motion of relativistic particles, first their kinematics and then their dynamics in external fields. Discussion of the difficult problem of charged particles acting as the sources of fields and being acted on by those same fields will be deferred to Chapter 17.

12.1   Momentum and Energy of a Particle

In nonrelativistic mechanics a particle of mass m and velocity $\mathbf{v}$ has a momentum $\mathbf{p} = m\mathbf{v}$ and a kinetic energy $T = \frac{1}{2}mv^2$. Newton's equation of motion relates the time rate of change of momentum to the applied force. For a charged particle the force is the Lorentz force. Since we have discussed the Lorentz transformation properties of the Lorentz force density in Section 11.11, we can immediately deduce the behavior of a charged particle's momentum under Lorentz transformations. For neutral particles with no detectable electromagnetic interactions it is clearly impossible to obtain their relativistic transformation properties in this way, but there is ample experimental evidence that all particles behave kinematically in the same way, whether charged or neutral.

A charged particle can be thought of as a very localized distribution of charge and mass. To find the force acting on such a particle we integrate the Lorentz force density f_μ (11.129) over the volume of the charge. If the total charge is e and the velocity of the particle is $\mathbf{v}$, then the volume integral of (11.129) is

$$\int f_\mu \, d^3x = \frac{e}{c} F_{\mu\nu} v_\nu \qquad (12.1)$$

where $v_\nu = (\mathbf{v}, ic)$, and $F_{\mu\nu}$ is interpreted as the average field acting on the particle. The left-hand side of (12.1) is now to be equated to the time rate of change of the momentum and energy of the particle, just as in Section 11.11. Thus

$$\frac{dp_\mu}{dt} = \int f_\mu \, d^3x \qquad (12.2)$$

where we have written p_k as the kth component of the particle's momentum and $p_4 = iE/c$ as proportional to the particle's energy. That p_μ is indeed a 4-vector follows immediately from (12.2). If we integrate both sides with respect to time, then the left-hand side becomes the momentum or energy of the particle while the right-hand side is the four-dimensional integral of f_μ. Since d^4x is a Lorentz invariant quantity, it follows that p_μ must have the same transformation properties as f_μ. Therefore the momentum $\mathbf{p}$ and the energy E of a particle form a 4-vector p_μ:

$$p_\mu = \left(\mathbf{p}, \frac{iE}{c} \right) \qquad (12.3)$$

The transformation of momentum and energy from one Lorentz frame K to another K' moving with a velocity $\mathbf{v}$ parallel to the z axis is

$$\left. \begin{aligned} p_1 &= p_1', \qquad p_2 = p_2' \\ p_3 &= \gamma\left(p_3' + \beta \frac{E'}{c} \right) \\ E &= \gamma(E' + \beta c p_3') \end{aligned} \right\} \qquad (12.4)$$

where $\beta = v/c$ and $\gamma = (1 - \beta^2)^{-\frac{1}{2}}$. The inverse transformation is obtained by changing $\beta \to -\beta$ and interchanging the primed and unprimed variables.

The length of the 4-vector p_μ is a Lorentz invariant quantity which is characteristic of the particle:

$$(p \cdot p) = (p' \cdot p') = -\frac{\lambda^2}{c^2} \tag{12.5}$$

In the rest frame of the particle ($\mathbf{p'} = 0$) the scalar product (12.5) gives the energy of the particle at rest:

$$E' = \lambda \tag{12.6}$$

To determine λ we consider the Lorentz transformation (12.4) of p_μ from the rest frame of the particle to the frame K in which the particle is moving in the z direction with a velocity $\mathbf{v}$. Then the momentum and energy are

$$\mathbf{p} = \frac{\gamma \boldsymbol{\beta} \lambda}{c} = \gamma \left(\frac{\lambda}{c^2}\right)\mathbf{v} \tag{12.7}$$

$$E = \gamma\lambda$$

From the nonrelativistic expression for momentum $\mathbf{p} = m\mathbf{v}$ we find that the invariant constant $\lambda = mc^2$. The nonrelativistic limit of the energy is

$$E = \gamma mc^2 \simeq mc^2 + \tfrac{1}{2}mv^2 + \cdots \tag{12.8}$$

This shows that E is the total energy of the particle, consisting of two parts: the rest energy (mc^2) and the kinetic energy. Even for a relativistic particle we can speak of the kinetic energy T, defined as the difference between the total and the rest energies:

$$T = E - mc^2 = (\gamma - 1)mc^2 \tag{12.9}$$

In summary, a free particle with mass m moving with a velocity $\mathbf{v}$ in a reference frame K has a momentum and energy in that frame:

$$\left.\begin{array}{l} \mathbf{p} = \gamma m\mathbf{v} \\[2mm] E = \gamma mc^2 \end{array}\right\} \tag{12.10}$$

From (12.5) it is evident that the energy E can be expressed in terms of the momentum as

$$E = (c^2p^2 + m^2c^4)^{\frac{1}{2}} \tag{12.11}$$

The velocity of the particle can likewise be expressed in terms of its momentum and energy:

$$\mathbf{v} = \frac{c^2\mathbf{p}}{E} \tag{12.12}$$

In dealing with relativistic-particle kinematics it is convenient to adopt a consistent, simple notation and set of units in which to express momenta and energies. In the formulas above we see that the velocity of light appears often. To suppress various powers of c and so simplify the notation we will adopt the convention that all momenta, energies, and masses will be measured in energy units, while velocities are measured in units of the velocity of light. All powers of c will be suppressed. Consequently in what follows, the symbols

$$\left.\begin{array}{c} p \\ E \\ m \\ v \end{array}\right\} \quad \text{stand for} \quad \left\{\begin{array}{c} cp \\ E \\ mc^2 \\ v/c \end{array}\right. \tag{12.13}$$

As energy units, the ev (electron volt), the Mev (million electron volt), and the Bev (10^9 ev) are convenient. One electron volt is the energy gained by a particle with electronic charge when it falls through a potential difference of one volt (1 ev $= 1.602 \times 10^{-12}$ erg).

12.2 Kinematics of Decay Products of an Unstable Particle

As a first illustration of relativistic kinematics which follow immediately from the 4-vector character of the momentum and energy of particles, we consider the two-body decay of an unstable particle at rest. Such decay processes are common among the unstable particles. Some examples are the following.

1. Charged pi meson decays into a mu meson and a neutrino with a lifetime $\tau = 2.6 \times 10^{-8}$ sec:

$$\pi \rightarrow \mu + \nu$$

The pi-meson rest energy is $M = 139.6$ Mev, while that of the mu meson is $m_\mu = 105.7$ Mev. The neutrino has zero rest mass, $m_\nu = 0$. There is, therefore, an energy release of 33.9 Mev in pi-meson decay.

2. Charged K meson sometimes decays into two pi mesons with a lifetime $\tau = 1.2 \times 10^{-8}$ sec:

$$K^\pm \rightarrow \pi^\pm + \pi^0$$

The charged K meson has a rest energy $M = 494$ Mev, while the two pi mesons have rest energies, $m_\pm = 139.6$ Mev, $m_0 = 135.0$ Mev. Thus the energy release is 219 Mev.

3. Lambda hyperon decays into a neutron or a proton and a pi meson with a lifetime $\tau = 2.9 \times 10^{-10}$ sec:

$$\Lambda \rightarrow \begin{cases} p + \pi^- \\ n + \pi^0 \end{cases}$$

The rest energy of the lambda hyperon is $M = 1115$ Mev; that of the proton $m_p = 938.5$ Mev, and of the neutron $m_n = 939.8$ Mev. With the pi-meson masses given above, we find that the energy release in lambda decay is 37 Mev in the charged mode and 40 Mev in the neutral mode.

The transformation of a system of mass M at rest into two particles of mass m_1 and m_2

$$M \rightarrow m_1 + m_2 \qquad (12.14)$$

can occur if the initial mass is greater than the sum of the final masses. We define the mass excess ΔM:

$$\Delta M = M - m_1 - m_2 \qquad (12.15)$$

The sum of the kinetic energies of the two particles must be equal to ΔM. Since the initial system had zero momentum, the two particles must have equal and opposite momenta, $\mathbf{p}_1 = -\mathbf{p}_2 = \mathbf{p}$. From (12.11) the conservation of energy can be written

$$\sqrt{p^2 + m_1{}^2} + \sqrt{p^2 + m_2{}^2} = M \qquad (12.16)$$

From this equation it is a straightforward matter to find the magnitude of the momentum $\mathbf{p}$ and the individual particle energies, E_1 and E_2.

Rather than solve (12.16) we wish to obtain our answers by illustrating a useful technique which exploits the Lorentz invariance of the scalar product of two 4-vectors. The conservation of energy and momentum in the two-body decay can be written as a 4-vector equation:

$$P = p_1 + p_2 \qquad (12.17)$$

where the 4-vector subscript μ on each symbol has been suppressed. The squares of the 4-vector momenta are the invariants:

$$(P \cdot P) = -M^2, \qquad (p_1 \cdot p_1) = -m_1{}^2, \qquad (p_2 \cdot p_2) = -m_2{}^2 \qquad (12.18)$$

In (12.18) we have written the squares of the 4-vectors as self-scalar products in order to distinguish the square of a vectorial quantity as a three-space self-scalar product (e.g., $p^2 = \mathbf{p} \cdot \mathbf{p}$). Using (12.17), we form the square of the 4-vector p_2:

$$(p_2 \cdot p_2) = (P - p_1) \cdot (P - p_1)$$

or

$$-m_2{}^2 = -M^2 - m_1{}^2 - 2(P \cdot p_1) \qquad (12.19)$$

The scalar product $(P \cdot p_1)$ is Lorentz invariant. In the frame in which the system M is at rest its space part vanishes, and it has the value:

$$(P \cdot p_1) = -ME_1 \qquad (12.20)$$

Therefore the total energy of the particle with mass m_1 is

$$E_1 = \frac{M^2 + m_1{}^2 - m_2{}^2}{2M} \qquad (12.21)$$

Similarly

$$E_2 = \frac{M^2 + m_2{}^2 - m_1{}^2}{2M} \qquad (12.22)$$

Often it is more convenient to have expressions for the kinetic energies than for the total energies. Using (12.15), it is easy to show that

$$T_i = \Delta M \left(1 - \frac{m_i}{M} - \frac{\Delta M}{2M} \right), \qquad i = 1, 2 \qquad (12.23)$$

where ΔM is the mass excess. The term $\Delta M/2M$ is a relativistic correction absent in the nonrelativistic result. Although it may not have obvious relativistic origin, a moment's thought shows that, if $\Delta M/2M$ is appreciable compared to unity, then necessarily the outgoing particles must be treated relativistically.

As a numerical illustration we consider the first example listed above, the decay of the pi meson. The mass excess is 33.9 Mev, while $M = 139.6$ Mev, $m_\mu = 105.7$ Mev, $m_\nu = 0$. Consequently the mu-meson and neutrino kinetic energies are

$$T_\mu = 33.9 \left(1 - \frac{105.7}{139.6} - \frac{33.9}{2(139.6)} \right) = 4.1 \text{ Mev}$$

$$T_\nu = 33.9 - T_\mu = 29.8 \text{ Mev}$$

The unique energy of 4.1 Mev for the mu meson was the characteristic of pi-meson decay at rest which led to its discovery in 1947 by Powell and coworkers from observations in photographic emulsions.

The lambda particle was first observed in flight by its charged decay products $(p + \pi^-)$ in cloud chambers. The charged particle tracks appear as shown in Fig. 12.1. The particles' initial momenta and identities can be inferred from their ranges and their curvatures in a magnetic field (or by other techniques, such as grain counting, in emulsions). The opening angle θ between the tracks provides the other datum required to determine the unseen particle's mass. Consider the square of (12.17):

$$(P \cdot P) = (p_1 + p_2) \cdot (p_1 + p_2) \qquad (12.24)$$

This becomes

$$-M^2 = -m_1{}^2 - m_2{}^2 + 2(p_1 \cdot p_2) \qquad (12.25)$$

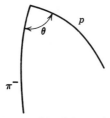

Fig. 12.1   Decay of lambda particle in flight.

If the scalar product $(p_1 \cdot p_2)$ is evaluated in the laboratory frame, we find

$$M^2 = m_1{}^2 + m_2{}^2 + 2E_1E_2 - 2p_1p_2 \cos \theta \qquad (12.26)$$

where p_1 and p_2 are the magnitudes of the three-dimensional momenta.

In a three- or more body decay process the particles do not have unique momenta, but are distributed in energy in some way. These energy spectra have definite upper end points which can be determined from the kinematics in ways similar to those used here (see Problem 12.2).

12.3   Center of Momentum Transformation and Reaction Thresholds

A common problem in nuclear or high-energy physics is the collision of two particles. Particle 1 (the projectile), with mass m_1, momentum $\mathbf{p}_1 = \mathbf{p}$, and energy E_1, is incident on particle 2 (the target) of mass m_2 at rest in the laboratory. The collision may involve elastic scattering,

$$① + ② \rightarrow ①' + ②' \qquad (12.27)$$

where the primes mean that the directions of the particles are in general different. The collision may, on the other hand, be a reaction

$$① + ② \rightarrow ③ + ④ + \cdots \qquad (12.28)$$

in which two or more particles are produced, at least one of which is different from the incident particles. Elastic scattering is always possible, but reactions may or may not be energetically possible, depending on the differences in masses of the particles and the incident energy. To determine the energetics involved and to see the processes in their simplest form kinematically it is convenient to transform to a coordinate frame K', where the projectile and the target have equal and oppositely directed momenta. This frame is called the *center of momentum* system (sometimes, loosely, the center of mass system) and is denoted by CM system. The scattered particles (or reaction products in a two-body reaction) have equal and opposite momenta making an angle θ' with the initial momenta.

Figure 12.2 shows the momentum vectors involved in elastic scattering or a two-body reaction. For elastic scattering, $|\mathbf{p}'| = |\mathbf{q}'|$, but for a reaction the magnitude of $\mathbf{q}'$ must be determined from conservation of total energy (including rest energies) in the CM system.

To relate the incident energy and momentum in the laboratory to the CM variables we can either make a direct Lorentz transformation to K', determining the transformation velocity $\mathbf{v}_{\mathrm{CM}}$ from the requirement that $\mathbf{p}_1' = \mathbf{p}' = -\mathbf{p}_2'$, or we can use the invariance of scalar products. Adopting the latter procedure, we consider the invariant scalar product

$$(p_1 + p_2) \cdot (p_1 + p_2) = (p_1' + p_2') \cdot (p_1' + p_2') \qquad (12.29)$$

The left-hand side is to be evaluated in the laboratory, where $\mathbf{p}_2 = 0$, and the right-hand side in the CM system, where $\mathbf{p}_1' + \mathbf{p}_2' = 0$. Consequently we obtain

$$p^2 - (E_1 + m_2)^2 = -(E_1' + E_2')^2 \qquad (12.30)$$

Using $E_1^2 = p^2 + m_1^2$, we find that the total energy in the CM system is

$$E' = E_1' + E_2' = (m_1^2 + m_2^2 + 2E_1 m_2)^{1/2} \qquad (12.31)$$

The separate energies E_1' and E_2' can be found by considering scalar products like

$$p_1 \cdot (p_1 + p_2) = p_1' \cdot (p_1' + p_2') \qquad (12.32)$$

This gives

$$E_1' = \frac{E'^2 + m_1^2 - m_2^2}{2E'} \left. \right\}$$

Similarly (12.33)

$$E_2' = \frac{E'^2 + m_2^2 - m_1^2}{2E'}$$

We note the similarity of these expressions to (12.21) and (12.22). The magnitude of the momentum $\mathbf{p}'$ can be obtained from (12.33):

$$\mathbf{p}' = \frac{m_2 \mathbf{p}}{E'} \qquad (12.34)$$

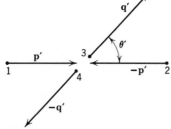

Fig. 12.2 Momentum vectors in the center of momentum frame for elastic scattering or a two-body reaction.

The Lorentz transformation parameters $\mathbf{v}_{CM}$ and γ_{CM} can be found by noting that $\mathbf{p}_2' = -\gamma_{CM}m_2\mathbf{v}_{CM} = -\mathbf{p}'$ and $E_2' = \gamma_{CM}m_2$. This gives

$$\mathbf{v}_{CM} = \frac{\mathbf{p}}{E_1 + m_2}, \qquad \gamma_{CM} = \frac{E_1 + m_2}{E'} \qquad (12.35)$$

For nonrelativistic motion the kinetic energy in the CM system reduces to

$$T' = E' - (m_1 + m_2) \simeq \left(\frac{m_2}{m_1 + m_2}\right)\tfrac{1}{2}m_1v_1^2 \qquad (12.36)$$

Similarly the CM velocity and the momentum in the CM system are

$$\mathbf{v}_{CM} = \left(\frac{m_2}{m_1 + m_2}\right)\mathbf{v}_1, \qquad \mathbf{p}' = \left(\frac{m_1m_2}{m_1 + m_2}\right)\mathbf{v}_1 \qquad (12.37)$$

We see that we can recover the familiar nonrelativistic results from our completely relativistic expressions. In the extreme of ultrarelativistic motion ($E_1 \gg m_1$ *and* m_2) the various quantities take on the approximate limiting values:

$$\left. \begin{aligned} E' &\simeq (2E_1m_2)^{\frac{1}{2}} \\[4pt] v_{CM} &\simeq 1 - \frac{m_2}{E_1} \\[4pt] p' &\simeq \tfrac{1}{2}E' \end{aligned} \right\} \qquad (12.38)$$

The energy available in the CM system is seen to increase only as the square root of the incident energy. This means that it is very difficult to obtain ultrahigh energies in the CM frame when bombarding stationary targets. The highest-energy accelerators presently existing (at CERN, near Geneva, Switzerland, and at Brookhaven, N.Y.) produce protons of approximately 30 Bev. If the target is a stationary nucleon, this means about 7 Bev total CM energy. To have 30 Bev available in the CM frame it would be necessary to bombard a stationary nucleon with protons of over 470 Bev! Considerable effort is being put into designs for so-called colliding or clashing beam accelerators so that no energy is wasted in CM motion.

In a reaction the initial particles of mass m_1 and m_2 are transformed into two or more particles with masses m_i ($i = 3,4, \ldots$). Let ΔM be the difference between the sum of masses finally and the sum of masses initially:*

$$\Delta M = (m_3 + m_4 + \cdots) - (m_1 + m_2) \qquad (12.39)$$

If ΔM is positive, the reaction will not occur below a certain incident kinetic energy T_{th}, called the *threshold* for the reaction. The criterion for the reaction just to occur is that there be enough energy available in the

---

* Note that this definition of ΔM is the negative of the one used in Section 12.2 for decay processes.

CM system that the particles can be created with no kinetic energy. This means that

$$(E')_{th} = m_1 + m_2 + \Delta M \tag{12.40}$$

Using (12.31), it is easy to show that the incident kinetic energy of the projectile at threshold is

$$T_{th} = \Delta M \left(1 + \frac{m_1}{m_2} + \frac{\Delta M}{2m_2}\right) \tag{12.41}$$

The first two terms in the parentheses are the nonrelativistic terms, while the last is a relativistic contribution. To illustrate the reaction-threshold formula we consider the calculation of the threshold energy for photo-production of neutral pi mesons from protons:

$$\gamma + p \rightarrow p + \pi^0$$

Since the photon has no rest mass, the mass difference is $\Delta M = m_{\pi^0} = 135.0$ Mev, while the target mass is $m_2 = m_p = 938.5$ Mev. Then the threshold energy is

$$T_{th} = 135.0\left[1 + \frac{135.0}{2(938.5)}\right] = 135.0(1.072) = 144.7 \text{ Mev}$$

As another example consider the production of a proton-antiproton pair in proton-proton collisions:

$$p + p \rightarrow p + p + p + \bar{p}$$

The mass difference is $\Delta M = 2m_p = 1.877$ Bev. From (12.41) we find

$$T_{th} = 2m_p(1 + 1 + 1) = 6m_p = 5.62 \text{ Bev}$$

In this example we find a factor-of-3 increase over the actual mass difference, whereas in the photoproduction example the increase was only 7.2 per cent. Other threshold calculations are left to Problem 12.1.

12.4 Transformation of Scattering or Reaction Momenta and Energies from CM to Laboratory System

In Fig. 12.2 the various CM momenta for a two-body collision are shown. The initial momenta and energies $(\mathbf{p_1}' = -\mathbf{p_2}' = \mathbf{p}', E_1', E_2')$ have already been calculated, (12.33) and (12.34). The final CM momenta and energies $(\mathbf{p_3}' = -\mathbf{p_4}' = \mathbf{q}', E_3', E_4')$ can be calculated similarly. Since energy and momentum are conserved, the 4-vector momenta satisfy

$$p_1' + p_2' = p_3' + p_4' \tag{12.42}$$

Then it is easy to show that the energies of the outgoing particles are

$$
\left.
\begin{aligned}
E_3' &= \frac{E'^2 + m_3{}^2 - m_4{}^2}{2E'} \\[2mm]
E_4' &= \frac{E'^2 + m_4{}^2 - m_3{}^2}{2E'}
\end{aligned}
\right\}
\tag{12.43}
$$

where E' is given by (12.31). The obvious symmetry with (12.33) should be noted. The CM momentum of the outgoing particles is

$$
q' = \frac{E'}{2} \left\{ \left[1 - \left(\frac{m_3 + m_4}{E'} \right)^2 \right] \left[1 - \left(\frac{m_3 - m_4}{E'} \right)^2 \right] \right\}^{\frac{1}{2}}
\tag{12.44}
$$

An alternative form of this result is

$$
q' = \frac{m_2}{E'} \left[\Delta E_1 \left(\Delta E_1 + \frac{2m_3 m_4}{m_2} \right) \right]^{\frac{1}{2}}
\tag{12.45}
$$

where ΔE_1 is the incident projectile's energy in the laboratory above the threshold energy (12.41):

$$
\Delta E_1 = T_1 - T_{\text{th}}
\tag{12.46}
$$

For elastic scattering where $m_3 = m_1$, $m_4 = m_2$, (12.45) obviously reduces to (12.34).

Since the scattering or reaction is actually observed in the laboratory, it is necessary to transform back from the CM frame to the laboratory. Figure 12.3 shows the initial momentum $\mathbf{p}$ and the final momenta $\mathbf{p}_3$ and $\mathbf{p}_4$ in the laboratory. The CM momenta in Fig. 12.2 have been thrown forward by the Lorentz transformation. We can express the laboratory energy E_3 in terms of CM quantities by the Lorentz transformation v_{CM}, using (12.35) and (12.4). If θ' is the CM angle of $\mathbf{p}_3'$ with respect to the incident direction, we find

$$
E_3 = \gamma_{\text{CM}}(E_3' + v_{\text{CM}} q' \cos \theta')
\tag{12.47}
$$

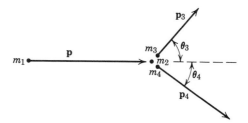

Fig. 12.3 Momentum vectors in laboratory for a two-body process.

Then an explicit expression is

$$E_3 = \tfrac{1}{2}(E_1 + m_2)\left(1 + \frac{m_3{}^2 - m_4{}^2}{E'^2}\right)$$

$$+ \frac{p}{2}\left\{\left[1 - \left(\frac{m_3 + m_4}{E'}\right)^2\right]\left[1 - \left(\frac{m_3 - m_4}{E'}\right)^2\right]\right\}^{\frac{1}{2}}\cos\theta' \quad (12.48)$$

where E' is given by (12.31). To obtain E_4 we merely interchange m_3 and m_4 and change θ' into $\pi - \theta'$ ($\cos\theta' \to -\cos\theta'$).

The relation between angles θ' and θ_3 can be obtained from the expression

$$\tan\theta_3 = \frac{p_{3\perp}}{p_{3\parallel}} = \frac{q'\sin\theta'}{\gamma_{\rm CM}(q'\cos\theta' + v_{\rm CM}E_3')} \quad (12.49)$$

Therefore we find

$$\tan\theta_3 = \frac{E'\sin\theta'}{(E_1 + m_2)(\cos\theta' + \alpha)} \quad (12.50)$$

where

$$\alpha = \frac{v_{\rm CM}E_3'}{q'} = \frac{p}{E_1 + m_2}\cdot\frac{1 + \dfrac{m_3{}^2 - m_4{}^2}{E'^2}}{\left\{\left[1 - \left(\dfrac{m_3 + m_4}{E'}\right)^2\right]\left[1 - \left(\dfrac{m_3 - m_4}{E'}\right)^2\right]\right\}^{\frac{1}{2}}}$$

$$(12.51)$$

We note that α is the ratio of the CM velocity to the velocity of particle 3 in the CM system. Just above threshold, α will be large compared to unity. This means that, as θ' ranges over all values from $0 \to \pi$ in the CM system, θ_3 will be confined to some forward cone, $0 \le \theta_3 \le \theta_{\max}$. Figure 12.4 shows the general behavior when $\alpha > 1$. The laboratory angle θ_3 is double valued if $\alpha > 1$, with particles emitted forwards and backwards in the CM system appearing at the same laboratory angle. The two types of particles can be distinguished by their energies. From (12.48) it is evident that the particles emitted forwards in the CM frame will be of higher energy than those emitted backwards. For $\alpha < 1$, it is evident that the denominator in (12.50) can vanish for some $\theta' > (\pi/2)$, implying $\theta_3 = (\pi/2)$, and is negative for large θ'. This means that θ_3 varies over the full range ($0 \le \theta_3 \le \pi$) and is a single-valued function of θ'. Such a curve is shown in Fig. 12.4.

Although it is not difficult to relate θ' and θ_3 through (12.50) and so obtain E_3 as a function of θ_3 from (12.48), it is sometimes convenient to have an explicit expression for this relationship. Using conservation of energy and momentum in the laboratory,

$$p_1 + p_2 = p_3 + p_4 \quad (12.52)$$

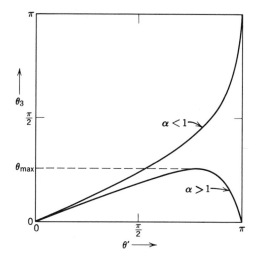

Fig. 12.4 Laboratory angle θ_3 of particle 3 versus center of momentum angle θ' for $\alpha < 1$ and $\alpha > 1$.

it is a straightforward, although tedious, matter to obtain the result:

$$E_3 = \frac{\left\{ \begin{array}{c} (E_1 + m_2)\left(m_2 E_1 + \dfrac{m_1{}^2 + m_2{}^2 + m_3{}^2 - m_4{}^2}{2} \right) \\[2mm] \pm p \cos \theta_3 \left[\left(m_2 E_1 + \dfrac{m_1{}^2 + m_2{}^2 - m_3{}^2 - m_4{}^2}{2} \right)^2 - m_3{}^2 m_4{}^2 - p^2 m_3{}^2 \sin^2 \theta_3 \right]^{1/2} \end{array} \right\}}{(E_1 + m_2)^2 - p^2 \cos^2 \theta_3}$$

(12.53)

Only the values of (12.53) greater than m_3 have physical significance. Both roots are allowed when $\alpha > 1$ in (12.50), but only one when $\alpha < 1$, as can be readily verified. To obtain E_4 we merely interchange m_3 and m_4 and replace θ_3 by θ_4.

For elastic scattering with $m_3 = m_1$, $m_4 = m_2$, the above relationships simplify considerably. The scattering angle in the laboratory is given by (12.50) with

$$\alpha = \frac{m_1}{m_2} \frac{\left(\dfrac{m_2}{m_1} E_1 + m_1 \right)}{(E_1 + m_2)}$$

(12.54)

In the nonrelativistic limit this reduces to the well-known result $\alpha = m_1/m_2$. The energy lost by the incident particle is $\Delta E = T_4 = E_4 - m_4$. From (12.48) we can obtain ΔE in terms of the CM scattering angle:

$$\Delta E = \frac{m_2 p^2 (1 - \cos \theta')}{2 m_2 E_1 + m_1{}^2 + m_2{}^2}$$

(12.55)

An alternative expression for ΔE in terms of the laboratory angle of recoil θ_4 can be found from (12.53):

$$\Delta E = \frac{2m_2 p^2 \cos^2 \theta_4}{2m_2 E_1 + m_1{}^2 + m_2{}^2 + p^2 \sin^2 \theta_4} \qquad (12.56)$$

For a head-on collision both expressions take on the maximum value

$$\Delta E_{\text{max}} = \frac{2m_2 p^2}{2m_2 E_1 + m_1{}^2 + m_2{}^2} \qquad (12.57)$$

The nonrelativistic value of ΔE_{max} is

$$\Delta E_{\text{max}} \simeq \frac{4m_1 m_2}{(m_1 + m_2)^2} (\tfrac{1}{2} m_1 v_1{}^2) \qquad (12.58)$$

showing that all the incident kinetic energy can be transferred in a head-on collision if $m_1 = m_2$ (true relativistically as well).

An important example of energy transfer occurs in collisions between incident charged particles and atomic electrons. These electrons can be treated as essentially at rest. If the incident particle is not an electron, $m_1 \gg m_2$. Then the maximum energy transfer can be written approximately as

$$\Delta E_{\text{max}} \simeq 2m_2 \left(\frac{p}{m_1}\right)^2 = 2m_2 \gamma^2 \beta^2 \qquad (12.59)$$

where γ, β are characteristic of the incident particle. Equation (12.59) is valid, provided the incident energy is not too large:

$$E_1 \ll \left(\frac{m_1}{m_2}\right) m_1 \qquad (12.60)$$

For mu mesons this limit is 20 Bev; for protons it is nearly 2000 Bev. For electron-electron collisions ($m_1 = m_2 = m$), the maximum energy transfer is

$$\Delta E_{\text{max}}^{(e)} = (\gamma - 1)m \qquad (12.61)$$

12.5 Covariant Lorentz Force Equation; Lagrangian and Hamiltonian for a Relativistic Charged Particle

In Section 12.1 we considered the Lorentz force equation as a method of establishing the Lorentz transformation properties of the momentum and energy of a particle, but we did not explicitly examine the equation as a covariant equation of motion for a particle moving in external fields. We now want to establish that covariance and discuss the associated Lagrangian, canonical momenta, and Hamiltonian. From equations (12.2) and and (11.129) we see that we can write the force equation in the form:

$$\frac{dp_\mu}{dt} = \frac{1}{c} \int F_{\mu\nu} J_\nu \, d^3x \qquad (12.62)$$

where the volume integral is over the extent of the charge. If the particle's velocity is **v** and its total charge e, then

$$\frac{dp_\mu}{dt} = \frac{e}{c} F_{\mu\nu} v_\nu \qquad (12.63)$$

where $v_\nu = v_k$ for $\nu = k = 1, 2, 3$ and $v_4 = ic$. This is not yet a covariant form for the equation, since v_ν is *not* a 4-vector, and dp_μ/dt is not one either. This deficiency can be remedied by writing a derivative with respect to proper time τ (11.62) rather than t. Since $dt = \gamma \, d\tau$, we obtain

$$\frac{dp_\mu}{d\tau} = \frac{e}{c} F_{\mu\nu} \gamma v_\nu \qquad (12.64)$$

But now $\gamma v_\nu = p_\nu/m$ is a 4-vector (sometimes called the 4-*velocity*). Consequently we arrive at the obviously covariant force equation for a particle:

$$\frac{dp_\mu}{d\tau} = \frac{e}{mc} F_{\mu\nu} p_\nu \qquad (12.65)$$

This is the counterpart for a discrete particle of the Lorentz force-density equation (11.129) for continuously distributed charge and current. Having established its covariance, it is often simplest to revert to the space-time forms:

$$\left.\begin{aligned}
\frac{d\mathbf{p}}{dt} &= e\left(\mathbf{E} + \frac{\mathbf{v} \times \mathbf{B}}{c}\right) \\
\frac{dE}{dt} &= e\mathbf{v} \cdot \mathbf{E}
\end{aligned}\right\} \qquad (12.66)$$

in any convenient reference frame. Equation (12.65) shows that, as long as all the different quantities are transformed according to their separate transformation laws, the noncovariant forms will be valid in any Lorentz frame.

Although the force equation (12.65) or (12.66) is sufficient to describe the general motion of a charged particle in external electromagnetic fields, it is sometimes convenient to use the ideas and formalism of Lagrangian and Hamiltonian mechanics. In order to see how to obtain an appropriate Lagrangian for the Lorentz force equation, we start with a free, but relativistic, particle. Since the Lagrangian must be a function of velocities and coordinates, we write the free-particle equation of motion as

$$\frac{d}{dt}(\gamma m \mathbf{v}) = 0 \tag{12.67}$$

where $\gamma = [1 - (v^2/c^2)]^{-\frac{1}{2}}$. At the least sophisticated level we know that the Lagrangian L must be chosen strategically so that the Euler-Lagrange equations of motion,

$$\frac{d}{dt}\left(\frac{\partial L}{\partial \dot{q}_i}\right) - \frac{\partial L}{\partial q_i} = 0 \tag{12.68}$$

are the same as Newton's equations of motion. Only a moment's consideration shows that a suitable Lagrangian for a free particle is

$$L_f = -mc^2\left(1 - \frac{v^2}{c^2}\right)^{\frac{1}{2}} \tag{12.69}$$

Evidently this form yields (12.67) when substituted into (12.68).

To obtain the free-particle Lagrangian in a more elegant way we consider Hamilton's principle or the principle of least action. This principle states that the motion of a mechanical system is such that in going from one configuration a at time t_1 to another configuration b at time t_2, the action integral A, defined as the time integral of the Lagrangian along the path of the system,

$$A = \int_a^b L \, dt \tag{12.70}$$

is an extremum (actually a minimum). By considering small variations of the path taken and demanding $\delta A = 0$, one obtains the Euler-Lagrange equations of motion (12.68). We now appeal to the Lorentz invariance of the action in order to determine the free-particle Lagrangian. That the action is a Lorentz scalar follows the first postulate of relativity, since the requirement that it be an extremum determines the mechanical equations of motion. If we introduce the proper time through $dt = \gamma \, d\tau$, the action integral becomes:

$$A = \int_a^b (\gamma L) \, d\tau \tag{12.71}$$

Since proper time is Lorentz invariant, the condition that A be also Lorentz invariant forces γL to be Lorentz invariant. This is a general condition on the Lagrangian. For a free particle L_f can be a function of only the velocity of the particle (and perhaps its mass). The only Lorentz invariant quantities involving the velocity are functions of the 4-vector scalar product $(1/m^2)(p \cdot p)$, where p_μ is the 4-momentum of the particle.

Since $(p \cdot p) = -m^2$, we see that for a free particle γL_f is a constant,

$$\gamma L_f = -\lambda \tag{12.72}$$

Then the action is proportional to the integral of the proper time over the path from the initial space-time point a to the final space-time point b. This integral is Lorentz invariant, but depends on the path taken. For purposes of calculation, consider a reference frame in which the particle is initially at rest. From definition (11.62) of proper time it is clear that, if the particle stays at rest in that frame, the integral over proper time will be larger than if it moves with a nonzero velocity along its path. Consequently we see that a straight world line joining the initial and final points of the path gives the maximum integral over proper time, or, with the negative sign in (12.72), a minimum for the action integral. Comparison with Newton's equation for nonrelativistic motion shows that $\lambda = mc^2$, yielding the free-particle Lagrangian (12.69).

The general requirement that γL be Lorentz invariant allows us to determine the Lagrangian for a relativistic charged particle in external electromagnetic fields, provided we know something about the Lagrangian (or equations of motion) for nonrelativistic motion in static fields. A slowly moving charged particle is influenced predominantly by the electric field which is derivable from the scalar potential Φ. The potential energy of interaction is $V = e\Phi$. Since the nonrelativistic Lagrangian is $(T - V)$, the interaction part L_{int} of the relativistic Lagrangian must reduce in the nonrelativistic limit to

$$L_{int} \to L_{int}^{NR} = -e\Phi \tag{12.73}$$

Our problem thus becomes that of finding a Lorentz invariant expression for γL_{int} which reduces to (12.73) for nonrelativistic velocities. Since Φ is the fourth component of the 4-vector potential A_μ, we anticipate that γL_{int} will involve the scalar product of A_μ with some 4-vector. The only other 4-vectors available are the momentum and position vectors of the particle. Since gamma times the Lagrangian must be translationally invariant as well as Lorentz invariant, it cannot involve the coordinates explicitly. Hence the interaction Lagrangian must be*

$$L_{int} = \frac{1}{\gamma}\frac{e}{mc}(p \cdot A) = \frac{e\mathbf{v} \cdot \mathbf{A}}{c} - e\Phi \tag{12.74}$$

---

* Without appealing to the nonrelativistic limit this form of L_{int} can be written down by demanding that γL_{int} be a Lorentz invariant which is (1) linear in the charge of the particle, (2) linear in the electromagnetic potentials, (3) translationally invariant, and (4) a function of no higher than the first time derivative of the particle coordinates. The reader may consider the possiblity of an interaction Lagrangian satisfying these conditions, but linear in the field strengths $F_{\mu\nu}$, rather than the potentials A_μ.

where the coefficient of the scalar product $(p \cdot A)$ is chosen to yield (12.73) in the limit $\mathbf{v} \to 0$.

The combination of (12.69) and (12.74) yields the complete relativistic Lagrangian for a charged particle:

$$L = \begin{cases} \dfrac{1}{\gamma}\left[-mc^2 + \dfrac{e}{mc}(p \cdot A) \right] \\[4mm] -mc^2\sqrt{1 - \dfrac{v^2}{c^2}} + \dfrac{e}{c}\mathbf{v} \cdot \mathbf{A} - e\Phi \end{cases} \tag{12.75}$$

where the upper (lower) line gives L in 4-vector (explicit space-time) form. Verification that (12.75) does indeed lead to the Lorentz force equation will be left as an exercise for the reader. Use must be made of the convective derivative $[d/dt = (\partial/\partial t) + \mathbf{v} \cdot \nabla]$ and the standard definitions of the fields in terms of the potentials.

The canonical momentum $\mathbf{P}$ conjugate to the position coordinate $\mathbf{x}$ is obtained by the definition,

$$P_i \equiv \frac{\partial L}{\partial v_i} = \gamma m v_i + \frac{e}{c} A_i \tag{12.76}$$

Thus the conjugate momentum is

$$\mathbf{P} = \mathbf{p} + \frac{e}{c}\mathbf{A} \tag{12.77}$$

where $\mathbf{p} = \gamma m \mathbf{v}$ is the momentum in the absence of fields. The Hamiltonian H is a function of the coordinate $\mathbf{x}$ and its conjugate momentum $\mathbf{P}$ and is a constant of the motion if the Lagrangian is not an explicit function of time. The Hamiltonian is defined in terms of the Lagrangian as

$$H = \mathbf{P} \cdot \mathbf{v} - L \tag{12.78}$$

The velocity $\mathbf{v}$ must be eliminated from (12.78) in favor of $\mathbf{P}$ and $\mathbf{x}$. From (12.76) or (12.77) we find that

$$\mathbf{v} = \frac{c\mathbf{P} - e\mathbf{A}}{\sqrt{\left(\mathbf{P} - \dfrac{e\mathbf{A}}{c}\right)^2 + m^2 c^2}} \tag{12.79}$$

When this is substituted into (12.78) and into L (12.75), the Hamiltonian takes on the form:

$$H = \sqrt{(c\mathbf{P} - e\mathbf{A})^2 + m^2 c^4} + e\Phi \tag{12.80}$$

Again the reader may verify that Hamilton's equations of motion can be combined to yield the Lorentz force equation. Equation (12.80) is an expression for the total energy W of the particle. It differs from the free-particle energy by the addition of the potential energy $e\Phi$ and by the replacement $\mathbf{p} \rightarrow [\mathbf{P} - (e/c)\mathbf{A}]$. These two modifications are actually only one 4-vector change. This can be seen by transposing $e\Phi$ in (12.80) and squaring both sides. Then

$$(c\mathbf{P} - e\mathbf{A})^2 - (W - e\Phi)^2 = -(mc^2)^2 \qquad (12.81)$$

This is just the 4-vector scalar product

$$(p \cdot p) = -(mc)^2 \qquad (12.82)$$

where
$$p_\mu \equiv \left(\mathbf{p}, \frac{iE}{c}\right) = \left[\left(\mathbf{P} - \frac{e\mathbf{A}}{c}\right), \frac{i}{c}(W - e\Phi)\right] \qquad (12.83)$$

We see that in some sense the total energy W is the fourth component of a canonically conjugate 4-momentum of which (12.77) is the space part. An alternative formulation with a relativistically invariant Lagrangian which is a function of the 4-velocity $u_\mu = p_\mu/m$ is discussed in Problem 12.5. There the canonical 4-momentum arises naturally.

The Lagrangian and Hamiltonian formulation of the dynamics of a charged particle has been outlined for several reasons. One is that the concept of Lorentz invariance, coupled with other physical requirements, was shown to be a powerful tool in the systematic construction of a Lagrangian which yields dynamic equations of motion. Another is that the Lagrangian is often a convenient starting point in discussing particle dynamics. Finally, the concepts and ideas of conjugate variables, etc., are useful even when one proceeds to solve the force equation directly.

12.6 Lowest-Order Relativistic Corrections to the Lagrangian for Interacting Charged Particles

In the previous section we discussed the general Lagrangian formalism for a relativistic particle in external electromagnetic fields described by the vector and scalar potentials, $\mathbf{A}$ and Φ. The appropriate interaction Lagrangian was given by (12.74). If we now consider the problem of a Lagrangian description of the interaction of two or more charged particles with each other, we see that it is possible only at nonrelativistic velocities. The Lagrangian is supposed to be a function of the instantaneous velocities and coordinates of all the particles. When the finite velocity of propagation of electromagnetic fields is taken into account, this is no longer possible, since the values of the potentials at one particle due to the other particles depend on their state of motion at "retarded" times. Only when retardation effects can be neglected is a Lagrangian description of the system of particles alone possible. In view of this one might think that a Lagrangian could be formulated only in the static limit, i.e., to zeroth order in (v/c). We will now show, however, that lowest-order relativistic corrections can be included, giving an approximate Lagrangian for interacting particles, correct to the order of $(v/c)^2$ inclusive.

It is sufficient to consider two interacting particles with charges q_1 and q_2, masses m_1 and m_2, and coordinates $\mathbf{x}_1$ and $\mathbf{x}_2$. The relative separation is $\mathbf{r} = \mathbf{x}_1 - \mathbf{x}_2$. The interaction Lagrangian in the static limit is just the negative of the electrostatic potential energy,

$$L_{\text{int}}^{\text{NR}} = - \frac{q_1 q_2}{r} \tag{12.84}$$

If attention is directed to the first particle, this can be viewed as the negative of the product of q_1 and the scalar potential Φ_{12} due to the second particle at the position of the first. This is of the same form as (12.73). If we wish to generalize beyond the static limit, we must, according to (12.74), determine both Φ_{12} and $\mathbf{A}_{12}$, at least to some degree of approximation. In general there will be relativistic corrections to both Φ_{12} and $\mathbf{A}_{12}$. But in the *Coulomb gauge*, the scalar potential is given correctly to all orders in v/c by the instantaneous Coulomb potential. Thus, if we calculate in that gauge, the scalar-potential contribution Φ_{12} is already known. All that needs to be considered is the vector potential $\mathbf{A}_{12}$.

If only the lowest-order relativistic corrections are desired, retardation effects can be neglected in computing $\mathbf{A}_{12}$. The reason is that the vector potential enters the Lagrangian (12.74) in the combination $q_1(\mathbf{v}_1/c) \cdot \mathbf{A}_{12}$. Since $\mathbf{A}_{12}$ itself is of the order of v_2/c, greater accuracy in calculating $\mathbf{A}_{12}$ is unnecessary. Consequently, we have the magnetostatic expression,

$$\mathbf{A}_{12} \simeq \frac{1}{c} \int \frac{\mathbf{J}_t(\mathbf{x}') \, d^3 x'}{|\mathbf{x}_1 - \mathbf{x}'|} \tag{12.85}$$

where $\mathbf{J}_t$ is the transverse part of the current due to the second particle, as discussed in Section 6.5. From equations (6.46)–(6.50) it is easy to see that the transverse current is

$$\mathbf{J}_t(\mathbf{x}') = q_2 \mathbf{v}_2 \, \delta(\mathbf{x}' - \mathbf{x}_2) - \frac{q_2}{4\pi} \boldsymbol{\nabla}' \left(\frac{\mathbf{v}_2 \cdot (\mathbf{x}' - \mathbf{x}_2)}{|\mathbf{x}' - \mathbf{x}_2|^3} \right) \tag{12.86}$$

When this is inserted in (12.85), the first term can be integrated immediately. Thus

$$\mathbf{A}_{12} \simeq \frac{q_2 \mathbf{v}_2}{cr} - \frac{q_2}{4\pi c} \int \frac{1}{|\mathbf{x}' - \mathbf{x}_1|} \boldsymbol{\nabla}' \left(\frac{\mathbf{v}_2 \cdot (\mathbf{x}' - \mathbf{x}_2)}{|\mathbf{x}' - \mathbf{x}_2|^3} \right) d^3 x' \tag{12.87}$$

By changing variables to $\mathbf{y} = \mathbf{x}' - \mathbf{x}_2$ and integrating by parts, the integral can be put in the form,

$$\mathbf{A}_{12} \simeq \frac{q_2\mathbf{v}_2}{cr} - \frac{q_2}{4\pi c}\boldsymbol{\nabla}_r \int \frac{\mathbf{v}_2 \cdot \mathbf{y}}{y^3} \frac{1}{|\mathbf{y} - \mathbf{r}|} d^3y \tag{12.88}$$

The integral can now be done in a straightforward manner to yield

$$\mathbf{A}_{12} \simeq \frac{q_2}{c}\left[\frac{\mathbf{v}_2}{r} - \tfrac{1}{2}\boldsymbol{\nabla}_r\left(\frac{\mathbf{v}_2 \cdot \mathbf{r}}{r}\right)\right] \tag{12.89}$$

The differentiation of the second term leads to the final result

$$\mathbf{A}_{12} \simeq \frac{q_2}{2cr}\left[\mathbf{v}_2 + \frac{\mathbf{r}(\mathbf{v}_2 \cdot \mathbf{r})}{r^2}\right] \tag{12.90}$$

With expression (12.90) for the vector potential due to the second particle at the position of the first, the interaction Lagrangian for two charged particles, including lowest-order relativistic effects, is

$$L_{\text{int}} = \frac{q_1 q_2}{r}\left\{-1 + \frac{1}{2c^2}\left[\mathbf{v}_1 \cdot \mathbf{v}_2 + \frac{(\mathbf{v}_1 \cdot \mathbf{r})(\mathbf{v}_2 \cdot \mathbf{r})}{r^2}\right]\right\} \tag{12.91}$$

This interaction form was first obtained by Darwin in 1920. It is of importance in a quantum-mechanical discussion of relativistic corrections in two-electron atoms. In the quantum-mechanical problem the velocity vectors are replaced by their corresponding quantum-mechanical operators (Dirac $\boldsymbol{\alpha}$'s). Then the interaction is known as the Breit interaction (1930).

12.7 Motion in a Uniform, Static, Magnetic Field

As a first important example of the dynamics of charged particles in electromagnetic fields we consider the motion in a uniform, static, magnetic induction **B**. The equations of motion (12.66) are

$$\frac{d\mathbf{p}}{dt} = \frac{e}{c}\mathbf{v} \times \mathbf{B}, \qquad \frac{dE}{dt} = 0 \tag{12.92}$$

Since the energy is constant in time, the magnitude of the velocity is constant and so is γ. Then the first equation can be written

$$\frac{d\mathbf{v}}{dt} = \mathbf{v} \times \boldsymbol{\omega}_B \tag{12.93}$$

where

$$\boldsymbol{\omega}_B = \frac{e\mathbf{B}}{\gamma mc} = \frac{ec\mathbf{B}}{E} \tag{12.94}$$

is the gyration or precession frequency. The motion described by (12.93) is a circular motion perpendicular to **B** and a uniform translation parallel to **B**. The solution for the velocity is easily shown to be

$$\mathbf{v}(t) = v_{\parallel}\boldsymbol{\epsilon}_3 + \omega_B a(\boldsymbol{\epsilon}_1 - i\boldsymbol{\epsilon}_2)e^{-i\omega_B t} \tag{12.95}$$

where $\boldsymbol{\epsilon}_3$ is a unit vector parallel to the field, $\boldsymbol{\epsilon}_1$ and $\boldsymbol{\epsilon}_2$ are the other orthogonal unit vectors, $v_{\parallel}$ is the velocity component along the field, and a is the gyration radius. The convention is that the real part of the equation is to be taken. Then one can see that (12.95) represents a counterclockwise rotation (for positive charge e) when viewed in the direction of **B**. Another integration yields the displacement of the particle,

$$\mathbf{x}(t) = \mathbf{X}_0 + v_{\parallel}t\boldsymbol{\epsilon}_3 + ia(\boldsymbol{\epsilon}_1 - i\boldsymbol{\epsilon}_2)e^{-i\omega_B t} \tag{12.96}$$

The path is a helix of radius a and pitch angle $\alpha = \tan^{-1}(v_{\parallel}/\omega_B a)$. The magnitude of the gyration radius a depends on the magnetic induction **B** and the transverse momentum $\mathbf{p}_{\perp}$ of the particle. From (12.94) and (12.95) it is evident that

$$cp_{\perp} = eBa$$

This form is convenient for the determination of particle momenta. The radius of curvature of the path of a charged particle in a known B allows the determination of its momentum. For particles with charge the same in magnitude as the electronic charge, the momentum can be written numerically as

$$p_{\perp}\,(\text{Mev}/c) = 3.00 \times 10^{-4}Ba\,\text{(gauss-cm)} \tag{12.97}$$

12.8   Motion in Combined, Uniform, Static Electric and Magnetic Fields

We now consider a charged particle moving in a combination of electric and magnetic fields **E** and **B**, both uniform and static, but in general not parallel. As an important special case, *perpendicular fields* will be treated first. The force equation (12.66) shows that the particle's energy is not constant in time. Consequently we cannot obtain a simple equation for the velocity, as was done for a static magnetic field. But an appropriate Lorentz transformation simplifies the equations of motion. Consider a Lorentz transformation to a coordinate frame K' moving with a velocity **u** with respect to the original frame. Then the Lorentz force equation for the particle in K' is

$$\frac{d\mathbf{p}'}{dt'} = e\left(\mathbf{E}' + \frac{\mathbf{v}' \times \mathbf{B}'}{c}\right)$$

where the primed variables are referred to the system K'. The fields E' and B' are given by relations (11.115) with v replaced by u, where $\parallel$ and $\perp$ refer to the direction of u. Let us first suppose that $|E| < |B|$. If u is now chosen perpendicular to the orthogonal vectors E and B,

$$\mathbf{u} = c\,\frac{(\mathbf{E} \times \mathbf{B})}{B^2} \tag{12.98}$$

we find the fields in K' to be

$$\left.\begin{aligned}
\mathbf{E}_{\parallel}' &= 0, & \mathbf{E}_{\perp}' &= \gamma\left(\mathbf{E} + \frac{\mathbf{u}}{c} \times \mathbf{B}\right) = 0 \\[2mm]
\mathbf{B}_{\parallel}' &= 0, & \mathbf{B}_{\perp}' &= \frac{1}{\gamma}\,\mathbf{B} = \left(\frac{B^2 - E^2}{B^2}\right)^{\!1/2}\mathbf{B}
\end{aligned}\right\} \tag{12.99}$$

In the frame K' the only field acting is a static magnetic field B' which points in the same direction as B, but is weaker than B by a factor γ^{-1}. Thus the motion in K' is the same as that considered in the previous section, namely a spiraling around the lines of force. As viewed from the original coordinate system, this gyration is accompanied by a uniform "drift" u perpendicular to E and B given by (12.98). This drift is sometimes called the $E \times B$ *drift*. It has already been considered for a conducting fluid in another context in Section 10.3. The drift can be understood qualitatively by noting that a particle which starts gyrating around B is accelerated by the electric field, gains energy, and so moves in a path with a larger radius for roughly half of its cycle. On the other half, the electric field decelerates it, causing it to lose energy and so move in a tighter arc. The combination of arcs produces a translation perpendicular to E and B as shown in Fig. 12.5. The direction of drift is independent of the sign of the charge of the particle.

The drift velocity u (12.98) has physical meaning only if it is less than the velocity of light, i.e., only if $|E| < |B|$. If $|E| > |B|$, the electric field

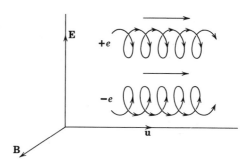

Fig. 12.5 E × B drift of charged
particles in crossed fields.

is so strong that the particle is continually accelerated in the direction of **E** and its average energy continues to increase with time. To see this we consider a Lorentz transformation from the original frame to a system K'' moving with a velocity

$$\mathbf{u}' = c \frac{(\mathbf{E} \times \mathbf{B})}{E^2} \qquad (12.100)$$

relative to the first. In this frame the electric and magnetic fields are

$$\left. \begin{array}{ll} \mathbf{E}_{\parallel}'' = 0, & \mathbf{E}_{\perp}'' = \dfrac{1}{\gamma'} \mathbf{E} = \left(\dfrac{E^2 - B^2}{E^2} \right)^{1/2} \mathbf{E} \\[4mm] \mathbf{B}_{\parallel}'' = 0, & \mathbf{B}_{\perp}'' = \gamma' \left(\mathbf{B} - \dfrac{\mathbf{u}' \times \mathbf{E}}{c} \right) = 0 \end{array} \right\} \qquad (12.101)$$

Thus in the system K'' the particle is acted on by a purely electrostatic field which causes hyperbolic motion with ever-increasing velocity (see Problem 12.7).

The fact that a particle can move through crossed **E** and **B** fields with the uniform velocity $u = cE/B$ provides the possibility of selecting charged particles according to velocity. If a beam of particles having a spread in velocities is normally incident on a region containing uniform crossed electric and magnetic fields, only those particles with velocities equal to cE/B will travel without deflection. Suitable entrance and exit slits will then allow only a very narrow band of velocities around cE/B to be transmitted, the resolution depending on the geometry, the velocities desired, and the field strengths. When combined with momentum selectors, such as a deflecting magnet, these $E \times B$ velocity selectors can separate a very pure and monoenergetic beam of particles of a definite mass from a mixed beam of particles with different masses and momenta. Large-scale devices of this sort are commonly used to provide experimental beams of particles produced in very high-energy accelerators.

If **E** has a component parallel to **B**, the behavior of the particle cannot be understood in such simple terms as above. The scalar product $\mathbf{E} \cdot \mathbf{B}$ is a Lorentz invariant quantity (see Problem 11.10), as is $(B^2 - E^2)$. When the fields were perpendicular ($\mathbf{E} \cdot \mathbf{B} = 0$), it was possible to find a Lorentz frame where $\mathbf{E} = 0$ if $|\mathbf{B}| > |\mathbf{E}|$, or $\mathbf{B} = 0$ if $|\mathbf{E}| > |\mathbf{B}|$. In those coordinate frames the motion was relatively simple. If $\mathbf{E} \cdot \mathbf{B} \neq 0$, electric and magnetic fields will exist simultaneously in all Lorentz frames, the angle between the fields remaining acute or obtuse depending on its value in the original coordinate frame. Consequently motion in combined fields must be considered. When the fields are static and uniform, it is a straightforward matter to obtain a solution for the motion in cartesian components. This will be left for Problem 12.10.

12.9 Particle Drifts in Nonuniform, Static Magnetic Fields

In astrophysical and thermonuclear applications it is of considerable interest to know how particles behave in magnetic fields which vary in space. Often the variations are gentle enough that a perturbation solution to the motion, first given by Alfvén, is an adequate approximation. "Gentle enough" generally means that the distance over which **B** changes appreciably in magnitude or direction is large compared to the gyration radius a of the particle. Then the lowest-order approximation to the motion is a spiraling around the lines of force at a frequency given by the local value of the magnetic induction. In the next approximation, slow changes occur in the orbit which can be described as a drifting of the guiding center.

The first type of spatial variation of the field to be considered is a gradient perpendicular to the direction of **B**. Let the gradient at the point of interest be in the direction of the unit vector **n**, with $\mathbf{n} \cdot \mathbf{B} = 0$. Then, to first order, the gyration frequency can be written

$$\boldsymbol{\omega}_B(\mathbf{x}) = \frac{e}{\gamma mc}\,\mathbf{B}(\mathbf{x}) \simeq \boldsymbol{\omega}_0\left[1 + \left(\frac{1}{B_0}\frac{\partial B}{\partial \xi}\right)\mathbf{n} \cdot \mathbf{x}\right] \qquad (12.102)$$

In (12.102) ξ is the coordinate in the direction **n**, and the expansion is about the origin of coordinates where $\omega_B = \omega_0$. Since the direction of **B** is unchanged, the motion parallel to **B** remains a uniform translation. Consequently we consider only modifications in the transverse motion. Writing $\mathbf{v}_\perp = \mathbf{v}_0 + \mathbf{v}_1$, where $\mathbf{v}_0$ is the uniform-field transverse velocity and $\mathbf{v}_1$ is a small correction term, we can substitute (12.102) into the force equation

$$\frac{d\mathbf{v}_\perp}{dt} = \mathbf{v}_\perp \times \boldsymbol{\omega}_B(\mathbf{x}) \qquad (12.103)$$

and, keeping only first-order terms, obtain the approximate result,

$$\frac{d\mathbf{v}_1}{dt} \simeq \left[\mathbf{v}_1 + \mathbf{v}_0(\mathbf{n} \cdot \mathbf{x}_0)\left(\frac{1}{B_0}\frac{\partial B}{\partial \xi}\right)\right] \times \boldsymbol{\omega}_0 \qquad (12.104)$$

From (12.95) and (12.96) it is easy to see that for a uniform field the transverse velocity $\mathbf{v}_0$ and coordinate $\mathbf{x}_0$ are related by

$$\left.\begin{aligned}
\mathbf{v}_0 &= -\boldsymbol{\omega}_0 \times (\mathbf{x}_0 - \mathbf{X}) \\[2mm]
(\mathbf{x}_0 - \mathbf{X}) &= \frac{1}{\omega_0{}^2}(\boldsymbol{\omega}_0 \times \mathbf{v}_0)
\end{aligned}\right\} \qquad (12.105)$$

where **X** is the center of gyration of the unperturbed circular motion ($\mathbf{X} = 0$ here). If $(\boldsymbol{\omega}_0 \times \mathbf{v}_0)$ is eliminated in (12.104) in favor of $\mathbf{x}_0$, we obtain

$$\frac{d\mathbf{v}_1}{dt} \simeq \left[\mathbf{v}_1 - \left(\frac{1}{B_0}\frac{\partial B}{\partial \xi}\right)\boldsymbol{\omega}_0 \times \mathbf{x}_0(\mathbf{n}\cdot\mathbf{x}_0)\right] \times \boldsymbol{\omega}_0 \qquad (12.106)$$

This shows that, apart from oscillatory terms, $\mathbf{v}_1$ has a non zero average value,

$$\mathbf{v}_G \equiv \langle\mathbf{v}_1\rangle = \left(\frac{1}{B_0}\frac{\partial B}{\partial \xi}\right)\boldsymbol{\omega}_0 \times \langle(\mathbf{x}_0)_\perp(\mathbf{n}\cdot\mathbf{x}_0)\rangle \qquad (12.107)$$

To determine the average value of $(\mathbf{x}_0)_\perp(\mathbf{n}\cdot\mathbf{x}_0)$, it is necessary only to observe that the rectangular components of $(\mathbf{x}_0)_\perp$ oscillate sinusoidally with peak amplitude a and a phase difference of $90°$. Hence only the component of $(\mathbf{x}_0)_\perp$ parallel to **n** contributes to the average, and

$$\langle(\mathbf{x}_0)_\perp(\mathbf{n}\cdot\mathbf{x}_0)\rangle = \frac{a^2}{2}\mathbf{n} \qquad (12.108)$$

Thus the gradient drift velocity is given by

$$\mathbf{v}_G = \frac{a^2}{2}\frac{1}{B_0}\frac{\partial B}{\partial \xi}(\boldsymbol{\omega}_0 \times \mathbf{n}) \qquad (12.109)$$

An alternative form, independent of coordinates, is

$$\frac{\mathbf{v}_G}{\omega_B a} = \frac{a}{2B^2}(\mathbf{B} \times \boldsymbol{\nabla}_\perp B) \qquad (12.110)$$

From (12.110) it is evident that, if the gradient of the field is such that a

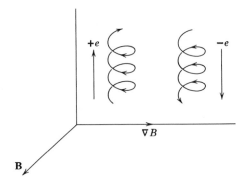

Fig. 12.6 Drift of charged particles due to transverse gradient of magnetic field.

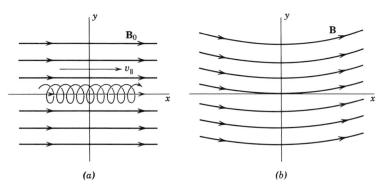

(a)   *(b)*

Fig. 12.7 (*a*) Particle moving in helical path along lines of uniform, constant magnetic induction. (*b*) Curvature of lines of magnetic induction will cause drift perpendicular to the (x, y) plane.

$|\nabla B/B| \ll 1$, the drift velocity is small compared to the orbital velocity ($\omega_B a$). The particle spirals rapidly while its center of rotation moves slowly perpendicular to both **B** and ∇B. The sense of the drift for positive particles is given by (12.110). For negatively charged particles the sign of the drift velocity is opposite; the sign change comes from the definition of ω_B. The gradient drift can be understood qualitatively from considera-tion of the variation of gyration radius as the particle moves in and out of regions of larger than average and smaller than average field strength. Figure 12.6 shows this qualitative behavior for both signs of charge.

Another type of field variation which causes a drifting of the particle's guiding center is curvature of the lines of force. Consider the two-dimensional field shown in Fig. 12.7. It is locally independent of z. On the left-hand side of the figure is a constant, uniform magnetic induction **B$_0$**, parallel to the x axis. A particle spirals around the field lines with a gyration radius a and a velocity $\omega_B a$, while moving with a uniform velocity $v_{\parallel}$ along the lines of force. We wish to treat that motion as a zero-order approximation to the motion of the particle in the field shown on the right-hand side of the figure, where the lines of force are curved with a local radius of curvature R which is large compared to a.

The first-order motion can be understood as follows. The particle tends to spiral around a field line, but the field line curves off to the side. As far as the motion of the guiding center is concerned, this is equivalent to a centrifugal acceleration of magnitude $v_{\parallel}^2/R$. This acceleration can be viewed as arising from an effective electric field,

$$\mathbf{E}_{\text{eff}} = \frac{\gamma m}{e} \frac{\mathbf{R}}{R^2} v_{\parallel}^2 \qquad (12.111)$$

in addition to the magnetic induction $\mathbf{B}_0$. From (12.98) we see that the combined effective electric field and the magnetic induction cause a *curvature drift velocity*,

$$\mathbf{v}_C \simeq c \frac{\gamma m}{e} v_\parallel^2 \frac{\mathbf{R} \times \mathbf{B}_0}{R^2 B_0^2} \tag{12.112}$$

With the definition of $\omega_B = eB_0/\gamma mc$, the curvature drift can be written

$$\mathbf{v}_C = \frac{v_\parallel^2}{\omega_B R} \left(\frac{\mathbf{R} \times \mathbf{B}_0}{R B_0} \right) \tag{12.113}$$

The direction of drift is specified by the vector product, in which $\mathbf{R}$ is the radius vector *from* the effective center of curvature *to* the position of the charge. The sign in (12.113) is appropriate for positive charges and is independent of the sign of $v_\parallel$. For negative particles the opposite sign arises from ω_B.

A more straightforward, although pedestrian, derivation of (12.113) can be given by solving the Lorentz force equation directly. If we use cylindrical coordinates (ρ, ϕ, z) appropriate to Fig. 12.7b with origin at the center of curvature, the magnetic induction has only a ϕ component, $B_\phi = B_0$. Then the force equation can be easily shown to give the three equations,

$$\ddot{\rho} - \rho\dot{\phi}^2 = -\omega_B \dot{z}$$
$$\rho\ddot{\phi} + 2\dot{\rho}\dot{\phi} = 0 \tag{12.114}$$
$$\ddot{z} = \omega_B \dot{\rho}$$

If the zero-order trajectory is a helix with radius a small compared to the radius of curvature R, then, to lowest order, $\dot{\phi} \simeq v_\parallel/R$, while $\rho \simeq R$. Thus the first equation of (12.114) yields an approximate result for $\dot{z}$:

$$\dot{z} \simeq \frac{v_\parallel^2}{\omega_B R} \tag{12.115}$$

This is just the curvature drift given by (12.113).

For regions of space in which there are no currents the gradient drift $\mathbf{v}_G$ (12.110) and the curvature drift $\mathbf{v}_C$ (12.113) can be combined into one simple form. This follows from the fact that $\nabla \times \mathbf{B} = 0$ implies

$$\frac{\nabla_\perp B}{B} = -\frac{\mathbf{R}}{R^2} \tag{12.116}$$

Evidently then the sum of $\mathbf{v}_G$ and $\mathbf{v}_C$ is a general drift velocity,

$$\mathbf{v}_D = \frac{1}{\omega_B R} (v_\parallel^2 + \tfrac{1}{2} v_\perp^2) \left(\frac{\mathbf{R} \times \mathbf{B}}{RB} \right) \tag{12.117}$$

where $v_\perp = \omega_B a$ is the transverse velocity of gyration. For singly charged nonrelativistic particles in thermal equilibrium, the magnitude of the drift velocity is

$$v_D(\text{cm/sec}) = \frac{172\ T(^\circ\text{K})}{R(\text{m})\ B(\text{gauss})} \tag{12.118}$$

The particle drifts implied by (12.117) are troublesome in certain types of thermonuclear machines designed to contain hot plasma. A possible configuration is a toroidal tube with a strong axial field supplied by solenoidal windings around the torus. With typical parameters of $R = 1$ meter, $B = 10^3$ gauss, particles in a 1-ev plasma ($T \simeq 10^{4\circ}\text{K}$) will have drift velocities $v_D \sim 1.8 \times 10^3$ cm/sec. This means that they will drift out to the walls in a small fraction of a second. For hotter plasmas the drift rate is correspondingly greater. One way to prevent this first-order drift in toroidal geometries is to twist the torus into a figure eight. Since the particles generally make many circuits around the closed path before drifting across the tube, they feel no net curvature or gradient of the field. Consequently they experience no net drift, at least to first order in $1/R$. This method of eliminating drifts due to spatial variations of the magnetic field is used in the Stellarator type of thermonuclear machine, in which containment is attempted with a strong, externally produced, axial magnetic field, rather than a pinch (see Sections 10.5–10.7).

12.10 Adiabatic Invariance of Flux through Orbit of Particle

The various motions discussed in the previous sections have been perpendicular to the lines of magnetic force. These motions, caused by electric fields or by the gradient or curvature of the magnetic field, arise because of the peculiarities of the magnetic-force term in the Lorentz force equation. To complete our general survey of particle motion in magnetic fields we must consider motion parallel to the lines of force. It turns out that for slowly varying fields a powerful tool is the concept of adiabatic invariants. In celestial mechanics and in the old quantum theory adiabatic invariants were useful in discussing perturbations on the one hand, and in deciding what quantities were to be quantized on the other. Our discussion will resemble most closely the celestial mechanical problem, since we are interested in the behavior of a charged particle in slowly varying fields which can be viewed as small departures from the simple, uniform, static field considered in Section 12.7.

The concept of adiabatic invariance is introduced by considering the action integrals of a mechanical system. If q_i and p_i are the generalized canonical coordinates and momenta, then, for each coordinate which is periodic, the action integral J_i is defined by

$$J_i = \oint p_i \, dq_i \tag{12.119}$$

The integration is over a complete cycle of the coordinate q_i. For a given mechanical system with specified initial conditions the action integrals J_i are constants. If now the properties of the system are changed in some way (e.g., a change in spring constant or mass of some particle), the question arises as to how the action integrals change. It can be proved* that, if the change in property is slow compared to the relevant periods of motion and is not related to the periods (such a change is called an *adiabatic change*), the action integrals are invariant. This means that, if we have a certain mechanical system in some state of motion and we make an adiabatic change in some property so that after a long time we end up with a different mechanical system, the final motion of that different system will be such that the action integrals have the same values as in the initial system. Clearly this provides a powerful tool in examining the effects of small changes in properties.

For a charged particle in a uniform, static, magnetic induction **B** the transverse motion is periodic. The action integral for this transverse motion is

$$J = \oint \mathbf{P}_{\perp} \cdot d\mathbf{l}, \tag{12.120}$$

where $\mathbf{P}_{\perp}$ is the transverse component of the canonical momentum (12.77) and $d\mathbf{l}$ is a directed line element along the circular path of the particle. From (12.77) we find that

$$J = \oint \gamma m \mathbf{v}_{\perp} \cdot d\mathbf{l} + \frac{e}{c} \oint \mathbf{A} \cdot d\mathbf{l} \tag{12.121}$$

Since $\mathbf{v}_{\perp}$ is parallel to $d\mathbf{l}$, we find

$$J = \oint \gamma m \omega_B a^2 \, d\theta + \frac{e}{c} \oint \mathbf{A} \cdot d\mathbf{l} \tag{12.122}$$

Applying Stokes's theorem to the second integral and integrating over θ in the first integral, we obtain

$$J = 2\pi \gamma m \omega_B a^2 + \frac{e}{c} \int_S \mathbf{B} \cdot \mathbf{n} \, da \tag{12.123}$$

---

* See, for example, M. Born, *The Mechanics of the Atom*, Bell, London (1927).

Since the line element $d\mathbf{l}$ in (12.120) is in a counterclockwise sense relative to $\mathbf{B}$, the unit vector $\mathbf{n}$ is antiparallel to $\mathbf{B}$. Hence the integral over the circular orbit subtracts from the first term. This gives

$$J = \gamma m \omega_B \pi a^2 = \frac{e}{c}(B\pi a^2) \tag{12.124}$$

making use of $\omega_B = eB/\gamma mc$. The quantity $B\pi a^2$ is the flux through the particle's orbit.

If the particle moves through regions where the magnetic field strength varies slowly in space or time, the adiabatic invariance of J means that the flux linked by the particle's orbit remains constant. If B increases, the radius a will decrease so that $B\pi a^2$ remains unchanged. This constancy of flux linked can be phrased in several ways involving the particle's orbit radius, its transverse momentum, its magnetic moment. These different statements take the forms:

$$\left.\begin{array}{c} Ba^2 \\ p_\perp^2/B \\ \gamma\mu \end{array}\right\} \text{ are adiabatic invariants} \tag{12.125}$$

where $\mu = (e\omega_B a^2/2c)$ is the magnetic moment of the current loop of the particle in orbit. If there are only static magnetic fields present, the speed of the particle is constant and its total energy does not change. Then the magnetic moment μ is itself an adiabatic invariant. In time-varying fields or with static electric fields, μ is an adiabatic invariant only in the nonrelativistic limit.

Let us now consider a simple situation in which a static magnetic field $\mathbf{B}$ acts mainly in the z direction, but has a small positive gradient in that direction. Figure 12.8 shows the general behavior of the lines of force. In addition to the z component of field there is a small radial component due to the curvature of the lines of force. For simplicity we assume cylindrical symmetry. Suppose that a particle is spiraling around the z axis in an

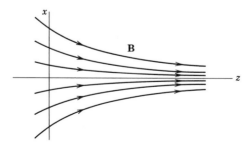

Fig. 12.8

orbit of small radius with a transverse velocity $\mathbf{v}_{\perp 0}$ and a component of velocity $v_{\parallel 0}$ parallel to $\mathbf{B}$ at $z = 0$, where the axial field strength is B_0. The speed of the particle is constant so that at any position along the z axis

$$v_{\parallel}^2 + v_{\perp}^2 = v_0^2 \tag{12.126}$$

where $v_0^2 = v_{\perp 0}^2 + v_{\parallel 0}^2$ is the square of the speed at $z = 0$. If we assume that the flux linked is a constant of the motion, then (12.125) allows us to write

$$\frac{v_{\perp}^2}{B} = \frac{v_{\perp 0}^2}{B_0} \tag{12.127}$$

where B is the axial magnetic induction. Then we find the parallel velocity at any position along the z axis given by

$$v_{\parallel}^2 = v_0^2 - v_{\perp 0}^2 \frac{B(z)}{B_0} \tag{12.128}$$

Equation (12.128) for the velocity of the particle in the z direction is equivalent to the first integral of Newton's equation of motion for a particle in a one-dimensional potential*

$$V(z) = \tfrac{1}{2} m \frac{v_{\perp 0}^2}{B_0} B(z)$$

If $B(z)$ increases enough, eventually the right-hand side of (12.128) will vanish at some point $z = z_0$. This means that the particle spirals in an ever-tighter orbit along the lines of force, converting more and more translational energy into energy of rotation, until its axial velocity vanishes. Then it turns around, still spiraling in the same sense, and moves back in the negative z direction. The particle is reflected by the magnetic field, as is shown schematically in Fig. 12.9.

Equation (12.128) is a consequence of the assumption that $p_{\perp}^2/B$ is invariant. To show that at least to first order this invariance follows directly from the Lorentz force equation, we consider an explicit solution of the equations of motion. If the magnetic induction along the axis is $B(z)$, there will be a radial component of the field near the axis given by the divergence equation as

$$B_\rho(\rho, z) \simeq -\tfrac{1}{2}\rho \frac{\partial B(z)}{\partial z} \tag{12.129}$$

where ρ is the radius out from the axis. The z component of the force equation is

$$\ddot{z} = \frac{e}{\gamma m c}(-\rho\dot{\phi}B_\rho) \simeq \frac{e}{2\gamma m c}\rho^2\dot{\phi}\frac{\partial B(z)}{\partial z} \tag{12.130}$$

---

* Note, however, that our discussion is fully relativistic. The analogy with one-dimensional nonrelativistic mechanics is only a formal one.

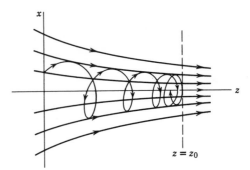

Fig. 12.9 Reflection of charged
particle out of region of high field
strength.

where $\dot{\phi}$ is the angular velocity around the z axis. This can be written, correct to first order in the small variation of $B(z)$, as

$$\ddot{z} \simeq -\frac{v_{\perp 0}^2}{2B_0} \frac{\partial B(z)}{\partial z} \tag{12.131}$$

where we have used $\rho^2 \dot{\phi} \simeq -(a^2 \omega_B)_0 = -(v_{\perp 0}^2/\omega_{B0})$. Equation (12.131) has as its first integral equation (12.128), showing that to first order in small quantities the constancy of flux linking the orbit follows directly from the equations of motion.

The adiabatic invariance of the flux linking an orbit is useful in discussing particle motions in all types of spatially varying magnetic fields. The simple example described above illustrates the principle of the "magnetic mirror": charged particles are reflected by regions of strong magnetic field. This mirror property formed the basis of a theory of Fermi for the acceleration of cosmic-ray particles to very high energies in interstellar space by collisions with moving magnetic clouds. The mirror principle can be applied to the containment of a hot plasma for thermonuclear energy production. A magnetic bottle can be constructed with an axial field produced by solenoidal windings over some region of space, and additional coils at each end to provide a much higher field towards the ends. The lines of force might appear as shown in Fig. 12.10. Particles created or injected into the field in the central region will spiral along the axis, but will be reflected by the magnetic mirrors at each end. If the ratio of maximum field B_m in the mirror to the field B in the central region is very large, only particles with a very large component of velocity parallel to the axis can penetrate through the ends. From (12.128) is it evident that the criterion for trapping is

$$\left|\frac{v_{\parallel 0}}{v_{\perp 0}}\right| < \left(\frac{B_m}{B} - 1\right)^{\frac{1}{2}} \tag{12.132}$$

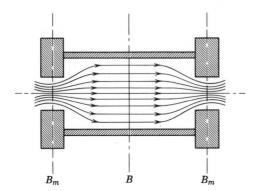

Fig. 12.10 Schematic diagram of "mirror" machine for the containment of a hot plasma.

B_m B B_m

If the particles are injected into the apparatus, it is easy to satisfy requirement (12.132). Then the escape of particles is governed by the rate at which they are scattered by residual gas atoms, etc., in such a way that their velocity components violate (12.132).

Another area of application of these principles is to terrestrial and stellar magnetic fields. The motion of charged particles in the magnetic dipole fields of the sun or earth can be understood in terms of the adiabatic invariant discussed here and the drift velocities of Section 12.9. Some aspects of this topic are left to Problems 12.11 and 12.12 on the trapped particles around the earth (the Van Allen belts).

REFERENCES AND SUGGESTED READING

The applications of relativistic kinematics, apart from precision work in low-energy nuclear physics, all occur in the field of high-energy physics. In books on that field, the relativity is taken for granted, and calculations of kinematics are generally omitted or put in appendices. One exception is the book by

Baldin, Gol'danskii, and Rozenthal

which covers the subject exhaustively with many graphs.

The Lagrangian and Hamiltonian formalism for relativistic charged particles is treated in every advanced mechanics textbook, as well as in books on electrodynamics. Some useful references are

Corben and Stehle, Chapter 16,

Goldstein, Chapter 6,

Landau and Lifshitz, *Classical Theory of Fields*, Chapters 2 and 3.

The motion of charged particles in external electromagnetic fields, especially inhomogeneous magnetic fields, is an increasingly important topic in geophysics, solar physics, and thermonuclear research. The classic reference for these problems is

Alfvén,

but the basic results are also presented by

Chandrasekhar, Chapters 2 and 3,

Linhart, Chapter 1,

Spitzer, Chapter 1.

Another important application of relativistic charged-particle dynamics is to high-energy accelerators. An introduction to the physics problems of this field will be found in

Corben and Stehle, Chapter 17,

Livingston.

For a more complete and technical discussion, with references, consult

E. D. Courant and H. S. Snyder, *Ann. Phys.*, **3**, 1 (1958).

PROBLEMS

12.1 Use the transformation to center of momentum coordinates to determine the threshold kinetic energies in Mev for the following processes:

(a) pi-meson production in nucleon-nucleon collisions ($m_\pi/M = 0.15$),

(b) pi-meson production in pi meson-nucleon collisions,

(c) pi-meson pair production in nucleon-nucleon collisions,

(d) nucleon-pair production in electron-electron collisions.

12.2 If a system of mass M decays or transforms at rest into a number of particles, the sum of whose masses is less than M by an amount ΔM,

(a) show that the maximum kinetic energy of the *i*th particle (mass m_i) is

$$(T_i)_{max} = \Delta Mc^2\left(1 - \frac{m_i}{M} - \frac{\Delta M}{2M}\right)$$

(b) determine the maximum kinetic energies in Mev and also the ratios to ΔMc^2 of each of the particles in the following decays or transformations of particles at rest:

$$\mu \to e + \nu + \bar{\nu}$$
$$K^+ \to \pi^+ + \pi^- + \pi^+$$
$$K^\pm \to e^\pm + \pi^0 + \nu$$
$$K^\pm \to \mu^\pm + \pi^0 + \nu$$
$$p + \bar{p} \to 2\pi^+ + 2\pi^- + \pi^0$$
$$p + \bar{p} \to K^+ + K^- + 3\pi^0$$

12.3 A pi meson ($m_1c^2 = 140$ Mev) collides with a proton ($m_2c^2 = 938$ Mev) at rest to create a K meson ($m_3c^2 = 494$ Mev) and a lambda hyperon ($m_4c^2 = 1115$ Mev). Use conservation of energy and momentum, plus relativistic kinematics, to find

(a) the kinetic energy in Mev of the incident pi meson at threshold for production of K mesons, and compare this with the Q value of the reaction;

(b) the kinetic energy of the pi meson in Mev in order to create K mesons at 90° in the laboratory;

(c) the kinetic energy of K mesons emerging at 0° in the laboratory when the kinetic energy of the pi meson is 20 per cent greater than in (b);

(d) the kinetic energy of K mesons at 90° in the laboratory when the incident pi meson has a kinetic energy of 1500 Mev.

12.4 It is a well-established fact that Newton's equation of motion

$$m\mathbf{a}' = e\mathbf{E}'$$

holds for a small charged body of mass m and charge e in a coordinate system K' where the body is momentarily at rest. Show that the Lorentz force equation

$$\frac{d\mathbf{p}}{dt} = e\left(\mathbf{E} + \frac{\mathbf{v}}{c} \times \mathbf{B}\right)$$

follows directly from the Lorentz transformation properties of accelerations and electromagnetic fields.

12.5 An alternative approach to the Lagrangian formalism for a relativistic charged particle is to treat the 4-vector of position x_μ and the 4-velocity $u_\mu = (\gamma\mathbf{v}, i\gamma c)$ as Lagrangian coordinates. Then the Euler-Lagrange equations have the obviously covariant form,

$$\frac{d}{d\tau}\left(\frac{\partial L}{\partial u_\mu}\right) - \frac{\partial L}{\partial x_\mu} = 0$$

where L is a Lorentz invariant Lagrangian and τ is the proper time.
  (a) Show that

$$L = \frac{1}{2} m u_\mu u_\mu + \frac{q}{c} u_\mu A_\mu$$

gives the correct relativistic equations of motion for a particle interacting with an external field described by the 4-vector potential A_μ.
  (b) Define the canonical momenta and write out the Hamiltonian in both covariant and space-time form. The Hamiltonian is a Lorentz invariant. What is its value?

12.6 (a) Show from Hamilton's principle that Lagrangians which differ only by a total time derivative of some function of the coordinates and time are equivalent in the sense that they yield the same Euler-Lagrange equations of motion.
  (b) Show explicitly that the gauge transformation $A_\mu \rightarrow A_\mu + (\partial\Lambda/\partial x_\mu)$ of the potentials in the charged-particle Lagrangian (12.75) merely generates another equivalent Lagrangian.

12.7 A particle with mass m and charge e moves in a uniform, static, electric field $\mathbf{E}_0$.
  (a) Solve for the velocity and position of the particle as explicit functions of time, assuming that the initial velocity $\mathbf{v}_0$ was perpendicular to the electric field.
  (b) Eliminate the time to obtain the trajectory of the particle in space. Discuss the shape of the path for short and long times (define "short" and "long" times).

12.8 It is desired to make an $E \times B$ velocity selector with uniform, static, crossed, electric and magnetic fields over a length L. If the entrance and exit slit widths are Δx, discuss the interval Δu of velocities around the mean value $u = cE/B$, which is transmitted by the device as a function of the mass, the momentum or energy of the incident particles, the field strengths, the length of the selector, and any other relevant variables. Neglect fringing effects at the ends. Base your discussion on the practical facts that $L \sim$ few meters, $E_{max} \sim 3 \times 10^4$ volts/cm, $\Delta x \sim 10^{-1}$ to 10^{-2} cm, $u \sim 0.5$ to $0.995c$.

12.9 A particle of mass m and charge e moves in the laboratory in crossed, static, uniform, electric and magnetic fields. **E** is parallel to the x axis; **B** is parallel to the y axis.

(a) For $|\mathbf{E}| < |\mathbf{B}|$ make the necessary Lorentz transformation described in Section 12.8 to obtain explicitly parametric equations for the particle's trajectory.

(b) Repeat the calculation of (a) for $|\mathbf{E}| > |\mathbf{B}|$.

12.10 Static, uniform electric and magnetic fields, **E** and **B**, make an angle of θ with respect to each other.

(a) By a suitable choice of axes, solve the force equation for the motion of a particle of charge e and mass m in rectangular coordinates.

(b) For **E** and **B** parallel, show that with appropriate constants of integration, etc., the parametric solution can be written

$$x = AR \sin \phi, \qquad y = AR \cos \phi, \qquad z = \frac{R}{\rho} \sqrt{1 + A^2} \cosh(\rho\phi)$$

$$ct = \frac{R}{\rho} \sqrt{1 + A^2} \sinh(\rho\phi).$$

where $R = (mc^2/eB)$, $\rho = (E/B)$, A is an arbitrary constant, and ϕ is the parameter [actually c/R times the proper time].

12.11 The magnetic field of the earth can be represented approximately by a magnetic dipole of magnetic moment $M = 8.1 \times 10^{25}$ gauss-cm³. Consider the motion of energetic electrons in the neighborhood of the earth under the action of this dipole field (Van Allen electron belts).

(a) Show that the equation for a line of magnetic force is $r = r_0 \sin^2 \theta$, where θ is the usual polar angle (colatitude) measured from the axis of the dipole, and find an expression to the magnitude of B along any line of force as a function of θ.

(b) A positively charged particle spirals around a line of force in the equatorial plane with a gyration radius a and a mean radius R ($a \ll R$). Show that the particle's azimuthal position (longitude) changes approximately linearly in time according to

$$\phi(t) = \phi_0 + \frac{3}{2}\left(\frac{a}{R}\right)^2 \omega_B(t - t_0)$$

where ω_B is the frequency of gyration at radius R.

(c) If, in addition to its circular motion of (b), the particle has a small component of velocity parallel to the lines of force, show that it undergoes small oscillations in θ around $\theta = \pi/2$ with a frequency $\Omega = (3/\sqrt{2})(a/R)\omega_B$. Find the change in longitude per cycle of oscillation in latitude.

(d) For an electron of 10 Mev at a mean radius $R = 3 \times 10^9$ cm, find ω_B and a, and so determine how long it takes to drift once around the earth and how long it takes to execute one cycle of oscillation in latitude. Calculate these same quantities for an electron of 10 Kev at the same radius.

12.12 A charged particle finds itself instantaneously in the equatorial plane of the earth's magnetic field (assumed to be a dipole field) at a distance R from the center of the earth. Its velocity vector at that instant makes an angle α with the equatorial plane ($v_\parallel/v_\perp = \tan \alpha$). Assuming that the particle spirals along the lines of force with a gyration radius $a \ll R$, and that the flux linked by the orbit is a constant of the motion, find an equation for the maximum magnetic latitude λ reached by the particle as a function of the angle α. Plot a graph (*not a sketch*) of λ versus α. Mark parametrically along the curve the values of α for which a particle at radius R in the equatorial plane will hit the earth (radius R_0) for $R/R_0 = 1.5, 2.0, 2.5, 3, 4, 6, 8, 10$.

13

Collisions between Charged Particles, Energy Loss, and Scattering

In this chapter collisions between swiftly moving, charged particles are considered, with special emphasis on the exchange of energy between collision partners and on the accompanying deflections from the incident direction. A fast charged particle incident on matter makes collisions with the atomic electrons and nuclei. If the particle is heavier than an electron (mu or pi meson, K meson, proton, etc.), the collisions with electrons and with nuclei have different consequences. The light electrons can take up appreciable amounts of energy from the incident particle without causing significant deflections, whereas the massive nuclei absorb very little energy but because of their greater charge cause scattering of the incident particle. Thus loss of energy by the incident particle occurs almost entirely in collisions with electrons. The deflection of the particle from its incident direction results, on the other hand, from essentially elastic collisions with the atomic nuclei. The scattering is confined to rather small angles, so that a heavy particle keeps a more or less straight-line path while losing energy until it nears the end of its range. For incident electrons both energy loss and scattering occur in collisions with the atomic electrons. Consequently the path is much less straight. After a short distance, electrons tend to diffuse into the material, rather than go in a rectilinear path.

The subject of energy loss and scattering is an important one and is discussed in several books* where numerical tables and graphs are

* See references at the end of the chapter.

presented. Consequently our discussion will emphasize the physical ideas involved, rather than the exact numerical formulas. Indeed, a full quantum-mechanical treatment is needed to obtain exact results, even though all the essential features are classical or semiclassical in origin. The order of magnitude of the quantum effects are all easily derivable from the uncertainty principle, as will be seen in what follows.

We will begin by considering the simple problem of energy transfer to a free electron by a fast heavy particle. Then the effects of a binding force on the electron are explored, and the classical Bohr formula for energy loss is obtained. Quantum modifications and the effect of the polarization of the medium are described, followed by a discussion of energy loss in an electronic plasma. Then the elastic scattering of incident particles by nuclei and multiple scattering are presented. Finally, a discussion is given of the electrical resistivity of a plasma caused by screened Coulomb collisions.

13.1 Energy Transfer in a Coulomb Collision

A swift particle of charge ze and mass M collides with an electron in an atom. If the particle moves rapidly compared to the characteristic velocity of the electron in its orbit, during the collision the electron can be treated as free and initially at rest. As further approximations we will assume that the momentum transfer Δp is sufficiently small that the incident particle is essentially undeflected from its straight-line path, and that the recoiling electron does not move appreciably during the collision. Then to find the energy transfer during the collision we need only calculate the momentum impulse caused by the electric field of the incident particle at the position of the electron. The particle's magnetic field is of negligible importance if the electron is essentially at rest.

Figure 13.1 shows the geometry of the collision. The incident particle has a velocity v and an energy $E = \gamma Mc^2$. It passes the electron of charge e and mass $m \ll M$ at an impact parameter b. At the position of the electron the fields of the incident particle are given by (11.118) with $q = ze$. Only the transverse electric field E_1 has a nonvanishing time integral.

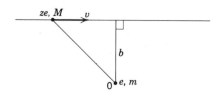

Fig. 13.1

Consequently the momentum impulse Δp is in the transverse direction and has the magnitude

$$\Delta p = \int_{-\infty}^{\infty} eE_1(t) \, dt = \frac{2ze^2}{bv} \tag{13.1}$$

It should be noted that Δp is independent of γ, as discussed in Section 11.10 below Eq. (11.119). The energy transferred to the electron is

$$\Delta E(b) = \frac{(\Delta p)^2}{2m} = \frac{2z^2 e^4}{mv^2} \left(\frac{1}{b^2}\right) \tag{13.2}$$

The angular deflection of the incident particle is given by $\theta \simeq \Delta p/p$, provided $\Delta p \ll p$. Thus, for small deflections,

$$\theta \simeq \frac{2ze^2}{pvb} \tag{13.3}$$

This result can be compared with the well-known exact expression for the Rutherford scattering of a nonrelativistic particle of charge ze by a Coulomb force field of charge $z'e$:

$$2 \tan \frac{\theta}{2} = \frac{2zz'e^2}{pvb} \tag{13.4}$$

We see that for small angles the two expressions agree.*

The energy transfer $\Delta E(b)$ given by (13.2) has several interesting features. It depends only on the charge and velocity of the incident particle, not on its mass. It varies inversely as the square of the impact parameter so that close collisions involve very large energy transfers. There is, of course, an upper limit on the energy transfer, corresponding to a head-on collision. Our method of calculation is really valid only for large values of b. We can obtain a lower limit b_{min} on the impact parameter for which our approximate calculation is valid by equating (13.2) to the maximum allowable energy transfer (12.59):

$$\Delta E(b_{min}) = \Delta E_{max} = 2m\gamma^2 v^2 \tag{13.5}$$

This yields the lower bound,

$$b_{min} = \frac{ze^2}{\gamma m v^2} \tag{13.6}$$

* Actually there is a question of reference frames in comparing (13.3) and (13.4). Since (13.4) holds for a fixed center of force (or the CM system), we should compare it with the result for the deflection of the light electron in the frame where the heavy incident particle is at rest. Then (13.3) holds with $p \simeq \gamma m v$ as the electron momentum in that frame. The reader may verify that (13.3) and (13.4) are also consistent in the frame in which the electron is at rest by using (12.50) and (12.54) to transform angles from the CM system to the laboratory.

below which our approximate result (13.2) must be replaced by a more exact expression which tends to (13.5) as $b \to 0$. It can be shown (Problem 13.1) that a proper treatment gives the more accurate result,

$$\Delta E(b) \simeq \frac{2z^2 e^4}{mv^2}\left(\frac{1}{b_{\min}^2 + b^2}\right) \tag{13.7}$$

Equation (13.7) exhibits the proper limiting behavior as $b \to 0$ and reduces to (13.2) for $b \gg b_{\min}$.

The lower limit on b can be obtained by another argument. Equation (13.2) was derived under the assumption that the electron did not move appreciably during the collision. As long as the distance d it actually moves is small compared to b, we may expect that (13.2) will be correct. An estimate of d can be obtained by saying that $\Delta p/2m$ is an average velocity of the electron during the collision, and that the time of collision is given by (11.120). Hence the distance traveled during the collision is of the order of

$$d \sim \frac{\Delta p}{2m} \times \Delta t = \frac{ze^2}{\gamma m v^2} = b_{\min} \tag{13.8}$$

As long as $b \gg d$, (13.2) should hold. This is exactly the condition implied by (13.7).

At the other extreme of very distant collisions the approximate result (13.2) for $\Delta E(b)$ is in error because of the binding of the atomic electrons. We assumed that the electrons were free, whereas they are actually bound in atoms. As long as the collision time (11.120) is short compared to the orbital period of motion, it may be expected that the collision will be sudden enough that the electron may be treated as free. If, on the other hand, the collision time (11.120) is very long compared to the orbital period, the electron will make many cycles of motion as the incident particle passes slowly by and will be influenced adiabatically by the fields with no net transfer of energy. The dividing point comes at impact parameter $b_{\max}$, where the collision time (11.120) and the orbital period are comparable. If ω is a characteristic atomic frequency of motion, this condition is

or

$$\left. \begin{aligned} \Delta t(b_{\max}) &\sim \frac{1}{\omega} \\[2mm] b_{\max} &= \frac{\gamma v}{\omega} \end{aligned} \right\} \tag{13.9}$$

For impact parameters greater than $b_{\max}$ it can be expected that the energy transfer falls below (13.2), going rapidiy to zero for $b \gg b_{\max}$.

The general behavior of $\Delta E(b)$ as a function of b is shown in Fig. 13.2. The dotted curve represents the approximate form (13.2), while the solid

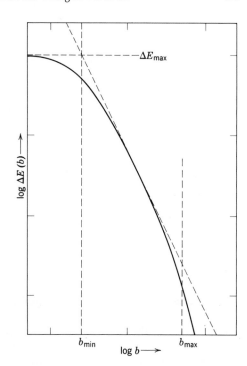

Fig. 13.2 Energy transfer as a function of impact parameter.

curve is a representation of the correct result. In the interval $b_{\min} < b < b_{\max}$ the energy transfer is given approximately by (13.2). But for impact parameters outside that interval, the energy transfer is considerably less.

A fast particle passing through matter "sees" electrons at various distances from its path. If there are N atoms per unit volume with Z electrons per atom, the number of electrons located at impact parameters between b and $(b + db)$ in a thickness dx of matter is

$$dn = NZ\, 2\pi b\, db\, dx \qquad (13.10)$$

To find the energy lost per unit distance by the incident particle we multiply (13.10) by the energy transfer $\Delta E(b)$ and integrate over all impact parameters. Thus the energy loss is

$$\frac{dE}{dx} = 2\pi NZ \int \Delta E(b) b\, db \qquad (13.11)$$

In view of the behavior of $\Delta E(b)$ shown in Fig. 13.2 we may use approximation (13.2) and integrate between $b_{\min}$ and $b_{\max}$. Then we find the result

$$\frac{dE}{dx} \simeq 4\pi NZ \frac{z^2 e^4}{mv^2} \int_{b_{\min}}^{b_{\max}} \frac{1}{b^2} b\, db \qquad (13.12)$$

or

$$\frac{dE}{dx} \simeq 4\pi NZ \frac{z^2 e^4}{mv^2} \ln B \tag{13.13}$$

where

$$B = \frac{b_{\max}}{b_{\min}} \simeq \frac{\gamma^2 m v^3}{z e^2 \omega} \tag{13.14}$$

This approximate expression for the energy loss exhibits all the essential features of the classical result due to Bohr (1915). The method of handling the lower limit of integration in (13.12) is completely equivalent to using (13.7) for $\Delta E(b)$. The cutoff at $b = b_{\max}$ is only approximate. Consequently B is uncertain by a factor of the order of unity. Because B appears in the logarithm, this factor is of negligible importance numerically. In any event, a proper treatment of binding effects is given in the next section. Discussion of (13.13) as a function of energy and its comparison with experiment will be deferred until Section 13.3.

13.2 Energy Transfer to a Harmonically Bound Charge

In order to justify the plausible value $b_{\max}$ (13.9) of the impact parameter which divides the Coulomb collisions for $b < b_{\max}$ with the free-energy transfer (13.2) and essentially adiabatic collisions for $b > b_{\max}$ with negligible energy transfer, we consider the problem of the energy lost by a massive charged particle with charge ze and velocity v passing a harmonically bound charge of mass m and charge e. This will serve as a simplified model for energy loss of particles passing through matter. As before, we will assume that the massive particle is deflected only slightly in the encounter so that its path can be approximated by a straight line. It passes by the bound particle at an impact parameter b, measured from the origin O of the binding force, as shown in Fig. 13.3. Since we are primarily interested in large impact parameters where binding effects are important,

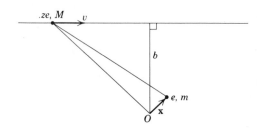

Fig. 13.3

we may assume that the energy transfer is not large, that the motion of the bound particle is nonrelativistic throughout the collision, and that its initial and final amplitudes of oscillation about the origin O are small compared to b. Then only the electric field of the incident particle need be included in the force equation. Furthermore, its variation over the position of the bound particle may be neglected, and its effective value can be taken as that at the origin O. This is sometimes called the *dipole approximation*, by analogy with the corresponding problem of absorption of radiation.

With these approximations the force equation for the harmonically bound charge can be written as

$$\ddot{\mathbf{x}} + \Gamma\dot{\mathbf{x}} + \omega_0^2\mathbf{x} = \frac{e}{m}\,\mathbf{E}(t) \tag{13.15}$$

where $\mathbf{E}(t)$ is the electric field at O due to the charge ze, its components being given by (11.118), ω_0 is the characteristic frequency of the binding, and Γ is a small damping constant. The damping factor is not essential, but it is present to at least some degree in actual physical systems and serves to remove certain ambiguities which would arise in its absence. To solve (13.15) we Fourier-analyze both $\mathbf{E}(t)$ and $\mathbf{x}(t)$:

$$\mathbf{x}(t) = \frac{1}{\sqrt{2\pi}} \int_{-\infty}^{\infty} \mathbf{x}(\omega)e^{-i\omega t}\,d\omega \tag{13.16}$$

$$\mathbf{E}(t) = \frac{1}{\sqrt{2\pi}} \int_{-\infty}^{\infty} \mathbf{E}(\omega)e^{-i\omega t}\,d\omega \tag{13.17}$$

Since both $\mathbf{x}(t)$ and $\mathbf{E}(t)$ are real, the positive and negative frequency parts of their transforms are related by

$$\mathbf{x}(-\omega) = \mathbf{x}^*(\omega)$$
$$\mathbf{E}(-\omega) = \mathbf{E}^*(\omega) \tag{13.18}$$

When the Fourier integral forms are substituted into the force equation, we find

$$\mathbf{x}(\omega) = \frac{e}{m}\,\frac{\mathbf{E}(\omega)}{\omega_0^2 - i\omega\Gamma - \omega^2} \tag{13.19}$$

With the known form of $\mathbf{E}(t)$ the Fourier amplitude $\mathbf{E}(\omega)$ can be determined. Then $\mathbf{x}(t)$ can be found from (13.16), using (13.19). The problem is solved, provided one can do the integrals.

The quantity of immediate interest is not the detailed motion of the bound particle, but the energy transfer in the collision. This can be found by considering the work done by the incident particle on the bound one.

The rate of doing work is given by

$$\frac{dE}{dt} = \int \mathbf{E} \cdot \mathbf{J} \, d^3x' \tag{13.20}$$

Thus the total work done by the particle passing by is

$$\Delta E = \int_{-\infty}^{\infty} dt \int d^3x' \, \mathbf{E} \cdot \mathbf{J} \tag{13.21}$$

The current density is $\mathbf{J} = e\mathbf{v} \, \delta[\mathbf{x}' - \mathbf{x}(t)]$ for the bound charge. Consequently

$$\Delta E = e \int_{-\infty}^{\infty} \mathbf{v} \cdot \mathbf{E} \, dt \tag{13.22}$$

where $\mathbf{v} = \dot{\mathbf{x}}$, and in the dipole approximation $\mathbf{E}$ is the field of the incident particle at the origin O. Using the Fourier representations (13.16) and (13.17), as well as that for a delta function (2.52), and the reality conditions (13.18), the energy transfer can be written

$$\Delta E = 2e \, \mathrm{Re} \int_{0}^{\infty} -i\omega \mathbf{x}(\omega) \cdot \mathbf{E}^*(\omega) \, d\omega \tag{13.23}$$

If now the result (13.19) for $\mathbf{x}(\omega)$ is inserted, this becomes

$$\Delta E = \frac{e^2}{m} \int_{0}^{\infty} |\mathbf{E}(\omega)|^2 \frac{2\omega^2 \Gamma}{(\omega_0^2 - \omega^2)^2 + \omega^2 \Gamma^2} \, d\omega \tag{13.24}$$

For small Γ the integrand peaks sharply around $\omega = \omega_0$ in an approximately Lorentzian line shape. Consequently the factor involving the electric field can be approximated by its value at $\omega = \omega_0$. Then (13.24) becomes

$$\Delta E = \frac{2e^2}{m} |\mathbf{E}(\omega_0)|^2 \int_{0}^{\infty} \frac{x^2 \, dx}{\left[\dfrac{\omega_0^2}{\Gamma^2} - x^2\right]^2 + x^2} \tag{13.25}$$

The integral has the value $\pi/2$, independent of ω_0/Γ. Thus the energy transfer is

$$\Delta E = \frac{\pi e^2}{m} |\mathbf{E}(\omega_0)|^2 \tag{13.26}$$

Equation (13.26) is a very general result for energy transfer to a non-relativistic oscillator by an external electromagnetic field. In the present application the field is produced by a passing charged particle. But a pulse of radiation or any combination of external fields will serve as well.

For a particle with charge ze passing by the origin O at an impact parameter b with a velocity v, the electromagnetic fields at the origin are given by (11.118) with $q = ze$. To illustrate the determination of the Fourier transform (13.17) we consider $E_1(t)$. Its transform $E_1(\omega)$ is defined to be

$$E_1(\omega) = \frac{zeb\gamma}{\sqrt{2\pi}} \int_{-\infty}^{\infty} \frac{e^{i\omega t} \, dt}{(b^2 + \gamma^2 v^2 t^2)^{3/2}} \tag{13.27}$$

By changing integration variable to $x = \gamma vt/b$, (13.27) can be written as

$$E_1(\omega) = \frac{ze}{\sqrt{2\pi}bv} \int_{-\infty}^{\infty} \frac{e^{i\omega bx/\gamma v}}{(1 + x^2)^{3/2}} \, dx \tag{13.28}$$

From a table of Fourier transforms* we find that the integral is proportional to a modified Bessel function of the order of unity [see (3.101)]. Thus

$$E_1(\omega) = \frac{ze}{bv} \left(\frac{2}{\pi}\right)^{1/2} \left[\frac{\omega b}{\gamma v} K_1\left(\frac{\omega b}{\gamma v}\right)\right] \tag{13.29}$$

Similarly $E_3(t)$ given by (11.118) has the Fourier transform:

$$E_3(\omega) = -i \frac{ze}{\gamma vb} \left(\frac{2}{\pi}\right)^{1/2} \left[\frac{\omega b}{\gamma v} K_0\left(\frac{\omega b}{\gamma v}\right)\right] \tag{13.30}$$

The energy transfer (13.26) to the harmonically bound charge can now be evaluated explicitly. Using (13.29) and (13.30), we find

$$\Delta E(b) = \frac{2z^2 e^4}{mv^2} \left(\frac{1}{b^2}\right) \left[\xi^2 K_1^2(\xi) + \frac{1}{\gamma^2} \xi^2 K_0^2(\xi)\right] \tag{13.31}$$

where

$$\xi = \frac{\omega_0 b}{\gamma v} \tag{13.32}$$

The factor multiplying the square bracket is just the approximate result (13.2). For small and large ξ, the limiting forms (3.103) and (3.104) show that the square bracket in (13.31) has the limiting values:

$$[\quad] = \begin{cases} 1, & \text{for } \xi \ll 1 \\ \left(1 + \dfrac{1}{\gamma^2}\right)\dfrac{\pi}{2} \xi e^{-2\xi} & \text{for } \xi \gg 1 \end{cases} \tag{13.33}$$

Since $\xi = b/b_{\max}$, we see that for $b \ll b_{\max}$ the energy transfer is essentially the approximate result (13.2), while for $b \gtrsim b_{\max}$ it falls off exponentially to zero. This justifies the qualitative arguments of the previous section on the upper limit $b_{\max}$.

---

* See, for example, Magnus and Oberhettinger, Chapter VIII, or Bateman Manuscript Project, *Tables of Integral Transforms*, Vol. I, Chapters I–III.

Classical Electrodynamics

13.3 Classical and Quantum-Mechanical Energy-Loss Formulas

The energy transfer (13.31) to a harmonically bound charge can be used to calculate a classical energy loss per unit length for a fast, heavy particle passing through matter. We suppose that there are N atoms per unit volume with Z electrons per atom. The Z electrons can be divided into groups specified by the index j, with f_j electrons having the same harmonic binding frequency ω_j. The number f_j is called the *oscillator strength* of the jth oscillator. The oscillator strengths satisfy the obvious sum rule, $\sum_j f_j = Z$. By a trivial extension of the arguments leading to (13.11) and (13.12) we find the energy loss to be

$$\frac{dE}{dx} = 2\pi N \sum_j f_j \int_{b_{\min}}^{\infty} \Delta E_j(b) b \, db \tag{13.34}$$

where $\Delta E_j(b)$ is given by (13.31) with $\xi = \omega_j b / \gamma v$, and a lower limit of $b_{\min}$ is specified, consistent with (13.7). No upper limit is necessary, since (13.31) falls rapidly to zero for large b. The integral over the modified Bessel functions can be done in closed form, leading to the result,

$$\frac{dE}{dx} = \frac{4\pi N z^2 e^4}{mv^2} \sum_j f_j \left[\xi_{\min} K_1(\xi_{\min}) K_0(\xi_{\min}) - \frac{v^2}{2c^2} \xi_{\min}^2 (K_1{}^2(\xi_{\min}) - K_0{}^2(\xi_{\min})) \right] \tag{13.35}$$

where $\xi_{\min} = \omega_j b_{\min} / \gamma v$. In general, $\xi_{\min} \ll 1$. This means that the limiting forms (3.103) may be used to simplify (13.35). This final expression for classical energy loss is

$$\frac{dE_c}{dx} = 4\pi N Z \frac{z^2 e^4}{mv^2} \left[\ln B_c - \frac{v^2}{2c^2} \right] \tag{13.36}$$

where the argument of the logarithm is

$$B_c = \frac{1.123 \gamma v}{\langle \omega \rangle b_{\min}} = \frac{1.123 \gamma^2 m v^3}{z e^2 \langle \omega \rangle} \tag{13.37}$$

The average frequency $\langle \omega \rangle$ appearing in B_c is a geometric mean defined by

$$Z \ln \langle \omega \rangle = \sum_j f_j \ln \omega_j \tag{13.38}$$

The result (13.36)–(13.38) is that obtained by Bohr in his classic paper on energy loss (1915). Our approximate expression (13.13) is in agreement with (13.36) in all its essentials, since the added $-v^2/2c^2$ is a small correction even at high velocities.

Bohr's formula (13.36) gives a reasonable description of the energy loss of relatively slow alpha particles and heavier nuclei. But for electrons, mesons, protons, and even fast alphas, it overestimates the energy loss considerably. The reason is that for the lighter particles quantum-mechanical modifications cause a breakdown of the classical result. The important quantum effects are (1) discreteness of the possible energy transfers, and (2) limitations due to the wave nature of the particles and the uncertainty principle.

The problem of the discrete nature of the energy transfer can be illustrated by calculating the classical energy transfer (13.2) at $b \simeq b_{\max}$. This is roughly the smallest energy transfer that is of importance in the energy-loss process. Assuming only one binding frequency ω_0 for simplicity, we find

$$\Delta E(b_{\max}) \simeq \frac{2}{\gamma^2} z^2 \left(\frac{v_0}{v}\right)^4 \hbar \omega_0 \qquad (13.39)$$

where $v_0 = c/137$ is the orbital velocity of an electron in the ground state of hydrogen. Since $\hbar \omega_0$ is of the order of the ionization potential of the atom, we see that for a fast particle ($v \gg v_0$) the classical energy transfer (13.39) is very small compared to the ionization potential, or even to the smallest excitation energy in the atom. But we know that energy must be transferred in definite quantum jumps. A tiny amount of energy like (13.39) simply cannot be absorbed by the atom. We conclude that our classical calculation fails in this domain. We might argue that only if our classical formula (13.2) gives an energy transfer *large* compared to typical atomic excitation energies would we expect it to be correct. This would set quite a different upper limit on the impact parameters. Fortunately the classical result can be applied in a statistical sense if we reinterpret its meaning. Quantum considerations show that the classical result of the transfer of a small amount of energy in every collision is incorrect. But if we consider a large number of collisions, we find that on the average a small amount of energy *is* transferred. It is not transferred in every collision, however. In most collisions no energy is transferred. But in a few collisions an appreciable excitation occurs, yielding a small average value over many collisions. In this statistical sense the quantum mechanism for discrete energy transfers and the classical process with a continuum of possible energy transfers can be reconciled. The detailed numerical agreement stems from the quantum-mechanical definitions of the oscillator strengths f_j and resonant frequencies ω_j.

The other important quantum modification arises from the wave nature of the particles. The uncertainty principle sets certain limits on the range of validity of classical orbit considerations. If we try to construct wave packets to give approximate meaning to a classical trajectory, we know that the path can be defined only to within an uncertainty $\Delta x \gtrsim \hbar/p$. For impact parameters b less than this uncertainty, classical concepts fail. Since the wave nature of the particles implies a smearing out in some sense over distances of the order of Δx, we anticipate that the correct quantum-mechanical energy loss will correspond to much smaller energy transfers than given by (13.2) for $b < \Delta x$. Thus $\Delta x \sim \hbar/p$ is a quantum analog of the minimum impact parameter (13.6). In the collision of two particles each one has a wave nature. For a given relative velocity the limiting uncertainty will come from the lighter of the two. For a heavy incident particle colliding with an electron, the momentum of the electron in the coordinate frame where the incident particle is at rest (almost the CM frame) is $p' = \gamma m v$, where m is the mass of the electron. Therefore the quantum-mechanical minimum impact parameter is

$$b_{\min}^{(q)} = \frac{\hbar}{\gamma m v} \qquad (13.40)$$

For electrons incident on electrons we must take more care and consider the CM momentum (12.34) for equal masses. Then for electrons we obtain the minimum impact parameter,

$$[b_{\min}^{(q)}]_{\text{electrons}} = \frac{\hbar}{mc} \sqrt{\frac{2}{\gamma - 1}} \qquad (13.41)$$

In a given situation the larger of the two minimum impact parameters (13.6) and (13.40) must be used to define argument B (13.14) of the logarithm in dE/dx. The ratio of the classical to quantum value of $b_{\min}$ is

$$\eta = \frac{ze^2}{\hbar v} \qquad (13.42)$$

If $\eta > 1$, the classical Bohr formula must be used. We see that this occurs for slow, highly charged, incident particles, in accord with observation. If $\eta < 1$, the quantum minimum impact parameter is larger than the classical one. Then quantum modifications appear in the energy-loss formula. The argument of the logarithm in (13.13) becomes

$$B_q = \frac{b_{\max}}{b_{\min}^{(q)}} = \eta B = \frac{\gamma^2 m v^2}{\hbar \langle \omega \rangle} \qquad (13.43)$$

Equation (13.13) with the quantum-mechanical B_q (13.43) in the logarithm is a good approximation to a quantum-theoretical result of Bethe (1930). Bethe's formula, including the effects of close collisions, is

$$\frac{dE_q}{dx} = 4\pi N Z \frac{z^2 e^4}{m v^2} \left[\ln \left(\frac{2\gamma^2 m v^2}{\hbar \langle \omega \rangle} \right) - \frac{v^2}{c^2} \right] \qquad (13.44)$$

Apart from the small correction term $-v^2/c^2$ and a factor of 2 in the argument of the logarithm, this is just our approximate expression.

For electrons the quantum effects embodied in (13.41) lead to a modified quantum-mechanical argument for the logarithm:

$$B_{\text{el}} \simeq (\gamma - 1)\sqrt{\frac{\gamma + 1}{2}}\,\frac{mc^2}{\hbar\langle\omega\rangle} \to \frac{\gamma^{3/2}}{\sqrt{2}}\,\frac{mc^2}{\hbar\langle\omega\rangle} \tag{13.45}$$

where the last expression is valid at high energies. Even though there are other quantum effects for electrons, such as spin and exchange effects, the dominant modifications are included in (13.45).

The general behavior of both the classical and quantum-mechanical energy-loss formulas is shown in Fig. 13.4. At low energies, the main energy variation is as v^{-2}, since the logarithm changes slowly. But at high energies where $v \to c$ the variation is upwards again, going as $\ln \gamma$ for $\gamma \gg 1$. Bethe's formula is in good agreement with experiment for all fast particles with $\eta < 1$, provided the energy is not too high (see the next section).

It is worth while to note the physical origins of the two powers of γ which appear in B_q (13.43). One power of γ comes from the increase of the maximum energy (13.5) which can be transferred in a head-on collision. The other power comes from the relativistic change in shape of the electromagnetic fields (11.118) of a fast particle with the consequent shortening of the collision time (11.120) and increase of b_{max} (13.9). The fields are effective in transferring energy at larger distances for a relativistic particle than for a nonrelativistic one.

Sometimes it is of interest to know the energy loss per unit distance due to collisions in which less than some definite amount ϵ of energy is transferred per collision. In photographic emulsions, for example, ejected

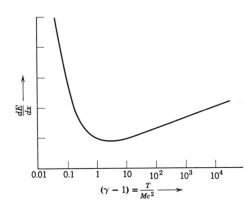

Fig. 13.4 Energy loss as a function of kinetic energy.

$$(\gamma - 1) = \frac{T}{Mc^2} \longrightarrow$$

electrons of more than about 10 Kev energy have a range greater than the average linear dimensions of the silver bromide grains. Consequently the energy dissipated in blackening of the grains corresponds to collisions where the energy transfer is less than about 10 Kev. Classically, the desired energy-loss formula can be obtained from the Bohr formula (13.35) with a minimum impact parameter $b_{min}(\epsilon)$ chosen so that (13.2) is equal to ϵ. Thus

$$b_{min}(\epsilon) = \frac{2ze^2}{v(2m\epsilon)^{\frac{1}{2}}} \tag{13.46}$$

This leads to a formula of the form of (13.36), but with an argument in the logarithm,

$$B_c(\epsilon) = \frac{1.123\gamma v^2 (2m\epsilon)^{\frac{1}{2}}}{2ze^2\langle\omega\rangle} \tag{13.47}$$

Since quantum-mechanical energy-loss formulas are obtained from classical ones by the replacement [see (13.43)],

$$B_q = \eta B_c = \frac{ze^2}{\hbar v} B_c \tag{13.48}$$

we expect that the quantum-mechanical formula for energy-loss per unit distance due to collisions with energy transfer less than ϵ will be

$$\frac{dE_q}{dx}(\epsilon) = 4\pi NZ \frac{z^2 e^4}{mv^2} \left[\ln B_q(\epsilon) - \frac{v^2}{2c^2} \right] \tag{13.49}$$

where

$$B_q(\epsilon) = \lambda \frac{\gamma v(2m\epsilon)^{\frac{1}{2}}}{\hbar\langle\omega\rangle} \tag{13.50}$$

The constant λ is a numerical factor of the order of unity that cannot be determined without detailed quantum-mechanical calculations. Bethe's calculations (1930) give the value $\lambda = 1$. The quantum-mechanical $B_q(\epsilon)$ can be written as

$$B_q(\epsilon) = \frac{b_{max}}{b_{min}^{(q)}(\epsilon)} \tag{13.51}$$

where b_{max} is given by (13.9), and the minimum impact parameter is

$$b_{min}^{(q)}(\epsilon) \simeq \frac{\hbar}{\Delta p} \simeq \frac{\hbar}{(2m\epsilon)^{\frac{1}{2}}} \tag{13.52}$$

The implication of this formula is that the classical trajectory must be ill defined by an amount at least as great as (13.52) in order that the uncertainty in transverse momentum Δp be less than the momentum transfer in the collision. Otherwise we would be unable to be certain that an energy transfer of less than ϵ had actually occurred. Hence (13.52) forms a natural quantum-mechanical lower limit on the classical orbit picture in this case.

13.4 Density Effect in Collision Energy Loss

For particles which are not too relativistic the observed energy loss is given accurately by (13.44) [or by (13.36) if $\eta > 1$] for all kinds of particles in all types of media. For ultrarelativistic particles, however, the observed energy loss is less than predicted by (13.44), especially for dense substances. In terms of Fig. 13.4 of (dE/dx), the observed energy loss increases beyond the minimum with a slope of roughly one-half that of the theoretical curve, corresponding to only one power of γ in the argument of the logarithm in (13.44) instead of two. In photographic emulsions the energy loss, as measured from grain densities, barely increases above the minimum to a plateau extending to the highest known energies. This again corresponds to a reduction of one power of γ, this time in $B_q(\epsilon)$ (13.50).

This reduction in energy loss, known as the density effect, was first treated theoretically by Fermi (1940). In our discussion so far we have tacitly made one assumption that is not valid in dense substances. We have assumed that it is legitimate to calculate the effect of the incident particle's fields on one electron in one atom at a time, and then sum up incoherently the energy transfers to all the electrons in all the atoms with $b_{min} < b < b_{max}$. Now b_{max} is very large compared to atomic dimensions, especially for large γ. Consequently in dense media there are many atoms lying between the incident particle's trajectory and the typical atom in question if b is comparable to b_{max}. These atoms, influenced themselves by the fast particle's fields, will produce perturbing fields at the chosen atom's position, modifying its response to the fields of the fast particle. Said in another way, in dense media the dielectric polarization of the material alters the particle's fields from their free-space values to those characteristic of macroscopic fields in a dielectric. This modification of the fields due to polarization of the medium must be taken into account in calculating the energy transferred in distant collisions. For close collisions the incident particle interacts with only one atom at a time. Then the free-particle calculation without polarization effects will apply. The dividing impact parameter between close and distant collisions is of the order of atomic dimensions. Since the joining of two logarithms is involved in calculating the sum, the dividing value of b need not be specified with great precision.

We will determine the energy loss in distant collisions ($b \geq a$), assuming that the fields in the medium can be calculated in the continuum approximation of a macroscopic dielectric constant $\epsilon(\omega)$. If a is of the order of atomic dimensions, this approximation will not be good for the closest of the distant collisions, but will be valid for the great bulk of the collisions.

The problem of finding the electric field in the medium due to the incident fast particle moving with constant velocity can be solved most readily by Fourier transforms. If the potentials $A_\mu(x)$ and source density $J_\mu(x)$ are transformed in space and time according to the general rule,

$$F(\mathbf{x}, t) = \frac{1}{(2\pi)^2} \int d^3k \int d\omega \, F(\mathbf{k}, \omega) e^{i\mathbf{k}\cdot\mathbf{x} - i\omega t} \tag{13.53}$$

then the transformed wave equations become

$$\left.\begin{aligned}
\left[k^2 - \frac{\omega^2}{c^2} \epsilon(\omega) \right] \Phi(\mathbf{k}, \omega) &= \frac{4\pi}{\epsilon(\omega)} \rho(\mathbf{k}, \omega) \\
\left[k^2 - \frac{\omega^2}{c^2} \epsilon(\omega) \right] \mathbf{A}(\mathbf{k}, \omega) &= \frac{4\pi}{c} \mathbf{J}(\mathbf{k}, \omega)
\end{aligned}\right\} \tag{13.54}$$

The dielectric constant $\epsilon(\omega)$ appears characteristically in positions dictated by the presence of **D** in Maxwell's equations. The Fourier transforms of

and
$$\left.\begin{aligned}
\rho(\mathbf{x}, t) &= ze\delta(\mathbf{x} - \mathbf{v}t) \\
\mathbf{J}(\mathbf{x}, t) &= \mathbf{v}\rho(\mathbf{x}, t)
\end{aligned}\right\} \tag{13.55}$$

are readily found to be

$$\left.\begin{aligned}
\rho(\mathbf{k}, \omega) &= \frac{ze}{2\pi} \delta(\omega - \mathbf{k} \cdot \mathbf{v}) \\
\mathbf{J}(\mathbf{k}, \omega) &= \mathbf{v}\rho(\mathbf{k}, \omega)
\end{aligned}\right\} \tag{13.56}$$

From (13.54) we see that the Fourier transforms of the potentials are

and
$$\left.\begin{aligned}
\Phi(\mathbf{k}, \omega) &= \frac{2ze}{\epsilon(\omega)} \cdot \frac{\delta(\omega - \mathbf{k} \cdot \mathbf{v})}{k^2 - \frac{\omega^2}{c^2} \epsilon(\omega)} \\
\mathbf{A}(\mathbf{k}, \omega) &= \epsilon(\omega) \frac{\mathbf{v}}{c} \Phi(\mathbf{k}, \omega)
\end{aligned}\right\} \tag{13.57}$$

From the definitions of the electromagnetic fields in terms of the potentials we obtain their Fourier transforms:

$$\left.\begin{aligned}
\mathbf{E}(\mathbf{k}, \omega) &= i\left(\frac{\omega\epsilon(\omega)}{c} \frac{\mathbf{v}}{c} - \mathbf{k} \right) \Phi(\mathbf{k}, \omega) \\
\mathbf{B}(\mathbf{k}, \omega) &= i\epsilon(\omega) \, \mathbf{k} \times \frac{\mathbf{v}}{c} \Phi(\mathbf{k}, \omega)
\end{aligned}\right\} \tag{13.58}$$

In calculating the energy loss it is apparent from (13.23) that we want the Fourier transform in time of the electromagnetic fields at a perpendicular distance b from the path of the particle moving along the z axis. Thus the required electric field is

$$\mathbf{E}(\omega) = \frac{1}{(2\pi)^{3/2}} \int d^3k \, \mathbf{E}(\mathbf{k}, \omega) \, e^{ibk_1} \tag{13.59}$$

where the observation point has coordinates $(b, 0, 0)$. To illustrate the determination of $\mathbf{E}(\omega)$ we consider the calculation of $E_3(\omega)$, the component of $\mathbf{E}$ parallel to $\mathbf{v}$. Inserting the explicit forms from (13.57) and (13.58), we obtain

$$E_3(\omega) = \frac{2ize}{\epsilon(\omega)} \frac{1}{(2\pi)^{3/2}} \int d^3k \, e^{ibk_1} \left(\frac{\omega\epsilon(\omega)v}{c^2} - k_3 \right) \frac{\delta(\omega - vk_3)}{k^2 - \dfrac{\omega^2}{c^2}\epsilon(\omega)} \quad (13.60)$$

The integral over dk_3 can be done immediately. Then

$$E_3(\omega) = -\frac{2ize\omega}{(2\pi)^{3/2} v^2} \left(\frac{1}{\epsilon(\omega)} - \beta^2 \right) \int_{-\infty}^{\infty} dk_1 e^{ibk_1} \int_{-\infty}^{\infty} \frac{dk_2}{k_2{}^2 + k_1{}^2 + \lambda^2}$$
$$(13.61)$$

where

$$\lambda^2 = \frac{\omega^2}{v^2} - \frac{\omega^2}{c^2}\epsilon(\omega) = \frac{\omega^2}{v^2}(1 - \beta^2\epsilon(\omega)) \quad (13.62)$$

The integral over dk_2 has the value $\pi/(\lambda^2 + k_1{}^2)^{1/2}$, so that $E_3(\omega)$ can be written

$$E_3(\omega) = -\frac{ize\omega}{\sqrt{2\pi}\,v^2}\left(\frac{1}{\epsilon(\omega)} - \beta^2 \right) \int_{-\infty}^{\infty} \frac{e^{ibk_1}}{(\lambda^2 + k_1{}^2)^{1/2}} dk_1 \quad (13.63)$$

The remaining integral is of the same general structure as (13.28). The result is

$$E_3(\omega) = -\frac{ize\omega}{v^2}\left(\frac{2}{\pi} \right)^{1/2}\left(\frac{1}{\epsilon(\omega)} - \beta^2 \right) K_0(\lambda b) \quad (13.64)$$

where the square root of (13.62) is chosen so that λ lies in the fourth quadrant. A similar calculation yields the other fields:

$$\left. \begin{aligned} E_1(\omega) &= \frac{ze}{v}\left(\frac{2}{\pi} \right)^{1/2} \frac{\lambda}{\epsilon(\omega)} K_1(\lambda b) \\ B_2(\omega) &= \epsilon(\omega)\beta E_1(\omega) \end{aligned} \right\} \quad (13.65)$$

In the limit $\epsilon(\omega) \to 1$ it is easily seen that fields (13.64) and (13.65) reduce to the earlier results (13.30) and (13.29).

To find the energy transferred to the atom at impact parameter b we merely write down the generalization of (13.23):

$$\Delta E(b) = 2e \sum_j f_j \, \text{Re} \int_0^{\infty} -i\omega\mathbf{x}_j(\omega) \cdot \mathbf{E}^*(\omega) \, d\omega \quad (13.66)$$

where $\mathbf{x}_j(\omega)$ is the amplitude of the jth type of electron in the atom. Rather than use (13.19) for $\mathbf{x}_j(\omega)$ we express the sum of dipole moments in terms of the molecular polarizability and so the dielectric constant:

$$e \sum_j f_j \mathbf{x}_j(\omega) = \frac{1}{4\pi N}(\epsilon(\omega) - 1)\mathbf{E}(\omega) \quad (13.67)$$

where N is the number of atoms per unit volume. Then the energy transfer can be written

$$\Delta E(b) = \frac{1}{2\pi N} \operatorname{Re} \int_0^\infty -i\omega \, \epsilon(\omega) \, |\mathbf{E}(\omega)|^2 \, d\omega \qquad (13.68)$$

The energy loss per unit distance in collisions with impact parameter $b \geq a$ is evidently

$$\left(\frac{dE}{dx}\right)_{b>a} = 2\pi N \int_a^\infty \Delta E(b)b \, db \qquad (13.69)$$

If fields (13.64) and (13.65) are inserted into (13.68) and (13.69), we find, after some calculation, the expression due to Fermi,

$$\left(\frac{dE}{dx}\right)_{b>a} = \frac{2}{\pi} \frac{(ze)^2}{v^2} \operatorname{Re} \int_0^\infty i\omega \, \lambda^* a K_1(\lambda^* a) K_0(\lambda a) \left(\frac{1}{\epsilon(\omega)} - \beta^2\right) d\omega \qquad (13.70)$$

where λ is given by (13.62). This result can be obtained more elegantly by calculating the electromagnetic energy radiated through a cylinder of radius a around the path of the incident particle. By conservation of energy this is the energy lost per unit time by the incident particle. Thus

$$\left(\frac{dE}{dx}\right)_{b>a} = \frac{1}{v} \frac{dE}{dt} = -\frac{c}{4\pi v} \int_{-\infty}^\infty 2\pi a B_2 E_3 \, dz \qquad (13.71)$$

The integral over dz at one instant of time is equivalent to an integral at one point on the cylinder over all time. Using $dz = v \, dt$, we have

$$\left(\frac{dE}{dx}\right)_{b>a} = -\frac{ca}{2} \int_{-\infty}^\infty B_2(t)E_3(t) \, dt \qquad (13.72)$$

In the standard way this can be converted into a frequency integral,

$$\left(\frac{dE}{dx}\right)_{b>a} = -ca \operatorname{Re} \int_0^\infty B_2^*(\omega)E_3(\omega) \, d\omega \qquad (13.73)$$

With fields (13.64) and (13.65) this gives the Fermi result (13.70).

The Fermi expression (13.70) bears little resemblance to our previous results for energy loss, such as (13.35). But under conditions where polarization effects are unimportant it yields the same results as before. For example, for nonrelativistic particles ($\beta \ll 1$) it is clear from (13.62) that $\lambda \simeq \omega/v$, independent of $\epsilon(\omega)$. Then in (13.70) the modified Bessel functions are real. Only the imaginary part of $1/\epsilon(\omega)$ contributes to the integral. If we neglect the Lorentz polarization correction (4.67) to the internal field at an atom, the dielectric constant can be written

$$\epsilon(\omega) \simeq 1 + \frac{4\pi N e^2}{m} \sum_j \frac{f_j}{\omega_j^2 - \omega^2 - i\omega \Gamma_j} \qquad (13.74)$$

where we have used the dipole moment expression (13.19). Assuming that the second term is small, the imaginary part of $1/\epsilon(\omega)$ can be readily calculated and substituted into (13.70). Then the integral over $d\omega$ can be performed in the same approximation as used in (13.24)–(13.26) to yield the nonrelativistic form of (13.35). If the departure of λ from $\omega/\gamma v$ is neglected, but no other approximations are made, then (13.70) yields precisely the Bohr result (13.35).

The density effect evidently comes from the presence of complex arguments in the modified Bessel functions, corresponding to taking into account $\epsilon(\omega)$ in (13.62). Since $\epsilon(\omega)$ there is multiplied by β^2, it is clear that the density effect can be really important only at high energies. The detailed calculations for all energies with some explicit expression such as (13.74) for $\epsilon(\omega)$ are quite complicated and not particularly informative. We will content ourselves with the extreme relativistic limit $(\beta \simeq 1)$. Furthermore, since the important frequencies in the integral over $d\omega$ are optical frequencies and the radius a is of the order of atomic dimensions, $|\lambda a| \sim (\omega a/c) \ll 1$. Consequently we can approximate the Bessel functions by their small argument limits (3.103). Then in the relativistic limit the Fermi expression (13.70) is

$$\left(\frac{dE}{dx}\right)_{b>a} \simeq \frac{2}{\pi}\frac{(ze)^2}{c^2}\,\mathrm{Re}\,\int_0^\infty i\omega\left(\frac{1}{\epsilon(\omega)}-1\right)$$
$$\times\left[\ln\left(\frac{1.123c}{\omega a}\right)-\frac{1}{2}\ln\left(1-\epsilon(\omega)\right)\right]d\omega \quad (13.75)$$

It is worth while right here to point out that the argument of the second logarithm is actually $[1-\beta^2\epsilon(\omega)]$. In the limit $\epsilon=1$, this log term gives a factor γ in the combined logarithm, corresponding to the old result (13.36). Provided $\epsilon(\omega)\neq 1$, we can write this factor as $[1-\epsilon(\omega)]$, thereby removing one power of γ from the logarithm, in agreement with experiment.

The integral in (13.75) with $\epsilon(\omega)$ given by (13.74) can be performed most easily by using Cauchy's theorem to change the integral over positive real ω to one over positive imaginary ω. Then $\epsilon(\omega)$ is purely real on the path of integration. Consequently contributions to the integral arise only from the phase of the logarithms. With the approximation that $\Gamma_j \ll \omega_j$ the result of this integration can be written in the simple form:

$$\left(\frac{dE}{dx}\right)_{b>a}=\frac{(ze)^2\omega_p^2}{c^2}\ln\left(\frac{1.123c}{a\omega_p}\right) \quad (13.76)$$

where ω_p is the electronic plasma frequency

$$\omega_p^2=\frac{4\pi NZe^2}{m} \quad (13.77)$$

The corresponding relativistic expression without the density effect is, from (13.36),

$$\left(\frac{dE}{dx}\right)_{b>a} = \frac{(ze)^2\omega_p^2}{c^2}\left[\ln\left(\frac{1.123\gamma c}{a\langle\omega\rangle}\right) - \frac{1}{2}\right] \qquad (13.78)$$

We see that the density effect produces a simplification in that the asymptotic energy loss no longer depends on the details of atomic structure through $\langle\omega\rangle$ (13.38), but only on the number of electrons per unit volume through ω_p. Two substances having very different atomic structures will produce the same energy loss for ultrarelativistic particles provided their densities are such that the density of electrons is the same in each.

Since there are numerous calculated curves of energy loss based on Bethe's formula (13.44), it is often convenient to tabulate the decrease in energy loss due to the density effect. This is just the difference between (13.78) and (13.76):

$$\lim_{\beta\to 1}\Delta\left(\frac{dE}{dx}\right) = \frac{(ze)^2\omega_p^2}{c^2}\left[\ln\left(\frac{\gamma\omega_p}{\langle\omega\rangle}\right) - \frac{1}{2}\right] \qquad (13.79)$$

For photographic emulsions, the relevant energy loss is given by (13.49) and (13.50) with $\epsilon \simeq 10$ Kev. With the density correction applied, this becomes constant at high energies with the value,

$$\frac{dE(\epsilon)}{dx} \to \frac{(ze)^2\omega_p^2}{2c^2}\ln\left(\frac{2mc^2\epsilon}{\hbar^2\omega_p^2}\right) \qquad (13.80)$$

For silver bromide, $\hbar\omega_p \simeq 48$ ev. Then for singly charged particles (13.80), divided by the density, has the value of approximately 1.02 Mev-cm²/gm. This energy loss is in good agreement with experiment, and corresponds to an increase above the minimum value of less than 10 per cent. Figure 13.5 shows total energy loss and loss from transfers of less than 10 Kev for a typical substance. The dotted curve is the Bethe curve for total energy loss without correction for density effect.

There is an interesting connection between the Fermi expression (13.70) for energy loss and the emission of Cherenkov radiation. Equation (13.70) represents energy transferred to the medium at distances greater than a. If we let $a \to \infty$ we can find out whether any of the energy escapes to infinity. Such energy would be properly described as radiation. For $a \to \infty$, the asymptotic forms (3.104) of the K functions can be used. Then (13.70) takes the form:

$$\lim_{a\to\infty}\left(\frac{dE}{dx}\right)_{b>a} = \frac{(ze)^2}{v^2}\,\text{Re}\int_0^\infty i\omega\left(\frac{1}{\epsilon(\omega)} - \beta^2\right)\left(\frac{\lambda^*}{\lambda}\right)^{1/2} e^{-(\lambda+\lambda^*)a}\,d\omega \qquad (13.81)$$

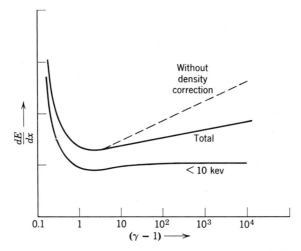

Fig. 13.5 Energy loss, including the density effect. The dotted curve is the total energy loss without density correction. The solid curves have the density effect incorporated, the upper one being the total energy loss and the lower one the energy loss due to individual energy transfers of less than 10 Kev.

If λ has a real part, the exponential factor causes the energy loss to go rapidly to zero at large distances. From (13.62) it is evident that this will always occur if the medium is absorbent, since then $\epsilon(\omega)$ has a positive imaginary part. But if $\epsilon(\omega)$ is real, λ can be pure imaginary for certain ω. This occurs whenever $\beta^2 > 1/\epsilon(\omega)$, i.e., whenever the velocity of the particle is greater than the phase velocity of light in the medium. This is the criterion for Cherenkov radiation. For such frequencies, $\lambda = -i\,|\lambda|$. Then the exponential equals unity, and we find

$$\lim_{a \to \infty} \left(\frac{dE}{dx}\right)_{b>a} = \frac{(ze)^2}{c^2} \int_{\epsilon(\omega)>(1/\beta^2)} \omega\left(1 - \frac{1}{\beta^2\epsilon(\omega)}\right) d\omega \qquad (13.82)$$

Since this expression is independent of the cylinder radius a, it represents true radiation. It is just the Frank-Tamm (1937) result for the total energy per unit distance emitted as Cherenkov radiation. A more detailed discussion of Cherenkov radiation as a radiative process will be given in Section 14.9.

For media in which the density effect is an important feature of the energy-loss process the absorption is almost always sufficiently great that the incipient Cherenkov radiation is absorbed very close to the path of the particle.

13.5   Energy Loss in an Electronic Plasma

The loss of energy by a nonrelativistic particle passing through a plasma can be treated in a manner similar to the density effect for a relativistic particle. As was discussed in Section 10.10, the length scale in a plasma is divided into two regions. For dimensions large compared to the Debye screening distance k_D^{-1} (10.106), the plasma acts as a continuous medium in which the charged particles participate in collective behavior such as plasma oscillations. For dimensions small compared to k_D^{-1}, individual-particle behavior dominates and the particles interact by the two-body screened potential (10.113). This means that in calculating energy loss the Debye screening distance plays the same role here as the atomic dimension *a* played in the density-effect calculation. For close collisions collective effects can be ignored, and the two-body screened potential can be used to evaluate this contribution to the energy loss. This is left as an exercise for the reader (Problem 13.3). For the distant collisions at impact parameters $bk_D > 1$ the collective effects can be calculated by utilizing Fermi's formula (13.70) with an appropriate dielectric constant for a plasma. The loss in distant collisions corresponds to the excitation of plasma oscillations in the medium.

For a nonrelativistic particle (13.70) yields the following expression for the energy loss to distances $b > k_D^{-1}$:

$$\left(\frac{dE}{dx}\right)_{k_D b > 1} \simeq \frac{2}{\pi} \frac{(ze)^2}{v^2} \operatorname{Re} \int_0^\infty \frac{i\omega}{\epsilon(\omega)} \left[\frac{\omega}{k_D v} K_1\left(\frac{\omega}{k_D v}\right) K_0\left(\frac{\omega}{k_D v}\right) \right] d\omega \quad (13.83)$$

Since the important frequencies in the integral turn out to be $\omega \sim \omega_p$, the relevant argument of the Bessel functions is

$$\frac{\omega_p}{k_D v} = \frac{\langle u^2 \rangle^{1/2}}{v} \quad (13.84)$$

For particles incident with velocities v less than thermal velocities this argument is large compared to unity. Because of the exponential fall-off of the Bessel functions for large argument, the energy loss in exciting plasma oscillations by such particles is negligible. Whatever energy is lost is in close binary collisions. If the velocity is comparable with or greater than thermal speeds, then the particle can lose appreciable amounts of energy in exciting collective oscillations. It is evident that this energy of oscillation is deposited in the neighborhood of the path of the particle, out to distances of the order of $(v/\langle u^2 \rangle^{1/2}) k_D^{-1}$.

For a particle moving rapidly compared to thermal speeds we may use the familiar small argument forms for the modified Bessel functions. Then (13.83) becomes

$$\left(\frac{dE}{dx}\right)_{k_Db>1} \simeq \frac{2}{\pi}\frac{(ze)^2}{v^2}\int_0^\infty \mathrm{Re}\left(\frac{i\omega}{\epsilon(\omega)}\right)\ln\left(\frac{1.123k_Dv}{\omega}\right)d\omega \qquad (13.85)$$

We shall take the simple dielectric constant (7.93), augmented by some damping:

$$\epsilon(\omega) = 1 - \frac{\omega_p^{\,2}}{\omega^2 + i\omega\Gamma} \qquad (13.86)$$

The damping constant Γ will be assumed small compared to ω_p. The necessary combination,

$$\mathrm{Re}\left(\frac{i\omega}{\epsilon(\omega)}\right) = \omega_p^{\,2}\frac{\omega^2\Gamma}{(\omega^2 - \omega_p^{\,2})^2 + \omega^2\Gamma^2} \qquad (13.87)$$

has the standard resonant character seen in (13.24), for example. In the limit $\Gamma \ll \omega_p$ the integral in (13.85) leads to the simple result,

$$\left(\frac{dE}{dx}\right)_{k_Db>1} \simeq \frac{(ze)^2}{v^2}\,\omega_p^{\,2}\ln\left(\frac{1.123k_Dv}{\omega_p}\right) \qquad (13.88)$$

This can be combined with the results of Problem 13.3 to give an expression for the total energy loss of a particle passing through a plasma. The presence of ω_p in the logarithm implies that the energy losses occur in quantum jumps of $\hbar\omega_p$, in the same way as the mean frequency $\langle\omega\rangle$ in (13.44) is indicative of the typical quantum jumps in atoms. Electrons passing through thin metal foils show this discreteness in their energy loss. The phenomenon can be used to determine the effective plasma frequency in metals.

13.6   Elastic Scattering of Fast Particles by Atoms

In the preceding sections we have been concerned with the energy loss of particles passing through matter. In these considerations it was assumed that the trajectory of the particle was a straight line. Actually this approximation is not rigorously true. As was discussed in Section 13.1, any momentum transfer between collision partners leads to a deflection in angle. In the introductory remarks at the beginning of the chapter it was pointed out that collisions with electrons determine the energy loss, whereas collisions with atoms determine the scattering. If the screening of the nuclear Coulomb field by the atomic electrons is neglected, a fast particle of momentum $p = \gamma M v$ and charge ze, passing a heavy nucleus of charge Ze at impact parameter b, will suffer an angular deflection,

$$\theta \simeq \frac{2zZe^2}{pvb} \qquad (13.89)$$

according to (13.3).

The differential scattering cross section $d\sigma/d\Omega$ (with dimensions of area per unit solid angle per atom) is defined by the relation,

$$nb\,db\,d\phi = n\,\frac{d\sigma}{d\Omega}\sin\theta\,d\theta\,d\phi \qquad (13.90)$$

where n is the number of particles incident on the atom per unit area per unit time. The left-hand side of (13.90) is the number of particles per unit time incident at azimuthal angles between ϕ and $(\phi + d\phi)$ and impact parameters between b and $(b + db)$. The right-hand side is the number of scattered particles per unit time emerging at polar angles (θ, ϕ) in the element of solid angle $d\Omega = \sin\theta\,d\theta\,d\phi$. Equation (13.90) is merely a statement of conservation of particles, since b and θ are functionally related. The classical differential scattering cross section can therefore be written.

$$\frac{d\sigma}{d\Omega} = \frac{b}{\sin\theta}\left|\frac{db}{d\theta}\right| \qquad (13.91)$$

The absolute value sign is put on, since db and $d\theta$ can in general have opposite signs, but the cross section is by definition positive definite. If b is a multiple-valued function of θ, then the different contributions must be added in (13.91).

With relation (13.89) between b and θ we find the small-angle nuclear Rutherford scattering cross section per atom,

$$\frac{d\sigma}{d\Omega} \simeq \left(\frac{2zZe^2}{pv}\right)^2 \frac{1}{\theta^4} \qquad (13.92)$$

We note that the Z electrons in each atom give a contribution Z^{-1} times the nuclear one. Hence the electrons can be ignored, except for their screening action. The small-angle Rutherford law (13.92) for nuclear scattering is found to be true quantum mechanically, independent of the spin nature of the incident particles. At wide angles spin effects enter, but for nonrelativistic particles the classical Rutherford formula,

$$\frac{d\sigma}{d\Omega} = \left(\frac{zZe^2}{2Mv^2}\right)^2 \operatorname{cosec}^4\frac{\theta}{2} \qquad (13.93)$$

which follows from (13.4), holds quantum mechanically as well.

Since most of the scattering occurs for $\theta \ll 1$, and even at $\theta = \pi/2$ the small-angle result (13.92) is within 30 per cent of the Rutherford expression, it is sufficiently accurate to employ (13.92) at all angles for which the unscreened point Coulomb-field description is valid.

Departures from the point Coulomb-field approximation come at large and small angles, corresponding to small and large impact parameters. At large b the screening effects of the atomic electrons cause the potential to fall off more rapidly than $(1/r)$. On the Fermi-Thomas model the potential can be approximated roughly by the form:

$$V(r) \simeq \frac{zZe^2}{r} \exp(-r/a) \qquad (13.94)$$

where the atomic radius a is

$$a \simeq 1.4a_0 Z^{-1/3} \qquad (13.95)$$

The length $a_0 = \hbar^2/me^2$ is the hydrogenic Bohr radius. For impact parameters of the order of, or greater than, a the rapid decrease of the potential (13.94) will cause the scattering angle to vanish much more rapidly with increasing b than is given by (13.89). This implies that the scattering cross section will flatten off at small angles to a finite value at $\theta = 0$, rather than increasing as θ^{-4}. A simple calculation with a cutoff Coulomb potential shows that the cross section has the general form:

$$\frac{d\sigma}{d\Omega} \sim \left(\frac{2zZe^2}{pv}\right)^2 \frac{1}{(\theta^2 + \theta_{min}^2)^2} \qquad (13.96)$$

where θ_{min} is a cutoff angle. The minimum angle θ_{min} below which the cross section departs appreciably from the simple result (13.92) can be determined either classically or quantum mechanically. As with b_{min} in the energy-loss calculations, the larger of the two angles is the correct one to employ. Classically θ_{min} can be estimated by putting $b = a$ in (13.89). This gives

$$\theta_{min}^{(c)} \simeq \frac{zZe^2}{pva} \qquad (13.97)$$

Quantum mechanically, the finite size of the scatterer implies that the approximately classical trajectory must be localized to within $\Delta x < a$; the incident particle must have a minimum uncertainty in transverse momentum $\Delta p \gtrsim \hbar/a$. For collisions in which the momentum transfer (13.1) is large compared to $\hbar/a$ the classical Rutherford formula will apply. But for smaller momentum transfers we expect the quantum-mechanical smearing out to flatten off the cross section. This leads to a quantum mechanical θ_{min}:

$$\theta_{min}^{(q)} \simeq \frac{\hbar}{pa} \qquad (13.98)$$

We note that the ratio of the classical to quantum-mechanical angles θ_{min} is $Zze^2/\hbar v$ in agreement with the ratio (13.42) of the classical and quantum values of b_{min}. For fast particles in all but the highest Z substances ($Zze^2/\hbar v$) is less than unity. Then the quantum value (13.98) will be used for θ_{min}. With value (13.95) for the screening radius a, (13.98) becomes

$$\theta^{(q)}_{min} \simeq \frac{Z^{\frac{1}{3}}}{192} \left(\frac{mc}{p}\right) \tag{13.99}$$

where p is the incident momentum ($p = \gamma Mv$), and m is the electronic mass.

At comparatively large angles the cross section departs from (13.92) because of the finite size of the nucleus. For electrons and mu mesons the influence of nuclear size is a purely electromagnetic effect, but for pi mesons protons, etc., there are specific effects of a nuclear-force nature as well. Since the gross overall effect is to lower the cross section below that predicted by (13.92) for whatever reason, we will consider only the electromagnetic aspect. The charge distribution of the atomic nucleus can be crudely approximated by a uniform volume distribution inside a sphere of radius R, falling rapidly to zero outside R. This means that the electrostatic potential inside the nucleus is not $1/r$, but rather parabolic in shape with a finite value at $r = 0$:

$$V(r) = \begin{cases} \dfrac{3}{2}\dfrac{zZe^2}{R}\left(1 - \dfrac{r^2}{3R^2}\right), & \text{for } r < R \\[3mm] \dfrac{zZe^2}{r}, & \text{for } r > R \end{cases} \tag{13.100}$$

It is a peculiarity of the point-charge Coulomb field that the quantum-mechanical cross section is the classical Rutherford formula. Thus for a point nucleus there is no need to consider a division of the angular region into angles corresponding to impact parameters less than, or greater than, the quantum-mechanical impact parameter $b^{(q)}_{min}$ (13.40). For a nucleus of finite size, however, the de Broglie wavelength of the incident particle does enter. When we consider wave packets incident on the relatively constant (inside $r = R$) potential (13.100), there will be appreciable departures from the simple formula (13.92). The situation is quite analogous to the diffraction of waves by a spherical object, considered in Chapter 9. The scattering is all confined to angles less than $\sim(\lambda/R)$, where λ is the wavelength (divided by 2π) of the waves involved. For wider angles the wavelets from different parts of the scatterer interfere, causing a rapid decrease in the scattering or perhaps subsidiary maxima and

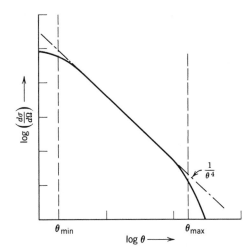

Fig. 13.6 Atomic scattering, including effects of electronic screening at small angles and finite nuclear size at large angles.

minima. Since the particle wavelength is $\lambda = \hbar/p$, the maximum scattering angle, beyond which the scattering cross section falls significantly below the θ^{-4} law, is

$$\theta_{max} \simeq \frac{\hbar}{pR} \tag{13.101}$$

Using the simple estimate $R \simeq \frac{1}{2}(e^2/mc^2) A^{1/3} = 1.4 A^{1/3} \times 10^{-13}$ cm, this has the numerical value,

$$\theta_{max} \simeq \frac{274}{A^{1/3}} \left(\frac{mc}{p}\right) \tag{13.102}$$

We note that, for all values of Z and A, $\theta_{max} \gg \theta_{min}$. If the incident momentum is so small that $\theta_{max} \gtrsim 1$, the nuclear size has no appreciable effect on the scattering. For an aluminum target $\theta_{max} = 1$ when $p \sim 50$ Mev/c, corresponding to ~ 50 Mev, 12 Mev, and 1.3 Mev kinetic energies for electrons, mu mesons, and protons, respectively. Only at higher energies than these are nuclear-size effects important in the scattering. At this momentum value $\theta_{min}^{(q)} \sim 10^{-4}$ radian.

The general behavior of the cross section is shown in Fig. 13.6. The dotted curve is the small-angle Rutherford approximation (13.92), while the solid curve shows the qualitative behavior of the cross section, including screening and finite nuclear size. The total scattering cross section can be obtained by integrating (13.96) over all solid angle:

$$\sigma = \int \frac{d\sigma}{d\Omega} \sin\theta \; d\theta \; d\phi \simeq 2\pi \left(\frac{2zZe^2}{pv}\right)^2 \int_0^\infty \frac{\theta \; d\theta}{(\theta_{min}^2 + \theta^2)^2} \tag{13.103}$$

This yields

$$\sigma \simeq \pi \left(\frac{2zZe^2}{pv}\right)^2 \frac{1}{\theta_{min}^2} = \pi a^2 \left(\frac{2zZe^2}{\hbar v}\right)^2 \tag{13.104}$$

where the final form is obtained by using $\theta_{\min}^{(q)}$ (13.98). It shows that at high velocities the total cross section can be far smaller than the classical value of geometrical area πa^2.

13.7 Mean Square Angle of Scattering and the Angular Distribution of Multiple Scattering

Rutherford scattering is confined to very small angles even for a point Coulomb field, and for fast particles $\theta_{\max}$ is small compared to unity. Thus there is a very large probability for small-angle scattering. A particle traversing a finite thickness of matter will undergo very many small-angle deflections and will generally emerge at a small angle which is the cumulative statistical superposition of a large number of deflections. Only rarely will the particle be deflected through a large angle; since these events are infrequent, such a particle will have made only one such collision. This circumstance allows us to divide the angular range into two regions—one region at comparatively large angles which contains only the single scatterings, and one region at very small angles which contains the multiple or compound scatterings. The complete distribution in angle can be approximated by considering the two regions separately. The intermediate region of so-called plural scattering must allow a smooth transition from small to large angles.

The important quantity in the multiple-scattering region, where there is a large succession of small-angle deflections symmetrically distributed about the incident direction, is the mean square angle for a single scattering. This is defined by

$$\langle \theta^2 \rangle = \frac{\displaystyle\int \theta^2 \frac{d\sigma}{d\Omega} \, d\Omega}{\displaystyle\int \frac{d\sigma}{d\Omega} \, d\Omega} \tag{13.105}$$

With the approximations of Section 13.6 we obtain

$$\langle \theta^2 \rangle = 2\theta_{\min}^2 \ln\left(\frac{\theta_{\max}}{\theta_{\min}}\right) \tag{13.106}$$

If the quantum value (13.99) of $\theta_{\min}$ is used along with $\theta_{\max}$ (13.102), then (13.106) has the numerical form:

$$\langle \theta^2 \rangle \simeq 4\theta_{\min}^2 \ln\left(210 Z^{-\frac{1}{3}}\right) \tag{13.107}$$

If nuclear size is unimportant (generally only of interest for electrons, and perhaps other particles at very low energies), $\theta_{\max}$ should be put equal to

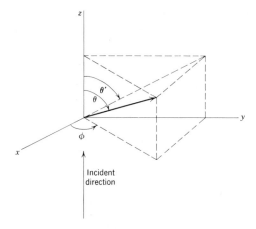

Fig. 13.7

unity in (13.106). Then the argument of the logarithm in (13.107) becomes $\left(\dfrac{192}{Z^{1/3}}\dfrac{p}{mc}\right)^{1/2}$, instead of $(210Z^{-1/3})$.

It is often desirable to use the projected angle of scattering θ', the projection being made on some convenient plane such as the plane of a photographic emulsion or a bubble chamber, as shown in Fig. 13.7. For small angles it is easy to show that

$$\langle \theta'^2 \rangle = \tfrac{1}{2}\langle \theta^2 \rangle \tag{13.108}$$

In each collision the angular deflections obey the Rutherford formula (13.92) suitably cut off at $\theta_{\min}$ and $\theta_{\max}$, with average value zero (when viewed relative to the forward direction, or as a projected angle) and mean square angle $\langle \theta^2 \rangle$ given by (13.106). Since the successive collisions are independent events, the central-limit theorem of statistics can be used to show that for a large number n of such collisions the distribution in angle will be approximately Gaussian around the forward direction with a mean square angle $\langle \Theta^2 \rangle = n \langle \theta^2 \rangle$. The number of collisions occurring as the particle traverses a thickness t of material containing N atoms per unit volume is

$$n = N\sigma t \simeq \pi N \left(\frac{2zZe^2}{pv}\right)^2 \frac{t}{\theta_{\min}^2} \tag{13.109}$$

This means that the mean square angle of the Gaussian is

$$\langle \Theta^2 \rangle \simeq 2\pi N \left(\frac{2zZe^2}{pv}\right)^2 \ln\left(\frac{\theta_{\max}}{\theta_{\min}}\right) t \tag{13.110}$$

Or, using (13.107) for $\langle \theta^2 \rangle$,

$$\langle \Theta^2 \rangle \simeq 4\pi N \left(\frac{2zZe^2}{pv}\right)^2 \ln\left(210Z^{-\frac{1}{3}}\right) t \tag{13.111}$$

The mean square angle increases linearly with the thickness t. But for reasonable thicknesses such that the particle does not lose appreciable energy, the Gaussian will still be peaked at very small forward angles.

The multiple-scattering distribution for the projected angle of scattering is

$$P_M(\theta') \, d\theta' = \frac{1}{\sqrt{\pi \langle \Theta^2 \rangle}} \exp\left(-\frac{\theta'^2}{\langle \Theta^2 \rangle}\right) d\theta' \tag{13.112}$$

where both positive and negative values of θ' are considered. The small-angle Rutherford formula (13.92) can be expressed in terms of the projected angle as

$$\frac{d\sigma}{d\theta'} = \frac{\pi}{2}\left(\frac{2zZe^2}{pv}\right)^2 \frac{1}{\theta'^3} \tag{13.113}$$

This gives a single-scattering distribution for the projected angle:

$$P_S(\theta') \, d\theta' = Nt \frac{d\sigma}{d\theta'} \, d\theta' = \frac{\pi}{2} Nt \left(\frac{2zZe^2}{pv}\right)^2 \frac{d\theta'}{\theta'^3} \tag{13.114}$$

The single-scattering distribution is valid only for angles large compared to $\langle \Theta^2 \rangle^{\frac{1}{2}}$, and contributes a tail to the Gaussian distribution.

If we express angles in terms of the relative projected angle,

$$\alpha = \frac{\theta'}{\langle \Theta^2 \rangle^{\frac{1}{2}}} \tag{13.115}$$

the multiple- and single-scattering distributions can be written

$$P_M(\alpha) \, d\alpha = \frac{1}{\sqrt{\pi}} e^{-\alpha^2} \, d\alpha$$

$$P_S(\alpha) \, d\alpha = \frac{1}{8 \ln\left(210Z^{-\frac{1}{3}}\right)} \frac{d\alpha}{\alpha^3} \tag{13.116}$$

where (13.111) has been used for $\langle \Theta^2 \rangle$. We note that the relative amounts of multiple and single scatterings are independent of thickness in these units, and depend only on Z. Even this Z dependence is not marked. The factor $8 \ln (210Z^{-\frac{1}{3}})$ has the value 36.0 for $Z = 13$ (aluminum) and the value 31.0 for $Z = 82$ (lead). Figure 13.8 shows the general behavior of the scattering distributions as a function of α. The transition from multiple to single scattering occurs in the neighborhood of $\alpha \simeq 2.5$. At this point the Gaussian has a value of 1/600 times its peak value. Thus the single-scattering distribution gives only a very small tail on the multiple-scattering curve.

There are two things which cause departures from the simple behavior shown in Fig. 13.8. The Gaussian shape is the limiting form of the

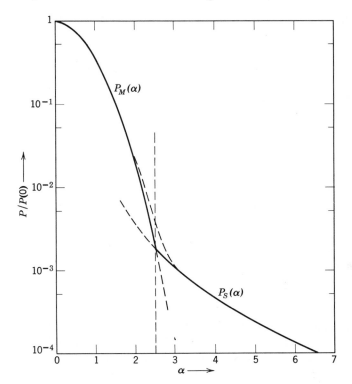

Fig. 13.8 Multiple and single scattering distributions of projected angle. In the region of plural scattering ($\alpha \sim 2$–3) the dotted curve indicates the smooth transition from the small-angle multiple scattering (approximately Gaussian in shape) to the wide-angle single scattering (proportional to α^{-3}).

angular distribution for very large n. If the thickness t is such that n (13.109) is not very large (i.e., $n \lesssim 100$), the distribution follows the single-scattering curve to smaller angles than $\alpha \simeq 2.5$, and is somewhat more sharply peaked at zero angle than a Gaussian. On the other hand, if the thickness is great enough, the mean square angle $\langle \Theta^2 \rangle$ becomes comparable with the angle $\theta_{\max}$ (13.102) which limits the angular width of the single-scattering distribution. For greater thicknesses the multiple-scattering curve extends in angle beyond the single-scattering region, so that there is no single-scattering tail on the distribution (see Problem 13.5).

13.8 Electrical Conductivity of a Plasma

The considerations of multiple scattering can be applied rather directly to the seemingly different problem of the electrical conductivity of a plasma. For simplicity we will consider the so-called Lorentz gas, which consists of N fixed ions of charge Ze per unit volume and NZ free electrons per unit volume. Furthermore electron-electron interactions will be ignored. The approximation of fixed ions is a reasonable one, at least for plasmas with electrons and ions at roughly the same kinetic temperatures. The effects of electron-electron collisions will be mentioned later.

The simple Drude theory of electrical conductivity, described briefly in Section 7.8, is based on the single electron equation,

$$m \frac{d\mathbf{v}}{dt} = e\mathbf{E} - m\nu\mathbf{v} \qquad (13.117)$$

where ν is the collision frequency. The low-frequency electrical conductivity σ due to electron motion is

$$\sigma = \frac{NZe^2}{m\nu} \qquad (13.118)$$

The problem of calculating the proper collision frequency can be approached by noting that the term $m\nu\mathbf{v}$ in (13.117) really represents the rate of decrease of forward momentum because of Coulomb collisions with the ions as the electron moves under the action of the applied electric field. If the scattering angle in a single elastic collision is θ, as indicated in Fig. 13.9, the forward momentum lost by a particle of momentum p is $p(1 - \cos \theta)$. The average value of this quantity multiplied by the number of collisions per unit distance is the loss in forward momentum per unit distance, namely, $m\nu$. Thus

$$m\nu = N\sigma p \langle 1 - \cos \theta \rangle \qquad (13.119)$$

where σ here is the total cross section (13.104). Since all the Coulomb scattering is at very small angles, $\langle 1 - \cos \theta \rangle \simeq \tfrac{1}{2} \langle \theta^2 \rangle$. Then the forward-momentum loss per unit distance is

$$m\nu \simeq \tfrac{1}{2}N\sigma p\langle\theta^2\rangle = 4\pi N \frac{(Ze^2)^2}{m v^3} \ln \left(\frac{\theta_{\max}}{\theta_{\min}}\right) \qquad (13.120)$$

Equation (13.106) has been used for $\langle\theta^2\rangle$. When (13.120) is inserted in (13.118), we obtain a conductivity,

$$\sigma(v) \simeq \frac{m v^3}{4\pi(Ze^2) \ln (\theta_{\max}/\theta_{\min})} \qquad (13.121)$$

This result holds for electrons of velocity v.

We now want to average over a thermal distribution. The variation with v in (13.121) comes mainly from the factor v^3. The argument of the logarithm can be evaluated at the mean velocity without introducing appreciable error. At energies appropriate to even the hottest plasmas nuclear-size effects are negligible. Consequently we put $\theta_{\max} = 1$. The value of $\theta_{\min}$ requires some discussion. For the screened atomic potential

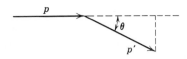

Fig. 13.9

the result (13.97) or (13.98) was appropriate, with the atomic radius a given by (13.95). For electron-ion collisions in a plasma the interaction is the Debye-Hückel screened potential (10.113). Consequently the Debye length k_D^{-1} plays the role of the atomic radius a in the formulas for θ_{min}. Either (13.97) or (13.98) is used, depending on which is larger. With these substitutions the argument of the logarithm in (13.121) can be written

$$\Lambda = \frac{\theta_{max}}{\theta_{min}} = \begin{cases} \dfrac{12\pi N}{k_D^{3}} \\[2ex] \dfrac{Ze^2}{\sqrt{3}\hbar\langle u^2\rangle^{\frac{1}{2}}}\dfrac{12\pi N}{k_D^{3}} \end{cases} \tag{13.122}$$

where k_D is given by (10.106) or (10.112), and $\langle u^2\rangle^{\frac{1}{2}} = kT/m$. The upper (lower) value of Λ is to be used when the mean electron energy $\frac{3}{2}kT$ is less (greater) than $13.6Z^2$ electron volts.

The average value of the nth power of the magnitude of velocity for a Maxwellian distribution is

$$\langle v^n\rangle = \left(\frac{2kT}{m}\right)^{n/2}\frac{\Gamma\left(\dfrac{n+3}{2}\right)}{\Gamma(\tfrac{3}{2})} \tag{13.123}$$

Consequently the value of the conductivity (13.121), averaged over a Maxwellian velocity distribution, is

$$\sigma \simeq \frac{m}{Ze^2\ln\Lambda}\left(\frac{2\,kT}{\pi\,m}\right)^{3/2} \tag{13.124}$$

This approximate result, obtained in a rather simple-minded way with the elementary Drude theory, is within a factor of 2 of the correct value found from an application of the Boltzmann equation. The physical mechanism is the same in both calculations, but the more rigorous treatment involves an averaging over v^5 rather than v^3.* Thus the two results differ by a factor $\langle v^5\rangle/\langle v^2\rangle\langle v^3\rangle = 2$.

* The added power of v^2 can be understood as follows. In the presence of the electric field the formerly spherically symmetric velocity distribution tends to become distorted in velocity directions parallel to the field. The amount of distortion determines the current and, through Ohm's law, the conductivity. The distorted distribution results from a balancing of the anisotropic electric force and the tendency towards isotropy produced by the collisions. Since the scattering cross section varies as v^{-2}, the anisotropic part of the distribution has more high-velocity components than normal by a factor v^2.

When electron-electron collisions are included, the forward-momentum loss is increased and so the conductivity is decreased from its value of twice (13.124). The relative decrease depends on Z roughly as $Z/(1 + Z)$, ranging from 0.58 for $Z = 1$ to 1.0 for $Z \to \infty$. Consequently (13.124) as it stands can be used as a good approximation to the conductivity for a hydrogen or deuterium plasma, including effects of electron-electron collisions. If the classical (low-energy) value of Λ (13.122) is used, (13.124) can be written in the instructive form:

$$\sigma \simeq \frac{1}{3} \left(\frac{2}{\pi} \right)^{3/2} \frac{\Lambda}{\ln \Lambda} \omega_p \qquad (13.125)$$

Since Λ is of the order of 10^4 for a typical hydrogen plasma ($n_e \sim 10^{15}$ cm^{-3}, $T \sim 10^{5\circ}$K), σ is $\sim 200 \omega_p \simeq 4 \times 10^{14}$ sec^{-1}. This is not quite as large as metallic conductivities ($\sim 10^{16}$ sec^{-1}), but is sufficiently large that the infinite conductivity approximation used in Chapter 10 is quite adequate in discussing the penetration of fields into a plasma.

REFERENCES AND SUGGESTED READING

The problems of the penetration of particles through matter have interested Niels Bohr all his life. A lovely presentation of the whole subject, with characteristic emphasis on the interplay of classical and quantum-mechanical effects, appears in his comprehensive review article of 1948:

Bohr.

Numerical tables and graphs of energy-loss data, as well as key formulas, are given by

Rossi, Chapter 2,

Segre, article by H. A. Bethe and J. Ashkin.

Rossi also gives a semiclassical treatment of energy loss and scattering similar to ours.

The density effect on the energy loss by extremely relativistic particles is discussed, with numerous results for different substances in graphical form, by

R. M. Sternheimer, *Phys. Rev.*, **88**, 851 (1952); **91**, 256 (1953).

The correct calculation of the conductivity of a plasma is outlined by

Spitzer, Chapter 5.

PROBLEMS

13.1 A heavy particle of charge ze, mass M, and nonrelativistic velocity v collides with a free electron of charge e and mass m initially at rest. With no approximations, other than that of nonrelativistic motion and $M \gg m$, show that the energy transferred to the electron in this Coulomb collision, as a function of the impact parameter b, is

$$\Delta E(b) = \frac{2(ze^2)^2}{mv^2} \frac{1}{b^2 + (ze^2/mv^2)^2}$$

13.2   (a) Taking $\hbar\langle\omega\rangle = 12Z$ ev in the quantum-mechanical energy-loss formula, calculate the rate of energy loss (in Mev/cm) in air at NTP, aluminum, copper, lead for a proton and a mu meson, each with kinetic energies of 10, 100, 1000 Mev.

(b) Convert your results to energy loss in units of Mev-cn^2/gm and compare the values obtained in different materials. Explain why all the energy losses in Mev-cm^2/gm are within a factor of 2 of each other, whereas the values in Mev/cm differ greatly.

13.3 Consider the energy loss by close collisions of a fast, but nonrelativistic, heavy particle of charge ze passing through an electronic plasma. Assume that the screened Coulomb interaction (10.113) acts between the electrons and the incident particle.

(a) Show that the energy transfer in a collision at impact parameter b is given approximately by

$$\Delta E(b) \simeq \frac{2(ze^2)^2}{mv^2} k_D^2 K_1^2(k_D b)$$

where m is the electron mass, v is the velocity of the incident particle, and k_D is the Debye wave number (10.112).

(b) Determine the energy loss per unit distance traveled for collisions with impact parameter greater than b_{min}. Assuming $k_D b_{min} \ll 1$, write down your result with both the classical and quantum-mechanical values of b_{min}.

13.4 With the same approximations as were used to discuss multiple scattering, show that the *projected* transverse displacement y (see Fig. 13.7) of an incident particle is described approximately by a Gaussian distribution,

$$P(y)\, dy = A \exp\left[\frac{-y^2}{2\langle y^2\rangle}\right] dy$$

where the mean square displacement is $\langle y^2\rangle = (x^2/6)\langle\Theta^2\rangle$, x being the thickness of the material traversed and $\langle\Theta^2\rangle$ the mean square angle of scattering.

13.5 If the finite size of the nucleus is taken into account in the "single-scattering" tail of the multiple-scattering distribution, there is a critical thickness x_c beyond which the single-scattering tail is absent.

(a) Define x_c in a reasonable way and calculate its value (in cm) for aluminum and lead, assuming that the incident particle is relativistic.

(b) For these thicknesses calculate the number of collisions which occur and determine whether the Gaussian approximation is valid.

14

Radiation by Moving Charges

It is well known that accelerated charges emit electromagnetic radiation. In Chapter 9 we discussed examples of radiation by macroscopic time-varying charge and current densities, which are fundamentally charges in motion. We will return to such problems in Chapter 16 where multipole radiation is treated in a systematic way. But there is a class of radiation phenomena where the source is a moving point charge or a small number of such charges. In these problems it is useful to develop the formalism in such a way that the radiation intensity and polarization are related directly to properties of the charge's trajectory and motion. Of particular interest are the total radiation emitted, the angular distribution of radiation, and its frequency spectrum. For nonrelativistic motion the radiation is described by the well-known Larmor result (see Section 14.2). But for relativistic particles a number of unusual and interesting effects appear. It is these relativistic aspects which we wish to emphasize. In the present chapter a number of general results are derived and applied to examples of charges undergoing prescribed motions, especially in external force fields. Chapter 15 deals with radiation emitted in atomic or nuclear collisions.

14.1 Liénard-Wiechert Potentials and Fields for a Point Charge

In Chapter 6 it was shown that for localized charge and current distributions without boundary surfaces the scalar and vector potentials can be written as

$$A_\mu(\mathbf{x}, t) = \frac{1}{c} \int \int \frac{J_\mu(\mathbf{x}', t')}{R} \delta\left(t' + \frac{R}{c} - t\right) d^3x' \, dt' \qquad (14.1)$$

where $\mathbf{R} = (\mathbf{x} - \mathbf{x}')$, and the delta function provides the retarded behavior demanded by causality. For a point charge e with velocity $c\boldsymbol{\beta}(t)$ at the point $\mathbf{r}(t)$ the charge-current density is

$$J_\mu(\mathbf{x}, t) = ec\beta_\mu \, \delta[\mathbf{x} - \mathbf{r}(t)] \tag{14.2}$$

where $\beta_\mu = (\boldsymbol{\beta}, i)$. With this source density the spatial integration in (14.1) can be done immediately, yielding

$$A_\mu(\mathbf{x}, t) = e \int \frac{\beta_\mu(t')}{R(t')} \delta\left[t' + \frac{R(t')}{c} - t\right] dt' \tag{14.3}$$

where now $R(t') = |\mathbf{x} - \mathbf{r}(t')|$. Although (14.3) is a convenient form to utilize in calculating the fields, the integral over dt' can be performed, provided we recall from Section 1.2 that when the argument of the delta function is a function of the variable of integration the standard results are modified as follows:

$$\int g(x) \, \delta[f(x) - \alpha] \, dx = \left[\frac{g(x)}{df/dx}\right]_{f(x)=\alpha} \tag{14.4}$$

The function $f(t') = t' + [R(t')/c]$ has a derivative

$$\frac{df}{dt'} \equiv \kappa = 1 + \frac{1}{c}\frac{dR}{dt'} = 1 - \mathbf{n} \cdot \boldsymbol{\beta} \tag{14.5}$$

where $c\boldsymbol{\beta}$ is the instantaneous velocity of the particle, and $\mathbf{n} = \mathbf{R}/R$ is a unit vector directed from the position of the charge to the observation point. With (14.5) in (14.4) and (14.3) the potentials of the point charge, called the *Liénard-Wiechert potentials*, are

$$\left.\begin{aligned}
\Phi(\mathbf{x}, t) &= e\left[\frac{1}{\kappa R}\right]_{\text{ret}} \\[2mm]
\mathbf{A}(\mathbf{x}, t) &= e\left[\frac{\boldsymbol{\beta}}{\kappa R}\right]_{\text{ret}}
\end{aligned}\right\} \tag{14.6}$$

The square bracket with subscript ret means that the quantity in brackets is to be evaluated at the retarded time, $t' = t - [R(t')/c]$. We note that, for nonrelativistic motion, $\kappa \to 1$. Then the potentials (14.6) reduce to the well-known nonrelativistic results.

To determine the fields $\mathbf{E}$ and $\mathbf{B}$ from the potentials A_μ it is possible to perform the specified differential operations directly on (14.6). But this is a more tedious procedure than working with the form (14.3). We note that in (14.3) the only dependence on the spatial coordinates $\mathbf{x}$ of the observation point is through R. Hence the gradient operation is equivalent to

$$\boldsymbol{\nabla} \to \boldsymbol{\nabla}R\,\frac{\partial}{\partial R} = \mathbf{n}\,\frac{\partial}{\partial R} \tag{14.7}$$

Consequently the electric and magnetic fields can be written as

$$
\left.
\begin{aligned}
\mathbf{E}(\mathbf{x}, t) &= e \int \left[\frac{\mathbf{n}}{R^2} \delta\left(t' + \frac{R}{c} - t \right) + \frac{1}{cR} (\boldsymbol{\beta} - \mathbf{n}) \delta'\left(t' + \frac{R}{c} - t \right) \right] dt' \\
\mathbf{B}(\mathbf{x}, t) &= e \int (\mathbf{n} \times \boldsymbol{\beta}) \left[-\frac{\delta\left(t' + \frac{R}{c} - t \right)}{R^2} + \frac{1}{cR} \delta'\left(t' + \frac{R}{c} - t \right) \right] dt'
\end{aligned}
\right\}
\tag{14.8}
$$

The primes on the delta functions mean differentiation with respect to their arguments. If the variable of integration is changed to $f(t') = t' + [R(t')/c]$, we can integrate by parts on the derivative of the delta function. Then we find readily

$$
\left.
\begin{aligned}
\mathbf{E}(\mathbf{x}, t) &= e \left[\frac{\mathbf{n}}{\kappa R^2} + \frac{1}{c\kappa} \frac{d}{dt'} \left(\frac{\mathbf{n} - \boldsymbol{\beta}}{\kappa R} \right) \right]_{\text{ret}} \\
\mathbf{B}(\mathbf{x}, t) &= e \left[\frac{\boldsymbol{\beta} \times \mathbf{n}}{\kappa R^2} + \frac{1}{c\kappa} \frac{d}{dt'} \left(\frac{\boldsymbol{\beta} \times \mathbf{n}}{\kappa R} \right) \right]_{\text{ret}}
\end{aligned}
\right\}
\tag{14.9}
$$

It is convenient to perform first the differentiation of the unit vector $\mathbf{n}$. It is evident from Fig. 14.1 that the rate of change of $\mathbf{n}$ with time is the negative of the ratio of the perpendicular component of $\mathbf{v}$ to R. Thus

$$
\frac{d\mathbf{n}}{c\,dt'} = \frac{\mathbf{n} \times (\mathbf{n} \times \boldsymbol{\beta})}{R}
\tag{14.10}
$$

When we perform the differentiation of $\mathbf{n}$ wherever it appears explicitly, we obtain

$$
\mathbf{E}(\mathbf{x}, t) = e \left[\frac{\mathbf{n}}{\kappa^2 R^2} + \frac{\mathbf{n}}{c\kappa} \frac{d}{dt'} \left(\frac{1}{\kappa R} \right) - \frac{\boldsymbol{\beta}}{\kappa^2 R^2} - \frac{1}{c\kappa} \frac{d}{dt'} \left(\frac{\boldsymbol{\beta}}{\kappa R} \right) \right]_{\text{ret}}
$$

$$
\mathbf{B}(\mathbf{x}, t) = e \left[\left(\frac{\boldsymbol{\beta}}{\kappa^2 R^2} + \frac{1}{c\kappa} \frac{d}{dt'} \left(\frac{\boldsymbol{\beta}}{\kappa R} \right) \right) \times \mathbf{n} \right]_{\text{ret}}
\tag{14.11}
$$

We observe at this point that the magnetic induction is related simply to

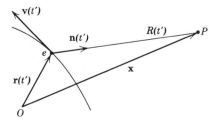

Fig. 14.1

the electric field by the relation,

$$\mathbf{B} = \mathbf{n} \times \mathbf{E} \qquad (14.12)$$

where the equation is understood to be in terms of the retarded quantities in square brackets.

The remaining derivatives needed in (14.11) are

$$\left.\begin{array}{l} \dfrac{d}{dt'}\boldsymbol{\beta} = \dot{\boldsymbol{\beta}} \\[3mm] \dfrac{1}{c}\dfrac{d}{dt'}(\kappa R) = \beta^2 - \boldsymbol{\beta} \cdot \mathbf{n} - \dfrac{R}{c}\,\mathbf{n} \cdot \dot{\boldsymbol{\beta}} \end{array}\right\} \qquad (14.13)$$

Then the electric field can be written

$$\mathbf{E}(\mathbf{x}, t) = e\left[\frac{(\mathbf{n} - \boldsymbol{\beta})(1 - \beta^2)}{\kappa^3 R^2}\right]_{\mathrm{ret}} + \frac{e}{c}\left[\frac{\mathbf{n}}{\kappa^3 R} \times \{(\mathbf{n} - \boldsymbol{\beta}) \times \dot{\boldsymbol{\beta}}\}\right]_{\mathrm{ret}} \qquad (14.14)$$

while the magnetic induction is given by (14.12). Fields (14.12) and (14.14) divide themselves naturally into "velocity fields," which are independent of acceleration, and "acceleration fields," which depend linearly on $\dot{\boldsymbol{\beta}}$. The velocity fields are essentially static fields falling off as R^{-2}, whereas the acceleration fields are typical radiation fields, both $\mathbf{E}$ and $\mathbf{B}$ being transverse to the radius vector and varying as R^{-1}.

For a particle in uniform motion the velocity fields must be the same as those obtained in Section 11.10 by means of a Lorentz transformation on the static Coulomb field. For example, the transverse electric field E_1 at a point a perpendicular distance b from the straight line path of the charge was found to be

$$E_1(t) = \frac{e\gamma b}{(b^2 + \gamma^2 v^2 t^2)^{3/2}} \qquad (14.15)$$

The origin of the time t is chosen so that the charge is closest to the observation point at $t = 0$. The electric field $E_1(t)$ given by (14.15) bears little resemblance to the velocity field in (14.14). The reason for this apparent difference is that field (14.15) is expressed in terms of the present position of the charge rather than its retarded position. To show the equivalence of the two expressions we consider the geometrical configuration shown in Fig. 14.2. Here O is the observation point, and the points P and P' are the present and apparent or retarded positions of the charge at time t. The distance $P'Q$ is $\beta R \cos \theta = (\mathbf{n} \cdot \boldsymbol{\beta})R$. Therefore the distance OQ is κR. But from triangles OPQ and $PP'Q$ we find

$$(\kappa R)^2 = r^2 - (PQ)^2 = r^2 - \beta^2(R \sin \theta)^2$$

Then from triangle OMP' we have $R \sin \theta = b$, so that

$$(\kappa R)^2 = b^2 + v^2 t^2 - \beta^2 b^2 = \frac{1}{\gamma^2}(b^2 + \gamma^2 v^2 t^2) \qquad (14.16)$$

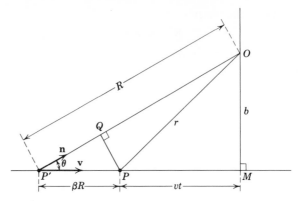

Fig. 14.2 Present and retarded positions of a charge in uniform motion.

The transverse component of the velocity field in (14.14) is

$$E_1(t) = e \left[\frac{b}{\gamma^2(\kappa R)^3} \right]_{\text{ret}} \tag{14.17}$$

With substitution (14.16) for κR in terms of the charge's present position, we find that (14.17) is equal to (14.15). The other components of **E** and **B** come out similarly.

14.2 Total Power Radiated by an Accelerated Charge—Larmor's Formula and Its Relativistic Generalization

If a charge is accelerated but is observed in a reference frame where its velocity is small compared to that of light, then in that coordinate frame the acceleration field in (14.14) reduces to

$$\mathbf{E}_a = \frac{e}{c} \left[\frac{\mathbf{n} \times (\mathbf{n} \times \dot{\boldsymbol{\beta}})}{R} \right]_{\text{ret}} \tag{14.18}$$

The instantaneous energy flux is given by the Poynting's vector,

$$\mathbf{S} = \frac{c}{4\pi} \mathbf{E} \times \mathbf{B} = \frac{c}{4\pi} |\mathbf{E}_a|^2 \mathbf{n} \tag{14.19}$$

This means that the power radiated per unit solid angle is*

$$\frac{dP}{d\Omega} = \frac{c}{4\pi} |R\mathbf{E}_a|^2 = \frac{e^2}{4\pi c} |\mathbf{n} \times (\mathbf{n} \times \dot{\boldsymbol{\beta}})|^2 \tag{14.20}$$

---

* In writing angular distributions of radiation we will always exhibit the polarization explicitly by writing the absolute square of a vector which is proportional to the electric field.

If Θ is the angle between the acceleration $\dot{\mathbf{v}}$ and $\mathbf{n}$, as shown in Fig. 14.3, then the power radiated can be written

$$\frac{dP}{d\Omega} = \frac{e^2}{4\pi c^3} \dot{v}^2 \sin^2 \Theta \qquad (14.21)$$

This exhibits the characteristic $\sin^2 \Theta$ angular dependence which is a well-known result. We note from (14.18) that the radiation is polarized in the plane containing $\dot{\mathbf{v}}$ and $\mathbf{n}$. The total instantaneous power radiated is obtained by integrating (14.21) over all solid angle. Thus

$$P = \frac{2}{3} \frac{e^2 \dot{v}^2}{c^3} \qquad (14.22)$$

This is the familiar Larmor result for a nonrelativistic, accelerated charge.

Larmor's formula (14.22) can be generalized by arguments about covariance under Lorentz transformations to yield a result which is valid for arbitrary velocities of the charge. Radiated electromagnetic energy behaves under Lorentz transformation like the fourth component of a 4-vector (see Problem 11.13). Since $dE_{\mathrm{rad}} = P\,dt$, this means that the power P is a Lorentz invariant quantity. If we can find a Lorentz invariant which reduces to the Larmor formula (14.22) for $\beta \ll 1$, then we have the desired generalization. There are, of course, many Lorentz invariants which reduce to the desired form when $\beta \to 0$. But from (14.14) it is evident that the general result must involve only $\boldsymbol{\beta}$ and $\dot{\boldsymbol{\beta}}$. With this restriction on the order of derivatives which can appear the result is unique. To find the appropriate generalization we write Larmor's formula in the suggestive form:

$$P = \frac{2}{3} \frac{e^2}{m^2 c^3} \left(\frac{d\mathbf{p}}{dt} \cdot \frac{d\mathbf{p}}{dt} \right) \qquad (14.23)$$

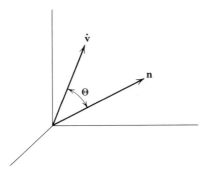

Fig. 14.3

where m is the mass of the charge, and $\mathbf{p}$ its momentum. The Lorentz invariant generalization is clearly

$$P = \frac{2}{3} \frac{e^2}{m^2 c^3} \left(\frac{dp_\mu}{d\tau} \frac{dp_\mu}{d\tau} \right) \qquad (14.24)$$

where $d\tau = dt/\gamma$ is the proper time element, and p_μ is the charged particle's momentum-energy 4-vector.* To check that (14.24) reduces properly to (14.23) as $\beta \to 0$ we evaluate the 4-vector scalar product,

$$\frac{dp_\mu}{d\tau} \frac{dp_\mu}{d\tau} = \left(\frac{d\mathbf{p}}{d\tau} \right)^2 - \frac{1}{c^2} \left(\frac{dE}{d\tau} \right)^2 = \left(\frac{d\mathbf{p}}{d\tau} \right)^2 - \beta^2 \left(\frac{dp}{d\tau} \right)^2 \qquad (14.25)$$

If (14.24) is expressed in terms of the velocity and acceleration by means of $E = \gamma mc^2$ and $\mathbf{p} = \gamma m\mathbf{v}$, we obtain the Liénard result (1898):

$$P = \frac{2}{3} \frac{e^2}{c} \gamma^6 [(\dot{\boldsymbol{\beta}})^2 - (\boldsymbol{\beta} \times \dot{\boldsymbol{\beta}})^2] \qquad (14.26)$$

One area of application of the relativistic expression for radiated power is that of charged-particle accelerators. Radiation losses are sometimes the limiting factor in the maximum practical energy attainable. For a given applied force (i.e., a given rate of change of momentum) the radiated power (14.24) depends inversely on the square of the mass of the particle involved. Consequently these radiative effects are largest for electrons. We will restrict our discussion to them.

In a linear accelerator the motion is one dimensional. From (14.25) it is evident that in that case the radiated power is

$$P = \frac{2}{3} \frac{e^2}{m^2 c^3} \left(\frac{dp}{dt} \right)^2 \qquad (14.27)$$

The rate of change of momentum is equal to the change in energy of the particle per unit distance. Consequently

$$P = \frac{2}{3} \frac{e^2}{m^2 c^3} \left(\frac{dE}{dx} \right)^2 \qquad (14.28)$$

showing that for linear motion the power radiated depends only on the external forces which determine the rate of change of particle energy with distance, not on the actual energy or momentum of the particle. The ratio

---

* That (14.24) is unique can be seen by noting that a Lorentz invariant is formed by taking scalar products of 4-vectors or higher-rank tensors. The available 4-vectors are p_μ and $dp_\mu/d\tau$. Only form (14.24) reduces to the Larmor formula for $\beta \to 0$. Contraction of higher-rank tensors such as $p_\mu (dp_\nu/d\tau)$ can be shown to vanish, or to give results proportional to (14.24) or m^2.

of power radiated to power supplied by the external sources is

$$\frac{P}{(dE/dt)} = \frac{2}{3}\frac{e^2}{m^2c^3}\frac{1}{v}\frac{dE}{dx} \rightarrow \frac{2}{3}\frac{(e^2/mc^2)}{mc^2}\frac{dE}{dx} \tag{14.29}$$

where the last form holds for relativistic particles ($\beta \rightarrow 1$). Equation (14.29) shows that the radiation loss will be unimportant unless the gain in energy is of the order of $mc^2 = 0.511$ Mev in a distance of $e^2/mc^2 = 2.82 \times 10^{-13}$ cm, or of the order of 2×10^{14} Mev/meter! Typical energy gains are less than 10 Mev/meter. Radiation losses are completely negligible in linear accelerators.

Circumstances change drastically in circular accelerators like the synchrotron or betatron. In such machines the momentum **p** changes rapidly in direction as the particle rotates, but the change in energy per revolution is small. This means that

$$\left|\frac{d\mathbf{p}}{d\tau}\right| = \gamma\omega\,|\mathbf{p}| \gg \frac{1}{c}\frac{dE}{d\tau} \tag{14.30}$$

Then the radiated power (14.24) can be written approximately

$$P = \frac{2}{3}\frac{e^2}{m^2c^3}\gamma^2\omega^2\,|\mathbf{p}|^2 = \frac{2}{3}\frac{e^2c}{\rho^2}\beta^4\gamma^4 \tag{14.31}$$

where we have used $\omega = (c\beta/\rho)$, ρ being the orbit radius. This result was first obtained by Liénard in 1898. The radiative-energy loss per revolution is

$$\delta E = \frac{2\pi\rho}{c\beta}P = \frac{4\pi}{3}\frac{e^2}{\rho}\beta^3\gamma^4 \tag{14.32}$$

For high-energy electrons ($\beta \simeq 1$) this has the numerical value,

$$\delta E\,(\text{Mev}) = 8.85 \times 10^{-2}\frac{[E\,(\text{Bev})]^4}{\rho\,(\text{meters})} \tag{14.33}$$

For a typical low-energy synchroton, $\rho \simeq 1$ meter, $E_{\max} \simeq 0.3$ Bev. Hence, $\delta E_{\max} \sim 1$ Kev per revolution. This is less than, but not negligible compared to, the energy gain of a few kilovolts per turn. In the largest electron synchrotrons, the orbit radius is of the order of 10 meters and the maximum energy is 5 Bev. Then the radiative loss is ~5.5 Mev per revolution. Since it is extremely difficult to generate radiofrequency power at levels high enough to produce energy increments much greater than this amount per revolution, it appears that 5–10 Bev is an upper limit on the maximum energy of circular electron accelerators.

The power radiated in circular accelerators can be expressed numerically as

$$P \text{ (watts)} = \frac{10^6}{2\pi} \frac{\delta E \text{ (Mev)}}{\rho \text{ (meters)}} J \text{ (amp)} \qquad (14.34)$$

where J is the circulating beam current. This equation is valid if the emission of radiation from the different electrons in the circulating beam is incoherent. In the largest electron synchrotrons the radiated power amounts to 0.1 watt per microampere of beam. Although this power dissipation is very small the radiated energy is readily detected and has some interesting properties which will be discussed in Section 14.6.

14.3 Angular Distribution of Radiation Emitted by an Accelerated Charge

For an accelerated charge in nonrelativistic motion the angular distribution shows a simple $\sin^2 \Theta$ behavior, as given by (14.21), where Θ is measured relative to the direction of acceleration. For relativistic motion the acceleration fields depend on the velocity as well as the acceleration. Consequently the angular distribution is more complicated. From (14.14) the radial component of Poynting's vector can be calculated to be

$$[\mathbf{S} \cdot \mathbf{n}]_{\text{ret}} = \frac{e^2}{4\pi c} \left[\frac{1}{\kappa^6 R^2} |\mathbf{n} \times [(\mathbf{n} - \boldsymbol{\beta}) \times \dot{\boldsymbol{\beta}}]|^2 \right]_{\text{ret}} \qquad (14.35)$$

It is evident that there are two types of relativistic effect present. One is the effect of the specific spatial relationship between $\boldsymbol{\beta}$ and $\dot{\boldsymbol{\beta}}$, which will determine the detailed angular distribution. The other is a general, relativistic effect arising from the transformation from the rest frame of the particle to the observer's frame and manifesting itself by the presence of the factors κ (14.5) in the denominator of (14.35). For ultrarelativistic particles the latter effect dominates the whole angular distribution.

In (14.35) $\mathbf{S} \cdot \mathbf{n}$ is the energy per unit area per unit time detected at an observation point at time t due to radiation emitted by the charge at time $t' = t - R(t')/c$. If we wanted to calculate the energy radiated during a finite period of acceleration, say from $t' = T_1$ to $t' = T_2$, we would write

$$W = \int_{t=T_1+[R(T_1)/c]}^{t=T_2+[R(T_2)/c]} [\mathbf{S} \cdot \mathbf{n}]_{\text{ret}} \, dt = \int_{t'=T_1}^{t'=T_2} (\mathbf{S} \cdot \mathbf{n}) \frac{dt}{dt'} \, dt' \qquad (14.36)$$

Thus we see that the useful and meaningful quantity is $(\mathbf{S} \cdot \mathbf{n}) (dt/dt')$, the power radiated per unit area in terms of the charge's own time. We

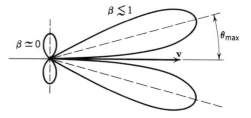

Fig. 14.4 Radiation pattern for charge accelerated in its direction of motion. The two patterns are not to scale, the relativistic one (appropriate for $\gamma \sim 2$) having been reduced by a factor $\sim 10^2$ for the same acceleration.

therefore define the power radiated per unit solid angle to be

$$\frac{dP(t')}{d\Omega} = R^2 (\mathbf{S} \cdot \mathbf{n}) \frac{dt}{dt'} = \kappa R^2 \mathbf{S} \cdot \mathbf{n} \qquad (14.37)$$

If we imagine the charge to be accelerated only for a short time during which $\boldsymbol{\beta}$ and $\dot{\boldsymbol{\beta}}$ are essentially constant in direction and magnitude, and we observe the radiation far enough away from the charge that $\mathbf{n}$ and R change negligibly during the acceleration interval, then (14.37) is proportional to the angular distribution of the energy radiated. With (14.35) for the Poynting's vector, the angular distribution is

$$\frac{dP(t')}{d\Omega} = \frac{e^2}{4\pi c} \frac{|\mathbf{n} \times \{(\mathbf{n} - \boldsymbol{\beta}) \times \dot{\boldsymbol{\beta}}\}|^2}{(1 - \mathbf{n} \cdot \boldsymbol{\beta})^5} \qquad (14.38)$$

The simplest example of (14.38) is linear motion in which $\boldsymbol{\beta}$ and $\dot{\boldsymbol{\beta}}$ are parallel. If θ is the angle of observation measured from the common direction of $\boldsymbol{\beta}$ and $\dot{\boldsymbol{\beta}}$, then (14.38) reduces to

$$\frac{dP(t')}{d\Omega} = \frac{e^2 \dot{v}^2}{4\pi c^3} \frac{\sin^2 \theta}{(1 - \beta \cos \theta)^5} \qquad (14.39)$$

For $\beta \ll 1$, this is the Larmor result (14.21). But as $\beta \to 1$, the angular distribution is tipped forward more and more and increases in magnitude, as indicated schematically in Fig. 14.4. The angle θ_{max} for which the intensity is a maximum is

$$\theta_{max} = \cos^{-1}\left[\frac{1}{3\beta}(\sqrt{1 + 15\beta^2} - 1)\right] \to \frac{1}{2\gamma} \qquad (14.40)$$

where the last form is the limiting value for $\beta \to 1$. In this same limit the peak intensity is proportional to γ^8. Even for $\beta = 0.5$, corresponding to electrons of ~ 80 Kev kinetic energy, $\theta_{max} = 38.2°$. For relativistic particles, θ_{max} is very small, being of the order of the ratio of the rest energy of the particle to its total energy. Thus the angular distribution is confined to a very narrow cone in the direction of motion. For such small angles the angular distribution (14.39) can be written approximately

$$\frac{dP(t')}{d\Omega} \simeq \frac{8}{\pi} \frac{e^2 \dot{v}^2}{c^3} \gamma^8 \frac{(\gamma\theta)^2}{(1 + \gamma^2\theta^2)^5} \tag{14.41}$$

The natural angular unit is evidently γ^{-1}. The angular distribution is shown in Fig. 14.5 with angles measured in these units. The peak occurs at $\gamma\theta = \frac{1}{2}$, and the half-power points at $\gamma\theta = 0.23$ and $\gamma\theta = 0.91$. The root mean square angle of emission of radiation in the relativistic limit is

$$\langle\theta^2\rangle^{\frac{1}{2}} = \frac{1}{\gamma} = \frac{mc^2}{E} \tag{14.42}$$

This is typical of the relativistic radiation patterns, regardless of the vectorial relation between $\boldsymbol{\beta}$ and $\dot{\boldsymbol{\beta}}$. The total power radiated can be obtained by integrating (14.39) over all angles. Thus

$$P(t') = \frac{2}{3}\frac{e^2}{c^3}\dot{v}^2\gamma^6 \tag{14.43}$$

in agreement with (14.26) and (14.27).

Another example of angular distribution of radiation is that for a charge in instantaneously circular motion with its acceleration $\dot{\boldsymbol{\beta}}$ perpendicular to its velocity $\boldsymbol{\beta}$. We choose a coordinate system such that instantaneously $\boldsymbol{\beta}$ is in the z direction and $\dot{\boldsymbol{\beta}}$ is in the x direction. With the customary polar angles θ, ϕ defining the direction of observation, as shown in Fig. 14.6, the general formula (14.38) reduces to

$$\frac{dP(t')}{d\Omega} = \frac{e^2 \dot{v}^2}{4\pi c^3} \frac{1}{(1 - \beta\cos\theta)^3} \left[1 - \frac{\sin^2\theta\cos^2\phi}{\gamma^2(1 - \beta\cos\theta)^2}\right] \tag{14.44}$$

We note that, although the detailed angular distribution is different from the linear acceleration case, the same characteristic relativistic peaking at forward angles is present. In the relativistic limit ($\gamma \gg 1$), the angular

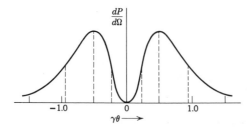

Fig. 14.5 Angular distribution of radiation for relativistic particle.

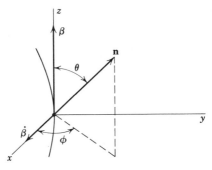

Fig. 14.6

distribution can be written approximately

$$\frac{dP(t')}{d\Omega} \simeq \frac{2}{\pi} \frac{e^2 \dot{v}^2}{c^3} \gamma^6 \frac{1}{(1 + \gamma^2\theta^2)^3} \left[1 - \frac{4\gamma^2\theta^2 \cos^2 \phi}{(1 + \gamma^2\theta^2)^2} \right] \qquad (14.45)$$

The root mean square angle of emission in this approximation is given by (14.42), just as for one-dimensional motion. The total power radiated can be found by integrating (14.44) over all angles or from (14.26):

$$P(t') = \frac{2}{3} \frac{e^2 \dot{v}^2}{c^3} \gamma^4 \qquad (14.46)$$

It is instructive to compare the power radiated for acceleration parallel to the velocity (14.43) or (14.27) with the power radiated for acceleration perpendicular to the velocity (14.46) for the same magnitude of applied force. For circular motion, the magnitude of the rate of change of momentum (which is equal to the applied force) is $\gamma m \dot{v}$. Consequently, (14.46) can be written

$$P_{\text{circular}}(t') = \frac{2}{3} \frac{e^2}{m^2 c^3} \gamma^2 \left(\frac{d\mathbf{p}}{dt}\right)^2 \qquad (14.47)$$

When this is compared to the corresponding result (14.27) for rectilinear motion, we find that for a given magnitude of applied force the radiation emitted with a transverse acceleration is a factor of γ^2 larger than with a parallel acceleration.

14.4 Radiation Emitted by a Charge in Arbitrary, Extreme Relativistic Motion

For a charged particle undergoing arbitrary, extreme relativistic motion the radiation emitted at any instant can be thought of as a coherent superposition of contributions coming from the components of acceleration parallel to and perpendicular to the velocity. But we have just seen that for comparable parallel and perpendicular forces the radiation from the parallel component is negligible (of order $1/\gamma^2$) compared to that from the perpendicular component. Consequently we may neglect the parallel component of acceleration and approximate the radiation intensity by that due to the perpendicular component alone. In other words, the radiation emitted by a charged particle in arbitrary, extreme relativistic motion is approximately the same as that emitted by a particle moving instantaneously along the arc of a circular path whose radius of curvature ρ is given by

$$\rho = \frac{v^2}{\dot{v}_\perp} \simeq \frac{c^2}{\dot{v}_\perp} \tag{14.48}$$

where $\dot{v}_\perp$ is the perpendicular component of acceleration. The form of the angular distribution of radiation is (14.44) or (14.45). It corresponds to a narrow cone or searchlight beam of radiation directed along the instantaneous velocity vector of the charge.

For an observer with a frequency-sensitive detector the confinement of the radiation to a narrow pencil parallel to the velocity has important consequences. The radiation will be visible only when the particle's velocity is directed towards the observer. For a particle in arbitrary motion the observer will detect a pulse or burst of radiation of very short time duration (or a succession of such bursts if the particle is in periodic motion), as sketched in Fig. 14.7. Since the angular width of the beam is of the order of γ^{-1}, the particle will illuminate the observer only for a time interval

$$\Delta t' \sim \frac{\rho}{c\gamma}$$

in terms of its own time, where ρ is the radius of curvature (14.48). The observer sees, however, a time interval,

$$\Delta t \sim \left\langle \frac{dt}{dt'} \right\rangle \Delta t'$$

where $\langle dt/dt' \rangle = \langle \kappa \rangle \sim (1/\gamma^2)$. Consequently the duration of the burst of radiation at the detector is

$$\Delta t \sim \frac{1}{\gamma^3} \frac{\rho}{c} \tag{14.49}$$

A pulse of this duration will contain, according to general arguments about Fourier integrals (see Section 7.3), appreciable frequency components up to a critical frequency, ω_c, of the order of

$$\omega_c \sim \frac{1}{\Delta t} \sim \left(\frac{c}{\rho}\right)\gamma^3 \tag{14.50}$$

For circular motion c/ρ is the angular frequency of rotation ω_0, and even

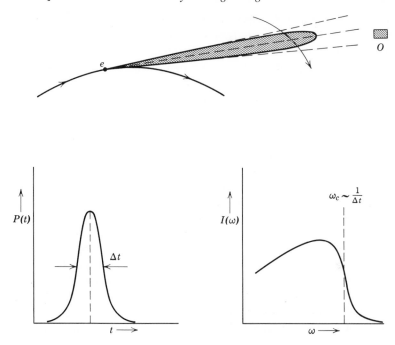

Fig. 14.7 Radiating particle illuminates the detector at O only for a time Δt. The frequency spectrum thus contains frequencies up to a maximum $\omega_c \sim (\Delta t)^{-1}$.

for arbitrary motion it plays the role of a fundamental frequency of motion. Equation (14.50) shows that a relativistic particle emits a broad spectrum of frequencies if $E \gg mc^2$, up to γ^3 times the fundamental frequency. In a 200-Mev synchrotron, $\gamma_{\max} \sim 400$. Therefore $\omega_c \sim 6 \times 10^7 \omega_0$. Since the rotation frequency is $\omega_0 \sim 3 \times 10^8 \sec^{-1}$, the frequency spectrum of emitted radiation extends up to $\sim 2 \times 10^{16} \sec^{-1}$. This represents a wavelength of 1000 angstroms. Hence the spectrum extends beyond the visible, even though the fundamental frequency is in the 100-Mc range. In Section 14.6 we will discuss in detail the angular distribution of the different frequency components, as well as the total energy radiated as a function of frequency.

14.5 Distribution in Frequency and Angle of Energy Radiated by Accelerated Charges

The qualitative arguments of the previous section show that for relativistic motion the radiated energy is spread over a wide range of frequencies.

The range of the frequency spectrum was estimated by appealing to properties of Fourier integrals. The argument can be made precise and quantitative by the use of Parseval's theorem of Fourier analysis.

The general form of the power radiated per unit solid angle is

$$\frac{dP(t)}{d\Omega} = |\mathbf{A}(t)|^2 \tag{14.51}$$

where

$$\mathbf{A}(t) = \left(\frac{c}{4\pi}\right)^{\frac{1}{2}} [R E]_{\text{ret}} \tag{14.52}$$

$\mathbf{E}$ being the electric field (14.14). In (14.51) the instantaneous power is expressed in the observer's time (contrary to the definition in Section 14.3), since we wish to consider a frequency spectrum in terms of the observer's frequencies. For definiteness we think of the acceleration occurring for some finite interval of time, or at least falling off for remote past and future times, so that the total energy radiated is finite. Furthermore, the observation point is considered far enough away from the charge that the spatial region spanned by the charge while accelerated subtends a small solid angle element at the observation point.

The total energy radiated per unit solid angle is the time integral of (14.51):

$$\frac{dW}{d\Omega} = \int_{-\infty}^{\infty} |\mathbf{A}(t)|^2 \, dt \tag{14.53}$$

This can be expressed alternatively as an integral over a frequency spectrum by use of Fourier transforms. We introduce the Fourier transform $\mathbf{A}(\omega)$ of $\mathbf{A}(t)$,

$$\mathbf{A}(\omega) = \frac{1}{\sqrt{2\pi}} \int_{-\infty}^{\infty} \mathbf{A}(t) e^{i\omega t} \, dt \tag{14.54}$$

and its inverse,

$$\mathbf{A}(t) = \frac{1}{\sqrt{2\pi}} \int_{-\infty}^{\infty} \mathbf{A}(\omega) e^{-i\omega t} \, d\omega \tag{14.55}$$

Then (14.53) can be written

$$\frac{dW}{d\Omega} = \frac{1}{2\pi} \int_{-\infty}^{\infty} dt \int_{-\infty}^{\infty} d\omega \int_{-\infty}^{\infty} d\omega' \, \mathbf{A}^*(\omega') \cdot \mathbf{A}(\omega) e^{i(\omega' - \omega)t} \tag{14.56}$$

Interchanging the orders of time and frequency integration, we see that the time integral is just a Fourier representation of the delta function $\delta(\omega' - \omega)$. Consequently the energy radiated per unit solid angle becomes

$$\frac{dW}{d\Omega} = \int_{-\infty}^{\infty} |\mathbf{A}(\omega)|^2 \, d\omega \tag{14.57}$$

The equality of (14.57) and (14.53), with suitable mathematical restrictions on the function $\mathbf{A}(t)$, is a special case of Parseval's theorem. It is customary to integrate only over positive frequencies, since the sign of the frequency has no physical meaning. Then the relation,

$$\frac{dW}{d\Omega} = \int_0^\infty \frac{dI(\omega)}{d\Omega} \, d\omega \tag{14.58}$$

defines a quantity $dI(\omega)/d\Omega$ which is the energy radiated per unit solid angle per unit frequency interval:

$$\frac{dI(\omega)}{d\Omega} = |\mathbf{A}(\omega)|^2 + |\mathbf{A}(-\omega)|^2 \tag{14.59}$$

If $\mathbf{A}(t)$ is real, from (14.55) it is evident that $\mathbf{A}(-\omega) = \mathbf{A}^*(\omega)$. Then

$$\frac{dI(\omega)}{d\Omega} = 2\,|\mathbf{A}(\omega)|^2 \tag{14.60}$$

This result relates in a quantitative way the behavior of the power radiated as a function of time to the frequency spectrum of the energy radiated.

By using (14.14) for the electric field of an accelerated charge we can obtain a general expression for the energy radiated per unit solid angle per unit frequency interval in terms of an integral over the trajectory of the particle. We must calculate the Fourier transform (14.54) of $\mathbf{A}(t)$ given by (14.52). Using (14.14), we find

$$\mathbf{A}(\omega) = \left(\frac{e^2}{8\pi^2 c}\right)^{1/2} \int_{-\infty}^{\infty} e^{i\omega t}\left[\frac{\mathbf{n} \times [(\mathbf{n} - \boldsymbol{\beta}) \times \dot{\boldsymbol{\beta}}]}{\kappa^3}\right]_{\text{ret}} dt \tag{14.61}$$

where ret means evaluated at $t' + [R(t')/c] = t$. We change the variable of integration from t to t', thereby obtaining the result:

$$\mathbf{A}(\omega) = \left(\frac{e^2}{8\pi^2 c}\right)^{1/2} \int_{-\infty}^{\infty} e^{i\omega(t' + [R(t')/c])}\frac{\mathbf{n} \times [(\mathbf{n} - \boldsymbol{\beta}) \times \dot{\boldsymbol{\beta}}]}{\kappa^2} dt' \tag{14.62}$$

Since the observation point is assumed to be far away from the region of space where the acceleration occurs, the unit vector $\mathbf{n}$ is sensibly constant in time. Furthermore the distance $R(t')$ can be approximated as

$$R(t') \simeq x - \mathbf{n} \cdot \mathbf{r}(t') \tag{14.63}$$

where x is the distance from an origin O to the observation point P, and $\mathbf{r}(t')$ is the position of the particle relative to O, as shown in Fig. 14.8. Then, apart from an overall phase factor, (14.62) becomes

$$\mathbf{A}(\omega) = \left(\frac{e^2}{8\pi^2 c}\right)^{1/2} \int_{-\infty}^{\infty} e^{i\omega(t - [\mathbf{n} \cdot \mathbf{r}(t)/c])}\frac{\mathbf{n} \times [(\mathbf{n} - \boldsymbol{\beta}) \times \dot{\boldsymbol{\beta}}]}{\kappa^2} dt \tag{14.64}$$

The primes on the time variable have been omitted for brevity. The

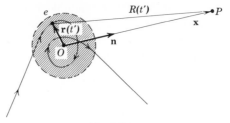

Fig. 14.8

energy radiated per unit solid angle per unit frequency interval (14.60) is accordingly

$$\frac{dI(\omega)}{d\Omega} = \frac{e^2}{4\pi^2 c} \left| \int_{-\infty}^{\infty} \frac{\mathbf{n} \times [(\mathbf{n} - \boldsymbol{\beta}) \times \dot{\boldsymbol{\beta}}]}{(1 - \boldsymbol{\beta} \cdot \mathbf{n})^2} e^{i\omega(t - [\mathbf{n} \cdot \mathbf{r}(t)/c])} dt \right|^2 \quad (14.65)$$

For a specified motion $\mathbf{r}(t)$ is known, $\boldsymbol{\beta}(t)$ and $\dot{\boldsymbol{\beta}}(t)$ can be computed, and the integral can be evaluated as a function of ω and the direction of $\mathbf{n}$. If accelerated motion of more than one charge is involved, a coherent sum of amplitudes $\mathbf{A}_j(\omega)$, one for each charge, must replace the single amplitude in (14.65) (see Problems 14.11, 15.2, and 15.3).

Even though (14.65) has the virtue that the time interval of integration is explicitly shown to be confined to times for which the acceleration is different from zero, a simpler expression for some purposes can be obtained by an integration by parts in (14.64). It is easy to demonstrate that the integrand in (14.64), excluding the exponential, is a perfect differential:

$$\frac{\mathbf{n} \times [(\mathbf{n} - \boldsymbol{\beta}) \times \dot{\boldsymbol{\beta}}]}{\kappa^2} = \frac{d}{dt}\left[\frac{\mathbf{n} \times (\mathbf{n} \times \boldsymbol{\beta})}{\kappa} \right] \quad (14.66)$$

Then an integration by parts leads to the intensity distribution:

$$\frac{dI(\omega)}{d\Omega} = \frac{e^2 \omega^2}{4\pi^2 c} \left| \int_{-\infty}^{\infty} \mathbf{n} \times (\mathbf{n} \times \boldsymbol{\beta}) e^{i\omega(t - [\mathbf{n} \cdot \mathbf{r}(t)/c])} dt \right|^2 \quad (14.67)$$

It should be observed that in (14.67) and (14.65) the polarization of the emitted radiation is specified by the direction of the vector integral in each. The intensity of radiation of a certain fixed polarization can be obtained by taking the scalar product of the appropriate unit polarization vector with the vector integral before forming the absolute square.

For a number of charges e_j in accelerated motion the integrand in (14.67) involves the replacement,

$$e\boldsymbol{\beta} e^{-i(\omega/c)\mathbf{n} \cdot \mathbf{r}(t)} \rightarrow \sum_{j=1}^{N} e_j \boldsymbol{\beta}_j e^{-i(\omega/c)\mathbf{n} \cdot \mathbf{r}_j(t)} \quad (14.68)$$

In the limit of a continuous distribution of charge in motion the sum over j becomes an integral over the current density $\mathbf{J(x, }t)$:

$$e\boldsymbol{\beta}e^{-i(\omega/c)\mathbf{n}\cdot\mathbf{r}(t)} \rightarrow \frac{1}{c}\int d^3x\ \mathbf{J(x, }t)e^{-i(\omega/c)\mathbf{n}\cdot\mathbf{x}} \qquad (14.69)$$

Then the intensity distribution becomes

$$\frac{dI(\omega)}{d\Omega} = \frac{\omega^2}{4\pi^2c^3}\left|\int dt\int d^3x\ \mathbf{n} \times [\mathbf{n} \times \mathbf{J(x, }t)]e^{i\omega[t-(\mathbf{n}\cdot\mathbf{x})/c]}\right|^2 \qquad (14.70)$$

a result which can be obtained from the direct solution of the inhomogeneous wave equation for the vector potential (14.1).

Of some interest is the radiation associated with a moving magnetic moment. This can be most easily expressed by recalling from Chapter 5 that a magnetization density $\mathcal{M}(\mathbf{x}, t)$ is equivalent to a current,

$$\mathbf{J}_M = c\boldsymbol{\nabla} \times \mathcal{M} \qquad (14.71)$$

Then substitution into (14.70) yields

$$\frac{dI_M(\omega)}{d\Omega} = \frac{\omega^4}{4\pi^2c^3}\left|\int dt\int d^3x\ \mathbf{n} \times \mathcal{M}(\mathbf{x}, t)e^{i\omega[t-(\mathbf{n}\cdot\mathbf{x})/c]}\right|^2 \qquad (14.72)$$

If the magnetization is a point magnetic moment $\boldsymbol{\mu}(t)$ at the point $\mathbf{r}(t)$, then

$$\mathcal{M}(\mathbf{x}, t) = \boldsymbol{\mu}(t)\ \delta[\mathbf{x} - \mathbf{r}(t)] \qquad (14.73)$$

and the energy radiated per unit solid angle per unit frequency interval is

$$\frac{dI_\mu(\omega)}{d\Omega} = \frac{\omega^4}{4\pi^2c^3}\left|\int dt\ \mathbf{n} \times \boldsymbol{\mu}(t)e^{i\omega(t-\mathbf{n}\cdot\mathbf{r}(t)/c)}\right|^2 \qquad (14.74)$$

We note that there is a characteristic difference of a factor ω^2 between the radiated intensity from a magnetic dipole and an accelerated charge, apart from the frequency dependence of the integrals.

The general formulas developed in this section, especially (14.65) and (14.67), will be applied in this chapter and subsequent ones to various problems involving the emission of radiation. The magnetic-moment formula (14.74) will be applied to the problem of radiation emitted in orbital-electron capture by nuclei in Chapter 15.

14.6 Frequency Spectrum of Radiation Emitted by a Relativistic Charged Particle in Instantaneously Circular Motion

In Section 14.4 we saw that the radiation emitted by an extremely relativistic particle subject to arbitrary accelerations is equivalent to that

Classical Electrodynamics

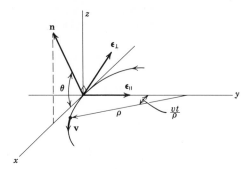

Fig. 14.9

emitted by a particle moving instantaneously at constant speed on an appropriate circular path. The radiation is beamed in a narrow cone in the direction of the velocity vector, and is seen by the observer as a short pulse of radiation as the searchlight beam sweeps across the observation point.

To find the distribution of energy in frequency and angle it is necessary to calculate the integral in (14.67). Because the duration of the pulse $\Delta t' \sim (\rho/c\gamma)$ is very short, it is necessary to know the velocity $\boldsymbol{\beta}$ and position $\mathbf{r}(t)$ over only a small arc of the trajectory whose tangent points in the general direction of the observation point. Figure 14.9 shows an appropriate coordinate system. The segment of trajectory lies in the x-y plane with instantaneous radius of curvature ρ. Since an integral will be taken over the path, the unit vector $\mathbf{n}$ can be chosen without loss of generality to lie in the x-z plane, making an angle θ (the colatitude) with the x axis. Only for very small θ will there be appreciable radiation intensity. The origin of time is chosen so that at $t = 0$ the particle is at the origin of coordinates.

The vector part of the integrand in (14.67) can be written

$$\mathbf{n} \times (\mathbf{n} \times \boldsymbol{\beta}) = \beta\left[-\boldsymbol{\epsilon}_{\shortparallel} \sin\left(\frac{vt}{\rho}\right) + \boldsymbol{\epsilon}_{\perp} \cos\left(\frac{vt}{\rho}\right) \sin\theta\right] \quad (14.75)$$

where $\boldsymbol{\epsilon}_{\shortparallel} = \boldsymbol{\epsilon}_2$ is a unit vector in the y direction, corresponding to polarization in the plane of the orbit; $\boldsymbol{\epsilon}_{\perp} = \mathbf{n} \times \boldsymbol{\epsilon}_2$ is the orthogonal polarization vector corresponding approximately to polarization perpendicular to the orbit plane (for θ small). The argument of the exponential is

$$\omega\left(t - \frac{\mathbf{n} \cdot \mathbf{r}(t)}{c}\right) = \omega\left[t - \frac{\rho}{c} \sin\left(\frac{vt}{\rho}\right) \cos\theta\right] \quad (14.76)$$

Since we are concerned with small angles θ and comparatively short times around $t = 0$, we can expand both trigonometric functions in (14.76) to obtain

$$\omega\left(t - \frac{\mathbf{n} \cdot \mathbf{r}(t)}{c}\right) \simeq \frac{\omega}{2}\left[\left(\frac{1}{\gamma^2} + \theta^2\right)t + \frac{c^2}{3\rho^2}t^3\right] \tag{14.77}$$

where β has been put equal to unity wherever possible. Using the time estimate $\rho/c\gamma$ for t and the estimate $\langle\theta^2\rangle^{1/2}$ (14.42) for θ, it is easy to see that neglected terms in (14.77) are of the order of γ^{-2} times those kept.

With the same type of approximations in (14.75) as led to (14.77), the radiated-energy distribution (14.67) can be written

$$\frac{dI(\omega)}{d\Omega} = \frac{e^2\omega^2}{4\pi^2 c}\left|-\boldsymbol{\epsilon}_{\|}A_{\|}(\omega) + \boldsymbol{\epsilon}_{\perp}A_{\perp}(\omega)\right|^2 \tag{14.78}$$

where the amplitudes are*

$$\left.\begin{aligned}
A_{\|}(\omega) &\simeq \frac{c}{\rho}\int_{-\infty}^{\infty} t \exp\left\{i\frac{\omega}{2}\left[\left(\frac{1}{\gamma^2} + \theta^2\right)t + \frac{c^2 t^3}{3\rho^2}\right]\right\} dt \\[2mm]
A_{\perp}(\omega) &\simeq \theta\int_{-\infty}^{\infty} \exp\left\{i\frac{\omega}{2}\left[\left(\frac{1}{\gamma^2} + \theta^2\right)t + \frac{c^2 t^3}{3\rho^2}\right]\right\} dt
\end{aligned}\right\} \tag{14.79}$$

A change of variable to $x = \left[ct/\rho\left(\frac{1}{\gamma^2} + \theta^2\right)^{1/2}\right]$ and introduction of the parameter ξ,

$$\xi = \frac{\omega\rho}{3c}\left(\frac{1}{\gamma^2} + \theta^2\right)^{3/2} \tag{14.80}$$

allows us to transform the integrals in $A_{\|}(\omega)$ and $A_{\perp}(\omega)$ into the form:

$$\left.\begin{aligned}
A_{\|}(\omega) &= \frac{\rho}{c}\left(\frac{1}{\gamma^2} + \theta^2\right)\int_{-\infty}^{\infty} x \exp\left[i\tfrac{3}{2}\xi(x + \tfrac{1}{3}x^3)\right] dx \\[2mm]
A_{\perp}(\omega) &= \frac{\rho}{c}\,\theta\left(\frac{1}{\gamma^2} + \theta^2\right)^{1/2}\int_{-\infty}^{\infty} \exp\left[i\tfrac{3}{2}\xi(x + \tfrac{1}{3}x^3)\right] dx
\end{aligned}\right\} \tag{14.81}$$

---

* The fact that the limits of integration in (14.79) are $t = \pm\infty$ may seem to contradict the approximations made in going from (14.76) to (14.77). The point is that for most frequencies the phase of the integrands in (14.79) oscillates very rapidly and makes the integrands effectively zero for times much smaller than those necessary to maintain the validity of (14.77). Hence the upper and lower limits on the integrals can be taken as infinite without error. Only for frequencies of the order of $\omega \sim (c/\rho) \sim \omega_0$ does the approximation fail. But we have seen in Section 14.4 that for relativistic particles essentially all the frequency spectrum is at much higher frequencies.

The integrals in (14.81) are identifiable as Airy integrals, or alternatively as modified Bessel functions:

$$\left.\begin{aligned}
\int_0^\infty x \sin\left[\tfrac{3}{2}\xi(x + \tfrac{1}{3}x^3)\right] dx &= \frac{1}{\sqrt{3}} K_{2/3}(\xi) \\
\int_0^\infty \cos\left[\tfrac{3}{2}\xi(x + \tfrac{1}{3}x^3)\right] dx &= \frac{1}{\sqrt{3}} K_{1/3}(\xi)
\end{aligned}\right\} \quad (14.82)$$

Consequently the energy radiated per unit frequency interval per unit solid angle is

$$\frac{dI(\omega)}{d\Omega} = \frac{e^2}{3\pi^2 c} \left(\frac{\omega\rho}{c}\right)^2 \left(\frac{1}{\gamma^2} + \theta^2\right)^2 \left[K_{2/3}^2(\xi) + \frac{\theta^2}{(1/\gamma^2) + \theta^2} K_{1/3}^2(\xi) \right] \quad (14.83)$$

The first term in the square bracket corresponds to radiation polarized in the plane of the orbit, and the second to radiation polarized perpendicular to that plane.

We now proceed to examine this somewhat complex result. First we integrate over all frequencies and find that the distribution of energy in angle is

$$\int_0^\infty \frac{dI(\omega)}{d\Omega} d\omega = \frac{7}{16} \frac{e^2}{\rho} \frac{1}{\left(\frac{1}{\gamma^2} + \theta^2\right)^{5/2}} \left[1 + \frac{5}{7} \frac{\theta^2}{(1/\gamma^2) + \theta^2}\right] \quad (14.84)$$

This shows the characteristic behavior seen in Section 14.3. Equation (14.84) can be obtained directly, of course, by integrating a slight generalization of the circular-motion power formula (14.44) over all times. As in (14.83), the first term in (14.84) corresponds to polarization parallel to the orbital plane, and the second to perpendicular polarization. Integrating over all angles, we find that seven times as much energy is radiated with parallel polarization as with perpendicular polarization. The radiation from a relativistically moving charge is very strongly, but not completely, polarized in the plane of motion.

The properties of the modified Bessel functions summarized in (3.103) and (3.104) show that the intensity of radiation is negligible for $\xi \gg 1$. From (14.80) we see that this will occur at large angles; the greater the frequency, the smaller the critical angle beyond which there will be negligible radiation. This shows that the radiation is largely confined to the plane containing the motion, as shown by (14.84), being more so confined the higher the frequency relative to c/ρ. If ω gets too large, however, we see that ξ will be large at *all* angles. Then there will be negligible total energy emitted at that frequency. The critical frequency

ω_c beyond which there is negligible radiation at any angle can be defined by $\xi = 1$ for $\theta = 0$. Then we find

$$\omega_c = 3\gamma^3 \left(\frac{c}{\rho}\right) = 3\left(\frac{E}{mc^2}\right)^3 \frac{c}{\rho} \qquad (14.85)$$

This critical frequency is seen to agree with our qualitative estimate (14.50) of Section 14.4. If the motion of the charge is truly circular, then c/ρ is the fundamental frequency of rotation, ω_0. Then we can define a critical harmonic frequency $\omega_c = n_c\omega_0$, with harmonic number,

$$n_c = 3\left(\frac{E}{mc^2}\right)^3 \qquad (14.86)$$

Since the radiation is predominantly in the orbital plane for $\gamma \gg 1$, it is instructive to evaluate the angular distribution (14.83) at $\theta = 0$. For frequencies well below the critical frequency ($\omega \ll \omega_c$), we find

$$\left.\frac{dI(\omega)}{d\Omega}\right|_{\theta=0} \simeq \frac{e^2}{c}\left[\frac{\Gamma(\frac{2}{3})}{\pi}\right]^2 \left(\frac{3}{4}\right)^{1/3}\left(\frac{\omega\rho}{c}\right)^{2/3} \qquad (14.87)$$

For the opposite limit of $\omega \gg \omega_c$, the result is

$$\left.\frac{dI(\omega)}{d\Omega}\right|_{\theta=0} \simeq \frac{3}{2\pi}\frac{e^2}{c}\gamma^2\frac{\omega}{\omega_c}e^{-2\omega/\omega_c} \qquad (14.88)$$

These limiting forms show that the spectrum at $\theta = 0$ increases with frequency roughly as $\omega^{2/3}$ well below the critical frequency, reaches a maximum in the neighborhood of ω_c, and then drops exponentially to zero above that frequency.

The spread in angle at a fixed frequency can be estimated by determining the angle θ_c at which $\xi(\theta_c) \simeq \xi(0) + 1$. In the low-frequency range ($\omega \ll \omega_c$), $\xi(0)$ is very small, so that $\xi(\theta_c) \simeq 1$. This gives

$$\theta_c \simeq \left(\frac{3c}{\omega\rho}\right)^{1/3} = \frac{1}{\gamma}\left(\frac{\omega_c}{\omega}\right)^{1/3} \qquad (14.89)$$

We note that the low-frequency components are emitted at much wider angles than the average, $\langle\theta^2\rangle^{1/2} \sim \gamma^{-1}$. In the high-frequency limit ($\omega > \omega_c$), $\xi(0)$ is large compared to unity. Then the intensity falls off in angle approximately as

$$\frac{dI(\omega)}{d\Omega} \sim \left.\frac{dI(\omega)}{d\Omega}\right|_{\theta=0} \cdot e^{-3\omega\gamma^2\theta^2/\omega_c} \qquad (14.90)$$

Thus the critical angle, defined by the $1/e$ point, is

$$\theta_c \simeq \frac{1}{\gamma}\left(\frac{\omega_c}{3\omega}\right)^{\!\frac{1}{3}} \tag{14.91}$$

This shows that the high-frequency components are confined to an angular range much smaller than average. Figure 14.10 shows qualitatively the angular distribution for frequencies small compared with, of the order of, and much larger than, ω_c. The natural unit of angle $\gamma\theta$ is used.

The frequency distribution of the total energy emitted as the particle passes by can be found by integrating (14.83) over angles:

$$I(\omega) = 2\pi \int_{-\pi/2}^{\pi/2} \frac{dI(\omega)}{d\Omega} \cos\theta\, d\theta \simeq 2\pi \int_{-\infty}^{\infty} \frac{dI(\omega)}{d\Omega}\, d\theta \tag{14.92}$$

(remember that θ is the colatitude). We can estimate the integral for the low-frequency range by using the value of the angular distribution (14.87) at $\theta = 0$ and the critical angle θ_c (14.89). Then we obtain

$$I(\omega) \sim 2\pi\theta_c \frac{dI(\omega)}{d\Omega}\bigg|_{\theta=0} \sim \frac{e^2}{c}\left(\frac{\omega\rho}{c}\right)^{\!\frac{1}{3}} \tag{14.93}$$

showing that the spectrum increases as $\omega^{\frac{1}{3}}$ for $\omega \ll \omega_c$. This gives a very broad, flat spectrum at frequencies below ω_c. For the high-frequency limit where $\omega \gg \omega_c$ we can integrate (14.90) over angles to obtain the reasonably accurate result,

$$I(\omega) \simeq \sqrt{3\pi}\,\frac{e^2}{c}\,\gamma\left(\frac{\omega}{\omega_c}\right)^{\!\frac{1}{2}} e^{-2\omega/\omega_c} \tag{14.94}$$

A proper integration of (14.83) over angles yields the expression,

$$I(\omega) = 2\sqrt{3}\,\frac{e^2}{c}\,\gamma\,\frac{\omega}{\omega_c} \int_{2\omega/\omega_c}^{\infty} K_{\frac{5}{3}}(x)\, dx \tag{14.95}$$

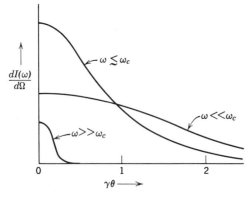

Fig. 14.10 Differential frequency spectrum as a function of angle. For frequencies comparable to the critical frequency ω_c, the radiation is confined to angles of the order of γ^{-1}. For much smaller (larger) frequencies, the angular spread is larger (smaller).

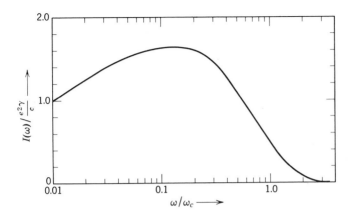

Fig. 14.11 Synchrotron radiation spectrum (energy radiated per unit frequency interval) as a function of frequency. The intensity is measured in units of $\gamma e^2/c$, while the frequency is expressed in units of ω_c (14.85).

In the limit $\omega \ll \omega_c$ this reduces to the form (14.93) with a numerical coefficient 3.25, while for $\omega \gg \omega_c$ it is equal to (14.94). The behavior of $I(\omega)$ as a function of frequency is shown in Fig. 14.11. The peak intensity is of the order of $e^2\gamma/c$, and the total energy is of the order of $e^2\gamma\omega_c/c = 3e^2\gamma^4/\rho$. This is in agreement with the value of $4\pi e^2\gamma^4/3\rho$ for the radiative loss per revolution (14.32) in circular accelerators.

The radiation represented by (14.83) and (14.95) is called *synchrotron radiation* because it was first observed in electron synchrotrons (1948). The theoretical results are much older, however, having been obtained for circular motion by Schott (1912). For periodic circular motion the spectrum is actually discrete, being composed of frequencies which are integral multiples of the fundamental frequency $\omega_0 = c/\rho$. Since the charged particle repeats its motion at a rate of $c/2\pi\rho$ revolutions per second, it is convenient to talk about the angular distribution of power radiated into the nth multiple of ω_0 instead of the energy radiated per unit frequency interval per passage of the particle. To obtain the harmonic power expressions we merely multiply $I(\omega)$ (14.95) or $dI(\omega)/d\Omega$ (14.83) by the repetition rate $c/2\pi\rho$ to convert energy to power, and by $\omega_0 = c/\rho$ to convert per unit frequency interval to per harmonic. Thus

$$\left.\begin{aligned} \frac{dP_n}{d\Omega} &= \frac{1}{2\pi}\left(\frac{c}{\rho}\right)^2 \frac{dI(\omega)}{d\Omega}\bigg|_{\omega=n\omega_0} \\[2mm] P_n &= \frac{1}{2\pi}\left(\frac{c}{\rho}\right)^2 I(\omega = n\omega_0) \end{aligned}\right\} \qquad (14.96)$$

These results have been compared with experiment in some detail.* For this purpose it is necessary to average the spectra over the acceleration cycle of the machine, since the electron's energy increases continually (see Problem 14.13). With 80 Mev maximum energy, the spectrum extends from the fundamental frequency of $\omega_0 \simeq 10^9$ sec^{-1} to $\omega_c \simeq 10^{16}$ sec^{-1}, or $\lambda \sim 1700$ angstroms. The radiation covers the visible region and is bluish white in color. Careful measurements are in full agreement with theory.

Synchrotron radiation has been observed in the astronomical realm associated with sunspots, the Crab nebula, and perhaps the $\sim 10^3$ Mc/sec radiation from Jupiter. For the Crab nebula the radiation spectrum extends over a frequency range from radiofrequencies into the extreme ultraviolet, and shows very strong polarization. From detailed observations it can be concluded that electrons with energies ranging up to 10^{12} ev are emitting synchrotron radiation while moving in circular or helical orbits in a magnetic induction of the order of 10^{-4} gauss (see Problem 14.15). The radio emission from Jupiter apparently comes from electrons trapped in Van Allen belts at distances several radii from Jupiter's surface. Whether these are relativistic electrons emitting synchrotron radiation, or nonrelativistic electrons emitting so-called cyclotron radiation as they spiral in the planet's magnetic field, is not clear at present. In any event, the radiation is strongly polarized parallel to the equator of Jupiter, as expected for particles trapped in a dipole field and spiraling around lines of force.

14.7 Thomson Scattering of Radiation

If a plane wave of monochromatic electromagnetic radiation is incident on a free particle of charge e and mass m, the particle will be accelerated and so emit radiation. This radiation will be emitted in directions other than that of the incident plane wave, but for nonrelativistic motion of the particle it will have the same frequency as the incident radiation. The whole process may be described as scattering of the incident radiation.

The instantaneous power radiated by a particle of charge e in nonrelativistic motion is given by Larmor's formula (14.21),

$$\frac{dP}{d\Omega} = \frac{e^2}{4\pi c^3} \dot{v}^2 \sin^2 \Theta \qquad (14.97)$$

where Θ is the angle between the observation direction and the acceleration. The acceleration is provided by the incident plane electromagnetic

* F. R. Elder, R. V. Langmuir, and H. C. Pollock, *Phys. Rev.*, **74**, 52 (1948); and especially D. H. Tomboulain and P. L. Hartman, *Phys. Rev.*, **102**, 1423 (1956).

wave. If the propagation vector is **k**, and the polarization vector **ε**, the electric field can be written

$$\mathbf{E}(\mathbf{x}, t) = \boldsymbol{\epsilon} E_0 e^{i\mathbf{k}\cdot\mathbf{x} - i\omega t} \qquad (14.98)$$

Then, from the force equation for nonrelativistic motion, we have the acceleration,

$$\dot{\mathbf{v}}(t) = \boldsymbol{\epsilon} \frac{e}{m} E_0 e^{i\mathbf{k}\cdot\mathbf{x} - i\omega t} \qquad (14.99)$$

If we assume that the charge moves a negligible part of a wavelength during one cycle of oscillation, the time average of $\dot{v}^2$ is $\frac{1}{2}\text{Re}\,(\dot{\mathbf{v}}\cdot\dot{\mathbf{v}}^*)$. Then the average power per unit solid angle can be expressed as

$$\left\langle \frac{dP}{d\Omega} \right\rangle = \frac{c}{8\pi} |E_0|^2 \left(\frac{e^2}{mc^2}\right)^2 \sin^2 \Theta \qquad (14.100)$$

Since the process is most simply viewed as a scattering, it is convenient to introduce a scattering cross section, defined by

$$\frac{d\sigma}{d\Omega} = \frac{\text{Energy radiated/unit time/unit solid angle}}{\text{Incident energy flux in energy/unit area/unit time}} \qquad (14.101)$$

The incident energy flux is just the time-averaged Poynting's vector for the plane wave, namely, $c\,|E_0|^2/8\pi$. Thus from (14.100) we obtain the differential scattering cross section,

$$\frac{d\sigma}{d\Omega} = \left(\frac{e^2}{mc^2}\right)^2 \sin^2 \Theta \qquad (14.102)$$

If the wave is incident along the z axis with its polarization vector making an angle of ψ with the x axis, as shown in Fig. 14.12, the angular distribution is

$$\sin^2 \Theta = 1 - \sin^2 \theta \cos^2 (\phi - \psi) \qquad (14.103)$$

For unpolarized radiation the cross section is given by averaging over the angle ψ. Thus

$$\frac{d\sigma}{d\Omega} = \left(\frac{e^2}{mc^2}\right)^2 \cdot \tfrac{1}{2}(1 + \cos^2 \theta) \qquad (14.104)$$

This is called the *Thomson formula* for scattering of radiation by a free charge, and is appropriate for the scattering of X-rays by electrons or gamma rays by protons. The angular distribution is as shown in Fig. 14.13 by the solid curve. The total scattering cross section, called the *Thomson cross section*, is

$$\sigma_T = \frac{8\pi}{3} \left(\frac{e^2}{mc^2}\right)^2 \qquad (14.105)$$

The Thomson cross section is equal to 0.665×10^{-24} cm² for electrons. The unit of length, $e^2/mc^2 = 2.82 \times 10^{-13}$ cm, is called the *classical electron radius*, since a classical distribution of charge totaling the electronic charge must have a radius of this order if its electrostatic self-energy is to equal the electron mass (see Chapter 17).

The classical Thomson result is valid only at low frequencies. For electrons quantum-mechanical effects enter importantly when the frequency ω becomes comparable to $mc^2/\hbar$, i.e, when the photon energy $\hbar\omega$ is comparable with, or larger than, the particle's rest energy mc^2. Another way of looking at this criterion is that we expect quantum effects to appear if the wavelength of the radiation is of the order of, or smaller than, the Compton wavelength $\hbar/mc$ of the particle. At these higher frequencies the angular distribution becomes peaked in the forward direction as shown in Fig. 14.13 by the dotted curves, always having, however, the Thomson value at zero degrees. The total cross section falls below the Thomson cross section (14.105). The process is then known as *Compton scattering*, and for electrons is described theoretically by the Klein-Nishina formula. For reference purposes we quote the asymptotic forms of the total cross section, as given by the Klein-Nishina formula:

$$\sigma_{KN} = \left(\frac{e^2}{mc^2}\right)^2 \begin{cases} \dfrac{8\pi}{3}\left(1 - \dfrac{2\hbar\omega}{mc^2} + \cdots\right), & \hbar\omega \ll mc^2 \\[2ex] \pi\dfrac{mc^2}{\hbar\omega}\left[\ln\left(\dfrac{2\hbar\omega}{mc^2}\right) + \tfrac{1}{2}\right], & \hbar\omega \gg mc^2 \end{cases} \qquad (14.106)$$

For protons the departures from the Thomson formula occur at photon

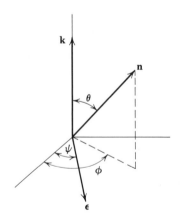

Fig. 14.12

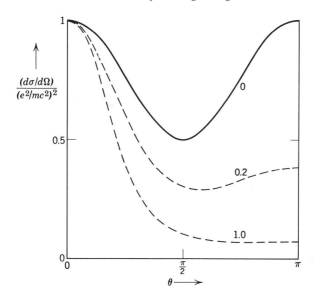

Fig. 14.13 Differential scattering cross section of unpolarized radiation by a free electron. The solid curve is the classical Thomson result. The dotted curves are given by the quantum-mechanical Klein-Nishina formula. The numbers on the curves refer to values of $\hbar\omega/mc^2$.

energies above about 100 Mev. This is far below the critical energy $\hbar\omega \sim Mc^2 \sim 1$ Bev which would be expected in analogy with the electron Compton effect. The reason is that a proton is not a point particle like the electron with nothing but electromagnetic interactions, but is a complex entity having a spread-out charge distribution with a radius of the order of 0.8×10^{-13} cm caused by strong interactions with pi mesons. The departure (a rapid increase in cross section) from Thomson scattering occurs at photon energies of the order of the rest energy of the pi meson (140 Mev).

14.8 Scattering of Radiation by Quasi-free Charges; Coherent and Incoherent Scattering

In the scattering of X-rays by atoms the angular distribution (14.104) is observed at wide angles, at least in light elements. But in the forward direction the scattering per electron increases rapidly to quite large values compared to the Thomson cross section. The reason is a coherent addition of the amplitudes from all electrons. From (14.18) it can be seen that the radiation field from a number of free charged particles will be

$$\mathbf{E}_a = \frac{1}{c} \sum_j e_j \left[\frac{\mathbf{n} \times (\mathbf{n} \times \dot{\boldsymbol{\beta}}_j)}{R_j} \right]_{\text{ret}} \tag{14.107}$$

With (14.99) for the acceleration of the typical particle, we find

$$\mathbf{E}_a = \frac{E_0}{c^2} \mathbf{n} \times (\mathbf{n} \times \boldsymbol{\epsilon}) \sum_j \frac{e_j^2}{m_j} \frac{\exp\left[i\mathbf{k} \cdot \mathbf{x}_j - i\omega\left(t - \frac{R_j}{c} \right) \right]}{R_j} \tag{14.108}$$

In calculating the radiation it is sufficient to approximate R_j in the exponent by the form (14.63). Then, in complete analogy with the steps from (14.97) to (14.102), we find the scattering cross section,

$$\frac{d\sigma}{d\Omega} = \left| \sum_j \frac{e_j^2}{m_j c^2} e^{i\mathbf{q} \cdot \mathbf{x}_j} \right|^2 \sin^2 \Theta \tag{14.109}$$

where

$$\mathbf{q} = \frac{\omega}{c} \mathbf{n} - \mathbf{k} \tag{14.110}$$

is the vectorial change in wave number in the scattering.

Equation (14.109) applies to free charged particles instantaneously at positions $\mathbf{x}_j$. Electrons in atoms, for example, are not free. But if the frequency of the incident radiation is large compared to the characteristic frequencies of binding, the particles can be treated as free while being accelerated by a pulse of finite duration. Thus (14.109) can be applied to the scattering of high-frequency (compared to binding frequencies) radiation by bound charged particles. The only thing that remains before comparison with experiment is to average (14.109) over the positions of all the particles in the bound system. Thus the observable cross section for scattering is

$$\frac{d\sigma}{d\Omega} = \left\langle \left| \sum_j \frac{e_j^2}{m_j c^2} e^{i\mathbf{q} \cdot \mathbf{x}_j} \right|^2 \right\rangle \sin^2 \Theta \tag{14.111}$$

where the symbol $\langle \ \rangle$ means average over all possible values of $\mathbf{x}_j$.

The cross section (14.111) shows very different behavior, depending on the value of $|\mathbf{q}|$. The coordinates $\mathbf{x}_j$ have magnitudes of the order of the linear dimensions of the bound system. If we call this dimension a, then the behavior of the cross section is very different in the two regions, $qa \ll 1$ and $qa \gg 1$. If the scattering angle is θ, the magnitude of q is $2k \sin(\theta/2)$. Thus the dividing line between the two domains occurs for angles such that

$$2ka \sin \frac{\theta}{2} \sim 1 \qquad (14.112)$$

If the frequency is low enough so that $ka \ll 1$, then the limit $qa \ll 1$ will apply at all angles. But for frequencies where $ka \gg 1$, there will be a region of forward angles less than

$$\theta_c \sim \frac{1}{ka} \qquad (14.113)$$

where the limit $qa \ll 1$ holds, and a region of wider angles where the limit $qa \gg 1$ applies.

For $qa \ll 1$, the arguments of exponents in (14.111) are all so small that the exponential factors can be approximated by unity. Then the differential cross section becomes

$$\lim_{qa \to 0} \frac{d\sigma}{d\Omega} = \left| \sum_j \frac{e_j^2}{m_j c^2} \right|^2 \sin^2 \Theta = Z^2 \left(\frac{e^2}{mc^2} \right)^2 \sin^2 \Theta \qquad (14.114)$$

where the last form is appropriate for electrons in an atom of atomic number Z. This shows the coherent effect of all the particles, giving an intensity corresponding to the *square* of the number of particles times the intensity for a single particle.

In the opposite limit of $qa \gg 1$ the arguments of the exponents are large and widely different in value. Consequently the cross terms in the square of the sum will average to zero. Only the absolute squared terms will survive. Then the cross section takes the form:

$$\lim_{qa \to \infty} \frac{d\sigma}{d\Omega} = \sum_j \left(\frac{e_j^2}{m_j c^2} \right)^2 \sin^2 \Theta = Z \left(\frac{e^2}{mc^2} \right)^2 \sin^2 \Theta \qquad (14.115)$$

where again the final form is for electrons in an atom. This result corresponds to the incoherent superposition of scattering from the individual particles.

For the scattering of X-rays by atoms the critical angle (14.113) can be estimated, using (13.95) as the atomic radius. Then one finds the numerical value,

$$\theta_c \sim \frac{Z^{1/3}}{\hbar\omega \, (\text{kev})} \qquad (14.116)$$

For angles less than θ_c the cross section rises rapidly to a value of the order of (14.114), while at wide angles it is given by Z times the Thomson result, (14.115), or for high-frequency X-rays or gamma rays by the Klein-Nishina formula, shown in Fig. 14.13.

14.9   Cherenkov Radiation

A charged particle in uniform motion in a straight line in free space does not radiate. But a particle moving with constant velocity through a material medium can radiate if its velocity is greater than the phase velocity of light in the medium. Such radiation is called *Cherenkov radiation*, after its discoverer, P. A. Cherenkov (1937). The emission of Cherenkov radiation is a cooperative phenomenon involving a large number of atoms of the medium whose electrons are accelerated by the fields of the passing particle and so emit radiation. Because of the collective aspects of the process it is convenient to use the macroscopic concept of a dielectric constant ϵ rather than the detailed properties of individual atoms.

A qualitative explanation of the effect can be obtained by considering the fields of the fast particle in the dielectric medium as a function of time. We denote the velocity of light in the medium by c and the particle velocity by v. Figure 14.14 shows a succession of spherical field wavelets for $v < c$ and for $v > c$. Only for $v > c$ do the wavelets interfere constructively to form a wake behind the particle. The normal to the wake makes an angle

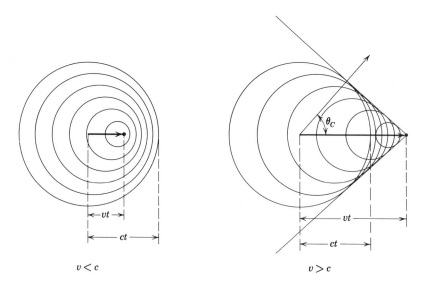

$v < c$ $\qquad\qquad\qquad\qquad\qquad\qquad$ $v > c$

Fig. 14.14 Cherenkov radiation. Spherical wavelets of fields of a particle traveling less than, and greater than, the velocity of light in the medium. For $v > c$, an electromagnetic "shock" wave appears, moving in the direction given by the Cherenkov angle θ_C.

θ_C with the velocity direction, where

$$\cos \theta_C = \frac{c}{v} \qquad (14.117)$$

This is the direction of emission of the Cherenkov radiation.

Although we have already found the fields appropriate to the Cherenkov-radiation problem in Section 13.4, and have even given an expression (13.82) for the energy emitted as Cherenkov radiation, it is instructive to look at the problem from the point of view of the Liénard-Wiechert potentials. We will make use of Section 13.4 to the extent of noting that, for a nonpermeable medium, we may discuss the fields and energy radiated as if the particle moved in free space with a velocity $v > c$, provided at the end of the calculation we make the replacements,

$$c \rightarrow \frac{c}{\sqrt{\epsilon}}, \qquad e \rightarrow \frac{e}{\sqrt{\epsilon}} \qquad (14.118)$$

where ϵ is the dielectric constant.* We will simplify the analysis by assuming that ϵ is independent of frequency. But our final results will be for individual frequency components and so will be easily generalized.

For a point charge in arbitrary motion the Liénard-Wiechert potentials were obtained in Section 14.1. It was tacitly assumed there that the particle velocity was less than the velocity of light. Then the potentials (14.6) at a given point in space-time depended on the behavior of the particle at *one* earlier point in space-time, the retarded position. This situation corresponds in the left side of Fig. 14.14 to the fact that a given point lies on only one circle. When $v > c$, however, we see from the right side of the figure that *two* retarded positions contribute to the field at a given point in space-time. The scalar potential in (14.6) is replaced by

$$\Phi(\mathbf{x}, t) = e\left[\frac{1}{\kappa R}\right]_1 + e\left[\frac{1}{\kappa R}\right]_2 \qquad (14.119)$$

where the indices 1 and 2 indicate the two retarded times t_1' and t_2'.

To determine the two times t_1' and t_2' we consider the vanishing of the argument of the delta function in (14.3):

$$t' + \frac{|\mathbf{x} - \mathbf{r}(t')|}{c} - t = 0 \qquad (14.120)$$

---

* From (13.54) it is evident that we are dealing in this way with potentials $\Phi' = \sqrt{\epsilon}\,\Phi$, $\mathbf{A}' = \mathbf{A}$, and fields $\mathbf{E}' = \sqrt{\epsilon}\,\mathbf{E}$, $\mathbf{B}' = \mathbf{B}$. Then, for example, Poynting's vector is

$$\mathbf{S}' = \frac{c}{4\pi}(\mathbf{E}' \times \mathbf{B}') \rightarrow \frac{c}{4\pi\sqrt{\epsilon}}(\mathbf{E}' \times \mathbf{B}') = \mathbf{S}$$

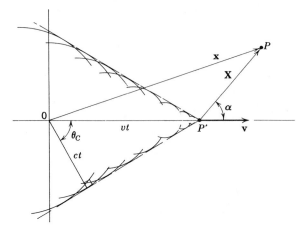

Fig. 14.15

For a particle with constant velocity $\mathbf{v}$, we can take $\mathbf{r}(t') = \mathbf{v}t'$. With $\mathbf{X} = (\mathbf{x} - \mathbf{v}t)$ as the vector distance from the *present* position P' of the particle to the observation point P, (14.120) becomes

$$(t - t') = \frac{1}{c}|\mathbf{X} + \mathbf{v}(t - t')| \tag{14.121}$$

The solution of this quadratic is

$$(t - t') = \frac{-\mathbf{X}\cdot\mathbf{v} \pm \sqrt{(\mathbf{X}\cdot\mathbf{v})^2 - (v^2 - c^2)X^2}}{(v^2 - c^2)} \tag{14.122}$$

Only roots that are real and positive have physical meaning. For $v < c$, the square root is real and larger than $|\mathbf{X}\cdot\mathbf{v}|$. Hence there is only one valid root for $(t - t')$, as already noted. But for $v > c$, there are other possibilities. First we note that even when the square root is real (as it is for directions more or less parallel or antiparallel to $\mathbf{v}$) it is smaller in magnitude than $|\mathbf{X}\cdot\mathbf{v}|$. Consequently there is no root for $(t - t')$ when $\mathbf{X}$ and $\mathbf{v}$ have an acute angle between them; the fields do not get ahead of the particle. If α is the angle between $\mathbf{X}$ and $\mathbf{v}$, as shown in Fig. 14.15, we see furthermore that the square root is imaginary for $\cos^2\alpha < [1 - (c^2/v^2)]$. But for backward angles, such that

$$\cos^{-1}\left[-\left(1 - \frac{c^2}{v^2}\right)^{1/2}\right] < \alpha < \pi$$

there are two real, positive values of $(t - t')$ as solutions of (14.121). Thus the potentials exist only inside the Cherenkov cone defined by $\cos\alpha = -[1 - (c^2/v^2)]^{1/2}$.

The values of κR corresponding to the two roots (14.122) are easily shown to be

$$\kappa R = \mp \frac{1}{c} [(\mathbf{X} \cdot \mathbf{v})^2 - (v^2 - c^2)X^2]^{\frac{1}{2}} \qquad (14.123)$$

Actually in (14.119) the absolute values are required, because of the sign inherent in the Jacobian derivative in (14.4). Thus the two terms add, and the potentials can be written

$$\left. \begin{aligned} \Phi(\mathbf{x}, t) &= \frac{2e}{X\sqrt{1 - (v^2/c^2) \sin^2 \alpha}} \\ \mathbf{A}(\mathbf{x}, t) &= \frac{\mathbf{v}}{c} \Phi(\mathbf{x}, t) \end{aligned} \right\} \qquad (14.124)$$

These potentials are valid inside the Cherenkov cone, become singular on its surface, and vanish outside the cone. They represent a wave front traveling at velocity c in the direction θ_C (14.117). The singularity is not a physical reality, of course. It comes from our assumption that the velocity of light in the medium is independent of frequency. For high enough frequencies (short enough wavelengths) the phase velocity of light in the medium will approach the velocity of light *in vacuo*. This variation with frequency will cause a smoothing at short distances which will eliminate singularities.

The potentials (14.124) are special cases of the potentials whose Fourier transforms are given by (13.57). The fields which can be found from (14.124) are similarly the Fourier transforms of the fields (13.64) and (13.65), assuming $\epsilon(\omega)$ is a real constant. The calculation of energy radiated proceeds exactly as in Section 13.4 with the integration of the Poynting's vector over a cylinder, as in (13.71), yielding the final expression (13.82) for energy radiated per unit distance.

The discussion presented so far, with the appearance of a Cherenkov "shock wave" for $v > c$, is the proper macroscopic description of the origin of Cherenkov radiation. If, however, one is interested only in the angular and frequency distribution of the radiation and not in the mechanism, it is possible to give a simple, nonrigorous derivation, using the substitutions (14.118). The angular and frequency distribution of radiation emitted by a charged particle in motion is given by (14.67):

$$\frac{dI(\omega)}{d\Omega} = \frac{e^2 \omega^2}{4\pi^2 c^3} \left| \int_{-\infty}^{\infty} \mathbf{n} \times (\mathbf{n} \times \mathbf{v}) e^{i\omega\left(t - \frac{\mathbf{n} \cdot \mathbf{r}(t)}{c}\right)} dt \right|^2 \qquad (14.125)$$

For a particle moving in a nonpermeable, dielectric medium transformation (14.118) yields

$$\frac{dI(\omega)}{d\Omega} = \frac{e^2 \omega^2}{4\pi^2 c^3} \epsilon^{\frac{1}{2}} \left| \int_{-\infty}^{\infty} \mathbf{n} \times (\mathbf{n} \times \mathbf{v}) e^{i\omega\left(t - \frac{\epsilon^{\frac{1}{2}} \mathbf{n} \cdot \mathbf{r}(t)}{c}\right)} dt \right|^2 \qquad (14.126)$$

For a uniform motion in a straight line, $\mathbf{r}(t) = \mathbf{v}t$. Then we obtain

$$\frac{dI(\omega)}{d\Omega} = \frac{e^2 \epsilon^{\frac{1}{2}}}{c^3}|\mathbf{n} \times \mathbf{v}|^2 \left| \frac{\omega}{2\pi} \int_{-\infty}^{\infty} e^{i\omega t\left(1 - \frac{\epsilon^{\frac{1}{2}}}{c}\mathbf{n}\cdot\mathbf{v}\right)} dt \right|^2 \quad (14.127)$$

The integral is a Dirac delta function. Then

$$\frac{dI(\omega)}{d\Omega} = \frac{e^2 \epsilon^{\frac{1}{2}}\beta^2 \sin^2 \theta}{c}|\delta(1 - \epsilon^{\frac{1}{2}}\beta \cos \theta)|^2 \quad (14.128)$$

where θ is measured relative to the velocity $\mathbf{v}$. The presence of the delta function guarantees that the radiation is emitted only at the Cherenkov angle θ_C:

$$\cos \theta_C = \frac{1}{\beta \epsilon^{\frac{1}{2}}} \quad (14.129)$$

The presence of the square of a delta function in angles in (14.128) means that the total energy radiated per unit frequency interval is infinite. This infinity occurs because the particle has been moving through the medium forever. To obtain a meaningful result we assume that the particle passes through a slab of dielectric in a time interval $2T$. Then the infinite integral in (14.127) is replaced by

$$\frac{\omega}{2\pi} \int_{-T}^{T} e^{i\omega t\left(1 - \frac{\epsilon^{\frac{1}{2}}}{c}\mathbf{n}\cdot\mathbf{v}\right)} dt = \frac{\omega T}{\pi} \frac{\sin\left[\omega T(1 - \epsilon^{\frac{1}{2}}\beta \cos \theta)\right]}{\left[\omega T(1 - \epsilon^{\frac{1}{2}}\beta \cos \theta)\right]} \quad (14.130)$$

The absolute square of this function peaks sharply at the angle θ_C, provided $\omega T \gg 1$. Assuming that $\beta > \frac{1}{\epsilon^{\frac{1}{2}}}$, so that the angle θ_C exists, the integral over angles is

$$\int d\Omega \left(\frac{\omega T}{\pi}\right)^2 \frac{\sin^2\left[\omega T(1 - \epsilon^{\frac{1}{2}}\beta \cos \theta)\right]}{\left[\omega T(1 - \epsilon^{\frac{1}{2}}\beta \cos \theta)\right]^2} = \frac{2\omega T}{\beta \epsilon^{\frac{1}{2}}} \quad (14.131)$$

showing that the amount of radiation is proportional to the time interval. From (14.128) we find that the total energy radiated per unit frequency interval in passing through the slab is

$$I(\omega) = \frac{e^2 \omega}{c^2} \sin^2 \theta_C (2c\beta T) \quad (14.132)$$

This can be converted into energy radiated per unit frequency interval per unit path length by dividing by $2c\beta T$. Then, with (14.129) for θ_C, we obtain

$$\frac{dI(\omega)}{dx} = \frac{e^2 \omega}{c^2}\left[1 - \frac{1}{\beta^2 \epsilon(\omega)}\right] \quad (14.133)$$

where ω is such that $\epsilon(\omega) > (1/\beta^2)$, in agreement with (13.82).

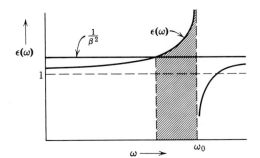

Fig. 14.16 Cherenkov band. Radiation is emitted only in shaded frequency range where $\epsilon(\omega) > \beta^{-2}$.

The properties of Cherenkov radiation can be utilized to measure velocities of fast particles. If the particles of a given velocity pass through a medium of known dielectric constant ϵ, the light is emitted at the Cherenkov angle (14.129). Thus a measurement of the angle allows determination of the velocity. Since the dielectric constant of a medium in general varies with frequency, light of different colors is emitted at somewhat different angles. Figure 14.16 shows a typical curve of $\epsilon(\omega)$, with a region of anomalous dispersion at the upper end of the frequency interval. The shaded region indicates the frequency range of the Cherenkov light. Since the dielectric medium is strongly absorbent at the region of anomalous dispersion, the escaping light will be centered somewhat below the resonance. Narrow band filters may be employed to select a small interval of frequency and so improve the precision of velocity measurement. For very fast particles ($\beta \lesssim 1$) a gas may be used to provide a dielectric constant differing only slightly from unity and having ($\epsilon - 1$) variable over wide limits by varying the gas pressure. Counting devices using Cherenkov radiation are employed extensively in high-energy physics, as instruments for velocity measurements, as mass analyzers when combined with momentum analysis, and as discriminators against unwanted slow particles.

REFERENCES AND SUGGESTED READING

The radiation by accelerated charges is at least touched on in all electrodynamics textbooks, although the emphasis varies considerably. The relativistic aspects are treated in more or less detail in
  Iwanenko and Sokolow, Sections 39–43,
  Landau and Lifshitz, *Classical Theory of Fields*, Chapters 8 and 9,
  Panofsky and Phillips, Chapters 18 and 19,
  Sommerfeld, *Electrodynamics*, Sections 29 and 30.
Extensive calculations of the radiation emitted by relativistic particles, anticipating many results rederived in the period 1940–1950, are presented in the interesting monograph by
  Schott.

The scattering of radiation by charged particles is presented clearly by
Landau and Lifshitz, *Classical Theory of Fields*, Sections 9.11–9.13, and
Electrodynamics of Continuous Media, Chapters XIV and XV.

PROBLEMS

14.1 Verify by explicit calculation that the Liénard-Wiechert expressions for *all*
components of **E** and **B** for a particle moving with constant velocity agree
with the ones obtained in the text by means of a Lorentz transformation.
Follow the general method at the end of Section 14.1.

14.2 Using the Liénard-Wiechert fields, discuss the time-average power radiated
per unit solid angle in nonrelativistic motion of a particle with charge e,
moving
 (*a*) along the z axis with instantaneous position $z(t) = a \cos \omega_0 t$,
 (*b*) in a circle of radius R in the x-y plane with constant angular
frequency ω_0.
Sketch the angular distribution of the radiation and determine the total
power radiated in each case.

14.3 A *nonrelativistic* particle of charge ze, mass m, and kinetic energy E makes
a *head-on* collision with a fixed central force field of finite range. The
interaction is repulsive and described by a potential $V(r)$, which becomes
greater than E at close distances.
 (*a*) Show that the total energy radiated is given by

$$\Delta W = \frac{4}{3} \frac{z^2 e^2}{m^2 c^3} \sqrt{\frac{m}{2}} \int_{r_{\min}}^{\infty} \left| \frac{dV}{dr} \right|^2 \frac{dr}{\sqrt{V(r_{\min}) - V(r)}}$$

where $r_{\min}$ is the closest distance of approach in the collision.
 (*b*) If the interaction is a Coulomb potential $V(r) = zZe^2/r$, show that
the total energy radiated is

$$\Delta W = \frac{8}{45} \frac{zmv_0^5}{Zc^3}$$

where v_0 is the velocity of the charge at infinity.

14.4 A particle of mass m, charge q, moves in a plane perpendicular to a
uniform, static, magnetic induction B.
 (*a*) Calculate the total energy radiated per unit time, expressing it in
terms of the constants already defined and the ratio γ of the particle's
total energy to its rest energy.
 (*b*) If at time $t = 0$ the particle has a total energy $E_0 = \gamma_0 mc^2$, show
that it will have energy $E = \gamma mc^2 < E_0$ at a time t, where

$$t \simeq \frac{3m^3 c^5}{2q^4 B^2} \left(\frac{1}{\gamma} - \frac{1}{\gamma_0} \right)$$

provided $\gamma \gg 1$.
 (*c*) If the particle is initially nonrelativistic and has a *kinetic* energy ϵ_0
at $t = 0$, what is its kinetic energy at time t?
 (*d*) If the particle is actually trapped in the magnetic dipole field of the
earth and is spiraling back and forth along a line of force, does it radiate

more energy while near the equator, or while near its turning points? Why? Make quantitative statements if you can.

14.5 As in Problem 14.2a a charge e moves in simple harmonic motion along the z axis, $z(t') = a \cos(\omega_0 t')$.

(a) Show that the instantaneous power radiated per unit solid angle is:

$$\frac{dP(t')}{d\Omega} = \frac{e^2 c \beta^4}{4\pi a^2} \frac{\sin^2 \theta \cos^2(\omega t')}{(1 + \beta \cos \theta \sin \omega_0 t')^5}$$

where $\beta = a\omega_0/c$.

(b) By performing a time averaging, show that the average power per unit solid angle is:

$$\frac{dP}{d\Omega} = \frac{e^2 c \beta^4}{32\pi a^2} \left[\frac{4 + \beta^2 \cos^2 \theta}{(1 - \beta^2 \cos^2 \theta)^{7/2}} \right] \sin^2 \theta$$

(c) Make rough sketches of the angular distribution for nonrelativistic and relativistic motion.

14.6 Show explicitly by use of the Poisson sum formula or other means that, if the motion of a radiating particle repeats itself with periodicity T, the continuous frequency spectrum becomes a discrete spectrum containing frequencies that are integral multiples of the fundamental. Show that a general expression for the power radiated per unit solid angle in each multiple m of the fundamental frequency $\omega_0 = 2\pi/T$ is:

$$\frac{dP_m}{d\Omega} = \frac{e^2 \omega_0^4 m^2}{(2\pi c)^3} \left| \int_0^{2\pi/\omega_0} \mathbf{v}(t) \times \mathbf{n} \exp\left[im\omega_0 \left(t - \frac{\mathbf{n} \cdot \mathbf{x}(t)}{c} \right) \right] dt \right|^2$$

14.7 (a) Show that for the simple harmonic motion of a charge discussed in Problem 14.5 the average power radiated per unit solid angle in the mth harmonic is:

$$\frac{dP_m}{d\Omega} = \frac{e^2 c \beta^2}{2\pi a^2} m^2 \tan^2 \theta J_m^2(m\beta \cos \theta)$$

(b) Show that in the nonrelativistic limit the total power radiated is all in the fundamental and has the value:

$$P \simeq \frac{2}{3} \frac{e^2}{c^3} \omega_0^4 \overline{a^2}$$

where $\overline{a^2}$ is the mean square amplitude of oscillation.

14.8 A particle of charge e moves in a circular path of radius R in the x-y plane with constant angular velocity ω_0.

(a) Show that the exact expression for the angular distribution of power radiated into the mth multiple of ω_0 is:

$$\frac{dP_m}{d\Omega} = \frac{e^2 \omega_0^4 R^2}{2\pi c^3} m^2 \left[\left(\frac{dJ_m(m\beta \sin \theta)}{d(m\beta \sin \theta)} \right)^2 + \frac{\cot^2 \theta}{\beta^2} J_m^2(m\beta \sin \theta) \right]$$

where $\beta = \omega_0 R/c$, and $J_m(x)$ is the Bessel function of order m.

(b) Assume nonrelativistic motion and obtain an approximate result for $dP_m/d\Omega$. Show that the results of Problem 14.2b are obtained in this limit.

(c) Assume extreme relativistic motion and obtain the results found in the text for a relativistic particle in instantaneously circular motion. (Watson, pp. 79, 249, may be of assistance to you.)

14.9 Bohr's correspondence principle states that in the limit of large quantum numbers the classical power radiated in the fundamental is equal to the product of the quantum energy ($\hbar\omega_0$) and the reciprocal mean lifetime of the transition from principal quantum number n to $(n - 1)$.

(a) Using nonrelativistic approximations, show that in a hydrogen-like atom the transition probability (reciprocal mean lifetime) for a transition from a circular orbit of principal quantum number n to $(n - 1)$ is given classically by

$$\frac{1}{\tau} = \frac{2}{3}\frac{e^2}{\hbar c}\left(\frac{Ze^2}{\hbar c}\right)^4 \frac{mc^2}{\hbar}\frac{1}{n^5}$$

(b) For hydrogen compare the classical value from (a) with the correct quantum-mechanical results for the transitions $2p \rightarrow 1s$ (1.6×10^{-9} sec), $4f \rightarrow 3d$ (7.3×10^{-8} sec), $6h \rightarrow 5g$ (6.1×10^{-7} sec).

14.10 Periodic motion of charges gives rise to a discrete frequency spectrum in multiples of the basic frequency of the motion. Appreciable radiation in multiples of the fundamental can occur because of relativistic effects (Problems 14.7 and 14.8) even though the components of velocity are truly *sinusoidal*, or it can occur if the components of velocity are not sinusoidal, even though periodic. An example of this latter motion is an electron undergoing nonrelativistic elliptic motion in a hydrogen atom.

The orbit can be specified by the parametric equations

$$x = a(\cos u - \epsilon)$$
$$y = a\sqrt{1 - \epsilon^2}\sin u$$

where

$$\omega_0 t = u - \epsilon \sin u$$

a is the semimajor axis, ϵ is the eccentricity, ω_0 is the orbital frequency, and u is the angle in the plane of the orbit. In terms of the binding energy B and angular momentum L, the various constants are

$$a = \frac{e^2}{2B}, \qquad \epsilon = \sqrt{1 - \frac{2BL^2}{me^4}}, \qquad \omega_0^2 = \frac{8B^3}{me^4}$$

(a) Show that the power radiated in the kth multiple of ω_0 is

$$P_k = \frac{4e^2}{3c^3}(k\omega_0)^4 a^2 \left\{\frac{1}{k^2}\left[(J_k'(k\epsilon))^2 + \left(\frac{1 - \epsilon^2}{\epsilon^2}\right)J_k^2(k\epsilon)\right]\right\}$$

where $J_k(x)$ is a Bessel function of order k.

(b) Verify that for circular orbits the general result (a) agrees with part (a) of Problem 14.9.

14.11 Instead of a single charge e moving with constant velocity ω_0 in a circular path of radius R, as in Problem 14.8, a set of N such charge moves with fixed relative positions around the same circle.

(a) Show that the power radiated into the mth multiple of ω_0 is

$$\frac{dP_m(N)}{d\Omega} = \frac{dP_m(1)}{d\Omega}F_m(N)$$

where $dP_m(1)/d\Omega$ is the result of part (a) in Problem 14.8, and

$$F_m(N) = \left| \sum_{j=1}^{N} e^{im\theta_j} \right|^2$$

θ_j being the angular position of the jth charge at $t = t_0$.

(b) Show that, if the charges are uniformly spaced around the circle, energy is radiated only into multiples of $N\omega_0$, but with an intensity N^2 times that for a single charge. Give a qualitative explanation of these facts.

(c) Without detailed calculations show that for nonrelativistic motion the dependence on N of the total power radiated is dominantly as β^{2N}, so that in the limit $N \to \infty$ no radiation is emitted.

(d) By arguments like those of (c) show that for relativistic particles the radiated power varies with N mainly as $\exp(-2N/3\gamma^3)$ for $N \gg \gamma^3$, so that again in the limit $N \to \infty$ no radiation is emitted.

(e) What relevance have the results of (c) and (d) to the radiation properties of a steady current in a loop?

14.12 As an idealization of steady-state currents flowing in a circuit, consider a system of N identical charges q moving with constant *speed* v (but subject to accelerations) in an arbitrary closed path. Successive charges are separated by a constant small interval Δ.

Starting with the Liénard-Wiechert fields for each particle, and making no assumptions concerning the speed v relative to the velocity of light, show that, in the limit $N \to \infty$, $q \to 0$, and $\Delta \to 0$, but $Nq = $ constant and $q/\Delta = $ constant, no radiation is emitted by the system and the electric and magnetic fields of the system are the usual static values.

(Note that for a real circuit the stationary positive ions in the conductors will produce an electric field which just cancels that due to the moving charges.)

14.13 Assume that the instantaneous power spectrum radiated by an electron in a synchrotron is given by

$$P(\omega, t) \simeq \frac{2}{\pi} \frac{e^2}{\rho} \gamma(t) \left(\frac{\omega}{\omega_c} \right)^{1/3} e^{-2\omega/\omega_c}$$

where $\omega_c = 3\omega_0 \gamma^3(t)$.

(a) If the electrons increase their energy approximately linearly during one cycle of operation, show that the power spectrum, averaged over one cycle of operation, is

$$\langle P(\omega, t) \rangle \simeq \frac{2^{2/3}}{3\pi} \frac{e^2}{\rho} \gamma_{max} x^{2/3} \int_x^{\infty} \frac{e^{-y}}{y^{1/3}} \, dy$$

where $x = 2\omega/\omega_{c\,max}$.

(b) Determine limiting forms for the spectrum when $x \ll 1$ and $x \gg 1$.

(c) By finding tables of the integral (it is an incomplete gamma function) or by graphical integration for $x = 0.1, 0.5, 1.0, 1.5$, determine numerically the spectrum, plot it as a function of $\log[\omega/\omega_{c\,max}]$, and compare it with the curves given by Elder, Langmuir, and Pollock, *Phys. Rev.*, **74**, 52 (1948), Fig. 1.

14.14 (a) Within the framework of approximations of Section 14.6, show that, for a relativistic particle moving in a path with instantaneous radius of curvature ρ, the frequency-angle spectra of radiations with positive and negative helicity are

$$\frac{dI_{\pm}(\omega)}{d\Omega} = \frac{e^2}{6\pi^2 c}\left(\frac{\omega\rho}{c}\right)^2\left(\frac{1}{\gamma^2} + \theta^2\right)^2 \left| K_{2/3}(\xi) \pm \frac{\theta}{\left(\frac{1}{\gamma^2} + \theta^2\right)^{1/2}} K_{1/3}(\xi) \right|^2$$

(b) From the formulas of Section 14.6 and (a) above, discuss the polarization of the total radiation emitted as a function of frequency and angle. In particular, determine the state of polarization at (1) high frequencies ($\omega > \omega_c$) for all angles, (2) intermediate and low frequencies ($\omega < \omega_c$) for large angles, (3) intermediate and low frequencies at very small angles.

(c) See the paper by P. Joos, *Phys. Rev. Letters*, **4**, 558 (1960), for experimental comparison.

14.15 Consider the synchrotron radiation from the Crab nebula. Electrons with energies up to at least 10^{12} ev move in a magnetic field of the order of 10^{-4} gauss.

(a) For $E = 10^{12}$ ev, $B = 10^{-4}$ gauss, calculate the orbit radius ρ, the fundamental frequency $\omega_0 = c/\rho$, and the critical frequency ω_c.

(b) Show that for a relativistic electron of energy E in a constant magnetic field the power spectrum of synchrotron radiation can be written

$$P(E, \omega) = \text{const} \left(\frac{\omega}{E^2}\right)^{1/3} f\left(\frac{\omega}{\omega_c}\right)$$

where $f(x)$ is a cutoff function having the value unity at $x = 0$ and vanishing rapidly for $x \gg 1$ [e.g., $f \simeq \exp(-2\omega/\omega_c)$, as in Problem 14.13], and $\omega_c = (3eB/mc)(E/mc^2)^2 \cos\theta$, where θ is the pitch angle of the helical path.

(c) If electrons are distributed in energy according to the spectrum $N(E)\,dE \sim E^{-n}\,dE$, show that the synchrotron radiation has the power spectrum

$$\langle P(\omega)\rangle\,d\omega \sim \omega^{-\alpha}\,d\omega$$

where $\alpha = (n - 1)/2$.

(d) Observations on the radiofrequency and optical continuous spectrum from the Crab nebula show that on the frequency interval from $\omega \sim 10^8$ sec^{-1} to $\omega \sim 6 \times 10^{15}$ sec^{-1} the constant $\alpha \simeq 0.35$. At higher frequencies the spectrum of radiation falls steeply with $\alpha \gtrsim 1.5$. Determine the index n for the electron-energy spectrum, as well as an upper cutoff for that spectrum. Is this cutoff consistent with the numbers of part (a)?

(e) From the result of Problem 14.4b find a numerical value for the time taken by an electron to decrease in energy from infinite energy to 10^{12} ev in a field of 10^{-4} gauss. How does this compare with the known lifetime of the Crab nebula?

14.16 Assuming that Plexiglas or Lucite has an index of refraction of 1.50 in the visible region, compute the angle of emission of visible Cherenkov radiation for electrons and protons as a function of their energies in Mev. Determine how many quanta with wavelengths between 4000 and 6000 angstroms are emitted per centimeter of path in Lucite by a 1-Mev electron, a 500-Mev proton, a 5-Bev proton.

15

Bremsstrahlung, Method of Virtual Quanta, Radiative Beta Processes

In Chapter 14 radiation by accelerated charges was discussed in a general way, formulas were derived for frequency and angular distributions, and examples of radiation by both nonrelativistic and relativistic charged particles in external fields were treated. The present chapter is devoted to problems of emission of electromagnetic radiation by charged particles in atomic and nuclear processes.

Particles passing through matter are scattered and lose energy by collisions, as described in detail in Chapter 13. In these collisions the particles undergo acceleration; hence they emit electromagnetic radiation. The radiation emitted during atomic collisions is customarily called *bremsstrahlung* (braking radiation) because it was first observed when high-energy electrons were stopped in a thick metallic target. For nonrelativistic particles the loss of energy by radiation is negligible compared with the collisional energy loss, but for ultrarelativistic particles radiation can be the dominant mode of energy loss. Our discussion of bremsstrahlung and related topics will begin with the nonrelativistic, classical situation. Then semiclassical arguments will be used, as in Chapter 13, to obtain plausible quantum-mechanical modifications. Relativistic effects, which produce significant changes in the results, will then be presented.

The creation or annihilation of charged particles is another process in which radiation is emitted. Such processes are purely quantum mechanical in origin. There can be no attempt at a classical explanation of the basic phenomena. But given that the process does occur, we may legitimately ask about the spectrum and intensity of electromagnetic radiation accompanying it. The sudden creation of a fast electron in nuclear beta decay, for example, can be viewed for our purposes as the violent acceleration of a charged particle initially at rest to some final velocity in a very short time interval, or, alternatively, as the sudden switching on of the charge of the moving particle in the same short interval. We will discuss nuclear beta decay and orbital-electron capture in these terms in Sections 15.7 and 15.8.

In radiation problems, such as the emission of bremsstrahlung or radiative beta decay, the wave nature of the charged particles involved produces quantum-mechanical modifications very similar to those appearing in our earlier energy-loss considerations. These can be taken into account in a relatively simple way. But there is a more serious deficiency which occurs only in radiation problems. It is very difficult to take into account the effects on the trajectory of the particle of the energy and momentum carried off by radiation. This is not only because radiation reaction effects are relatively hard to include (see Chapter 17), but also because of the discrete quantum nature of the photons emitted. Thus, even when modifications are made to describe the quantum-mechanical nature of the particles, our results are limited in validity by the restriction that the emitted photon have an energy small compared to the total energy available. At the upper end of the frequency spectrum our semiclassical expressions will generally have only qualitative validity.

15.1 Radiation Emitted during Collisions

If a charged particle makes a collision, it undergoes acceleration and emits radiation. If its collision partner is also a charged particle, they both emit radiation and a coherent superposition of the radiation fields must be made. Since the amplitude of the radiation fields depends (nonrelativistically) on the charge times the acceleration, the lighter particle will radiate more, provided the charges are not too dissimilar. In many applications the mass of one collision partner is much greater than the mass of the other. Then for the emission of radiation it is sufficient to treat the collision as the interaction of the lighter of the two particles with a fixed field of force. We will consider only this situation, and will leave more involved cases to the problems at the end of the chapter.

From (14.65) we see that a nonrelativistic particle with charge e and acceleration $c\dot{\boldsymbol{\beta}}(t)$ radiates energy with an intensity per unit solid angle per unit frequency interval,

$$\frac{dI(\omega)}{d\Omega} = \frac{e^2}{4\pi^2 c} \left| \int \mathbf{n} \times (\mathbf{n} \times \dot{\boldsymbol{\beta}}) e^{i\omega\left(t - \frac{\mathbf{n}\cdot\mathbf{r}(t)}{c}\right)} dt \right|^2 \qquad (15.1)$$

The position vector $\mathbf{r}(t)$ has the order of magnitude $\langle v \rangle t$, relative to a suitable origin, where $\langle v \rangle$ is a typical velocity of the problem. This means that the second term in the exponential in (15.1) is of the order of $\langle v \rangle / c$ times the first. For nonrelativistic motion, it can be neglected. Its neglect is sometimes called the *dipole approximation*, in analogy with the multipole expansion of Section 9.2. Then we find the approximate expression,

$$\frac{dI(\omega)}{d\Omega} = \frac{e^2}{4\pi^2 c} \left| \int \mathbf{n} \times (\mathbf{n} \times \dot{\boldsymbol{\beta}}) e^{i\omega t} \, dt \right|^2 \tag{15.2}$$

If we consider a collision process, the acceleration caused by the field of force exists only for a limited time τ, namely, the collision time:

$$\tau \simeq \frac{a}{v} \tag{15.3}$$

where a is a characteristic distance over which the force is appreciable. Then the integral in (15.2) is over a time interval of order τ. This means that τ provides a natural parameter with which to divide the frequencies of the radiation emitted into low frequencies ($\omega\tau \ll 1$) and high frequencies ($\omega\tau \gg 1$). In the low-frequency limit, the exponential in (15.2) is sensibly constant over the period of acceleration. Then the integration can be performed immediately:

$$\int \dot{\boldsymbol{\beta}}(t) e^{i\omega t} \, dt \simeq \int \dot{\boldsymbol{\beta}}(t) \, dt = \boldsymbol{\beta}_2 - \boldsymbol{\beta}_1 \equiv \Delta\boldsymbol{\beta} \tag{15.4}$$

where $c\boldsymbol{\beta}_1$ and $c\boldsymbol{\beta}_2$ are the initial and final velocities, and $\Delta\boldsymbol{\beta}$ is the vectorial change. Then the energy radiated is

$$\frac{dI(\omega)}{d\Omega} \simeq \frac{e^2}{4\pi^2 c} |\Delta\boldsymbol{\beta}|^2 \sin^2 \Theta, \qquad \omega\tau \ll 1 \tag{15.5}$$

where Θ is measured relative to the direction of $\Delta\boldsymbol{\beta}$. The total energy radiated per unit frequency interval in this limit is

$$I(\omega) \simeq \frac{2}{3\pi} \frac{e^2}{c} |\Delta\boldsymbol{\beta}|^2, \qquad \omega\tau \ll 1 \tag{15.6}$$

In the high-frequency limit ($\omega\tau \gg 1$) the exponential in (15.2) oscillates very rapidly compared to the variation of $\dot{\boldsymbol{\beta}}(t)$ in time. Consequently the integrand has a very small average value, and the energy radiated is negligible. The frequency spectrum will appear qualitatively as shown in Fig. 15.1. It will be convenient sometimes to make the approximation that the spectrum is given by a step function:

$$I(\omega) = \begin{cases} \dfrac{2e^2}{3\pi c} |\Delta\boldsymbol{\beta}|^2, & \omega\tau < 1 \\[2mm] 0, & \omega\tau > 1 \end{cases} \qquad (15.7)$$

For a single encounter with a definite $\Delta\boldsymbol{\beta}$ this is not a very good approximation, but if an average over many collisions with various $\Delta\boldsymbol{\beta}$ is wanted the approximation is adequate.

The angular distribution (15.5) includes all polarizations of the emitted radiation. Sometimes it is of interest to exhibit the intensity for a definite state of polarization. In collision problems it is usual that the direction of the incident particle is known and the direction of the radiation is known, but the deflected particle's direction, and consequently that of $\Delta\boldsymbol{\beta}$, are not known. Consequently the plane containing the incident beam direction and the direction of the radiation is a natural one with respect to which one specifies the state of polarization of the radiation.

For simplicity we consider a small angle deflection so that $\Delta\boldsymbol{\beta}$ is approximately perpendicular to the incident direction. Figure 15.2 shows the vectorial relationships. Without loss of generality $\mathbf{n}$, the observation direction, is chosen in the x-z plane, making an angle θ with the incident beam. The change in velocity $\Delta\boldsymbol{\beta}$ lies in the x-y plane, making an angle ϕ with the x axis. Since the direction of the scattered particle is not observed, we will average over ϕ. The unit vectors $\boldsymbol{\epsilon}_{\parallel}$ and $\boldsymbol{\epsilon}_{\perp}$ are polarization vectors parallel and perpendicular to the plane containing $\boldsymbol{\beta}_1$ and $\mathbf{n}$.

The direction of polarization of the radiation is given by the vector $\mathbf{n} \times (\mathbf{n} \times \Delta\boldsymbol{\beta})$. This is perpendicular to $\mathbf{n}$ (as it must be) and can be resolved into components along $\boldsymbol{\epsilon}_{\parallel}$ and $\boldsymbol{\epsilon}_{\perp}$. Thus

$$\mathbf{n} \times (\mathbf{n} \times \Delta\boldsymbol{\beta}) = \Delta\beta[\cos\theta\cos\phi\,\boldsymbol{\epsilon}_{\parallel} - \sin\phi\,\boldsymbol{\epsilon}_{\perp}] \qquad (15.8)$$

The absolute squares of the components in (15.8), averaged over ϕ, give

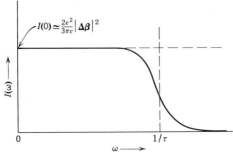

Fig. 15.1 Frequency spectrum of radiation emitted in a collision of duration τ with velocity change $\Delta\beta$.

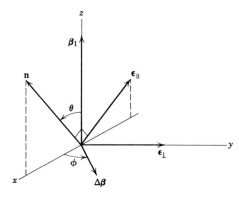

Fig. 15.2

the intensities of radiation for the two polarization states. The results are

$$
\left.
\begin{aligned}
\frac{dI_{\parallel}(\omega)}{d\Omega} &= \frac{e^2}{8\pi^2 c} |\Delta\boldsymbol{\beta}|^2 \cos^2\theta \\[2mm]
\frac{dI_{\perp}(\omega)}{d\Omega} &= \frac{e^2}{8\pi^2 c} |\Delta\boldsymbol{\beta}|^2
\end{aligned}
\right\}
\tag{15.9}
$$

These angular distributions are valid for all types of nonrelativistic small-angle collisions. They have been verified in detail for the continuous X-ray spectrum produced by electrons of kinetic energies in the kilovolt range. It is evident that the sum of the two intensities is consistent with (15.5) and yields a total radiated intensity equal to (15.6).

15.2 Bremsstrahlung in Nonrelativistic Coulomb Collisions

The most common situation where a continuum of radiation is emitted is the collision of a fast particle with an atom. As a model for this process we will consider first the collision of a fast, but nonrelativistic, particle of charge ze, mass M, and velocity v with a fixed point charge Ze. For simplicity we will assume that the deflection of the incident particle is small. Then the same arguments as were used in Chapter 13 on the limits of impact parameters will be involved. In fact, much of the discussion can be transplanted bodily from the treatment of energy loss.

For a small deflection in a point Coulomb field of charge Ze the momentum change is transverse and is given by (13.1) times Z. Thus the net change of velocity of the incident particle passing at impact parameter b has the magnitude

$$
\Delta v = \frac{2zZe^2}{Mvb}
\tag{15.10}
$$

The frequency spectrum will be given approximately by (15.7) (times z^2) with a collision time (15.3) $\tau \simeq b/v$. Thus the spectrum extends from $\omega = 0$ to $\omega_{max} \simeq v/b$:

$$I(\omega, b) \simeq \begin{cases} \dfrac{8}{3\pi}\left(\dfrac{z^2e^2}{Mc^2}\right)^2 \dfrac{Z^2e^2}{c}\left(\dfrac{c}{v}\right)^2 \dfrac{1}{b^2}, & \omega < \dfrac{v}{b} \\[4mm] 0, & \omega > \dfrac{v}{b} \end{cases} \tag{15.11}$$

Just as in the energy-loss process, the useful physical quantity is a cross section obtained by integrating over all possible impact parameters. Accordingly we define the *radiation cross section* $\chi(\omega)$, with dimensions (area-energy/frequency),

$$\chi(\omega) = \int I(\omega, b)2\pi b \; db \tag{15.12}$$

The classical limits on the impact parameters can be found by arguments analogous to those of Section 13.1. The classical minimum impact parameter is [see (13.5)–(13.7)]:

$$b_{min}^{(c)} = \frac{zZe^2}{Mv^2} \tag{15.13}$$

while the maximum value is governed by the cutoff in the spectrum (15.11). If we fix our attention on a given frequency ω in $\chi(\omega)$, it is evident that only for impact parameters less than

$$b_{max} \simeq \frac{v}{\omega} \tag{15.14}$$

will the accelerations be violent and rapid enough to produce significant radiation at that frequency. With these limits on b the radiation cross section is

$$\chi_c(\omega) \simeq \frac{16}{3}\frac{Z^2e^2}{c}\left(\frac{z^2e^2}{Mc^2}\right)^2\left(\frac{c}{v}\right)^2 \ln\left(\frac{\lambda Mv^3}{zZe^2\omega}\right) \tag{15.15}$$

where λ is a numerical factor of the order of unity which takes into account our uncertainties in the exact limits on impact parameters. This result is valid only for frequencies where the argument of the logarithm is large compared to unity, corresponding to $b_{max} \gg b_{min}$. This means that there is a classical upper limit $\omega_{max}^{(c)}$ to the frequency spectrum given by

$$\omega_{max}^{(c)} \sim \frac{Mv^3}{zZe^2} \tag{15.16}$$

For highly charged, massive, slow particles the classical radiation cross section will be valid, but just as in the energy-loss phenomena the wave nature of particles enters importantly for lightly charged, swift particles. The quantum modifications are very similar to those discussed in Section 13.3. The wave nature of the incident particle sets a quantum-mechanical lower limit on the impact parameters,

$$b_{\min}^{(q)} \simeq \frac{\hbar}{Mv} \tag{15.17}$$

This means that the radiation cross section is approximately

$$\chi_q(\omega) \simeq \frac{16}{3} \frac{Z^2 e^2}{c} \left(\frac{z^2 e^2}{Mc^2}\right)^2 \left(\frac{c}{v}\right)^2 \ln\left(\frac{\lambda M v^2}{\hbar \omega}\right) \tag{15.18}$$

instead of (15.15). We note that the arguments of the logarithm differ by the factor η (13.42) (times Z to give the product of the charges). The same rules about domains of validity of the classical and quantum-mechanical formulas apply here as for the energy loss. The frequency spectrum of the quantum cross section extends up to a maximum frequency $\omega_{\max}^{(q)}$ of the order of

$$\omega_{\max}^{(q)} \sim \frac{Mv^2}{\hbar} \tag{15.19}$$

We note that this is approximately the conservation of energy limit, $\omega_{\max} = Mv^2/2\hbar$. Since the classical result holds only when $\eta \gg 1$, we see that

$$\omega_{\max}^{(c)} \simeq \frac{1}{\eta}\, \omega_{\max}^{(q)} \ll \omega_{\max}^{(q)} \tag{15.20}$$

This shows that the classical frequency spectrum is always confined to very low frequencies compared to the maximum allowed by conservation of energy. Thus the classical domain is of little interest. In what follows we will concentrate on the quantum-mechanical results.

Although the upper limit (15.19) is in rough accord with conservation of energy, the quantum radiation cross section has only qualitative validity at the upper end of the frequency spectrum. As was discussed in the introduction to this chapter, the reason is the discrete quantum nature of the photons emitted. For soft photons with energies far from the maximum the discrete nature is unimportant because the energy and momentum carried off are negligible. But for hard photons near the end point of the spectrum the effects are considerable. One obvious and plausible way to include the conservation of energy requirement is to argue that the impact parameters (15.14) and (15.17) should involve an *average* velocity,

$$\langle v \rangle = \tfrac{1}{2}(v_i + v_f) = \frac{1}{\sqrt{2M}} (\sqrt{E} + \sqrt{E - \hbar\omega}) \tag{15.21}$$

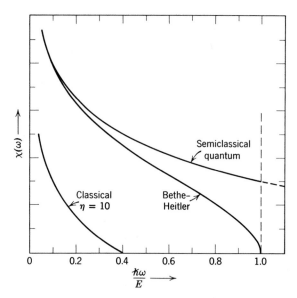

Fig. 15.3 Radiation cross section (energy × area/unit frequency) for Coulomb collisions as a function of frequency in units of the maximum frequency ($E/\hbar$). The classical spectrum is confined to very low frequencies. The curve marked "Bethe-Heitler" is the quantum-mechanical Born approximation result, while the "semi-classical quantum" curve is (15.18).

where $E = \frac{1}{2}Mv^2$ is the initial kinetic energy of the particle, and $\hbar\omega$ is the energy of the photon emitted. With this average velocity in place of v in (15.18), we obtain

$$\chi_q(\omega) \simeq \frac{16}{3}\frac{Z^2e^2}{c}\left(\frac{z^2e^2}{Mc^2}\right)^2\left(\frac{c}{v}\right)^2 \ln\left[\frac{\lambda}{2}\frac{(\sqrt{E}+\sqrt{E-\hbar\omega})^2}{\hbar\omega}\right] \quad (15.22)$$

If $\lambda = 2$, this cross section is exactly the quantum-mechanical result in the Born approximation, first calculated by Bethe and Heitler (1934). The argument of the logarithm evidently equals unity when $\hbar\omega = \frac{1}{2}Mv^2$, so that the conservation of energy requirement is properly satisfied. Figure 15.3 shows the shape of the radiation cross section as a function of frequency. The Bethe-Heitler formula (15.22) is compared with our classical and semiclassical quantum formulas (15.15) and (15.18) with $\lambda = 2$ and $\eta = 10$.

The bremsstrahlung spectrum is sometimes expressed as a cross section for photon emission with dimensions of area/unit photon energy. Thus

$$\hbar\omega\sigma_{\text{brems}}(\hbar\omega)\,d(\hbar\omega) = \chi(\omega)\,d\omega \quad (15.23)$$

The bremsstrahlung photon cross section is evidently

$$\sigma_{\text{brems}}(\hbar\omega) \simeq \frac{16}{3} \frac{Z^2 e^2}{\hbar c} \left(\frac{z^2 e^2}{Mc^2}\right)^2 \left(\frac{c}{v}\right)^2 \frac{\ln (\ \ )}{\hbar\omega} \tag{15.24}$$

where the argument of the logarithm is that of (15.15) or (15.22). Since the logarithm varies relatively slowly with photon energy, the main dependence of the cross section on photon energy is as $(\hbar\omega)^{-1}$, known as the typical bremsstrahlung spectrum.

The radiation cross section $\chi(\omega)$ depends on the properties of the particles involved in the collision as $Z^2 z^4/M^2$, showing that the emission of radiation is most important for electrons in materials of high atomic number. The total energy lost in radiation by a particle traversing unit thickness of matter containing N fixed charges Ze (atomic nuclei) per unit volume is

$$\frac{dE_{\text{rad}}}{dx} = N \int_0^{\omega_{\text{max}}} \chi(\omega) \, d\omega \tag{15.25}$$

Using (15.22) for $\chi(\omega)$ and converting to the variable of integration $x = (\hbar\omega/E)$, we can write the radiative energy loss as

$$\frac{dE_{\text{rad}}}{dx} = \frac{16}{3} NZ \left(\frac{Ze^2}{\hbar c}\right) \frac{z^4 e^4}{Mc^2} \int_0^1 \ln \left(\frac{1 + \sqrt{1-x}}{\sqrt{x}}\right) dx \tag{15.26}$$

The dimensionless integral has the value unity. For comparison we write the ratio of radiative energy loss to collision energy loss (13.13) or (13.44):

$$\frac{dE_{\text{rad}}}{dE_{\text{coll}}} \simeq \frac{4}{3\pi} z^2 \frac{Z}{137} \frac{m}{M} \left(\frac{v}{c}\right)^2 \frac{1}{\ln B_q} \tag{15.27}$$

For nonrelativistic particles ($v \ll c$) the radiative loss is completely negligible compared to the collision loss. The fine structure constant ($e^2/\hbar c = 1/137$) enters characteristically whenever there is emission of radiation as an additional step beyond the basic process (here the deflection of the particle in the nuclear Coulomb field). The factor m/M appears because the radiative loss involves the acceleration of the incident particle, while the collision loss involves the acceleration of an electron.

15.3   Relativistic Bremsstrahlung

For relativistic particles making collisions with atomic nuclei there are characteristic modifications in the radiation emitted. Our first thought would be that the nonrelativistic discussion of the previous sections would not be valid at all, and that a full relativistic treatment would be necessary.

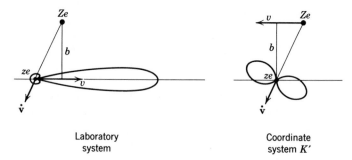

Laboratory
system

Coordinate
system K'

Fig. 15.4 Radiation emitted during relativistic collisions viewed from the laboratory (nucleus at rest) and the frame K' (incident particle essentially at rest).

But it is one of the great virtues of the special theory of relativity (aside from being correct and necessary) that it allows us to choose a convenient reference frame for our calculation and then transform to the laboratory at the end. Thus we will find that all but the final steps of the relativistic bremsstrahlung calculation can be done nonrelativistically.

There are two aspects. First of all, we know that radiation emitted by a highly relativistic particle is confined to a narrow cone of half-angle of the order of Mc^2/E, where E is its total energy. Thus, unless we are interested in very fine details, it is sufficient to consider only the total energy radiated at a given frequency. The second point is that except for very close collisions the incident particle is deflected only slightly in an encounter and loses only a very small amount of energy. In the reference frame K', where the incident particle is at rest initially and the nucleus moves by with velocity $v \sim c$, the corresponding motion of the incident particle is non-relativistic throughout the collision. This means that in the frame K' the radiation process can be treated entirely nonrelativistically. The connection between the radiation process as viewed in the laboratory and in the coordinate frame K' is sketched in Fig. 15.4.

Almost all the arguments previously presented in Sections 15.1 and 15.2 apply. The only modifications are in the limits on the impact parameters. The relativistic contraction of the fields (see Section 11.10) makes the collision time (11.120) smaller by a factor $\gamma = E/Mc^2$. This means that the maximum impact parameter is increased from (15.14) to

$$b_{\text{max}} \simeq \frac{\gamma v}{\omega'} \tag{15.28}$$

where ω' is the emitted frequency in the coordinate system K'. The minimum impact parameter for these radiation problems is *not* the expected $\hbar/p = \hbar/\gamma Mv$, even though this is the magnitude of "smearing out" of the particle due to quantum effects. The proper value is still (15.17), without factors of γ, as can be seen from the following argument. In the emission of radiation all parts of an extended charge distribution must experience the same acceleration at the same time. Otherwise interference effects will greatly reduce the amount of radiation. This means that appreciable radiation will occur only if the width of the pulse of accelera- tion due to the passing nuclear field is large compared to the "smearing out" of the charge. The pulse width is of the order of b/γ, while the transverse smearing-out distance is of the order of $\hbar/\gamma Mv$. This sets a lower limit on impact parameters equal to (15.17), even for relativistic motion. With (15.11), (15.12), and these revised impact parameters, the radiation cross section $\chi'(\omega')$ in the system K' is

$$\chi'(\omega') \simeq \frac{16}{3} \frac{Z^2 e^2}{c} \left(\frac{z^2 e^2}{Mc^2}\right)^2 \left(\frac{c}{v}\right)^2 \ln\left(\frac{\lambda \gamma Mv^2}{\hbar\omega'}\right) \qquad (15.29)$$

To transform this result to the (unprimed) laboratory frame we need to know the transformation properties of the radiation cross section and the frequency. The radiation cross section has the dimensions of (cross- sectional area) $\cdot$ (energy) $\cdot$ (frequency)$^{-1}$. Since energy and frequency transform in the same way under Lorentz transformations, while transverse dimensions are invariant, the radiation cross section is a Lorentz invariant:

$$\chi(\omega) = \chi'(\omega') \qquad (15.30)$$

The transformation of frequencies is according to the relativistic Doppler- shift formula (11.38):

$$\omega = \gamma\omega'(1 + \beta \cos \theta') \qquad (15.31)$$

where θ' is the angle of emission in the frame K'. The cross section $\chi'(\omega')$ is the *total* cross section, integrated over angles in K'. Since the accelera- tion is predominantly transverse in that frame, the radiation is emitted essentially symmetrically about $\theta' = \pi/2$. Consequently on the average we have $\omega = \gamma\omega'$.* With this substitution for ω' in (15.29) we obtain the radiation cross section in the laboratory:

$$\chi(\omega) \simeq \frac{16}{3} \frac{Z^2 e^2}{c} \left(\frac{z^2 e^2}{Mc^2}\right)^2 \left(\frac{c}{v}\right)^2 \ln\left(\frac{\lambda \gamma^2 Mv^2}{\hbar\omega}\right) \qquad (15.32)$$

The only change from the nonrelativistic result (15.18) is the factor γ^2 in the argument of the logarithm. Conservation of energy requires that this

---

* This result can be obtained from the original transformation (11.37), $\omega' = \gamma\omega(1 - \beta \cos \theta)$, by noting that, for $\gamma \gg 1$ and $\theta \ll 1$, we have $\omega' \simeq (\omega/2\gamma)(1 + \gamma^2\theta^2)$. Since the average value of $\gamma^2\theta^2$ in the laboratory is of the order of unity, we obtain $\omega' \simeq \omega/\gamma$.

expression be used only for frequencies such that $0 < \hbar\omega < (\gamma - 1)Mc^2 \simeq \gamma Mc^2$. We note that quanta with laboratory energies in the range $Mc^2 \ll \hbar\omega \ll \gamma Mc^2$ come from quanta with $\hbar\omega' \ll Mc^2$ in the transformed frame K'.

The above derivation of $\chi(\omega)$ in the laboratory is somewhat casual in that the dependence of transformed frequency on angle was not treated rigorously. We should actually consider the *differential* cross section in the frame K':

$$\frac{d\chi'(\omega')}{d\Omega'} = A \ln\left(\frac{\lambda\gamma Mv^2}{\hbar\omega'}\right)\left[\frac{3}{16\pi}(1 + \cos^2\theta')\right] \qquad (15.33)$$

where A is the coefficient of the logarithm in (15.32). The square-bracketed angular factor comes from the sum of the two terms in (15.9) and is normalized to a unit integral over solid angles. When transformed according to (11.38), (15.33) becomes in the laboratory

$$\frac{d\chi(\omega)}{d\Omega} \simeq A \ln\left(\frac{2\lambda\gamma^2 Mv^2}{\hbar\omega(1 + \gamma^2\theta^2)}\right)\frac{3}{2\pi}\frac{\gamma^2(1 + \gamma^4\theta^4)}{(1 + \gamma^2\theta^2)^4} \qquad (15.34)$$

The angular distribution is peaked sharply in the forward direction. The angular factor falls off as $(\gamma\theta)^{-4}$ for $\gamma\theta \gg 1$. Of course, (15.34) is not valid for angles $\theta \gtrsim 1$. But the order of magnitude is correct, the intensity of radiation being a factor γ^{-4} smaller at backward angles than in the forward direction, and approaching the limiting value (at $\theta = \pi$):

$$\lim_{\theta \to \pi}\frac{d\chi(\omega)}{d\Omega} = \frac{3}{32\pi}\frac{A}{\gamma^2}\ln\left(\frac{\lambda Mv^2}{2\hbar\omega}\right) \qquad (15.35)$$

Since almost all the radiation is confined to angles $\theta \ll 1$, we may approximate the solid angle element $d\Omega \simeq 2\pi\theta\, d\theta = (\pi/\gamma^2)\, d(\gamma^2\theta^2)$, and integrate over the interval $0 < \gamma^2\theta^2 < \infty$ with little error. This yields the total radiation cross section,

$$\chi(\omega) = A\left[\ln\left(\frac{2\lambda\gamma^2 Mv^2}{\hbar\omega}\right) - \frac{13}{12}\right] \qquad (15.36)$$

which differs insignificantly from the previous result (15.32).

15.4 Screening Effects; Relativistic Radiative Energy Loss

In the treatment of bremsstrahlung so far we have ignored the effects of the atomic electrons. As direct contributors to the acceleration of the incident particle they can be safely ignored, since their contribution per atom is of the order of Z^{-1} times the nuclear one. But they have an indirect effect through their screening of the nuclear charge. The potential energy of the incident particle in the field of the atom can be approximated by the form (13.94). This means that there will be negligible radiation emitted for collisions at impact parameters greater than the atomic radius (13.95).

We can include this approximately in our previous calculations by defining a maximum impact parameter due to screening by the atomic electrons,

$$b^{(s)}_{max} \sim a \simeq 1.4 \frac{a_0}{Z^{\frac{1}{3}}} \tag{15.37}$$

Then we must use the smaller of the two values (15.28) and (15.37) for b_{max} in the argument of the logarithm. The ratio is

$$\frac{b^{(s)}_{max}}{b_{max}} \simeq \frac{192}{Z^{\frac{1}{3}}} \left(\frac{v}{c}\right) \frac{\hbar\omega}{\gamma^2 m v^2} \tag{15.38}$$

where m is the electronic mass and we have used the average value $\omega' = \omega/\gamma$. This shows that for low enough frequencies the screening value is always smaller than (15.28). The limiting frequency ω_s below which we must use $b^{(s)}_{max}$ is, in units of the spectrum end point $\omega_{max} = \frac{1}{\hbar}(\gamma - 1)Mc^2$,

$$\frac{\omega_s}{\omega_{max}} \simeq \frac{Z^{\frac{1}{3}}}{192} \left(\frac{m}{M}\right) \gamma \left(\frac{\gamma+1}{\gamma-1}\right)^{\frac{1}{2}} \simeq \begin{cases} \dfrac{2Z^{\frac{1}{3}}}{192} \dfrac{m}{M} \left(\dfrac{c}{v}\right) \\[2mm] \dfrac{Z^{\frac{1}{3}}}{192} \dfrac{m}{M} \gamma \end{cases} \tag{15.39}$$

where the upper (lower) line is the nonrelativistic (relativistic) limiting form. When $\omega < \omega_s$ the argument of the logarithm in the radiation cross section (15.32) becomes independent of frequency:

$$\frac{b^{(s)}_{max}}{b_{min}} \simeq \lambda \frac{192}{Z^{\frac{1}{3}}} \frac{M}{m} \frac{v}{c} \tag{15.40}$$

This makes the radiation cross section approach the constant value,

$$\chi(\omega) \simeq \frac{16}{3} \frac{Z^2 e^2}{c} \left(\frac{z^2 e^2}{Mc^2}\right)^2 \left(\frac{c}{v}\right)^2 \ln\left(\lambda \frac{192}{Z^{\frac{1}{3}}} \frac{M}{m} \frac{v}{c}\right) \tag{15.41}$$

for $\omega \ll \omega_s$. Then the energy radiated per unit frequency interval at low frequencies is finite, rather than logarithmically divergent. This is the same type of effect as the screening produces in making the small-angle scattering (13.96) finite, rather than divergent as θ^{-4} for a pure Coulomb field.

Except for extremely low velocities the screening frequency ω_s is very small compared to ω_{max} in the nonrelativistic limit. A typical figure is $\omega_s/\omega_{max} \simeq 0.07$ for electrons of 100-Kev kinetic energy incident on a gold target $(Z = 79)$. For heavier nonrelativistic particles the ratio is even smaller. This means that the spectrum shown in Fig. 15.3 is altered only at very low frequencies for nonrelativistic bremsstrahlung.

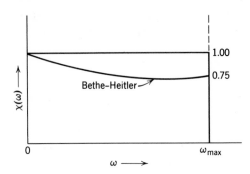

Fig. 15.5 Radiation cross section in the complete screening limit. The constant value is the semi-classical result. The curve marked "Bethe-Heitler" is the quantum-mechanical Born approximation.

For extremely relativistic particles the screening can be "complete." Complete screening occurs when $\omega_s > \omega_{max}$. This occurs at energies greater than the critical value,

$$E_s = \left(\frac{192M}{Z^{1/3}m}\right) Mc^2 \qquad (15.42)$$

For electrons, $E_s \simeq 42$ Mev in aluminum ($Z = 13$) and 23 Mev in lead ($Z = 82$). The corresponding values for mu mesons are 2×10^6 Mev and 10^6 Mev. Because of the factor M/m, screening is important only for electrons. When $E > E_s$, the radiation cross section is given by the constant value (15.41) for all frequencies. Figure 15.5 shows the radiation cross section (15.41) in the limit of complete screening, as well as the corresponding Bethe-Heitler result. Their proper quantum treatment involves a slowly varying factor which changes from unity at $\omega = 0$ to 0.75 at $\omega = \omega_{max}$. For cosmic-ray electrons and for electrons from most high-energy electron accelerators, the bremsstrahlung is in the complete screening limit. Thus the photon spectrum shows a typical $(\hbar\omega)^{-1}$ behavior.

The radiative energy loss was considered in the nonrelativistic limit in Section 15.2 and was found to be negligible compared to the energy loss by collisions. For ultrarelativistic particles, especially electrons, this is no longer true. The radiative energy loss is given approximately in the limit $\gamma \gg 1$ by

$$\frac{dE_{rad}}{dx} \simeq \frac{16}{3} N \frac{Z^2 e^2}{c} \left(\frac{z^2 e^2}{Mc^2}\right)^2 \int_0^{\gamma Mc^2/\hbar} \ln\left(\frac{b_{max}}{b_{min}}\right) d\omega \qquad (15.43)$$

where the argument of the logarithm depends on whether $\omega < \omega_s$ or $\omega > \omega_s$. For negligible screening ($\omega_s \ll \omega_{max}$) we find approximately

$$\frac{dE_{rad}}{dx} \simeq \frac{16}{3} N \frac{Z^2 e^2}{\hbar c} \left(\frac{z^2 e^2}{Mc^2}\right)^2 \ln(\lambda\gamma) \gamma Mc^2 \qquad (15.44)$$

For higher energies where complete screening occurs this is modified to

$$\frac{dE_{\text{rad}}}{dx} \sim \left[\frac{16}{3} N \frac{Z^2 e^2}{\hbar c} \left(\frac{z^2 e^2}{Mc^2} \right)^2 \ln \left(\frac{\lambda 192 M}{Z^{1/3} m} \right) \right] \gamma M c^2 \qquad (15.45)$$

showing that eventually the radiative loss is proportional to the particle's energy.

The comparison of radiative loss to collision loss now becomes

$$\frac{dE_{\text{rad}}}{dE_{\text{coll}}} \sim \frac{4}{3\pi} \left(\frac{Zz^2}{137} \right) \frac{m}{M} \frac{\ln \left(\frac{\lambda 192 M}{Z^{1/3} m} \right)}{\ln B_q} \gamma \qquad (15.46)$$

The value of γ for which this ratio is unity depends on the particle and on Z. For electrons it is $\gamma \sim 200$ for air and $\gamma \sim 20$ for lead. At higher energies, the radiative energy loss is larger than the collision loss and for ultrarelativistic particles is the dominant loss mechanism.

At energies where the radiative energy loss is dominant the complete screening result (15.45) holds. Then it is useful to introduce a unit of length X_0, called the *radiation length*, which is the distance a particle travels while its energy falls to e^{-1} of its initial value. By conservation of energy, we may rewrite (15.45) as

$$\frac{dE}{dx} = -\frac{E}{X_0}$$

with solution, $$E(x) = E_0 e^{-x/X_0} \qquad (15.47)$$

where the radiation length X_0 is

$$X_0 = \left[\frac{16}{3} N \frac{Z^2 e^2}{\hbar c} \left(\frac{z^2 e^2}{Mc^2} \right)^2 \ln \left(\frac{\lambda 192 M}{Z^{1/3} m} \right) \right]^{-1} \qquad (15.48)$$

For electrons, some representative values of X_0 are 32 gm/cm² (270 meters) in air at NTP, 19 gm/cm² (7.2 cm) in aluminum, and 4.4 gm/cm² (0.39 cm) in lead.* In studying the passage of cosmic-ray or man-made high-energy particles through matter, the radiation length X_0 is a convenient unit to employ, since not only the radiative energy loss is governed by it, but also the production of negaton-positon pairs by the radiated photons, and so the whole development of the electronic cascade shower.

---

* These numerical values differ by $\sim$20–30 per cent from those given by Rossi, p. 55, because he uses a more accurate coefficient of 4 instead of 16/3 and $Z(Z + 1)$ instead of Z^2 in (15.48).

15.5   Weizsäcker-Williams Method of Virtual Quanta

The emission of bremsstrahlung and other processes involving the electromagnetic interaction of relativistic particles can be viewed in a way that is very helpful in providing physical insight into the processes. This point of view is called the *method of virtual quanta*. It exploits the similarity between the fields of a rapidly moving charged particle and the fields of a pulse of radiation (see Section 11.10) and correlates the effects of the collision of the relativistic charged particle with some system with the corresponding effects produced by the interaction of radiation (the virtual quanta) with the same system. The method was developed independently by C. F. Weizsäcker and E. J. Williams in 1934.

In any given collision there are an "incident particle" and a "struck system." The perturbing fields of the incident particle are replaced by an equivalent pulse of radiation which is analyzed into a frequency spectrum of virtual quanta. Then the effects of the quanta (either scattering or absorption) on the struck system are calculated. In this way the charged-particle interaction is correlated with the photon interaction. The table lists a few typical correspondences and specifies the incident particle and

Correspondences between charged particle interactions and photon interactions

Particle Process	Incident Particle	Struck System	Radiative Process	b_{min}
Bremsstrahlung in electron (light particle)-nucleus collision	Nucleus	Electron (light particle)	Scattering of virtual photons of nuclear Coulomb field by the electron (light particle)	h/Mv
Collisional ionization of atoms (in distant collisions)	Incident particle	Atom	Photoejection of atomic electrons by virtual quanta	a
Electron disintegration of nuclei	Electron	Nucleus	Photodisintegration of nuclei by virtual quanta	Larger of $h/\gamma mv$ and R
Production of pions in electron-nuclear collisions	Electron	Nucleus	Photoproduction of pions by virtual quanta interactions with nucleus	

struck system. From the table we see that the struck system is not always the target in the laboratory. For bremsstrahlung the struck system is the

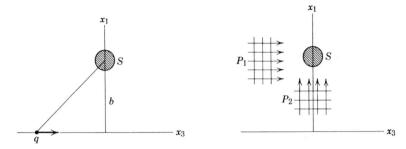

Fig. 15.6 Relativistic charged particle passing the struck system S and the equivalent
pulses of radiation.

lighter of the two collision partners, since its radiation scattering power is
greater. For bremsstrahlung in electron-electron collision it is necessary
from symmetry to take the sum of two contributions where each electron
in turn is the struck system at rest initially in some reference frame.

The chief assumption in the method of virtual quanta is that the effects
of the various frequency components of equivalent radiation add incoherently. This will be true provided the perturbing effect of the fields is
small, and is related to our assumption in Section 15.2 that the struck
particle moves only slightly during the collision.

The spectrum of equivalent radiation for an incident particle of charge
q, velocity $v \simeq c$, passing a struck system S at impact parameter b, can be
found from the fields of Section 11.10:

$$\left.\begin{aligned}
E_1(t) &= q\,\frac{\gamma b}{(b^2 + \gamma^2 v^2 t^2)^{3/2}}\\[4pt]
B_2(t) &= \beta E_1(t)\\[4pt]
E_3(t) &= -q\,\frac{\gamma v t}{(b^2 + \gamma^2 v^2 t^2)^{3/2}}
\end{aligned}\right\}
\tag{15.49}$$

For $\beta \simeq 1$ the fields $E_1(t)$ and $B_2(t)$ are completely equivalent to a pulse of
plane polarized radiation P_1 incident on S in the x_3 direction, as shown in
Fig. 15.6. There is no magnetic field to accompany $E_3(t)$ and so form a
proper pulse of radiation P_2 incident along the x_1 direction, as shown.
Nevertheless, if the motion of the charged particles in S is nonrelativistic
in this coordinate frame, we can add the necessary magnetic field to create
the pulse P_2 without affecting the physical problem because the particles
in S respond only to electric forces. Even if the particles in S are influenced
by magnetic forces, the additional magnetic field implied by replacing $E_3(t)$
by the radiation pulse P_2 is not important, since the pulse P_2 will be seen
to be of minor importance anyway.

From the discussion of Section 14.5, especially equations (14.51), (14.52), and (14.60), it is evident that the equivalent pulse P_1 has a frequency spectrum (energy per unit area per unit frequency interval) $I_1(\omega, b)$ given by

$$I_1(\omega, b) = \frac{c}{2\pi} |E_1(\omega)|^2 \qquad (15.50)$$

where $E_1(\omega)$ is the Fourier transform (14.54) of $E_1(t)$ in (15.49). Similarly the pulse P_2 has the frequency spectrum,

$$I_2(\omega, b) = \frac{c}{2\pi} |E_3(\omega)|^2 \qquad (15.51)$$

The Fourier integrals have already been calculated in Chapter 13 and are given by (13.29) and (13.30). The two frequency spectra are

$$\left.\begin{aligned} I_1(\omega, b) \\ I_2(\omega, b) \end{aligned}\right\} = \frac{1}{\pi^2} \frac{q^2}{c} \left(\frac{c}{v}\right)^2 \frac{1}{b^2} \left\{\begin{aligned} \left(\frac{\omega b}{\gamma v}\right)^2 K_1^2\left(\frac{\omega b}{\gamma v}\right) \\ \frac{1}{\gamma^2}\left(\frac{\omega b}{\gamma v}\right)^2 K_0^2\left(\frac{\omega b}{\gamma v}\right) \end{aligned}\right. \qquad (15.52)$$

We note that the intensity of the pulse P_2 involves a factor γ^{-2} and so is of little importance for ultrarelativistic particles. The shapes of these spectra are shown qualitatively in Fig. 15.7. The behavior is easily understood if one recalls that the fields of pulse P_1 are bell-shaped in time with a width $\Delta t \sim b/\gamma v$. Thus the frequency spectrum will contain all frequencies up to a maximum of order $\omega_{max} \sim 1/\Delta t$. On the other hand, the fields of pulse P_2 are similar to one cycle of a sine wave of frequency $\omega \sim \gamma v/b$.

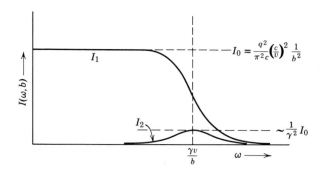

Fig. 15.7 Frequency spectra of the two equivalent pulses of radiation.

Consequently its spectrum will contain only a narrow range of frequencies centered around $\gamma v/b$.

In collision problems we must sum the frequency spectra (15.52) over the various possible impact parameters. This gives the energy per unit frequency interval present in the equivalent radiation field. As always in such problems we must specify a minimum impact parameter b_{min}. The method of virtual quanta will be useful only if b_{min} can be so chosen that for impact parameters greater than b_{min} the effects of the incident particle's fields can be represented accurately by the effects of equivalent pulses of radiation, while for small impact parameters the effects of the particle's fields can be neglected or taken into account by other means. Setting aside for the moment how we choose the proper value of b_{min}, we can write down the frequency spectrum integrated over possible impact parameters,

$$I(\omega) = 2\pi \int_{b_{min}} [I_1(\omega, b) + I_2(\omega, b)]b \, db \qquad (15.53)$$

where we have combined the contributions of pulses P_1 and P_2. This integral has already been done in Section 13.3, equation (13.35). The result is

$$I(\omega) = \frac{2}{\pi}\frac{q^2}{c}\left(\frac{c}{v}\right)^2\left[xK_0(x)K_1(x) - \frac{v^2}{2c^2}x^2(K_1^{\,2}(x) - K_0^{\,2}(x))\right] \qquad (15.54)$$

where

$$x = \frac{\omega b_{min}}{\gamma v} \qquad (15.55)$$

For low frequencies ($\omega \ll \gamma v/b_{min}$) the energy per unit frequency interval reduces to

$$I(\omega) \simeq \frac{2}{\pi}\frac{q^2}{c}\left(\frac{c}{v}\right)^2\left[\ln\left(\frac{1.123\gamma v}{\omega b_{min}}\right) - \frac{v^2}{2c^2}\right] \qquad (15.56)$$

whereas for high frequencies ($\omega \gg \gamma v/b_{min}$) the spectrum falls off exponentially as

$$I(\omega) \simeq \frac{q^2}{c}\left(\frac{c}{v}\right)^2\left(1 - \frac{v^2}{2c^2}\right)e^{-(2\omega b_{min}/\gamma v)} \qquad (15.57)$$

Figure 15.8 shows an accurate plot of $I(\omega)$ (15.54) for $v \simeq c$, as well as the low-frequency approximation (15.56). We see that the energy spectrum consists predominantly of low-frequency quanta with a tail extending up to frequencies of the order of $2\gamma v/b_{min}$.

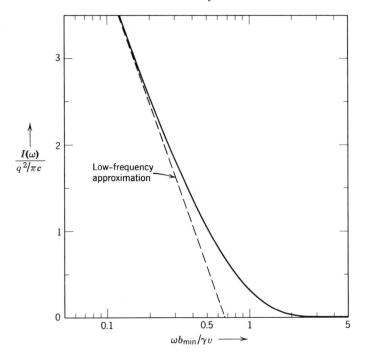

Fig. 15.8 Frequency spectrum of virtual quanta for a relativistic particle, with the energy per unit frequency $I(\omega)$ in units of $q^2/\pi c$ and the frequency in units of $\gamma v/b_{min}$. The number of virtual quanta per unit energy interval is obtained by dividing by $\hbar^2\omega$.

The number spectrum of virtual quanta $N(\hbar\omega)$ is obtained by using the relation,

$$I(\omega)\,d\omega = \hbar\omega N(\hbar\omega)\,d(\hbar\omega) \tag{15.58}$$

Thus the number of virtual quanta per unit energy interval in the low-frequency limit is

$$N(\hbar\omega) \simeq \frac{2}{\pi}\left(\frac{q^2}{\hbar c}\right)\left(\frac{c}{v}\right)^2 \frac{1}{\hbar\omega}\left[\ln\left(\frac{1.123\gamma v}{\omega b_{min}}\right) - \frac{v^2}{2c^2}\right] \tag{15.59}$$

The choice of minimum impact parameter b_{min} must be considered. In bremsstrahlung, $b_{min} = \hbar/Mv$, where M is the mass of the lighter particle, as discussed in Section 15.3. For collisional ionization of atoms, $b_{min} \simeq a$, the atomic radius, closer impacts being treated as collisions between the incident particle and free electrons. In electron disintegration of nuclei or electron production of mesons from nuclei, $b_{min} = \hbar/\gamma Mv$ or $b_{min} = R$, the nuclear radius, whichever is larger. The values are summarized in the table on p. 520.

15.6   Bremsstrahlung as the Scattering of Virtual Quanta

The emission of bremsstrahlung in a collision between an incident relativistic particle of charge ze and mass M and an atomic nucleus of charge Ze can be viewed as the scattering of the virtual quanta in the nuclear Coulomb field by the incident particle in the coordinate system K', where the incident particle is at rest. The spectrum of virtual quanta $I(\omega')$ is given by (15.54) with $q = Ze$. The minimum impact parameter is $\hbar/Mv$, so that the frequency spectrum extends up to $\omega' \sim \gamma Mc^2/\hbar$.

The virtual quanta are scattered by the incident particle (the struck system in K') according to the Thomson cross section (14.105) at low frequencies and the Klein-Nishina formula (14.106) at photon energies $\hbar\omega' \gtrsim Mc^2$. Thus, in the frame K', for frequencies small compared to $Mc^2/\hbar$, the radiation cross section $\chi'(\omega')$ is given by

$$\chi'(\omega') \simeq \frac{8\pi}{3}\left(\frac{z^2 e^2}{Mc^2}\right)^2 I(\omega') \tag{15.60}$$

Since the spectrum of virtual quanta extends up to $\gamma Mc^2/\hbar$, the approximation (15.56) can be used for $I(\omega')$ in the region ($\omega' < Mc^2/\hbar$). Thus the radiation cross section becomes

$$\chi'(\omega') \simeq \frac{16}{3}\frac{Z^2 e^2}{c}\left(\frac{z^2 e^2}{Mc^2}\right)^2\left[\ln\left(\frac{\lambda\gamma Mc^2}{\hbar\omega'}\right) - \tfrac{1}{2}\right] \tag{15.61}$$

where extreme relativistic motion ($v \simeq c$) has been assumed.

This is essentially the same cross section as (15.29), and can be transformed to the laboratory in the same way as was done in Section 15.3. Equations (15.60) and (15.61), involving the Thomson cross section, are valid only for quanta in K' with frequencies $\omega' \lesssim Mc^2/\hbar$. For frequencies $\omega' \gtrsim Mc^2/\hbar$, we must replace the constant Thomson cross section (14.105) with the quantum-mechanical Klein-Nishina formula (14.106), which falls off rapidly with increasing frequency. This shows that in K' the bremsstrahlung quanta are confined to a frequency range $0 < \omega' \lesssim Mc^2/\hbar$, even though the spectrum of virtual quanta in the nuclear Coulomb field extends to much higher frequencies. The restricted spectrum in K' is required physically by conservation of energy, since in the laboratory system where $\omega = \gamma\omega'$ the frequency spectrum is limited to $0 < \omega < (\gamma Mc^2/\hbar)$. A detailed treatment using the angular distribution of scattering from the Klein-Nishina formula yields a bremsstrahlung cross section in complete agreement with the Bethe-Heitler formulas (Weizsäcker, 1934).

The effects of screening on the bremsstrahlung spectrum can be discussed in terms of the Weizsäcker-Williams method. For a screened

Coulomb potential the spectrum of virtual quanta is modified from (15.56). The argument of the logarithm is changed to a constant, as was discussed in Section 15.4.

Further applications of the method of virtual quanta to such problems as collisional ionization of atoms and electron disintegration of nuclei are deferred to the problems at the end of the chapter.

15.7 Radiation Emitted during Beta Decay

In the process of beta decay an unstable nucleus with atomic number Z transforms spontaneously into another nucleus of atomic number $(Z \pm 1)$ while emitting an electron ($\mp e$) and a neutrino. The process is written symbolically as

$$Z \rightarrow (Z \pm 1) + e^{\mp} + \nu \qquad (15.62)$$

The energy released in the decay is shared almost entirely by the electron and the neutrino, with the recoiling nucleus getting a completely negligible share because of its very large mass. Even without knowledge of why or how beta decay takes place, we can anticipate that the sudden creation of a rapidly moving charged particle will be accompanied by the emission of radiation. As mentioned in the introduction, either we can think of the electron initially at rest and being accelerated violently during a short time interval to its final velocity, or we can imagine that its charge is suddenly turned on in the same short time interval. The heavy nucleus receives a negligible acceleration and so does not contribute to the radiation.

For the purposes of calculation we can assume that at $t = 0$ an electron is created at the origin with a constant velocity $\mathbf{v} = c\boldsymbol{\beta}$. Then, according to (14.67), the intensity distribution in frequency and angle of the radiation emitted is

$$\frac{dI(\omega)}{d\Omega} = \frac{e^2\omega^2}{4\pi^2 c} \left| \int_0^\infty \mathbf{n} \times (\mathbf{n} \times \boldsymbol{\beta}) e^{i\omega\left(t - \frac{\mathbf{n}\cdot\mathbf{r}(t)}{c}\right)} \, dt \right|^2 \qquad (15.63)$$

Since $\boldsymbol{\beta}$ is constant, the position of the electron is $\mathbf{r}(t) = c\boldsymbol{\beta}t$. Then the intensity distribution is

$$\frac{dI(\omega)}{d\Omega} = \frac{e^2\omega^2}{4\pi^2 c} \beta^2 \sin^2\theta \left| \int_0^\infty e^{i\omega(1 - \beta\cos\theta)t} \, dt \right|^2 \qquad (15.64)$$

where θ is measured from the direction of motion of the emerging electron. Thus the angular distribution is

$$\frac{dI(\omega)}{d\Omega} = \frac{e^2}{4\pi^2 c} \beta^2 \frac{\sin^2\theta}{(1 - \beta\cos\theta)^2} \qquad (15.65)$$

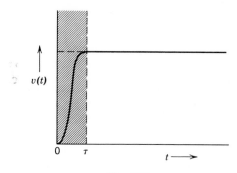

Fig. 15.9

while the total intensity per unit frequency interval is

$$I(\omega) = \frac{e^2}{\pi c}\left[\frac{1}{\beta}\ln\left(\frac{1+\beta}{1-\beta}\right) - 2\right] \qquad (15.66)$$

For $\beta \ll 1$, (15.66) reduces to $I(\omega) \simeq 2e^2\beta^2/3\pi c$, showing that for low-energy beta particles the radiated intensity is negligible.

The intensity distribution (15.66) is a typical bremsstrahlung spectrum with number of photons per unit energy range given by

$$N(\hbar\omega) = \frac{e^2}{\pi\hbar c}\left(\frac{1}{\hbar\omega}\right)\left[\frac{1}{\beta}\ln\left(\frac{1+\beta}{1-\beta}\right) - 2\right] \qquad (15.67)$$

It sometimes bears the name "inner bremsstrahlung" to distinguish it from bremsstrahlung emitted by the same beta particle in passing through matter. It appears that the spectrum extends to infinity, thereby violating conservation of energy. We can obtain qualitative agreement with conservation of energy by appealing to the uncertainty principle. Figure 15.9 shows a qualitative sketch of the electron velocity as a function of time. Our calculation is based on a step function with the acceleration time τ vanishingly small. From the uncertainty principle, however, we know that for a given uncertainty in energy ΔE the uncertainty in time Δt cannot be smaller than $\Delta t \sim \hbar/\Delta E$. In the act of creation of the beta particle, $\Delta E = E = \gamma mc^2$, so that the acceleration time τ must be of the order of $\tau \sim \hbar/E$. When this is transformed into frequency, the well-known arguments show that the frequency spectrum will not extend appreciably beyond $\omega_{max} \sim E/\hbar$, thereby satisfying the conservation of energy requirement at least qualitatively.

The total energy radiated is approximately

$$E_{rad} = \int_0^{\omega_{max}} I(\omega)\,d\omega \simeq \frac{e^2}{\pi\hbar c}\left[\frac{1}{\beta}\ln\left(\frac{1+\beta}{1-\beta}\right) - 2\right]E \qquad (15.68)$$

For very fast beta particles, the ratio of energy going into radiation to the particle energy is

$$\frac{E_{\text{rad}}}{E} \simeq \frac{2}{\pi} \frac{e^2}{\hbar c} \left[\ln \left(\frac{2E}{mc^2} \right) - 1 \right] \qquad (15.69)$$

This shows that the radiated energy is a very small fraction of the total energy released in beta decay, even for the most energetic beta processes ($E_{\text{max}} \sim 30mc^2$). Nevertheless, the radiation can be observed, and provides useful information for nuclear physicists.

In the actual beta process the energy release is shared by the electron and the neutrino so that the electron has a whole spectrum of energies up to some maximum. Then the radiation spectrum (15.66) must be averaged over the energy distribution of the beta particles. Furthermore, a quantum-mechanical treatment leads to modifications near the upper end of the photon spectrum. These are important details for quantitative comparison with experiment. But the origins of the radiation and its semiquantitative description are given adequately by our classical calculation.

15.8 Radiation Emitted in Orbital-Electron Capture—Disappearance of Charge and Magnetic Moment

In beta emission the sudden creation of a fast electron gives rise to radiation. In orbital-electron capture the sudden disappearance of an electron does likewise. Orbital-electron capture is the process whereby an orbital electron around an unstable nucleus of atomic number Z is captured by the nucleus, transforming it into another nucleus with atomic number $(Z - 1)$, with the simultaneous emission of a neutrino which carries off the excess energy. The process can be written symbolically as

$$Z + e^- \rightarrow (Z - 1) + \nu \qquad (15.70)$$

Since a virtually undetectable neutrino carries away the decay energy if there is no radiation, the spectrum of photons accompanying orbital electron capture is of great importance in yielding information about the energy release.

As a simplified model we consider an electron moving in a circular atomic orbit of radius a with a constant angular velocity ω_0. The orbit lies in the x-y plane, as shown in Fig. 15.10, with the nucleus at the center. The observation direction $\mathbf{n}$ is defined by the polar angle θ and lies in the x-z plane. The velocity of the electron is

$$\mathbf{v}(t) = -\boldsymbol{\epsilon}_1 \omega_0 a \sin (\omega_0 t + \alpha) + \boldsymbol{\epsilon}_2 \omega_0 a \cos (\omega_0 t + \alpha) \qquad (15.71)$$

where α is an arbitrary phase angle. If the electron vanishes at $t = 0$, the

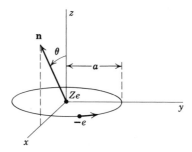

Fig. 15.10

frequency spectrum of emitted radiation (14.67) is approximately

$$\frac{dI(\omega)}{d\Omega} = \frac{e^2\omega^2}{4\pi^2 c^3} \left| \int_{-\infty}^{0} \mathbf{n} \times [\mathbf{n} \times \mathbf{v}(t)] e^{i\omega t}\, dt \right|^2 \qquad (15.72)$$

where we have assumed that $(\omega a/c) \ll 1$ (dipole approximation) and put the retardation factor equal to unity. The integral in (15.72) can be written

$$\int_{-\infty}^{0} dt = -\omega_0 a(\boldsymbol{\epsilon}_\perp I_1 + \boldsymbol{\epsilon}_\| \cos \theta I_2) \qquad (15.73)$$

where

$$\left. \begin{aligned} I_1 &= \int_{-\infty}^{0} \cos(\omega_0 t + \alpha) e^{i\omega t}\, dt \\[2mm] I_2 &= \int_{-\infty}^{0} \sin(\omega_0 t + \alpha) e^{i\omega t}\, dt \end{aligned} \right\} \qquad (15.74)$$

and $\boldsymbol{\epsilon}_\perp$, $\boldsymbol{\epsilon}_\|$ are unit polarization vectors perpendicular and parallel to the plane containing $\mathbf{n}$ and the z axis. The integrals are elementary and lead to an intensity distribution,

$$\frac{dI(\omega)}{d\Omega} = \frac{e^2\omega^2}{4\pi^2 c^3} \frac{\omega_0{}^2 a^2}{(\omega^2 - \omega_0{}^2)^2}$$
$$\times \left[(\omega^2 \cos^2 \alpha + \omega_0{}^2 \sin^2 \alpha) + \cos^2 \theta (\omega^2 \sin^2 \alpha + \omega_0{}^2 \cos^2 \alpha) \right] \quad (15.75)$$

Since the electron can be captured from any position around the orbit, we average over the phase angle α. Then the intensity distribution is

$$\frac{dI(\omega)}{d\Omega} = \frac{e^2}{4\pi^2 c} \left(\frac{\omega_0 a}{c} \right)^2 \frac{\omega^2 (\omega^2 + \omega_0{}^2)}{(\omega^2 - \omega_0{}^2)^2} \tfrac{1}{2}(1 + \cos^2 \theta) \qquad (15.76)$$

The total energy radiated per unit frequency interval is

$$I(\omega) = \frac{2}{3\pi} \frac{e^2}{c} \left(\frac{\omega_0 a}{c} \right)^2 \left[\frac{\omega^2 (\omega_0{}^2 + \omega^2)}{(\omega^2 - \omega_0{}^2)^2} \right] \qquad (15.77)$$

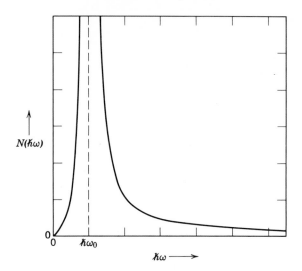

Fig. 15.11 Spectrum of photons emitted in orbital electron capture because of disappearance of the charge of the electron.

while the number of photons per unit energy interval is

$$N(\hbar\omega) = \frac{2}{3\pi}\left(\frac{e^2}{\hbar c}\right)\left(\frac{\omega_0 a}{c}\right)^2\left[\frac{\omega^2(\omega_0^2 + \omega^2)}{(\omega^2 - \omega_0^2)^2}\right]\frac{1}{\hbar\omega} \qquad (15.78)$$

For $\omega \gg \omega_0$ the square-bracketed quantity approaches unity. Then the spectrum is a typical bremsstrahlung spectrum. But for $\omega \simeq \omega_0$ the intensity is very large (infinite in our approximation). The behavior of the photon spectrum is shown in Fig. 15.11. The singularity at $\omega = \omega_0$ may seem alarming, but it is really quite natural and expected. If the electron were to keep orbiting forever, the radiation spectrum would be a sharp line at $\omega = \omega_0$. The sudden termination of the periodic motion produces a broadening of the spectrum in the neighborhood of the characteristic frequency.

Quantum mechanically the radiation arises when an $l = 1$ electron (mainly from the $2p$ orbit) makes a virtual radiative transition to an $l = 0$ state, from which it can be absorbed by the nucleus. Thus the frequency ω_0 must be identified with the frequency of the characteristic $2p \to 1s$ X-ray, $\hbar\omega_0 \simeq (3Z^2 e^2/8a_0)$. Similarly the orbit radius is actually a transitional dipole moment. With the estimate $a \simeq a_0/Z$, where a_0 is the Bohr radius, the photon spectrum (15.78) is

$$N(\hbar\omega) \simeq \frac{3}{32\pi}Z^2\left(\frac{e^2}{\hbar c}\right)^3\frac{1}{\hbar\omega}\left[\frac{\omega^2(\omega^2 + \omega_0^2)}{(\omega^2 - \omega_0^2)^2}\right] \qquad (15.79)$$

The essential characteristics of this spectrum are its strong peaking at the X-ray energy and its dependence on atomic number as Z^2.

So far we have considered the radiation which accompanies the disappearance of the charge of an orbital electron in the electron-capture process. An electron possesses a magnetic moment as well as a charge. The disappearance of the magnetic moment also gives rise to radiation, but with a spectrum of quite different character. The intensity distribution in angle and frequency for a point magnetic moment in motion is given by (14.74). The electronic magnetic moment can be treated as a constant vector in space until its disappearance at $t = 0$. Then, in the dipole approximation, the appropriate intensity distribution is

$$\frac{dI(\omega)}{d\Omega} = \frac{\omega^4}{4\pi^2 c^3} \left| \int_{-\infty}^{0} \mathbf{n} \times \boldsymbol{\mu} e^{i\omega t} \, dt \right|^2 \tag{15.80}$$

This gives

$$\frac{dI(\omega)}{d\Omega} = \frac{\omega^2}{4\pi^2 c^3} \mu^2 \sin^2 \Theta \tag{15.81}$$

where Θ is the angle between $\boldsymbol{\mu}$ and the observation direction $\mathbf{n}$.

In a semiclassical sense the electronic magnetic moment can be thought of as having a magnitude $\mu = \sqrt{3}(e\hbar/2mc)$, but being observed only through its projection $\mu_z = \pm(e\hbar/2mc)$ on an arbitrary axis. The moment can be thought of as precessing around the axis at an angle $\alpha = \tan^{-1}\sqrt{2}$, so that on the average only the component of the moment along the axis survives. It is easy to show that on averaging over this precession $\sin^2 \Theta$ in (15.81) becomes equal to its average value of $\frac{2}{3}$, independent of observation direction. Thus the angular and frequency spectrum becomes

$$\frac{dI(\omega)}{d\Omega} = \frac{e^2}{8\pi^2 c} \left(\frac{\hbar\omega}{mc^2} \right)^2 \tag{15.82}$$

The total energy radiated per unit frequency interval is

$$I(\omega) = \frac{e^2}{2\pi c} \left(\frac{\hbar\omega}{mc^2} \right)^2 \tag{15.83}$$

while the corresponding number of photons per unit energy interval is

$$N(\hbar\omega) = \frac{e^2}{2\pi\hbar c} \frac{\hbar\omega}{(mc^2)^2} \tag{15.84}$$

These spectra are very different in their frequency dependence from a bremsstrahlung spectrum. They increase with increasing frequency, apparently without limit. Of course, we have been forewarned that our classical results are valid only in the low-frequency limit. We can imagine that some sort of uncertainty-principle argument such as was used in Section 15.7 for radiative beta decay holds here and that conservation of energy, at least, is guaranteed. Actually, modifications arise because a neutrino is always emitted in the electron-capture process. The probability of emission of the neutrino can be shown to depend on the square of its energy E_ν. When no photon is emitted, the neutrino has the full decay energy $E_\nu = E_0$. But when a photon of energy $\hbar\omega$ accompanies it, the neutrino's energy is reduced to $E_\nu' = E_0 - \hbar\omega$. Then the probability of neutrino emission is reduced by a factor,

$$\left(\frac{E_\nu'}{E_\nu}\right)^2 = \left(1 - \frac{\hbar\omega}{E_0}\right)^2 \tag{15.85}$$

This means that our classical spectra (15.83) and (15.84) must be corrected by multiplication with (15.85) to take into account the kinematics of the neutrino emission. The modified classical photon spectrum is

$$N(\hbar\omega) = \frac{e^2}{2\pi\hbar c}\frac{\hbar\omega}{(mc^2)^2}\left(1 - \frac{\hbar\omega}{E_0}\right)^2 \tag{15.86}$$

This is essentially the correct quantum-mechanical result. A comparison of the corrected distribution (15.86) and the classical one (15.84) is shown in Fig. 15.12. Evidently the neutrino-emission probability is crucial in obtaining the proper behavior of the photon energy spectrum. For the

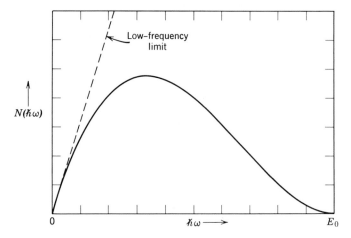

Fig. 15.12 Spectrum of photons emitted in orbital electron capture because of disappearance of the magnetic moment of the electron.

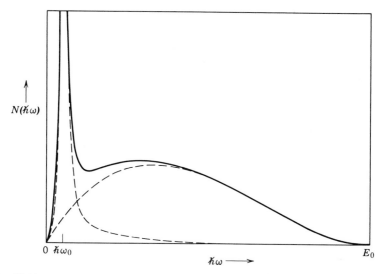

Fig. 15.13 Typical photon spectrum for radiative orbital electron capture with energy release E_0, showing the contributions from the disappearance of the electronic charge and magnetic moment.

customary bremsstrahlung spectra such correction factors are less important because the bulk of the radiation is emitted in photons with energies much smaller than the maximum allowable value.

The total radiation emitted in orbital electron capture is the sum of the contributions from the disappearance of the electric charge and of the magnetic moment. From the different behaviors of (15.79) and (15.86) we see that the upper end of the spectrum will be dominated by the magnetic-moment contribution, unless the energy release is very small, whereas the lower end of the spectrum will be dominated by the electric-charge term, especially for high Z. Figure 15.13 shows a typical combined spectrum for $Z \sim 20$–30. Observations on a number of nuclei confirm the general features of the spectra and allow determination of the energy release E_0.

REFERENCES AND SUGGESTED READING

Classical bremsstrahlung is discussed briefly by
  Landau and Lifshitz, *Classical Theory of Fields*, Section 9.4,
  Panofsky and Phillips, Section 19.6.
A semiclassical discussion, analogous to ours, but much briefer, appears in
  Rossi, Section 2.12.

Bremsstrahlung can be described accurately only by a proper quantum-mechanical treatment. The standard reference is
   Heitler.
The method of virtual quanta (Weizsäcker-Williams method) has only one proper reference, the classic article by
   Williams.
Short discussions appear in
   Heitler, Appendix 6,
   Panofsky and Phillips, Section 18.5.
Among the quantum-mechanical treatments of radiative beta processes, having comparisons with experiment in some cases, are those by
   C. S. W. Chang and D. L. Falkoff, *Phys. Rev.*, **76**, 365 (1949),
   P. C. Martin and R. J. Glauber, *Phys. Rev.*, **109**, 1307 (1958),
   Siegbahn, Chapter XX (III) by C. S. Wu.

PROBLEMS

15.1 A nonrelativistic particle of charge e and mass m collides with a fixed, smooth, hard sphere of radius R. Assuming that the collision is elastic, show that in the dipole approximation (neglecting retardation effects) the classical differential cross section for the emission of photons per unit solid angle per unit energy interval is

$$\frac{d^2\sigma}{d\Omega\, d(\hbar\omega)} = \frac{R^2}{12\pi}\frac{e^2}{\hbar c}\left(\frac{v}{c}\right)^2\frac{1}{\hbar\omega}(2 + 3\sin^2\theta)$$

where θ is measured relative to the incident direction. Sketch the angular distribution. Integrate over angles to get the total bremsstrahlung cross section. Qualitatively what factor (or factors) govern the upper limit to the frequency spectrum?

15.2 Two particles with charges q_1 and q_2 and masses m_1 and m_2 collide under the action of electromagnetic (and perhaps other) forces. Consider the angular and frequency distributions of the radiation emitted in the collision.

(*a*) Show that for nonrelativistic motion the energy radiated per unit solid angle per unit frequency interval in the center of mass coordinate system is given by

$$\frac{dI(\omega, \Omega)}{d\Omega} = \frac{\mu^2}{4\pi^2 c^3}$$

$$\times \left| \int e^{-i\omega t}\ddot{\mathbf{x}} \times \mathbf{n}\left(\frac{q_1}{m_1} e^{i(\omega/c)(\mu/m_1)\mathbf{n}\cdot\mathbf{x}(t)} - \frac{q_2}{m_2} e^{-i(\omega/c)(\mu/m_2)\mathbf{n}\cdot\mathbf{x}(t)}\right) dt \right|^2$$

where $\mathbf{x} = (\mathbf{x}_1 - \mathbf{x}_2)$ is the relative coordinate, $\mathbf{n}$ is a unit vector in the direction of observation, and $\mu = m_1 m_2/(m_1 + m_2)$ is the reduced mass.

(*b*) By expanding the retardation factors show that, if the two particles have the same charge to mass ratio (e.g., a deuteron and an alpha particle), the leading (dipole) term vanishes and the next-order term gives

$$\frac{dI(\omega, \Omega)}{d\Omega} = \frac{\omega^2}{4\pi^2 c^5}\left(\frac{q_1\mu^2}{m_1^2} + \frac{q_2\mu^2}{m_2^2}\right)^2 \left|\int e^{-i\omega t}(\mathbf{n}\cdot\mathbf{x})(\ddot{\mathbf{x}}\times\mathbf{n})\, dt\right|^2$$

(c) Relate result (b) to the multipole expansion of Sections 9.1–9.3.

15.3 Two identical point particles of charge q and mass m interact by means of a short-range repulsive interaction which is equivalent to a hard sphere of radius R in their relative separation. Neglecting the electromagnetic *inter-action* between the two particles, determine the radiation cross section in the center of mass system for a collision between these identical particles to the lowest nonvanishing approximation. Show that the differential cross section for emission of photons per unit solid angle per unit energy interval is

$$\frac{d^2\sigma}{d(\hbar\omega)\,d\Omega} = \frac{R^2}{60\pi}\left(\frac{q^2}{\hbar c}\right)\frac{v^2}{c^2}\left(\frac{R}{\hbar c}\right)^2\hbar\omega[1 + \tfrac{5}{14}P_2(\cos\theta) - \tfrac{3}{28}P_4(\cos\theta)]$$

where θ is measured relative to the incident direction. Compare this result with that of Problem 15.1 as to frequency dependence, relative magnitude, etc.

15.4 A particle of charge ze, mass m, and nonrelativistic velocity v is deflected in a screened Coulomb field, $V(r) = Zze^2e^{-\alpha r}/r$, and consequently emits radiation. Discuss the radiation with the approximation that the particle moves in a straight-line trajectory past the force center.

(a) Show that, if the impact parameter is b, the energy radiated per unit frequency interval is

$$I(\omega, b) = \frac{8}{3\pi}\frac{Z^2e^2}{c}\left(\frac{z^2e^2}{mc^2}\right)^2\left(\frac{c}{v}\right)^2\alpha^2K_1^2(\alpha b)$$

for $\omega \ll v/b$, and negligible for $\omega \gg v/b$.

(b) Show that the radiation cross section is

$$\chi(\omega) \simeq \frac{16}{3}\frac{Z^2e^2}{c}\left(\frac{z^2e^2}{mc^2}\right)^2\left(\frac{c}{v}\right)^2\left[\frac{x^2}{2}\left(K_0^2(x) - K_1^2(x) + \frac{2K_0(x)K_1(x)}{x}\right)\right]_{x_1}^{x_2}$$

where $x_1 = \alpha b_{\min}$, $x_2 = \alpha b_{\max}$.

(c) With $b_{\min} = \hbar/mv$, $b_{\max} = v/\omega$, and $\alpha^{-1} = 1.4a_0Z^{-\frac{1}{3}}$, determine the radiation cross section in the two limits, $x_2 \ll 1$ and $x_2 \gg 1$. Compare your results with the "screening" and "no screening" limits of the text.

15.5 A particle of charge ze, mass m, and velocity v is deflected in a hyperbolic path by a fixed repulsive Coulomb potential, $V(r) = Zze^2/r$. In the non-relativistic dipole approximation (but with no further approximations),

(a) show that the energy radiated per unit frequency interval by the particle when initially incident at impact parameter b is:

$$I(\omega, b) = \frac{8}{3\pi}\frac{(zea\omega)^2}{c^3}e^{-(\pi\omega/\omega_0)}\left\{\left[K'_{i\omega/\omega_0}\left(\frac{\omega\epsilon}{\omega_0}\right)\right]^2 + \frac{\epsilon^2-1}{\epsilon^2}\left[K_{i\omega/\omega_0}\left(\frac{\omega\epsilon}{\omega_0}\right)\right]^2\right\}$$

(b) show that the radiation cross section is:

$$\chi(\omega) = \frac{16}{3}\frac{(zeav)^2}{c^3}e^{-(\pi\omega/\omega_0)}\frac{\omega}{\omega_0}K_{i\omega/\omega_0}\left(\frac{\omega}{\omega_0}\right)\left[-K'_{i\omega/\omega_0}\left(\frac{\omega}{\omega_0}\right)\right]$$

(c) Prove that the radiation cross section reduces to that obtained in the text for classical bremsstrahlung for $\omega \ll \omega_0$. What is the limiting form for $\omega \gg \omega_0$?

(d) What modifications occur for an attractive Coulomb interaction? The hyperbolic path may be described by

$$x = a(\epsilon + \cosh \xi), \quad y = -b \sinh \xi, \quad \omega_0 t = (\xi + \epsilon \sinh \xi)$$

where $a = Zze^2/mv^2$, $\epsilon = \sqrt{1 + (b/a)^2}$, $\omega_0 = v/a$.

15.6 Using the method of virtual quanta, discuss the relationship between the cross section for photodisintegration of a nucleus and electrondisintegration of a nucleus.

(a) Show that, for electrons of energy $E = \gamma mc^2 \gg mc^2$, the electron-disintegration cross section is approximately:

$$\sigma_{\text{el}}(E) \simeq \frac{2}{\pi} \frac{e^2}{\hbar c} \int_{\omega_T}^{E/\hbar} \sigma_{\text{photo}}(\omega) \ln \left(\frac{k\gamma^2 mc^2}{\hbar \omega} \right) \frac{d\omega}{\omega}$$

where $\hbar \omega_T$ is the threshold energy for the process.

(b) Assuming that $\sigma_{\text{photo}}(\omega)$ has the resonance shape:

$$\sigma_{\text{photo}}(\omega) \simeq \frac{A}{2\pi} \frac{e^2}{Mc} \frac{\Gamma}{(\omega - \omega_0)^2 + (\Gamma/2)^2}$$

where the width Γ is small compared to $(\omega_0 - \omega_T)$, sketch the behavior of $\sigma_{\text{el}}(E)$ as a function of E and show that for $E \gg \hbar \omega_0$,

$$\sigma_{\text{el}}(E) \simeq \frac{2}{\pi} \left(\frac{e^2}{\hbar c} \right) \frac{Ae^2}{Mc} \frac{1}{\omega_0} \ln \left(\frac{kE^2}{mc^2 \hbar \omega_0} \right)$$

(c) Discuss the experimental comparison between activities produced by bremsstrahlung spectra and monoenergetic electrons as presented by Brown and Wilson, *Phys. Rev.*, **93**, 443 (1954), and show that the quantity defined as $F_{\text{exp}}(Z, E)$ has the approximate value $8\pi/3$ at high energies if the Weizsäcker-Williams spectrum is used to describe both processes and the photodisintegration cross section has a resonance shape.

15.7 A fast particle of charge ze, mass M, velocity v, collides with a hydrogen-like atom with one electron of charge $-e$, mass m, bound to a nuclear center of charge Ze. The collisions can be divided into two kinds: close collisions where the particle passes through the atom ($b < d$), and distant collisions where the particle passes by outside the atom ($b > d$). The atomic "radius" d can be taken as a_0/Z. For the close collisions the interaction of the incident particle and the electron can be treated as a two-body collision and the energy transfer calculated from the Rutherford cross section. For the distant collisions the excitation and ionization of the atom can be considered the result of the photoelectric effect by the virtual quanta of the incident particle's fields.

For simplicity assume that for photon energies Q greater than the ionization potential I the photoelectric cross section is

$$\sigma_\gamma(Q) = \frac{8\pi^2}{137} \left(\frac{a_0}{Z} \right)^2 \left(\frac{I}{Q} \right)^3$$

(This obeys the empirical $Z^4\lambda^3$ law for X-ray absorption and has a coefficient adjusted to satisfy the dipole sum rule, $\int \sigma_\gamma(Q)\, dQ = 2\pi^2 e^2\hbar/mc$.)

(*a*) Calculate the cross sections for energy transfer Q for close and distant collisions (write them as functions of Q/I as far as possible and in units of $2\pi z^2 e^4/mv^2 I^2$). *Plot* the two distributions for $Q/I > 1$ for non-relativistic motion of the incident particle and $\frac{1}{2}mv^2 = 10^3 I$.

(*b*) Show that the number of distant collisions is much larger than the number of close collisions, but that the energy transfer per collision is much smaller. Show that the energy loss is divided approximately equally between the two kinds of collisions, and verify that your total energy loss is in essential agreement with Bethe's result (13.44).

15.8 In the decay of a pi meson at rest a mu meson and a neutrino are created. The total kinetic energy available is $(m_\pi - m_\mu)c^2 = 34$ Mev. The mu meson has a kinetic energy of 4.1 Mev. Determine the number of quanta emitted per unit energy interval because of the sudden creation of the moving mu meson. Assuming that the photons are emitted perpendicular to the direction of motion of the mu meson (actually it is a $\sin^2\theta$ distribution), show that the maximum photon energy is 17 Mev. Find how many quanta are emitted with energies greater than one-tenth of the maximum, and compare your result with the observed ratio of radiative pi-mu decays. [W. F. Fry, *Phys. Rev.*, **86**, 418 (1952); H. Primakoff, *Phys. Rev.*, **84**, 1255 (1951).]

15.9 In internal conversion, the nucleus makes a transition from one state to another and an orbital electron is ejected. The electron has a kinetic energy equal to the transition energy minus its binding energy. For a conversion line of 1 Mev determine the number of quanta emitted per unit energy because of the sudden ejection of the electron. What fraction of the electrons will have energies less than 99 per cent of the total energy? Will this low-energy tail on the conversion line be experimentally observable?

16

Multipole Fields

In Chapters 3 and 4 on electrostatics the spherical harmonic expansion of the scalar potential was used extensively for problems possessing some symmetry property with respect to an origin of coordinates. Not only was it useful in handling boundary-value problems in spherical coordinates, but with a source present it provided a systematic way of expanding the potential in terms of multipole moments of the charge density. For time-varying electromagnetic fields the scalar spherical harmonic expansion can be generalized to an expansion in vector spherical waves. These vector spherical waves are convenient for electromagnetic boundary-value problems possessing spherical symmetry properties and for the discussion of multipole radiation from a localized source distribution. In Chapter 9 we have already considered the simplest radiating multipole systems. In the present chapter we present a systematic development.

16.1 Basic Spherical Wave Solutions of the Scalar Wave Equation

As a prelude to the vector spherical wave problem, we consider the scalar wave equation. A scalar field $\psi(\mathbf{x}, t)$ satisfying the source-free wave equation,

$$\nabla^2\psi - \frac{1}{c^2}\frac{\partial^2\psi}{\partial t^2} = 0 \tag{16.1}$$

can be Fourier-analyzed in time as

$$\psi(\mathbf{x}, t) = \int_{-\infty}^{\infty} \psi(\mathbf{x}, \omega)e^{-i\omega t}\, d\omega \tag{16.2}$$

with each Fourier component satisfying the Helmholtz wave equation,

$$(\nabla^2 + k^2)\psi(\mathbf{x}, \omega) = 0 \tag{16.3}$$

with $k^2 = \omega^2/c^2$. For problems possessing symmetry properties about some origin it is convenient to have fundamental solutions appropriate to spherical coordinates. The representation of the Laplacian operator in spherical coordinates is given in equation (3.1). The separation of the angular and radial variables follows the well-known expansion,

$$\psi(\mathbf{x}, \omega) = \sum_{l,m} f_l(r) Y_{lm}(\theta, \phi) \tag{16.4}$$

where the spherical harmonics Y_{lm} are defined by (3.53). The radial functions $f_l(r)$ satisfy the radial equation,

$$\left[\frac{d^2}{dr^2} + \frac{2}{r}\frac{d}{dr} + k^2 - \frac{l(l+1)}{r^2} \right] f_l(r) = 0 \tag{16.5}$$

With the substitution,

$$f_l(r) = \frac{1}{r^{1/2}} u_l(r) \tag{16.6}$$

equation (16.5) is transformed into

$$\left[\frac{d^2}{dr^2} + \frac{1}{r}\frac{d}{dr} + k^2 - \frac{(l+\frac{1}{2})^2}{r^2} \right] u_l(r) = 0 \tag{16.7}$$

This equation is just Bessel's equation (3.75) with $\nu = l + \frac{1}{2}$. Thus the solutions for $f_l(r)$ are

$$f_l(r) \simeq \frac{1}{r^{1/2}} J_{l+1/2}(kr), \qquad \frac{1}{r^{1/2}} N_{l+1/2}(kr) \tag{16.8}$$

It is customary to define *spherical Bessel and Hankel functions*, denoted by $j_l(x)$, $n_l(x)$, $h_l^{(1,2)}(x)$, as follows:

$$\left. \begin{aligned} j_l(x) &= \left(\frac{\pi}{2x}\right)^{1/2} J_{l+1/2}(x) \\[2mm] n_l(x) &= \left(\frac{\pi}{2x}\right)^{1/2} N_{l+1/2}(x) \\[2mm] h_l^{(1,2)}(x) &= \left(\frac{\pi}{2x}\right)^{1/2} [J_{l+1/2}(x) \pm i N_{l+1/2}(x)] \end{aligned} \right\} \tag{16.9}$$

For real x, $h_l^{(2)}(x)$ is the complex conjugate of $h_l^{(1)}(x)$. From the series expansions (3.82) and (3.83) one can show that

$$\left. \begin{aligned} j_l(x) &= (-x)^l \left(\frac{1}{x}\frac{d}{dx}\right)^l \left(\frac{\sin x}{x}\right) \\[2mm] n_l(x) &= -(-x)^l \left(\frac{1}{x}\frac{d}{dx}\right)^l \left(\frac{\cos x}{x}\right) \end{aligned} \right\} \tag{16.10}$$

For the first few values of l the explicit forms are:

$$j_0(x) = \frac{\sin x}{x}, \qquad n_0(x) = -\frac{\cos x}{x}, \qquad h_0^{(1)}(x) = \frac{e^{ix}}{ix}$$

$$j_1(x) = \frac{\sin x}{x^2} - \frac{\cos x}{x}, \qquad n_1(x) = -\frac{\cos x}{x^2} - \frac{\sin x}{x}$$

$$h_1^{(1)}(x) = -\frac{e^{ix}}{x}\left(1 + \frac{i}{x}\right)$$

$$j_2(x) = \left(\frac{3}{x^3} - \frac{1}{x}\right)\sin x - \frac{3\cos x}{x^2}, \qquad n_2(x) = -\left(\frac{3}{x^3} - \frac{1}{x}\right)\cos x - 3\frac{\sin x}{x^2}$$

$$h_2^{(1)}(x) = \frac{ie^{ix}}{x}\left(1 + \frac{3i}{x} - \frac{3}{x^2}\right)$$

$$(16.11)$$

From the asymptotic forms (3.89)–(3.91) it is evident that the small argument limits are

$$x \ll l$$

$$j_l(x) \rightarrow \frac{x^l}{(2l+1)!!}$$

$$n_l(x) \rightarrow -\frac{(2l-1)!!}{x^{l+1}}$$

$$(16.12)$$

where $(2l+1)!! = (2l+1)(2l-1)(2l-3)\cdots(5)\cdot(3)\cdot(1)$. Similarly the large argument limits are

$$x \gg l$$

$$j_l(x) \rightarrow \frac{1}{x}\sin\left(x - \frac{l\pi}{2}\right)$$

$$n_l(x) \rightarrow -\frac{1}{x}\cos\left(x - \frac{l\pi}{2}\right)$$

$$h_l^{(1)}(x) \rightarrow (-i)^{l+1}\frac{e^{ix}}{x}$$

$$(16.13)$$

The spherical Bessel functions satisfy the recursion formulas,

$$\frac{2l+1}{x}z_l(x) = z_{l-1}(x) + z_{l+1}(x)$$

$$z_l'(x) = \frac{1}{2l+1}\left[lz_{l-1}(x) - (l+1)z_{l+1}(x)\right]$$

$$(16.14)$$

where $z_l(x)$ is any one of the functions $j_l(x)$, $n_l(x)$, $h_l^{(1)}(x)$, $h_l^{(2)}(x)$. The Wronskians of the various pairs are

$$W(j_l, n_l) = \frac{1}{i} W(j_l, h_l^{(1)}) = -W(n_l, h_l^{(1)}) = \frac{1}{x^2} \qquad (16.15)$$

The general solution of (16.3) in spherical coordinates can be written

$$\psi(\mathbf{x}) = \sum_{l,m} [A_{lm}^{(1)} h_l^{(1)}(kr) + A_{lm}^{(2)} h_l^{(2)}(kr)] Y_{lm}(\theta, \phi) \qquad (16.16)$$

where the coefficients $A_{lm}^{(1)}$ and $A_{lm}^{(2)}$ will be determined by the boundary conditions.

For reference purposes we present the spherical wave expansion for the outgoing wave Green's function $G(\mathbf{x}, \mathbf{x}')$, which is appropriate to the equation,

$$(\nabla^2 + k^2) G(\mathbf{x}, \mathbf{x}') = -\delta(\mathbf{x} - \mathbf{x}') \qquad (16.17)$$

in the infinite domain. The closed form for this Green's function, as was shown in Chapter 9, is

$$G(\mathbf{x}, \mathbf{x}') = \frac{e^{ik|\mathbf{x}-\mathbf{x}'|}}{4\pi|\mathbf{x} - \mathbf{x}'|} \qquad (16.18)$$

The spherical wave expansion for $G(\mathbf{x}, \mathbf{x}')$ can be obtained in exactly the same way as was done in Sections 3.8 and 3.10 for Poisson's equation [see especially equation (3.117) and below, and (3.138) and below]. An expansion of the form,

$$G(\mathbf{x}, \mathbf{x}') = \sum_{l,m} g_l(r, r') Y_{lm}^*(\theta', \phi') Y_{lm}(\theta, \phi) \qquad (16.19)$$

substituted into (16.17) leads to an equation for $g_l(r, r')$:

$$\left[\frac{d^2}{dr^2} + \frac{2}{r} \frac{d}{dr} + k^2 - \frac{l(l+1)}{r^2} \right] g_l = -\frac{1}{r^2} \delta(r - r') \qquad (16.20)$$

The solution which satisfies the boundary conditions of finiteness at the origin and outgoing waves at infinity is

$$g_l(r, r') = A j_l(kr_<) h_l^{(1)}(kr_>) \qquad (16.21)$$

The correct discontinuity in slope is assured if $A = ik$. Thus the expansion of the Green's function is

$$\frac{e^{ik|\mathbf{x}-\mathbf{x}'|}}{4\pi|\mathbf{x} - \mathbf{x}'|} = ik \sum_{l=0}^{\infty} j_l(kr_<) h_l^{(1)}(kr_>) \sum_{m=-l}^{l} Y_{lm}^*(\theta', \phi') Y_{lm}(\theta, \phi) \qquad (16.22)$$

Our emphasis so far has been on the radial functions appropriate to the scalar wave equation. We now re-examine the angular functions in order to introduce some concepts of use in considering the vector wave equation. The basic angular functions are the spherical harmonics $Y_{lm}(\theta, \phi)$ (3.53), which are solutions of the equation,

$$-\left[\frac{1}{\sin\theta}\frac{\partial}{\partial\theta}\left(\sin\theta\frac{\partial}{\partial\theta}\right) + \frac{1}{\sin^2\theta}\frac{\partial^2}{\partial\phi^2}\right]Y_{lm} = l(l+1)Y_{lm} \quad (16.23)$$

As is well known in quantum mechanics, this equation can be written in the form:

$$L^2 Y_{lm} = l(l+1)Y_{lm} \quad (16.24)$$

The differential operator $L^2 = L_x{}^2 + L_y{}^2 + L_z{}^2$, where

$$\mathbf{L} = \frac{1}{i}(\mathbf{r} \times \nabla) \quad (16.25)$$

is the orbital angular-momentum operator of wave mechanics.

The components of $\mathbf{L}$ can be written conveniently in the combinations,

$$\left.\begin{array}{l}
L_+ = L_x + iL_y = e^{i\phi}\left(\dfrac{\partial}{\partial\theta} + i\cot\theta\dfrac{\partial}{\partial\phi}\right) \\[3mm]
L_- = L_x - iL_y = e^{-i\phi}\left(-\dfrac{\partial}{\partial\theta} + i\cot\theta\dfrac{\partial}{\partial\phi}\right) \\[3mm]
L_z = -i\dfrac{\partial}{\partial\phi}
\end{array}\right\} \quad (16.26)$$

We note that $\mathbf{L}$ operates only on angular variables and is independent of r. From definition (16.25) it is evident that

$$\mathbf{r} \cdot \mathbf{L} = 0 \quad (16.27)$$

holds as an operator equation. From the explicit forms (16.26) it is easy to verify that L^2 is equal to the operator on the left side of (16.23).

From the explicit forms (16.26) and recursion relations for Y_{lm} the following useful relations can be established:

$$\left.\begin{array}{l}
L_+ Y_{lm} = \sqrt{(l-m)(l+m+1)}\,Y_{l,m+1} \\[2mm]
L_- Y_{lm} = \sqrt{(l+m)(l-m+1)}\,Y_{l,m-1} \\[2mm]
L_z Y_{lm} = m Y_{lm}
\end{array}\right\} \quad (16.28)$$

Finally we note the following operator equations concerning the commutation properties of $\mathbf{L}$, L^2, and ∇^2:

$$\left.\begin{aligned}
L^2\mathbf{L} &= \mathbf{L}L^2 \\[4pt]
\mathbf{L} \times \mathbf{L} &= i\mathbf{L} \\[4pt]
L_j\nabla^2 &= \nabla^2 L_j
\end{aligned}\right\} \tag{16.29}$$

where

$$\nabla^2 = \frac{1}{r}\frac{\partial^2}{\partial r^2}(r) - \frac{L^2}{r^2} \tag{16.30}$$

16.2 Multipole Expansion of the Electromagnetic Fields

In a source-free region Maxwell's equations are

$$\left.\begin{aligned}
\nabla \times \mathbf{E} &= -\frac{1}{c}\frac{\partial \mathbf{B}}{\partial t}, & \nabla \times \mathbf{B} &= \frac{1}{c}\frac{\partial \mathbf{E}}{\partial t} \\[4pt]
\nabla \cdot \mathbf{E} &= 0, & \nabla \cdot \mathbf{B} &= 0
\end{aligned}\right\} \tag{16.31}$$

With the assumption of a time dependence, $e^{-i\omega t}$, these equations become

$$\left.\begin{aligned}
\nabla \times \mathbf{E} &= ik\mathbf{B}, & \nabla \times \mathbf{B} &= -ik\mathbf{E} \\[4pt]
\nabla \cdot \mathbf{E} &= 0, & \nabla \cdot \mathbf{B} &= 0
\end{aligned}\right\} \tag{16.32}$$

If $\mathbf{E}$ is eliminated between the two curl equations, we obtain the following equations,

$$(\nabla^2 + k^2)\mathbf{B} = 0, \qquad \nabla \cdot \mathbf{B} = 0$$

and the defining relation,

$$\mathbf{E} = \frac{i}{k}\nabla \times \mathbf{B} \tag{16.33}$$

Alternatively $\mathbf{B}$ can be eliminated to yield

$$(\nabla^2 + k^2)\mathbf{E} = 0, \qquad \nabla \cdot \mathbf{E} = 0$$

plus

$$\mathbf{B} = \frac{-i}{k}\nabla \times \mathbf{E} \tag{16.34}$$

Either (16.33) or (16.34) is a set of three equations which is equivalent to Maxwell's equations (16.32).

We now wish to determine multipole solutions for $\mathbf{E}$ and $\mathbf{B}$. From (16.33) it is evident that each *rectangular* component of $\mathbf{B}$ satisfies the Helmholtz wave equation (16.3). Hence each component of $\mathbf{B}$ can be represented by the general solution (16.16). These can be combined to yield the vectorial result:

$$\mathbf{B} = \sum_{l,m} [\mathbf{A}_{lm}^{(1)} h_l^{(1)}(kr) + \mathbf{A}_{lm}^{(2)} h_l^{(2)}(kr)] Y_{lm}(\theta, \phi) \qquad (16.35)$$

where $\mathbf{A}_{lm}$ are arbitrary constant vectors.

The coefficients $\mathbf{A}_{lm}$ in (16.35) are not completely arbitrary. The divergence condition $\nabla \cdot \mathbf{B} = 0$ must be satisfied. Since the radial functions are linearly independent, the condition $\nabla \cdot \mathbf{B} = 0$ must hold for the two sets of terms in (16.35) separately. Thus we require the coefficients $\mathbf{A}_{lm}$ to be so chosen that

$$\nabla \cdot \sum_{l,m} h_l(kr) \mathbf{A}_{lm} Y_{lm}(\theta, \phi) = 0 \qquad (16.36)$$

The gradient operator can be written in the form:

$$\nabla = \frac{\mathbf{r}}{r} \frac{\partial}{\partial r} - \frac{i}{r^2} \mathbf{r} \times \mathbf{L} \qquad (16.37)$$

where $\mathbf{L}$ is the operator (16.25). When this is applied in (16.36), we obtain the requirement,

$$\mathbf{r} \cdot \sum_l \left[\frac{\partial h_l}{\partial r} \sum_m \mathbf{A}_{lm} Y_{lm} - \frac{ih_l}{r} \mathbf{L} \times \sum_m \mathbf{A}_{lm} Y_{lm} \right] = 0 \qquad (16.38)$$

From recursion formulas (16.14) it is evident that in general the coefficients $\mathbf{A}_{lm}$ for a given l will be coupled with those for $l' = l \pm 1$. This will happen unless the $(2l + 1)$ vector coefficients for each l value are such that

$$\mathbf{r} \cdot \sum_m \mathbf{A}_{lm} Y_{lm} = 0 \qquad (16.39)$$

For this special circumstance, the second term in (16.38) shows that the final condition on the coefficients is

$$\mathbf{r} \cdot (\mathbf{L} \times \sum_m \mathbf{A}_{lm} Y_{lm}) = 0 \qquad (16.40)$$

The assumption (16.39) that the field is transverse to the radius vector, together with (16.40), is sufficient to determine a unique set of vector angular functions of order l, one for each m value. These can be found in a straightforward manner from (16.39) and (16.40), and the properties of the Y_{lm}'s. But it is expedient, and not too damaging at this point, to observe that the appropriate angular solution is

$$\sum_{m'} \mathbf{A}_{lm'} Y_{lm'} = \sum_m a_{lm} \mathbf{L} Y_{lm} \qquad (16.41)$$

From (16.27) it is clear that the transversality condition (16.39) is satisfied. Similarly, from the second commutation relation in (16.29) and (16.27), the final condition (16.40) is obeyed. That the functions $[f_l(r)\mathbf{L}\,Y_{lm}]$ satisfy the wave equation (16.3) follows from the last commutation relation in (16.29).

By assumption (16.39) we have found a special set of electromagnetic multipole fields,

$$\left. \begin{array}{l} \mathbf{B}_{lm} = f_l(kr)\mathbf{L}Y_{lm}(\theta,\ \phi) \\[2mm] \mathbf{E}_{lm} = \dfrac{i}{k}\,\boldsymbol{\nabla}\times\mathbf{B}_{lm} \end{array} \right\} \qquad (16.42)$$

where

$$f_l(kr) = A_l^{(1)}h_l^{(1)}(kr) + A_l^{(2)}h_l^{(2)}(kr) \qquad (16.43)$$

Any linear combination of these fields, summed over l and m, satisfies the set of equations (16.33). They have the characteristic that the magnetic induction is perpendicular to the radius vector $(\mathbf{r}\cdot\mathbf{B}_{lm}=0)$. They therefore do not represent a general solution to equations (16.33). They are, in fact, the spherical equivalent of the *transverse magnetic* (TM), or *electric*, cylindrical fields of Chapter 8.

If we had started with the set of equations (16.34) instead of (16.33), we would have obtained an alternative set of multipole fields in which $\mathbf{E}$ is transverse to the radius vector:

$$\left. \begin{array}{l} \mathbf{E}_{lm} = f_l(kr)\mathbf{L}Y_{lm}(\theta,\ \phi) \\[2mm] \mathbf{B}_{lm} = \dfrac{-i}{k}\,\boldsymbol{\nabla}\times\mathbf{E}_{lm} \end{array} \right\} \qquad (16.44)$$

These are the spherical wave analogs of the *transverse electric* (TE), or *magnetic*, cylindrical fields of Chapter 8.

Just as for the cylindrical wave-guide case, the two sets of multipole fields (16.42) and (16.44) can be shown to form a complete set of vector solutions to Maxwell's equations. The terminology electric and magnetic multipole fields will be used, rather than TM and TE, since the sources of each type of field will be seen to be the electric-charge density and the magnetic-moment density, respectively. Since the vector spherical harmonic, $\mathbf{L}Y_{lm}$, plays an important role, it is convenient to introduce the normalized form,*

$$\mathbf{X}_{lm}(\theta,\ \phi) = \frac{1}{\sqrt{l(l+1)}}\,\mathbf{L}Y_{lm}(\theta,\ \phi) \qquad (16.45)$$

with the orthogonality property,

$$\int \mathbf{X}_{l'm'}^{*}\cdot\mathbf{X}_{lm}\,d\Omega = \delta_{ll'}\delta_{mm'} \qquad (16.46)$$

---

* $\mathbf{X}_{lm}$ is defined to be identically zero for $l=0$. Spherically symmetric solutions to the source-free Maxwell's equations exist only in the static limit $k\to 0$.

By combining the two types of fields we can write the general solution to Maxwell's equations (16.32):

$$\left.\begin{aligned}
\mathbf{B} &= \sum_{l,m} \left[a_E(l,m) f_l(kr) \mathbf{X}_{lm} - \frac{i}{k} a_M(l,m) \boldsymbol{\nabla} \times g_l(kr) \mathbf{X}_{lm} \right] \\
\mathbf{E} &= \sum_{l,m} \left[\frac{i}{k} a_E(l,m) \boldsymbol{\nabla} \times f_l(kr) \mathbf{X}_{lm} + a_M(l,m) g_l(kr) \mathbf{X}_{lm} \right]
\end{aligned}\right\} \tag{16.47}$$

where the coefficients $a_E(l,m)$ and $a_M(l,m)$ specify the amounts of electric (l,m) multipole and magnetic (l,m) multipole fields. The radial functions $f_l(kr)$ and $g_l(kr)$ are of form (16.43). The coefficients $a_E(l,m)$ and $a_M(l,m)$, as well as the relative proportions in (16.43), will be determined by the sources and boundary conditions.

16.3 Properties of Multipole Fields; Energy and Angular Momentum of Multipole Radiation

Before considering the connection between the general solution (16.47) and a localized source distribution, we examine the properties of the individual multipole fields (16.42) and (16.44). In the near zone ($kr \ll 1$) the radial function $f_l(kr)$ is proportional to n_l, given by (16.12), unless its coefficient vanishes identically. Excluding this possibility, the limiting behavior of the magnetic induction for an electric (l,m) multipole is

$$\mathbf{B}_{lm} \to - \frac{k}{l} \mathbf{L} \frac{Y_{lm}}{r^{l+1}} \tag{16.48}$$

where the proportionality coefficient is chosen for later convenience. To find the electric field we must take the curl of the right-hand side. A useful operator identity is

$$i\boldsymbol{\nabla} \times \mathbf{L} = \mathbf{r}\nabla^2 - \boldsymbol{\nabla}\left(1 + r\frac{\partial}{\partial r}\right) \tag{16.49}$$

The electric field (16.42) is

$$\mathbf{E}_{lm} \to \frac{-i}{l} \boldsymbol{\nabla} \times \mathbf{L}\left(\frac{Y_{lm}}{r^{l+1}}\right) \tag{16.50}$$

Since (Y_{lm}/r^{l+1}) is a solution of Laplace's equation, the first term in (16.49) vanishes. The second term merely gives a factor l. Consequently the electric field at close distances for an electric (l,m) multipole is

$$\mathbf{E}_{lm} \to - \boldsymbol{\nabla}\left(\frac{Y_{lm}}{r^{l+1}}\right) \tag{16.51}$$

This is exactly the electrostatic multipole field of Section 4.1. We note that the magnetic induction $\mathbf{B}_{lm}$ is smaller in magnitude than $\mathbf{E}_{lm}$ by a factor kr. Hence, in the near zone, the magnetic induction of an electric multipole is always much smaller than the electric field. For the magnetic multipole fields (16.44) evidently the roles of $\mathbf{E}$ and $\mathbf{B}$ are interchanged according to the transformation,

$$\mathbf{E}_E \rightarrow -\mathbf{B}_M, \qquad \mathbf{B}_E \rightarrow \mathbf{E}_M \qquad (16.52)$$

In the far or radiation zone ($kr \gg 1$) the multipole fields depend on the boundary conditions imposed. For definiteness we consider the example of outgoing waves, appropriate to radiation by a localized source. Then the radial function $f_l(kr)$ is proportional to the spherical Hankel function $h_l^{(1)}(kr)$. From the asymptotic form (16.13) we see that in the radiation zone the magnetic induction for an electric (l, m) multipole goes as

$$\mathbf{B}_{lm} \rightarrow (-i)^{l+1} \frac{e^{ikr}}{kr} \mathbf{L} Y_{lm} \qquad (16.53)$$

Then the electric field can be written

$$\mathbf{E}_{lm} = \frac{(-i)^l}{k^2} \left[\nabla \left(\frac{e^{ikr}}{r} \right) \times \mathbf{L} Y_{lm} + \frac{e^{ikr}}{r} \nabla \times \mathbf{L} Y_{lm} \right] \qquad (16.54)$$

Since we have already used the asymptotic form of the spherical Hankel function, we are not justified in keeping higher powers in $(1/r)$ than the first. With this restriction and use of the identity (16.49) we find

$$\mathbf{E}_{lm} = -(-i)^{l+1} \frac{e^{ikr}}{kr} \left[\mathbf{n} \times \mathbf{L} Y_{lm} - \frac{1}{k} (\mathbf{r}\nabla^2 - \nabla) Y_{lm} \right] \qquad (16.55)$$

where $\mathbf{n} = (\mathbf{r}/r)$ is a unit vector in the radial direction. The second term is evidently $1/kr$ times some dimensionless function of angles and can be omitted in the limit $kr \gg 1$. Then we find that the electric field in the radiation zone is

$$\mathbf{E}_{lm} = \mathbf{B}_{lm} \times \mathbf{n} \qquad (16.56)$$

where $\mathbf{B}_{lm}$ is given by (16.53). These fields are typical radiation fields, transverse to the radius vector and falling off as r^{-1}. For magnetic multipoles we merely make the interchanges (16.52).

The multipole fields of a radiating source can be used to calculate the energy and angular momentum carried off by the radiation. For definiteness we consider an electric (l, m) multipole and, following (16.47), write the fields as

$$\left. \begin{aligned} \mathbf{B}_{lm} &= a_E(l, m) h_l^{(1)}(kr) \mathbf{X}_{lm} e^{-i\omega t} \\ \mathbf{E}_{lm} &= \frac{i}{k} \nabla \times \mathbf{B}_{lm} \end{aligned} \right\} \qquad (16.57)$$

For harmonically varying fields the time-averaged energy density is

$$u = \frac{1}{16\pi} (\mathbf{E} \cdot \mathbf{E}^* + \mathbf{B} \cdot \mathbf{B}^*) \qquad (16.58)$$

In the radiation zone the two terms are equal. Consequently the energy in a spherical shell between r and $(r + dr)$ (for $kr \gg 1$) is

$$dU = \frac{|a_E(l, m)|^2}{8\pi} |h_l^{(1)}(kr)|^2 r^2 \, dr \int \mathbf{X}_{lm}^* \cdot \mathbf{X}_{lm} \, d\Omega \qquad (16.59)$$

With the orthogonality integral (16.46) and the asymptotic form (16.13) of the spherical Hankel function, this becomes

$$\frac{dU}{dr} = \frac{|a_E(l, m)|^2}{8\pi k^2} \qquad (16.60)$$

independent of the radius. For a magnetic (l, m) multipole we merely replace $a_E(l, m)$ by $a_M(l, m)$.

The time-averaged angular-momentum density is

$$\mathbf{m} = \frac{1}{8\pi c} \operatorname{Re} [\mathbf{r} \times (\mathbf{E} \times \mathbf{B}^*)] \qquad (16.61)$$

The triple cross product can be expanded and the electric field (16.57) substituted to yield, for electric multipoles,

$$\mathbf{m} = \frac{1}{8\pi\omega} \operatorname{Re} [\mathbf{B}^*(\mathbf{L} \cdot \mathbf{B})] \qquad (16.62)$$

Then the angular momentum in a spherical shell between r and $(r + dr)$ is

$$d\mathbf{M} = \frac{|a_E(l, m)|^2}{8\pi\omega} |h_l^{(1)}(kr)|^2 r^2 \, dr \int \operatorname{Re} [\mathbf{X}_{lm}^* \mathbf{L} \cdot \mathbf{X}_{lm}] \, d\Omega \qquad (16.63)$$

With the explicit form (16.45) for $\mathbf{X}_{lm}$, (16.63) becomes in the radiation zone

$$\frac{d\mathbf{M}}{dr} = \frac{|a_E(l, m)|^2}{8\pi\omega k^2} \int \operatorname{Re} [Y_{lm}^* \mathbf{L} Y_{lm}] \, d\Omega \qquad (16.64)$$

From the properties of $\mathbf{L} Y_{lm}$ listed in (16.28) and the orthogonality of the spherical harmonics we see that only the z-component of $d\mathbf{M}$ exists. It has the value,

$$\frac{dM_z}{dr} = \frac{m}{\omega} \frac{|a_E(l, m)|^2}{8\pi k^2} \qquad (16.65)$$

Comparison with the energy radiated (16.60) shows that the ratio of z component of angular momentum to energy is

$$\frac{M_z}{U} = \frac{m}{\omega} = \frac{m\hbar}{\hbar\omega} \qquad (16.66)$$

This has the obvious quantum interpretation that the radiation from a multipole of order (l, m) carries off $m\hbar$ units of z component of angular momentum per photon of energy $\hbar\omega$. In further analogy with quantum mechanics we would expect the ratio of the magnitude of the angular momentum to the energy to have the value,

$$\frac{M^{(q)}}{U} = \frac{(M_x^2 + M_y^2 + M_z^2)_q^{1/2}}{U} = \frac{\sqrt{l(l + 1)}}{\omega} \qquad (16.67)$$

But from (16.64) and (16.65) the classical result is

$$\frac{M^{(c)}}{U} = \frac{|M_z|}{U} = \frac{|m|}{\omega} \qquad (16.68)$$

The reason for this difference lies in the quantum nature of the electromagnetic fields for a single photon. If the z component of angular momentum of a single photon is known precisely, the uncertainty principle requires that the other components be uncertain, with mean square values such that (16.67) holds. On the other hand, for a state of the radiation field containing many photons (the classical limit) the mean square values of the transverse components of angular momentum can be made negligible compared to the square of the z component. Then the classical limit (16.68) applies.*

The quantum-mechanical interpretation of the radiated angular momentum per photon for multipole fields contains the selection rules for multipole transitions between quantum states. A multipole transition of order (l, m) will connect an initial quantum state specified by total angular momentum J and z component M to a final quantum state with J' in the range $|J - l| \leq J' \leq J + l$ and $M' = M - m$. Or, alternatively, with two states (J, M) and (J', M'), possible multipole transitions have (l, m) such that $|J - J'| \leq l \leq J + J'$ and $m = M - M'$.

To complete the quantum-mechanical specification of a multipole transition it is necessary to state whether the parities of the initial and final states are the same or different. The parity of the initial state is equal to the product of the parities of the final state and the multipole field. To determine the parity of a multipole field we merely examine the behavior of the magnetic induction $\mathbf{B}_{lm}$ under the parity transformation of inversion through the origin $(\mathbf{r} \rightarrow -\mathbf{r})$. One way of seeing that $\mathbf{B}_{lm}$ specifies the parity of a multipole field is to recall that the interaction of a charged particle and the electromagnetic field is proportional to $(\mathbf{v} \cdot \mathbf{A})$. If $\mathbf{B}_{lm}$ has

---

* For a detailed discussion of this point, see C. Morette De Witt and J. H. D. Jensen, *Z. Naturforsch.*, **8a**, 267 (1953). They show that for a multipole field containing N photons the square of the angular momentum is equal to $[N^2m^2 + Nl(l + 1) - m^2]\hbar^2$.

a certain parity (even or odd) for a multipole transition, then the corresponding $\mathbf{A}_{lm}$ will have the opposite parity, since the curl operation changes parity. Then, because $\mathbf{v}$ is a polar vector with odd parity, the states connected by the interaction operator $(\mathbf{v} \cdot \mathbf{A})$ will differ in parity by the parity of the magnetic induction $\mathbf{B}_{lm}$.

For electric multipoles the magnetic induction is given by (16.57). The parity transformation $(\mathbf{r} \to -\mathbf{r})$ is equivalent to $(r \to r,\ \theta \to \pi - \theta,\ \phi \to \phi + \pi)$ in spherical coordinates. The operator $\mathbf{L}$ is invariant under inversion. Consequently the parity properties of $\mathbf{B}_{lm}$ for electric multipoles are specified by the transformation of $Y_{lm}(\theta, \phi)$. From (3.53) and (3.50) it is evident that the parity of Y_{lm} is $(-1)^l$. Thus we see that the *parity* of fields of *an electric multipole of order* (l, m) *is* $(-1)^l$. Specifically, the magnetic induction $\mathbf{B}_{lm}$ has parity $(-1)^l$, while the electric field $\mathbf{E}_{lm}$ has parity $(-1)^{l+1}$, since $\mathbf{E}_{lm} \sim \nabla \times \mathbf{B}_{lm}$.

For a *magnetic multipole of order* (l, m) the *parity is* $(-1)^{l+1}$. In this case the electric field $\mathbf{E}_{lm}$ is of the same form as $\mathbf{B}_{lm}$ for electric multipoles. Hence the parities of the fields are just opposite to those of an electric multipole of the same order.

Correlating the parity changes and angular-momentum changes in quantum transitions, we see that only certain combinations of multipole transitions can occur. For example, if the states have $J = \frac{1}{2}$ and $J' = \frac{3}{2}$, the allowed multipole orders are $l = 1, 2$. If the parities of the two states are the same, we see that parity conservation restricts the possibilities, so that only magnetic dipole and electric quadruple transitions occur. If the states differ in parity, then electric dipole and magnetic quadrupole radiation can be emitted or absorbed.

16.4 Angular Distribution of Multipole Radiation

For a general localized source distribution the fields in the radiation zone are given by the superposition,

$$\mathbf{B} \to \frac{e^{ikr - i\omega t}}{kr} \sum_{l,m} (-i)^{l+1}[a_E(l, m)\mathbf{X}_{lm} + a_M(l, m)\mathbf{n} \times \mathbf{X}_{lm}] \quad (16.69)$$

$$\mathbf{E} \to \mathbf{B} \times \mathbf{n}.$$

The coefficients $a_E(l, m)$ and $a_M(l, m)$ will be related to the properties of the source in the next section. The time-averaged power radiated per unit solid angle is

$$\frac{dP}{d\Omega} = \frac{c}{8\pi k^2} \left| \sum_{l,m} (-i)^{l+1}[a_E(l, m)\mathbf{X}_{lm} \times \mathbf{n} + a_M(l, m)\mathbf{X}_{lm}] \right|^2 \quad (16.70)$$

Within the absolute value signs the polarization of the radiation is specified by the directions of the vectors. We note that electric and magnetic multipoles of a given (l, m) have the same angular dependence, but have polarizations at right angles to one another. Thus the multipole order can be determined by measurement of the angular distribution of radiated power, but the character of the radiation (electric or magnetic) can be determined only by a polarization measurement.

For pure multipole of order (l, m) the angular distribution (16.70) reduces to a single term,

$$\frac{dP(l, m)}{d\Omega} = \frac{c}{8\pi k^2} |a(l, m)|^2 |\mathbf{X}_{lm}|^2 \qquad (16.71)$$

From definition (16.45) of $\mathbf{X}_{lm}$ and properties (16.28), this can be transformed into the explicit form:

$$\frac{dP(l, m)}{d\Omega} = \frac{c\,|a(l, m)|^2}{8\pi k^2 l(l+1)} \left\{ \begin{array}{l} \tfrac{1}{2}(l - m)(l + m + 1)\,|Y_{l,m+1}|^2 \\ + \tfrac{1}{2}(l + m)(l - m + 1)\,|Y_{l,m-1}|^2 + m^2|Y_{lm}|^2 \end{array} \right\} (16.72)$$

The table lists some of the simpler angular distributions.

$$|\mathbf{X}_{lm}(\theta, \phi)|^2$$

l	m		
	0	± 1	± 2
1 Dipole	$\dfrac{3}{8\pi} \sin^2 \theta$	$\dfrac{3}{16\pi} (1 + \cos^2 \theta)$	
2 Quadrupole	$\dfrac{15}{8\pi} \sin^2 \theta \cos^2 \theta$	$\dfrac{5}{16\pi} (1 - 3\cos^2 \theta + 4\cos^4 \theta)$	$\dfrac{5}{16\pi} (1 - \cos^4 \theta)$

The dipole distributions are seen to be those of a dipole oscillating parallel to the z axis ($m = 0$) and of two dipoles, one along the x axis and one along the y axis, 90° out of phase ($m = \pm 1$). The dipole and quadrupole angular distributions are plotted as polar intensity diagrams in Fig. 16.1. These are representative of $l = 1$ and $l = 2$ multipole angular distributions, although a general multipole distribution of order l will involve a coherent superposition of the $(2l + 1)$ amplitudes for different m, as shown in (16.70).

552   *Classical Electrodynamics*

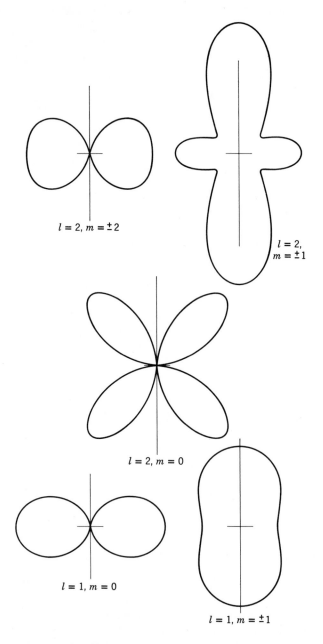

Fig. 16.1   Dipole and quadrupole radiation patterns for pure (l, m) multipoles.

It can be shown by means of (3.69) that the absolute squares of the vector spherical harmonics obey the sum rule,

$$\sum_{m=-l}^{l} |\mathbf{X}_{lm}(\theta, \phi)|^2 = \frac{2l+1}{4\pi} \tag{16.73}$$

Hence the radiation distribution will be isotropic from a source which consists of a set of multipoles of order *l*, with coefficients $a(l, m)$ independent of *m*, superposed incoherently. This situation usually prevails in atomic and nuclear radiative transitions unless the initial state has been prepared in a special way.

The total power radiated by a pure multipole of order (l, m) is given by the integral of (16.71) over all angles. Since the $\mathbf{X}_{lm}$ are normalized to unity, the power radiated is

$$P(l, m) = \frac{c}{8\pi k^2} |a(l, m)|^2 \tag{16.74}$$

For a general source the angular distribution is given by the coherent sum (16.70). On integration over angles it is easy to show that the interference terms do not contribute. Hence the total power radiated is just an incoherent sum of contributions from the different multipoles:

$$P = \frac{c}{8\pi k^2} \sum_{l,m} [|a_E(l, m)|^2 + |a_M(l, m)|^2] \tag{16.75}$$

16.5 Sources of Multipole Radiation; Multipole Moments

Having discussed the properties of multipole fields, the radiation patterns, and the angular momentum and energy carried off, we now turn to the connection of the fields with the sources which generate them. We assume that there exists a localized distribution of charge $\rho(\mathbf{x}, t)$, current $\mathbf{J}(\mathbf{x}, t)$, and intrinsic magnetization $\mathcal{M}(\mathbf{x}, t)$. Furthermore, we assume that the time dependence can be analyzed into its Fourier components, and we consider only harmonically varying sources,

$$\rho(\mathbf{x})e^{-i\omega t}, \quad \mathbf{J}(\mathbf{x})e^{-i\omega t}, \quad \mathcal{M}(\mathbf{x})e^{-i\omega t} \tag{16.76}$$

where it is understood that we take the real part of such complex quantities. A more general time dependence can be obtained by linear superposition.

Since we are considering a magnetization density, we must preserve the distinction between **B** and **H**. Maxwell's equations in the presence of harmonically varying sources can be written

$$\mathbf{\nabla} \times \mathbf{E} = ik\mathbf{B}, \qquad \mathbf{\nabla} \cdot \mathbf{B} = 0$$
$$\mathbf{\nabla} \times \mathbf{H} + ik\mathbf{E} = \frac{4\pi}{c}\mathbf{J}, \qquad \mathbf{\nabla} \cdot \mathbf{E} = 4\pi\rho \tag{16.77}$$

with the continuity equation,

$$i\omega\rho = \mathbf{\nabla} \cdot \mathbf{J} \tag{16.78}$$

In order to make use of the structure of the general solution (16.47) for the source-free Maxwell's equations, we use as field variables,

$$\mathbf{B} \quad \text{and} \quad \mathbf{E}' = \mathbf{E} + \frac{4\pi i}{\omega}\mathbf{J} \tag{16.79}$$

Then the two sets of equations equivalent to (16.33) and (16.34) are

$$(\nabla^2 + k^2)\mathbf{B} = -\frac{4\pi}{c}[\mathbf{\nabla} \times \mathbf{J} + c\mathbf{\nabla} \times (\mathbf{\nabla} \times \mathscr{M})]$$
$$\mathbf{\nabla} \cdot \mathbf{B} = 0, \qquad \mathbf{E}' = \frac{i}{k}(\mathbf{\nabla} \times \mathbf{B} - 4\pi\mathbf{\nabla} \times \mathscr{M}) \tag{16.80}$$

and

$$(\nabla^2 + k^2)\mathbf{E}' = -\frac{4\pi ik}{c}\left[c\mathbf{\nabla} \times \mathscr{M} + \frac{1}{k^2}\mathbf{\nabla} \times (\mathbf{\nabla} \times \mathbf{J})\right]$$
$$\mathbf{\nabla} \cdot \mathbf{E}' = 0, \qquad \mathbf{B} = \frac{-i}{k}\left(\mathbf{\nabla} \times \mathbf{E}' - \frac{4\pi i}{\omega}\mathbf{\nabla} \times \mathbf{J}\right) \tag{16.81}$$

Evidently, outside the source these sets of equations reduce to (16.33) and (16.34). Consequently the general solution for **B** and **E**' outside the source is given by (16.47). Furthermore, even inside the source both fields still have vanishing divergence. Thus the general structure of (16.47) will be preserved, with modifications arising only in the form of the radial functions $f_l(r)$ and $g_l(r)$, in a way that is familiar in scalar field problems such as electrostatics or wave mechanics.

Consider, for example, the magnetic induction,

$$\mathbf{B} = \sum_{l,m}\left[f_{lm}(r)\mathbf{X}_{lm} - \frac{i}{k}\mathbf{\nabla} \times g_{lm}(r)\mathbf{X}_{lm}\right] \tag{16.82}$$

where, outside the source, to conform to the notation of (16.57) and (16.69),

$$f_{lm}(r) \to a_E(l, m)h_l^{(1)}(kr)$$
$$g_{lm}(r) \to a_M(l, m)h_l^{(1)}(kr) \tag{16.83}$$

To determine the equation satisfied by the electric multipole function $f_{lm}(r)$ inside the source, we substitute (16.82) into the first equation of (16.80), take the scalar product of both sides with a typical $\mathbf{X}_{lm}^*$, and integrate over all angles. All the terms on the left-hand side of the equation involving $g_{lm}(r)$ vanish because of orthogonality, and only one term involving an $f_{lm}(r)$ survives:

$$\left[\frac{d^2}{dr^2} + \frac{2}{r}\frac{d}{dr} + k^2 - \frac{l(l+1)}{r^2}\right] f_{lm}(r) =$$

$$- \frac{4\pi}{c}\int \mathbf{X}_{lm}^* \cdot [\mathbf{\nabla} \times \mathbf{J} + c\mathbf{\nabla} \times (\mathbf{\nabla} \times \mathcal{M})]\, d\Omega \quad (16.84)$$

By substituting the equivalent expansion for $\mathbf{E}'$ into the first equation of (16.81) and carrying out the same manipulations, we obtain an equation for $g_{lm}(r)$:

$$\left[\frac{d^2}{dr^2} + \frac{2}{r}\frac{d}{dr} + k^2 - \frac{l(l+1)}{r^2}\right] g_{lm}(r)$$

$$= -4\pi ik \int \mathbf{X}_{lm}^* \cdot \left[\mathbf{\nabla} \times \mathcal{M} + \frac{1}{ck^2}\mathbf{\nabla} \times (\mathbf{\nabla} \times \mathbf{J})\right] d\Omega \quad (16.85)$$

These inhomogeneous equations for $f_{lm}(r)$ and $g_{lm}(r)$ can be solved by means of the Green's function technique. The appropriate Green's function is (16.21), which satisfies (16.20). Denoting the right-hand side of (16.84) as $-K_E(r)$, we can write the solution for $f_{lm}(r)$ in the form:

$$f_{lm}(r) = ik \int_0^\infty r'^2 j_l(kr_<)h_l^{(1)}(kr_>)K_E(r')\, dr' \quad (16.86)$$

Outside the source $r_< = r'$ and $r_> = r$. Then

$$f_{lm}(r) \rightarrow ikh_l^{(1)}(kr) \int_0^\infty r'^2 j_l(kr')K_E(r')\, dr' \quad (16.87)$$

Comparison with (16.83) allows us to identify the electric multipole coefficient $a_E(l, m)$. With the explicit form of $K_E(r)$ from the right side of (16.84), we have

$$a_E(l, m) = \frac{4\pi ik}{c}\int j_l(kr)\mathbf{X}_{lm}^* \cdot [\mathbf{\nabla} \times \mathbf{J} + c\mathbf{\nabla} \times (\mathbf{\nabla} \times \mathcal{M})]\, d^3x$$

$$(16.88)$$

Similarly the magnetic multipole coefficient $a_M(l, m)$ is

$$a_M(l, m) = -4\pi k^2 \int j_l(kr)\mathbf{X}_{lm}^* \cdot \left[\mathbf{\nabla} \times \mathcal{M} + \frac{1}{ck^2}\mathbf{\nabla} \times (\mathbf{\nabla} \times \mathbf{J})\right] d^3x$$

$$(16.89)$$

Results (16.88) and (16.89) can be transformed into more useful forms by means of the following identity:

$$\int j_l(kr)\mathbf{X}_{lm}^* \cdot (\nabla \times \mathbf{A}) \, d^3x$$

$$= \frac{i}{\sqrt{l(l+1)}} \int Y_{lm}^* \left\{ (\nabla \cdot \mathbf{A}) \frac{\partial}{\partial r} [rj_l(kr)] - k^2 \mathbf{r} \cdot \mathbf{A} j_l(kr) \right\} d^3x \quad (16.90)$$

$\mathbf{A}$ is any well-behaved vector field which vanishes at infinity faster than r^{-2}. The proof of (16.90) involves integration by parts to cast the curl operator over on $\mathbf{X}_{lm}$, then use of the operator relation (16.49), and another integration by parts. With $\mathbf{A}$ equal to $\mathbf{J}, \mathcal{M}, \nabla \times \mathbf{J}, \nabla \times \mathcal{M}$, the various terms in (16.88) and (16.89) can be transformed to yield the expressions:

$$a_E(l, m) = \frac{4\pi k^2}{i\sqrt{l(l+1)}} \int Y_{lm}^* \left\{ \begin{array}{l} \rho \dfrac{\partial}{\partial r} [rj_l(kr)] + \dfrac{ik}{c} (\mathbf{r} \cdot \mathbf{J}) j_l(kr) \\[2mm] -ik\nabla \cdot (\mathbf{r} \times \mathcal{M}) j_l(kr) \end{array} \right\} d^3x$$

$$(16.91)$$

and

$$a_M(l, m) = \frac{4\pi k^2}{i\sqrt{l(l+1)}} \int Y_{lm}^* \left\{ \begin{array}{l} \nabla \cdot \left(\dfrac{\mathbf{r} \times \mathbf{J}}{c}\right) j_l(kr) + \nabla \cdot \mathcal{M} \dfrac{\partial}{\partial r} [rj_l(kr)] \\[2mm] -k^2 (\mathbf{r} \cdot \mathcal{M}) j_l(kr) \end{array} \right\} d^3x$$

$$(16.92)$$

These results are the exact multipole coefficients, valid for arbitrary frequency and source size.

For many applications in atomic and nuclear physics the source dimensions are very small compared to a wavelength ($kr_{max} \ll 1$). Then the multipole coefficients can be simplified considerably. The small argument limit (16.12) can be used for the spherical Bessel functions. Keeping only the lowest powers in kr for terms involving ρ or $\mathbf{J}$ and $\mathcal{M}$, we find the approximate electric multipole coefficient,

$$a_E(l, m) \simeq \frac{4\pi k^{l+2}}{i(2l+1)!!} \left(\frac{l+1}{l}\right)^{1/2} (Q_{lm} + Q_{lm}') \quad (16.93)$$

where the multipole moments are

$$Q_{lm} = \int r^l Y_{lm}^* \rho \, d^3x$$

and

$$Q_{lm}' = \frac{-ik}{l+1} \int r^l Y_{lm}^* \nabla \cdot (\mathbf{r} \times \mathcal{M}) \, d^3x$$

$$(16.94)$$

The moment Q_{lm} is seen to be the same in form as the electrostatic multipole moment q_{lm} (4.3). The moment Q_{lm}' is an induced electric multipole moment due to the magnetization. It is generally at least a factor kr smaller than the normal moment Q_{lm}. For the magnetic multipole coefficient $a_M(l, m)$ the corresponding long-wavelength approximation is

$$a_M(l, m) \simeq \frac{4\pi i k^{l+2}}{(2l+1)!!}\left(\frac{l+1}{l}\right)^{\!\!1/2}(M_{lm} + M_{lm}') \qquad (16.95)$$

where the magnetic multipole moments are

$$M_{lm} = -\frac{1}{l+1}\int r^l Y_{lm}^* \boldsymbol{\nabla}\cdot\left(\frac{\mathbf{r}\times\mathbf{J}}{c}\right)d^3x$$

and $\left.\vphantom{\begin{array}{c}a\\a\\a\\a\end{array}}\right\}$ (16.96)

$$M_{lm}' = -\int r^l Y_{lm}^* \boldsymbol{\nabla}\cdot\boldsymbol{\mathscr{M}}\,d^3x$$

In contrast to the electric multipole moments Q_{lm} and Q_{lm}', for a system with intrinsic magnetization the magnetic moments M_{lm} and M_{lm}' are generally of the same order of magnitude.

In the long-wavelength limit we see clearly the fact that electric multipole fields are related to the electric-charge density ρ, while the magnetic multipole fields are determined by the magnetic-moment densities, $(1/2c)(\mathbf{r}\times\mathbf{J})$ and $\boldsymbol{\mathscr{M}}$.

16.6 Multipole Radiation in Atomic and Nuclear Systems

Although a full discussion involves a proper quantum-mechanical treatment of the states involved,* the essential features of multipole radiation in atoms and nuclei can be presented with simple arguments. From (16.74) and the multipole coefficients (16.93) and (16.95), the total power radiated by a multipole of order (l, m) is

$$P_E(l, m) = \frac{2\pi c}{[(2l+1)!!]^2}\left(\frac{l+1}{l}\right)k^{2l+2}\,|Q_{lm} + Q_{lm}'|^2$$

$$\left.\vphantom{\begin{array}{c}a\\a\\a\\a\end{array}}\right\} (16.97)$$

$$P_M(l, m) = \frac{2\pi c}{[(2l+1)!!]^2}\left(\frac{l+1}{l}\right)k^{2l+2}\,|M_{lm} + M_{lm}'|^2$$

---

* See Blatt and Weisskopf, pp. 597–599, for the quantum-mechanical definitions of the multipole moments. Beware of factors of 2 between our moments and theirs, due to their definitions, (3.1) and (3.2) on p. 590, of the source densities, as compared to our (16.76).

In quantum-mechanical terms we are interested in the transition probability (reciprocal mean life), defined as the power divided by the energy of a photon:

$$\frac{1}{\tau} = \frac{P}{\hbar\omega} \tag{16.98}$$

Since we are concerned only with order-of-magnitude estimates, we make the following schematic model of the source. The oscillating charge density is assumed to be

$$\rho(\mathbf{x}) = \begin{cases} \dfrac{3e}{a^3} Y_{lm}(\theta, \phi), & r < a \\ 0, & r > a \end{cases} \tag{16.99}$$

Then an estimate of the electric multipole moment Q_{lm} is

$$Q_{lm} \simeq \frac{3}{l + 3} ea^l \tag{16.100}$$

independent of m. Similarly for the divergences of the magnetizations we assume the schematic form:

$$\nabla \cdot \mathscr{M} + \frac{1}{l + 1} \nabla \cdot \left(\frac{\mathbf{r} \times \mathbf{J}}{c} \right) = \begin{cases} \dfrac{2g}{a^3} Y_{lm}(\theta, \phi) \left(\dfrac{e\hbar}{mcr} \right), & r < a \\ 0, & r > a \end{cases} \tag{16.101}$$

where g is the effective g factor for the magnetic moments of the particles in the atomic or nuclear system, and $e\hbar/mc$ is twice the Bohr magneton for those particles. Then an estimate of the sum of magnetic multipole moments is

$$M_{lm} + M_{lm}' \simeq - \frac{2}{l + 2} ea^l \left(\frac{g\hbar}{mca} \right) \tag{16.102}$$

From the definition of Q_m' (16.94) we see that

$$Q_{lm}' \sim g \left(\frac{\hbar\omega}{mc^2} \right) Q_{lm} \tag{16.103}$$

Since the energies of radiative transitions in atoms and nuclei are always very small compared to the rest energies of the particles involved, Q_{lm}' is always completely negligible compared to Q_{lm}.

For electric multipole transitions of order l, estimate (16.100) leads to a transition probability (16.98):

$$\frac{1}{\tau_E(l)} \simeq \left(\frac{e^2}{\hbar c} \right) \frac{2\pi}{[(2l + 1)!!]^2} \left(\frac{l + 1}{l} \right) \left(\frac{3}{l + 3} \right)^2 (ka)^{2l} \omega \tag{16.104}$$

Apart from factors of the order of unity, the transition probability for magnetic multipoles is, according to (16.102),

$$\frac{1}{\tau_M(l)} \simeq \left(\frac{g\hbar}{mca}\right)^2 \frac{1}{\tau_E(l)} \tag{16.105}$$

The presence of the factor $(ka)^{2l}$ in the transition probability (16.104) means that in the long-wavelength limit ($ka \ll 1$) the transition rate falls off rapidly with increasing multipole order, for a fixed frequency. Consequently in an atomic or nuclear transition the lowest nonvanishing multipole will generally be the only one of importance. The ratio of transition probabilities for successive orders of either electric or magnetic multipoles of the same frequency is

$$\frac{[\tau(l+1)]^{-1}}{[\tau(l)]^{-1}} \sim \frac{(ka)^2}{4l^2} \tag{16.106}$$

where we have omitted numerical factors of relative order $(1/l)$.

In atomic systems the electrons are the particles involved in the radiation process. The dimensions of the source can be taken as $a \sim (a_0/Z_{\rm eff})$, where a_0 is the Bohr radius and $Z_{\rm eff}$ is an effective nuclear charge ($Z_{\rm eff} \sim 1$ for transitions by valence electrons; $Z_{\rm eff} \lesssim Z$ for X-ray transitions). To estimate ka we note that the atomic transition energy is generally of the order

$$\hbar\omega \lesssim Z_{\rm eff}^2 \frac{e^2}{a_0} \tag{16.107}$$

so that

$$ka \lesssim \frac{Z_{\rm eff}}{137} \tag{16.108}$$

From (16.106) we see that successive multipoles will be in the ratio $(Z_{\rm eff}/137)^2$. The ratio of magnetic to electric multipole transition rates can be estimated from (16.105). The g factor is of the order of unity for electrons. With $a \sim a_0/Z_{\rm eff} = 137(\hbar/mcZ_{\rm eff})$, we see that the magnetic lth multipole rate is a factor $(Z_{\rm eff}/137)^2$ smaller than the corresponding electric multipole rate. We conclude that in atoms electric dipole transitions will be most intense, with electric quadrupole and magnetic dipole transitions a factor $(Z_{\rm eff}/137)^2$ weaker. Only for X-ray transitions in heavy elements is there the possibility of competition from other than the lowest-order electric multipole.

We now turn to the question of radiative transitions in atomic nuclei. Because nuclear radiative transition energies vary greatly (from ~ 10 Kev to several Mev), the values of ka cover a wide range. This means that for a given multipole order the transition probabilities (or mean lifetimes)

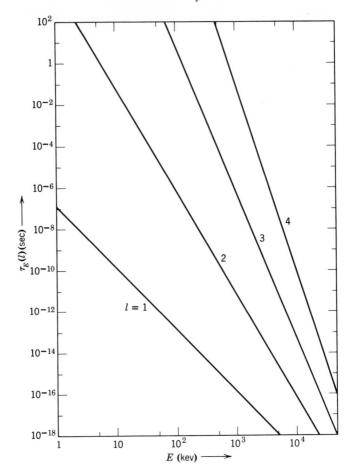

Fig. 16.2 Estimated lifetimes of excited nuclear states against emission of electric multipole radiation as a function of the photon energy for $l = 1, 2, 3, 4$.

will range over many powers of 10, depending on the energy release, overlapping the multipoles on either side. In spite of this, rough estimates (16.104) and (16.105) are useful in cataloging nuclear multipole transitions, because for a *fixed energy release* the estimates for different multipoles differ greatly.

Figure 16.2 shows a log-log plot of estimate (16.104) for lifetimes of electric multipole transitions, using e as the protonic charge and $a \simeq 5.6 \times 10^{-13}$ cm. This is a nuclear radius appropriate to mass number $A \sim 100$. We see that, although the curves tend to converge at high energies, the lifetimes for different multipoles at the same energy differ by factors typically of order 10^5. This means that the actual multipole moments in individual transitions can deviate widely from our simple estimates without vitiating the usefulness of those estimates as a guide in assigning multipole orders. Experimentally, the lifetime-energy diagram shows broad, but well-defined, bands lying in the vicinity of the straight lines in Fig. 16.2. There is a general tendency for estimate (16.104) to serve as a lower bound on the lifetime, corresponding to (16.100) being an upper bound on the multipole moment, but for certain so-called "enhanced" electric quadrupole transitions the lifetimes can be as much as 100 times shorter than shown in Fig. 16.2.

Magnetic and electric multipoles of the same order can be compared using (16.105). For nucleons the effective g factor is typically of the order of $g \sim 3$ because of their anomalous magnetic moments. Then, with the source size estimate $a \sim R = 1.2A^{\frac{1}{3}} \times 10^{-13}$ cm, we find

$$\frac{1}{\tau_M(l)} \sim \frac{0.3}{A^{\frac{2}{3}}} \frac{1}{\tau_E(l)} \tag{16.109}$$

The numerical factor ranges from 4×10^{-2} to 0.8×10^{-2} for $20 < A < 250$. We thus anticipate that for a given multipole order electric transitions will be 25–120 times as intense as magnetic transitions. For most multipoles this is generally true. But for $l = 1$ there are special circumstances in nuclei (strongly attractive, charge-independent forces) which inhibit electric dipole transitions (at least at low energies). Then estimate (16.109) fails; magnetic dipole transitions are far commoner and just as intense as electric dipole transitions.

In Section 16.3 the parity and angular-momentum selection rules were discussed, and it was pointed out that in a transition between two quantum states a mixture of multipoles, such as magnetic $l, (l + 2), \ldots$ pole and electric $(l + 1), (l + 3), \ldots$ pole, could occur, In the long-wavelength limit we need consider only the lowest multipole of each type. Ratios (16.105) and (16.106) can be combined to yield the relative transition rates of electric $(l + 1)$ pole to magnetic l pole (most commonly used for $l = 1$),

$$\frac{[\tau_E(l + 1)]^{-1}}{[\tau_M(l)]^{-1}} \sim \left(\frac{A^{\frac{1}{3}}E}{200l}\right)^2 \tag{16.110}$$

where E is the photon energy in Mev. For energetic transitions in heavy elements the electric quadrupole *amplitude* is ~ 5 per cent of the magnetic dipole amplitude. If, as actually occurs in the rare earth and transuranic elements, there is an enhancement of the effective quadrupole moment by a factor of 10, the electric quadrupole transition competes favorably with the magnetic dipole transition.

For a mixture of magnetic $(l + 1)$ pole and electric l pole, the ratio of transition rates is

$$\frac{[\tau_M(l + 1)]^{-1}}{[\tau_E(l)]^{-1}} \sim \left(\frac{E}{600l}\right)^2 \tag{16.111}$$

Even for energetic transitions, a magnetic $(l + 1)$ pole never comes close to competing with an electric l pole.

16.7   Radiation from a Linear, Center-fed Antenna

As an illustration of the use of a multipole expansion for a source whose dimensions are comparable to a wavelength, we consider the radiation from a thin, linear, center-fed antenna, as shown in Fig. 16.3. We have already given in Chapter 9 a direct solution for the fields when the current distribution is sinusoidal. This will serve as a basis of comparison to test the convergence of the multipole expansion. We assume the antenna to lie along the z axis from $-(d/2) \leq z \leq (d/2)$, and to have a small gap at its center so that it can be suitably excited. The current along the antenna vanishes at the end points and is an even function of z. For the moment we will not specify it more than to write

$$I(z, t) = I(|z|)e^{-i\omega t}, \qquad I\left(\frac{d}{2}\right) = 0 \tag{16.112}$$

Since the current flows radially, $(\mathbf{r} \times \mathbf{J}) = 0$. Furthermore there is no

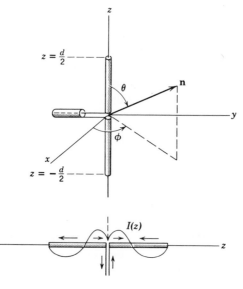

Fig. 16.3   Linear, center-fed antenna.

intrinsic magnetization. Consequently all magnetic multipole coefficients $a_M(l, m)$ vanish. To calculate the electric multipole coefficient $a_E(l, m)$ (16.91) we need expressions for charge and current densities. The current density **J** is a radial current, confined to the z axis. In spherical coordinates this can be written for $r < (d/2)$

$$\mathbf{J}(\mathbf{x}) = \boldsymbol{\epsilon}_r \frac{I(r)}{2\pi r^2} [\delta(\cos\theta - 1) - \delta(\cos\theta + 1)] \qquad (16.113)$$

where the delta functions cause the current to flow only upwards (or downwards) along the z axis. From the continuity equation (16.78) we find the charge density,

$$\rho(\mathbf{x}) = \frac{1}{i\omega} \frac{dI(r)}{dr} \left[\frac{\delta(\cos\theta - 1) - \delta(\cos\theta + 1)}{2\pi r^2} \right] \qquad (16.114)$$

These expressions for **J** and ρ can be inserted into (16.91) to give

$$a_E(l, m) = \frac{2k^2}{\sqrt{l(l + 1)}} \int_0^{d/2} dr \left\{ \frac{k}{c} rj_l(kr)I(r) - \frac{1}{\omega} \frac{dI}{dr} \frac{d}{dr} [rj_l(kr)] \right\}$$

$$\times \int d\Omega\, Y_{lm}^*[\delta(\cos\theta - 1) - \delta(\cos\theta + 1)] \qquad (16.115)$$

The integral over angles is

$$\int d\Omega = 2\pi \delta_{m,0} [Y_{l0}(0) - Y_{l0}(\pi)] \qquad (16.116)$$

showing that only $m = 0$ multipoles occur. This is obvious from the cylindrical symmetry of the antenna. The Legendre polynomials are even (odd) about $\theta = \pi/2$ for l even (odd). Hence, the only nonvanishing multipoles have l odd. Then the angular integral has the value,

$$\int d\Omega = \sqrt{4\pi(2l + 1)}, \qquad l\ \text{odd},\ m = 0 \qquad (16.117)$$

With slight manipulation (16.115) can be written

$$a_E(l, 0) = \frac{2k}{c} \left[\frac{4\pi(2l + 1)}{l(l + 1)} \right]^{\frac{1}{2}} \int_0^{d/2} \left\{ -\frac{d}{dr} \left[rj_l(kr) \frac{dI}{dr} \right] \right.$$

$$\left. + rj_l(kr) \left(\frac{d^2I}{dr^2} + k^2I \right) \right\} dr \qquad (16.118)$$

To evaluate (16.118) we must specify the current $I(z)$ along the antenna. If no radiation occurred, the sinusoidal variation in time at frequency ω would imply a sinusoidal variation in space with wave number $k = \omega/c$.

The emission of radiation modifies this somewhat. To find the correct spatial variation of the current along the antenna we would have to consider the whole boundary-value problem of antenna plus radiation fields. This is a difficult task which must be faced if one wishes precise answers to antenna problems. Fortunately neglect of the effects of the radiation on the current distribution is not serious. Reasonably good answers are obtained with the sinusoidal approximation. Accordingly we assume

$$I(z) = I \sin\left(\frac{kd}{2} - k\,|z|\right) \qquad (16.119)$$

where I is the peak current, and the phase is so chosen that the current vanishes at the ends of the antenna. With a sinusoidal current the second part of the integrand in (16.118) vanishes. The first part is a perfect differential. Consequently we immediately obtain, with $I(z)$ from (16.119),

$$a_E(l, 0) = \frac{4I}{cd}\left[\frac{4\pi(2l+1)}{l(l+1)}\right]^{\frac{1}{2}}\left[\left(\frac{kd}{2}\right)^2 j_l\left(\frac{kd}{2}\right)\right], \qquad l \text{ odd} \quad (16.120)$$

Since we wish to test the multipole expansion when the source dimensions are comparable to a wavelength, we consider the special cases of a half-wave antenna $(kd = \pi)$ and a full-wave antenna $(kd = 2\pi)$. For these two values of kd the $l = 1$ coefficient is tabulated, along with the relative values for $l = 3, 5$. From the table it is evident that (*a*) the coefficients decrease

kd	$a_E(1, 0)$	$a_E(3, 0)/a_E(1, 0)$	$a_E(5, 0)/a_E(1, 0)$
π	$4\sqrt{6\pi}\,\dfrac{I}{cd}$	4.95×10^{-2}	1.02×10^{-3}
2π	$4\pi\sqrt{6\pi}\,\dfrac{I}{cd}$	0.325	3.09×10^{-2}

rapidly in magnitude as l increases, and (*b*) higher l coefficients are more important the larger the source dimensions. But even for the full-wave antenna it is probably adequate to keep only $l = 1$ and $l = 3$ in the angular distribution and certainly adequate for the total power (which involves the squares of the coefficients).

With only dipole and octupole terms in the angular distribution we find that the power radiated per unit solid angle (16.70) is

$$\frac{dP}{d\Omega} = \frac{c\,|a_E(1, 0)|^2}{16\pi k^2}\left|\mathbf{L}Y_{1,0} - \frac{a_E(3, 0)}{\sqrt{6}\,a_E(1, 0)}\mathbf{L}Y_{3,0}\right|^2 \qquad (16.121)$$

The various factors in the absolute square are

$$|\mathbf{L}Y_{1,0}|^2 = \frac{3}{4\pi} \sin^2 \theta$$

$$|\mathbf{L}Y_{3,0}|^2 = \frac{63}{16\pi} \sin^2 \theta\, (5 \cos^2 \theta - 1)^2 \qquad (16.122)$$

$$(\mathbf{L}Y_{1,0})^* \cdot (\mathbf{L}Y_{3,0}) = \frac{3\sqrt{21}}{8\pi} \sin^2 \theta(5 \cos^2 \theta - 1)$$

With these angular factors (16.121) becomes

$$\frac{dP}{d\Omega} = \lambda\, \frac{12I^2}{\pi^2 c} \left(\frac{3}{8\pi} \sin^2 \theta\right) \left| 1 - \sqrt{\frac{7}{8}} \frac{a_E(3,0)}{a_E(1,0)} (5 \cos^2 \theta - 1) \right|^2 \quad (16.123)$$

where the factor λ is equal to 1 for the half-wave antenna and $(\pi^2/4)$ for the full wave. The coefficient of $(5 \cos^2 \theta - 1)$ in (16.123) is 0.0463 and 0.304 for the half-wave and full-wave antenna, respectively.

From Chapter 9 the exact angular distributions (for sinusoidal driving currents) are

$$\frac{dP}{d\Omega} = \frac{I^2}{2\pi c} \begin{cases} \dfrac{\cos^2\left(\dfrac{\pi}{2}\cos\theta\right)}{\sin^2\theta}, & kd=\pi \\[4mm] 4\dfrac{\cos^4\left(\dfrac{\pi}{2}\cos\theta\right)}{\sin^2\theta}, & kd=2\pi \end{cases} \qquad (16.124)$$

A numerical comparison of the exact and approximate angular distributions is shown in Fig. 16.4. The solid curves are the exact results, the dashed curves the two-term multipole expansions. For the half-wave case (Fig. 16.4a) the simple dipole result [first term in (16.123)] is also shown as a dotted curve. The two-term multipole expansion is almost indistinguishable from the exact result for $kd = \pi$. Even the lowest-order approximation is not very far off in this case. For the full-wave antenna (Fig. 16.4b) the dipole approximation is evidently quite poor. But the two-term multipole expansion is reasonably good, differing by less than 5 per cent in the region of appreciable radiation.

The total power radiated is, according to (16.75),

$$P = \frac{c}{8\pi k^2} \sum_{l\,\text{odd}} |a_E(l,0)|^2 \qquad (16.125)$$

For the half-wave antenna the coefficients in the table on p. 564 show that the power radiated is a factor 1.00245 times larger than the simple dipole

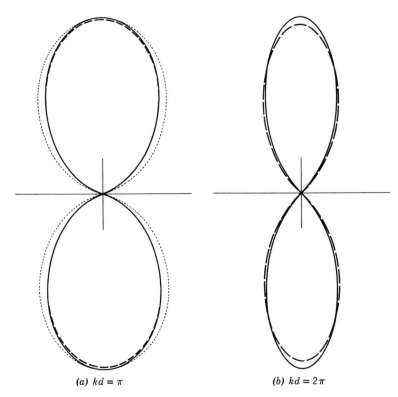

(a) $kd = \pi$ (b) $kd = 2\pi$

Fig. 16.4 Comparison of exact radiation patterns (solid curves) for half-wave ($kd = \pi$) and full-wave ($kd = 2\pi$) center-fed antennas with two-term multipole expansions (dashed curves). For the half-wave pattern, the dipole approximation (dotted curve) is also shown. The agreement between the exact and two-term multipole results is excellent, especially for $kd = \pi$.

result, $(12I^2/\pi^2c)$. For the full-wave antenna, the power is a factor 1.114 times larger than the dipole form $(3I^2/c)$.

16.8   Spherical Wave Expansion of a Vector Plane Wave

In discussing the scattering or absorption of electromagnetic radiation by spherical objects, or localized systems in general, it is useful to have an expansion of a plane electromagnetic wave in spherical waves.

For a scalar field $\psi(\mathbf{x})$ satisfying the wave equation the necessary expansion can be obtained by using the orthogonality properties of the basic spherical solutions $j_l(kr)Y_{lm}(\theta, \phi)$. An alternative derivation makes use of the spherical wave expansion (16.22) of the Green's function $(e^{ikR}/4\pi R)$. We let $|\mathbf{x}'| \to \infty$ on both sides of (16.22). Then we can put $|\mathbf{x} - \mathbf{x}'| \simeq r' - \mathbf{n} \cdot \mathbf{x}$ on the left-hand side, where $\mathbf{n}$ is a unit vector in the direction of $\mathbf{x}'$. On the right side $r_> = r'$ and $r_< = r$. Furthermore we can use the asymptotic form (16.13) for $h_l^{(1)}(kr')$. Then we find

$$\frac{e^{ikr'}}{4\pi r'}e^{-ik\mathbf{n}\cdot\mathbf{x}} = ik\frac{e^{ikr'}}{kr'}\sum_{l,\,m}(-i)^{l+1}j_l(kr)Y_{lm}^*(\theta',\phi')Y_{lm}(\theta,\phi) \quad (16.126)$$

Canceling the factor $e^{ikr'}/r'$ on either side and taking the complex conjugate, we have the expansion of a plane wave,

$$e^{i\mathbf{k}\cdot\mathbf{x}} = 4\pi\sum_{l=0}^{\infty}i^l j_l(kr)\sum_{m=-l}^{l}Y_{lm}^*(\theta,\phi)Y_{lm}(\theta',\phi') \quad (16.127)$$

where $\mathbf{k}$ is the wave vector with spherical coordinates k, θ', ϕ'. The addition theorem (3.62) can be used to put this in a more compact form,

$$e^{i\mathbf{k}\cdot\mathbf{x}} = \sum_{l=0}^{\infty}i^l(2l+1)j_l(kr)P_l(\cos\gamma) \quad (16.128)$$

where γ is the angle between $\mathbf{k}$ and $\mathbf{x}$. With (3.57) for $P_l \cos(\gamma)$, this can also be written as

$$e^{i\mathbf{k}\cdot\mathbf{x}} = \sum_{l=0}^{\infty}i^l\sqrt{4\pi(2l+1)}j_l(kr)Y_{l,0}(\gamma) \quad (16.129)$$

We now wish to make an equivalent expansion for a circularly polarized plane wave incident along the z axis,

$$\left.\begin{array}{l}\mathbf{E}(\mathbf{x}) = (\boldsymbol{\epsilon}_1 \pm i\boldsymbol{\epsilon}_2)e^{ikz} \\[2mm] \mathbf{B}(\mathbf{x}) = \boldsymbol{\epsilon}_3 \times \mathbf{E} = \mp i\mathbf{E}\end{array}\right\} \quad (16.130)$$

Since the plane wave is finite everywhere, we can write its multipole expansion (16.47) involving only the regular radial functions $j_l(kr)$:

$$\left.\begin{array}{l}\mathbf{E}(\mathbf{x}) = \displaystyle\sum_{l,\,m}\left[a_\pm(l,m)j_l(kr)\mathbf{X}_{lm} + \frac{i}{k}b_\pm(l,m)\boldsymbol{\nabla} \times j_l(kr)\mathbf{X}_{lm}\right] \\[4mm] \mathbf{B}(\mathbf{x}) = \displaystyle\sum_{l,\,m}\left[\frac{-i}{k}a_\pm(l,m)\boldsymbol{\nabla} \times j_l(kr)\mathbf{X}_{lm} + b_\pm(l,m)j_l(kr)\mathbf{X}_{lm}\right]\end{array}\right\} \quad (16.131)$$

To determine the coefficients $a_\pm(l,m)$ and $b_\pm(l,m)$ we utilize the orthogonality properties of the vector spherical harmonics $\mathbf{X}_{lm}$. For reference purposes we summarize the basic relation (16.46), as well as some other useful relations:

$$
\left.
\begin{aligned}
&\int [f_l(r)\mathbf{X}_{l'm'}]^* \cdot [g_l(r)\mathbf{X}_{lm}] \, d\Omega = f_l^* g_l \, \delta_{ll'} \delta_{mm'} \\[2mm]
&\int [f_l(r)\mathbf{X}_{l'm'}]^* \cdot [\boldsymbol{\nabla} \times g_l(r)\mathbf{X}_{lm}] \, d\Omega = 0 \\[2mm]
&\frac{1}{k^2} \int [\boldsymbol{\nabla} \times f_l(r)\mathbf{X}_{l'm'}]^* \cdot [\boldsymbol{\nabla} \times g_l(r)\mathbf{X}_{lm}] \, d\Omega \\[2mm]
&\quad = \delta_{ll'}\delta_{mm'} \left\{ f_l^* g_l + \frac{1}{k^2 r^2} \frac{\partial}{\partial r} \left[r f_l^* \frac{\partial}{\partial r} (r g_l) \right] \right\}
\end{aligned}
\right\}
\qquad (16.132)
$$

In these relations $f_l(r)$ and $g_l(r)$ are linear combinations of spherical Bessel functions, satisfying (16.5). The second and third relations can be proved using the operator identity (16.49), the representation (16.37) for the gradient operator, and the radial differential equation (16.5).

To determine the coefficients $a_\pm(l, m)$ and $b_\pm(l, m)$ we take the scalar product of both sides of (16.131) with $\mathbf{X}_{lm}^*$ and integrate over angles. Then with the first and second orthogonality relations in (16.132) we obtain

$$
a_\pm(l, m) j_l(kr) = \int \mathbf{X}_{lm}^* \cdot \mathbf{E}(\mathbf{x}) \, d\Omega \qquad (16.133)
$$

and

$$
b_\pm(l, m) j_l(kr) = \int \mathbf{X}_{lm}^* \cdot \mathbf{B}(\mathbf{x}) \, d\Omega \qquad (16.134)
$$

With (16.130) for the electric field, (16.133) becomes

$$
a_\pm(l, m) j_l(kr) = \int \frac{(L_\mp Y_{lm})^*}{\sqrt{l(l + 1)}} \, e^{ikz} \, d\Omega \qquad (16.135)
$$

where the operators $L_\pm$ are defined by (16.26), and the results of their operating by (16.28). Thus we obtain

$$
a_\pm(l, m) j_l(kr) = \frac{\sqrt{(l \pm m)(l \mp m + 1)}}{\sqrt{l(l + 1)}} \int Y_{l, m \mp 1}^* e^{ikz} \, d\Omega \qquad (16.136)
$$

If expansion (16.129) for e^{ikz} is inserted, the orthogonality of the Y_{lm}'s evidently leads to the result,

$$
a_\pm(l, m) = i^l \sqrt{4\pi(2l + 1)} \, \delta_{m, \pm 1} \qquad (16.137)
$$

From (16.134) and (16.130) it is clear that

$$
b_\pm(l, m) = \mp i a_\pm(l, m) \qquad (16.138)
$$

Then the multipole expansion of the plane wave (16.130) is

$$\mathbf{E}(\mathbf{x}) = \sum_{l=1}^{\infty} i^l \sqrt{4\pi(2l+1)} \left[j_l(kr)\mathbf{X}_{l,\,\pm 1} \pm \frac{1}{k}\boldsymbol{\nabla} \times j_l(kr)\mathbf{X}_{l,\,\pm 1} \right]$$

$$(16.139)$$

$$\mathbf{B}(\mathbf{x}) = \sum_{l=1}^{\infty} i^l \sqrt{4\pi(2l+1)} \left[\frac{-i}{k}\boldsymbol{\nabla} \times j_l(kr)\mathbf{X}_{l,\,\pm 1} \mp ij_l(kr)\mathbf{X}_{l,\,\pm 1} \right]$$

For such a circularly polarized wave the m values of $m = \pm 1$ have the obvious interpretation of ± 1 unit of angular momentum per photon parallel to the propagation direction. This has already been established in Problem 6.12.

16.9 Scattering of Electromagnetic Waves by a Conducting Sphere

If a plane wave of electromagnetic radiation is incident on a spherical obstacle, as indicated schematically in Fig. 16.5, it is scattered so that far away from the scatterer the fields are represented by a plane wave plus outgoing spherical waves. There may be absorption by the obstacle as well as scattering. Then the total energy flow away from the obstacle will be less than the total energy flow towards it, the difference being absorbed. We will consider the simple example of scattering by a sphere of radius a and infinite conductivity.

The fields outside the sphere can be written as a sum of incident and scattered waves:

$$\left.\begin{aligned} \mathbf{E}(\mathbf{x}) &= \mathbf{E}_{\text{inc}} + \mathbf{E}_{\text{sc}} \\ \mathbf{B}(\mathbf{x}) &= \mathbf{B}_{\text{inc}} + \mathbf{B}_{\text{sc}} \end{aligned}\right\} \qquad (16.140)$$

where $\mathbf{E}_{\text{inc}}$ and $\mathbf{B}_{\text{inc}}$ are given by (16.139). Since the scattered fields are outgoing waves at infinity, their expansions must be of the form,

$$\mathbf{E}_{\text{sc}} = \frac{1}{2}\sum_{l=1}^{\infty} i^l \sqrt{4\pi(2l+1)} \left[\alpha_{\pm}(l)h_l^{(1)}(kr)\mathbf{X}_{l,\,\pm 1} \right.$$

$$\left. \pm \frac{\beta_{\pm}(l)}{k}\boldsymbol{\nabla} \times h_l^{(1)}(kr)\mathbf{X}_{l,\,\pm 1} \right]$$

$$(16.141)$$

$$\mathbf{B}_{\text{sc}} = \frac{1}{2}\sum_{l=1}^{\infty} i^l \sqrt{4\pi(2l+1)} \left[\frac{-i\alpha_{\pm}(l)}{k}\boldsymbol{\nabla} \times h_l^{(1)}(kr)\mathbf{X}_{l,\,\pm 1} \right.$$

$$\left. \mp i\beta_{\pm}(l)h_l^{(1)}(kr)\mathbf{X}_{l,\,\pm 1} \right]$$

The coefficients $\alpha_{\pm}(l)$ and $\beta_{\pm}(l)$ will be determined by the boundary conditions on the surface of the sphere.

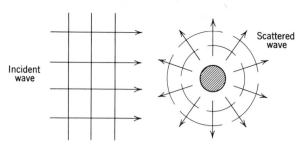

Fig. 16.5 Scattering of radiation by a localized object.

For a perfectly conducting sphere the boundary conditions at the surface $r = a$ are

$$\mathbf{n} \times \mathbf{E} = 0, \qquad \mathbf{n} \cdot \mathbf{B} = 0 \tag{16.142}$$

In order to apply these boundary conditions we must know the vectorial character of the two types of terms in (16.141). We already know that $\mathbf{X}_{lm}$ is perpendicular to the radius vector. The other type of term is

$$\nabla \times f_l(r)\mathbf{X}_{lm} = \frac{i n \sqrt{l(l + 1)}}{r} f_l(r) Y_{lm} + \frac{1}{r} \frac{\partial}{\partial r} [r f_l(r)] \mathbf{n} \times \mathbf{X}_{lm} \tag{16.143}$$

where f_l is any spherical Bessel function satisfying (16.5). Applying the boundary conditions (16.142) to the *total* fields (16.140), we find conditions on the coefficients $\alpha_{\pm}(l)$ and $\beta_{\pm}(l)$:

$$\left. \begin{array}{r} \tfrac{1}{2}\alpha_{\pm}(l)h_l^{(1)}(ka) + j_l(ka) = 0 \\[2mm] \left\{ \tfrac{1}{2}\beta_{\pm}(l) \dfrac{\partial}{\partial r}\left[rh_l^{(1)}(kr)\right] + \dfrac{\partial}{\partial r}\left[rj_l(kr)\right] \right\}_{r=a} = 0 \end{array} \right\} \tag{16.144}$$

We note that the coefficients are the same for both states of circular polarization. Since $2j_l(kr) = h_l^{(1)}(kr) + h_l^{(2)}(kr)$, the coefficients can be written

$$\left. \begin{array}{l} \alpha_{\pm}(l) = -\left[\dfrac{h_l^{(2)}(ka)}{h_l^{(1)}(ka)} + 1\right] \\[6mm] \beta_{\pm}(l) = -\left\{ \left. \dfrac{\dfrac{d}{dr}\left[rh_l^{(2)}(kr)\right]}{\dfrac{d}{dr}\left[rh_l^{(1)}(kr)\right]} \right|_{r=a} + 1 \right\} \end{array} \right\} \tag{16.145}$$

The ratios are ratios of complex conjugate quantities and so are complex numbers of absolute value unity. It is convenient to define two angles, called *phase shifts*, as follows:

$$
\left.
\begin{aligned}
e^{2i\delta_l} &= -\frac{h_l^{(2)}(ka)}{h_l^{(1)}(ka)} \\[2ex]
e^{2i\delta_{l'}} &= -\left.\frac{\dfrac{d}{dr}\left[rh_l^{(2)}(kr)\right]}{\dfrac{d}{dr}\left[rh_l^{(1)}(kr)\right]}\right|_{r=a}
\end{aligned}
\right\}
\tag{16.146}
$$

or alternatively,

$$
\left.
\begin{aligned}
\tan\delta_l &= \frac{j_l(ka)}{n_l(ka)} \\[2ex]
\tan\delta_{l'} &= \left.\frac{\dfrac{d}{dr}\left[rj_l(kr)\right]}{\dfrac{d}{dr}\left[rn_l(kr)\right]}\right|_{r=a}
\end{aligned}
\right\}
\tag{16.147}
$$

Then the coefficients are

$$
\alpha_{\pm}(l) = (e^{2i\delta_l} - 1), \qquad \beta_{\pm}(l) = (e^{2i\delta_{l'}} - 1) \tag{16.148}
$$

The asymptotic forms (16.12) and (16.13) can be used to find limiting values of these phase shifts for $ka \ll 1$ and $ka \gg 1$:

$ka \ll 1$:

$$
\left.
\begin{aligned}
\delta_l &\to -\frac{(ka)^{2l+1}}{(2l+1)[(2l-1)!\,!]^2} \\[2ex]
\delta_{l'} &\to -\left(\frac{l+1}{l}\right)\delta_l
\end{aligned}
\right\}
\tag{16.149}
$$

$ka \gg 1$:

$$
\left.
\begin{aligned}
\delta_l &\to \frac{l\pi}{2} - ka \\[2ex]
\delta_{l'} &\to (l+1)\frac{\pi}{2} - ka
\end{aligned}
\right\}
\tag{16.150}
$$

The coefficient $\alpha(l)$ and the phase shift δ_l can be termed *magnetic* parameters since they relate to the magnetic multipole fields in (16.141). Similarly $\beta(l)$ and $\delta_{l'}$ are *electric* parameters.

With coefficients (16.148) the magnetic induction of the scattered wave therefore becomes

$$
\mathbf{B}_{sc} = \sum_{l=1}^{\infty} i^l\sqrt{4\pi(2l+1)}\left[\frac{e^{i\delta_l}\sin\delta_l}{k}\nabla \times h_l^{(1)}\mathbf{X}_{l,\pm1} \pm e^{i\delta_{l'}}\sin\delta_{l'}h_l^{(1)}\mathbf{X}_{l,\pm1}\right]
\tag{16.151}
$$

with the asymptotic form $(kr \to \infty)$,

$$\mathbf{B}_{sc} \to \frac{e^{ikr}}{kr} \sum_{l=1}^{\infty} \sqrt{4\pi(2l+1)} \left[e^{i\delta_l} \sin \delta_l (\mathbf{n} \times \mathbf{X}_{l,\pm 1}) \mp i e^{i\delta_l'} \sin \delta_l' \mathbf{X}_{l,\pm 1} \right]$$

(16.152)

The scattered field (16.152) corresponds in general to elliptically polarized radiation. Only if the electric and magnetic phases were equal would it represent circularly polarized radiation. This means that, if linearly polarized radiation is incident, the scattered radiation will be elliptically polarized; and, if the incident radiation is unpolarized, the scattered radiation will exhibit partial polarization depending on the angle of observation.

In discussing the scattered intensity it is convenient to use the concept of a scattering cross section. This has already been defined in (14.101). The scattered power per unit solid angle is

$$\frac{dP_{sc}}{d\Omega} = \frac{c}{8\pi} |r\mathbf{B}_{sc}|^2$$

(16.153)

The incident flux is

$$\mathbf{S} = \frac{c}{8\pi} \operatorname{Re} (\mathbf{E}_{inc} \times \mathbf{B}_{inc}^*) = \frac{c}{4\pi} \boldsymbol{\epsilon}_3$$

(16.154)

Consequently the scattering cross section is

$$\frac{d\sigma}{d\Omega} = \frac{2\pi}{k^2} \left| \sum_{l=1}^{\infty} \sqrt{2l+1} \left[e^{i\delta_l} \sin \delta_l (\mathbf{n} \times \mathbf{X}_{l,\pm 1}) \mp i e^{i\delta_l'} \sin \delta_l' \mathbf{X}_{l,\pm 1} \right] \right|^2$$

(16.155)

This angular distribution is rather complicated, except in the long-wavelength limit (see below). But the total cross section can be calculated directly. From the second orthogonality relation in (16.132) and (16.143) it is evident that the cross terms in (16.155) vanish on integration over angles. Then the total cross section is easily found to be

$$\sigma = \frac{2\pi}{k^2} \sum_{l=1}^{\infty} (2l+1)[\sin^2 \delta_l + \sin^2 \delta_l']$$

(16.156)

The electric and magnetic multipole parts of the wave contribute incoherently to the total cross section.

In the long-wavelength limit $(ka \ll 1)$ the scattering cross section becomes relatively simple because the phase shifts (16.149) decrease rapidly with increasing l. Keeping only $l = 1$ terms in the expansion, we find

$$\frac{d\sigma}{d\Omega} \simeq \frac{2\pi}{3} a^2(ka)^4 |\mathbf{n} \times \mathbf{X}_{1,\pm 1} \pm 2i\mathbf{X}_{1,\pm 1}|^2$$

(16.157)

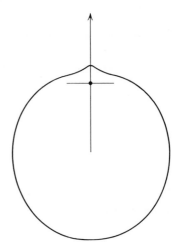

Fig. 16.6 Angular distribution of radiation scattered by a perfectly conducting sphere in the long-wavelength limit ($ka \ll 1$).

From the table on p. 551 we obtain the absolute squared terms,

$$|\mathbf{n} \times \mathbf{X}_{1, \pm 1}|^2 = |\mathbf{X}_{1, \pm 1}|^2 = \frac{3}{16\pi} (1 + \cos^2 \theta) \qquad (16.158)$$

The cross terms can be easily worked out:

$$\text{Re} \left[\pm i(\mathbf{n} \times \mathbf{X}_{1, \pm 1})^* \cdot \mathbf{X}_{1, \pm 1} \right] = \frac{-3}{8\pi} \cos \theta \qquad (16.159)$$

Thus the long-wavelength limit of the differential scattering cross section is

$$\frac{d\sigma}{d\Omega} \simeq a^2 (ka)^4 \left[\tfrac{5}{8}(1 + \cos^2 \theta) - \cos \theta \right] \qquad (16.160)$$

independent of the state of polarization of the incident radiation. The angular distribution of scattered radiation is shown in Fig. 16.6. The scattering is predominantly backwards, the marked asymmetry about 90° being caused by the electric dipole-magnetic dipole interference term.

The total cross section in the long-wavelength limit is

$$\sigma = \frac{10\pi}{3} a^2 (ka)^4 \qquad (16.161)$$

This is a well-known result, first obtained by Mie and Debye (1908–1909). The dependence of the cross section on frequency as ω^4 is known as Rayleigh's law, and is characteristic of all systems which possess a dipole moment.

16.10 Boundary-Value Problems with Multipole Fields

The scattering of radiation by a conducting sphere is an example of a boundary-value problem with multipole fields. Other examples are the free oscillations of a conducting sphere, the spherical resonant cavity, and scattering by a dielectric sphere. The possibility of resistive losses in conductors adds problems such as Q values of cavities and absorption cross sections to the list. The general techniques for handling these problems are the same ones as have been met in Section 16.9 and in Chapter 8. The necessary mathematical apparatus has been developed in the present chapter. We will leave the discussion of these examples to the problems at the end of the chapter.

REFERENCES AND SUGGESTED READING

The theory of vector spherical harmonics and multipole vector fields is discussed thoroughly by
  Blatt and Weisskopf, Appendix B,
  Morse and Feshbach, Section 13.3.
Applications to nuclear multipole radiation are given in
  Blatt and Weisskopf, Chapter XII,
  Siegbahn, Chapter XIII by S. A. Moszkowski and
          Chapter XVI (II) by M. Goldhaber and A. W. Sunyar.
A number of books on antennas were cited at the end of Chapter 9. None of them discusses multipole expansions in a rigorous way, however.
  The scattering of radiation by a perfectly conducting sphere is treated briefly by
  Morse and Feshbach, pp. 1882–1886,
  Panofsky and Phillips, Section 12.9.
Much more elaborate discussions, with arbitrary dielectric and conductive properties for the sphere, are given by
  Born and Wolf, Section 13.5,
  Stratton, Section 9.25.
  Mathematical information on spherical Bessel functions, etc., will be found in
  Morse and Feshbach, pp. 1573–1898.

PROBLEMS

16.1 Three charges are located along the z axis, a charge $+2q$ at the origin and charges $-q$ at $z = \pm a \cos \omega t$. Determine the lowest nonvanishing multipole moments, the angular distribution of radiation, and the total power radiated. Assume that $ka \ll 1$.

16.2 An almost spherical surface defined by

$$R(\theta) = R_0[1 + \beta P_2(\cos \theta)]$$

has inside of it a uniform volume distribution of charge totaling Q. The small parameter β varies harmonically in time at frequency ω. This corresponds to surface waves on a sphere. Keeping only lowest-order terms in β and making the long-wavelength approximation, calculate the nonvanishing multipole moments, the angular distribution of radiation, and the total power radiated.

16.3 The uniform charge density of Problem 16.2 is replaced by a uniform density of intrinsic magnetization parallel to the z axis and having total magnetic moment M. With the same approximations as above calculate the nonvanishing radiation multipole moments, the angular distribution of radiation, and the total power radiated.

16.4 An antenna consists of a circular loop of wire of radius a located in the x-y plane with its center at the origin. The current in the wire is

$$I = I_0 \cos \omega t = \text{Re } I_0 e^{-i\omega t}$$

(a) Find the expressions for $\mathbf{E}$, $\mathbf{B}$ in the radiation zone without approximations as to the magnitude of ka. Determine the power radiated per unit solid angle.

(b) What is the lowest nonvanishing multipole moment ($Q_{l,m}$ or $M_{l,m}$)? Evaluate this moment in the limit $ka \ll 1$.

16.5 Two fixed electric dipoles of dipole moment p are located in a plane a distance $2a$ apart, their axes parallel and perpendicular to the plane, but their moments directed oppositely. The dipoles rotate with constant angular velocity ω about a parallel axis located halfway between them ($\omega \ll c/a$).

(a) Calculate the components of the quadrupole moment.

(b) Show that the angular distribution of radiation is proportional to $(1 - 3\cos^2\theta + 4\cos^4\theta)$, and that the total power radiated is

$$P = \frac{2cp^2a^2}{5}\left(\frac{\omega}{c}\right)^6.$$

16.6 In the long-wavelength limit evaluate the nonvanishing electric multipole moments for the charge distribution:

$$\rho = Cr^3 e^{-5r/6} Y_{1,1}(\theta, \phi) Y_{2,0}(\theta, \phi) e^{-i\omega_0 t}$$

and determine the angular distribution and total power radiated for each multipole. This charge distribution is appropriate to a transition between the states $n = 3$, $l = 2$ $(3d)$ and $n = 2$, $l = 1$ $(2p)$ in a hydrogen atom.

16.7 The fields representing a transverse magnetic wave propagating in a cylindrical wave guide of radius R are:

$$E_z = J_m(\gamma r)e^{im\phi}e^{i\beta z - i\omega t}, \qquad H_z = 0$$

$$E_\phi = \frac{-m\beta}{\gamma^2}\frac{E_z}{r}, \qquad H_r = -\frac{k}{\beta}E_\phi$$

$$E_r = \frac{i\beta}{\gamma^2}\frac{\partial E_z}{\partial r}, \qquad H_\phi = \frac{k}{\beta}E_r$$

where m is the index specifying the angular dependence, β is the propagation constant, $\gamma^2 = k^2 - \beta^2$ $(k = \omega/c)$, where γ is such that $J_m(\gamma R) = 0$. Calculate the ratio of the z component of the electromagnetic angular momentum to the energy in the field. It may be advantageous to perform some integrations by parts, and to use the differential equation satisfied by E_z, in order to simplify your calculations.

16.8 A spherical hole of radius a in a conducting medium can serve as an electromagnetic resonant cavity.

(a) Assuming infinite conductivity, determine the transcendental equations for the characteristic frequencies ω_{ln} of the cavity for TE and TM modes.

(b) Calculate numerical values for the wavelength λ_{ln} in units of the radius a for the four lowest modes for TE and TM waves.

(c) Calculate explicitly the electric and magnetic fields inside the cavity for the lowest TE and lowest TM mode.

16.9 The spherical resonant cavity of Problem 16.8 has nonpermeable walls of large, but finite, conductivity. In the approximation that the skin depth δ is small compared to the cavity radius a, show that the Q of the cavity, defined by equation (8.82), is given by

$$Q = \frac{a}{2\pi\delta}, \qquad \text{for all TE modes}$$

and

$$Q = \frac{a}{2\pi\delta}\left(1 - \frac{l(l+1)}{x_{ln}^2}\right), \qquad \text{for TM modes}$$

where $x_{ln} = (a/c)\omega_{ln}$ for TM modes.

16.10 Discuss the normal modes of oscillation of a perfectly conducting solid sphere of radius a in free space.

(a) Determine the characteristic equations for the eigenfrequencies for TE and TM modes of oscillation. Show that the roots for ω always have a negative imaginary part, assuming a time dependence of $e^{-i\omega t}$.

(b) Calculate the eigenfrequencies for the $l = 1$ and $l = 2$ TE and TM modes. Tabulate the wavelength (defined in terms of the real part of the frequency) in units of the radius a and the decay time (defined as the time taken for the *energy* to fall to e^{-1} of its initial value) in units of the transit time (a/c) for each of the modes.

16.11 A circularly polarized plane wave of radiation of frequency $\omega = ck$ is incident on a nonpermeable, conducting sphere of radius a.

(a) Assuming that the conductivity of the sphere is infinite, write down explicit expressions for the electric and magnetic fields near and at the surface of the sphere in the long-wavelength limit, $ka \ll 1$.

(b) Using the techniques of Chapter 8, calculate the power absorbed by the sphere from the incident wave, assuming that the conductivity is large but finite. Express your result as an absorption cross section in terms of the wave number k, the radius a, and the skin depth δ. Assume $ka \ll 1$.

16.12 Discuss the scattering of a plane wave of electromagnetic radiation by a nonpermeable, dielectric sphere of radius a and dielectric constant ϵ.

(a) By finding the fields inside the sphere and matching to the incident plus scattered wave outside the sphere, determine the multipole coefficients in the scattered wave. Define suitable phase shifts for the problem.

(b) Consider the long-wavelength limit $(ka \ll 1)$ and determine explicitly the differential and total scattering cross sections. Sketch the angular distribution for $\epsilon = 2$.

(c) In the limit $\epsilon \to \infty$ compare your results to those for the perfectly conducting sphere.

17

Radiation Damping, Self-Fields of a Particle, Scattering and Absorption of Radiation by a Bound System

17.1 Introductory Considerations

In the preceding chapters the problems of electrodynamics have been divided into two classes, one in which the sources of charge and current are specified and the resulting electromagnetic fields are calculated, and the other in which the external electromagnetic fields are specified and the motions of charged particles or currents are calculated. Antennas and radiation from multipole sources are examples of the first type of problem, while motion of charges in electric and magnetic fields and energy-loss phenomena are examples of the second type. Occasionally, as in the discussion of bremsstrahlung, the two problems are combined. But the treatment is a stepwise one—first the motion of the charged particle in an external field is determined, neglecting the emission of radiation; then the radiation is calculated from the trajectory as a given source distribution.

It is evident that this manner of handling problems in electrodynamics can be of only approximate validity. The motion of charged particles in external force fields necessarily involves the emission of radiation whenever the charges are accelerated. The emitted radiation carries off energy, momentum, and angular momentum and so must influence the subsequent motion of the charged particles. Consequently the motion of the sources of radiation is determined, in part, by the manner of emission of the radiation. A correct treatment must include the reaction of the radiation on the motion of the sources.

Why is it that we have taken so long in our discussion of electrodynamics to face this fact? Why is it that many answers calculated in an apparently erroneous way agree so well with experiment? A partial answer to the first question lies in the second. There *are* very many problems in electrodynamics which can be put with negligible error into one of the two categories described in the first paragraph. Hence it is worth while discussing them without the added and unnecessary complication of including reaction effects. The remaining answer to the first question is that a completely satisfactory treatment of the reactive effects of radiation does not exist. The difficulties presented by this problem touch one of the most fundamental aspects of physics, the nature of an elementary particle. Although partial solutions, workable within limited areas, can be given, the basic problem remains unsolved. One might hope that the transition from classical to quantum-mechanical treatments would remove the difficulties. While there is still hope that this may eventually occur, the present quantum-mechanical discussions are beset with even more elaborate troubles than the classical ones. It is one of the triumphs of comparatively recent years ($\sim$1948–1950) that the concepts of Lorentz covariance and gauge invariance were exploited sufficiently cleverly to circumvent these difficulties in quantum electrodynamics and so allow the calculation of very small radiative effects to extremely high precision, in full agreement with experiment. From a fundamental point of view, however, the difficulties still remain. In this chapter we will consider only the classical aspects, but will indicate some of the quantum-mechanical analogs along the way.

The question as to why many problems can apparently be handled neglecting reactive effects of the radiation has the obvious answer that such effects must be of negligible importance. To see qualitatively when this is so, and to obtain semiquantitative estimates of the ranges of parameters where radiative effects are or are not important, we need a simple criterion. One such criterion can be obtained from energy considerations. If an external force field causes a particle of charge e to have an acceleration of typical magnitude a for a period of time T, the energy radiated is of the order of

$$E_{\text{rad}} \sim \frac{2e^2a^2T}{3c^3} \tag{17.1}$$

from Larmor's formula (14.22). If this energy lost in radiation is negligible compared to the relevant energy E_0 of the problem, we can expect that radiative effects will be unimportant. But if $E_{\text{rad}} \gtrsim E_0$, the effects of radiation reaction will be appreciable. The criterion for the point when radiative effects begin to be important can thus be expressed by

$$E_{\text{rad}} \sim E_0 \qquad (17.2)$$

The specification of the relevant energy E_0 demands a little care. We will distinguish two apparently different situations, one in which the particle is initially at rest and is acted on by the applied force only for the finite interval T, and one where the particle undergoes continual acceleration, e.g., in quasiperiodic motion at some characteristic frequency ω_0. For the particle at rest initially a typical energy is evidently its kinetic energy after the period of acceleration. Thus

$$E_0 \sim m(aT)^2$$

Criterion (17.2) then becomes

$$\frac{2}{3} \frac{e^2 a^2 T}{c^3} \sim ma^2 T^2$$

or

$$T \sim \frac{2}{3} \frac{e^2}{mc^3}$$

It is useful to define the characteristic time in this relation as

$$\tau = \frac{2}{3} \frac{e^2}{mc^3} \qquad (17.3)$$

Then the conclusion is that for time T long compared to τ radiative effects are unimportant. Only when the force is applied so suddenly and for such a short time that $T \sim \tau$ will radiative effects modify the motion appreciably. It is useful to note that the longest characteristic time τ for charged particles is for electrons and that its value is $\tau = 6.26 \times 10^{-24}$ sec. This is of the order of the time taken for light to travel 10^{-13} cm. Only for phenomena involving such distances or times will we expect radiative effects to play a crucial role.

If the motion of the charged particle is quasi-periodic with a typical amplitude d and characteristic frequency ω_0, the mechanical energy of motion can be identified with E_0 and is of the order of

$$E_0 \sim m\omega_0^2 d^2$$

The accelerations are typically $a \sim \omega_0^2 d$, and the time interval $T \sim (1/\omega_0)$. Consequently criterion (17.2) is

$$\frac{2e^2 \omega_0^4 d^2}{3c^3 \omega_0} \simeq m\omega_0^2 d^2$$

or

$$\omega_0 \tau \sim 1 \qquad (17.4)$$

where τ is given by (17.3). Since ω_0^{-1} is a time appropriate to the mechanical motion, again we see that, if the relevant mechanical time interval is long compared to the characteristic time τ (17.3), radiative reaction effects on the motion will be unimportant.

The examples of the last two paragraphs show that the reactive effects of radiation on the motion of a charged particle can be expected to be important if the external forces are such that the motion changes appreciably in times of the order of τ or over distances of the order of $c\tau$. This is a general criterion within the framework of classical electrodynamics. For motions less violent, the reactive effects are sufficiently small that they have a negligible effect on the short-term motion. Their long-term, cumulative effects can be taken into account in an approximate way, as we will see immediately.

17.2 Radiative Reaction Force from Conservation of Energy

The question now arises as to how to include the reactive effects of radiation in the equations of motion for a charged particle. We begin with a simple plausibility argument based on conservation of energy for a nonrelativistic charged particle. A more fundamental derivation and the incorporation of relativistic effects will be deferred to later sections.

If the emission of radiation is neglected, a charged particle of mass m and charge e acted on by an external force $\mathbf{F}_{\text{ext}}$ moves according to Newton's equation of motion:

$$m\dot{\mathbf{v}} = \mathbf{F}_{\text{ext}} \tag{17.5}$$

Since the particle is accelerated, it emits radiation at a rate given by Larmor's power formula (14.22),

$$P(t) = \frac{2}{3}\frac{e^2}{c^3}(\dot{\mathbf{v}})^2 \tag{17.6}$$

To account for this radiative energy loss and its effect on the motion of the particle we modify Newton's equation (17.5) by adding a *radiative reaction force* $\mathbf{F}_{\text{rad}}$:

$$m\dot{\mathbf{v}} = \mathbf{F}_{\text{ext}} + \mathbf{F}_{\text{rad}} \tag{17.7}$$

While $\mathbf{F}_{\text{rad}}$ is not determined at this stage, we can see some of the requirements it "must" satisfy:

$\mathbf{F}_{\text{rad}}$ "must" (1) vanish if $\dot{\mathbf{v}} = 0$, since then there is no radiation;

(2) be proportional to e^2, since (*a*) the radiated power is proportional to e^2, and (*b*) the sign of the charge cannot enter in radiative effects;

(3) in fact involve the characteristic time τ (17.3), since that is apparently the only parameter of significance available.

We will determine the form of $\mathbf{F}_{\text{rad}}$ by demanding that the work done by this force on the particle in the time interval $t_1 < t < t_2$ be equal to the negative of the energy radiated in that time. Then energy will be conserved, at least over the interval (t_1, t_2). With the Larmor result (17.6), this requirement is

$$\int_{t_1}^{t_2} \mathbf{F}_{\text{rad}} \cdot \mathbf{v}\, dt = -\int_{t_1}^{t_2} \frac{2}{3} \frac{e^2}{c^3} \dot{\mathbf{v}} \cdot \dot{\mathbf{v}}\, dt$$

The second integral can be integrated by parts to yield

$$\int_{t_1}^{t_2} \mathbf{F}_{\text{rad}} \cdot \mathbf{v}\, dt = \frac{2}{3} \frac{e^2}{c^3} \int_{t_1}^{t_2} \ddot{\mathbf{v}} \cdot \mathbf{v}\, dt - \frac{2}{3} \frac{e^2}{c^3} (\dot{\mathbf{v}} \cdot \mathbf{v})\Big|_{t_1}^{t_2}$$

If the motion is periodic or such that $(\dot{\mathbf{v}} \cdot \mathbf{v}) = 0$ at $t = t_1$ and $t = t_2$, we may write

$$\int_{t_1}^{t_2} \left(\mathbf{F}_{\text{rad}} - \frac{2}{3} \frac{e^2}{c^3} \ddot{\mathbf{v}} \right) \cdot \mathbf{v}\, dt = 0$$

Then it is permissible to identify the radiative reaction force as

$$\mathbf{F}_{\text{rad}} = \frac{2}{3} \frac{e^2}{c^3} \ddot{\mathbf{v}} = m\tau\ddot{\mathbf{v}} \tag{17.8}$$

The modified equation of motion then reads

$$m(\dot{\mathbf{v}} - \tau\ddot{\mathbf{v}}) = \mathbf{F}_{\text{ext}} \tag{17.9}$$

Equation (17.9) is sometimes called the *Abraham-Lorentz equation of motion*. It can be considered as an equation which includes in some approximate and time-average way the reactive effects of the emission of radiation. The equation can be criticized on the grounds that it is second order in time, rather than first, and therefore runs counter to the well-known requirements for a dynamical equation of motion. This difficulty manifests itself immediately in the so-called "runaway" solutions. If the external force is zero, it is obvious that (17.9) has two possible solutions,

$$\dot{\mathbf{v}}(t) = \begin{cases} 0 \\ \mathbf{a}e^{t/\tau} \end{cases} \tag{17.10}$$

where $\mathbf{a}$ is the acceleration at $t = 0$. Only the first solution is reasonable. The method of derivation shows that the second solution is unacceptable, since $(\dot{\mathbf{v}} \cdot \mathbf{v}) \neq 0$ at t_1 and t_2. It is clear that the equation is useful only in the domain where the reactive term is a small correction. Then the radiative reaction can be treated as a perturbation producing slow or small changes in the state of motion of the particle. The problem of the "runaway" solutions can be avoided by replacing (17.9) by an integro-differential equation (see Section 17.7).

To illustrate the use of (17.9) to account for small radiative effects we consider a particle moving in an attractive, conservative, central force field. In the absence of radiation reaction, the particle's energy and angular momentum are conserved and determine the motion. The emission of radiation causes changes in these quantities. Provided the accelerations are not too violent, the energy and angular momentum will change appreciably only in a time interval long compared to the characteristic period of the motion. Thus the motion will instantaneously be essentially the same as in the absence of radiative reaction. The long-term changes can be described by averages over the particle's unperturbed orbit.

If the attractive, conservative, central force field is described by a potential $V(r)$, the acceleration, neglecting reactive effects, is

$$\dot{\mathbf{v}} = \frac{-1}{m}\left(\frac{dV}{dr}\right)\frac{\mathbf{r}}{r} \tag{17.11}$$

By conservation of energy the rate of change of the particle's total energy is given by the negative of the Larmor power:

$$\frac{dE}{dt} = -\frac{2}{3}\frac{e^2}{c^3}(\dot{\mathbf{v}})^2 = -\frac{2e^2}{3m^2c^3}\left(\frac{dV}{dr}\right)^2$$

With the definition of τ (17.3) this can be written

$$\frac{dE}{dt} = -\frac{\tau}{m}\left(\frac{dV}{dr}\right)^2 \tag{17.12}$$

Since the change in energy is assumed to be small in one cycle of the orbit, the right-hand side may be replaced by its time-averaged value in terms of the Newtonian orbit. Then we obtain

$$\frac{dE}{dt} \simeq -\frac{\tau}{m}\left\langle\left(\frac{dV}{dr}\right)^2\right\rangle \tag{17.13}$$

The secular change in angular momentum can be found by considering the vector product of (17.9) with the radius vector $\mathbf{r}$. Since the angular momentum is $\mathbf{L} = m\mathbf{r} \times \mathbf{v}$, we find

$$\frac{d\mathbf{L}}{dt} = \mathbf{r} \times \mathbf{F}_{\text{ext}} + m\tau\mathbf{r} \times \dddot{\mathbf{v}} \tag{17.14}$$

Since the external force is central, the applied torque vanishes. But the radiative torque term can be expressed as

$$m\tau\mathbf{r} \times \dddot{\mathbf{v}} = \tau\left(\frac{d^2\mathbf{L}}{dt^2} - m\mathbf{v} \times \dot{\mathbf{v}}\right) \tag{17.15}$$

The angular momentum is assumed to change slowly in time, certainly when time is measured in units of τ. Consequently it is consistent to omit in (17.15) the second derivative of **L** with respect to t and to substitute $\dot{\mathbf{v}}$ from the unperturbed equation of motion (17.11). Then the rate of change of angular momentum can be written as

$$\frac{d\mathbf{L}}{dt} \simeq -\frac{\tau}{m}\left\langle \frac{1}{r}\frac{dV}{dr}\right\rangle \mathbf{L} \qquad (17.16)$$

where a time average over the instantaneous orbit has been performed, as in (17.13).

Equations (17.13) and (17.16) determine how the particle orbit changes as a function of time because of radiative reaction. Although the detailed behavior depends on the specific law of force, some qualitative statements can be made. If the characteristic frequency of motion is ω_0, the average value in (17.16) can be written

$$\frac{\tau}{m}\left\langle \frac{1}{r}\frac{dV}{dr}\right\rangle \sim \frac{\tau}{m} m\omega_0^2 = \omega_0^2\tau$$

with some dimensionless numerical coefficient of the order of unity. This shows that the characteristic time over which the angular momentum changes is of the order of $1/(\omega_0\tau)\omega_0$. This time is very long compared to the orbital period $2\pi/\omega_0$, provided $\omega_0\tau \ll 1$. Similar arguments can be made with the energy equations.

These equations including radiative effects can be used to discuss practical problems such as the moderation time of a mu or pi meson in cascading from an orbit of very large quantum number around a nucleus down to the low-lying orbits. Over most of the time interval the quantum numbers are sufficiently large that the classical description of continuous motion is an adequate approximation. Discussion of examples of this kind will be left to the problems.

17.3 Abraham-Lorentz Evaluation of the Self-Force

The derivation of the radiation reaction force in the previous section, while plausible, is certainly not rigorous or fundamental. The problem is to give a satisfactory account of the reaction back on the charged particle of its own radiation fields. Thus any systematic discussion must consider the charge structure of the particle and its self-fields. Abraham (1903) and Lorentz (1904) made the first attempt at such a treatment by trying to make a purely electromagnetic model of a charged particle. Our discussion is patterned after that given by Lorentz in his book, *Theory of Electrons*, Note 18, p. 252.

Let us consider a single charged particle of total charge e with a sharply localized charge density $\rho(\mathbf{x})$ in the particle's rest frame. The particle is in external electromagnetic fields, $\mathbf{E}_{\text{ext}}(\mathbf{x}, t)$, $\mathbf{B}_{\text{ext}}(\mathbf{x}, t)$. We have seen in Sections 6.9 and 11.11 that the rate of change of mechanical momentum plus electromagnetic momentum in a given volume vanishes, provided there is no flow of momentum out of or into the volume. Abraham and Lorentz proposed that the apparently mechanical momentum of a charged particle is actually electromagnetic in origin. Then the momentum-conservation law can be phrased,

$$\frac{d\mathbf{G}}{dt} = 0$$

or equivalently in terms of the Lorentz force,

$$\int \left(\rho \mathbf{E} + \frac{1}{c} \mathbf{J} \times \mathbf{B} \right) d^3x = 0 \tag{17.17}$$

In this equation the fields are the *total* fields, and the integration is over the volume of the particle.

In order that (17.17) take on the form of Newton's equation of motion

$$\frac{d\mathbf{p}}{dt} = \mathbf{F}_{\text{ext}}$$

we decompose the total fields into the external fields and the self-fields $\mathbf{E}_s$, $\mathbf{B}_s$ due to the particle's own charge and current densities, ρ and $\mathbf{J}$:

$$\left. \begin{array}{l} \mathbf{E} = \mathbf{E}_{\text{ext}} + \mathbf{E}_s \\ \mathbf{B} = \mathbf{B}_{\text{ext}} + \mathbf{B}_s \end{array} \right\} \tag{17.18}$$

Then (17.17) can be written as Newton's equations of motion, with the external force as

$$\mathbf{F}_{\text{ext}} = \int \left(\rho \mathbf{E}_{\text{ext}} + \frac{1}{c} \mathbf{J} \times \mathbf{B}_{\text{ext}} \right) d^3x \tag{17.19}$$

and the rate of change of momentum of the particle as

$$\frac{d\mathbf{p}}{dt} = -\int \left(\rho \mathbf{E}_s + \frac{1}{c} \mathbf{J} \times \mathbf{B}_s \right) d^3x \tag{17.20}$$

Provided the external fields vary only slightly over the extent of the particle, the external force (17.19) becomes just the ordinary Lorentz force on a particle of charge e and velocity $\mathbf{v}$.

To calculate the self-force [the integral on the right side of (17.20)] it is necessary to have a model of the charged particle. We will assume for simplicity that:

(a) the particle is instantaneously at rest;

(b) the charge distribution is rigid and spherically symmetric.

Our results will then necessarily be restricted to nonrelativistic motions and will lack the proper Lorentz transformation properties. These deficiences can be remedied later.

For a particle instantaneously at rest (17.20) becomes

$$\frac{d\mathbf{p}}{dt} = -\int \rho(\mathbf{x}, t)\mathbf{E}_s(x, t)\, d^3x \tag{17.21}$$

The self-field can be expressed in terms of the self-potentials, $\mathbf{A}$ and Φ, so that

$$\frac{d\mathbf{p}}{dt} = \int \rho(\mathbf{x}, t)\left[\nabla\Phi(\mathbf{x}, t) + \frac{1}{c}\frac{\partial \mathbf{A}}{\partial t}(\mathbf{x}, t)\right] d^3x \tag{17.22}$$

The potentials are given by $A_\mu = (\mathbf{A}, i\Phi)$:

$$A_\mu(\mathbf{x}, t) = \frac{1}{c}\int \frac{[J_\mu(\mathbf{x}', t')]_{\text{ret}}}{R}\, d^3x' \tag{17.23}$$

with $J_\mu = (\mathbf{J}, ic\rho)$ and $\mathbf{R} = \mathbf{x} - \mathbf{x}'$.

In (17.23) the 4-current must be evaluated at the retarded time t'. This differs from the time t by a time of the order of $\Delta t \sim (a/c)$, where a is the dimension of the particle. For a highly localized charge distribution this time interval is extremely short. During such a short time the motion of the particle can be assumed to change only slightly. Consequently it is natural to make a Taylor series expansion in (17.23) around the time $t' = t$. Since $[\ ]_{\text{ret}}$ means evaluated at $t' = t - (R/c)$, any retarded quantity has the expansion,

$$[\ ]_{\text{ret}} = \sum_{n=0}^{\infty} \frac{(-1)^n}{n!}\left(\frac{R}{c}\right)^n \frac{\partial^n}{\partial t^n}[\ ] \tag{17.24}$$

With this expansion for the retarded 4-current in (17.23), expression (17.22) becomes

$$\frac{d\mathbf{p}}{dt} = \sum_{n=0}^{\infty} \frac{(-1)^n}{n!\, c^n}\int d^3x \int d^3x' \rho(\mathbf{x}, t)\frac{\partial^n}{\partial t^n}\left[\rho(\mathbf{x}', t)\nabla R^{n-1} + \frac{R^{n-1}}{c^2}\frac{\partial \mathbf{J}(\mathbf{x}', t)}{\partial t}\right]$$

Consider the $n = 0$ and $n = 1$ terms in the scalar potential part (the first term in the square bracket) of the right-hand side. For $n = 0$ the term is proportional to

$$\int d^3x \int d^3x' \rho(\mathbf{x}, t)\rho(\mathbf{x}', t)\nabla\left(\frac{1}{R}\right)$$

This is just the electrostatic self-force. For spherically symmetric charge distributions it vanishes. The $n = 1$ term is identically zero, since it involves ∇R^{n-1}. Thus the first nonvanishing contribution from the scalar potential part comes from $n = 2$. This means that we can change the summation indices so that the sum now reads

$$\frac{d\mathbf{p}}{dt} = \sum_{n=0}^{\infty} \frac{(-1)^n}{n!\, c^{n+2}} \int d^3x \int d^3x'\, \rho(\mathbf{x}, t) R^{n-1} \frac{\partial^{n+1}}{\partial t^{n+1}} \{\ \}$$

where

$$\{\ \} = \mathbf{J}(\mathbf{x}', t) + \frac{\partial \rho}{\partial t}(\mathbf{x}', t) \frac{\nabla R^{n+1}}{(n+1)(n+2)R^{n-1}}$$

$$(17.25)$$

With the continuity equation for charge and current densities, the curly bracket in (17.25) can be written

$$\{\ \} = \mathbf{J}(\mathbf{x}', t) - \frac{\mathbf{R}}{n+2} \nabla' \cdot \mathbf{J}(\mathbf{x}', t)$$

In the integral over d^3x' we can integrate the second term by parts. We then have

$$-\int d^3x'\, R^{n-1} \frac{\mathbf{R}}{n+2} \nabla' \cdot \mathbf{J} = +\frac{1}{n+2} \int d^3x'\, (\mathbf{J} \cdot \nabla') R^{n-1} \mathbf{R}$$

$$= \frac{-1}{n+2} \int d^3x'\, R^{n-1} \left(\mathbf{J} + (n-1) \frac{\mathbf{J} \cdot \mathbf{R}}{R^2} \mathbf{R} \right)$$

This means that the curly bracket in (17.25) is effectively equal to

$$\{\ \} = \left(\frac{n+1}{n+2} \right) \mathbf{J}(\mathbf{x}', t) - \left(\frac{n-1}{n+2} \right) \frac{(\mathbf{J} \cdot \mathbf{R})\mathbf{R}}{R^2} \qquad (17.26)$$

For a rigid charge distribution the current is

$$\mathbf{J}(\mathbf{x}', t) = \rho(\mathbf{x}', t)\mathbf{v}(t)$$

If the charge distribution is spherically symmetric, the only relevant direction in the problem is that of $\mathbf{v}(t)$. Consequently in the integration over d^3x and d^3x' only the component of (17.26) along the direction of $\mathbf{v}(t)$ survives. Hence (17.26) is equivalent to

$$\{\ \} = \rho(\mathbf{x}', t)\mathbf{v}(t) \left[\frac{n+1}{n+2} - \frac{n-1}{n+2} \left(\frac{\mathbf{R} \cdot \mathbf{v}}{Rv} \right)^2 \right]$$

Furthermore all directions of $\mathbf{R}$ are equally probable. This means that the second term above can be replaced by its average value of $\frac{1}{3}$. This leads to the final simple form for our curly bracket in (17.25):

$$\{\ \} = \tfrac{2}{3} \rho(\mathbf{x}', t)\mathbf{v}(t) \qquad (17.27)$$

With (17.27) in (17.25) the self-force becomes, apart from neglected non-linear terms in time derivatives of $\mathbf{v}$ (which appear for $n \geq 4$),

$$\frac{d\mathbf{p}}{dt} = \sum_{n=0}^{\infty} \frac{(-1)^n}{c^{n+2}} \frac{2}{3n!} \frac{\partial^{n+1} \mathbf{v}}{\partial t^{n+1}} \int d^3x' \int d^3x \, \rho(\mathbf{x}') R^{n-1} \rho(\mathbf{x}) \qquad (17.28)$$

To understand the meaning of (17.28) we consider the first few terms in the expansion:

$$\left. \begin{array}{l} \left(\dfrac{d\mathbf{p}}{dt}\right)_0 = \dfrac{2}{3c^2} \dot{\mathbf{v}} \displaystyle\int d^3x \int d^3x' \, \dfrac{\rho(\mathbf{x})\rho(\mathbf{x}')}{R} \\[12pt] \left(\dfrac{d\mathbf{p}}{dt}\right)_1 = \dfrac{-2}{3c^3} \ddot{\mathbf{v}} \displaystyle\int d^3x \int d^3x' \, \rho(\mathbf{x})\rho(\mathbf{x}') = -\dfrac{2e^2}{3c^3} \ddot{\mathbf{v}} \\[12pt] \left(\dfrac{d\mathbf{p}}{dt}\right)_n \sim \dfrac{e^2}{n! \, c^{n+2}} \overset{(n+1)}{\mathbf{v}} a^{n-1} \end{array} \right\} \qquad (17.29)$$

In the third expression a is a length characteristic of the extension of the charge distribution of the particle. We note that for $n \geq 2$ the terms in the expansion vanish in the limit of a point particle ($a \to 0$). Thus for very localized charge distributions we need only consider the $n = 0$ and $n = 1$ contributions. The $n = 1$ term is just the radiative reaction force already found in (17.9). It is independent of the structure of the particle, depending only on its total charge. Our present derivation can be considered as placing it on a much more fundamental footing than the treatment of Section 17.2.

The $n = 0$ term in (17.29) deserves special attention. The double integral is proportional to the electrostatic self-energy U of the charge distribution,

$$U = \frac{1}{2} \int d^3x \int d^3x' \, \frac{\rho(\mathbf{x})\rho(\mathbf{x}')}{R} \qquad (17.30)$$

Consequently the $n = 0$ term can be expressed as

$$\left(\frac{d\mathbf{p}}{dt}\right)_0 = \frac{4}{3} \frac{U}{c^2} \dot{\mathbf{v}} \qquad (17.31)$$

This has the general form required of a rate of change of momentum. The electrostatic self-energy divided by c^2 can be identified with the electromagnetic mass of the particle:

$$m_e = \frac{U}{c^2} \qquad (17.32)$$

Then Newton's equation of motion for the Abraham-Lorentz model takes the form,

$$(\tfrac{4}{3}m_e)\dot{\mathbf{v}} - \frac{2e^2}{3c^3}\ddot{\mathbf{v}} = \mathbf{F}_{\text{ext}} \qquad (17.33)$$

provided higher terms in expansion (17.28) are neglected. This is the same as (17.9), apart from the factor of $\tfrac{4}{3}$ multiplying the electromagnetic mass.

17.4 Difficulties with the Abraham-Lorentz Model

Although the Abraham-Lorentz approach is a significant step towards a fundamental description of a charged particle, it is deficient in several respects.

1. One obvious deficiency is the nonrelativistic nature of the model. For the reactive force term alone a relativistic generalization can be made easily (see Problem 17.4), but that in itself is not sufficient.

2. The electromagnetic mass enters with an incorrect coefficient in (17.33). This is a symptom of improper Lorentz covariance properties inherent in the model, as will become clearer in the following section.

3. If we wish to be able to ignore the higher terms in the self-force expansion, we must take $a \to 0$. But the electromagnetic mass is of the order $(e^2/c^2 a)$. Hence, in the limit $a \to 0$, the mass becomes infinite. If we wish to keep the mass of the order of the observed mass m of the particle, the extent of the charge distribution must be $a \sim r_0$, where

$$r_0 = \frac{e^2}{mc^2}$$

For electrons this distance, called the *classical electron radius*, is 2.82×10^{-13} cm. Although this is very small, motions can be envisioned as sufficiently violent that for such a finite extent the higher terms in the expansion would become significant.* Thus, if the particle has a finite extent, the truncated theory must be considered as only an approximate description.

4. The localized charge distribution must have forces of nonelectromagnetic character holding it stable. Thus the idea of a purely electromagnetic model for matter must be abandoned within the framework of Maxwell's equations and special relativity. We know of strong, nonelectromagnetic forces in nature. But at the present time the internal

---

* Successive terms in the expansion are in the ratio $\left(\dfrac{a}{c}\dfrac{d^{n+2}v}{dt^{n+2}} \middle/ \dfrac{d^{n+1}v}{dt^{n+1}} \right)$. This means that the motion must change appreciably in a time interval (a/c). With $a \sim e^2/mc^2$, this time interval is just τ, given by (17.3). We thus return to the same criterion as before.

structure of particles is largely unknown. The only exceptions are the electromagnetic structures of the neutron and proton. These have been explored by high-energy electron scattering, assuming that the electrons are point particles with no structure and that no changes occur in electrodynamics at the small distances involved. It is found that the distributions of charge and magnetic-moment densities extend over distances of the order of $(0.5\text{-}1.0) \times 10^{-13}$ cm.* This is somewhat smaller than, but of the same order as, the classical electron radius r_0. The reader is cautioned, however, against attributing any deep significance to this occurrence. The structure of an elementary particle is very much a quantum-mechanical phenomenon. Of much greater relevance as a length parameter for the neutron and proton size is the Compton wavelength $(\hbar/m_\pi c \simeq 1.4 \times 10^{-13}$ cm) of the pi meson, which acts as the quantum of the nuclear force field in the same way as the photon acts as the quantum of the electromagnetic field.†

Nonelectromagnetic forces imply a contribution m_0 to the mass of a particle from such forces. Within the framework of the Abraham-Lorentz model as discussed so far, this additional mass merely appears as an added coefficient of the acceleration in (17.33).

17.5 Lorentz Transformation Properties of the Abraham-Lorentz Model; Poincaré Stresses

The troublesome and puzzling factor of $\frac{4}{3}$ in the inertia of electromagnetic energy was first found by J. J. Thomson (1881). To see its origins clearly we will consider the electromagnetic self-energy and momentum of the Abraham-Lorentz model, rather than the equation of motion. In Sections 6.9 and 11.11 we considered the conservation laws of energy and momentum. There we interpreted the fourth column (or row) of the stress-energy tensor $T_{\mu\nu}$ (11.134) as the momentum and energy densities of the electromagnetic field. For our model of a charged particle it is therefore natural to identify the electromagnetic self-energy and momentum as the appropriate volume integral of the *self* stress-energy tensor $T_{\mu\nu}$ of the self-fields. Thus the particle's energy-momentum is defined as

$$P_\mu = \frac{i}{c} \int d^3x \, T_{\mu 4} \tag{17.34}$$

* For a discussion of these experiments, see the article by R. Hofstadter, *Ann. Rev. Nuclear Sci.*, **7**, 231 (1957).

† The fact that $(\hbar/m_\pi c) \simeq \frac{1}{2}(e^2/m_e c^2)$, corresponding to the pi-meson mass being 2×137 times the electron mass, is another of these numerological coincidences which may ultimately have some deep significance.

where the integral is taken over all space. Written out separately, the electromagnetic energy and momentum are

$$E_e = \int d^3x \, T_{44} = \int d^3x \, u_s$$

$$P_{ek} = \frac{i}{c} \int d^3x \, T_{k4} = \int d^3x \, g_{sk}$$

(17.35)

where u_s and $\mathbf{g}_s$ are the self-energy and momentum densities.

In the rest frame of the particle definitions (17.35) reduce to $\mathbf{p}_e = 0$ (since $\mathbf{g}_s = 0$ identically) and

$$E_e = \int d^3x^{(0)} \, T_{44}^{(0)} = U$$

(17.36)

The superscript (0) means rest frame of the particle; U is the electrostatic self-energy (17.30).

From these values of energy and momentum in the rest frame we wish to obtain the corresponding values in a different Lorentz frame and so exhibit the transformation properties. Let the electromagnetic energy-momentum in a reference frame moving with velocity $-\mathbf{v}$ relative to the rest frame be given by (17.34). In that Lorentz frame the charged particle is moving with velocity $\mathbf{v}$. In order to express (17.34) in terms of rest-frame quantities we must transform the integrand. Since $\gamma \, d^3x$ is a Lorentz invariant volume element, we have $\gamma \, d^3x = \gamma^{(0)} \, d^3x^{(0)} = d^3x^{(0)}$. The tensor $T_{\mu\nu}$ transforms according to (11.88). Thus (17.34) can be written

$$P_\mu = \frac{i}{\gamma c} a_{\mu\lambda} a_{4\sigma} \int d^3x^{(0)} \, T_{\lambda\sigma}^{(0)}$$

With the velocity $\mathbf{v}$ chosen parallel to the x_3 axis for convenience, the inverse of (11.75) can be used for $a_{\mu\nu}$. Then it is easy to show that the energy and momentum are

$$E_e = \gamma \int d^3x^{(0)} \, [T_{44}^{(0)} - \beta^2 T_{33}^{(0)}]$$

$$cp_e = \gamma\beta \int d^3x^{(0)} \, [T_{44}^{(0)} - T_{33}^{(0)}]$$

(17.37)

These results differ from the expected ones by the added terms involving the Maxwell self-stress $T_{33}^{(0)}$. Thus we reach the conclusion that the electromagnetic contributions to a particle's energy and momentum do not transform properly unless the self-stress vanishes in the rest frame. For the Abraham-Lorentz model it is evident that the self-stress does *not* vanish. In fact, with the assumption of spherical symmetry, the vanishing

of the trace of $T_{\mu\nu}$ (11.136) can be used to show that the self-stress contribution to the momentum in (17.37) is just one-third of the contribution from the self-energy, leading to the factor $\frac{4}{3}$ which appears in (17.31), even for nonrelativistic velocities.

The improper Lorentz transformation properties implied in (17.37) stem from the nonvanishing of the Maxwell self-stress. On the other hand, the existence of the self-stress is a consequence of the instability of a charge distribution interacting by electromagnetic forces alone. To say that the Maxwell stresses are nonvanishing is merely another way of saying that electrostatic forces tend to push apart the various segments of the localized charge distribution. A stable configuration of matter is one in which the *total* stress, due to all kinds of forces, vanishes.

In 1906 Poincaré saw that he could remove two of the difficulties of the Abraham-Lorentz model simultaneously by postulating appropriate nonelectromagnetic forces, called *Poincaré stresses*, which would compensate for the Maxwell stresses, producing stability of the charged particle and making the *total* self-stress vanish in the rest frame. Thus we can introduce a nonelectromagnetic stress-energy-momentum tensor $P_{\mu\nu}$, which is added to the electromagnetic tensor $T_{\mu\nu}$ to give the *total* stress-energy-momentum tensor,

$$S_{\mu\nu} = T_{\mu\nu} + P_{\mu\nu} \qquad (17.38)$$

Then the particle's energy-momentum 4-vector is defined by

$$P_\mu = \frac{i}{c} \int d^3x \, S_{\mu 4} \qquad (17.39)$$

instead of (17.34). This has the correct Lorentz transformation properties if, in the rest frame of the particle, the Poincaré tensor has the form,

$$P_{\mu\nu}^{(0)} = \begin{pmatrix} P_{11}^{(0)} & P_{12}^{(0)} & P_{13}^{(0)} & 0 \\ P_{21}^{(0)} & P_{22}^{(0)} & P_{23}^{(0)} & 0 \\ P_{31}^{(0)} & P_{32}^{(0)} & P_{33}^{(0)} & 0 \\ 0 & 0 & 0 & P_{44}^{(0)} \end{pmatrix} \qquad (17.40)$$

and furthermore that

$$\int P_{ij}^{(0)} \, d^3x^{(0)} = - \int T_{ij}^{(0)} \, d^3x^{(0)} \qquad (17.41)$$

This last statement is just the mathematical requirement that the attractive Poincaré stresses balance out the repulsive electrostatic forces to give stability.

The Poincaré model shows that it is not allowable to break up the self-energy or mass into an electromagnetic contribution and other

contributions because the separate parts behave differently under Lorentz transformations. Only the *total* self-energy or mass,

$$m = \frac{1}{c^2} \int [T_{44}^{(0)} + P_{44}^{(0)}] \, d^3x^{(0)} \tag{17.42}$$

has physical meaning.

The postulate of Poincaré simultaneously gives the proper Lorentz transformation properties to the particle's total energy and momentum and provides stability for the particle's charge distribution. It can therefore be thought an acceptable solution to the problem within the limitations of classical theory. The origins and fundamental nature of the Poincaré stresses are, of course, unknown. The stresses were merely postulated to meet the obvious experimental facts that charged particles do exist as stable entities with well-established transformation properties for their energies and momenta. Since classical electrodynamics cannot within itself provide a mechanism for vanishing self-stress, we are forced to go outside that framework.

In quantum electrodynamics (strictly, the theory of the interaction of photons with negatons and positons, rather than all charged particles) essentially the same difficulties appear, although in a rather different guise. The theory deals with point charged particles having "bare" charge $\pm e_0$ and "bare" mass m_0. The particles, as well as the electromagnetic field, are described by a quantized field. Thus the quantum theory describes the interaction of two fields, while the classical theory involves one field interacting with "matter."

When the electromagnetic self-energy of a particle is calculated, the result is infinite because of the assumption of point particles. But the singularity is logarithmic,

$$m_e^{(q)} \sim \frac{e^2}{\hbar c} \, m \ln\left(\frac{\hbar}{mca}\right)$$

compared to the linear singularity in the classical result,

$$m_e^{(c)} \sim \frac{e^2}{ac^2}$$

Here a is the length parameter which tends to zero for point particles. The reduction of the singularity from linear to logarithmic is a consequence of cancellations which arise from the different contributions involving the electromagnetic field and the negaton-positon field. In addition, the negaton-positon field gives rise to an effect not found in the classical theory—an infinite contribution to the charge of a particle. These contributions to the mass and charge of a particle are absorbed into the observed mass and charge by attributing suitable Lorentz transformation properties to the infinite integrals which represent them. This procedure is known as *renormalization*. It is successful because (*a*) it can be done in a Lorentz covariant manner, (*b*) it involves a finite number of quantities to be renormalized (actually three—mass, charge, and wave function), and (*c*) it leaves a well-defined theory with which to calculate. The calculated values of small radiative effects, such as the Lamb shift and the anomalous magnetic moment of the electron, are found to be in complete agreement with experiment. The accuracy involved is of the order of 1 part in 10^6 or 10^7.

In the quantum-mechanical calculation of the self-stress, ambiguities also arise because of the presence of infinite integrals. One method of handling this problem is to exploit the connection between the vanishing of the divergence of the stress tensor and the 4-vector character of energy-momentum (see Problem 11.13). Calculation of the electrodynamic contributions to the divergence of the stress tensor and to the self-stress shows that the same infinite integrals appear in each. Conservation of energy and momentum demands that the additions to the divergence of the stress tensor be zero. Hence the infinite integrals there must be formally put equal to zero. Since the *same* integrals appear in the self-stress, *it* may be said to vanish because of conservation of energy and momentum. An alternative method involves formally adding one or more vector fields in interaction with the particle in addition to the electromagnetic field, with the coupling constants (charges) chosen so that the contributions of the added fields to the self-stress and the divergence of the stress tensor cancel the electrodynamic parts.* Although rather different in detail, this method of making the self-stress vanish is in essence similar to the classical, mechanistic postulate of Poincaré.

17.6 Covariant Definition of the Electromagnetic Self-Energy and Momentum of a Charged Particle

The discussion of the previous section has one puzzling aspect. Classical electrodynamics is a properly covariant theory. Hence we might rightfully expect that a correct calculation of any quantity should not violate the requirements of Lorentz covariance. In the Abraham-Lorentz-Poincaré model we seem to have such a violation. A noncovariant electromagnetic contribution to the self-energy or momentum of a charged particle is balanced out by a noncovariant contribution from the Poincaré stresses, so that the result is properly covariant.

---

* For a discussion of these quantum-mechanical approaches, see S. Borowitz and W. Kohn, *Phys. Rev.*, **86**, 985 (1952).

One might argue, as we did in Section 17.5, that, if nonelectromagnetic cohesive forces are necessary to create a stable entity of localized charge and associated self-fields, then only the total forces and stresses have physical meaning. Nevertheless it is legitimate to ask whether the purely electromagnetic contributions to the self-energy and momentum *can be defined* to have the proper Lorentz transformation properties. This not only would be esthetically pleasing, but also would have the added virtue of separating, at least formally, the Lorentz transformation properties and the question of stability.

The task of defining a 4-vector representing the electromagnetic self-energy-momentum is straightforward. We merely create a 4-vector which reduces to the electrostatic self-energy (17.36) in the particle's rest frame.* Clearly we must take the scalar product $T_{\mu\nu}$ with some 4-vector n_ν in order to get a 4-vector. Then we must use the invariant volume element $\gamma\, d^3x$. Hence the covariant generalization of (17.36) is

$$P_{e\mu} = \frac{\gamma}{c} \int d^3x\, T_{\mu\nu} n_\nu \qquad (17.43)$$

where n_ν is a 4-vector which has components $(0, i)$ in the rest frame. From the Lorentz transformation (11.75) it is easy to see that the general form of n_ν is

$$n_\nu = \left(\gamma\frac{\mathbf{v}}{c}, i\gamma\right) \qquad (17.44)$$

Written out in terms of energy and momentum densities and the Maxwell stress tensor, the self-energy and momentum are

$$E_e = \gamma^2 \int d^3x(u - \mathbf{v}\cdot\mathbf{g})$$

$$\mathbf{p}_e = \gamma^2 \int d^3x\left(\mathbf{g} + \frac{\mathbf{v}\cdot\overset{\leftrightarrow}{\mathbf{T}}}{c^2}\right) \qquad (17.45)$$

These expressions differ from the Abraham-Lorentz forms (17.35) by the added terms in $-\mathbf{v}\cdot\mathbf{g}$ and $(\mathbf{v}\cdot\overset{\leftrightarrow}{\mathbf{T}})/c^2$, apart from factors of γ^2.

The physical meaning of these covariant definitions of E_e and p_e can be found in the following argument. Consider the self-energy. One may argue that the proper quantity to identify as the electromagnetic self-energy of a localized distribution of charge in motion is not the total field-energy content, but rather that energy diminished by the work done by the electromagnetic forces (Maxwell stresses) which are eventually balanced

---

* This seems to have been done first by B. Kwal, *J. phys. radium*, **10**, 103 (1949). See also F. Rohrlich, *Am. J. Phys.*, **28**, 639 (1960).

out by some nonelectromagnetic forces. In the rest frame of the particle
these electromagnetic forces do no work. Then the self-energy is given by
the Abraham-Lorentz result (17.36). But in the moving frame the rate of
doing work by these forces is $\int \mathbf{v} \cdot (\partial \mathbf{g}/\partial t)\, d^3x$, neglecting factors of γ which
are unimportant for the physical argument. The time integral of this rate
is just the term subtracted from the total field energy in (17.45). Similarly,
the Maxwell-stress term in the momentum (17.45) represents the negative
of the momentum contribution from the transport of purely electromagnetic stresses.

Since the energy-momentum (17.45) was constructed to be a 4-vector,
there is no need to make an explicit verification of that fact. But it is
nevertheless interesting to see the factor of $\frac{4}{3}$ in the momentum removed.
In the nonrelativistic limit, the self-magnetic field is given by

$$\mathbf{B}_s \simeq \frac{\mathbf{v}}{c} \times \mathbf{E}_s \tag{17.46}$$

Then the first term in the momentum (17.45) is

$$\mathbf{p}_{e1} \simeq \int \mathbf{g}\, d^3x \simeq \frac{1}{4\pi c^2} \int \mathbf{E}_s \times (\mathbf{v} \times \mathbf{E}_s)\, d^3x$$

or

$$\mathbf{p}_{e1} \simeq \frac{1}{4\pi c^2} \int \left[E_s^2 \mathbf{v} - (\mathbf{E}_s \cdot \mathbf{v})\mathbf{E}_s \right] d^3x \tag{17.47}$$

This is the Abraham-Lorentz momentum. If the field is assumed spherically
symmetric, the second term can be averaged over angles to give one-third
of the first term, leading to $\frac{4}{3}(U/c^2)\mathbf{v}$, as already discussed below (17.37).

The second term in the momentum is

$$\mathbf{p}_{e2} \simeq \frac{1}{c^2} \int \mathbf{v} \cdot \overleftrightarrow{\mathbf{T}}\, d^3x \simeq \frac{1}{4\pi c^2} \int [(\mathbf{v} \cdot \mathbf{E}_s)\mathbf{E}_s - \tfrac{1}{2}E_s^2\mathbf{v}]\, d^3x \tag{17.48}$$

With *no assumption* about spherical symmetry, the sum of (17.47) and
(17.48) is

$$\mathbf{p}_e = \mathbf{p}_{e1} + \mathbf{p}_{e2} = \frac{\mathbf{v}}{8\pi c^2} \int E_s^2\, d^3x = \frac{U}{c^2}\mathbf{v} \tag{17.49}$$

as expected from Lorentz covariance.

The modified definitions (17.45) allow one from a formal point of view
to obtain manifestly covariant expressions for the electromagnetic self-
energy and momentum, without reference to other types of forces or
questions of stability. We have seen, however, that the covariant forms
are obtained by omitting the work done or the momentum generated by the
electromagnetic forces which must eventually be canceled by the attractive
nonelectromagnetic forces necessary for stability. Hence, at least for an

extended classical charge distribution, it is a matter of taste whether the Poincaré approach is completely satisfactory or whether one demands that the various contributions to the self-energy and momentum be defined to be separately covariant.

An apparently different, but closely related, solution to the lack of covariance implied by the appearance of the factor $\frac{4}{3}$ in the Abraham-Lorentz force equation (17.33) was made by Fermi* in 1922, when he demonstrated that a covariant application of Hamilton's principle led to an appropriate modification of the self-force (17.20) so that a factor of unity is obtained, instead of $\frac{4}{3}$. A discussion similar in some respects to that given here was presented by Wilson.†

17.7 Integrodifferential Equation of Motion, Including Radiation Damping

In Section 17.2 the Abraham-Lorentz equation (17.9) was discussed qualitatively. It was pointed out that, if the radiative effects were considered as small in some sense, a scheme of successive approximations could be used to describe the motion. Nevertheless, the equation in its differential form contains unphysical behavior [e.g., solution (17.10)] because it is higher order in time differentiation than a mechanical equation of motion should be. It is desirable to have an equivalent equation of motion which is of the correct order, has no grossly unphysical solutions, and exhibits the successive approximations aspect in a natural manner. The discussion will be limited to nonrelativistic motion, although the generalization to fully relativistic motion is not difficult.

The guiding principle in converting (17.9) into an equivalent equation of motion is that the new equation should have solutions which evolve continuously into those for a neutral particle in the limit as the charge of the particle tends to zero. The smaller the particle's charge, the smaller the self-fields, and the smaller the radiative effects, other things being equal.

If the external force is thought of as a given function of time, (17.9) can be integrated once with respect to time by use of an integrating factor. We put

$$\dot{\mathbf{v}}(t) = e^{t/\tau} \mathbf{u}(t)$$

Then we find from (17.9) that

$$m\dot{\mathbf{u}} = -\frac{1}{\tau} e^{-t/\tau} \mathbf{F}(t)$$

---

* E. Fermi, *Physik. Z.*, **23**, 340 (1922), or *Atti accad. nazl. Lincei Rend.*, **31**, 184, 306 (1922).

† W. Wilson, *Proc. Phys. Soc.* (*London*), **A48**, 376 (1936).

The first integral is therefore

$$m\dot{\mathbf{v}}(t) = \frac{e^{t/\tau}}{\tau} \int_t^C e^{-t'/\tau} \mathbf{F}(t') \, dt' \qquad (17.50)$$

The minus sign of the preceding line has been absorbed by making the lower limit of the integral the indefinite one. The constant of integration C is to be determined on physical grounds.

The integrodifferential equation of motion (17.50) differs from customary mechanical equations of motion in that the acceleration of the particle at any time depends, not on the instantaneous value of the force acting, but on a weighted time average of the force. The presence of the factor $e^{-(t'-t)/\tau}$ means that only a small time interval of order τ is involved. Since $\tau \propto e^2$, that time interval becomes vanishingly small as $e^2 \to 0$. Then we demand that the equation of motion become Newton's equation, $m\dot{\mathbf{v}}(t) = \mathbf{F}(t)$. This is accomplished by choosing the upper limit on the integral in (17.50) as infinity. To see the behavior in detail, we introduce a new variable of integration,

$$s = \frac{1}{\tau}(t' - t)$$

Then (17.50) can be written

$$m\dot{\mathbf{v}}(t) = \int_0^\infty e^{-s} \mathbf{F}(t + \tau s) \, ds \qquad (17.51)$$

If the force is slowly varying in time (measured in units of τ), a Taylor's series expansion around $s = 0$ can be expected to converge rapidly. Thus we write

$$\mathbf{F}(t + \tau s) = \sum_{n=0}^\infty \frac{(\tau s)^n}{n!} \frac{d^n \mathbf{F}(t)}{dt^n} \qquad (17.52)$$

On substitution into (17.51) this gives

$$m\dot{\mathbf{v}}(t) = \sum_{n=0}^\infty \tau^n \frac{d^n \mathbf{F}(t)}{dt^n} \qquad (17.53)$$

In the limit $\tau \to 0$ only the $n = 0$ term in the series survives. Then one has the ordinary equation of motion of an uncharged particle. The higher terms represent radiative corrections for a charged particle, terms which are important only if the force varies in time sufficiently rapidly.

The integrodifferential equation (17.51) can be regarded as a physically reasonable replacement for the Abraham-Lorentz equation of motion (17.9). All solutions of (17.51) satisfy (17.9). But unphysical "runaway" solutions, such as (17.10), do not occur. Equation (17.51) still has certain peculiarities. The chief of these is its violation of the traditional concept

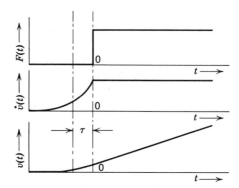

Fig. 17.1 "Preacceleration" of charged particle.

of causality. It is evident from (17.51) that the acceleration at time t depends on the force acting at times *later* than t. This is contrary to our ideas of cause and effect. Figure 17.1 shows a typical example of such acausal behavior. A constant force is applied to the particle for times $t > 0$. The equation of motion predicts "preacceleration" before the force is "actually" applied.

To understand whether such effects are in contradiction to known facts we must consider the time scale involved. The acausal effects are limited to time intervals of the order of $\tau \sim e^2/mc^3 \sim 10^{-24}$ sec. This is the time it takes light to travel a distance of the order of the "size" of elementary particles. Such a short time interval is impossible to detect by macroscopic means. Forces, for example, cannot be turned on and off at a particle with the rapidity indicated in the figure. Hence the lack of causality inherent in (17.51) cannot be observed in the laboratory. We describe this state of affairs by saying that, while (17.51) implies lack of microscopic causality, the model satisfies the requirements of macroscopic causality. Another important point is that the model is a classical one which surely fails at distances and times much greater than e^2/mc^2 and τ. As a consequence of the uncertainty principle the turning on of an external force in a time interval Δt is accompanied by uncertainties in energy of the order of $\Delta E \sim \hbar/\Delta t$. If these energy uncertainties are of the order of the rest energy mc^2 of the particle, the behavior will be far from classical. This sets a quantum-mechanical limit on the time intervals, $\tau_q \sim \hbar/mc^2 \sim 137\tau$. Since $\tau_q \gg \tau$, we reach the conclusion that in the domain where the classical equation is expected to hold the motions are sufficiently gentle that (1) acausal effects are of very minor importance, and (2) radiative reaction causes only small corrections to the motion.

If the applied force $\mathbf{F}$ is given as a function of position rather than time, the solution of the integrodifferential equation becomes somewhat more involved, although no different in principle.

17.8 Line Breadth and Level Shift of an Oscillator

The effects of radiative reaction are of great importance in the detailed behavior of atomic systems. Although a complete discussion involves the rather elaborate formalism of quantum electrodynamics, the qualitative features are apparent from a classical treatment. As a typical example we consider a charged particle bound by a one-dimensional linear restoring force with force constant $k = m\omega_0^2$. In the absence of radiation damping the particle oscillates with constant amplitude at the characteristic frequency ω_0. When the reactive effects are included, the amplitude of oscillation gradually decreases, since energy of motion is being converted into radiant energy.

If the displacement of the charged particle from equilibrium is $x(t)$, the equation of motion (17.51) for this problem is

$$\ddot{x}(t) + \omega_0^2 \int_0^\infty e^{-s} x(t + \tau s) \, ds = 0 \tag{17.54}$$

Since the solution when $\tau = 0$ is $x(t) \sim e^{-i\omega_0 t}$, it is natural to assume a solution of the form,

$$x(t) = x_0 e^{-\alpha t} \tag{17.55}$$

We anticipate on physical grounds that the imaginary part of α will be closely equal to ω_0, at least for $\omega_0 \tau \ll 1$, but that α will have a positive real part to describe the dissipative effect of the emission of radiation. When (17.55) is substituted into (17.54), there results a cubic equation for α:

$$\tau \alpha^3 + \alpha^2 + \omega_0^2 = 0 \tag{17.56}$$

There are three roots for α; two are complex conjugates and one is real. The real root is always negative and must be discarded [it corresponds to the "runaway" solutions of (17.9)]. The two physically meaningful roots can be exhibited in closed form for arbitrary τ and ω_0, but the formula is sufficiently involved that it is of little value except for numerical computation. We are interested in the range of parameters where $\omega_0 \tau \ll 1$. Then it is a simple matter to show directly from (17.56) that, correct to order $(\omega_0 \tau)^2$ inclusive, α is given by

$$\alpha = \frac{\Gamma}{2} \pm i(\omega_0 + \Delta\omega) \left.\vphantom{\begin{array}{c}1\\1\\1\end{array}}\right\}$$

where
$$\Gamma = \omega_0^2 \tau \qquad\qquad \left.\vphantom{\begin{array}{c}1\\1\\1\end{array}}\right\} \tag{17.57}$$

$$\Delta\omega = -\tfrac{5}{8}\omega_0^3 \tau^2 \left.\vphantom{\begin{array}{c}1\\1\\1\end{array}}\right.$$

The constant Γ is known as the *decay constant*, while $\Delta\omega$ is called the *level shift*.

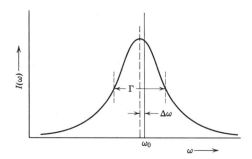

Fig. 17.2 Broadening and shifting of spectral line because of radiative reaction. The Lorentz line shape has width Γ. The level shift is $\Delta\omega$.

The energy of the oscillator decays exponentially as $e^{-\Gamma t}$ because of radiation damping. This means that the emitted radiation appears as a wave train with effective length of the order of c/Γ. Such a finite pulse of radiation is not exactly monochromatic but has a frequency spectrum covering an interval of order Γ. The exact shape of the frequency spectrum is given by the square of the Fourier transform of the electric field or the acceleration. Neglecting an initial transient (of duration τ), the amplitude of the spectrum is thus proportional to

$$E(\omega) \sim \int_0^\infty e^{-\alpha t} e^{i\omega t}\, dt = \frac{1}{\alpha - i\omega}$$

The energy radiated per unit frequency interval is therefore

$$I(\omega) = I_0 \frac{\Gamma}{2\pi} \frac{1}{(\omega - \omega_0 - \Delta\omega)^2 + (\Gamma/2)^2} \qquad (17.58)$$

where I_0 is the total energy radiated. This spectral distribution is called a *Lorentz line shape*. The width of the distribution at half-maximum intensity is called the *half-width* or *line breadth* and is equal to Γ. Shown in Fig. 17.2 is such a spectral line. Because of the reactive effects of radiation the line is *broadened and shifted* in frequency.

The classical line breadth for electronic oscillators is a universal constant when expressed in terms of wavelength:

$$\Delta\lambda = 2\pi \frac{c}{\omega_0^2} \Gamma = 2\pi c\tau = 1.2 \times 10^{-4}\,\text{Å}$$

Quantum mechanically the natural widths of spectral lines vary. In order to establish a connection with the classical treatment, the quantum-mechanical line width is sometimes written as

$$\Gamma_q = f_{ij}\Gamma$$

where f_{ij} is the "oscillator strength" of the transition $(i \to j)$. Oscillator strengths vary considerably, sometimes being nearly unity for strong single-electron transitions and sometimes much smaller.

The classical level shift $\Delta\omega$ is smaller than the line width Γ by a factor $\omega_0\tau \ll 1$. Quantum mechanically (and experimentally) this is not so. The reason is that in the quantum theory there is a different mechanism for the level shift, although still involving the electromagnetic field. Even in the absence of photons, the quantized radiation field has nonvanishing expectation values of the *squares* of the electromagnetic field strengths (vacuum fluctuations). These fluctuating fields (along with vacuum fluctuations in the negaton-positon field) act on the charged particle to cause a shift in its energy. The quantum-mechanical level shift for an oscillator is of the order of

$$\frac{\Delta\omega_q}{\omega_0} \sim \omega_0\tau \log\left(\frac{mc^2}{\hbar\omega_0}\right)$$

as compared to the classical shift due to emission of radiation,

$$\frac{|\Delta\omega_c|}{\omega_0} \sim (\omega_0\tau)^2$$

The quantum-mechanical level shift is seen to be comparable to, or greater than, the line width. The small radiative shift of energy levels of atoms was first observed by Lamb in 1947* and is called the *Lamb shift* in his honor. A readable account of the quantum-theoretical aspects of the problem, requiring only a rudimentary knowledge of quantum field theory, has been given by Weisskopf.†

17.9 Scattering and Absorption of Radiation by an Oscillator

The scattering of radiation by free charged particles has been discussed in Sections 14.7 and 14.8. We now wish to consider the scattering and absorption of radiation by bound charges. The first example chosen is the scattering of radiation of frequency ω by a single nonrelativistic particle of mass m and charge e bound by a spherically symmetric, linear, restoring force $m\omega_0^2\mathbf{x}$. Because we will be dealing with steady-state oscillations, it is allowable to employ the Abraham-Lorentz equation (17.9), rather than the integrodifferential form (17.51). Then the equation of motion is

$$m(\ddot{\mathbf{x}} - \tau\dddot{\mathbf{x}} + \omega_0^2\mathbf{x}) = \mathbf{F}(t)$$

If we wish to allow for other dissipative processes (corresponding quantum-mechanically to other modes of decay besides photon emission), we can add a resistive term $(m\Gamma'\dot{\mathbf{x}})$ to the left-hand side, Γ' being a decay constant

---

* W. E. Lamb and R. C. Retherford, *Phys. Rev.*, **72**, 241 (1947).
† V. F. Weisskopf, *Revs. Modern Phys.*, **21**, 305 (1949).

with dimensions of frequency which measures the strength of the non-electromagnetic dissipative effects. The incident electromagnetic field provides the driving force. In the dipole approximation the equation of motion then becomes

$$\ddot{\mathbf{x}} + \Gamma'\dot{\mathbf{x}} - \tau\dddot{\mathbf{x}} + \omega_0^2\mathbf{x} = \frac{e}{m}\boldsymbol{\epsilon}E_0 e^{-i\omega t} \tag{17.59}$$

where E_0 is the electric field at the center of force, and $\boldsymbol{\epsilon}$ is the incident polarization vector. The steady-state solution is

$$\mathbf{x} = \frac{e}{m}\frac{E_0 e^{-i\omega t}}{\omega_0^2 - \omega^2 - i\omega\Gamma_t}\boldsymbol{\epsilon} \tag{17.60}$$

where

$$\Gamma_t(\omega) = \Gamma' + \left(\frac{\omega}{\omega_0}\right)^2\Gamma \tag{17.61}$$

is called the *total decay constant* or *total width*. The radiative decay constant is $\Gamma = \omega_0^2\tau$.

The accelerated motion described by (17.60) gives rise to radiation fields. From (14.18) the radiation electric field is

$$\mathbf{E}_{\text{rad}} = \frac{e}{c^2}\frac{1}{r}[\mathbf{n} \times (\mathbf{n} \times \ddot{\mathbf{x}})]_{\text{ret}}$$

Consequently the radiation field with polarization $\boldsymbol{\epsilon}'$ is given by

$$\boldsymbol{\epsilon}' \cdot \mathbf{E}_{\text{rad}} = \frac{e^2}{mc^2}\omega^2\frac{E_0 e^{-i\omega t}e^{ikr}}{\omega_0^2 - \omega^2 - i\omega\Gamma_t}\left(\frac{\boldsymbol{\epsilon} \cdot \boldsymbol{\epsilon}'}{r}\right) \tag{17.62}$$

From definition (14.101) of differential-scattering cross section we find that the cross section for scattered radiation of frequency ω and polarization $\boldsymbol{\epsilon}'$ is

$$\frac{d\sigma(\omega, \boldsymbol{\epsilon}')}{d\Omega} = \left|\frac{r\boldsymbol{\epsilon}' \cdot \mathbf{E}_{\text{rad}}}{E_0}\right|^2 = \left(\frac{e^2}{mc^2}\right)^2(\boldsymbol{\epsilon} \cdot \boldsymbol{\epsilon}')^2\left[\frac{\omega^4}{(\omega_0^2 - \omega^2)^2 + \omega^2\Gamma_t^2}\right] \tag{17.63}$$

The factor multiplying the square bracket is just the Thomson cross section for scattering by a free particle.

For frequencies very small compared to the binding frequency ($\omega \ll \omega_0$) the cross section reduces to

$$\frac{d\sigma(\omega, \boldsymbol{\epsilon}')}{d\Omega} = \left(\frac{e^2}{mc^2}\right)^2(\boldsymbol{\epsilon} \cdot \boldsymbol{\epsilon}')^2\left(\frac{\omega}{\omega_0}\right)^4 \tag{17.64}$$

The scattering at long wavelengths is thus inversely proportional to the fourth power of the wavelength. This is the Rayleigh law of scattering. As mentioned in Section 16.9, it is a general property of all systems possessing an electric dipole polarizability.

For frequencies near the binding frequency ω_0 the scattering becomes very great, showing a typical resonance behavior. In the neighborhood of the resonance the cross section can be approximated by

$$\frac{d\sigma(\omega, \boldsymbol{\epsilon}')}{d\Omega} \simeq \frac{9}{16} \lambdabar_0{}^2 \frac{\Gamma^2}{(\omega - \omega_0)^2 + (\Gamma_t/2)^2} (\boldsymbol{\epsilon} \cdot \boldsymbol{\epsilon}')^2 \qquad (17.65)$$

where $\lambdabar_0 = (c/\omega_0)$ is the wavelength (divided by 2π) at resonance, $\Gamma = \omega_0{}^2\tau$ is the radiative decay constant, and $\Gamma_t \simeq \Gamma + \Gamma'$. If a sum is taken over scattered polarizations and an integration is made over all angles, there results a total scattering cross section,

$$\sigma_{\text{sc}}(\omega) \simeq \frac{3\pi}{2} \lambdabar_0{}^2 \frac{\Gamma^2}{(\omega - \omega_0)^2 + (\Gamma_t/2)^2} \qquad (17.66)$$

This exhibits the typical Lorentz line shape with half-width given by Γ_t and peak cross section,

$$\sigma_{\text{sc}}(\omega_0) = 6\pi\lambdabar_0{}^2 \left(\frac{\Gamma}{\Gamma_t}\right)^2 \qquad (17.67)$$

At high frequencies $(\omega \gg \omega_0)$ the cross section (17.63) approaches the Thomson free-particle value, apart from a factor $(1 + \omega^2\tau^2)^{-1}$ due to radiation damping. In the classical domain this factor can be taken as unity: $\omega\tau \sim 1$ corresponds to photons of energies $\hbar\omega \sim 137mc^2$. Quantum effects become important when $\hbar\omega \sim mc^2$, as discussed in Sections 14.7 and 17.7.

Figure 17.3 shows the scattering cross section over the whole classical range of frequencies.

The sharply resonant scattering at $\omega = \omega_0$ is called *resonance fluorescence*. Quantum mechanically it corresponds to the absorption of radiation by an atom, molecule, or nucleus in a transition from its ground

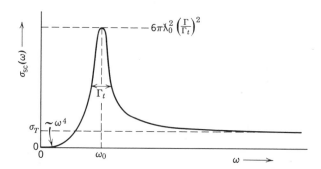

Fig. 17.3 Total cross section for the scattering of radiation by an oscillator as a function of frequency. σ_T is the Thomson free-particle scattering cross section.

state to an excited state with the subsequent re-emission of the radiation in other directions in the process of de-excitation. The factor $6\pi\lambda_0^2$ in the peak cross section is replaced quantum mechanically by the statistical factor,

$$6\pi\lambda_0^2 \rightarrow 4\pi\lambda_0^2 \frac{(2J_{ex} + 1)}{2(2J_g + 1)}$$

where J_g and J_{ex} are the angular momenta of the ground and excited states, and $4\pi\lambda_0^2$ is the maximum allowable scattering for any single quantum state. The remaining factors represent a sum over all final magnetic substates and an average over initial ones, the factor 2 being the statistical weight associated with the incident radiation's polarizations. The classical result corresponds to $J_g = 0$ and $J_{ex} = 1$.

The absorption of radiation, as distinct from scattering, has already been discussed for an oscillator in Section 13.2. The driving fields there were those of a swift, charged particle, but the treatment [from equation (13.15) to (13.24)] was general enough to allow direct transcription. The only differences are that the Γ of Section 13.2 is to be replaced by Γ_t (17.61), and the incident electric field is to be taken as essentially monochromatic. From (13.24) we find that in the dipole approximation the energy absorbed per unit frequency interval is

$$\frac{dE}{d\omega} = \frac{e^2}{m} |E_0(\omega)|^2 \frac{2\omega^2\Gamma_t}{(\omega_0^2 - \omega^2)^2 + \omega^2\Gamma_t^2} \tag{17.68}$$

The *absorption cross section* can be defined as the energy absorbed per unit frequency interval divided by the incident energy per unit area per unit frequency. The incident flux is $(c/2\pi)|E_0(\omega)|^2$. Consequently the absorption cross section is

$$\sigma_{abs}(\omega) = 4\pi \frac{e^2}{mc} \frac{\omega^2\Gamma_t}{(\omega_0^2 - \omega^2)^2 + \omega^2\Gamma_t^2} \tag{17.69}$$

Using the definition of $\Gamma = \omega_0^2\tau$, this can be written

$$\sigma_{abs}(\omega) = 6\pi\lambda_0^2 \frac{\omega^2\Gamma\Gamma_t}{(\omega_0^2 - \omega^2)^2 + \omega^2\Gamma_t^2} \tag{17.70}$$

In the three regions, $\omega \ll \omega_0$, $\omega \sim \omega_0$, $\omega \gg \omega_0$, the cross section can be approximated as

$$\sigma_{abs}(\omega) \simeq \begin{cases} 6\pi\lambda_0^2 \dfrac{\omega^2\Gamma\Gamma_t}{\omega_0^4}, & \omega \ll \omega_0 \\[2ex] \dfrac{3\pi}{2}\lambda_0^2 \dfrac{\Gamma\Gamma_t}{(\omega_0 - \omega)^2 + (\Gamma_t/2)^2}, & \omega \sim \omega_0 \\[2ex] 6\pi\lambda_0^2 \dfrac{\Gamma\Gamma_t}{\omega^2}, & \omega \gg \omega_0 \end{cases} \tag{17.71}$$

We see that near the resonant frequency ω_0 the absorption cross section has the same Lorentz shape as the scattering cross section, but is larger by a factor Γ_t/Γ. At high frequencies $\Gamma_t \to \omega^2\tau$, so that the absorption cross section approaches the constant Thomson value (we have again ignored $\omega\tau$ compared to unity).

The absorption cross section is sometimes called the *total* cross section because it includes all processes, scattering as well as other dissipative effects described by Γ'. To obtain the absorption cross section we merely calculated the energy absorbed by the oscillator, without asking whether it was re-emitted as radiation or dissipated in some other way. The cross section for processes other than scattering is called the *reaction* cross section, $\sigma_r(\omega)$. It can be calculated simply by taking the difference between the total cross section and the scattering cross section integrated over angles. All three cross sections can be written in a suggestive form based on (17.70):

$$\left.\begin{array}{c} \sigma_{sc}(\omega) \\[8pt] \sigma_r(\omega) \\[8pt] \sigma_{abs}(\omega) \end{array}\right\} = 6\pi c^2 \frac{(\omega^2/\omega_0^2)\Gamma}{(\omega_0^2 - \omega^2)^2 + \omega^2\Gamma_t^2} \left\{\begin{array}{c} \dfrac{\omega^2}{\omega_0^2}\Gamma \\[8pt] \Gamma' \\[8pt] \Gamma_t \end{array}\right. \tag{17.72}$$

The resonant denominator is the same for all three cross sections. The radiative process is proportional to $(\omega^2/\omega_0^2)\Gamma = \omega^2\tau$. The other dissipative processes (reactions) are proportional to Γ'. The common factor $(\omega^2/\omega_0^2)\Gamma$ represents the incident radiation. For scattering a second factor $(\omega^2/\omega_0^2)\Gamma$ appears, while for reactions a factor Γ' appears. The absorption or total cross section involves the total width Γ_t. This characteristic product of decay constants or widths appropriate to the initial and final states of the process also occurs quantum mechanically in the theory of resonance reactions.

The integral over all frequencies of the absorption cross section, neglecting the radiative scattering, yields a relation called the *dipole sum rule*. Neglect of the scattering is equivalent to the assumption that the width Γ_t in (17.70) is a constant, independent of frequency. The integral of $\sigma_{abs}(\omega)$ over all frequencies in this approximation is then easily shown to be

$$\int_0^\infty \sigma_{abs}(\omega)\,d\omega = \frac{2\pi^2 e^2}{mc} \tag{17.73}$$

We note that the sum rule depends on the charge and mass of the particle, but not on other detailed properties, such as ω_0 and Γ'. It is equivalent to expression (13.26) for the total energy absorbed by the system from the passing fields.

The dipole sum rule is a general statement that is true both classically and quantum mechanically, no matter how complicated the response of the system to the incident radiation as a function of frequency. It depends only on the two physical requirements that (*a*) the normal modes of oscillation of the system must decay in time (even if very slowly) because of ever-present resistive losses, and (*b*) at high frequencies binding effects are unimportant and the particle responds as if it were free (see Problem 17.8).

For a system of independent particles with charges e_j and masses m_j bound to a fixed center the sum rule has an obvious generalization,

$$\int_0^\infty \sigma_{abs}(\omega)\, d\omega = \frac{2\pi^2}{c} \sum_j \frac{e_j^2}{m_j} \tag{17.74}$$

If the particles are bound together by mutual interactions, the center of mass motion must be removed. It is easy to show that this is accomplished by subtracting from the sum in (17.74) a term (Q^2/M), where Q is the total charge of the system of particles and M the total mass. For a nucleus with Z protons and $N\,(= A - Z)$ neutrons, the sum rule then becomes

$$\int_0^\infty \sigma_{abs}(\omega)\, d\omega = \frac{2\pi^2 e^2}{mc} \left(\frac{NZ}{A}\right) \tag{17.75}$$

where e is the protonic charge, and m is the mass of one nucleon.*

REFERENCES AND SUGGESTED READING

The history of the attempts at classical models of charged particles and associated questions is treated in interesting detail by
Whittaker.
The ideas of Abraham, Lorentz, Poincaré, and others are presented in lucid and elegant fashion by
Lorentz, Sections 26–37, 179–184, Note 18.
Lorentz's discussion seems old fashioned by our standards, but is a model of clear physical thinking for its time.
Clear, if brief, treatments of self-energy effects and radiative reaction are given by
Abraham and Becker, Band II, Sections 13, 14, 66,
Landau and Lifshitz, *Classical Theory of Fields*, Section 9.9,
Panofsky and Phillips, Chapters 20 and 21,
Sommerfeld, *Electrodynamics*, Section 36.
A relativistic classical point electron theory was first developed by
P. A. M. Dirac, *Proc. Roy Soc.*, **A167**, 148 (1938).

* Actually the sum rule for the nuclear photoeffect has an added factor $(1 + x)$, where $x \sim \frac{1}{2}$ is the contribution due to exchange forces in the nucleus. A physical way to understand this increase is to think of the exchange forces as caused by the transfer of virtual charged pi mesons between nucleons. These virtual charged mesons contribute to the total nuclear current. Since their e/m ratio is larger than that of the nucleons, they cause an increase of the sum rule over its ordinary value (17.75).

The integrodifferential equation of motion (17.51) seems to have been first written down
by
    Iwanenko and Sokolow, Section 35.
Many aspects of the relativistic classical theory are discussed by
    Rohrlich,
    F. Rohrlich, *Am. J. Phys.*, **28**, 639 (1960).
Examples of the solution of the integrodifferential equation of motion are given by
    G. N. Plass, *Revs. Modern Phys.*, **33**, 37 (1961).

PROBLEMS

17.1 A nonrelativistic particle of charge e and mass m is bound by a linear,
isotropic, restoring force with force constant $m\omega_0^2$.
    Using (17.13) and (17.16) of Section 17.2, show that the energy and
angular momentum of the particle both decrease exponentially from their
initial values as $e^{-\Gamma t}$, where $\Gamma = \omega_0^2\tau$.

17.2 A nonrelativistic electron of charge $-e$ and mass m bound in an attractive
Coulomb potential $(-Ze^2/r)$ moves in a circular orbit in the absence of
radiation reaction.
    (*a*) Show that both the energy and angular-momentum equations (17.13)
and (17.16) lead to the solution for the slowly changing orbit radius,

$$r^3(t) = r_0^3 - 9Z(c\tau)^3 \frac{t}{\tau}$$

where r_0 is the value of $r(t)$ at $t = 0$.
    (*b*) For circular orbits in a Bohr atom the orbit radius and the principal
quantum number n are related by $r = n^2 a_0/Z$. If the transition probability
for transitions from $n \to (n-1)$ is defined as $-dn/dt$, show that the result
of (*a*) agrees with that found in Problem 14.9.
    (*c*) From (*a*) calculate the numerical value of the time taken for a mu
meson of mass $m = 207m_e$ to fall from a circular orbit with principal
quantum number $n_1 = 10$ to one with $n_2 = 4$, $n_2 = 1$. This is a reasonable
estimate of the time taken for a mu meson to cascade down to its lowest
orbit after capture by an isolated atom.

17.3 An electron moving in an attractive Coulomb field $(-Ze^2/r)$ with binding
energy ϵ and angular momentum L has an elliptic orbit,

$$\frac{1}{r} = \frac{Ze^2 m}{L}\left[1 + \sqrt{1 - \frac{2\epsilon L^2}{Z^2 e^4 m}}\cos(\theta - \theta_0)\right]$$

The eccentricity ξ of the ellipse is given by the square root multiplying the
cosine.
    (*a*) By performing the appropriate time averages over the orbit show that
the secular changes in energy and angular momentum are

$$\frac{d\epsilon}{dt} = \frac{2^{3/2}}{3}\frac{Z^3 e^8 m^{1/2}}{c^3}\frac{\epsilon^{3/2}}{L^5}\left(3 - \frac{2\epsilon L^2}{Z^2 e^4 m}\right)$$

$$\frac{dL}{dt} = -\frac{2^{5/2}}{3}\frac{Ze^4}{m^{1/2}c^3}\frac{\epsilon^{3/2}}{L^2}$$

(b) If the initial values of ϵ and L are ϵ_0 and L_0, show that

$$\epsilon(L) = \frac{Z^2 e^4 m}{2L^2}\left[1 - \left(\frac{L}{L_0}\right)^3\right] + \frac{\epsilon_0}{L_0} L$$

Calculate the eccentricity of the ellipse, and show that it decreases from its initial value as $(L/L_0)^{3/2}$, showing that the orbit tends to become circular as time goes on.

(c) Compare your results here to the special case of a circular orbit of Problem 17.2.

Hint: In performing the time averages make use of Kepler's law of equal areas $(dt = mr^2\,d\theta/L)$ to convert time integrals to angular integrals.

17.4 The Dirac (1938) relativistic theory of classical point electrons has as its equation of motion,

$$\frac{dp_\mu}{d\tau} = F_\mu^{\text{ext}} + F_\mu^{\text{rad}}$$

where p_μ is the particle's 4-momentum, τ is the particle's proper time, and F_μ^{rad} is the covariant generalization of the radiative reaction force (17.8). Using the requirement that any force must satisfy $F_\mu p_\mu = 0$, show that

$$F_\mu^{\text{rad}} = \frac{2e^2}{3mc^3}\left[\frac{d^2 p_\mu}{d\tau^2} - \frac{p_\mu}{m^2 c^2}\left(\frac{dp_\nu}{d\tau}\frac{dp_\nu}{d\tau}\right)\right]$$

17.5 (a) Show that for relativistic motion in one dimension the equation of motion of Problem 17.4 can be written in the form,

$$\dot{p} - \frac{2e^2}{3mc^3}\left(\ddot{p} - \frac{p\dot{p}^2}{p^2 + m^2 c^2}\right) = \sqrt{1 + \frac{p^2}{m^2 c^2}}\, f(\tau)$$

where p is the momentum in the direction of motion, a dot means differentiation with respect to proper time, and $f(\tau)$ is the ordinary Newtonian force as a function of proper time.

(b) Show that the substitution of $p = mc \sinh y$ reduces the relativistic equation to the Abraham-Lorentz form (17.9) in y and τ. Write down the general solution for $p(\tau)$, with the initial condition that

$$p(\tau) = p_0 \quad \text{at} \quad \tau = 0$$

17.6 A nonrelativistic particle of charge e and mass m is accelerated in one-dimensional motion across a gap of width d by a constant electric field. The mathematical idealization is that the particle has applied to it an external force $m\alpha$ while its coordinate lies in the interval $(0, d)$. Without radiation damping the particle, having initial velocity v_0, is accelerated uniformly for a time $T = (-v_0/\alpha) + \sqrt{(v_0^2/\alpha^2) + (2d/\alpha)}$, emerging at $x = d$ with a final velocity $v_1 = \sqrt{v_0^2 + 2\alpha d}$.

With radiation damping the motion is altered so that the particle takes a time T' to cross the gap and emerges with a velocity v_1'.

(a) Solve the integrodifferential equation of motion, including damping, assuming T and T' large compared to τ. Sketch a velocity versus time diagram for the motion with and without damping.

(b) Show that to lowest order in τ,

$$T' = T - \tau\left(1 - \frac{v_0}{v_1}\right)$$

$$v_1' = v_1 - \frac{\alpha^2\tau}{v_1} T$$

(c) Verify that the sum of the energy radiated and the change in the particle's kinetic energy is equal to the work done by the applied field.

17.7 A classical model for the description of collision broadening of spectral lines is that the oscillator is interrupted by a collision after oscillating for a time T so that the coherence of the wave train is lost.

(a) Taking the oscillator used in Section 17.8 and assuming that the probability that a collision will occur between time T and $(T + dT)$ is $(ve^{-vT} dT)$, where v is the mean collision frequency, show that the averaged spectral distribution is:

$$I(\omega) = \frac{I_0}{2\pi} \frac{\Gamma + 2v}{(\omega - \omega_0)^2 + \left(\dfrac{\Gamma}{2} + v\right)^2}$$

so that the breadth of the line is $(2v + \Gamma)$.

(b) For the sodium doublet at 5893 Å the oscillator strength is $f = 0.975$, so that the natural width is essentially the classical value, $\Delta\lambda = 1.2 \times 10^{-4}$ Å. Estimate the Doppler width of the line, assuming the sodium atoms are in thermal equilibrium at a temperature of 500°K, and compare it with the natural width. Assuming a collision cross section of 10^{-16} cm², determine the collision breadth of the sodium doublet as a function of the pressure of the sodium vapor. For what pressure is the collision breadth equal to the natural breadth? The Doppler breadth?

17.8 A single particle oscillator under the action of an applied electric field $E_0 e^{-i\omega t}$ has a dipole moment given by

$$\mathbf{p} = \alpha(\omega)\mathbf{E}_0 e^{-i\omega t}$$

(a) Show that the dipole absorption cross section can be written as

$$\sigma_{abs}(\omega) = \frac{2\pi}{c}[-i\omega\alpha(\omega) + \text{c.c.}]$$

(b) Using only the facts that all the normal modes of oscillation must have some damping and that the polarizability $\alpha(\omega)$ must approach the free-particle value $(-e^2/m\omega^2)$ at high frequencies, show that the absorption cross section satisfies the dipole sum rule,

$$\int_0^\infty \sigma_{abs}(\omega)\, d\omega = \frac{2\pi^2 e^2}{mc}$$

Appendix on Units and Dimensions

The question of units and dimensions in electricity and magnetism has exercised a great number of physicists and engineers over the years. This situation is in marked contrast with the almost universal agreement on the basic units of length (centimeter or meter), mass (gram or kilogram), and time (mean solar second). The reason perhaps is that the mechanical units were defined when the idea of "absolute" standards was a novel concept (just before 1800), and they were urged on the professional and commercial world by a group of scientific giants (Borda, Laplace, and others). By the time the problem of electromagnetic units arose there were (and still are) many experts. The purpose of this appendix is to add as little heat and as much light as possible without belaboring the issue.

1 Units and Dimensions; Basic Units and Derived Units

The *arbitrariness* in the *number* of fundamental units and in the *dimensions* of any physical quantity in terms of those units has been emphasized by Abraham, Planck, Bridgman,* Birge,† and others. The reader interested in units as such will do well to familiarize himself with the excellent series of articles by Birge.

The desirable features of a system of units in any field are convenience and clarity. For example, theoretical physicists active in relativistic quantum field theory and the theory of elementary particles find it convenient to *choose* the universal constants such as Planck's quantum of

* P. W. Bridgman, *Dimensional Analysis*, Yale University Press (1931).
† R. T. Birge, *Am. Phys. Teacher* (now *Am. J. Phys.*), **2**, 41 (1934); **3**, 102, 171 (1935).

action and the velocity of light in vacuum to be *dimensionless* and of *unit magnitude*. The resulting system of units (called "natural" units) has only *one* basic unit, customarily chosen to be length. All quantities, whether length or time or force or energy, etc., are expressed in terms of this one unit and have dimensions which are powers of its dimension. There is nothing contrived or less fundamental about such a system than one involving the meter, the kilogram, and the second as basic units (mks system). It is merely a matter of convenience.*

A word needs to be said about basic units or standards, considered as independent quantities, and derived units or standards, which are defined in both magnitude and dimension through theory and experiment in terms of the basic units. Tradition requires that mass (m), length (l), and time (t) be treated as basic units. But for electrical quantities there is as yet no compelling tradition. Consider, for example, the unit of current. The "international" ampere (for a long period the accepted practical unit of current) is defined in terms of the mass of silver deposited per unit time by electrolysis in a standard silver voltameter. Such a unit of current is properly considered a basic unit, independent of mass, length and time units, since the amount of current serving as the unit is found from a supposedly reproducible experiment in electrolysis.

On the other hand, the presently accepted standard of current, the "absolute" ampere, is defined as that current which when flowing in each of two infinitely long, parallel wires of negligible cross-sectional area, separated by a distance of 1 meter in vacuum, causes a transverse force per unit length of 2×10^{-7} newton/meter to act between the wires. This means that the "absolute" ampere is a derived unit, since its definition is in terms of the mechanical force between two wires through equation (A.4) below.† The "absolute" ampere is, by this definition, exactly one-tenth of the em unit of current, the abampere. Since 1948 the internationally accepted system of electromagnetic standards has been based on the meter, the kilogram, the second, and the above definition of the "absolute" ampere plus other derived units for resistance, voltage, etc. This seems to

---

* In quantum field theory, powers of the coupling constant play the role of other basic units in doing dimensional analysis.

† The proportionality constant k_2 in (A.4) is thereby given the magnitude $k_2 = 10^{-7}$ in the mks system. The *dimensions* of the "absolute" ampere, as distinct from its magnitude, depend on the dimensions assigned k_2. In the conventional mks system of electromagnetic units, electric charge (q) is arbitrarily chosen as a *fourth* basic unit. Consequently the ampere has dimensions of qt^{-1}, and k_2 has dimensions of mlq^{-2}. If k_2 is taken to be dimensionless, then current has the dimensions $m^{1/2}l^{1/2}t^{-1}$. The question of whether a fourth basic unit like charge is introduced or whether electromagnetic quantities have dimensions given by powers (sometimes fractional) of the three basic mechanical units is a purely subjective matter and has no fundamental significance.

be a desirable state of affairs. It avoids such difficulties as arose when, in 1894, by Act of Congress (based on recommendations of an international commission of engineers and scientists), independent basic units of current, voltage, and resistance were defined in terms of three independent experiments (silver voltameter, Clark standard cell, specified column of mercury).* Soon afterwards, because of systematic errors in the experiments outside the claimed accuracy, Ohm's law was no longer valid, by Act of Congress!

2 Electromagnetic Units and Equations

In discussing the units and dimensions of electromagnetism we will take as our starting point the traditional choice of length (l), mass (m), and time (t) as independent, basic units. Furthermore, we will make the commonly accepted definition of current as the time rate of change of charge ($I = dq/dt$). This means that the dimension of the ratio of charge and current is that of time.† The continuity equation for charge and current densities then takes the form:

$$\mathbf{V} \cdot \mathbf{J} + \frac{\partial \rho}{\partial t} = 0 \qquad (A.1)$$

To simplify matters we will initially consider only electromagnetic phenomena in free space, apart from the presence of charges and currents.

The basic physical law governing electrostatics is Coulomb's law on the force between two point charges q and q', separated by a distance r. In symbols this law is

$$F_1 = k_1 \frac{qq'}{r^2} \qquad (A.2)$$

The constant k_1 is a proportionality constant whose magnitude and dimensions *either* are determined by the equation if the magnitude and dimensions of the unit of charge have been specified independently *or* are chosen arbitrarily in order to define the unit of charge. Within our present framework all that is determined at the moment is that the product ($k_1 qq'$) has the dimensions (ml^3t^{-2}).

---

* See, for example, F. A. Laws, *Electrical Measurements*, McGraw-Hill, New York (1917), pp. 705–706.

† From the point of view of special relativity it would be more natural to give current the dimensions of charge divided by length. Then current density J and charge density ρ would have the same dimensions and would form a "natural" 4-vector. This is the choice made in a modified Gaussian system (see the footnote (p. 621) for Table 4).

The electric field $\mathbf{E}$ is a derived quantity, customarily defined to be the force per unit charge. A more general definition would be that the electric field be numerically proportional to the force per unit charge, with a proportionality constant which is a universal constant perhaps having dimensions such that the electric field is dimensionally different from force per unit charge, There is, however, nothing to be gained by this extra freedom in the definition of $\mathbf{E}$, since $\mathbf{E}$ is the first derived field quantity to be defined. Only when we define other field quantities may it be convenient to insert dimensional proportionality constants in the definitions in order to adjust the dimensions and magnitude of these fields relative to the electric field. Consequently, with no significant loss of generality the electric field of a point charge q may be defined from (A.2) as the force per unit charge,

$$E = k_1 \frac{q}{r^2} \qquad (A.3)$$

All systems of units known to the author use this definition of electric field.

For steady-state magnetic phenomena Ampère's observations form a basis for specifying the interaction and defining the magnetic induction. According to Ampère, the force per unit length between two infinitely long, parallel wires separated by a distance d and carrying currents I and I' is,

$$\frac{dF_2}{dl} = 2k_2 \frac{II'}{d} \qquad (A.4)$$

The constant k_2 is a proportionality constant akin to k_1 in (A.2). The dimensionless number 2 is inserted in (A.4) for later convenience in specifying k_2. Because of our choice of the dimensions of current and charge embodied in (A.1) the dimensions of k_2 relative to k_1 are determined. From (A.2) and (A.4) it is easily found that the ratio k_1/k_2 has the dimension of a velocity squared $(l^2 t^{-2})$. Furthermore, by comparison of the magnitude of the two mechanical forces (A.2) and (A.4) for known charges and currents, the magnitude of the ratio k_1/k_2 in free space can be found. The numerical value is closely given by the square of the velocity of light in vacuum. Therefore in symbols we can write

$$\frac{k_1}{k_2} = c^2 \qquad (A.5)$$

where c stands for the velocity of light in magnitude and dimension ($c = 2.997930 \pm 0.000003 \times 10^{10}$ cm/sec).

The magnetic induction $\mathbf{B}$ is derived from the force laws of Ampère as being numerically proportional to the force per unit current with a proportionality constant α which may have certain dimensions chosen for convenience. Thus for a long straight wire carrying a current I, the magnetic induction $\mathbf{B}$ at a distance d has the magnitude (and dimensions)

$$B = 2k_2\alpha \frac{I}{d} \tag{A.6}$$

The dimensions of the ratio of electric field to magnetic induction can be found from (A.1), (A.3), (A.5), and (A.6). The result is that (E/B) has the dimensions $(l/t\alpha)$.

The third and final relation in the specification of electromagnetic units and dimensions is Faraday's law of induction, which connects electric and magnetic phenomena. The observed law that the electromotive force induced around a circuit is proportional to the rate of change of magnetic flux through it takes on the differential form,

$$\nabla \times \mathbf{E} + k_3 \frac{\partial \mathbf{B}}{\partial t} = 0 \tag{A.7}$$

where k_3 is a constant of proportionality. Since the dimensions of $\mathbf{E}$ relative to $\mathbf{B}$ are established, the dimensions of k_3 can be expressed in terms of previously defined quantities merely by demanding that both terms in (A.7) have the same dimensions. Then it is found that k_3 has the dimensions of α^{-1}. Actually, k_3 is *equal* to α^{-1}. This is established on the basis of Galilean invariance in Section 6.1. But the easiest way to prove the equality is to write all Maxwell's equations in terms of the fields defined here:

$$\left. \begin{aligned} \nabla \cdot \mathbf{E} &= 4\pi k_1 \rho \\[4pt] \nabla \times \mathbf{B} &= 4\pi k_2 \alpha \mathbf{J} + \frac{k_2\alpha}{k_1} \frac{\partial \mathbf{E}}{\partial t} \\[4pt] \nabla \times \mathbf{E} + k_3 \frac{\partial \mathbf{B}}{\partial t} &= 0 \\[4pt] \nabla \cdot \mathbf{B} &= 0 \end{aligned} \right\} \tag{A.8}$$

Then for source-free regions the two curl equations can be combined into the wave equation,

$$\nabla^2 \mathbf{B} - k_3 \frac{k_2\alpha}{k_1} \frac{\partial^2 \mathbf{B}}{\partial t^2} = 0 \tag{A.9}$$

The velocity of propagation of the waves described by (A.9) is related to the combination of constants appearing there. Since this velocity is known to be that of light, we may write

$$\frac{k_1}{k_3 k_2 \alpha} = c^2 \tag{A.10}$$

Combining (A.5) with (A.10), we find

$$k_3 = \frac{1}{\alpha} \qquad \text{(A.11)}$$

an equality holding for both magnitude and dimensions.

3 Various Systems of Electromagnetic Units

The various systems of electromagnetic units differ in their choices of the magnitudes and dimensions of the various constants above. Because of relations (A.5) and (A.11) there are only two constants (e.g., k_1, k_3)

Table 1
Magnitudes and dimensions of the electromagnetic constants for various systems of units

The dimensions are given after the numerical values. The symbol c stands for the velocity of light in vacuum ($c = 2.998 \times 10^{10}$ cm/sec $= 2.998 \times 10^8$ m/sec). The first four systems of units use the centimeter, gram, and second as their fundamental units (l, m, t). The mks system uses the meter, kilogram, and second, plus charge (q) as a fourth unit.

System	k_1	k_2	α	k_3
Electrostatic (esu)	1	$c^{-2}(t^2 l^{-2})$	1	1
Electromagnetic (emu)	$c^2(l^2 t^{-2})$	1	1	1
Gaussian	1	$c^{-2}(t^2 l^{-2})$	$c(lt^{-1})$	$c^{-1}(tl^{-1})$
Heaviside-Lorentz	$\dfrac{1}{4\pi}$	$\dfrac{1}{4\pi c^2}(t^2 l^{-2})$	$c(lt^{-1})$	$c^{-1}(tl^{-1})$
Rationalized mks	$\dfrac{1}{4\pi\epsilon_0} = 10^{-7}c^2$ $(ml^3 t^{-2} q^{-2})$	$\dfrac{\mu_0}{4\pi} \equiv 10^{-7}$ (mlq^{-2})	1	1

that can (and must) be chosen arbitrarily. It is convenient, however, to tabulate all four constants (k_1, k_2, α, k_3) for the commoner systems of units. These are given in Table 1. We note that, apart from dimensions, the em units and mks units are very similar, differing only in various powers of 10 in their mechanical and electromagnetic units. The Gaussian and Heaviside-Lorentz systems differ only by factors of 4π. Only in the

Gaussian (and Heaviside-Lorentz) system does k_3 have dimensions. It is evident from (A.7) that, with k_3 having dimensions of a reciprocal velocity, **E** and **B** have the same dimensions. Furthermore, with $k_3 = c^{-1}$, (A.7) shows that for electromagnetic waves in free space **E** and **B** are equal in magnitude as well.

Only electromagnetic fields in free space have been discussed so far. Consequently only the two fundamental fields **E** and **B** have appeared. There remains the task of defining the macroscopic field variables **D** and **H**. If the averaged electromagnetic properties of a material medium are described by a macroscopic polarization **P** and a magnetization **M**, the general forms of the definitions of **D** and **H** are

$$\left. \begin{aligned} \mathbf{D} &= \epsilon_0 \mathbf{E} + \lambda \mathbf{P} \\ \mathbf{H} &= \frac{1}{\mu_0} \mathbf{B} - \lambda' \mathbf{M} \end{aligned} \right\} \tag{A.12}$$

where ϵ_0, μ_0, λ, λ' are proportionality constants. Nothing is gained by making **D** and **P** or **H** and **M** have different dimensions. Consequently λ and λ' are chosen as pure numbers ($\lambda = \lambda' = 1$ in rationalized systems, $\lambda = \lambda' = 4\pi$ in unrationalized systems). But there is the choice as to whether **D** and **P** will differ in dimensions from **E**, and **H** and **M** differ from **B**. This choice is made for convenience and simplicity, usually in order to make the macroscopic Maxwell's equations have a relatively simple, neat form. Before tabulating the choices made for different systems, we note that for linear, isotropic media the constitutive relations are always written

$$\left. \begin{aligned} \mathbf{D} &= \epsilon \mathbf{E} \\ \mathbf{B} &= \mu \mathbf{H} \end{aligned} \right\} \tag{A.13}$$

Thus in (A.12) the constants ϵ_0 and μ_0 are the vacuum values of ϵ and μ. The relative permittivity of a substance (often called the *dielectric constant*) is defined as the dimensionless ratio (ϵ/ϵ_0), while the relative permeability (often called the *permeability*) is defined as (μ/μ_0).

Table 2 displays the values of ϵ_0 and μ_0, the defining equations for **D** and **H**, the macroscopic forms of Maxwell's equations, and the Lorentz force equation in the five common systems of units of Table 1. For each system of units the continuity equation for charge and current is given by (A.1), as can be verified from the first pair of Maxwell's equations in the table in each case.* Similarly, in all systems the statement of Ohm's law is $\mathbf{J} = \sigma \mathbf{E}$, where σ is the conductivity.

---

* Some workers employ a modified Gaussian system of units in which current is defined by $I = (1/c)(dq/dt)$. Then the current density **J** in the table must be replaced by $c\mathbf{J}$, and the continuity equation is $\nabla \cdot \mathbf{J} + (1/c)(\partial \rho / \partial t) = 0$. See also the footnote below Table 4.

Table 2

Definitions of ϵ_0, μ_0, D, H, macroscopic Maxwell's equations, and Lorentz force equation in various systems of units

Where necessary the dimensions of quantities are given in parentheses. The symbol c stands for the velocity of light in vacuum with dimensions (lt^{-1}).

System	ϵ_0	μ_0	D, H	Macroscopic Maxwell's Equations				Lorentz Force per Unit charge
Electrostatic (esu)	1	c^{-2} $(t^2 l^{-2})$	$D = E + 4\pi P$ $H = c^2 B - 4\pi M$	$\nabla \cdot D = 4\pi\rho$	$\nabla \times H = 4\pi J + \dfrac{\partial D}{\partial t}$	$\nabla \times E + \dfrac{\partial B}{\partial t} = 0$	$\nabla \cdot B = 0$	$E + v \times B$
Electromagnetic (emu)	c^{-2} $(t^2 l^{-2})$	1	$D = \dfrac{1}{c^2} E + 4\pi P$ $H = B - 4\pi M$	$\nabla \cdot D = 4\pi\rho$	$\nabla \times H = 4\pi J + \dfrac{\partial D}{\partial t}$	$\nabla \times E + \dfrac{\partial B}{\partial t} = 0$	$\nabla \cdot B = 0$	$E + v \times B$
Gaussian	1	1	$D = E + 4\pi P$ $H = B - 4\pi M$	$\nabla \cdot D = 4\pi\rho$	$\nabla \times H = \dfrac{4\pi}{c} J + \dfrac{1}{c}\dfrac{\partial D}{\partial t}$	$\nabla \times E + \dfrac{1}{c}\dfrac{\partial B}{\partial t} = 0$	$\nabla \cdot B = 0$	$E + \dfrac{v}{c} \times B$
Heaviside-Lorentz	1	1	$D = E + P$ $H = B - M$	$\nabla \cdot D = \rho$	$\nabla \times H = \dfrac{1}{c}\left(J + \dfrac{\partial D}{\partial t}\right)$	$\nabla \times E + \dfrac{1}{c}\dfrac{\partial B}{\partial t} = 0$	$\nabla \cdot B = 0$	$E + \dfrac{v}{c} \times B$
Rationalized mks	$\dfrac{10^7}{4\pi c^2}$ $(q^2 t^2 m^{-1} l^{-3})$	$4\pi \times 10^{-7}$ (mlq^{-2})	$D = \epsilon_0 E + P$ $H = \dfrac{1}{\mu_0} B - M$	$\nabla \cdot D = \rho$	$\nabla \times H = J + \dfrac{\partial D}{\partial t}$	$\nabla \times E + \dfrac{\partial B}{\partial t} = 0$	$\nabla \cdot B = 0$	$E + v \times B$

Table 3

Conversion table for symbols and formulas

The symbols for mass, length, time, force, and other not specifically electromagnetic quantities are unchanged. To convert any equation in Gaussian variables to the corresponding equation in mks quantities, on both sides of the equation replace the relevant symbols listed below under "Gaussian" by the corresponding "mks" symbols listed on the right. The reverse transformation is also allowed. Since the length and time symbols are unchanged, quantities which differ dimensionally from one another only by powers of length and/or time are grouped together where possible.

Quantity	Gaussian	mks
Velocity of light	c	$(\mu_0\epsilon_0)^{-1/2}$
Electric field (potential, voltage)	$\mathbf{E}(\Phi, V)$	$\sqrt{4\pi\epsilon_0}\,\mathbf{E}(\Phi, V)$
Displacement	$\mathbf{D}$	$\sqrt{\dfrac{4\pi}{\epsilon_0}}\,\mathbf{D}$
Charge density (charge, current density, current, polarization)	$\rho(q, \mathbf{J}, I, \mathbf{P})$	$\dfrac{1}{\sqrt{4\pi\epsilon_0}}\rho(q, \mathbf{J}, I, \mathbf{P})$
Magnetic induction	$\mathbf{B}$	$\sqrt{\dfrac{4\pi}{\mu_0}}\,\mathbf{B}$
Magnetic field	$\mathbf{H}$	$\sqrt{4\pi\mu_0}\,\mathbf{H}$
Magnetization	$\mathbf{M}$	$\sqrt{\dfrac{\mu_0}{4\pi}}\,\mathbf{M}$
Conductivity	σ	$\dfrac{\sigma}{4\pi\epsilon_0}$
Dielectric constant	ϵ	$\dfrac{\epsilon}{\epsilon_0}$
Permeability	μ	$\dfrac{\mu}{\mu_0}$
Resistance (impedance)	$R(Z)$	$4\pi\epsilon_0 R(Z)$
Inductance	L	$4\pi\epsilon_0 L$
Capacitance	C	$\dfrac{1}{4\pi\epsilon_0}C$

Table 4

Conversion table for given amounts of a physical quantity

The table is arranged so that a given amount of some physical quantity, expressed as so many mks or Gaussian units of that quantity, can be expressed as an equivalent number of units in the other system . Thus the entries in each row stand for the same amount, expressed in different units. All factors of 3 (apart from exponents) should, for accurate work, be replaced by (2.997930 ± 0.000003), arising from the numerical value of the velocity of light. For example, in the row for displacement (D), the entry $(12\pi \times 10^5)$ is actually $(2.99793 \times 4\pi \times 10^5)$. Where a name for a unit has been agreed on or is in common usage, that name is given. Otherwise, one merely reads so many Gaussian units, or mks units.

Physical Quantity	Symbol	Rationalized mks		Gaussian
Length	l	1 meter (m)	10^2	centimeters (cm)
Mass	m	1 kilogram (kg)	10^3	grams (gm)
Time	t	1 second (sec)	1	second (sec)
Force	F	1 newton	10^5	dynes
Work / Energy	W / U	1 joule	10^7	ergs
Power	P	1 watt	10^7	ergs sec^{-1}
Charge	q	1 coulomb (coul)	3×10^9	statcoulombs
Charge density	ρ	1 coul m^{-3}	3×10^3	statcoul cm^{-3}
Current	I	1 ampere (coul sec^{-1})	3×10^9	statamperes
Current density	J	1 amp m^{-2}	3×10^5	statamp cm^{-2}
Electric field	E	1 volt m^{-1}	$\frac{1}{3} \times 10^{-4}$	statvolt cm^{-1}
Potential	Φ, V	1 volt	$\frac{1}{300}$	statvolt
Polarization	P	1 coul m^{-2}	3×10^5	dipole moment cm^{-3}
Displacement	D	1 coul m^{-2}	$12\pi \times 10^5$	statvolt cm^{-1} (statcoul cm^{-2})
Conductivity	σ	1 mho m^{-1}	9×10^9	sec^{-1}
Resistance	R	1 ohm	$\frac{1}{9} \times 10^{-11}$	sec cm^{-1}
Capacitance	C	1 farad	9×10^{11}	cm
Magnetic flux	ϕ, F	1 weber	10^8	gauss cm^2 or maxwells
Magnetic induction	B	1 weber m^{-2}	10^4	gauss
Magnetic field	H	1 ampere-turn m^{-1}	$4\pi \times 10^{-3}$	oersted
Magnetization	M	1 ampere m^{-1}	$\frac{1}{4\pi} \times 10^{-3}$	magnetic moment cm^{-3}
*Inductance	L	1 henry	$\frac{1}{9} \times 10^{-11}$	

4 Conversion of Equations and Amounts between Gaussian Units and mks Units

The two systems of electromagnetic units in most common use today are the Gaussian and rationalized mks systems. The mks system has the virtue of overall convenience in practical, large-scale phenomena, especially in engineering applications. The Gaussian system is more suitable for microscopic problems involving the electrodynamics of individual charged particles, etc. Since microscopic, relativistic problems are emphasized in this book, it has been found most convenient to use Gaussian units throughout. In Chapter 8 on wave guides and cavities an attempt has been made to placate the engineer by writing each key formula in such a way that omission of the factor in square brackets in the equation will yield the equivalent mks equation (provided all symbols are reinterpreted as mks variables).

Tables 3 and 4 are designed for general use in conversion from one system to the other. Table 3 is a conversion scheme for *symbols and equations* which allows the reader to convert any equation from the Gaussian system to the mks system and vice versa. Simpler schemes are available for conversion only *from* the mks system *to* the Gaussian system, and other general schemes are possible. But by keeping all mechanical quantities unchanged, the recipe in Table 3 allows the straightforward conversion of quantities which arise from an interplay of electromagnetic and mechanical forces (e.g., the fine structure constant $e^2/\hbar c$ and the plasma frequency $\omega_p{}^2 = 4\pi n e^2/m$) without additional considerations. Table 4 is a conversion table for units to allow the reader to express a given amount of any physical entity as a certain number of mks units or cgs-Gaussian units.

---

* There is some confusion prevalent about the unit of inductance in Gaussian units. This stems from the use by some authors of a modified system of Gaussian units in which current is measured in electromagnetic units, so that the connection between charge and current is $I_m = (1/c)(dq/dt)$. Since inductance is defined through the induced voltage $V = L(dI/dt)$ or the energy $U = \frac{1}{2}LI^2$, the choice of current defined in Section 2 means that our Gaussian unit of inductance is equal in magnitude and dimensions $(t^2 l^{-1})$ to the electrostatic unit of inductance. The electromagnetic current I_m is related to our Gaussian current I by the relation $I_m = (1/c)I$. From the energy definition of inductance we see that the electromagnetic inductance L_m is related to our Gaussian inductance L through $L_m = c^2 L$. Thus L_m has the dimensions of length. The modified Gaussian system generally uses the electromagnetic unit of inductance, as well as current. Then the voltage relation reads $V = (L_m/c)(dI_m/dt)$. The numerical connection between units of inductance is

$$1 \text{ henry} = \tfrac{1}{9} \times 10^{-11} \text{ Gaussian (es) unit} = 10^9 \text{ emu}$$

Bibliography

Abraham, M., and R. Becker, *Electricity and Magnetism*, Blackie, London (1937), trans. from 8th German ed. of *Theorie der Elektrizität*, Band I.

— — —, *Theorie der Elektrizität*, Band II, *Elektronentheorie*, Tuebner, Leipzig (1933).

Adler, R. B., L. J. Chu, and R. M. Fano, *Electromagnetic Energy Transmission and Radiation*, Wiley, New York (1960).

Aharoni, J. *The Special Theory of Relativity*, Oxford University Press (1959).

Alfvén, H. *Cosmical Electrodynamics*, Oxford University Press (1950).

Baker, B. B., and E. T. Copson, *Mathematical Theory of Huygens' Principle*, 2nd ed., Oxford University Press (1950).

Baldin, A. M., V. I. Gol'danskii, and I. L. Rozenthal, *Kinematics of Nuclear Reactions*, Pergamon Press, New York (1961).

Bateman Manuscript Project, *Higher Transcendental Functions*, 3 vols., ed. by A. Erdélyi, McGraw-Hill, New York (1953).

— — —, *Tables of Integral Transforms*, 2 vols., ed. by A. Erdéyli, McGraw-Hill, New York (1954).

Bergmann, P. G., *Introduction to the Theory of Relativity*, Prentice-Hall, Englewood Cliffs, N.J. (1942).

Blatt, J. M., and V. F. Weisskopf, *Theoretical Nuclear Physics*, Wiley, New York (1952).

Bohr, N., "Penetration of Atomic Particles through Matter," *Kgl. Danske Videnskab. Selskab Mat-fys. Medd.*, **XVIII**, No. 8 (1948).

Born, M., and E. Wolf, *Principles of Optics*, Pergamon Press, New York (1959).

Böttcher, C. J. F., *Theory of Electric Polarization*, Elsevier, New York (1952).

Brillouin, L., *Wave Propagation and Group Velocity*, Academic Press, New York (1960).

Byerly, W. E., *Fourier Series and Spherical Harmonics*, Ginn, Boston (1893); also Dover reprint.

Chandrasekhar, S., *Plasma Physics*, University of Chicago Press (1960).

Churchill, R. V., *Fourier Series and Boundary Value Problems*, McGraw-Hill, New York (1941).

Condon, E. U., and H. Odishaw, ed., *Handbook of Physics*, McGraw-Hill, New York (1958).

Corben, H. C., and P. Stehle, *Classical Mechanics*, 2nd. ed., Wiley, New York (1960).

Courant, R., and D. Hilbert, *Methoden der mathematischen Physik*, 2 vols., Springer, Berlin (1937). Volume 1 is available in English from Interscience, New York.

Cowling, T. G., *Magnetohydrodynamics*, Interscience, New York (1957).

Debye, P., *Polar Molecules*, Dover, New York (1945).

Durand, E., *Electrostatique et magnétostatique*, Masson, Paris (1953).

Einstein A., H. A. Lorentz, H. Minkowski, and H. Weyl, *The Principle of Relativity*, collected papers, with notes by A. Sommerfeld, Dover, New York (1952).

Friedman, B., *Principles and Techniques of Applied Mathematics*, Wiley, New York (1956).

Fröhlich, H., *Theory of Dielectrics*, Oxford University Press (1949).

Glasstone, S., and R. H. Lovberg, *Controlled Thermonuclear Reactions*, Van Nostrand, Princeton, N.J. (1960).

Goldstein, H., *Classical Mechanics*, Addison-Wesley, Reading, Mass. (1950).

Hadamard, J., *Lectures on Cauchy's Problem*, Yale University Press (1923); Dover reprint (1952).

Heitler, W., *Quantum Theory of Radiation*, 3rd ed., Oxford University Press (1954).

Hildebrand, F. B., *Advanced Calculus for Engineers*, Prentice-Hall, Englewood Cliffs, N.J. (1948).

Iwanenko, D., and A. Sokolow, *Klassische Feldtheorie*, Akademie-Verlag, Berlin (1953), trans. from the Russian ed. (1949).

Jahnke F., and E. Emde, *Tables of Functions*, 4th ed., Dover, New York (1945).

Jeans, J. H., *Mathematical Theory of Electricity and Magnetism*, 5th ed., Cambridge University Press (1948).

Jordan, E. C., *Electromagnetic Waves and Radiating Systems*, Prentice-Hall, Englewood Cliffs, N.J. (1950).

Kelvin, Lord (Sir W. Thomson), *Reprints of Papers on Electrostatics and Magnetism*, 2nd ed., Macmillan, London (1884).

King, R. W. P., and T. T. Wu, *Scattering and Diffraction of Waves*, Harvard University Press (1959).

Kraus, J. D., *Antennas*, McGraw-Hill, New York (1950).

Landau, L. D., and E. M. Lifshitz, *Classical Theory of Fields*, Addison-Wesley, Reading, Mass. (1951).

——— , *Electrodynamics of Continuous Media*, Addison-Wesley, Reading, Mass. (1960).

Lighthill, M. J., *Introduction to Fourier Analysis and Generalized Functions*, Cambridge University Press (1958).

Linhart, J. G., *Plasma Physics*, North-Holland, Amsterdam (1960).

Livingston, M. S., *High-Energy Accelerators*, Interscience, New York (1954).

Lorentz, H. A., *Theory of Electrons*, 2nd ed. (1915), Dover Publications, New York (1952).

Magnus, W., and F. Oberhettinger, *Special Functions of Mathematical Physics*, Chelsea, New York (1949).

Mason, M., and W. Weaver, *The Electromagnetic Field*, University of Chicago Press (1929); Dover reprint.

Maxwell, J. C., *Treatise on Electricity and Magnetism*, 3rd ed., 2 vols., reprint by Dover, New York (1954).

Møller, C., *Theory of Relativity*, Oxford University Press (1952).

Morse, P. M., and H. Feshbach, *Methods of Theoretical Physics*, 2 Pts., McGraw-Hill, New York (1953).

Panofsky, W. K. H., and M. Phillips, *Classical Electricity and Magnetism*, Addison-Wesley, Reading, Mass. (1955).

Pauli, W., *Theory of Relativity*, Pergamon Press, New York (1958), trans. from an article in the *Encyklopedia der mathematischen Wissenschaften*, Vol. V19, Tuebner, Leipzig (1921), with supplementary notes by the author (1956).

Rohrlich, F., in *Lectures in Theoretical Physics*, Vol II, ed. W. E. Brittin and B. W. Downs, Interscience, New York (1960).

Rose, D. J., and M. Clark, *Plasmas and Controlled Fusion*, M.I.T.–Wiley, New York (1961).

Rosenfeld, L., *Theory of Electrons*, North-Holland, Amsterdam (1951).

Rossi, B., *High-Energy Particles*, Prentice-Hall, Englewood Cliffs, N.J. (1952).

Rothe, R., F. Ollendorff, and K. Polhausen, *Theory of Functions as Applied to Engineering Problems*, Technology Press, Cambridge, Mass. (1933).

Schelkunoff, S. A., *Advanced Antenna Theory*, Wiley, New York (1952).

Schott, G. A., *Electromagnetic Radiation*, Cambridge University Press (1912).

Segre, E., ed., *Experimental Nuclear Physics*, Vol. 1, Wiley, New York (1953).

Siegbahn, K., ed., *Beta- and Gamma-Ray Spectroscopy*, North-Holland, Amsterdam, and Interscience, New York (1955).

Silver, S., ed., *Microwave Antenna Theory and Design*, M.I.T. Radiation Laboratory Series, Vol. 12, McGraw-Hill, New York (1949).

Simon, A., *Introduction to Thermonuclear Research*, Pergamon, New York (1959).

Slater, J. C., *Microwave Electronics*, McGraw-Hill, New York (1950).

— — —, and N. H. Frank, *Electromagnetism*, McGraw-Hill, New York (1947).

Smythe, W. R., *Static and Dynamic Electricity*, 2nd ed., McGraw-Hill, New York (1950).

Sommerfeld, A., *Electrodynamics*, Academic, New York (1952).

— — —, *Partial Differential Equations in Physics*, Academic, New York (1949).

Spitzer, L., *Physics of Fully Ionized Gases*, Interscience, New York (1956).

Stratton, J. A., *Electromagnetic Theory*, McGraw-Hill, New York (1941).

Titchmarsh, E. C., *Introduction to the Theory of Fourier Integrals*, 2nd ed., Oxford University Press (1948).

Tranter, C. J., *Integral Transforms in Mathematical Physics*, 2nd ed., Methuen, London (1956).

Van Vleck, J. H., *Theory of Electric and Magnetic Susceptibilities*, Oxford University Press (1932).

Watson, G. N., *Theory of Bessel Functions*, 2nd ed., Cambridge University Press (1952).

Whittaker, E. T., *History of the Theories of Aether and Electricity*, 2 vols., Nelson, London (1951, 1953).

Williams, E. J., "Correlation of Certain Collision Problems with Radiation Theory," *Kgl. Danske Videnskab. Selskab Mat-fys. Medd.*, **XIII**, No. 4 (1935).

Index